Fraction	Decimal
1/64	0.015625
1/32	0.03125
3/64	0.046875
1/16	0.0625
5/64	0.078125
3/32	0.09375
7/64	0.109375
1/8	0.125
9/64	0.140625
5/32	0.15625
11/64	0.171875
3/16	0.1875
13/64	0.203125
7/32	0.21875
15/64	0.234375
1/4	0.25
17/64	0.265625
9/32	0.28125
19/64	0.296875
5/16	0.3125
21/64	0.328125
11/32	0.34375
23/64	0.359375
3/8	0.375
25/64	0.390625
13/32	0.40625
27/64	0.421875
7/16	0.4375
29/64	0.453125
15/32	0.46875
31/64	0.484375
1/2	0.5
33/64	0.515625
17/32	0.53125
35/64	0.546875
9/16	0.5625
37/64	0.578125
19/32	0.59375
39/64	0.609375
5/8	0.625
41/64	0.640625
21/32	0.65625
43/64	0.671875
11/16	0.6875
45/64	0.703125
23/32	0.71875
47/64	0.734375
3/4	0.75
49/64	0.765625
25/32	0.78125
51/64	0.796875
13/16	0.8125
53/64	0.828125
27/32	0.84375
55/64	0.859375
7/8	0.875
57/64	0.890625
29/32	0.90625
59/64	0.921875
15/16	0.9375
61/64	0.953125
31/32	0.96875
63/64	0.984375
1	1

Decimal Equivalents of an Inch

Applied Manufacturing Process Planning
with Emphasis on Metal Forming and Machining

Donald H. Nelson
Manager Manufacturing Engineering
Chrysler Corporation, Retired

George Schneider, Jr., CMfgE
Professor Emeritus
Lawrence Technological University

Upper Saddle River, New Jersey
Columbus, Ohio

Library of Congress Cataloging-in-Publication Data

Nelson, Donald H.
Applied manufacturing process planning: with emphasis on metal forming and machining/Donald H. Nelson, George Schneider, Jr.–1st ed.
p. cm.
ISBN 0-13-532458-0
1. Manufacturing processes–Planning. 2. Metal-work. 3. Metal-cutting. I. Schneider, George. II. Title

TS183.3. N45 2001
658.5–dc21

00-029853
CIP

Vice President and Publisher: Dave Garza
Editor in Chief: Stephen Helba
Acquisitions Editor: Debbie Yarnell
Associate Editor: Michelle Churma
Production Editor: Louise N. Sette
Project Supervision: York Production Services
Design Coordinator: Robin G. Chukes
Cover Designer: Steve Penn
Cover photo: © EX-CELL-O Machine Tools, Inc.
Production Manager: Brian Fox
Marketing Manager: Chris Bracken

This book was set in Clearface by York Graphic Services, Inc. It was printed and bound by Courier Westford, Inc. The cover was printed by Phoenix Color Corp.

10 9 8 7 6 5 4 3 2 1
ISBN: 0-13-532458-0

About the Authors

Donald H. Nelson's education and hands-on experience provide him with unique qualifications to write this book. Mr. Nelson graduated from high school in Austin, Minnesota, and subsequently from Lawrence Institute of Technology (now Lawrence Technological University) with a Bachelor of Mechanical Engineering degree. During his undergraduate years, he was employed by General Motors, Hydromatic Division, in the Tool Engineering department.

At the outbreak of World War II, he was assigned to Wright Aeronautical Corporation in Paterson, New Jersey, and worked as a tool engineer on aircraft engines. After the war, Mr. Nelson joined the Chrysler Corporation as a manufacturing engineer at the Highland Park Plant and was promoted through the ranks to Manager, Manufacturing Engineering.

Highland Park was a unique production and service parts supplier that produced many different products in quantity, including torque converters, power steering pumps, brake wheel and master cylinders, rear axles, transmission gear sets, aluminum in an aluminium foundry and permanent-mold pistons, cold-headed parts, automatic screw machine parts, and also many different stampings in a press plant.

A transfer and assignment to the Indianapolis plant placed him in charge of Production, Production Control, and Manufacturing Engineering. Indianapolis produced starting motors, alternators, and power steering units.

A subsequent transfer to the Kokomo transmission plant provided responsibility for tool engineering, plant engineering, and maintenance during the introduction of a new model automatic transmission and expansion of the manufacturing facilities.

A final assignment to Chrysler corporate staff provided opportunity and responsibility for consulting on manufacturing plant problems and for review and approval of manufacturing plant's project requests for tooling and equipment.

Following retirement, Mr. Nelson served General Dynamics Land Systems Division as a consultant for three years and still serves as a consultant to private industry.

Mr. Nelson taught manufacturing engineering classes at Lawrence Technological University for several years. He is a life member of the Society of Automotive Engineers and served as program cochairman of the Indiana section. He is also a senior member of the Society of Manufacturing Engineers.

George Schneider, Jr. immigrated to the United States in 1955 from Germany, after leaving his birthplace in Transylvania at the age of 11 in 1950. He completed high school in Lorain, Ohio, and served in the U.S. Navy aboard the aircraft carriers *USS Leyte* and *USS Wasp*.

After working as a tool and die apprentice, Professor Schneider decided to go to college. He holds a BS degree in Industrial Production from Kent State University, an MBA degree in Industrial Management from Case Western Reserve University, and an AS degree in Computer Technology from Lawrence Technological University.

Professor Schneider spent many years in industry working for General Motors, Modco-Valenite, Ford Motor, the Carmet Company, and the Dawson Tool Company. In 1982 he joined Lawrence Technological University's engineering faculty, teaching manufacturing processes, materials, robotics, and senior project courses. Professor Schneider also served as Director of

Cooperative and Continuing Education and as Chairman of the Division of Corporate Credit and Cooperative Education.

Professor Schneider was inducted into Tau Beta Pi as an Eminent Engineer in 1985. He is an SME Certified Manufacturing Engineer (CMfgE) in Numerical Control; he has also been treasurer (1987–1988) and chairman (1988–1989) of the Society of Manufacturing Engineers Detroit Area Chapter ONE. From 1989 to 1990 Professor Schneider served as President of the International Executive Board of ASM's Society of Carbide and Tool Engineers (SCTE).

In 1998 Professor Schneider retired from Lawrence Technological University and was granted emeritus status by the university as Professor of Engineering Technology. He currently serves industry as a manufacturing consultant, and Lawrence Technological University as a global education consultant.

Preface

Applied Manufacturing Process Planning is unique in that it is application-oriented and follows actual manufacturing plan development closely. Some differences do exist between manufacturers, such as reporting relationships and product design, but basic plan development is common.

Chapters are organized in the sequence used to develop manufacturing plans in actual practice. Each chapter is illustrated to clarify the discussion of the subject involved, and examples are provided. Also, chapters are provided that serve as support and refresher updates to chapters that detail elements of a manufacturing plan. These pages are identified with shaded edges.

Machined parts, upset parts, and sheet metal parts have been selected as a representative cross section of manufactured parts for detailed discussion and illustration of manufacturing plan development. The planning functions discussed and illustrated in this book can be employed to develop manufacturing plans for most other manufactured products. Key parts of a manufacturing plan for machined parts are:

- **Processing:** Processing is determining the operations and sequence of operations required to manufacture a part. Machining dimensions, manufacturing tolerances, and locating surfaces are also determined.
- **Tolerance Charting:** Tolerance charting is mathematically checking, verifying, and graphically displaying (charting) the machining dimensions, manufacturing tolerances, and stock removal of each operation planned in processing—all from the specified locating points.
- **Workpiece Holding:** Workpieces must rest, during machining, on the identical locating surfaces specified in processing and confirmed with tolerance charting. Clamping must keep the workpiece in contact with the locating surfaces; supports are sometimes required to prevent distortion caused by clamping and/or tool pressures.

Processing, tolerance charting, and workpiece holding of a part must be developed to be in complete agreement; afterwards, ancillary, auxiliary, and support services, and costs are added to complete a manufacturing plan.

Chapters on manufacturing plan elements of special machines, machine selection, and group technology are provided. Developing a manufacturing plan for metal-worked parts follows the same general course as for machined parts; however, actual processes, tooling, and production materials are unique to that type of manufacture. Chapters in this area describe and illustrate processes and tooling used, as well as similarities and differences from machining processes.

The book is designed for use in manufacturing, mechanical, and industrial engineering courses in two- and four-year schools of engineering, technology, and skilled trades.

Flexibility in the use of the book exists in that individual chapters may be used to teach specific subjects—that is, tolerance charting, work holding, machining processes, or others may be studied in any sequence and in conjunction with other source material. However, where a complete manufacturing plan is to be studied and developed, part design analysis, manufacturing processing, and tolerance charting should be studied in that order.

The book will prove helpful to individuals involved in both mechanical and manufacturing engineering as well as individuals transiting from skilled trades to manufacturing engineering.

Newly graduated manufacturing engineers will find the text a valuable guide and reference in the application of theory to practice.

One proviso—the book user should have a working knowledge of basic machining processes, math, and materials to use it effectively.

The authors' careers in industry, including years of hands-on experience at levels from Process Engineer to Manager of Manufacturing Engineering at General Motors, Wright Aeronautical, Chrysler, General Dynamics, EX-CELL-O, Modco-Valenite, and Carmet companies, provide the first-hand knowledge required for a book of this kind.

Both authors have taught Manufacturing Engineering courses at Lawrence Institute of Technology (now Lawrence Technological University), and both are active as consultants.

Acknowledgments

I acknowledge the help received from a multitude of college professors, fellow workers, engineers, and machine-tool builders over the years of my career and in writing this book. I owe and thank them all.

My appreciation to Patsy LaFave for her invaluable assistance in word processing the manuscript.

My thanks to Athena Nelson, my wife, for her help, understanding, support, and patience during the several years it took to write and produce this book. My thanks to George Schneider, colleague and friend, for his superior and unfailing support of knowledge, experience, and on-the-mark suggestions. His enthusiasm has been indispensable to the completion of this book.

Many companies and individuals were of great aid supplying photographs and information. Three individuals went beyond the second mile to help: Steve Penn of French & Rogers, Inc., representing EX-CELL-O, ESAB, and Trumpf; Gorden Mikula of Lamb Technicon; and William Zink of Minster Machine Company.

Thank you all.

Donald H. Nelson

To have planned, researched, and written parts of *Applied Manufacturing Process Planning* has been a challenging opportunity. This accomplishment would not have been possible without help and support from many people. First of all I wish to thank my colleague and friend Don Nelson, and Maria, my wife of nearly 38 years.

I met Don at Lawrence Technological University, where we both taught a manufacturing systems course using a rather outdated textbook. We both augmented the text extensively with handouts of current materials. Don proposed a new textbook, and I have been privileged to have collaborated with him for over five years to publish this new text.

Without my wife Maria's help, my contribution to this text would have been next to impossible. Over the years my handwriting has deteriorated to the point where I have been accused of writing in hieroglyphics. My wife has learned to decipher my chicken scratch. I have not been blessed with a lot of patience; in fact I have been known to be short-tempered at times. Thanks, Hon, for your patience and support.

Many, many companies have contributed to this text by supplying technical information as well as photographic exhibits. Forty-three manufacturers are represented in my three parts of the text: "Metal Removal," "Single-Point Machining," and "Multipoint Machining." Three individuals have been of tremendous help: Kennametal's Bill Kennedy, Sandvik's Bob Kogan, and Clausing's Brad Coombs. As stated earlier, many others have helped with technical information and photo exhibits. Their contributions have made this textbook possible.

Then there is Prentice Hall's Technology Senior Editor, Steve Helba, who had confidence in Don and me to offer us a contract in 1995. Steve and his assistant, Michelle Churma, have been very generous with their time and advice and most helpful in preparing this text for publication. Thanks, Steve and Michelle.

George Schneider, Jr.

We would also like to thank the reviewers of this text:

Gerald W. Hieronymus, Central Piedmont Community College (NC)
James B. Higley, Purdue University–Calumet (IN)
Harvey Hoy, Milwaukee School of Engineering (WI)
Billy R. McElhaney, Nashville State Technical Institute (TN)
Lee Rosenthal
C. Sahay, State University of New York at Binghamton
Jeffrey Short, Southwestern Oklahoma State University
Marvin Simon, University of Dayton (OH)
Rob Speckert, Miami University (OH)
James Kevin Standiford, ITT Technical Institute of Little Rock (AR)

Special thanks to the following companies who have contributed to this textbook:

ABB I-R Waterjet Systems LLC.; American National Carbide Co.; Armstrong-Blum Mfg. Co.; Banner Engineering Corp.; Boehringer Gröppingen; Bridgeport Machine, Inc.; Chevalier Machinery, Inc.; Cincinnati Machine, A UNOVA Co.; Clausing Industries, Inc.; Cleveland Twist Drill/Greenfield Ind.; Danly Die Set; Detroit Broach Co.; Dorian Tool; Duramet Corporation; Engis Corporation; Erie Press Systems; ESAB Cutting Systems; EX-CELL-O Machine Tools, Inc.; George Fischer FMS Turning Technology; Greenfield Industries; Greenleaf Corporation; Hardinge, Inc.; Ingersoll Cutting Tool Co.; Ingersoll Milling Machine Co.; Iscar Metals, Inc.; K. O. Lee Co.; Kasto-Racine, Inc.; Kennametal, Inc.; Kitagawa Div. Sumikin Bussan Int'l Corp.; Komet of America, Inc.; Kurt Manufacturing; Lamb Technicon Machining Systems; Lyndex Corporation; The Minster Machine Company; The Monarch Machine Tool Co.; Moore Specialty Tool Co.; Morse Cutting Tools; National Acme Co.; Div. DeVlieg-Bullard, Inc.; Norton Company; Palmgren Steel Products, Inc.; Royal Products; Sandvik Coromant Co.; Star Cutter Co.; Stark Industrial, Inc.; Summit Machine Tool Mfg. Corp.; Sunnen Products Co.; Tapmatic Corporation; TechniDrill Systems, Inc.; TRUMPH, USA; U.S. Broach & Machine Co.; Valenite, Inc.; The Weldon Tool Co.-Talbot Holdings; and WMW Machinery Co., Inc.

Contents

PART VII Multipoint Machining

Introduction

All manufactured products, whether simple or complex, require a plan to translate them from design to actual production. A manufacturing plan depicts the engineering, facilities, manpower, and costs required to translate a product design into production. A good manufacturing plan guarantees that all parts will be produced to the part design; all specified dimensions, tolerances, and surface finishes will be held, and all other specified physical requirements will be met. Such a plan employs the most capable and cost-effective technology available to assure production of quality parts to scheduled requirements at competitive costs. Processing is central to the planning, but all related areas must be considered.

Manufacturing Engineering develops the plan—coordinating processing, industrial engineering, quality control, material handling, purchasing, and other areas of involvement. Manufacturing Engineering is responsible for providing the required cost estimates, tools, and facilities, and for coordinating all areas involved in the installation and debugging of the resulting line. Manufacturing Engineering is responsible from design release to start of production—and beyond; it will subsequently be responsible for implementing engineering changes, production rate changes, and cost reductions.

Applied Manufacturing Process Planning describes and illustrates the development of manufacturing planning in a seven-part, application-oriented text. The book is organized to include illustrations, explanations, and examples of process planning for machined parts and for metal-worked parts as separate studies. Basic process descriptions and refresher update materials are also provided. Pertinent tables and data are included in the text at points required. A glossary of terms used is provided.

Illustrated examples are included in areas of design for producibility, processing, and tolerance charting. Line drawings are used liberally, along with photographs, to clarify and accent points of discussion. Basic process drawings, used to illustrate principles, do not imply the level of machine sophistication. Basics are the same on a piece of toolroom equipment as they are on a complex transfer machine. Review questions and problems support each chapter and provide both additional review and practice for the student.

PART I

METAL REMOVAL

INTRODUCTION

The cutting process is as old as the earliest civilizations. Prehistoric people discovered some of the principles of cutting when they found that a sharp edge, properly inclined, allowed them to remove materials with less effort, and that a hard tool like flint lasted longer than softer ones. The art of cutting materials developed through the years, but little was done in early times that could be truly classified as "science." In 1798, Count Rumford reported that he measured the heat generated in metal cutting in an attempt to improve the boring of cannons. This is the first known scientific approach to machining problems.

Studies in cutting of materials began with the general use of powered machine tools about 1900. F. W. Taylor recognized that the knowledge of metal cutting was artistic and empirical up to that time, and he did extensive research to correct this condition. This research was the basis of a comprehensive paper for the American Society of Mechanical Engineers. Published in 1907, it was called "On the Art of Cutting Metals." Some of Taylor's concepts are still highly regarded by people working in the metal-cutting industry.

Between World Wars I and II, metal-cutting research turned from observation of the effects of machining to the physics of the chip formation process. Thus, the deformation of metal at the cutting-tool edge, the friction in cutting, and the type of chip formed were studied. Previous observations had been concerned only with the power used, the finish produced, and the tool life. In the United States, this work was spearheaded by O. W. Boston and H. Ernst. About the time of World War II, M. E. Merchant popularized the rational mechanics of metal cutting, combining the chip formation process and cutting forces into mathematical equations.

There has been rapid advancement in discovering the physical and chemical laws of cutting since World War II. A new understanding of cutting has been obtained, and its impact on shop practice is now being felt. While the general principles of cutting apply to nonmetallics as well as to metals, the chief interest in this book, because of the great commercial importance of metals, is the removal of metal chips.

Figure I.1a shows many of the tools used in machines such as the one shown in Figure I.1b to produce parts similar to the ones shown in Figure I.1c. Such parts in turn are assembled into everyday products such as planes, trains, and automobiles.

(a)

(b)

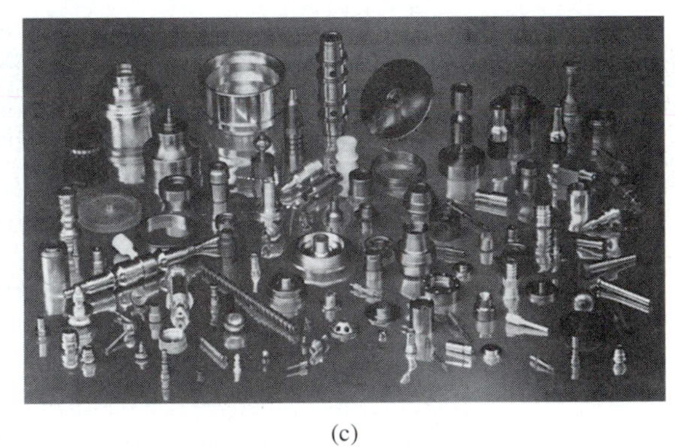

(c)

FIGURE I.1 a–c Tools, machines, and parts used to manufacture various products such as trains, planes, and automobiles. (a) Courtesy Valenite Inc. (b) Courtesy Monarch Machine Tool (c) Courtesy Hardinge Inc.

CHAPTER ONE

CUTTING-TOOL MATERIALS

CHAPTER OVERVIEW

- Introduction
- Tool Steels and Cast Alloys
- Cemented Tungsten Carbide
  - Manufacture of Carbide Products
  - Classification of Carbide Tools
  - Coated Carbide Tools
- Ceramic and Cermet Tools
- Diamond, CBN, and Whisker-Reinforced Tools
- Review Questions and Problems

1.1 INTRODUCTION

Many types of tool materials, ranging from **high-carbon steel** to ceramics and diamonds, are used as cutting tools in today's metalworking industry. It is important to be aware that differences do exist among tool materials, what these differences are, and the correct application for each type of material.

The various tool manufacturers assign names and numbers to their products. Many of these names and numbers may appear to be similar, but the applications of these tool materials may be entirely different. In most cases the tool manufacturers will provide tools made of the proper material for each given application. In some particular applications, a premium or higher-priced material will be justified.

This does not mean that the most expensive tool is always the best tool. Cutting-tool users cannot afford to ignore the constant changes and advancements that are being made in the field of tool material technology. When a tool change is needed or anticipated, a performance comparison should be made before selecting the tool for the job. The optimum tool is not necessarily the least expensive or the most expensive, and it is not always the same as the tool that was used for the job last time. The best tool is the one that has been carefully chosen to get the job done quickly, efficiently, and economically.

A cutting tool must have the following characteristics in order to produce good quality and economical parts:

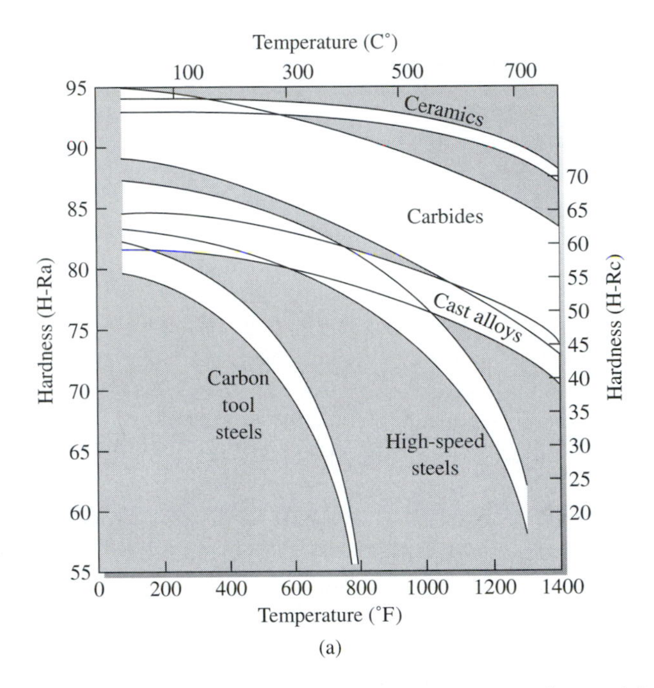

(a)

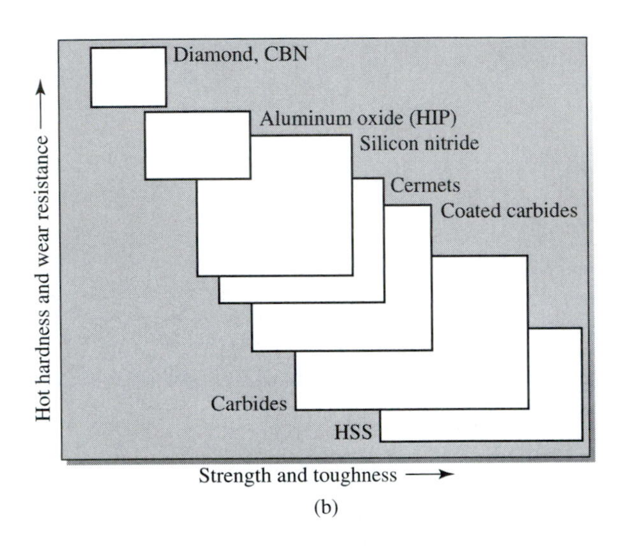

(b)

FIGURE 1.1 (a) Hardness of various cutting-tool materials as a function of temperature. (b) Ranges of properties of various groups of tool materials.

Hardness: Hardness and strength of the cutting tool must be maintained at elevated temperatures, also called hot hardness (Fig. 1.1).

Toughness: Toughness of cutting tools is needed so that tools don't chip or fracture, especially during interrupted cutting operations.

Wear Resistance: Wear resistance means the attainment of acceptable tool life before tools need to be replaced.

The materials from which cutting tools are made are all characteristically hard and strong. A wide range of tool materials are available for machining operations, and the general classification and use of these materials are of interest here.

1.2 TOOL STEELS AND CAST ALLOYS

Plain carbon tool steel is the oldest of the tool materials dating back hundreds of years. In simple terms it is a high-carbon steel (steel that contains about 1.05% carbon). This high carbon content allows the steel to be hardened, offering greater resistance to **abrasive wear**. Plain high-carbon steel served its purpose well for many years. However, because it is quickly overtempered (softened) at relatively low cutting temperatures (300 to 500° F), it is now rarely used as cutting-tool material except in files, saw blades, chisels, and so on. The use of plain high-carbon steel is limited to low-heat applications.

High-Speed Tool Steel The need for tool materials that could withstand increased cutting speeds and temperatures led to the development of high-speed tool steel (HSS). The major difference between high-speed tool steel and plain high-carbon steel is the addition of alloying elements to harden and strengthen the steel and make it more resistant to heat (hot hardness).

Some of the most commonly used alloying elements are manganese, chromium, tungsten, vanadium, molybdenum, cobalt, and niobium (columbium). While each of these elements will add certain specific desirable characteristics, it can be generally stated that they add deep hardening capability, high hot hardness, resistance to abrasive wear, and strength to high-speed tool steel. These characteristics allow relatively higher machining speeds and improved performance over plain high-carbon steel.

The most common high-speed steels used primarily as cutting tools are divided into the M and T series. The M series represents tool steels of the molybdenum type, and the T series represents those of the tungsten type. Although there seems to be a great deal of similarity among these high-speed steels, each one serves a specific purpose and offers significant benefits in its special application.

An important point to remember is that none of the alloying elements for either series of high-speed tool steels is in abundant supply and the cost of these elements is skyrocketing. In addition, U.S. manufacturers must rely on foreign countries for supply of these very important elements.

Some of the high-speed steels are now available in a **powdered metal** (PM) form. The difference between powdered and conventional metals is in the method by which they are made. The majority of conventional high-speed steel is poured into an ingot and then, either hot or cold, worked to the desired shape. Powdered metal is exactly what its name indicates. Basically the same elements that are used in conventional high-speed steel are prepared in a very fine powdered form. These powdered elements are carefully blended together, pressed into a die under extremely high pressure, and then **sintered** in an atmospherically controlled furnace. The PM method of manufacturing cutting tools is explained in Section 1.3.1, "Manufacture of Carbide Products."

HSS Surface Treatment Many surface treatments have been developed in an attempt to extend tool life, reduce power consumption, and control other factors that affect operating conditions and costs. Some of these treatments have been used for many years and have proven to have some value. For example, the black oxide coatings that commonly appear on drills and taps are of value as a deterrent to buildup on the tool. The black oxide is basically a "dirty" surface, which discourages the buildup of work material.

One of the more recent developments in coatings for high-speed steel is titanium nitride by the physical vapor deposition (PVD) method. Titanium nitride is deposited on the tool surface in one of several different types of furnace at relatively low temperatures, which does not significantly affect the heat treatment (hardness) of the tool being coated. This coating is known to extend the life of a cutting tool significantly or to allow the tool to be used at higher operating speeds. Tool life can be extended by as much as three times, or operating speeds can be increased by up to 50%.

Cast Alloys The use of alloying elements in high-speed steel, principally cobalt, chromium, and tungsten, improved the cutting properties sufficiently that metallurgical researchers developed the cast alloys, a family of these materials without iron.

A typical composition for this class of tool material was 45% cobalt, 32% chromium, 21% tungsten, and 2% carbon. The purpose of such alloying was to obtain a cutting tool with hot hardness superior to that of high-speed steel.

When applying cast alloy tools, their **brittleness** should be kept in mind, and sufficient support should be provided at all times. Cast alloys provide high abrasion resistance and are thus useful for cutting scaly materials or those with hard inclusions.

1.3 CEMENTED TUNGSTEN CARBIDE

Tungsten carbide was discovered by Henri Moissan in 1893 during a search for a method of making artificial diamonds. Charging sugar and tungsten oxide, he melted tungsten subcarbide in an arc furnace. The carbonized sugar reduced the oxide and carburized the tungsten. Moissan recorded that the tungsten carbide was extremely hard, approaching the hardness of diamond and exceeding that of sapphire. It was more than 16 times as heavy as water. The material proved to be extremely brittle, and this seriously limited its industrial use.

Commercial **tungsten carbide** with 6% cobalt binder was first produced and marketed in Germany in 1926. Production of the same carbide began in the United States in 1928 and in Canada in 1930. At this time, hard carbides consisted of the basic tungsten carbide system with cobalt **binders**. These carbides exhibited superior performance in the machining of cast iron, nonferrous, and nonmetallic materials, but were disappointing when used for the machining of steel.

Most of the subsequent developments in the hard carbides have been modifications of the original patents, principally involving replacement of part or all of the tungsten carbide with other carbides, especially **titanium carbide** and/or **tantalum carbide**. This led to the development of the modern multicarbide cutting-tool materials, permitting the high-speed machining of steel.

A new phenomenon was introduced with the development of the cemented carbides, again making higher speeds possible. Previous cutting-tool materials, products of molten metallurgy, depended largely upon heat treatment for their properties, and these properties could, in turn, be destroyed by further heat treatment. At high speeds, and consequently high temperatures, these products of molten metallurgy failed.

A different set of conditions exists with the cemented carbides. The hardness of the carbide is greater than that of most other tool materials at room temperature, and it has the ability to retain its hardness at elevated temperatures to a greater degree so greater speeds can be adequately supported.

1.3.1 Manufacture of Carbide Products

The term "tungsten carbide" describes a comprehensive family of hard carbide compositions used for metal-cutting tools, dies of various types, and wear parts. In general, these materials are composed of the carbides of tungsten, titanium, tantalum, or some combination of these, sintered or cemented in a matrix binder, usually cobalt.

Blending The first operation after reduction of the tungsten compounds to tungsten metal powder is the milling of tungsten and carbon prior to the carburizing operation. Here, 94 parts by weight of tungsten and 6 parts by weight of carbon, usually added in the form of lampblack, are blended together in a rotating mixer or ball mill. This operation must be performed under carefully controlled conditions in order to ensure optimum dispersion of the carbon in the tungsten. Carbide blending equipment, better known as a ball mill, is shown in Figure 1.2.

To provide the necessary strength, a binding agent, usually cobalt (Co) is added to the tungsten (WC) in powder form, and these two are ball milled together for a period of several days to form a very intimate mixture. Careful control of conditions, including time, must be exercised to obtain a uniform, homogeneous product. Blended tungsten carbide powder is shown in Figure 1.3.

Compacting The most common compacting method for grade powders involves the use of a die, made to the shape of the eventual product desired. The size of the die must be greater than the final product size to allow for dimensional shrinkage, which takes place in the final

FIGURE 1.2 Carbide blending equipment, better known as ball mill is used to ensure optimum dispersion of the carbon within the tungsten. (Courtesy American National Carbide Co.)

FIGURE 1.3 Blended tungsten carbide powder is produced by mixing tungsten carbide (WC) with a cobalt (Co) binder in a ball milling process. (Courtesy American National Carbide Co.)

sintering operation. These dies are expensive and usually made with tungsten carbide liners. Therefore sufficient numbers of the final product (compacts) are required to justify the expense involved in manufacturing a special die. Carbide compacting equipment, better known as a pill press, is shown in Figure 1.4. Various pill-pressed carbide parts are shown in Figure 1.5.

If the quantities are not high, a larger briquette, or **billet** may be pressed. This billet may then be cut up (usually after presintering) into smaller units and shaped or preformed to the required configuration, and again, allowance must be made to provide for shrinkage. Ordinarily pressures used in these cold compacting operations are in the neighborhood of 30,000 psi. Various carbide preformed parts are shown in Figure 1.6.

A second compacting method is the hot pressing of grade powders in graphite dies at the sintering temperature. After cooling, the part has attained full hardness. Because the graphite dies are expendable, this system is generally used only when the part to be produced is too large for cold pressing and sintering.

A third compacting method, usually used for large pieces, is **isostatic pressing**. Powders are placed into a closed, flexible container, which is then suspended in a liquid in a closed pressure vessel. Pressure in the liquid is built up to the point where the powders become properly compacted. This system is advantageous for pressing large pieces because the pressure acting on the powders operates equally from all directions, resulting in a compact of uniform pressed density.

FIGURE 1.4 Carbide compacting equipment, better known as a pill press, is used to produce carbide products in various shapes. (Courtesy American National Carbide Co.)

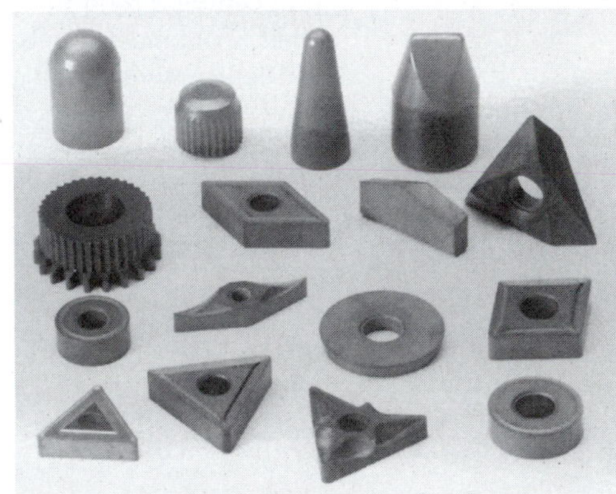

FIGURE 1.5 Various carbide compacts, which are produced with special dies mounted into pill presses. (Courtesy American National Carbide Co.)

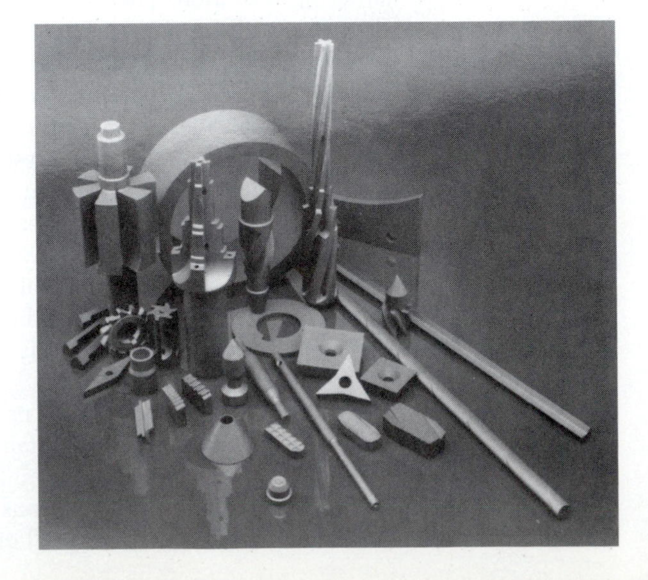

FIGURE 1.6 If quantities are not high, presintered billets are shaped or preformed into required shapes. (Courtesy Duramet Corp.)

FIGURE 1.7 Carbide parts are loaded into a sintering furnace, where they are heated to temperatures ranging from 2500 to 2900 degrees Fahrenheit. (Courtesy American National Carbide Co.)

Sintering Sintering of tungsten–cobalt (WC–Co) compacts is performed with the cobalt binder in liquid phase. The compact is heated in hydrogen atmosphere or vacuum furnaces to temperatures ranging from 2500 to 2900° F, depending on the composition. Both time and temperature must be carefully adjusted in combination to effect optimum control over properties and geometry. The compact will shrink by approximately 16% on linear dimensions, or by 40% in volume. The exact amount of shrinkage depends on several factors, including the particle size of the powders and the composition of the grade. Control of size and shape is most important, and least predictable, during the cooling cycle. This is particularly true with those grades of cemented carbides with higher cobalt contents.

With cobalt having a lesser density than tungsten, it occupies a greater part of the volume than would be indicated by the rated cobalt content of the grade, and because cobalt contents are generally a much higher percentage of the mass in liquid phase, extreme care is required to control and predict with accuracy the magnitude and direction of shrinkage. Figure 1.7 shows carbide parts being loaded into a sintering furnace. A more detailed schematic diagram of the cemented tungsten carbide manufacturing process is shown in Figure 1.8.

1.3.2 Classification of Carbide Tools

Cemented carbide products are classified into three major grades:

Wear Grades: Used primarily in dies, machine and tool guides, and such everyday items as the line guides on fishing rods and reels—anywhere good wear resistance is required.

Impact Grades: Also used for dies, particularly for stamping and forming, and in tools such as mining drill heads.

Cutting-Tool Grades: The cutting-tool grades of cemented carbides are divided into two groups depending on their primary application. If the carbide is intended for use on cast iron, which is a nonductile material, it is graded as a cast-iron carbide. If it is to be used to cut steel, a ductile material, it is graded as a steel-grade carbide.

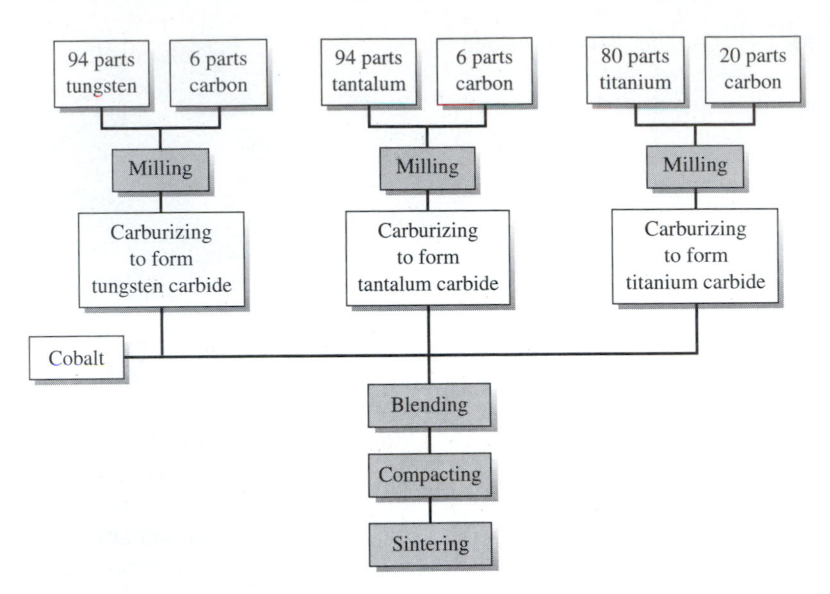

FIGURE 1.8 Schematic diagram of the cemeted tungsten carbide manufacturing process.

Cast-iron carbides must be more resistant to abrasive wear. Steel carbides require more resistance to **cratering** and heat. The tool-wear characteristics of various metals are different, thereby requiring different tool properties. The high abrasiveness of cast iron causes mainly edge wear to the tool. The long **chip** of steel, which flows across the tool at normally higher cutting speeds, causes mainly cratering and heat **deformation** to the tool. Tool-wear characteristics and chip formation will be discussed in Chapter 2.

It is important to choose and use the correct carbide grade for each job application. Several factors make one carbide grade different from another and therefore more suitable for a specific application. The carbide grades may appear to be similar, but the difference between the right and wrong carbide for the job can mean the difference between success and failure.

Figure 1.8 illustrates how carbide is manufactured, using pure tungsten carbide with a cobalt binder. The pure tungsten carbide makes up the basic carbide tool and is often used as such, particularly when machining cast iron. This is because pure tungsten carbide is extremely hard and offers the best resistance to abrasive wear.

Large amounts of tungsten carbide are present in all of the grades in the two cutting groups, and cobalt is always used as the binder. The more common alloying additions to the basic tungsten/cobalt material are tantalum carbide and titanium carbide.

Some of these alloys may be present in cast-iron grades of cutting tools, but they are primarily added to steel grades. Pure tungsten carbide is the most abrasive resistant and will work most effectively with the abrasive nature of cast iron. The addition of the alloying materials such as tantalum carbide and titanium carbide offers many benefits:

- The most significant benefit of **titanium carbide** is that it reduces cratering of the tool by reducing the tendency of the long steel chips to erode the surface of the tool.
- The most significant contribution of **tantalum carbide** is that it increases the hot hardness of the tool, which in turn reduces thermal deformation.

Varying the amount of cobalt binder in the tool material largely affects both the cast-iron and steel grades in three ways. Cobalt is far more sensitive to heat than the carbide around it. Cobalt is also more sensitive to abrasion and chip welding. Therefore, the more cobalt present, the softer the tool is, making it more sensitive to heat deformation, abrasive wear, and chip welding and leaching, which causes cratering. On the other hand, cobalt is stronger than carbide. Therefore, more cobalt improves the tool strength and resistance to shock. The strength of a carbide tool is expressed in terms of **transverse rupture strength** (TRS). Figure 1.9 shows how transverse rupture strength is measured.

The third difference between the cast-iron and steel-grade cutting tools is carbide grain size. The carbide grain size is controlled by the ball mill process. There are some exceptions, such as micrograin carbides, but generally the smaller the carbide grains, the harder the tool. Conversely, the larger the carbide grain, the stronger the tool. Carbide grain sizes at 1500× magnification are shown in Figures 1.10 and 1.11.

In the C- classification method (Fig. 1.12), grades C-1 through C-4 are for cast iron and grades C-5 through C-8 for steel. The higher the C- number in each group, the harder the grade, the lower the C- number, the stronger the grade. The harder grades are used for finish-cut applications; the stronger grades are used for rough-cut applications.

Many manufacturers produce and distribute charts showing a comparison of their carbide grades with those of other manufacturers. These are not equivalency charts, even though they may imply that one manufacturer's carbide is equivalent to that of another manufacturer. Each manufacturer knows his carbide best, and only the manufacturer of that specific carbide can

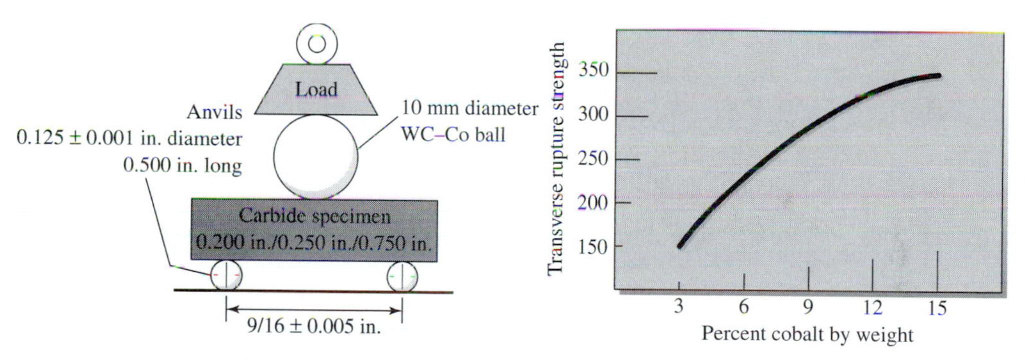

FIGURE 1.9 The method used to measure Transverse Rupture Strength (TRS) is shown as well as the relationship of TRS to cobalt (Co) content.

FIGURE 1.10 Carbide grain size (0.8 micron WC @ 1500×) consisting of 90% WC and 10% Co.

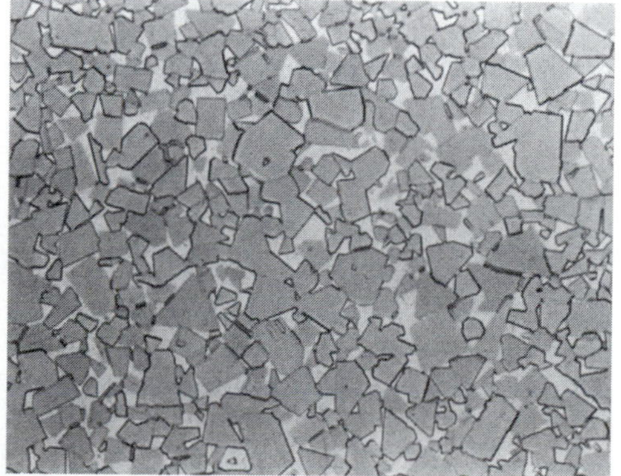

FIGURE 1.11 Carbide grain size (7 microns WC @ 1500×) consisting of 90% WC and 10% Co.

Classification Number	Materials to be Machined	Machining Operation	Type of Carbide	Characteristics Of		Typical Properties	
				Cut	Carbide	Hardness H-Ra	Transverse Rupture Strength (MPa)
C-1	Cast iron, nonferrous metals, and nonmetallic materials requiring abrasion resistance	Roughing cuts	Wear-resistant grades; generally straight WC–Co with varying grain sizes	Increasing cutting speed	Increasing hardness and wear resistance	89.0	2,400
C-2		General purpose				92.0	1,725
C-3		Finishing				92.5	1,400
C-4		Precision boring and fine finishing		Increasing feed rate	Increasing strength and binder content	93.5	1,200
C-5	Steels and steel-alloys requiring crater and deformation resistance	Roughing cuts	Crater-resistant grades; various WWC–Co compositions with TIC and/or TaC alloys	Increasing cutting speed	Increasing hardness and wear resistance	91.0	2,070
C-6		General purpose				92.0	1,725
C-7		Finishing				93.0	1,380
C-8		Precision boring and fine finishing		Increasing feed rate	Increasing strength and binder content	94.0	1,035

FIGURE 1.12 Classification, application, characteristics, and typical properties of metal-cutting carbide grades.

accurately place that carbide on the C- chart. Many manufacturers, especially those outside the United States, do not use the C- classification system for carbides. The placement of these carbides on a C- chart by a competing company is based upon similarity of application and is at best an educated guess. Tests have shown a marked difference in performance among carbide grades that manufacturers using the C- classification system have listed in the same category.

1.3.3 Coated-Carbide Tools

Coated carbides have been in existence since the late 1960s, but they did not reach their full potential until the mid-1970s. The first coated carbides were nothing more than standard carbide grades that were subjected to a coating process. As the manufacturers gained experience in producing coated carbides, they began to realize that the coating was only as good as the base carbide under the coating (known as the substrate).

It is advisable to consider coated carbides for most applications. When the proper coated carbide with the right edge preparation is used in the right application, it will generally outperform any uncoated grade. The microstructure of a coated carbide insert at 1500× magnification is shown in Figure 1.13.

Numerous types of coating materials are used, each for a specific application. It is important to observe the do's and don't's in the application of coated carbides. The most common coating materials are

- Titanium carbide
- Titanium nitride

FIGURE 1.13 Microstructure of a coated carbide insert at 1500× magnification. (Courtesy Kennametal Inc.)

- Ceramic coating
- Diamond coating
- Titanium carbonitride

In addition, multilayered combinations of these coating materials are used. The microstructure of a multilayered coated carbide insert at 1500× magnification is shown in Figure 1.14.

In general the coating process is accomplished by Chemical Vapor Deposition (CVD). The substrate is placed in an environmentally controlled chamber having an elevated temperature. The coating material is then introduced into the chamber as a chemical vapor. The coating material is drawn to and deposited on the substrate by a magnetic field around the substrate. It takes many hours in the chamber to achieve a coating of .0002 to .0003 in. on the substrate. Another process is Physical Vapor Deposition (PVD).

Titanium Carbide Coating: Of all the coatings, titanium carbide is the most widely used. Titanium carbide is used on many different substrate materials for cutting various materials under varying conditions. Titanium carbide coatings allow the use of higher cutting speeds because of their greater resistance to abrasive wear and cratering, and higher heat resistance.

FIGURE 1.14 Microstructure of a multilayered coated carbide insert at 1500× magnification. (Courtesy Kennametal Inc.)

Titanium Nitride Coating—Gold Color: Titanium nitride is used on many different substrate materials. The primary advantage of titanium nitride is its resistance to cratering. Titanium nitride also offers some increased abrasive wear resistance and a significant increase in heat resistance, permitting higher cutting speeds. It is also said that titanium nitride is more slippery, allowing chips to pass over it at the cutting interface with less friction.

Ceramic Coating—Black Color: Because **aluminum oxide** (ceramic) is extremely hard and brittle, it is not optimal for interrupted cuts, scaly cuts, and hard spots in the workpiece. This is not to say that it will never work under these conditions, but it may be more subject to failure by chipping. Even with these limitations, aluminum oxide is probably the greatest contributor to the coated carbides. Aluminum oxide ceramic allows much higher cutting speeds than other coated carbides because of its outstanding resistance to abrasive wear and its resistance to heat and chemical interaction.

Diamond Coating: A recent development concerns the use of diamond polycrystalline as coating for tungsten carbide cutting tools. Problems exist regarding adherence of the diamond film to the substrate and the difference in thermal expansion between diamond and substrate materials. Thin-film diamond-coated inserts are now available using either PVD (Physical Vapor Deposition) or CVD (Chemical Vapor Deposition) coating methods. Diamond-coated tools are effective in machining abrasive materials, such as aluminum alloys containing silicon, fiber-reinforced materials, and graphite. Improvements in tool life of as much as tenfold have been obtained over other coated tools.

Titanium Carbonitride—Black Color Multilayered Coatings: Titanium carbonitride normally appears as the intermediate layer of two or three phase coatings. The role of titanium carbonitride is one of neutrality, helping the other coating layers to bond into a sandwich-like structure (Fig. 1.14). Other multilayer coating combinations are being developed to effectively machine stainless steels and aerospace alloys. Chromium-based coatings such as chromium carbide have been found to be effective in machining softer metals such as aluminum, copper, and titanium.

There are a few important points to remember about using coated carbides. Coated carbides will not always outperform uncoated grades, but because of the benefits offered by coated carbides, they should always be a first consideration when selecting cutting tools.

When comparing cost between coated and uncoated carbides there will be little difference when the benefits of coated carbides are considered. Because coated carbides are more resistant to abrasive wear, cratering, and heat, and because they are more resistant to work-material buildup at lower cutting speeds, tool life is extended, reducing tool replacement costs. Coated carbides permit operation at higher speeds, reducing production costs.

All coated carbides have an edge **hone** to prevent coating buildup during the coating process. This is because the coating will generally seek sharp edges. The edge hone is usually very slight and actually extends tool life. However, a coated insert should *never* be reground or honed. If a special edge preparation is required, the coated carbides must be ordered that way. The only time the edge hone may be of any disadvantage is when making a very light finishing cut. Carbide insert edge preparations will be discussed in Chapter 2.

1.4 CERAMIC AND CERMET TOOLS

Ceramic aluminum oxide ($A1_2O_3$) material for cutting tools was first developed in Germany sometime around 1940. Although ceramics were slow to develop as tool materials, advancements made since the mid-1970s have greatly improved their usefulness. **Cermets** are basically a combination of ceramic and titanium carbide. The word "cermet" is derived from the words "ceramic" and "metal."

Ceramic Cutting Tools Ceramics are nonmetallic materials. This puts them in an entirely different category than HSS and carbide tool materials. The use of ceramics as a cutting-tool materials has distinct advantages and disadvantages. The application of ceramic cutting tools is limited because of their extreme brittleness. The Transverse Rupture Strength (TRS) is very low. This means that they will fracture more easily when making heavy or interrupted cuts. However, the strength of ceramics under compression is much higher than that of HSS and carbide tools.

There are two basic types of ceramic material; hot pressed and cold pressed. In hot-pressed ceramics, which are usually black or gray in color, the aluminum oxide grains are pressed together under extremely high pressure and at a very high temperature to form a billet. The billet is then cut to insert size. With cold-pressed ceramics, which are usually white in color, the aluminum oxide grains are pressed together, again under extremely high pressure but at a lower temperature. The billets are then sintered to achieve bonding. This procedure is similar to carbide manufacture, except no metallic binder material is used. Although both hot- and cold-pressed ceramics are similar in hardness, the cold-pressed ceramic is slightly harder. The hot-pressed ceramic has greater transverse rupture strength. Various shapes of both hot- and cold-pressed ceramic inserts are shown in Figure 1.15.

The brittleness, or relative strength, of ceramic materials is their greatest disadvantage when they are compared to HSS or carbide tools. Proper tool geometry and edge preparation play an important role in the application of ceramic tools and help to overcome their weakness. Some of the advantages of ceramic tools are

- High strength for light cuts on very hard work materials
- Extremely high resistance to abrasive wear and cratering
- Capability of running at speeds in excess of 2000 surface feet per minute. (SFPM)
- Extremely high hot hardness
- Low thermal conductivity

Ceramics may not be the all-around tool for the average shop, but they can be useful in certain applications. Ceramic tools have been alloyed with zirconium (about 15%) to increase their strength.

Many ceramic-tool manufacturers are recommending the use of ceramic tools for both rough-cutting and finishing operations. Practical shop experience indicates that these recommendations are somewhat optimistic. To use ceramic tools successfully, insert shape, work-material

FIGURE 1.15 Various sizes and shapes of hot- and cold-pressed ceramic inserts. (Courtesy Greenleaf Corp.)

condition, machine-tool capability, setup, and general machining conditions must all be correct. High rigidity of the machine tool and setup is also important for the application of ceramic tools. Ceramics are being developed to have greater strength (higher TRS). Some manufacturers are offering ceramic inserts with positive geometry and even formed chipbreaker grooves.

Cermet Cutting Tools The manufacturing process for cermets is similar to the process used for hot-pressed ceramics. The materials, approximately 70% ceramic and 30% titanium carbide, are pressed into billets under extremely high pressure and temperature. After sintering, the billets are sliced to the desired tool shapes. Subsequent grinding operations for final size and edge preparation complete the manufacturing process.

The strength of cermets is greater than that of hot-pressed ceramics. Therefore, cermets perform better on interrupted cuts. However, when compared to solid ceramics, the presence of the 30% titanium carbide in cermets decreases the hot hardness and resistance to abrasive wear. The hot hardness and resistance to abrasive wear of cermets are high when compared to HSS and carbide tools. The greater strength of cermets allows them to be available in a significantly larger selection of geometries and to be used in standard insert holders for a greater variety of applications. The geometries include many positive/negative, and chipbreaker configurations.

Silicon Nitride–Base Ceramics Developed in the 1970s, silicon nitride (SIN)–base ceramic tool materials consist of silicon nitride with various additions of aluminum oxide, yttrium oxide, and titanium carbide. These tools have high toughness, hot hardness and good thermal shock resistance. Sialon for example is recommended for machining cast irons and nickel base superalloys at intermediate cutting speeds.

1.5 DIAMOND, CBN, AND WHISKER-REINFORCED TOOLS

The materials described here are not commonly found in a heavy metalworking environment. They are most commonly used in high-speed automatic production systems for light finishing of precision surfaces. To complete the inventory of tool materials, it is important to note the characteristics and general applications of these specialty materials.

Diamond: The two types of **diamonds** being used as cutting tools are industrial-grade natural diamonds and synthetic polycrystalline diamonds. Because diamonds are pure carbon, they have an affinity for the carbon of **ferrous** metals. Therefore, they can only be used on **nonferrous** metals.

Some diamond cutting tools are made of a diamond crystal compaction (many small crystals pressed together) bonded to a carbide base (Fig. 1.16). These diamond cutting tools should only be used for light finishing cuts of precision surfaces. **Feeds** should be very light, and **speeds** are usually in excess of 5000 SFPM. Rigidity in the machine tool and the setup is very critical because of the extreme hardness and brittleness of diamond.

Cubic Boron Nitride: Cubic boron nitride (CBN) is similar to diamond in its polycrystalline structure and is also bonded to a carbide base. With the exception of titanium or titanium-alloyed materials, CBN will work effectively as a cutting tool on most common work materials. However, the use of CBN should be reserved for very hard and difficult-to-machine materials. CBN will run at lower speeds, around 600 SFPM, and will take heavier cuts with higher **lead angles** than diamond. Still, CBN should mainly be considered as a finishing tool material because of its extreme hardness and brittleness. Machine tool and setup rigidity for CBN, as with diamond, is critical.

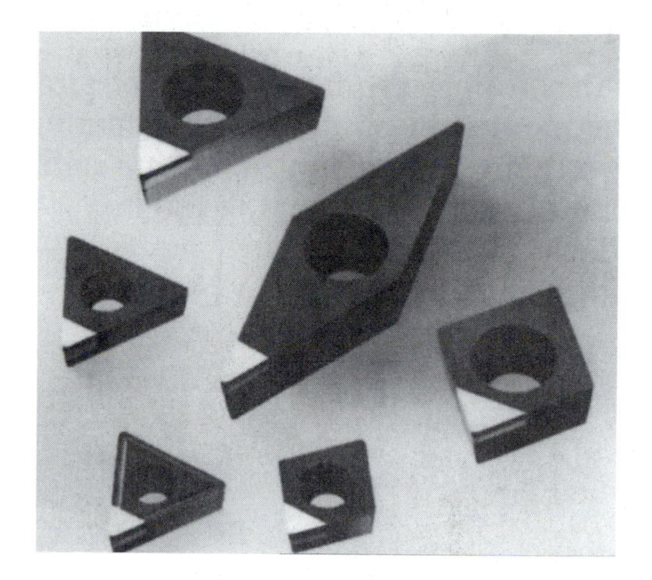

FIGURE 1.16 Polycrystalline diamond material bonded to a carbide base of various insert sizes and shapes. (Courtesy Sandvik Coromant Co.)

Whisker-Reinforced Materials: To further improve the performance and wear resistance of cutting tools to machine new work materials and composites, whisker-reinforced composite cutting-tool materials have been developed. Whisker-reinforced materials include silicon nitride–base tools and aluminum oxide–base tools, reinforced with silicon carbide (SiC) whiskers. Such tools are effective in machining composites and nonferrous materials, but are not suitable for machining irons and steels.

1.6 REVIEW QUESTIONS AND PROBLEMS

1. Name and describe the three cutting-tool characteristics.
2. What are the limitations of plain carbon tool steel?
3. What are the differences between plain carbon tool steel and high-speed steel (HSS)?
4. Explain the differences between M- and T-series high-speed steel.
5. Explain the composition and limitations of cast alloys.
6. What are the two major ingredients of basic tungsten carbide?
7. Name and describe the major steps in the manufacture of carbide products.
8. Name and describe the three grades of carbide tools.
9. Explain the benefits of titanium carbide and tantalum carbide.
10. Explain Transverse Rupture Strength (TRS).
11. What is the C- method of carbide classification?
12. Explain the carbide coating process.
13. What are some problems associated with diamond coating?
14. Discuss advantages and disadvantages of ceramic cutting tools.
15. Discuss the differences between hot- and cold-pressed ceramic inserts.
16. What are the advantages and disadvantages of Cermet cutting tools?
17. Why can diamond inserts not be used to machine steel parts?
18. What are the most common applications for CBN cutting tools?

CHAPTER TWO

METAL REMOVAL METHODS

CHAPTER OVERVIEW

- Introduction
- Cutting-Tool Forces
- Chip Formation and Tool Wear
  - Chip Formation
  - Cutting-Tool Wear
- Single-Point Cutting Tools
  - Cutting-Tool Geometry
  - Edge Preparation
  - Chip Breakers
- Indexable-Type Tooling
  - Indexable-Insert Shapes
  - Indexable Inserts—Classes and Sizes
  - Indexable-Insert Identification System
  - Mechanical Toolholders
- Review Questions and Problems

2.1 INTRODUCTION

The process of metal removal, a process in which a wedge-shaped tool engages a workpiece to remove a layer of material in the form of a chip, goes back many years. Even with all of the sophisticated equipment and techniques used in today's modern industry, the basic mechanics of forming a chip remain the same. As the cutting tool engages the workpiece, the material directly ahead of the tool is sheared and deformed under tremendous pressure. The deformed material then seeks to relieve its stressed condition by fracturing and flowing into the space above the tool in the form of a chip. A turning toolholder generating a chip is shown in Figure 2.1.

2.2 CUTTING-TOOL FORCES

The deformation of a work material means that enough force has been exerted by the tool to permanently reshape or fracture the work material. If a material is reshaped, it is said to have exceeded its plastic limit. A chip is a combination of reshaping and fracturing. The deformed

FIGURE 2.1 A turning toolholder insert generating a chip. (Courtesy Kennametal Inc.)

chip is separated from the parent material by fracture. The cutting action and the chip formation can be more easily analyzed if the edge of the tool is set perpendicular to the relative motion of the material, as shown in Figure 2.2. Here the undeformed chip thickness t_1 is the value of the depth of cut and t_2 is the thickness of the deformed chip after leaving the workpiece. The major deformation starts at the shear zone, and diameter determines the angle of shear.

A general discussion of the forces acting in metal cutting is presented by using the example of a typical turning operation. When a solid bar is turned, there are three forces acting on the cutting tool (Fig. 2.3):

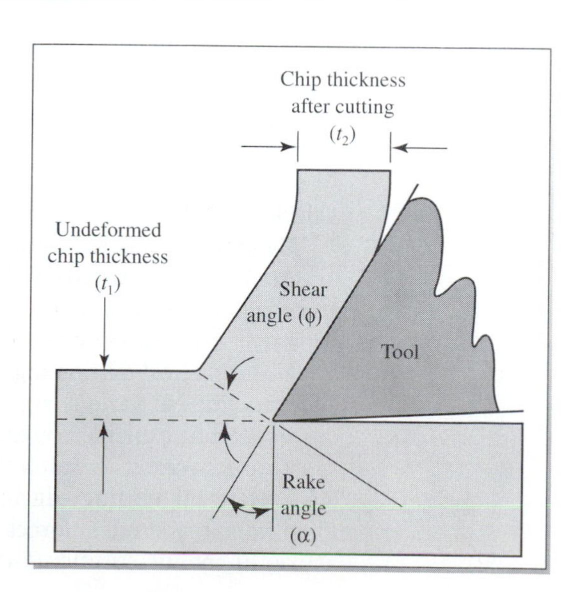

FIGURE 2.2 Chip formation showing the deformation of the material being machined.

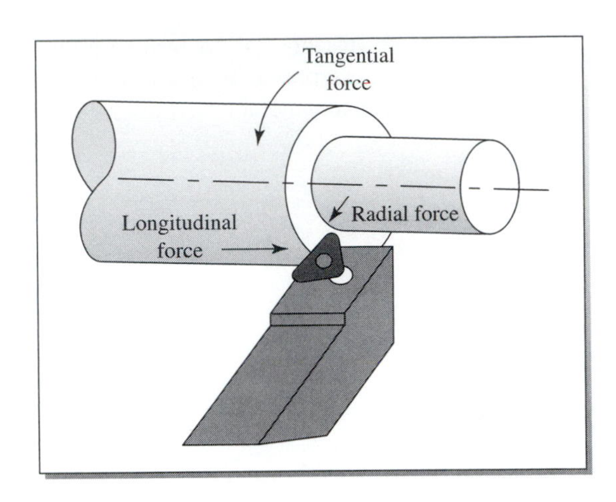

FIGURE 2.3 Typical turning operation showing the forces acting on the cutting tool.

Tangential Force: This acts in a direction tangential to the revolving workpiece and represents the resistance to the rotation of the workpiece. In a normal operation, tangential force is the highest of the three forces and accounts for 99% of the total power required by the operation.

Longitudinal Force: Longitudinal force acts in the direction parallel to the axis of the work and represents the resistance to the longitudinal feed of the tool. Longitudinal force is usually about 50% as great as tangential force. Since feed velocity is usually very low in relation to the velocity of the rotating workpiece, longitudinal force accounts for only about 1% of total power required.

Radial Force: Radial force acts in a radial direction from the centerline of the workpiece. The radial force is generally the smallest of the three, often about 50% as large as longitudinal force. Its effect on power requirements is very small because velocity in the radial direction is negligible.

2.3 CHIP FORMATION AND TOOL WEAR

Regardless of the tool being used or the metal being cut, the chip-forming process occurs by a mechanism called plastic deformation. This deformation can be visualized as shearing; that is, when a metal is subjected to a load exceeding its elastic limit, the crystals of the metal elongate through an action of slipping, or shearing, which takes place within the crystals and between adjacent crystals. This action, shown in Figure 2.4, is similar to the action that takes place when a deck of cards is given a push and sliding, or shearing, occurs between the individual cards.

Metals are composed of many crystals, and each crystal in turn is composed of atoms arranged into some definite pattern. Without getting into a complicated discussion on the atomic makeup and characteristics of metals, it should be noted that the slipping of the crystals takes place along a plane of greatest ionic density.

Most practical cutting operations, such as **turning** and **milling**, involve two or more cutting edges inclined at various angles to the direction of the cut. However, the basic mechanism of cutting can be explained by analyzing cutting done with a single cutting edge.

Chip formation is simplest when a continuous chip is formed in orthogonal cutting (Fig. 2.5a). Here the cutting edge of the tool is perpendicular to the line of tool travel; tangential, longitudinal, and radial forces are in the same plane; and only a single straight cutting edge is active. In oblique cutting, (Fig. 2.5b), a single straight cutting edge is inclined in the direction

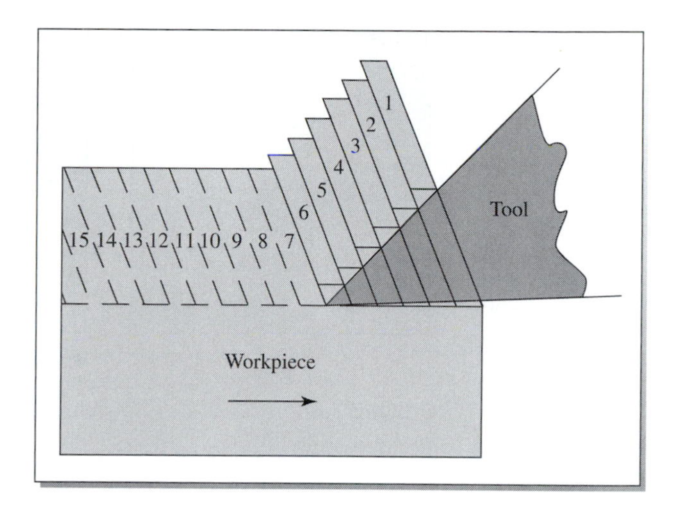

FIGURE 2.4 Chip formation compared to a sliding deck of cards.

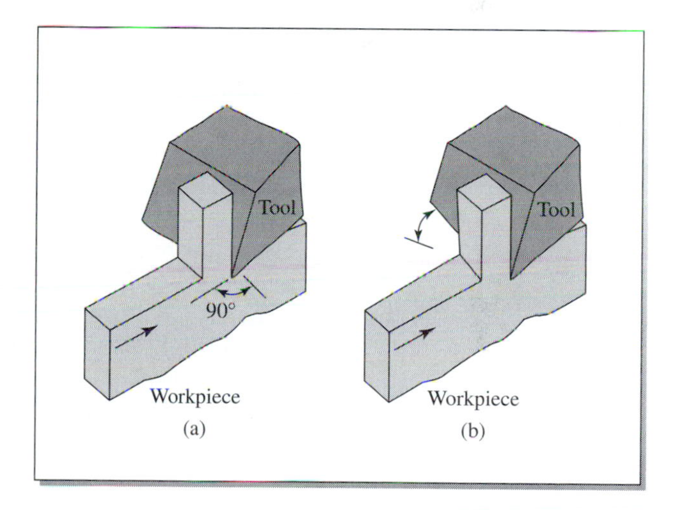

FIGURE 2.5 Chip formation showing both (a) orthogonal cutting and (b) oblique cutting.

of tool travel. This inclination causes changes in the direction of chip flow up the face of the tool. When the cutting edge is inclined, the chip flows across the tool face with a sideways movement that produces a helical form of chip.

2.3.1 Chip Formation

Metal cutting chips have been classified into three basic types:

- Discontinuous or segmented
- Continuous
- Continuous with a built-up edge

All three types of chips are shown in Figure 2.6.

Discontinuous Chip—Type 1 Discontinuous or segmented chips are produced when brittle metal such as cast iron and hard bronze are cut or when some ductile metals are cut under poor cutting conditions. As the point of the cutting tool contacts the metal, some compression occurs and the chip begins flowing along the chip–tool interface. As more stress is applied to brittle

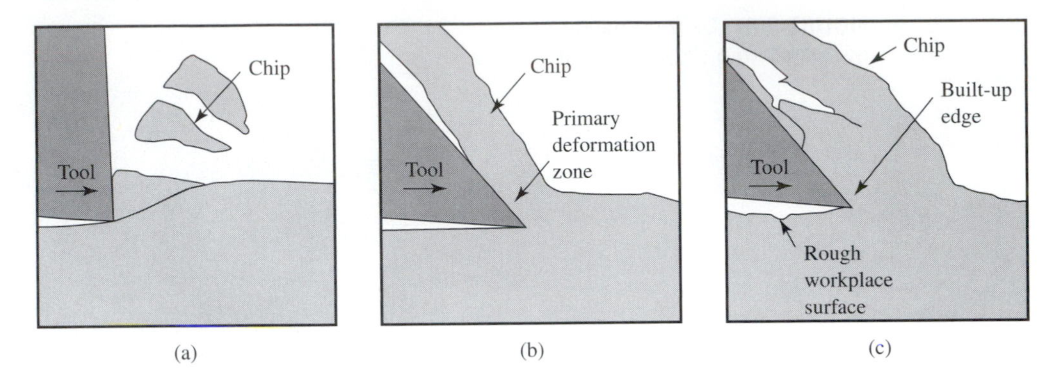

FIGURE 2.6 Types of chip formations: (a) discontinuous, (b) continuous, (c) continuous with built-up edge (BUE).

metal by the cutting action, the metal compresses until it reaches a point where rupture occurs and the chip separates from the unmachined portion. This cycle is repeated indefinitely during the cutting operation, with the rupture of each segment occurring on the **shear angle** or plane. Generally, as a result of these successive ruptures, a poor surface is produced on the workpiece.

Continuous Chip—Type 2 The Type 2 chip is a continuous ribbon produced when the flow of metal next to the tool face is not greatly restricted by a built-up edge or friction at the chip–tool interface. The continuous-ribbon chip is considered ideal for efficient cutting action because it results in better finishes.

Unlike the Type 1 chip, fractures or ruptures do not occur here, because of the ductile nature of the metal. The crystal structure of the ductile metal is elongated when it is compressed by the action of the cutting tool and as the chip separates from the metal. The process of chip formation occurs in a single plane, extending from the cutting tool to the unmachined work surface. The area where plastic deformation of the crystal structure and shear occurs is called the shear zone. The angle on which the chip separates from the metal is called the shear angle, as shown in Figure 2.2.

Continuous Chip with a Built-Up Edge (BUE)—Type 3 The metal ahead of the cutting tool is compressed and forms a chip, which begins to flow along the chip–tool interface. As a result of the high temperature, the high pressure, and the high frictional resistance against the flow of the chip along the chip–tool interface, small particles of metal begin adhering to the edge of the cutting tool while the chip shears away. As the cutting process continues, more particles adhere to the cutting tool and a larger buildup results, which affects the cutting action. The **built-up edge** increases in size and becomes more unstable. Eventually a point is reached where fragments are torn off. Portions of these fragments that break off stick to both the chip and the workpiece. The buildup and breakdown of the built-up edge occur rapidly during a cutting action and cover the machined surface with a multitude of built-up fragments. These fragments adhere to and score the machined surface, resulting in a poor surface finish.

Shear Angle Certain characteristics of continuous chips are determined by the shear angle. The shear angle is the plane where slip occurs, to begin chip formation (Fig. 2.2). In Figure 2.7 the distortion of the work material grains in the chip, as compared to the parent material, is visible. Each fracture line in the chip as it moves upward over the tool surface can be seen, as well as the distorted surface grains where the tool has already passed. In certain work materials, these distorted surface grains account for **work hardening**.

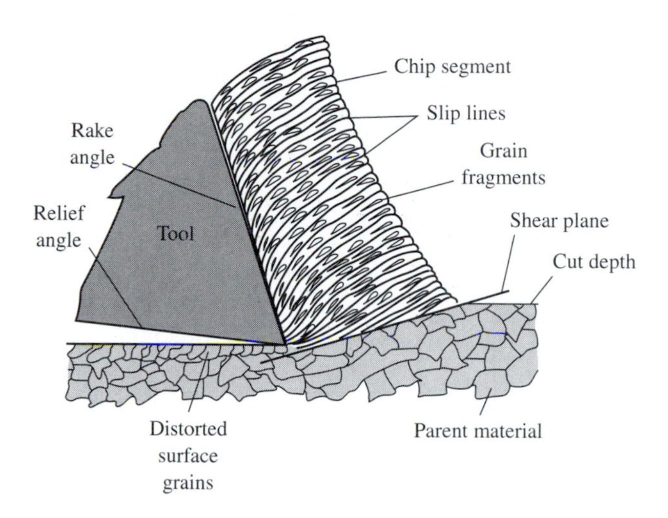

FIGURE 2.7 Distribution of work material during chip formation.

Regardless of the shear angle, the compressive deformation caused by the tool force against the chip will cause the chip to be thicker and shorter than the layer of workpiece material removed. The work or energy required to deform the material usually accounts for the largest portion of forces and power involved in a metal removing operation. For a layer of work material of given dimensions, the thicker the chip, the greater the force required to produce it.

Heat in Metal Cutting The mechanical energy consumed in the cutting area is converted into heat. The main sources of heat are the shear zone (the interface between the tool and the chip where the friction force generates heat) and the lower portion of the tool tip, which rubs against the machined surface. The interaction of these heat sources, combined with the geometry of the cutting area, results in a complex temperature distribution, as shown in Figure 2.8.

The temperature generated in the **shear plane** is a function of the shear energy and the specific heat of the material. Temperature increase on the tool face depends on the friction conditions at the interface. A low coefficient of friction is, of course, desirable. Temperature distribution will be a function of, among other factors, the thermal conductivities of the workpiece and the tool materials, the specific heat, the cutting speed, the depth of cut, and the use of a cutting

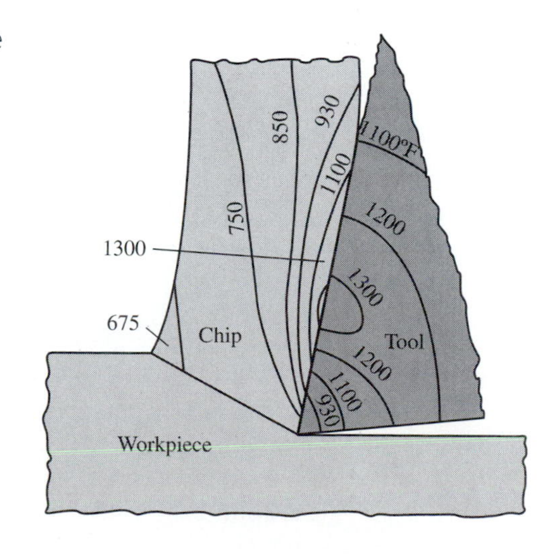

FIGURE 2.8 Typical temperature distribution in the cutting zone.

fluid. As cutting speed increases, there is little time for the heat to be dissipated away from the cutting area, so the proportion of the heat carried away by the chip increases. In Chapter 3, "Machinability of Metals," this topic is discussed in more detail.

2.3.2 Cutting-Tool Wear

Cutting-tool life is one of the most important economic considerations in metal cutting. In roughing operations, the tool material, the various tool angles, cutting speeds, and feedrates are usually chosen to give an economical tool life. Conditions giving a very short tool life will not be economical because tool-grinding, indexing, and tool-replacement costs will be high. On the other hand, the use of very low speeds and feeds to give long tool life will not be economical because of the low production rate. Clearly, any tool or work-material improvements that increase tool life without causing unacceptable drops in production will be beneficial. In order to form a basis for such improvements, efforts have been made to understand the behavior of the tool, how it physically wears, the wear mechanisms, and forms of tool failure.

While the tool is engaged in the cutting operation, wear may develop in one or more areas on and near the cutting edge:

Crater Wear: Typically, cratering occurs on the top face of the tool. It is essentially the erosion of an area parallel to the cutting edge. This erosion process takes place as the chip being cut rubs the top face of the tool. Under very high-speed cutting conditions and when machining tough materials, crater wear can be the factor that determines the life of the tool. Typical crater wear patterns are shown in Figures 2.9a and 2.10a. However, when tools are used under economical conditions, the edge wear and not the crater wear is more commonly the controlling factor in the life of the tool.

Edge Wear: Edge wear occurs on the clearance face of the tool and is mainly caused by the rubbing of the newly machined workpiece surface on the contact area of the tool edge. This type of wear occurs on all tools while cutting any type of work material. Edge wear begins along the lead cutting edge and generally moves downward, away from the cutting edge. Typical edge wear patterns are shown in Figures 2.9b and 2.10b. The edge wear is also commonly known as the **wearland**.

Nose Wear: Usually observed after a considerable cutting time, nose wear appears when the tool has already exhibited land and/or crater wear. Wear on the nose of the cutting edge usually affects the quality of the surface finish on the workpiece.

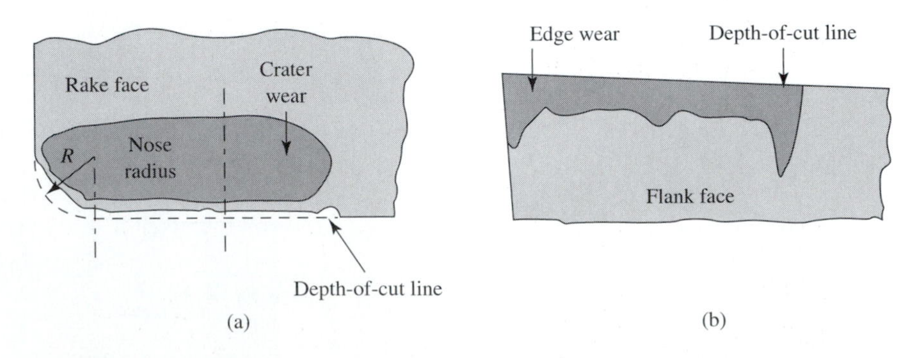

FIGURE 2.9 Carbide insert wear patterns: (a) crater wear, (b) edge wear.

(a) (b)

FIGURE 2.10 Carbide insert wear patterns: (a) crater wear, (b) edge wear. (Courtesy Kennametal Inc.)

Cutting-tool materials in general, and carbide tools in particular, exhibit different types of wear and/or failure:

Plastic Deformation: Edge depression and body bulging appear, due to excessive heat. The tool loses strength and consequently flows plastically.

Mechanical Breakage: Excessive force may cause immediate failure. Alternatively, the mechanical failure (**chipping**) may result from a fatigue-type failure. Thermal shock also causes mechanical failure.

Gradual Wear: The tool assumes a form of stability wear due to interaction between tool and work, resulting in crater wear. Four basic wear mechanisms affecting tool material have been categorized:

Abrasion: Because hard inclusions in the workpiece microstructure plow into the tool face and flank surfaces, abrasion wear predominates at relatively low cutting temperatures. The **abrasion resistance** of a tool material is proportional to its hardness.

Adhesion: Caused by formation and subsequent destruction of minute welded junctions, adhesion wear is commonly observed as a built-up edge (BUE) on the top face of the tool. This BUE may eventually disengage from the tool, causing a craterlike wear. Adhesion can also occur when minute particles of the tool surface are instantaneously welded to the chip surface at the tool–chip interface and carried away with the chip.

Diffusion: Because of high temperatures and pressures in diffusion wear, microtransfer on an atomic scale takes place. The rate of diffusion increases exponentially with increases in temperature.

Oxidation: At elevated temperatures, the oxidation of the tool material can cause high tool-wear rates. The oxides that are formed are easily carried away, leading to increased wear.

The different wear mechanisms, as well as the different phenomena contributing to the attritious wear of the cutting tool, depend on the multitude of cutting conditions, and especially on the cutting speeds and cutting fluids.

Aside from the sudden premature breakage of the cutting edge (tool failure), there are several indicators of the progression of physical wear. The machine operator can observe these factors prior to total rupture of the edge. The indicators are

- Increase in the flank wear size above a predetermined value
- Increase in the crater depth, width, or other parameter of the crater, in the rake face

- Increase in the power consumption or cutting forces required to perform the cut
- Failure to maintain the dimensional quality of the machined part within a specified tolerance limit
- Significant increase in the surface roughness of the machined part
- Change in the chip formation due to increased crater wear or excessive heat generation

2.4 SINGLE-POINT CUTTING TOOLS

The metal-cutting tool separates chips from the workpiece in order to cut the part to the desired shape and size. There is a great variety of metal-cutting tools, each of which is designed to perform a particular job or a group of metal-cutting operations in an efficient manner. For example, a twist drill is designed to drill a hole of a particular size, whereas a turning tool might be used to turn a variety of cylindrical shapes.

2.4.1 Cutting-Tool Geometry

The shape and position of the tool relative to the workpiece have an important effect on metal cutting. The most important geometric elements, relative to chip formation, are the location of the cutting edge and the orientation of the tool face with respect to the workpiece and the direction of cut. Other shape considerations are concerned primarily with relief or clearance (e.g., taper applied to tool surfaces to prevent rubbing or dragging against the work).

Terminology used to designate the surfaces, angles, and radii of single-point tools is shown in Figure 2.11. The tool shown here is a brazed-tip type, but the same definitions apply to indexable tools.

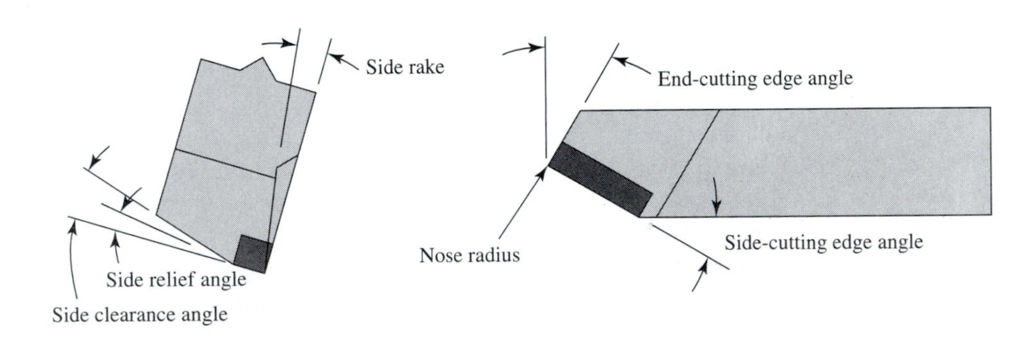

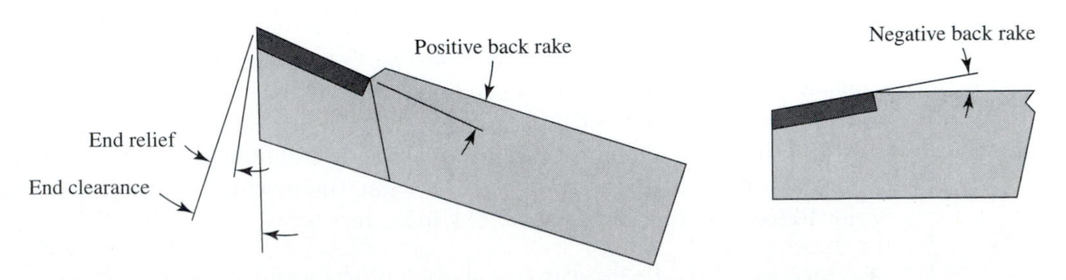

FIGURE 2.11 Terminology used to designate the surfaces, angles, and radii of single-point tools.

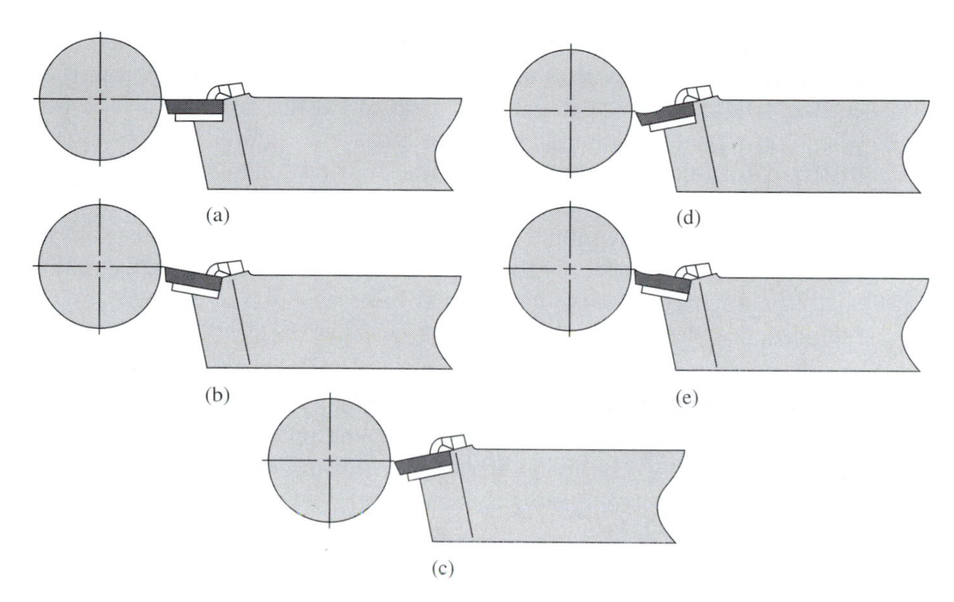

FIGURE 2.12 With the cutting tool on center, various back rake angles are shown: (a) neutral, (b) positive, (c) negative, (d) positive/negative, (e) double positive.

Rake Angle The basic tool geometry is determined by the rake angle of the tool as shown in Figure 2.12. The rake angle is always at the top side of the tool. With the tool tip at the centerline of the workpiece, the rake angle is determined by the angle of the tool as it goes away from the workpiece centerline location. The neutral, positive, and negative rakes are seen in Figure 2.12a, b, and c. The angle for these geometries is set by the position of the insert pocket in the tool holder. The positive/negative (d) and double positive (e) rake angles are set by a combination of the insert pocket in the tool holder and the insert shape itself.

There are two rake angles: back rake as shown in Figure 2.12 and side rake as shown in Figure 2.13. In most turning and **boring** operations, it is the **side rake** that is the most influential. This is because the side rake is in the direction of the cut.

Rake angle has two major effects during the metal-cutting process. One major effect of rake angle is its influence on tool strength. An insert with **negative rake** will withstand far more loading than an insert with **positive rake**. The cutting force and heat are absorbed by a greater mass of tool material, and the compressive strength of carbide is about two and one-half times greater than its transverse rupture strength.

The other major effect of rake angle is its influence on cutting pressure. An insert with a positive rake angle reduces cutting forces by allowing the chips to flow more freely across the rake surface.

Negative Rake. Negative-rake tools should be selected whenever workpiece and machine-tool stiffness and rigidity allow. Negative rake, because of its strength, offers great advantage during roughing, interrupted, scaly, and hard-spot cuts. Negative rake also offers more cutting edges for economy and often eliminates the need for a chip breaker. Negative rakes are recommended on insert grades that do not possess good toughness (low transverse rupture strength).

However, negative rake is not without some disadvantages. Negative rake requires more horsepower and maximum machine rigidity. It is more difficult to achieve good surface finishes with negative rake. Negative rake forces the chip into the workpiece, generates more heat into the tool and workpiece, and is generally limited to boring on larger diameters because of chip jamming.

Positive Rake. Positive-rake tools should be selected only when negative-rake tools can't get the job done. Some areas of cutting where positive rake may prove more effective are when cutting tough, alloyed materials that tend to "work harden," such as certain stainless steels; when cutting soft or gummy metals; or when low rigidity of the workpiece, tooling, machine tool, or fixture allows chatter to occur. The shearing action and free cutting of positive-rake tools will often eliminate problems in these areas.

One exception that should be noted when experiencing **chatter** with a positive rake is that at times the preload effect of the higher cutting forces of a negative-rake tool will dampen out chatter in a marginal situation. This may be especially true during lighter cuts, when tooling is extended, or when the machine tool has excessive backlash.

Neutral Rake. Neutral-rake tools are seldom used or encountered. When a negative-rake insert is used in a neutral-rake position, the end relief (between tool and workpiece) is usually inadequate. On the other hand, when a positive insert is used at a neutral rake, the tip of the insert is less supported, making the insert extremely vulnerable to breakage.

Positive/Negative Rake. The positive/negative rake is generally applied using the same guidelines as a positive rake. The major advantages of a positive/negative insert are that it can be used in a negative holder, it offers greater strength than a positive rake, and it doubles the number of cutting edges when using a two-sided insert.

The positive/negative insert has a 10° positive rake. It is mounted in the normal 5° negative pocket, which gives it an effective 5° positive rake when cutting. The positive/negative rake still maintains a cutting attitude that keeps the carbide under compression and offers more mass for heat dissipation. The positive/negative insert also aids in chip breaking on many occasions, as it tends to curl the chip.

Double-Positive Rake. The double-positive insert is the weakest of all inserts. It is free cutting, and generally used only when delicate, light cuts are required that exert minimum force against the workpiece, as in the case of thin-wall tubing, for example. Other uses of double-positive inserts are for very soft or gummy work materials, such as low-carbon steel, and for boring small-diameter holes when maximum clearance is needed.

Side-Rake Angles. In addition to the **back-rake** angles there are side-rake angles as shown in Figure 2.13. These angles are normally determined by the tool manufacturers. Each manufacturer's tools may vary slightly, but usually an insert from one manufacturer can be used in the toolholder from another. The advantage of positive and negative geometry that was discussed for back rake applies to side rake. When back rake is positive, so is side rake, and when back rake is negative, so is side rake.

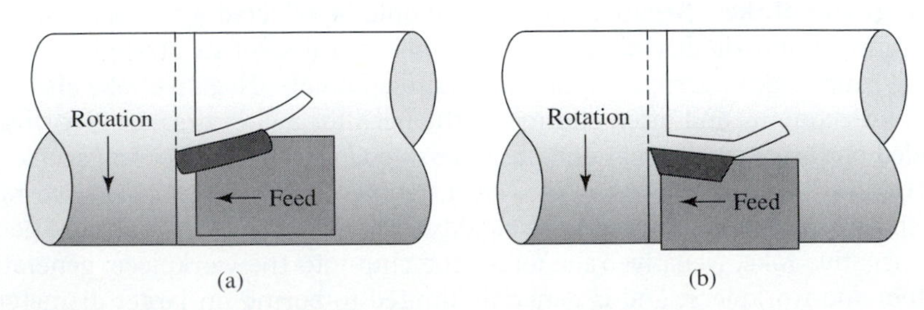

FIGURE 2.13 Side-rake-angle variations: (a) negative, (b) positive.

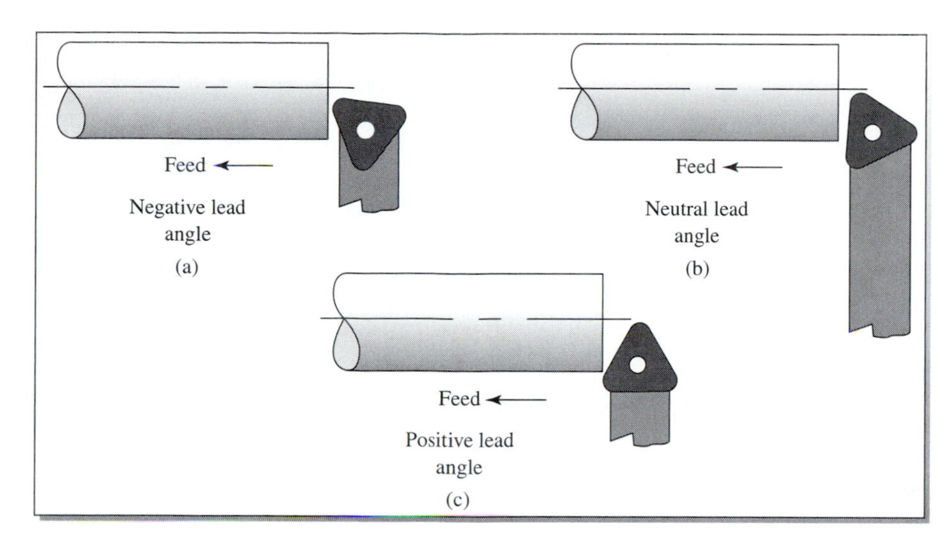

FIGURE 2.14 Lead-angle variations: (a) negative, (b) neutral, (c) positive.

Side and End Relief Angles. Relief angles are for the purpose of helping to eliminate tool breakage and to increase tool life. The included angle under the cutting edge must be made as large as practical. If the relief angle is too large, the cutting tool may chip or break. If the angle is too small, the tool will rub against the workpiece and generate excessive heat, and this will cause premature dulling of the cutting tool.

Small relief angles are essential when machining hard and strong materials, and they should be increased for the weaker and softer materials. A smaller angle should be used for interrupted cuts or heavy feeds, and a larger angle for semifinish and finish cuts.

Lead Angle

The lead angle (Fig. 2.14) is determined by the tool holder, which must be chosen for each particular job. The insert itself can be used in any appropriate holder for that particular insert shape, regardless of lead angle.

Lead angle is an important consideration when choosing a tool holder. A positive lead angle is the most commonly used and should be the choice for the majority of applications. A positive lead angle performs two main functions:

- It thins the chip.
- It protects the insert.

The undeformed chip thickness decreases when using a positive lead angle.

Positive lead angles vary, but the most common lead angles available on standard holders are 10, 15, 30, and 45°. As seen in Figure 2.15, the volume of chip material is about the same

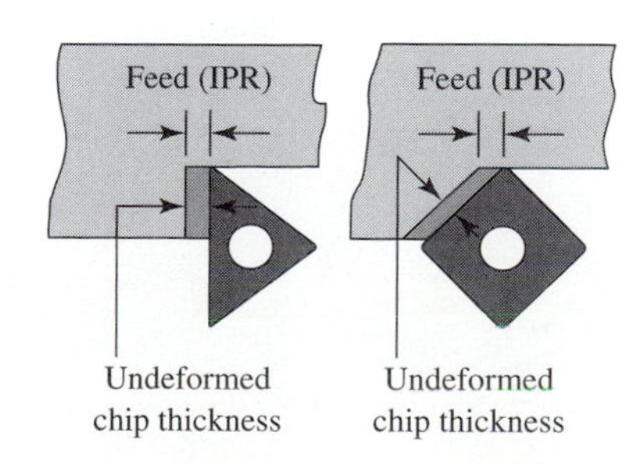

FIGURE 2.15 Lead angle vs. chip thickness. A positive lead angle thins the chip and protects the insert.

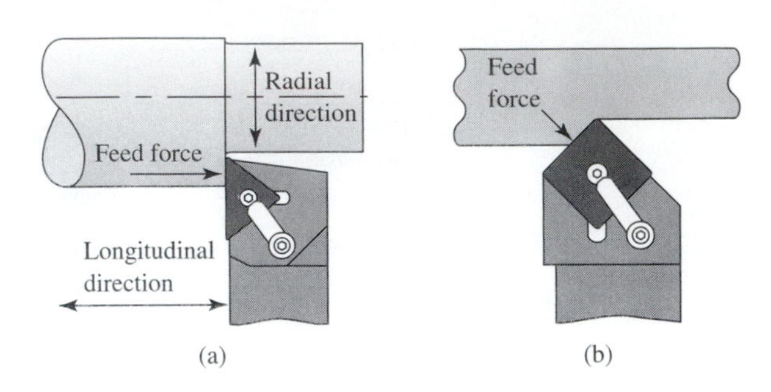

FIGURE 2.16 Lead angles and their effects on longitudinal and radial cutting-tool feed forces.

in each case but the positive lead angle distributes the cutting force over a greater area of the tool's edge. This allows a substantial increase in feedrate without reducing the tool life because of excessive loading. The greater the lead angle, the more the feedrate can be increased.

A positive lead angle also reduces the longitudinal force (direction of feed) on the workpiece, but it increases the radial force because the cutting force is always approximately perpendicular to the cutting edge (Fig. 2.16). This may become a problem when machining a workpiece that is not well supported. Care must be taken in cases where an end support, such as a tailstock center, is not used.

A heavy positive lead angle also has a tendency to induce chatter because of the greater tool contact area. This chatter is an amplification of tool or workpiece deflection resulting from the increased contact. In this situation it is appropriate to decrease the positive lead angle.

A positive lead angle protects the tool and promotes longer tool life. As shown in Figure 2.17, the tool comes in contact with the workpiece well away from the tool tip, which is the weakest point of the tool. As the tool progresses into the cut, the load against the tool gradually increases, rather than occurring as a sudden shock to the cutting edge. The positive lead angle also reduces the wear on the cutting edge caused by a layer of hardened material or scale, by thinning the layer and spreading it over a greater area. These advantages are extremely beneficial during interrupted cuts. Another way positive lead angle helps to extend tool life is by allowing intense heat buildup to dissipate more rapidly, because more of the tool is in contact with the workpiece.

Neutral- and negative-lead-angle tools also have some benefits. A neutral angle offers the least amount of tool contact, which will sometimes reduce the tendency to chatter and lowers longitudinal forces. This is important on less-stable workpieces or setups. Negative lead angles permit machining to a shoulder or a corner and are useful for **facing**. Cutting forces tend to pull the insert out of the seat, leading to erratic size control. Therefore, negative lead angles should be avoided if at all possible.

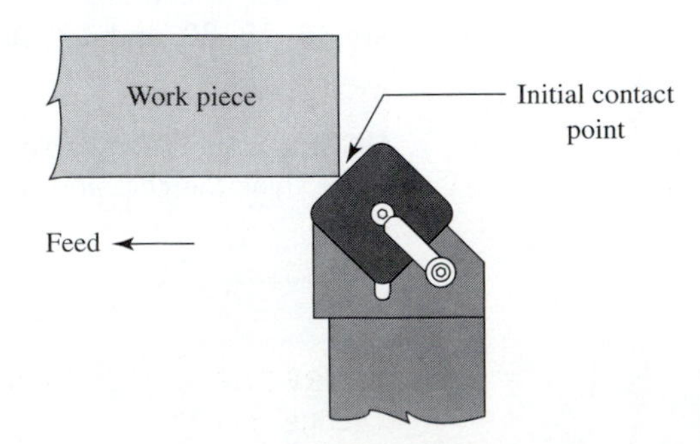

FIGURE 2.17 Gradual feed/workpiece contact, protects the cutting tool by slowly increasing the load.

2.4.2 Edge Preparation

Edge preparation is a step taken to prolong tool life or to enhance tool performance. There are four basic approaches to edge preparation:

- Edge hone
- Edge L land
- Edge chamfer
- Combinations of the above

Many inserts, including carbide, ceramic, and so on, are purchased with a standard edge preparation, normally an edge hone. The primary purpose of edge preparation is to increase the insert's resistance to chipping, breaking, and wear. Figure 2.18 illustrates the basic edge preparations.

Tool materials such as carbide and ceramic are very hard and brittle. Therefore, a sharp lead cutting edge on inserts made of these materials is extremely prone to chipping and breaking. Once a cutting edge is chipped, the wear rate is greatly accelerated or breakage occurs. A prepared edge eliminates the sharp edge and provides other benefits such as redistributing cutting forces.

Edge Hone The edge hone is by far the most commonly used edge preparation. Many inserts are automatically provided with an edge hone at the time of purchase, especially larger inserts that will be exposed to heavy cutting. An edge hone on a ground or precision insert must usually be specially requested. A standard light hone in the United States usually has a radius of .001 to .003 in.; a standard heavy hone has a radius of .004 to .007 in. Heavier hones are available on request. The heavier the hone, the more resistance an edge has to chipping and breaking, especially in heavy roughing cuts, interrupted cuts, hard-spot cuts, and scaly cuts.

It is standard practice of all manufacturers to hone inserts that are to be coated before the inserts are subjected to the coating process. The reason for this is that during the coating process, the coating material tends to build up on sharp edges. Therefore it is necessary to hone those edges to prevent buildup.

L Land The L-land edge preparation adds strength to the cutting edge of an insert. Essentially, the L land amplifies the advantages of negative rake by diverting a greater amount of cutting force into the body of the insert. The L land amplifies this condition because the included angle at the insert's edges is 110° as opposed to 90°. The L land is particularly beneficial when engaging severe scale, interruptions, and roughing.

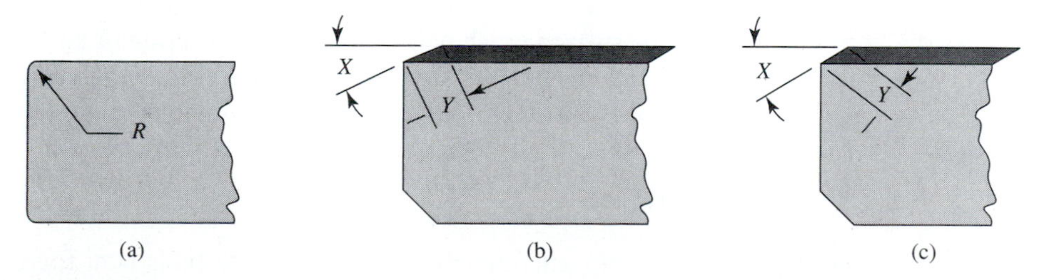

FIGURE 2.18 The three basic edge preparations are (a) edge hone, (b) L land, (c) edge chamfer.

The L land configuration is normally 20° by two thirds of the feedrate. The feedrate should exceed the land width by about one third. This is not a hard and fast rule, but it does serve as a good starting point. If the land width is greater than the feedrate, severe jamming of the chips, excessive high pressures, and high heat will likely occur, resulting in rapid tool failure.

Something other than a 20° land angle may be considered, with varying land width. Some experimentation may prove beneficial; however, if the land angle is varied from 20° it should probably be less rather than more than 20° to keep from jamming the chips.

An L land is normally used only on negative, flat-top inserts placed at a negative rake angle. To use an L land on a positive or a positive/negative insert would defeat the purpose of positive cutting action.

Chamfer A **chamfer** is a compromise between a heavy hone and an L land. A chamfer will also increase an insert's resistance to chipping and breaking. In a shop situation a chamfer is easier and quicker to apply than a heavy hone, because it can be applied with a grinder rather than a hand hone. When a chamfer is applied it should be very slight, 45° by .005 to .030 in.

Normally a chamfer presents a negative-cutting situation, which can result in some problems. The area of application for chamfers is limited, and caution must be exercised. A slight chamfer is often used on a hard and brittle tool for making a very light finishing cut on hard work material. In this instance, the chamfer will strengthen the cutting edge.

Combinations Anytime a sharp edge can be eliminated the life of an insert will likely be extended. When an L land or chamfer is put on an insert, it will make a dramatic improvement in performance, but the L land or chamfer will leave some semisharp corners. To get the maximum benefit from an L land or chamfer, it will help to add a slight hone to each semisharp corner. This will be of significant value in extending tool life, particularly when a large L land is used.

Nose Radius The **nose radius** of an insert has a great influence in the metal-cutting process. The primary function of the nose radius is to provide strength to the tip of the tool. Most of the other functions and the size of the nose radius are just as important. The choice of nose radius will affect the results of the cutting operation; however, inserts are provided with various standard radii and, in most cases, one of these will meet each specific cutting need.

The larger the radius, the stronger the tool tip will be. However, a large radius causes more contact with the work surface and can cause chatter. The cutting forces will increase with a large radius for the same reason, increased contact with the work surface. When taking a shallow cut, a depth approximately equal to the radius or less, the radius acts as a positive lead angle, thinning the chip. A large radius will allow the cutting heat to dissipate more quickly into the insert body, reducing the temperature buildup at the cutting edge.

One of the most important influences of a large radius is that of surface finish. The larger the radius, the better the surface finish will be at an equal feedrate. A larger radius will allow a faster feedrate and yet obtain a satisfactory finish. During a finishing cut, the feedrate should not exceed the radius if a reasonable surface finish is required.

2.4.3 Chip Breakers

Breaking the chip effectively when machining with carbide tools is of the utmost importance, not only from the production viewpoint, but also from the safety viewpoint. When steel is machined at efficient carbide cutting speeds, a continuous chip flows away from the work at high speed.

If this chip is allowed to continue, it may wrap around the toolpost, the workpiece, the **chuck**, and perhaps around the operator's arm. Not only is the operator in danger of receiving a nasty laceration but, if the chip winds around the workpiece and the machine, he or she must spend considerable time in removing it. A loss of production will be encountered. Therefore it is imperative that this chip be controlled and broken in some manner.

With the advent of NC machining and automatic chip-handling systems, the control of chips is becoming more important than ever. The control of chips on any machine tool, old or new, helps to avoid jam-ups with tooling and reduces safety hazards from flying chips. There is a great deal of research and development being conducted in chip control, much of which has been very successful.

Two basic types of chip control are being used with indexable insert tooling: the mechanical chip breaker, Figure 2.19, and the sintered chip breaker, Figure 2.20. Mechanical chip breakers are not as commonly used as sintered chip breakers. There are more parts involved with the mechanical chip breaker, which increases the cost, and the chip breaker hampers changing and indexing the insert. However, mechanical chip breakers are extremely effective in controlling chips during heavy metal-removing operations.

There are two groups of mechanical chip breakers, solid and adjustable, as shown in Figure 2.21. Solid chip breakers are available in various lengths and angles to suit each metal-cutting application. The adjustable chip breaker can eliminate the need for stocking various sizes of solid chip breakers.

Sintered chip breakers are available in many different configurations, some designed for light feeds, some for heavy feeds, and still others for handling both light and heavy feeds. Figure 2.22 shows examples of the various sintered chip-breaker configurations available from a single manufacturer. There are single-sided and double-sided designs of sintered chip breaker inserts.

FIGURE 2.19 Mechanical chip breaker.

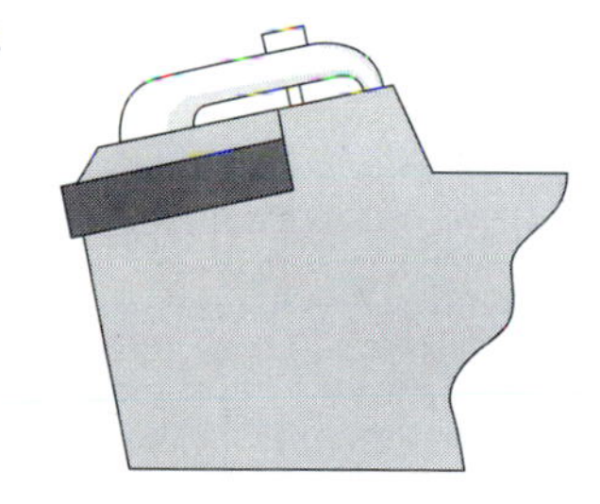

FIGURE 2.20 Sintered chip breaker.

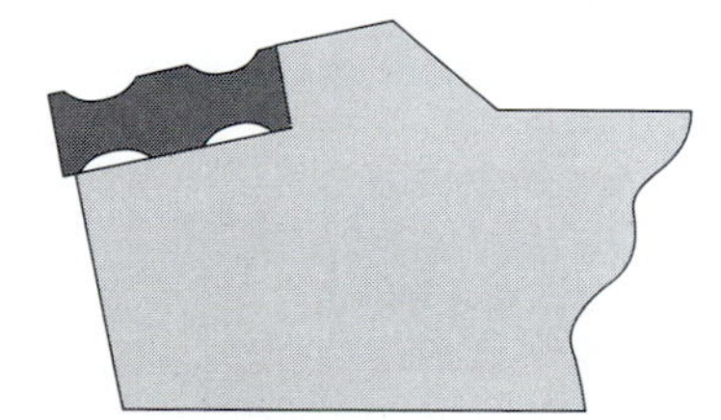

FIGURE 2.21 Solid and adjustable chip breakers.

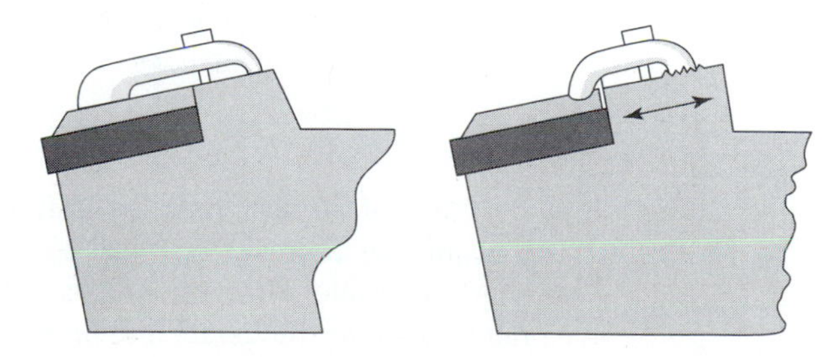

Double-Sided General-Purpose Groove Geometries

.004–.020 ipr feed range		Offers excellent mix of low cost per cutting edge and effective chip control. Designed for general-purpose use at low feed rates.
.005–.065 ipr feed range		Offers excellent mix of low cost per cutting edge and effective chip control. Designed for general-purpose use at medium feed rates
.012–.070 ipr feed range		Offers excellent mix of low cost per cutting edge and effective chip control. Designed for general-purpose use at high feed rates

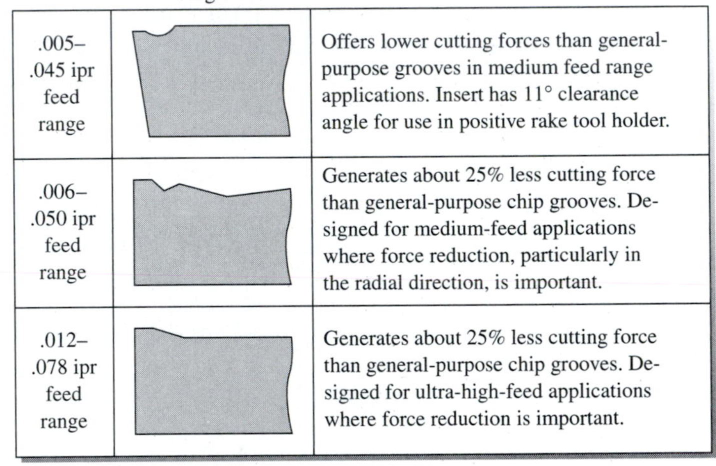

Single-Side Low Force Groove Geometries

.005–.045 ipr feed range		Offers lower cutting forces than general-purpose grooves in medium feed range applications. Insert has 11° clearance angle for use in positive rake tool holder.
.006–.050 ipr feed range		Generates about 25% less cutting force than general-purpose chip grooves. Designed for medium-feed applications where force reduction, particularly in the radial direction, is important.
.012–.078 ipr feed range		Generates about 25% less cutting force than general-purpose chip grooves. Designed for ultra-high-feed applications where force reduction is important.

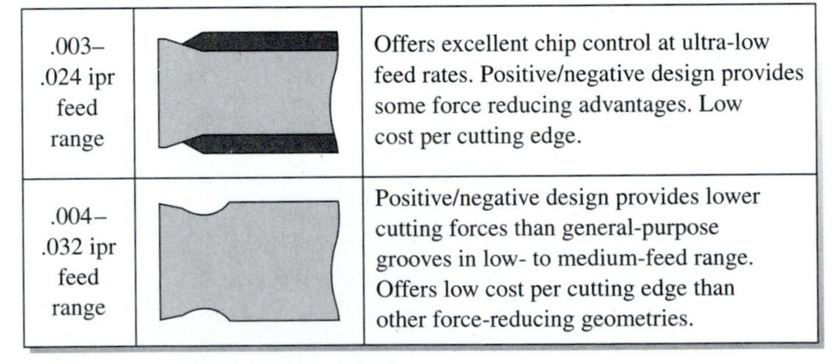

Double-Sided Low Feed Groove Geometries

.003–.024 ipr feed range		Offers excellent chip control at ultra-low feed rates. Positive/negative design provides some force reducing advantages. Low cost per cutting edge.
.004–.032 ipr feed range		Positive/negative design provides lower cutting forces than general-purpose grooves in low- to medium-feed range. Offers low cost per cutting edge than other force-reducing geometries.

FIGURE 2.22 Various sintered chip breaker configurations, with application recommendations.

Many of the designs will significantly reduce cutting forces as well as control chips. Normally it would be more economical to use a double-sided insert because of the additional cutting edges available. However, this is not always true. Although a double-sided insert is more economical under moderate and finish cutting conditions because of its additional cutting edges, a

FIGURE 2.23 Five common insert shapes with various sintered chip-breaker configurations. (Courtesy American National Carbide Co.)

single-sided design will justify itself from a cost standpoint through more effective chip control and reduced cutting forces in certain situations. Figure 2.23 shows five common insert styles with sintered chip breakers.

Figure 2.22 illustrates that a single-sided insert is flat on the bottom as compared to a double-sided insert. This flat bottom provides a single-sided insert with better support under the cutting edge in a severe cutting situation. The single-sided insert, because of its added support, has the ability to remove larger amounts of material with greater ease and efficiency, making it more economical to use. Another reason the single-sided insert may be more economical is that, under heavy machining conditions, it is rare that all of the cutting edges of a double-sided insert can be used. The intense thermal and mechanical shock to the insert will normally damage it to the point where the opposite cutting edge is not usable and, in a sense, wasted. Figure 2.24 shows two square inserts with special-purpose chip breakers.

Statistics have proven that under severe conditions a single-sided insert is more often the most economical choice because its higher efficiency will remove more metal in less time. Additionally, if half of the available cutting edges of a double-sided insert are unusable, for reasons stated before, then the more efficient single-sided insert, having essentially the same number of usable cutting edges, is the most economical insert to use.

There are many configurations of chip-breaker designs other than the ones shown in Figure 2.22. Each manufacturer has its own. The recommended application areas are generally listed in each manufacturer's catalog. However, for specific recommendations and special applications, it is best to consult the manufacturer.

Figure 2.25 shows the various types of chips that are encountered every day. Examining the chips that are coming off a workpiece will give a lot of information as to how well the job is going, how tool wear is progressing, and why premature tool failure or short tool life is occurring.

FIGURE 2.24 Two square inserts with one-sided special-purpose chip breakers. (Courtesy Iscar Metals, Inc.)

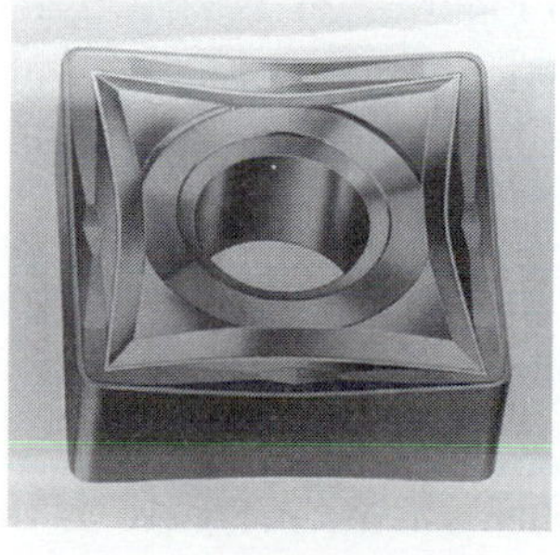

(a)

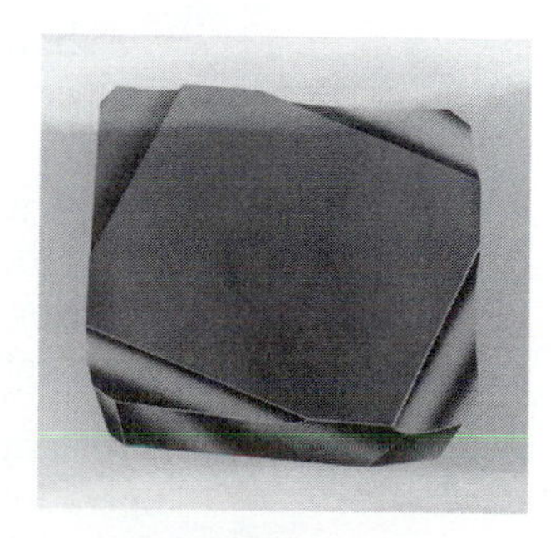

(b)

FIGURE 2.25 Various types of chip formations.

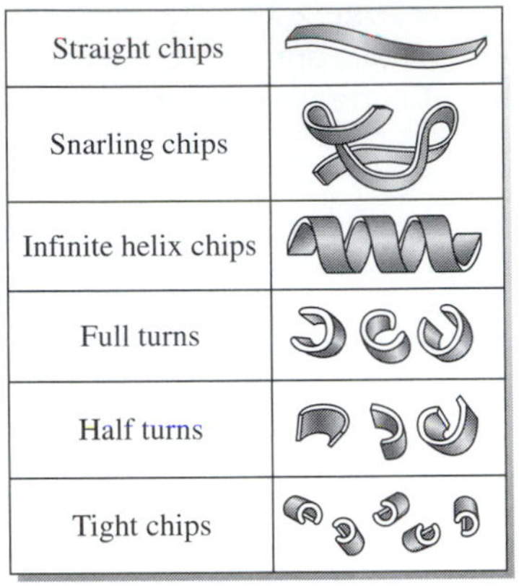

Straight chips	
Snarling chips	
Infinite helix chips	
Full turns	
Half turns	
Tight chips	

Straight Chips: Straight chips are usually the most troublesome. They string out all over the machine tool; they get snarled in the tool, workpiece, and fixturing; they cause tooling to break; they jam up chip-handling equipment; they are difficult to remove; and they are dangerous, especially when they begin to whip around. Soft, gummy, low-carbon, and tough steels usually cause this type of chip. One of the quickest ways to eliminate the straight chip is to increase the feedrate, because a thicker chip breaks more easily. Other ways to eliminate straight chips are to decrease the lead angle, which would also thicken the chip, increase the speed, use a negative-rake tool, or use a chipbreaker insert.

Snarling Chips: Snarling chips are continuous chips much the same as straight chips. They are generally caused by the same conditions as straight chips and create the same problems. It stands to reason, therefore, that to correct a snarling chip situation, the same methods would be used as with straight chips. In addition, cooling the chips with a flood or mist **coolant** as they come off the tool will frequently help to break them.

Infinite-Helix Chips: Infinite-helix chips are chips that are near the breaking point. The problems this type of chip creates are similar to those created by straight chips. Infinite-helix chips are common when machining very ductile material, such as leaded or resulfurized steels, and other soft materials. They will most often occur when making light cuts with positive-rake tools. Using a sintered chipbreaker insert, which will force the natural chip flow direction to change, is often effective in breaking the infinite-helix chip. An increase in feed or speed will also help break the chip.

Full-Turn Chips: Full-turn chips are not usually a problem so long as they are consistent and without occasional stringers. A consistent full-turn chip is near the ideal half-turn chip.

Half-Turn Chip: If there is such a thing as a perfect chip, it is the half-turn or 6-shape chip. This is the chip shape that the machinist strives for in the cutting operation. The half-turn chip, shown in Figure 2.26, is known as the classic chip form.

Tight Chips: Tight chips do not present a problem from a handling or interfacing point of view, but these tight chips are a sign that poor tool life or premature tool failure may occur. The tight chip is formed by very high pressure and causes intense heat, deflection of the tool and workpiece, and rapid tool failure. A tight chip is a jammed chip, meaning that its flow path is overly restricted. Causes include too high a feed rate, too negative a rake angle, improper chipbreaker selection, or setting, or a worn insert.

Many times a straight, snarled, or infinite-helix chip will be generated at the start of a cutting operation, when the insert is new. As the insert begins to wear, the chip gradually becomes well shaped and properly broken. It may even progress into a tight chip and eventually cause

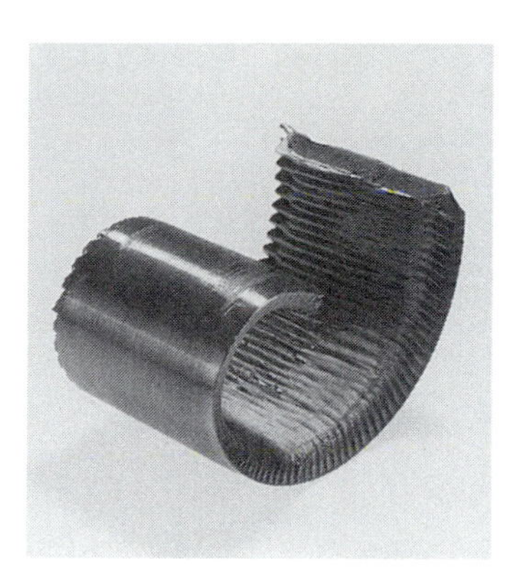

FIGURE 2.26 Half-turn chip or "perfect" chip. (Courtesy Kennametal Inc.)

catastrophic tool failure. This is caused by a type of insert wear known as cratering (see Figures 2.9a and 2.10a). In cratering, a groove is worn into the insert, causing a false chipbreaker groove to be formed. This is a definite sign of a problem, such as the insert is not of the correct carbide grade, is not the correct geometry, or the cutting speed may be too fast.

2.5 INDEXABLE-TYPE TOOLING

One of the more recent developments in cutting-tool design is the **indexable insert**, which is mechanically held in a **toolholder**. Inserts are available in several thicknesses and a variety of sizes and shapes. The round, square, triangle, and diamond account for the greatest percentage. Many other shapes, including the parallelogram, hexagon, and pentagon, are used to meet specific machining requirements. Each shape has its advantages and limitations because operational as well as economic factors must be considered in tooling selection. The most common insert shapes are shown in Figure 2.23.

2.5.1 Indexable-Insert Shapes

Indexable inserts have certainly established their position and potential in the metalworking industry. The elimination of regrinding, the accuracy of tool geometry, reduced inventory tool costs, and reduced downtime for tool changes are some of the advantages resulting from the use of this tooling.

There are four basic shapes and a variety of special shapes. Because approximately 95% of all machining is done with the four basic shapes, these are the ones of interest here. The four basic shapes are

- Square
- Triangle
- Diamond
- Round

These shapes are available in many different configurations for almost any job. Each shape can be obtained for positive, negative, or positive/negative rake, with or without chipbreaker grooves, with or without holes, with various edge preparations, in various tolerances, and in various radii and sizes. A variety of insert shapes and configurations are shown in Figure 2.27.

Choosing a particular shape or insert requires a great deal of planning and thought. The choice of insert shape must be based on such factors as the workpiece configuration and tolerance, workpiece material, amount of material to be removed, machine-tool capability, and economics.

FIGURE 2.27 Various insert shapes, with and without holes, with and without chip breakers. (Courtesy American National Carbide Co.)

The insert shape also has an influence on insert strength. As shown in Figure 2.28, the greater the included angle at the insert tip, the greater the strength. The round insert and the 100° corner of the first diamond-shaped insert are shown as the strongest. Because of the higher cutting forces and the possibility of chatter, these inserts are more limited in use than the square shape. Therefore, for practical purposes, the square insert is the strongest for general use. Triangle and diamond inserts should only be used when a square cannot be used, such as when machining to a corner or a shoulder.

The Round Insert Round or button inserts give a good finish at heavy feeds, and they are also ideal for forming inside corner radii. Their shape provides the greatest geometric strength, and they offer the maximum number of indexes when light cuts are being taken. The solid button type, which is held in place by means of a clamp, generally has edges at 90° to the surfaces for use in negative-rake holders, thereby providing cutting edges on both sides of the insert. The CDH button type is made in larger sizes and has a **counterbored** hole. This button has clearance and is normally held in the toolholder with neutral rake. A typical application is for tracing or **contouring**, where the tool must generate forms that require a large portion of the cutting edge to be in the cut.

Round inserts have their limitations, however, because the large nose radius thins the chips and increases the forces between the tool and workpiece for a given size cut. Very high radial forces are usually incurred as compared with normal cutting, particularly at normal feedrates. Chatter and deflection often result, especially when machining long-chip materials. For this reason, button inserts are applied with greater success on cast iron and the other short-chip, low-strength materials, although heavy feedrates will often improve the cutting action on ductile materials.

The Square Insert Square inserts provide four or eight cutting edges, depending on the design of the toolholder. Positive rakes mean that relief angles must be ground on the insert, thereby eliminating the use of one side.

Square inserts are preferred for most machining jobs, where the workpiece and tool design relationships allow their use. Their shape provides strength close to that of the round insert, but with the economy of four or eight cutting edges, and also permits a reduction in the side

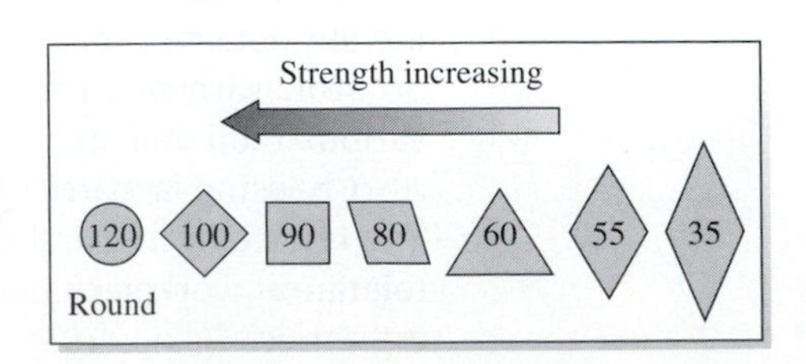

FIGURE 2.28 Various insert shapes as related to strength.

cutting-edge angle and the problem related to the chip-thinning action of the round. Economical tool application dictates the use of an insert shape that gives the maximum number of cutting edges and is compatible with the machining operation. If the operation requires machining to a square shoulder, the square insert would be eliminated because of the design of an A-style tool. The **end cutting-edge angle** (ECEA) must be such that the tool will clear the machined surface, so something less than a 90° included angle between the side and end of the tool is mandatory.

The Triangular Insert Owing to design and application requirements, one of which has just been pointed out, the triangular insert has assumed an important place in indexable tooling. The triangle provides three or six cutting edges, depending on whether relief angles are required on the insert for use in a positive-rake holder. The 60° included angle is not as strong as the 90° of the square or the radius of the button, yet many machining operations are performed satisfactorily with triangular inserts. Turning to a shoulder, plunging and contouring, and numerous other operations require a generous end cutting-edge angle, which the triangle can provide. The 60° included angle is also suitable for threading operations. Because of its fewer cutting edges and lower strength, the triangular insert and holder should only be used when other geometric shapes will not meet the job requirements.

The Pentagonal Insert A pentagonal or five-sided insert is a means of providing one or two more cutting edges per insert, and the extra edges are the main reason for this design. There is, of course, a strength advantage over the square and triangle in the 108° included angle. As in the case of the square, the pentagonal shape sets up certain design and application limitations. The tool must always cut with a **side cutting-edge angle** (SCEA), which thins the chip and improves tool life. However, an SCEA cannot always be used owing to the requirements of the finished part's shape or because the increased radial forces cause chatter and deflection of the workpiece. The minimum SCE angle that can be used is 24°. This then leaves 6° end cutting-edge (ECE) angle. An SCEA of 33° results in 15° of ECEA, which is the same as that used on standard B-style tools and is quite adequate.

The Diamond Insert The trend in lathe design is toward machines that generate the form on the workpiece. This is accomplished by guiding the tool so that it faces, plunges, turns, forms radii and chamfers, and machines other configurations. For a tool to satisfy the requirements of these complex maneuvers, it must meet certain design standards. Because the tool often plunges along an angle, a great amount of ECEA is needed. Back facing is also a common operation on such setups, and this requires a negative SCEA.

The diamond insert was developed specifically for tracing operations. The industry's standard marking system includes designations for diamond inserts with included angles of 86, 80, 55, and 35°. By far the most popular size is the 55° included angle diamond. This geometry apparently meets the requirements of most tracing operations. When the insert is positioned in the toolholder and tool block so that it cuts with a 3° negative SCEA, it will back face with depths of cut up to .020 in., and in most toolholders will be able to plunge at an angle of 20° with adequate clearance.

Holding the insert securely in the holder so that duplication of workpiece size to tolerances specified is achieved has been a problem. The tendency for the insert to twist in the pocket on turning and plunging operations, and to be pulled out of the pocket on back-facing operations, has resulted in design changes by some manufacturers. Diamond tracer inserts are made in regular and elongated shapes. The elongated diamond provides greater resistance to the twisting action set up by the cutting forces.

Further developments are still being made in tracer inserts and holders so that they will meet the exacting requirements of tracing operations better. In some designs the diamond-shaped insert, either regular or elongated, is locked into the pocket with an eccentric pin. This gives a positive holding action and locates the insert against the back walls of the pocket, minimizing the chances for movement during the contouring operations.

The selection of a tool for a tracing operation should begin with an analysis of the requirements of the contouring operations. The tool selected should be the one that provides the strongest geometric shape and still meets the contouring requirements. Many tracing jobs can be done satisfactorily with a triangular insert. If no back facing is included in the operations, no negative SCEA is needed and a standard A-style tool can be used. In some cases it is possible to use a tool designed to cut with an SCEA. Generally, better tool life will be realized with lower cost per cutting edge when tools without negative SCEA can be used.

The Parallelogram Insert The parallelogram-shaped insert provides some advantages, which make its use justified in certain applications. When a long side cutting edge is needed, it is sometimes more economical and advantageous from a machining standpoint to use a parallelogram rather than a square or triangle.

The parallelogram also permits the construction of an A-style tool with greater geometric strength than is possible with a triangular insert. A limitation of the parallelogram design is the number of usable cutting edges. A negative-rake insert can be used on two corners in a right- or left-hand holder. To use the remaining two cutting edges, the opposite-hand holder is required. Unless all four corners can be used, the use of the parallelogram insert may not be economically justifiable.

The Hexagonal Insert A versatile tool makes use of a hexagonal insert. Turning, facing, and chamfering can all be done from a number of positions. The hexagon's shape provides strong cutting edges as in the case of the pentagon, but also necessitates cutting with considerable SCEA. The number of usable cutting edges in this design makes it a most economical insert where it can be applied.

The On-Edge Insert The on-edge insert concept (Fig. 2.29) has only been in use for a short time but is becoming more common. The on-edge insert was first developed for milling operations. The main reason for its development was to provide the strength needed to withstand the constant interruption of milling cuts. The on-edge concept is now becoming more popular for turning inserts as well.

The main use of the on-edge insert is for rough cutting when cutting forces are high and the interruptions are often severe. The extra thickness of the on-edge insert offers more

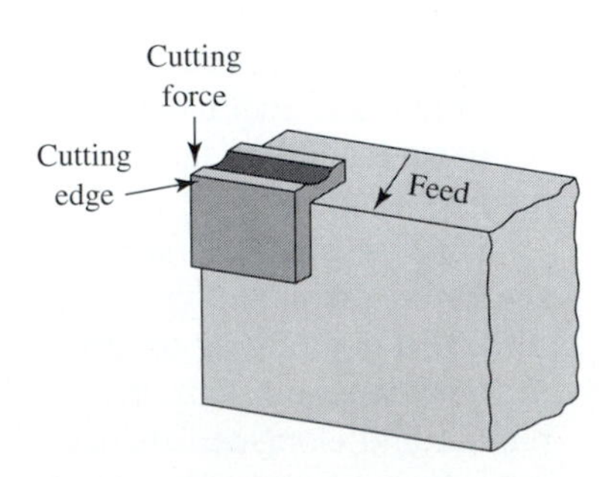

FIGURE 2.29 On-edge turning tool design.

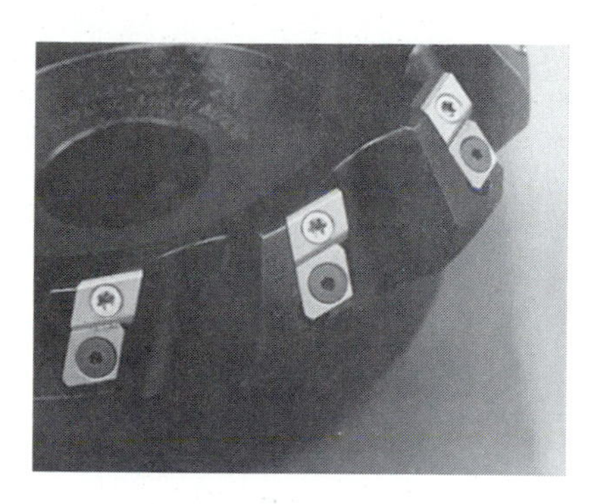

FIGURE 2.30 On-edge milling cutter section. (Courtesy Ingersoll Cutting Tool Co.)

protection from heat and shock damage to the opposite side cutting edge during heavy roughing, than is common with standard inserts. A milling cutter section with on-edge inserts is shown in Figure 2.30.

2.5.2 Indexable Inserts—Classes and Sizes

Inserts are commercially available with various degrees of dimensional tolerances, such as the inscribed circle of a triangle, the measurement across the flats of a square or elongated diamond, thickness, nose radii, and tangency. All these dimensions and several other factors contribute to the ability of an insert to be accurately indexed and to machine a given material to a specific size. The need for inserts with different tolerances depends not so much on the dimensional size of the finished part as on how the insert is to be used in the machining operation.

Unground Inserts: Through improved manufacturing techniques, many carbide producers can supply inserts that are to the required specifications, thus eliminating the grinding operation. Cutting edges produced by this method are not only metallurgically sound in structure, they are also honed to give them geometric increase in strength.

Utility Inserts: This type of insert is ground on the top and bottom faces only.

Precision Inserts: These are ground all over and to close tolerances.

Honed Inserts: The development of production honing techniques for inserts has made standard inserts available to the machining industry in the prehoned condition. These inserts have the advantage of having the cracked crystal layer removed not only from the cutting-edge area but also from the cutting-tool surfaces. Lighter finishing cuts taken with finishing grades of carbide should have small amounts of honing performed on the cutting edge. Roughing grades should, conversely, be honed heavily. Carbide-insert honing equipment is shown in Figure 2.31.

Insert Size: The size of an insert is determined by its inscribed circle (I.C.). Every insert has an I.C. regardless of the insert shape (Fig. 2.32). The I.C. is designated in fractions of an inch in the United States, normally in 1/8-in. increments. The thickness of the insert is designated by its actual thickness in increments of 1/16 in., and the nose radius is designated in increments of 1/64 in.

The thickness of the insert is usually standard to a particular I.C. Sometimes, however, a choice of thickness will be available. In these situations, the thickness that is appropriate to the amount of cutting force that will be applied is the optimum choice. If a thin insert is chosen, a thicker **shim** should be used to keep the cutting edge at the workpiece centerline.

FIGURE 2.31 Carbide-insert honing equipment. (Courtesy American National Carbide Co.)

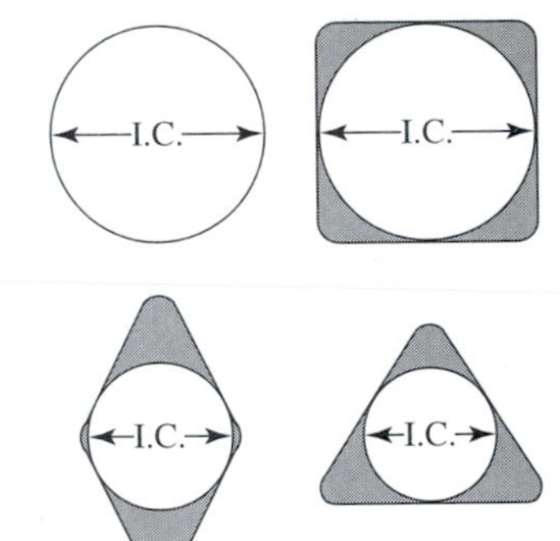

FIGURE 2.32 The size of an insert is determined by its inscribed circle (I.C.).

2.5.3 Indexable-Insert Identification System

A standard marking system, proposed by the Cemented Carbide Producers Association and approved by the American National Standards Institute (ANSI), has been adopted by the cemented carbide manufacturers. A new identification and numbering system became necessary, because of the addition of an expanded range of types and sizes of inserts incorporating a wide variety of detail. Under this new system, the insert number, with the manufacturer's grade of carbide, is all that is needed to describe the insert (see Fig. 2.33). The eight sequences of marking indexable inserts are

- Shape
- Thickness
- Cutting point
- Other conditions
- Size
- Clearance angle
- Class
- Type

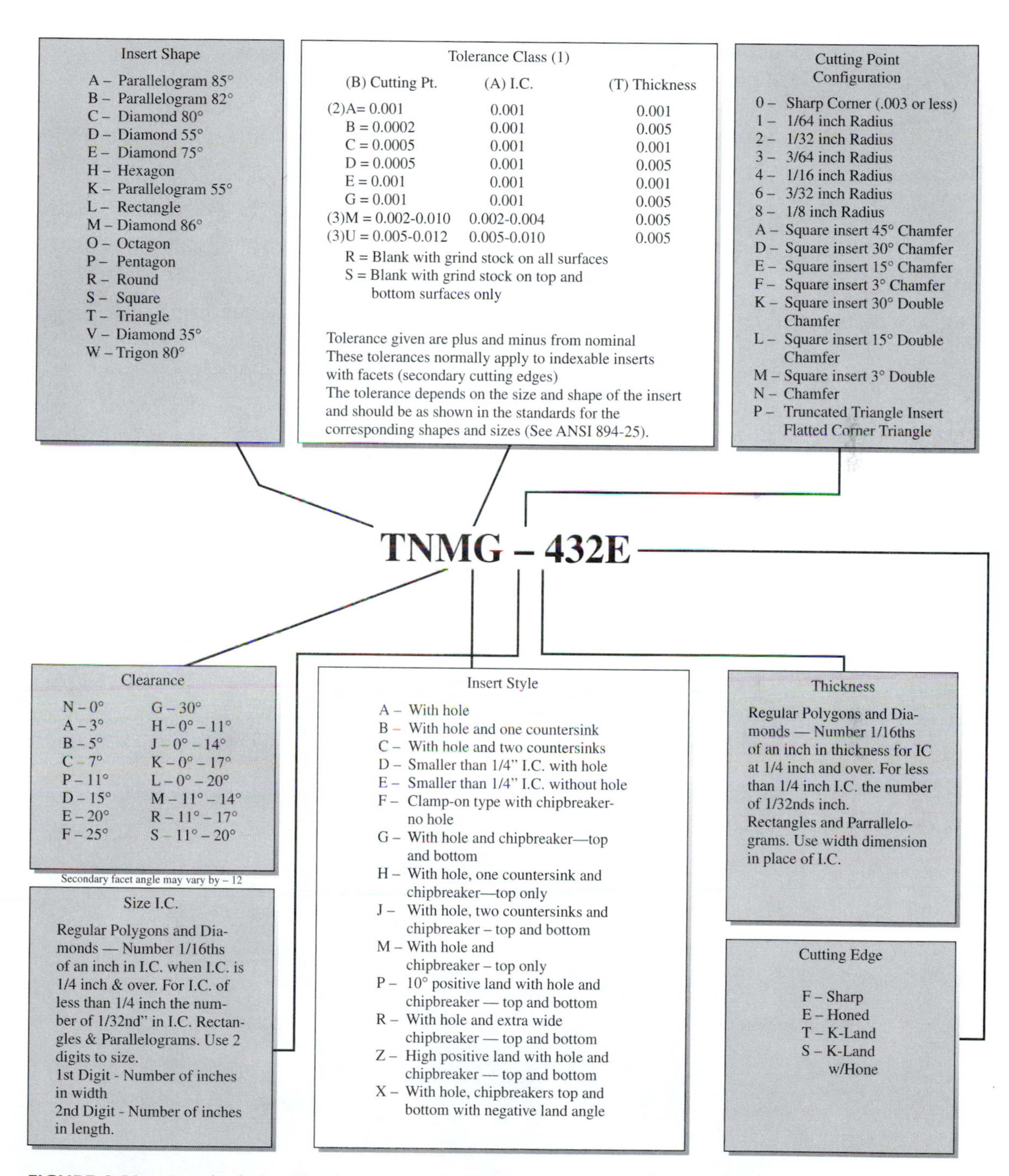

FIGURE 2.33 Standard Identification System for indexable inserts. (Courtesy Cemented Carbide Producers Association)

Insert Economics The cost of carbide and other tool materials as well as the cost of preparing these materials into cutting tools is relatively high and continuing to increase. Therefore, it is most important to choose tool inserts wisely. Here are some important things to consider when making the choice:

- Choose a shape that offers the most cutting edges.

Examples

A negative insert has twice as many cutting edges as a positive insert.

A square insert has 25% more cutting edges than a triangle insert.

A double-sided chip breaker insert has twice as many cutting edges as a single-sided insert.

- Choose an I.C. appropriate to the amount of material to be removed.

Example

A 1 in. I.C. square insert for a 1/4-in. depth of cut would be wasteful because a large piece of expensive carbide would be used where a smaller piece would achieve the same result.

- Choose an insert tolerance that is appropriate to the job being done. In most cases an unground utility grade will do the job. The closer the tolerance, the higher the cost. Tight insert tolerances are normally required only when the indexability of an insert is critical.

Example

A C-tolerance insert used for finishing to a workpiece tolerance of plus or minus .010 in. would not be necessary. An M- or even a U-tolerance insert would be satisfactory.

- Choose a single-sided insert when conditions make its efficiency more economical.

Example

A heavy roughing cut has made the second side of a less-efficient double-sided insert unusable because of heat and shock damage.

2.5.4 Mechanical Toolholders

The revolution of the indexable insert has resulted in the availability of a wide range and variety of toolholders. A number of toolholders with inserts are shown in Figure 2.34.

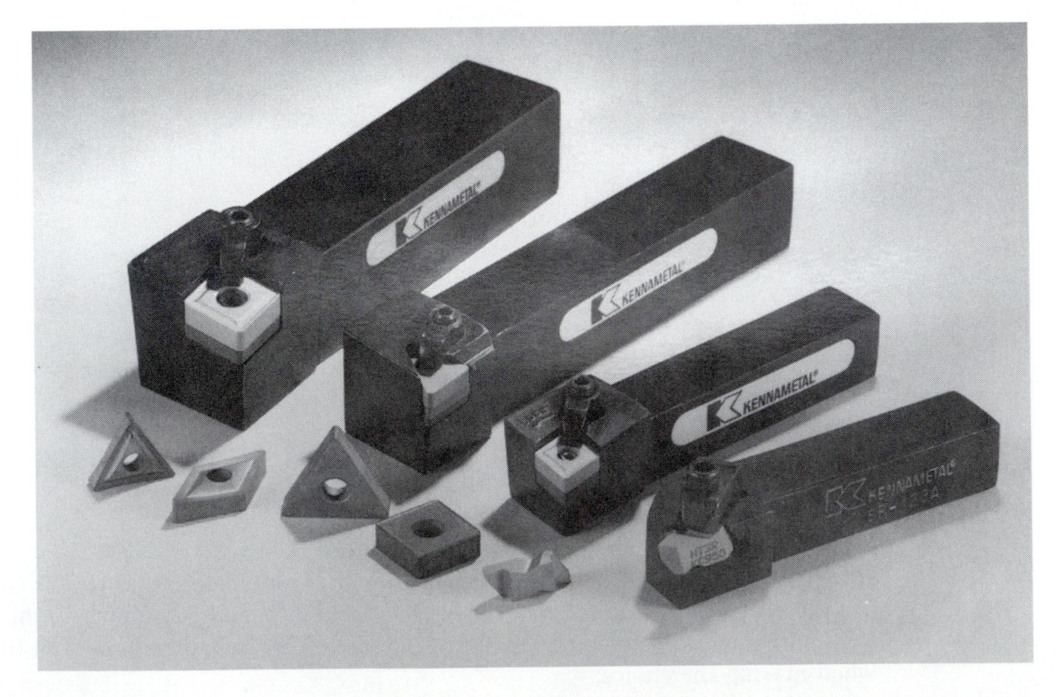

FIGURE 2.34 Four toolholders with various insert styles and sizes. (Courtesy Kennametal Inc.)

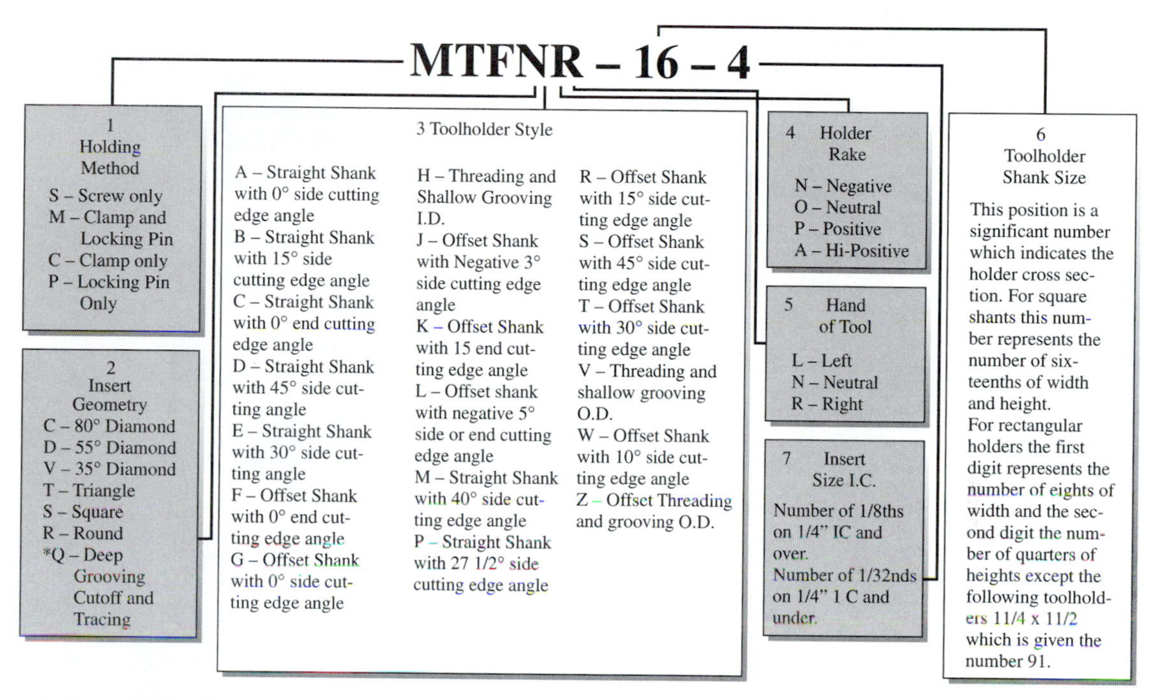

FIGURE 2.35 Standard identification system for turning tool holders. (Courtesy Cemented Carbide Producers Association)

To select or recommend the best holder for every machining application would be a formidable task. The practice in many manufacturing plants is to standardize on one or two designs so that a minimum of repair parts and accessories need to be carried in inventory. Three basic design and construction elements are common to all holders:

- The shank
- The seat
- The clamp or locking device

Turning toolholders have been standardized as shown in Figure 2.35.

The Shank The shank is the basic element of the toolholder, and its purpose is to hold and present the cutting edge to the workpiece. It usually has drilled and tapped holes, slots, and cutouts, and it must provide a firm support for the carbide cutting edge. Generally shanks are made of high-carbon or low-alloy steel, heat treated to give physical properties that will resist thread damage, chip erosion, and deformation under the toolblock clamping screws. Some designs and sizes that do not make use of a carbide seat are made of high-alloy steel to resist deformation under the insert.

The machined area for the seat and insert is one of the most critical areas and must be flat to provide the proper support for the carbide seat and insert. Common practice is to relieve the inside corner for seat and insert clearance. The intersections of the sides and bottom of the pocket usually have a small radius, because sharp corners may be the source of cracks during heat treatment. A tool shank with basic components is shown in Figure 2.36.

The Seat Most toolholders for indexable inserts use a carbide seat or pad as support for the insert. Cemented tungsten carbide has a high compressive strength, is hard, and can be ground

FIGURE 2.36 Tool shank with basic components. (Courtesy Kennametal Inc.)

to a smooth, flat surface. Although hardened steel has been used, and still is in some designs, a strong preference for carbide seats prevails.

The seats shown in Figure 2.37 are typical and will serve to illustrate the basic design. The periphery is chamfered at one face to clear any radius in the steel-shank-pocket area. If the seat or pad is held in place by a screw, the hole will be deeply countersunk so that the head of the screw will be well below the surface. If the screw head projects above the seat surface and the insert is clamped down on it, breakage of the latter could result.

The seat is attached to the shank only for convenience and to prevent its loss when inserts are removed and replaced, or if the holder is used vertically as in a vertical turret lathe or upside down, as in the rear tool post of a **turret lathe**.

Seat flatness is one of the most critical requirements of toolholders. Application tests have shown that an out-of-flatness condition of as little as .001 in. can result in insert breakage. Regardless of the design of the toolholder selected, the pocket and seat flatness specifications should be carefully examined and the highest standards insisted upon.

The Clamp or Locking Device Many clamping and locking arrangements have been developed for holding the insert in a toolholder, and there is probably no one best method or design because specific application requirements vary so greatly. However, a number of features and construction elements warrant consideration and should influence the selection of a toolholder (Fig. 2.38).

The main function of the clamping mechanism is to hold the insert securely in position, and many methods of doing so are in use. On normal turning and facing operations, the insert in most styles of toolholders is held in the pocket by the cutting pressures, and the load on the clamp is very light except as affected by the chip. Tracing and threading operations change the

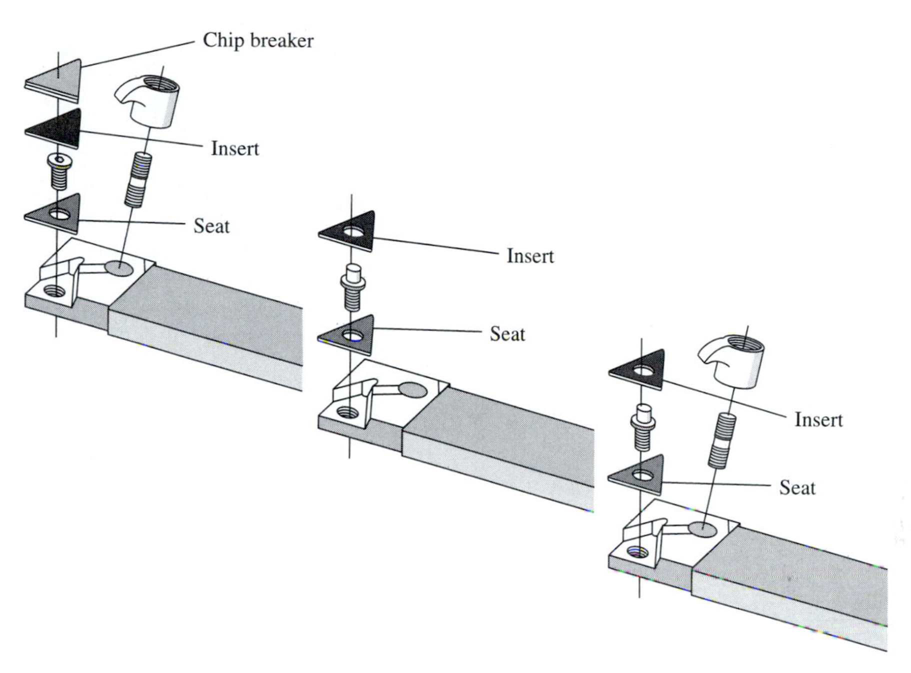

FIGURE 2.37 Schematic drawing of various insert locking options.

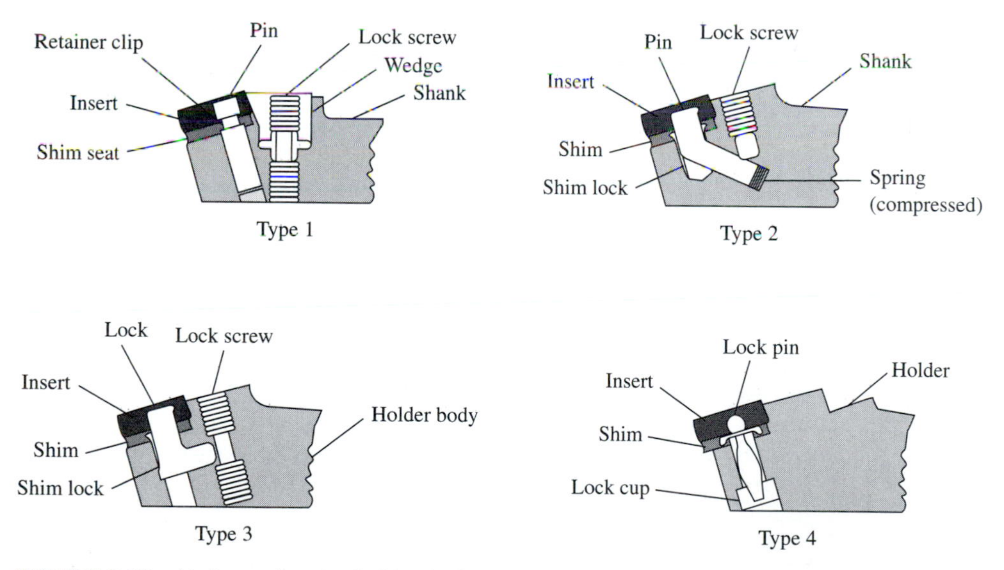

FIGURE 2.38 Various pin-type holder locking options.

direction and amount of the load applied to the insert, and there is more tendency to twist or pull the insert out of the pocket. The ability of the clamping mechanism to perform satisfactorily under such conditions should be carefully evaluated. The use of a pin or lever mechanism has been incorporated in some designs to give a more positive holding action against the insert.

The suitability of the clamp design to the machine toolholding blocks and to the workpiece configuration should be considered. Bulky club heads, high clamps on clamping screws, or intricate adjusting mechanisms may be in the way, especially when tools must be ganged up or when

FIGURE 2.39 Three tool holders in which inserts are held by both pins and clamps. (Courtesy Sandvik Coromant Corp.)

machine and workpiece clearances are small. A toolholder that is not easily accessible and must be removed from the machine so that the insert can be indexed or the chip breaker adjusted should not be considered suitable for the application. A number of toolholders are shown in Figure 2.39, with indexable inserts being held by both pins and clamps.

Tools that are positioned upside down should have a wrench socket in the lower end of the clamping screw so that it can be easily reached. Chip-breaker plates and clamp parts should be secured so that they will not be dropped in the chip pan when loosened for insert changing.

2.6 REVIEW QUESTIONS AND PROBLEMS

1. Explain the three forces acting on a cutting tool.
2. Explain the difference between orthogonal and oblique cutting.
3. Name and explain the three types of chip formation.
4. What problems are caused by edge buildup?
5. Explain the difference between crater and edge wear.
6. Discuss the four gradual wear mechanisms.
7. Discuss and explain the effect of rake angles on cutting tools.
8. Discuss the advantages of positive/negative rake inserts.
9. Explain the need for side and end relief angles.
10. Discuss the two benefits of a positive lead angle.
11. Discuss the pros and cons of various edge preparations.
12. Explain the effect a nose radius has on cutting performance and surface finish.
13. Name and describe the two types of chip breakers.
14. Explain what is meant by a "perfect chip."
15. What are the advantages of indexable inserts?
16. What effect do insert shapes have on the machining process?
17. Discuss the importance of diamond-shaped inserts.
18. Explain the advantages of the on-edge insert design concept.
19. What are the eight sequences of marking indexable inserts?
20. Explain the importance of shim seats in indexable toolholder design.

CHAPTER THREE

MACHINABILITY OF METALS

CHAPTER OVERVIEW

- Introduction
- Condition of Work Material
- Physical Properties of Work Materials
- Metal Machining
  - Cast Iron
  - Steel
  - Nonferrous Metals and Alloys
- Judging Machinability
- Review Questions and Problems

3.1 INTRODUCTION

The condition and physical properties of a work material have a direct influence on its **machinability**. The various conditions and characteristics described as "condition of work material," individually and in combinations, directly influence and determine the machinability. Operating conditions, tool material and geometry, and workpiece requirements exercise indirect effects on machinability and can often be used to overcome difficult conditions presented by the work material. On the other hand, they can create situations that increase machining difficulty if they are ignored. A thorough understanding of all of the factors affecting machinability and machining will help in selecting material and workpiece designs to achieve the optimum machining combinations critical to maximum productivity.

3.2 CONDITION OF WORK MATERIAL

The following eight factors determine the condition of the work material: microstructure, grain size, heat treatment, chemical composition, fabrication, hardness, yield strength, and tensile strength.

Microstructure The microstructure of a metal refers to its crystal or grain structure as shown through examination of etched and polished surfaces under a microscope. Metals whose microstructures are similar have like machining properties, but there can be variations in the microstructure of the same workpiece that will affect machinability.

Grain Size Grain size and structure of a metal serve as general indicators of its machinability. A metal with small, undistorted grains tends to cut easily and finish easily. Such a metal is ductile, but it is also "gummy." Metals of an intermediate grain size represent a compromise that permits both cutting and finishing machinability. Hardness of a metal must be correlated with grain size and it is generally used as an indicator of machinability.

Heat Treatment To provide desired properties in metals, they are sometimes put through a series of heating and cooling operations when in the solid state. A material may be treated to reduce brittleness, to remove stress, to obtain ductility or toughness, to increase strength, to obtain a definite microstructure, to change hardness, or to make other changes that affect machinability.

Chemical Composition The chemical composition of a metal is a major factor in determining its machinability. The effects of composition, though, are not always clear, because the elements that make up an alloy metal work both singly and collectively. Certain generalizations about chemical composition of steels in relation to machinability can be made, but nonferrous alloys are too numerous and varied to permit such generalizations.

Fabrication Whether a metal has been **hot-rolled**, **cold rolled**, **cold drawn**, cast, or **forged** will affect its grain size, ductility, strength, hardness, and structure—and therefore its machinability.

The term "wrought" refers to the hammering or forming of materials into premanufactured shapes that are readily altered into components or products using traditional manufacturing techniques. Wrought metals are materials that are mechanically shaped into bars, billets, rolls, sheets, plates, or tubing.

Casting involves pouring molten metal into a mold to arrive at a near-component shape that requires minimal or, in some cases, no machining. Molds for these operations are made from sand, plaster, metals, and a variety of other materials.

Hardness The textbook definition of hardness is the tendency for a material to resist deformation. Hardness is often measured using either the **Brinell** or **Rockwell** scale. The method used to measure hardness involves embedding a specific size and shape indenter into the surface of the test material, using a predetermined load or weight. The distance the indenter penetrates the material surface will correspond to a specific Brinell or Rockwell hardness reading. The greater the indenter surface penetration, the lower the ultimate Brinell or Rockwell number, and thus the lower the corresponding hardness level. Therefore, high Brinell or Rockwell numbers or readings represent a minimal amount of indenter penetration into the workpiece and thus, by definition, are an indication of an extremely hard part. Figure 3.1 shows how hardness is measured.

The Brinell hardness test involves embedding a steel ball of a specific diameter, using a kilogram load, in the surface of a test piece. The Brinell Hardness Number (BHN) is determined by dividing the kilogram load by the area (in square millimeters) of the circle created at the rim

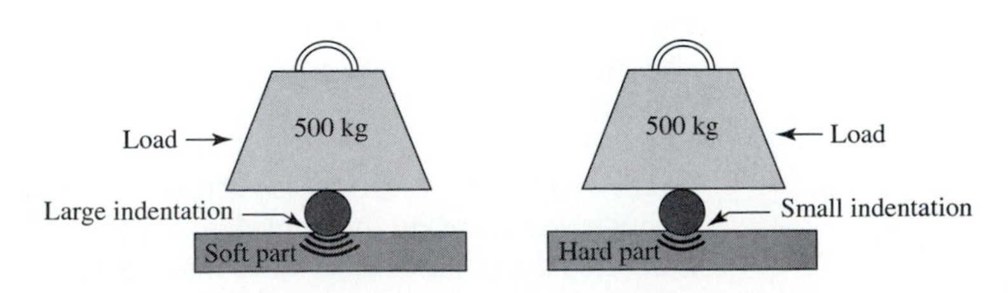

FIGURE 3.1 Hardness is measured by depth of indentations made.

of the dimple or impression left in the workpiece surface. This standardized approach provides a consistent method to make comparative tests between a variety of workpiece materials or a single material that has undergone various hardening processes.

The Rockwell test can be performed with various indenter sizes and loads. Several different scales exist for the Rockwell method or hardness testing. The three most popular are outlined here in terms of the actual application the test is designed to address:

Rockwell Scale	Testing Application
A	For tungsten carbide and other extremely hard materials, and thin, hard sheets.
B	For medium-hardness low- and medium-carbon steels in the **annealed** condition.
C	For materials less hard than Rockwell B 100.

Machining Considerations. In terms of general machining practice, low material hardness enhances productivity because cutting speed is often selected based on material hardness (the lower the hardness, the higher the speed). Tool life is adversely affected by an increase in workpiece hardness because the cutting loads and temperatures rise for a specific cutting speed with part hardness, thereby reducing tool life. In drilling and turning, the added cutting temperature is deterimental to tool life because it produces excess heat, causing accelerated edge wear. In milling, increased material hardness produces higher impact loads because inserts enter the cut, which often leads to a premature breakdown of the cutting edge.

Yield Strength

Tensile test work is used as a means of comparison of metal material conditions. These tests can establish the yield strength, **tensile strength**, and many other conditions of a material based on its heat treatment. In addition, these tests are used to compare different workpiece materials. The tensile test involves taking a cylindrical rod or shaft and pulling it from opposite ends with a progressively larger force in a hydraulic machine. Prior to the start of the test, two marks either 2 or 8 in. apart are made on the rod or shaft. As the rod is systematically subjected to increased loads, the marks begin to move farther apart. A material is in the "elastic zone" when the load can be removed from the rod and the marks return to their initial distance apart of either 2 or 8 in. If the test is allowed to progress, a point is reached where, when the load is removed, the marks will not return to their initial distance apart. At this point, permanent set or deformation of the test specimen has taken place. Figure 3.2 shows how yield strength is measured.

Yield strength is measured just before permanent deformation takes place. Yield strength is stated in pounds per square inch (psi) and is determined by dividing the load just prior to permanent deformation by the cross-sectional area of the test specimen. This material property has been referred to as a condition because it can be altered during heat treatment. Increased part hardness produces an increase in yield strength, so as a part becomes harder, it takes a larger

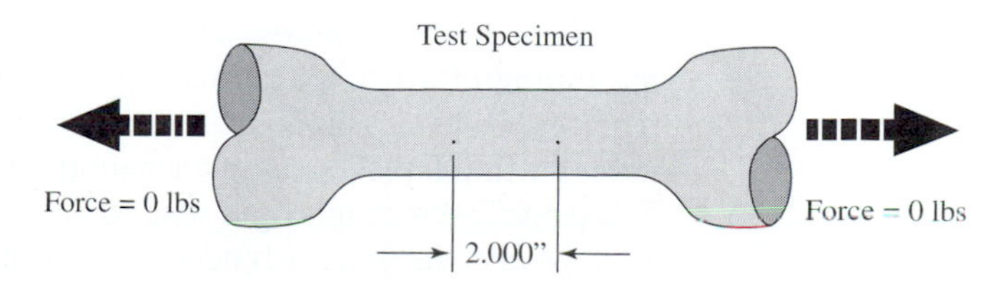

FIGURE 3.2 Yield strength is measured by pulling a test specimen as shown.

force to produce permanent deformation of the part. Yield strength should not be confused with fracture strength, cracking, or the actual breaking of the material into pieces because these properties are quite different and unrelated to the current subject.

Machining Considerations. By definition a material with high yield strength (force required per unit of area to create permanent deformation) requires a high level of force to initiate chip formation in a machining operation. This implies that as a material's yield strength increases, stronger insert shapes as well as less-positive cutting geometries are necessary to combat the additional load encountered in the cutting zone. Material hardness and yield strength increase simultaneously during heat treatment. Therefore, materials with relatively high yield strengths will be more difficult to machine and will reduce tool life when compared to materials with more moderate strengths.

Tensile Strength The tensile strength of a material increases along with yield strength as it is heat treated to greater hardness levels. This material condition is also established using a tensile test. Tensile strength (or ultimate strength) is defined as the maximum load that results during the tensile test, divided by the cross-sectional area of the test specimen. Therefore, tensile strength, like yield strength, is expressed in psi. This value is referred to as a material condition rather than a property because its level, just like yield strength and hardness, can be altered by heat treatment. Therefore, based on the material selected, distinct tensile and yield strength levels exist for each hardness reading.

Machining Considerations. Just as increased yield strength implied higher cutting forces during machining operations, the same could be said for increased tensile strength. Again, as the workpiece tensile strength is elevated, stronger cutting-edge geometries are required for productive machining and acceptable tool life.

3.3 PHYSICAL PROPERTIES OF WORK MATERIALS

Physical properties encompass these characteristics included in the individual material groups, such as the modulus of elasticity, thermal conductivity, thermal expansion, and work hardening.

Modulus of Elasticity The modulus of elasticity can be determined during a tensile test in the same manner as the previously mentioned conditions. However, unlike hardness, yield or tensile strength, the modulus of elasticity is a fixed material property and therefore is unaffected by heat treatment. This particular property is an indicator of the rate at which a material will deflect when subjected to an external force. This property is stated in psi, and typical values are several million psi for metals. A 2-in. × 4-in. × 8-ft. wood beam supported on either end, with a 200-lb weight hanging in the middle, will sag 17 times more than a beam of the same dimensions made out of steel and subjected to the same load. The difference is not because that steel is harder or stronger but that steel has a modulus of elasticity that is 17 times greater than wood.

Machining Considerations. General manufacturing practice dictates that productive machining of a workpiece material with a relatively moderate modulus of elasticity normally requires positive or highly positive raked cutting geometries. Positive cutting geometries produce lower cutting forces, so chip formation is enhanced on elastic material using these types of tools. Sharp positive cutting edges tend to bite and promote shearing of a material, whereas blunt negative geometries have a tendency to create large cutting forces, which impede chip formation by severely pushing or deflecting the part as the tool enters the cut.

Thermal Conductivity Materials are frequently labeled as being either heat conductors or insulators. Conductors tend to transfer heat from a hot or cold object at a high rate; insulators impede the flow of heat. Thermal conductivity is a measure of how efficiently a material transfers heat. Therefore, a material that has a relatively high thermal conductivity would be considered a conductor, whereas one with a relatively low level would be regarded as an insulator.

Machining Considerations. Metals that exhibit low thermal conductivities will not dissipate heat freely, and therefore, during the machining of these materials, the cutting tool and workpiece become extremely hot. This excess heat accelerates wear at the cutting edge and reduces tool life. The proper application of sufficient amounts of coolant directly in the cutting zone (between the cutting edge and workpiece) is essential to improving tool life in metals with low thermal conductivities.

Thermal Expansion Many materials, especially metals, tend to increase in dimensional size as their temperature rises. This physical property is referred to as **thermal expansion**. The rate at which metals expand varies, depending on the type or alloy of material under consideration. The rate at which a metal expands can be determined using the material's expansion coefficient. The greater the value of this coefficient, the more a material will expand when subjected to a temperature rise or contract when subjected to a temperature reduction. For example, a 100-in. bar of steel that encounters a 100° F rise in temperature would measure 100.065 in. A bar of aluminum exposed to the same set of test conditions would measure 100.125 in. In this case, the change in the aluminum bar length was nearly twice that of the steel bar. This is a clear indication of the significant difference in thermal expansion coefficients between these materials.

Machining Considerations. In terms of general machining practice, materials with large thermal expansion coefficients will make holding close finish tolerances extremely difficult because a small rise in workpiece temperature will result in dimensional change. The machining of these types of materials requires adequate coolant supplies for thermal and dimensional stability. In addition, the use of positive cutting geometries on these materials will also reduce machining temperatures.

Work Hardening Many metals exhibit a physical characteristic that produces dramatic increases in hardness due to cold work. Cold work involves changing the shape of a metal object by bending, shaping, rolling, or forming. As the metal is shaped, internal stresses develop that act to harden the part. The rate and magnitude of this internal hardening vary widely from one material to another. Heat also plays an important role in the work hardening of a material. When materials that exhibit work hardening tendencies are subjected to increased temperature, it acts like a catalyst to produce higher hardness levels in the workpiece.

Machining Considerations. The machining of workpiece materials with work-hardening properties should be undertaken with a generous amount of coolant. In addition, cutting speeds should correlate specifically to the material machined and should not be recklessly altered to meet a production rate. The excess heat created by unusually high cutting speeds could be extremely detrimental to the machining process by promoting work hardening of the workpiece. Low chip thicknesses should be avoided on these materials because this type of inefficient machining practice creates heat due to friction, which produces the same type of effect mentioned earlier. Positive low-force cutting geometries at moderate speeds and feeds are normally very effective on these materials.

3.4 METAL MACHINING

The term "machinability" is a relative measure of how easily a material can be machined when compared to 160 Brinell AISI B1112 free-machining low-carbon steel. The American Iron and Steel Institute (AISI) ran turning tests of this material at 180 surface ft and compared their results for B1112 against several other materials. If B1112 represents a 100% rating, then materials with a rating less than this level would be decidedly more difficult to machine, whereas those that exceed 100% would be easier to machine.

The machinability rating of a metal takes the normal cutting speed, surface finish, and tool life attained into consideration. These factors are weighted and combined to arrive at a final machinability rating. The following list shows a variety of materials and their specific machinability ratings:

Material	Hardness	Machinability Rating
6061-T aluminum	—	190%
7075-T aluminum	—	120%
B1112 steel	160 BHN	100%
416 Stainless steel	200 BHN	90%
1120 Steel	160 BHN	80%
1020 Steel	148 BHN	65%
8620 Steel	194 BHN	60%
304 Stainless steel	160 BHN	40%
Iconel X	360 BHN	15%
Rene 41	215 BHN	15%
Waspalloy	270 BHN	12%
Hastelloy X	197 BHN	9%

3.4.1 Cast Iron

All metals that contain iron (Fe) are known as ferrous materials. The word "ferrous" is defined as "relating to or containing iron." Ferrous materials include **cast iron**, pig iron, wrought iron, and low-carbon and alloy steels. The extensive use of cast-iron and steel workpiece materials can be attributed to the fact that iron is one of the most frequently occurring elements in nature.

When iron ore and carbon are metallurgically mixed, a wide variety of workpiece materials result, with a fairly unique set of physical properties. Carbon contents are altered in cast irons and steels to provide changes in hardness, yield, and tensile strengths. The physical properties of cast irons and steels can be modified by changing the amount of the iron–carbon mixtures in these materials as well as their manufacturing process.

Pig iron is created after iron ore is mixed with carbon in a series of furnaces. This material can be further changed into cast iron, steel, or wrought iron depending on the selected manufacturing process.

Cast iron is an iron–carbon mixture that is generally used to pour sand castings, as opposed to making billets or bar stock. It has excellent flow properties and therefore, when it is heated to extreme temperatures, is an ideal material for complex cast shapes and intricate molds. This material is often used for automotive engine blocks, cylinder heads, valve bodies, manifolds, heavy-equipment oil pans and machine bases.

Gray Cast Iron

Gray cast iron is an extremely versatile, very machinable, relatively low-strength cast iron used for pipe, automotive engine blocks, farm implements, and fittings. This material receives the dark gray color that gives it its name from the excess carbon in the form of graphite flakes.

Machining Considerations. Gray cast-iron workpieces have relatively low hardness and strength levels. However, because of their tendency to produce short discontinuous chips, double-negative or negative (axial)–positive (radial) rake-angle geometries are used to machine these materials. When this type of chip is produced during the machining of these workpieces, the entire cutting force is concentrated on a very narrow area of the cutting edge, so double-positive rake tools normally chip prematurely on these types of materials due to their lower edge strength.

White Cast Iron

White cast iron occurs when all of the carbon in the casting is combined with iron to form **cementite**. This is an extremely hard substance that results from the rapid cooling of the casting after it is poured. Because the carbon in this material is transformed into cementite, the resulting color of the material when chipped or fractured is a silvery white—thus the name white cast iron. However, white cast iron has almost no ductility, so when it is subjected to any type of bending or twisting loads, it fractures. The hard, brittle, white cast iron surface is desirable when a material with extreme abrasion resistance is required. Applications of this material would include plate rolls in a mill or rock crushers.

Machining Considerations. Due to the extreme hardness of white cast iron, it is very difficult to machine. Double-negative insert geometries are almost exclusively required for these materials, because their normal hardness is 450–600 Brinell. As stated earlier with gray iron, this class of cast material subjects the cutting edge to extremely concentrated loads, thus requiring added edge strength.

Malleable Cast Iron

When white cast iron castings are annealed (softened by heating to a controlled temperature for a specific length of time), malleable iron castings are formed. Malleable iron castings result when hard, brittle cementite in white iron castings is transformed into tempered carbon or graphite in the form of rounded nodules or aggregate. The resulting material is a strong, ductile, tough, and very machinable product that is used on a broad scope of applications.

Machining Considerations. Malleable cast irons are relatively easy to machine when compared to white iron castings. However, double-negative or negative (axial)–positive (radial) rake-angle geometries are also used to machine these materials as with gray iron, because of their tendency to produce short, discontinuous chips.

Nodular Cast Iron

Nodular or "ductile" iron is used to manufacture a wide range of automotive engine components including cam shafts, crank shafts, bearing caps, and cylinder heads. This material is also frequently used for heavy-equipment cast parts as well as heavy machinery faceplates and guides. Nodular iron is strong, ductile, tough, and extremely shock resistant.

Machining Considerations. Although nodular iron castings are very machinable when compared with gray iron castings of the same hardness, high-strength nodular iron castings can have relatively low machinability ratings. The cutting geometry selected for nodular iron castings is also dependent on the grade to be machined. However, double-negative or positive (radial) and negative (axial) rake angles are normally used.

3.4.2 Steel

Steel materials are composed mainly of iron and carbon, often with a modest mixture of alloying elements. The biggest difference between cast-iron materials and steel is the carbon content. Cast-iron materials are compositions of iron and carbon, with a minimum of 1.7% carbon to 4.5% carbon. Steel has a typical carbon content of .05% to 1.5%.

The commercial production of a significant number of steel grades is further evidence of the demand for this versatile material. Very soft steels are used in drawing applications for automobile fenders, hoods, and oil pans, and premium-grade high-strength steels are used for cutting tools. Steels are often selected for their electrical properties or resistance to corrosion. In other applications, nonmagnetic steels are selected for wristwatches and minesweepers.

Plain Carbon Steel

This category of steels includes materials that are a combination of iron and carbon with no alloying elements. As the carbon content in these materials is increased, the ductility (ability to stretch or elongate without breaking) of the material is reduced. Plain carbon steels are numbered in a four-digit code according to the AISI of SAE system (e.g., 10XX). The last two digits of the code indicate the carbon content of the material in hundredths of a percentage point. For example, a 1018 steel has a .18% carbon content.

Machining Considerations. The machinability of plain carbon steels is primarily dependent on the carbon content of the material and its heat treatment. Materials in the low-carbon category are extremely ductile, which creates problems in chip breaking on turning and drilling operations. As the carbon content of the material rises above 0.30%, reliable chip control is often attainable. These materials should be milled with a positive (radial) and negative (axial) rake-angle geometry. In turning and drilling operations on these materials, negative or neutral geometries should be used whenever possible. The plain carbon steels as a group are relatively easy to machine; they only present machining problems when their carbon content is very low (chip breaking or built-up edge), or when they have been heat treated to an extreme (wear, insert breakage, or depth of cut notching).

Alloy Steels

Plain carbon steels are made up primarily of iron and carbon, whereas alloy steels include these same elements with many other elemental additions. The purpose of alloying steel is to enhance the material's physical properties or its ultimate manufacturability. The physical property enhancements include improved toughness, tensile strength, hardenability (the relative ease with which a higher hardness level can be attained), ductility, and wear resistance. The use of alloying elements can alter the final grain size of a heat-treated steel, which often results in a lower machinability rating of the final product. The primary types of alloyed steel are nickel, chromium, manganese, vanadium, molybdenum, chrome–nickel, chrome–vanadium, chrome–molybdenum, and nickel–molybdenum. The following summaries detail some of the differences in these alloys in terms of their physical as well as mechanical properties for alloyed carbon steels:

- Nickel. This element is used to increase the hardness and ultimate strength of the steel without sacrificing ductility.
- Chromium. Chromium will extend the hardness and strength gains that can be realized with nickel. However, these gains are offset by a reduction in ductility.
- Manganese. This category of alloyed steels possesses a greater strength level than nickel-alloyed steels and improved toughness when compared to chromium-alloyed steels.

- **Vanadium.** Vanadium-alloyed steels are stronger, harder, and tougher than their manganese counterparts. However, this group of materials, loses a significant amount of its ductility when compared to the manganese group.
- **Molybdenum.** These alloyed steels benefit from increased strength and hardness without adversely affecting ductility. They are often considered very tough, with an impact strength that approaches the vanadium steels.
- **Chrome–Nickel.** The alloying elements present in the chrome–nickel steels produce a very ductile, tough, fine-grain, wear-resistant material. However, they are relatively unstable when heat treated and tend to distort, especially as their chromium and nickel content is increased.
- **Chrome–Vanadium.** This combination of alloying elements produces hardness, impact strength, and toughness properties that exceed those of the chrome–nickel steels. This alloyed steel has a very fine grain structure and therefore improved wear resistance.
- **Chrome–Molybdenum.** This alloyed steel has slightly different properties from a straight molybdenum alloy due to the chromium content of the alloy. The final hardness and wear resistance of this alloy exceed that of a normal molybdenum alloy steel.
- **Nickel-Molybdenum.** The properties of this material are similar to chrome–molybdenum–alloyed steels except for one, its increased toughness.

Machining Considerations. The machinability of alloy steels varies widely, depending on their hardness and chemical compositions. The correct geometry selection for these materials is often totally dependent on the hardness of the part. Double-positive milling or turning geometries should be selected for these materials only when the workpiece, machine, or fixturing lacks the necessary rigidity to use stronger, higher-force-generating geometries. In milling, positive (radial)–negative (axial) geometries are preferred on alloyed steels due to their strength and toughness. In turning operations, double-negative or neutral geometries should be used on softer alloy steels. Lead-angled tools should be used on these materials whenever possible to minimize the shock associated with cutter entry into the cut.

Tool Steels

This group of high-strength steels is often used in the manufacture of cutting tools for metals, wood, and other workpiece materials. In addition, these high-strength materials are used as die and punch materials due to their extreme hardness and wear resistance after heat treatment. The key to achieving the hardness, strength, and wear resistance desired for any tool steel is normally through careful heat treatment. These materials are available in a wide variety of grades with a substantial number of chemical compositions designed to satisfy specific as well as general application criteria.

Machining Considerations. Tool steels are highly alloyed and therefore quite tough. However, they can often be readily machined prior to heat treatment. Negative cutting geometries will extend tool life when machining these materials, provided the system (machine, part, and fixturing) is able to withstand the additional tool force.

Stainless Steels

As the name implies, this group of materials is designed to resist oxidation and other forms of corrosion, in addition to heat in some instances. These materials tend to have significantly greater corrosion resistance than their plain or alloy steel counterparts due to the substantial additions of chromium as an alloying element. **Stainless steels** are used extensively in the food processing, chemical, and petroleum industries to transfer corrosive liquids between processing and storage facilities. Stainless steels can be cold formed, forged, machined, welded, or **extruded**. This group of materials can attain relatively high strength levels when compared to plain carbon and alloy steels. Stainless steels are available in up to 150 different chemical compositions. The wide selection of these materials is designed to satisfy the broad range of physical properties required by potential customers and industries.

Machining Considerations. Stainless steels fall into four distinct metallurgical categories austenitic, ferritic, martensitic, and precipitation hardening. Austenitic (300-series) steels are generally difficult to machine. Chatter could be a problem, thus requiring machine tools with high stiffness. However, ferritic stainless steels (also 300-series) have good machinability. Martensitic (400-series) steels are abrasive and tend to form built-up edges, and require tool materials with high hot hardness and crater-wear resistance. Precipitation-hardening stainless steels are strong and abrasive, requiring hard and abrasion-resistant tool materials.

3.4.3 Nonferrous Metals and Alloys

Nonferrous metals and alloys cover a wide range of materials, from the more common metals such as aluminum, copper, and magnesium to high-strength, high-temperature alloys such as tungsten, tantalum, and molybdenum. Although more expensive than ferrous metals, nonferrous metals and alloys have important applications because of their numerous properties, such as corrosion resistance, high thermal and electrical conductivity, low density, and ease of fabrication.

Aluminum The relatively extensive use of **aluminum** as an industrial as well as consumer-based material revolves around its many unique properties. For example, aluminum is a very lightweight metal (one-third the density when compared to steel), yet it possesses great strength for its weight. Therefore, aluminum has been an excellent material for framing structures in military and commercial aircraft. The corrosive resistance of aluminum has made it a popular material selection for the soft drink industry (cans) and the residential building industry (windows and siding). In addition, most grades of aluminum are easily machined and yield greater tool life and productivity than many other metals.

Machining Considerations. Aluminum is a soft, machinable metal, and the limitations on speeds are governed by the capacity of the machine and good safety practices. Chips are of the continuous type, and frequently they are a limiting safety factor because they tend to bunch up. Aluminum has been machined at such high speeds that the chip becomes an oxide powder. To increase its strength and hardness, aluminum is alloyed with silicon, iron, manganese, nickel, chromium, and other metals. These materials should be machined with positive cutting geometries.

Copper Copper is a very popular material that is widely used for its superior electrical conductivity, corrosion resistance, and ease in formability. In addition, when alloyed properly, copper alloys can exhibit a vast array of strength levels and unique mechanical properties.

Several copper alloys are now in widespread commercial use, including copper nickels, **brasses**, **bronzes**, copper–nickel–zinc alloys, leaded copper, and many special alloys. Brass and bronze are the most popular copper alloys in use.

Machining Considerations. The machinability of copper and its alloys varies widely. Pure copper and high-copper alloys are very tough, abrasive, and prone to tearing. To limit and prevent tearing, these materials should be machined with positive cutting geometries. Positive geometries should also be used on bronze and bronze alloys due to their toughness and ductility.

Negative axial and positive radial rake-angle geometries should be used on brass alloys because they have greater levels of machinability and in a cast state their chip formation is similar to that of cast iron.

Nickel Nickel is often used as an alloying element to improve corrosion, heat resistance, and the strength of many materials. When nickel is alloyed or combined with copper (Monels), chromium (Inconels and Hastelloys), or chromium and cobalt (Waspalloys), it provides a vast array of alloys, which exhibit a wide range of physical properties. Other important alloys belonging to this group of materials include Rene, Astroloy, Udimet, Incoloys, and several Haynes alloys. The machinability of nickel-based alloys is generally quite low.

Machining Considerations. Most nickel-based alloys should be machined using positive cutting geometries. Since these materials are machined with carbide at 120 sfpm or less, positive rake-angle geometries are required to minimize cutting forces and heat generation. In the machining of most materials, increased temperature enhances chip flow and reduces the physical force on the cutting edge. Adequate clearance angles must be utilized on these materials because many of them are very ductile and prone to work hardening. When a tool is stopped and left to rub on the workpiece, hardening of the workpiece surface will often occur. To avoid this condition, care should be taken to ensure that as long as the cutting edge and part are touching, the tool is always feeding.

Titanium and Titanium Alloys Titanium is one of the earth's most abundant metals. Thus, its application is fairly widespread, from a cutting-tool material to the struts and framing members on jet aircraft. Titanium and its alloys are often selected to be used in aerospace applications due to their high strength-to-weight ratio and ductility.

Machining Considerations. The machining of titanium and its alloys involves the careful selection of cutting geometry and speed. Positive-rake tools are often preferred on these materials to minimize part deflection and to reduce cutting temperatures in the cutting zone. The generous use of coolants on titanium and its alloys is strongly advised to maintain thermal stability and thus avoid the disastrous effects of accelerated heat and temperature buildup, which leads to workpiece galling or tool breakage (drilling) and rapid edge wear. Type machinability rating for titanium and its alloys is approximately 30% or less.

Refractory Alloys The group of materials designated as refractory alloys includes those metals which contain high concentrations of either tungsten (W), tantalum (Ta), molybdenum (Mo), or columbium (Co). These materials are known for their heat-resistance properties, which allow them to operate in extreme thermal environments without permanent damage. In addition, they are known for their extremely high melting points and abrasiveness. Most of these materials are quite brittle, so they possess very low machinability ratings when their heat resistance and extreme melting properties are considered. The machining of this group of materials is characterized by extremely low cutting speeds and feedrates when utilizing carbide cutting tools.

Machining Considerations. Cast molybdenum has a machinability rating of approximately 30% whereas pure tungsten has a rating of only 5%. The machinability of tantalum and columbium is at a more moderate level and thus falls between these two figures. Generally speaking, these materials should be machined at moderate to low speeds at light depths of cut using positive-rake tools.

3.5 JUDGING MACHINABILITY

The factors affecting machinability have been explained; four methods used to judge machinability are discussed here.

Tool Life Metals that can be cut without rapid tool wear are generally thought of as being quite machinable and vice versa. A workpiece material with many small, hard inclusions may appear to have the same mechanical properties as a less-abrasive metal. It may require no greater power consumption during cutting, yet the machinability of this material would be lower because its abrasive properties are responsible for rapid wear on the tool, resulting in higher machining costs.

One problem arising from the use of tool life as a machinability index is its sensitivity to the other machining variables. Of particular importance is the effect of tool material. Machinability ratings based on tool life cannot be compared if a high-speed steel tool is used in one case and a sintered carbide tool in another. The superior life of the carbide tool would cause the machinability of the metal cut with the steel tool to appear unfavorable. Even if identical types of tool materials are used in evaluating the workpiece materials, meaningless ratings may still result. For example, cast-iron cutting grades of carbide will not hold up when cutting steel because of excessive cratering, and steel-cutting grades of carbide are not hard enough to give sufficient abrasion resistance when cutting cast iron.

Tool life may be defined as the period of time over which the cutting tool performs efficiently. Many variables—such as material to be machined, cutting-tool material, cutting-tool geometry, machine condition, cutting-tool clamping, cutting speed, feed, and depth of cut—make cutting-tool-life determination very difficult.

The first comprehensive tool-life data were reported by F. W. Taylor in 1907, and his work has been the basis for later studies. Taylor showed that the relationship between cutting speed and tool life can be expressed empirically by

$$VT^n = C$$

where V = cutting speed, in feet per minute.

T = tool life, in minutes.

C = a constant depending on work material, tool material, and other machine variables. Numerically it is the cutting speed that would give 1 min of tool life.

n = a constant depending on work and tool material.

This equation predicts that, when plotted on log–log scales, there is a linear relationship between tool life and cutting speed. The exponent n has values ranging from .125 for high-speed steel (HSS) tools, to .70 for ceramic tools.

Tool Forces and Power Consumption The use of tool forces or power consumption as criteria of machinability of the workpiece material comes about for two reasons. First, the concept of machinability as the ease with which a metal is cut implies that a metal through which a tool is easily pushed should have a good machinability rating. Second, the more practical concept of machinability in terms of minimum cost per part machined relates to forces and power consumption, and the overhead cost of a machine of proper capacity.

When using tool forces as a machinability rating, either the cutting force or the thrust force (feeding force) may be used. The cutting force is the more popular of the two because it is the force that pushes the tool through the workpiece and determines the power consumed. Although machinability ratings could be listed according to the cutting forces under a set of standard machining conditions, the data are usually presented in terms of specific energy. Workpiece materials having a high specific energy of metal removal are said to be less machinable than those with a lower specific energy.

The use of net power consumption during machining as an index of the machinability of the workpiece is similar to the use of cutting force. Again, the data are most useful in terms of specific energy. One advantage of using the specific energy of metal removal as an indication of machinability is that it is mainly a property of the workpiece material itself and is quite insensitive to tool material. By contrast, tool life is strongly dependent on tool material.

The metal-removal factor is the reciprocal of the specific energy and can be used directly as a machinability rating if forces or power consumption are used to define machinability. That is, metals with a high metal-removal factor could be said to have high machinability.

Cutting-tool forces were discussed in Chapter 2. Tool-force and power-consumption formulas and calculations are beyond the scope of this text; they are discussed in other textbooks that are more theoretical in their approach to discussing machinability of metals.

Surface Finish The quality of the surface finish left on the workpiece during a cutting operation is sometimes useful in determining the machinability rating of a metal. Some workpieces will not "take a good finish" as well as others. The fundamental reason for surface roughness is the formation and sloughing off of parts of the built-up edge on the tool. Soft, ductile materials tend to form a built-up edge rather easily. Stainless steels, gas turbine alloy, and other metals with high strain-hardening ability also tend to machine with built-up edges. Materials that machine with high shear-zone angles tend to minimize built-up edge effects. These include the aluminum alloys, cold-worked steels, free-machining steels, brass, and titanium alloys. If surface finish alone is the chosen index of machinability, these latter metals would rate higher than those in the first group.

In many cases, surface finish is a meaningless criterion of workpiece machinability. In roughing cuts, for example, no attention to finish is required. In many finishing cuts, the conditions producing the desired dimension on the part will inherently provide a good finish within the engineering specification.

Machinability figures based on surface finish measurements do not always agree with figures obtained by force or tool-life determinations. Stainless steels would have a low rating by any of these standards, while aluminum alloys would be rated high. Titanium alloys would have a high rating by finish measurements, low by tool-life tests, and intermediate by force readings.

The machinability ratings of various materials by surface finish are easily determined. Surface finish readings are taken with an appropriate instrument after standard workpieces of various materials are machined under controlled cutting conditions. The machinability rating varies inversely with the instrument reading. A low reading means good finish and thus high machinability. Relative ratings may be obtained by comparing the observed value of surface finish with that of a material chosen as the reference. Surface finishes will be discussed in Chapter 5.

Chip Form There have been machinability ratings based on the type of chip that is formed during the machining operation. The machinability might be judged by the ease of handling and disposing of chips. A material that produces long, stringy chips would receive a low rating, as would one that produces fine, powdery chips. Materials that inherently form nicely broken chips, a half or full turn of the normal chip helix, would receive top rating. Chip handling and disposal can be quite expensive. Stringy chips are a menace to the operator and to the finish on the freshly machined surface. However, chip formation is a function of the machine variables as well as the workpiece material, and the ratings obtained by this method could be changed by provision of a suitable chip breaker.

Ratings based on the ease of chip disposal are basically qualitative and would be judged by an individual, who might assign letter gradings of some kind. Wide use is not made of this method of interpreting machinability. It finds some application in drilling, where good chip formation action is necessary to keep the chips running up the **flutes**. However, the whipping action of long coils once they are clear of the hole is undesirable. Chip formation and tool wear were discussed in Chapter 2; Figure 3.3 shows ideal chips developed from a variety of common materials.

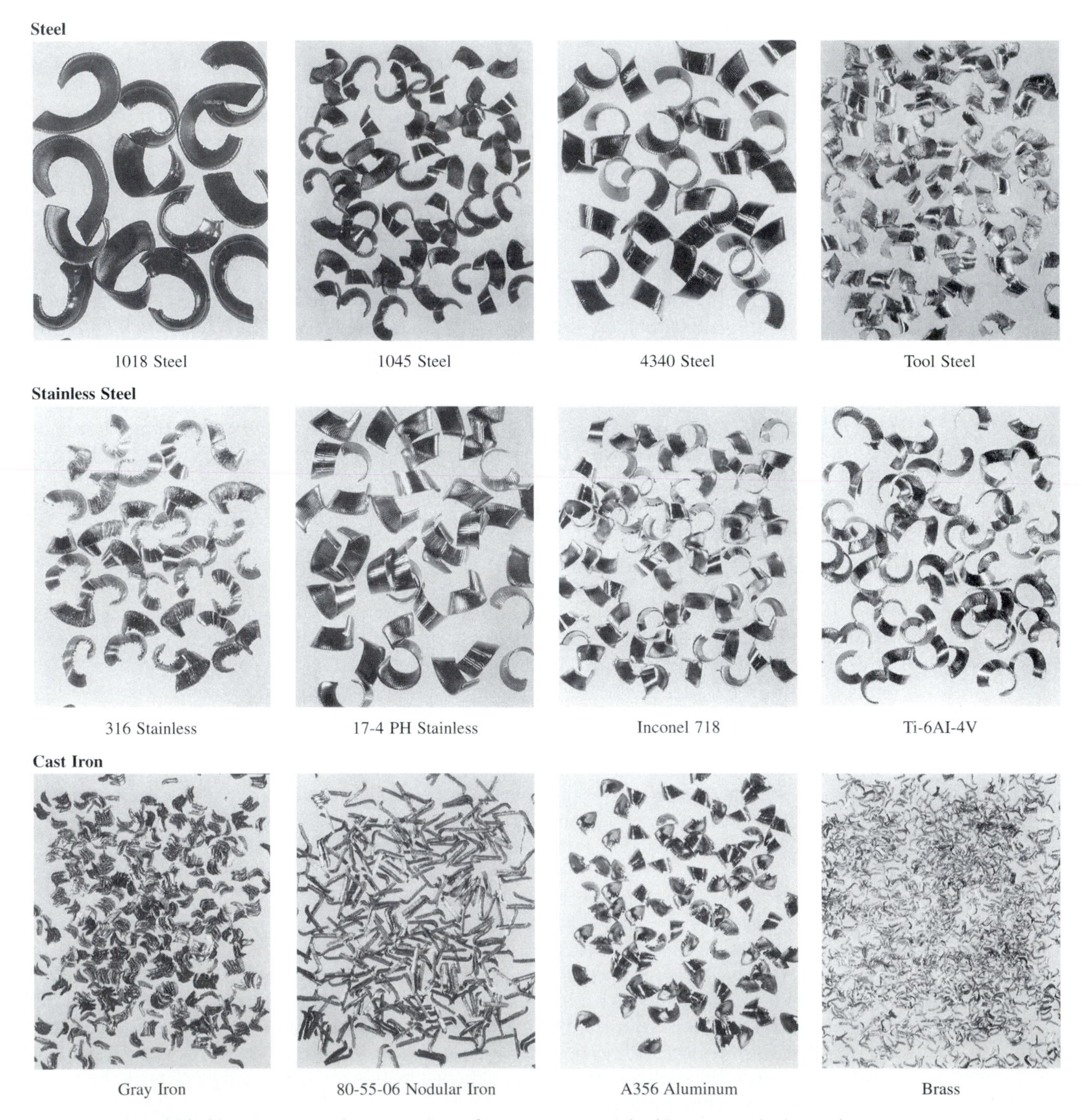

FIGURE 3.3 Ideal chips developed from a variety of common materials. (Courtesy Valenite Inc.)

3.6 REVIEW QUESTIONS AND PROBLEMS

1. List and discuss four steel fabrication methods.
2. Describe how hardness of metals is determined.
3. Describe how yield strength of metals is determined.
4. Explain how work hardening affects the machining process.
5. Describe the various types of cast iron and their uses.
6. How is malleable cast iron formed?
7. What is the difference between cast iron and steel?
8. How does carbon content change the property of steel?
9. What is the purpose of alloying steel?
10. What is the major alloying element in stainless steel, and how does it affect material properties?
11. What are the advantages of using aluminum as compared to steel?
12. List some copper alloys commonly in use.
13. Why are titanium and titanium alloys often used in the aerospace industry?
14. Why is it so difficult to determine tool life?
15. Explain how tool forces and power consumption affect machinability.
16. Why are surface finish and chip formation normally not good machinability indicators?

PART II

MACHINING PROCESS PLANNING

INTRODUCTION

All chapters of *Applied Manufacturing Process Planning* are designed to provide data, principles, and descriptive information required in the application of methods and techniques for the manufacture of a product. Chapters in Part II specifically address the development of a manufacturing plan to produce a specific product.

In Chapter 4, the problems of design features that effect producibility are addressed. Several areas of design producibility considerations are illustrated. Almost always, a product will be more producible at lower cost with higher quality when properly designed originally than when redesigned later with engineering changes and requiring reprocessing. Manufacturing Engineering must acquire all required information and criteria relating to the design of a product before a complete analysis of manufacturing requirements can be made.

Chapter 5 deals with the analysis of part design, including areas that must be checked for specifications, standards, notes, and other related data. Much required information in areas other than the actual part design is found in this chapter.

Processing, as defined in Chapter 6, is the determination of the processes and the sequence of processes that will produce a desired product. This chapter outlines actions that must be taken by Manufacturing Engineering to process a product. Several processing examples are provided.

However, as cautioned in the general introduction for this text, an engineer must be familiar with manufacturing processes to be able to process a part. There is no substitute for this knowledge and experience.

Tolerance charts, shown and explained in Chapter 7, are used to check the manufacturing dimensions and tolerances developed to process a part. A tolerance chart can guarantee that the properly developed processing, dimensioning, and tolerancing of a part will produce a part to the part design. Conversely, a tolerance chart will identify errors in dimensioning, tolerancing, and amounts of stock removed in an improperly developed processing of a part. Several examples of tolerance chart development are illustrated.

Workpiece holders, the design theory of which is described in Chapter 8, must be designed to accommodate workpieces in accordance with the processes, locating surfaces, and support areas determined during processing and tolerance charting.

Areas covered in this chapter are

- Stability of the workpiece
- Locating surfaces and locaters
- Clamping and supporting

All areas are illustrated.

CHAPTER 4

PART DESIGN FOR PRODUCIBILITY

CHAPTER OVERVIEW

4.1 INTRODUCTION

Producibility is inherent in a functional part when that part is so designed as to incorporate *all* elements required for the economical production of that part.

The time to achieve maximum producibility of a part is during the original product design. Decisions made here will increase or decrease the rate of production, can add or eliminate operations, and will greatly affect the cost of required manufacturing facilities.

The elements of a designed part that must be considered for economical production are, literally, every element of the part: materials and material form, design methods, dimensional tolerances, and surface finish restrictions. Each feature will be discussed in this chapter as a very significant preliminary to the development of a manufacturing plan.

Development of the manufacturing plan is the second step in manufacturing; it must produce parts that will satisfy the part design. Hence, the part design dictates much of the processing and manufacturing systems used to manufacture the part. Conversely, the manufacturing process can dictate much of the part design.

4.2 LEAD TIME

4.2.1 Definition

Lead time, to *Manufacturing Engineering*, is the time interval between completion and release of the product design and the required start-up date for manufacture. Every effort must be made to provide sufficient time to develop an optimum manufacturing plan and to obtain and install the required manufacturing facilities.

Insufficient lead time results in two different, usually cumulative, cost penalties:

1. Engineers, designers, and technicians must work on an overtime basis to develop a manufacturing plan that will satisfy the start-up date. Additionally, machine and tool suppliers will incur overtime costs to meet the required delivery date.
2. Insufficient lead time precludes exploring other manufacturing plan options that could result in lower piece costs, higher rates of production, and perhaps higher quality—less scrap. This cost can prove extremely high and is further compounded by the fact that the equipment, once purchased, will be used probably for the life of the product.

It must be recognized that Product Engineering, too, requires its own lead time to properly design, explore alternative designs, select and develop the final design, and officially release a design for production.

4.2.2 Ways to Reduce Lead Time

While products are becoming more sophisticated—requiring more total lead time to design and plan for manufacture—the realities of competition are actually compressing total lead time to bring a new product, or an improved product, to market sooner.

One solution to this enigma includes early collaboration between **Product** and **Manufacturing Engineering**. These two disciplines must collaborate as early on as possible, exchanging requirements and possibilities to steer the product design into a configuration that is functionally acceptable to the product engineer and producible from the viewpoint of the manufacturing engineer.

So much importance has been attached to this subject that almost all of the major automobile manufacturers have set up formal programs, known as 'SE'—'***simultaneous engineering***'—(sometimes called concurrent engineering) to address the problem. Some of the SE programs go beyond fostering Product Engineering and Manufacturing Engineering working together: they include purchasing, quality control, finance, and some suppliers—also Industrial Engineering.

It is beyond the scope of this book to discuss the SE programs now in practice. In general, however, they are comprised of teams or groups, in which product engineers, manufacturing engineers, and other required representatives work simultaneously on assigned car parts and assemblies.

The groups are given the responsibility for designing and tooling their assigned parts by a specified target date. They are supplied with help and varying degrees of decision-making authority.

SE has already become one of the most effective factors in increasing producibility because it has the backing of company management, which can encourage the simultaneous working and cooperating between the involved parties with a little pressure if need be.

The following examples are given to illustrate how concurrent working between Product Engineering and Manufacturing Engineering can increase producibility.

EXAMPLE 1

A **critical surface** of a new product design will require certain minimum characteristics. Rather than Product Engineering empirically determining "safe" dimensions, tolerances, and surface finish limits that are more restrictive (and costly) than is actually required—Manufacturing Engineering should be requested to provide recommendations and equipment capabilities. Manufacturing Engineering will study the part and provide a recommended processing along with the obtainable tolerances and finishes of that processing. Mutually satisfactory dimensions, tolerances, and surface finish specifications can then be settled on for the surface in question.

EXAMPLE 2

The foregoing approach is also feasible with intricate stampings: Product Engineering can supply the stamping material specifications, material thickness, and overall general outline and dimensions to Manufacturing Engineering, who study the part for processing options such as

- Forming the part in one piece.
- Manufacturing the part in several pieces and joining the pieces into the final part by welding, brazing, riveting, and so on.
- Manufacturing the part in another configuration that—when split—results in two or more of the desired parts. Manufacturing Engineering will then recommend the most producible form.

Again, a mutually satisfactory design can be agreed upon at the beginning.

The time to eliminate a manufacturing operation is before it is required—at the product design inception. If an operation cannot be entirely eliminated, careful consideration can very often reduce the production time and cost required.

For instance, where a surface can be used "as cast," it must, of course, be so specified. In some instances, the superior surfaces provided by die casting will dictate that process.

When a hole is required that cannot be cored, the drilling time can sometimes be reduced as shown in example Figure 4.1. The drilling time of the hole shown in Figure 4.1(a) is not only the longest of the three examples, but it will cause additional tool wear and breakage, particularly at the start of drilling when an interrupted-cut condition exists.

If the hole is for drainage, venting, or other such use, it might be possible to design the hole normal to the surface as shown in Figure 4.1(b). Here the hole depth is reduced from 2.5 times the drill diameter, as shown in Figure 4.1(a), to 1.0 drill diameter as shown in Figure 4.1(b).

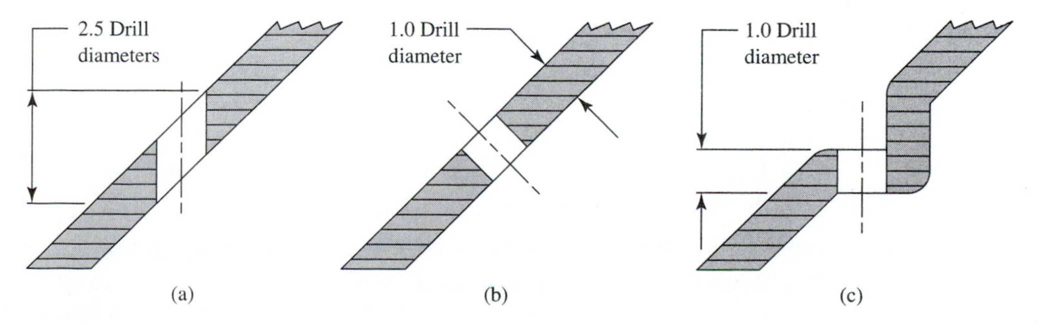

FIGURE 4.1 Drilling conditions.

FIGURE 4.2 Drilling paths.

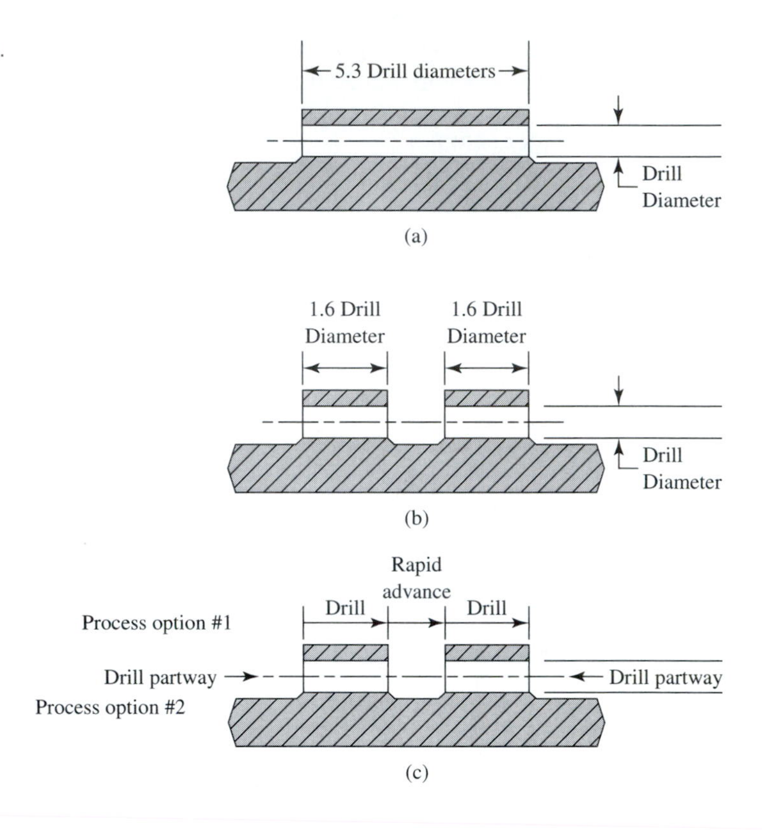

The interrupted-cut condition is also eliminated. Figure 4.1(c) is a composite in which the hole position remains as at Figure 4.1(a), but the drill starting conditions and drill starting and finishing times are improved. A change of the casting would be required.

Figure 4.2(a) shows a forged steel bracket containing a drilled hole with a length of five times the drill diameter. This is a costly design from the standpoint of drilling time. The cost of drilling a hole with a depth of up to three times the drill diameter is a straight-line function, whereas the cost accelerates beyond the three diameters' ratio due to the drill flutes loading with chips, difficulties in providing coolant, and tool wear (see Fig. 4.3). Special drills, special coolants and systems, and "drill and withdraw" drilling feeds to unload chips all increase the time required, and costs, to drill deep holes.

The requirement for some deep drilled holes can be eliminated "at the source"—the original part design. One example is shown in Figure 4.2. A redesign, as shown in Figure 4.2(b) provides two options:

1. The drilling depth is now 1.6 times the drill diameter. This depth can be drilled with no withdrawal of the drill to unload chips. The drill can be rapidly advanced to the second section, which can then also be drilled through with a continuous feed.
2. The part can be drilled from each end simultaneously, which will reduce drilling time by one half. If close alignment is a consideration, the hole probably will require **line reaming** or **line boring** in any case.

In summary, lead time can be shortened when Product and Manufacturing Engineering collaborate and work concurrently rather than successively.

Tooling, labor, and equipment costs can be significantly reduced by eliminating or simplifying manufacturing operations at the source—the original product design. Also, later, when

FIGURE 4.3 Drilled hole length to drill diameter ratio.

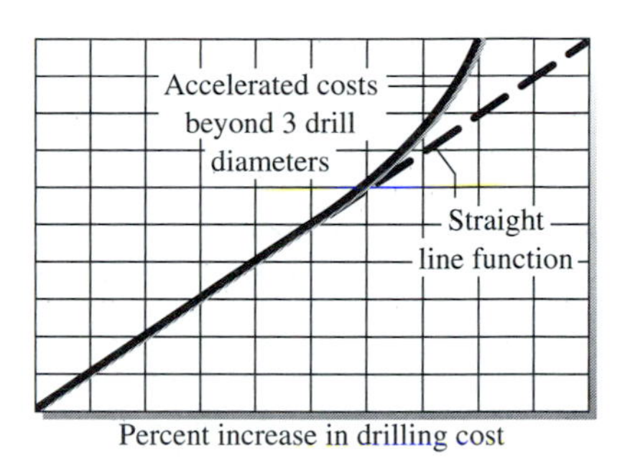

Product Engineering completes and releases final designs for production, Manufacturing Engineering will not lose time studying the prints and requesting information and/or changes. In fact, time will have been saved by determining equipment availability, machine delivery times, and other vital information concurrently with the product design time.

4.3 PRODUCT ENGINEERING FUNCTIONS

Product Engineering's functions are threefold:

- Product Engineering must design a functional part.
- Product Engineering must design a producible part.
- Product Engineering must meet target costs.

4.3.1 Functional Responsibility

Organizational charts and job descriptions spell out reporting structures, responsibilities, and functions of the different areas and departments of manufacturing companies. Product Engineering's functions include, but are not limited to,

1. Develop preliminary designs.
2. Build product prototypes.
3. Test.
4. Complete final designs including
   a. Part prints—detail and assembly drawings
   b. Material specifications
   c. Specifications of heat treat: hardness, toughness
   d. Specifications of coating: organic, inorganic.
5. Officially release part designs and specifications for manufacture to all involved areas.

Product Engineering must include many significant considerations in the product final design:

1. Useful life of the product as designed and as determined by testing program.
2. Ease of servicing and maintaining the product.
   a. Spare, or service parts' design must be determined. This may or may not be the same as the original product design.
   b. Service procedures must be determined and spelled out in service manuals.
3. Appearance.

4. Safety.
5. Packaging.
6. Cost.

Product Engineering is not usually responsible for styling. Styling is commonly set up as a separate function and administered by a defined styling department. Styling often consists of adding a shaped exterior skin, cover, or housing over mechanical, hydraulic, and/or electrical components. Some products have the desired shape or style inherent in the exterior surfaces of the mechanical components. In any case, Product Engineering must design the product to include the required shape or styling.

Components other than those containing exterior, visual surfaces must also be designed for customer acceptance and sales appeal. Mechanical, electrical, electronic, hydraulic, pneumatic, and all other such components can and must be designed to reflect clean, strong, well-balanced concepts.

Simplified designs that eliminate intricate detail and involved surface shapes not only look good, but will function as well or better. Such designs almost always cost less to produce.

Famed Ford trimotor airplane designer William B. Stout admonished his assistants to "design for strength and then add lightness" (by means of lightening holes, thinner sections with stiffening ribs, etc.). Perhaps we could add a corollary—"Design for function and then add simplicity."

4.3.2 Producibility

Product design is the first step in the manufacture of a product. The design dictates to a very great degree the processing and manufacturing systems that will be required to manufacture the product.

For instance, the specifications of

1. Finished diameters in soft material will require turning operations.
2. Finished flat surfaces in soft material will require milling or broaching operations.
3. Finished surfaces in hard material will require grinding operations.
4. Powdered metal parts can eliminate many operations.
5. Stamped parts versus machined parts will require entirely different tooling.
6. Bar stock and tubing will require lathe-type machines.
7. Castings and forgings likely will require surfaces machined on milling machines and broaches.

Close tolerances on *dimensions* add higher costs to a product through the requirement of slower feeds, longer time cycles, and added operations. Close tolerance on surface finishes has the same effect as restricting dimensional tolerances.

4.4 PART DESIGN

4.4.1 Coordinate Dimensioning

It is not within the scope of this book to cover drafting standards. Most major manufacturing companies have their own standards, which are usually used in conjunction with the American National Standard for Engineering Drawings (ANSI Y14.2M)—ASME 414.5M. However, certain techniques in the applications of such standards will aid in the development of a more producible design.

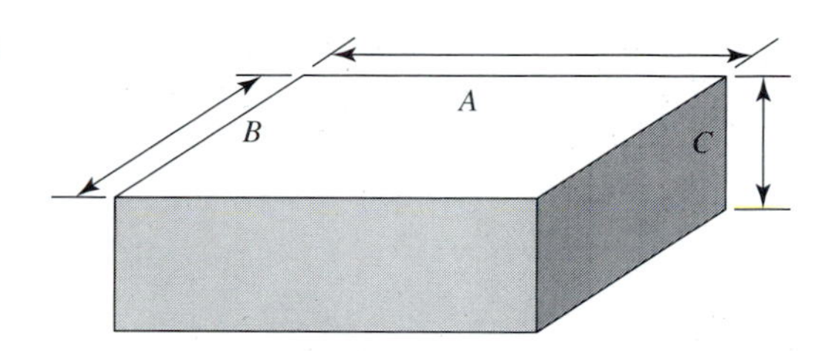

FIGURE 4.4 Coordinate dimensioning of surfaces *A*, *B*, and *C*.

Rectangular Parts Coordinate dimensioning is the conventional system used in designing the majority of parts. A rectangular part consisting of surfaces *A, B,* and *C* is dimensioned as shown in Figure 4.4.

As parts increase in complexity of design from the basic *A–B–C* form, features of the part should be related to (or located *from*) one of the basic surfaces as shown in Figure 4.5.

- Figure 4.5(a) shows a hole dimensioned from basic surfaces *A* and *B* (datums).
- Figure 4.5(b) shows a notch added and correctly dimensioned from common surfaces *A* and *B*.
- Figure 4.5(c) shows the hole and notch incorrectly dimensioned from other than common surfaces *A* and *B*. Such dimensioning reduces manufacturing tolerances and will be further discussed in the Sections 4.4.2 to 4.4.5 on baselines and tolerance charting.

Angular Parts Angular parts pose special problems when dimensioning, or measuring, as shown in Figure 4.6.

- Figure 4.6(a) shows the manufacturing tolerance applied directly to the angle. This dimensioning method is not desirable for the reasons shown in Figure 4.6(b).

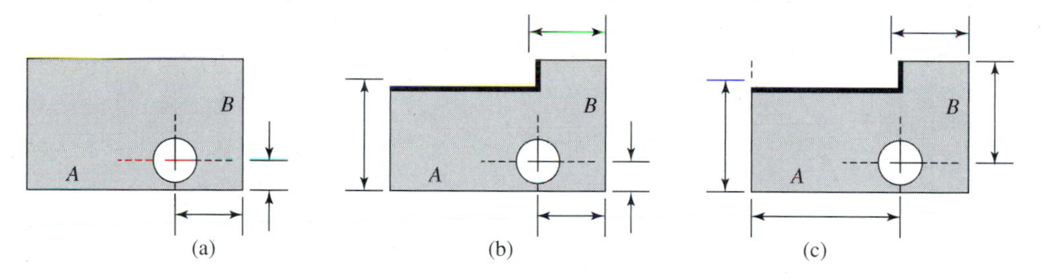

FIGURE 4.5 Correct dimensioning and Incorrect dimensioning.

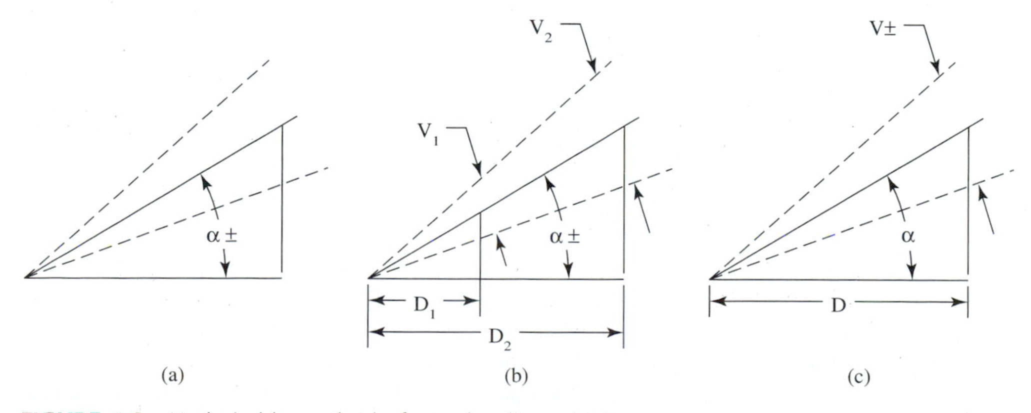

FIGURE 4.6 Undesirable method of angular dimensioning.

FIGURE 4.7 Desirable angular dimensioning.

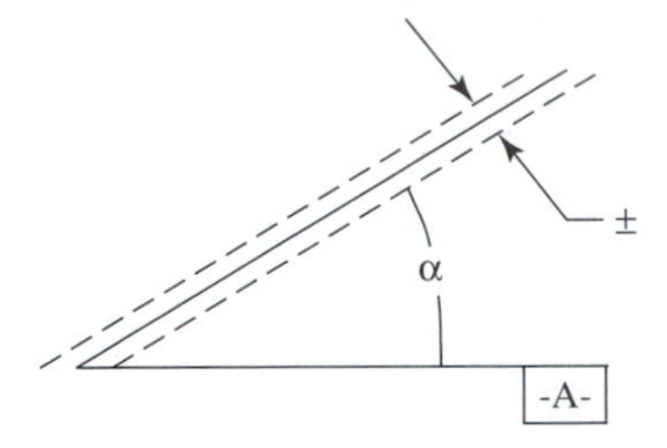

- Figure 4.6(b) shows that the *amount* of variation becomes greater as the distance from the vertex increases: V_1 at D_1 increases to V_2 at D_2.
- An angle can be measured by comparing it with a master, or the allowable amount of variations at a specific distance, *D*, from the vertex can be specified and so measured as shown in Figure 4.6(c).
- The preferred way to treat an angle and its tolerance is shown in Figure 4.7.
- Figure 4.7 shows the angle as a basic quantity with no tolerance and is drawn from line *A*. The manufacturing tolerance is applied to that entire line. This method of dimensioning is discussed more thoroughly in Section 4.4.5, "Geometric Dimensioning and Tolerancing."

Cylindrical Parts Since cylindrical parts have uniform sections, dimensioning them is relatively simple, consisting of radii, diameters, lengths, and notes. The design elements that affect producibility are

1. The *amounts* of allowable manufacturing tolerances specified for straightness, squareness, concentricity, run out, and dimensions. Of course, the more manufacturing tolerance that can be applied to a product design, and still retain a quality, functional part, the more producible the design will be.
2. The *surfaces* that are selected from which length dimensions are located. As in rectangular part dimensioning, distances between surfaces should be related to, or located from, one common surface as shown in Figure 4.8(a).

4.4.2 Baseline Dimensioning

Baseline dimensioning, as the term implies, is an arrangement whereby dimensioning is drawn from a common surface or base to the various related surfaces. This arrangement realizes higher producibility by providing for the part to be located on the baseline surface when machining the other surfaces related to it. Full manufacturing tolerances can be used, and most related surfaces can be machined at one time. Surfaces left unmachined will also retain their relationship during manufacture as will be shown later in Chapter 7, "Tolerance Charting." Examples of parts employing the baseline concept are shown in Figure 4.9.

It is not always feasible to dimension parts from one common *A–B–C* surface. This is increasingly true as the part design becomes more complex. However, producibility can still be greatly enhanced if a conscientious effort is made to design the part with, perhaps, no more than two or three related baselines. Examples of this are shown in Figure 4.10.

Tolerance Stacking Parts designed so that machined surfaces are not dimensioned from a common baseline may incur reduced producibility due to "tolerance stacking." A more thorough discussion of this condition is covered in Chapter 7, "Tolerance Charting." However, to clarify at this point what is meant by "tolerance stacking," refer to Figures 4.11(a) and 4.11(b).

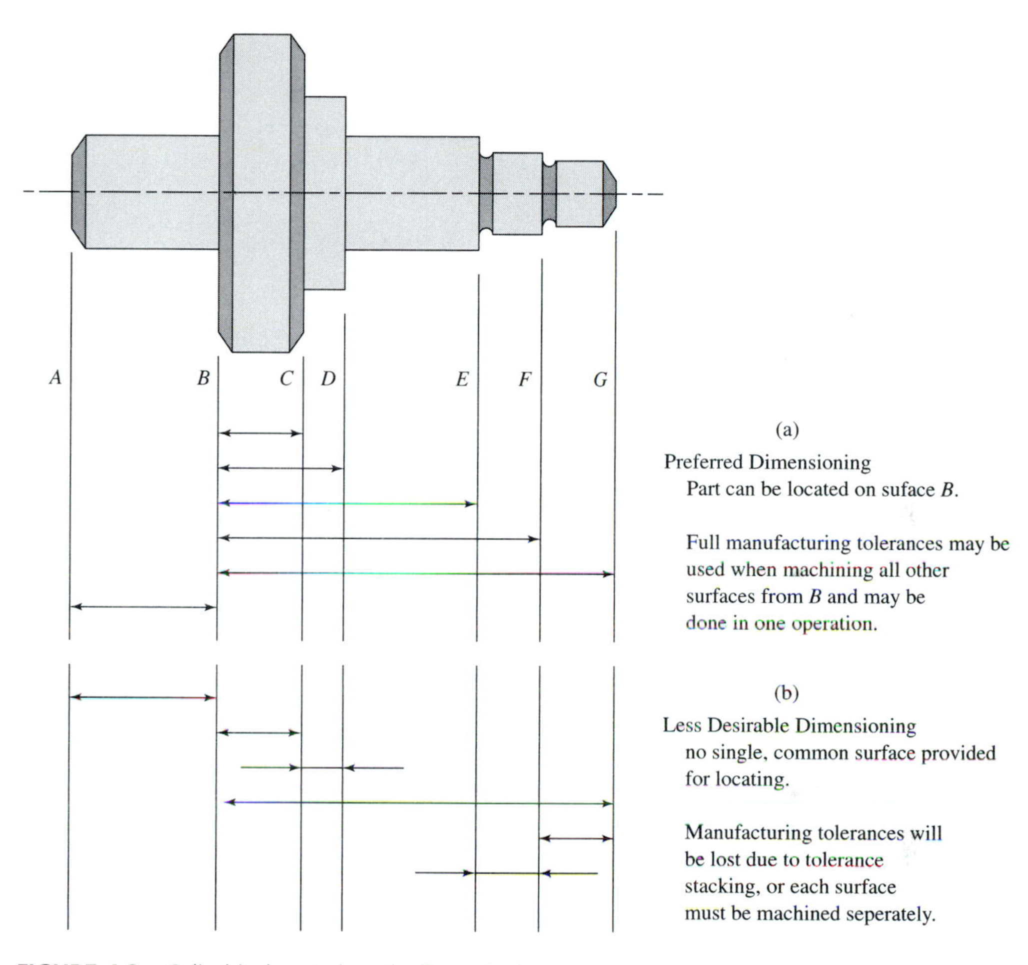

FIGURE 4.8 Cylindrical parts length dimensioning.

- With the baseline located at *A*—Figure 4.11(a), no tolerance stacking occurs when machining surface *D*.
- Surface *D* can be machined directly from the baseline surface, *A*, to whatever the full specified tolerance is for dimension *A–D*. There will be no effect on the tolerances for dimensions *A–B* or *A–C*.
- Also, dimensions *A–B* and *A–C* can be machined in the same operation as *A–D* because the baseline, or locating surface, is common to all three dimensions.

If the part, Figure 4.11(b), is processed and machined the way it is dimensioned, there will be a tolerance stack on the finished part:

- *A* is the baseline for dimension *A–B*. If the part is machined to the full manufacturing tolerance shown from *A*, *A–B* will have a tolerance of ±.010.
- *B* is the baseline for dimension *B–C*. If the part is machined to the full manufacturing tolerance shown from *B*, *B–C* will have a tolerance of ±.010.
- *C* is the baseline for dimension *C–D*. If the part is manufactured to the full manufacturing tolerance shown from *C*, *C–D* will have a tolerance of ±.010.

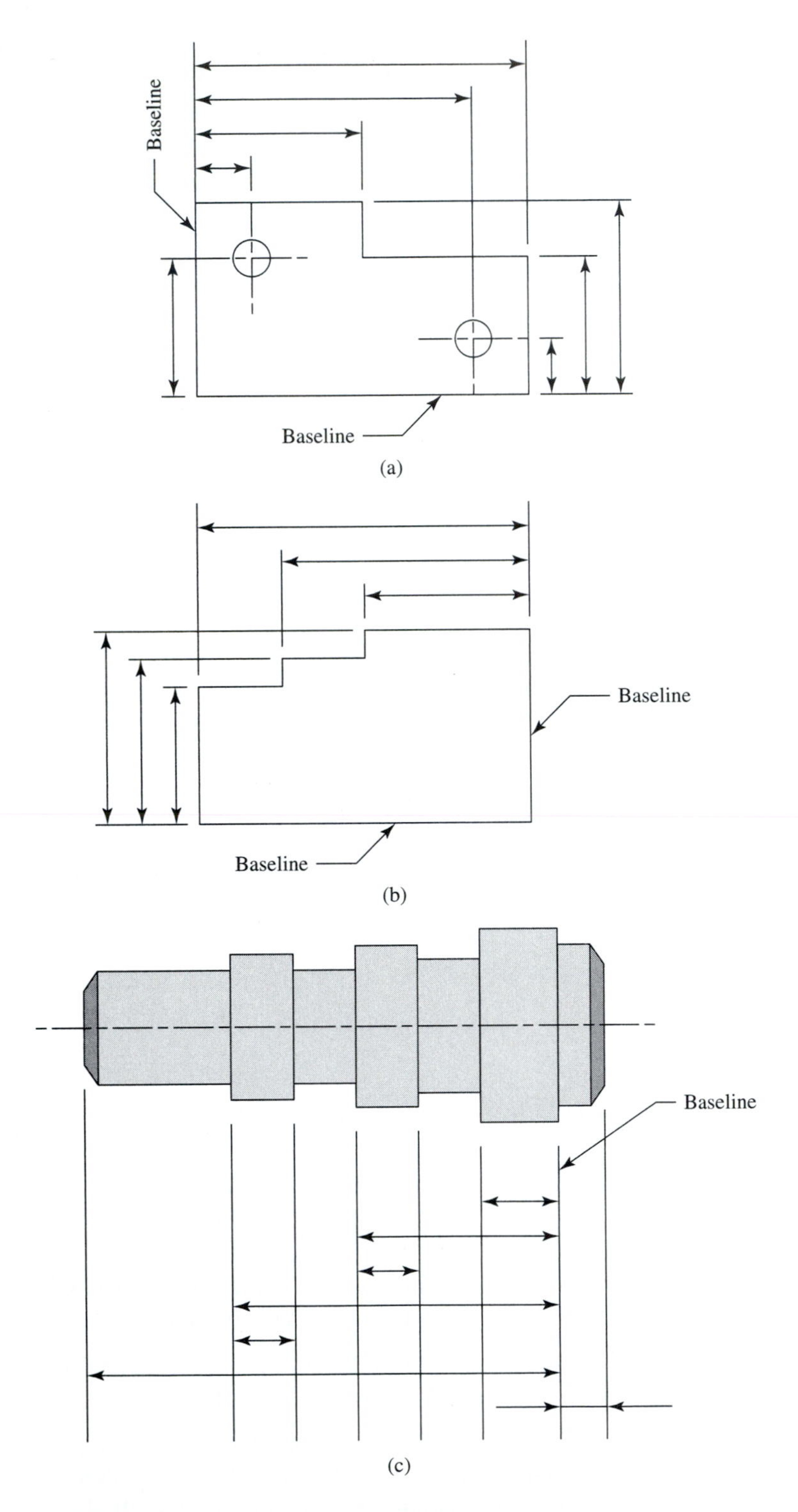

FIGURE 4.9 Baseline dimensioning. (a, b) Rectangular parts, (c) Cylindrical parts.

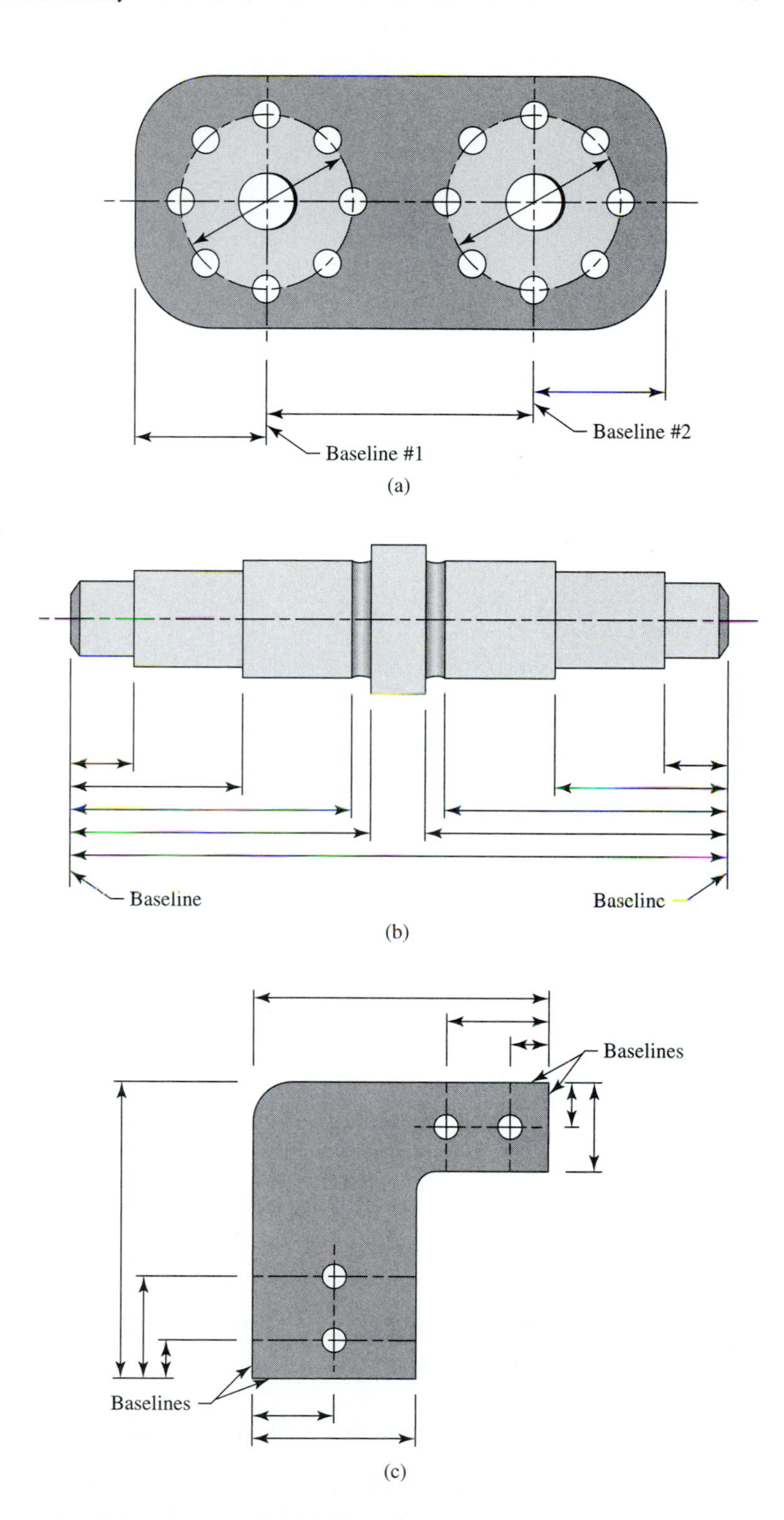

FIGURE 4.10 Baseline dimensioning of more complex parts.

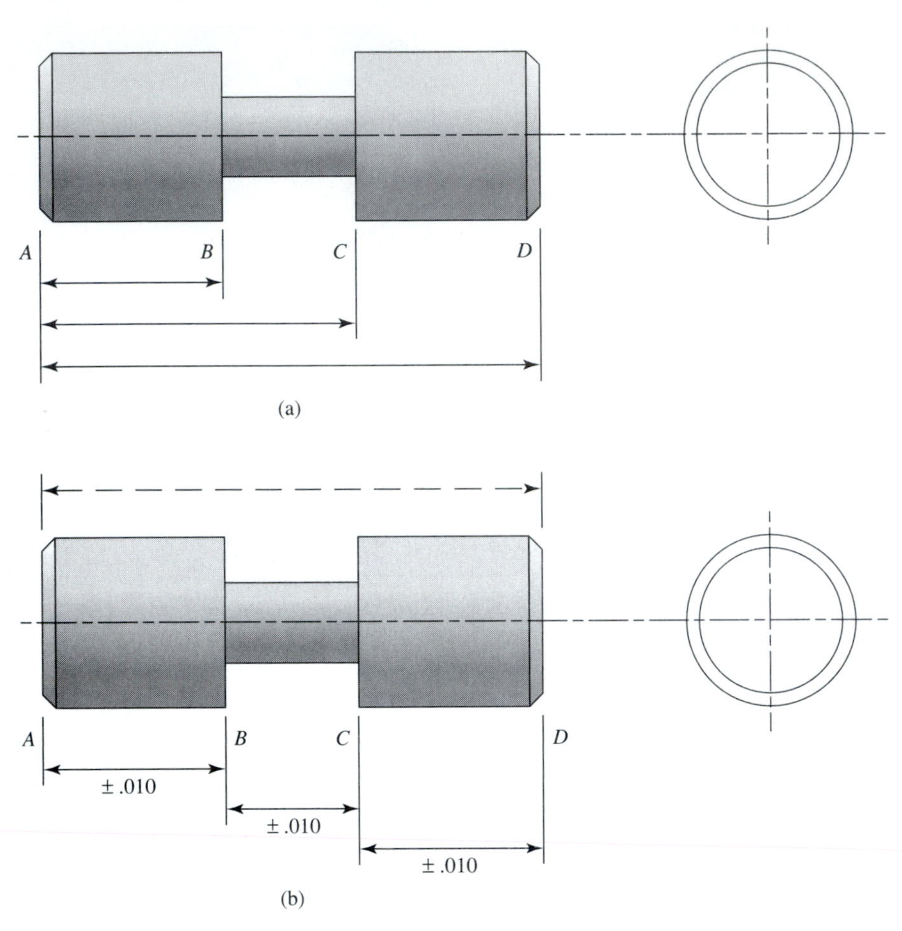

FIGURE 4.11 Tolerance stacking examples: (a) No tolerance stacking occurs for dimensions *A–D*. (b) Tolerance stacking does exist for dimension *A–D*.

- Tolerances always add. Therefore, the total tolerance, or *tolerance stack*, for the overall dimension *A–D* of (b) is:

$$\begin{aligned} \text{Tolerance } A\text{–}B &= \pm.010 \\ \text{Tolerance } B\text{–}C &= \pm.010 \\ \underline{\text{Tolerance } C\text{–}D} &= \underline{\pm.010} \\ \text{Tolerance stack of } A\text{–}D &= \pm.030 \end{aligned}$$

If the overall length tolerance must be reduced, the manufacturing tolerances of each of the contributing dimensional tolerances must also be reduced accordingly.

4.4.3 Tolerances—Dimensional

Dimensional tolerances have a direct effect on producibility. Tolerances are the amount of "leeway" or manufacturing error the designed part can tolerate and remain reliable and functional. This is the amount of latitude available for the manufacturing process.

More tolerance equates to greater producibility and less cost; a higher cutting speed and feeds with fewer qualifying operations can be used. Less tolerance reduces producibility and raises costs; slower cutting speeds and feeds with additional operations required are the probabilities.

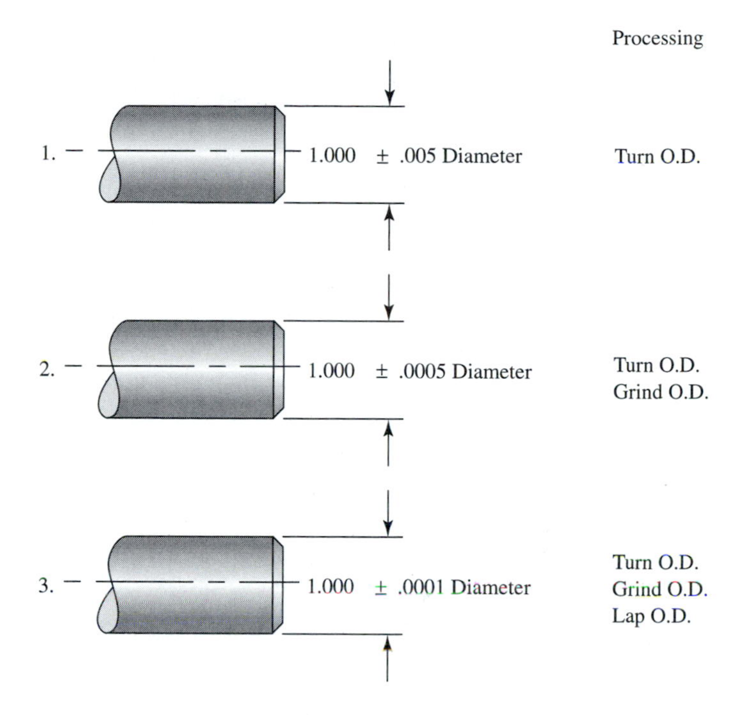

FIGURE 4.12 Process costs increase with more restrictive tolerances.

A general comparison of tolerances versus the manufacturing operation, or operations, probably required is included and discussed in the sections covering part print analysis in Chapter 5. Many attempts have been made to chart part processing and part costs versus specified dimensional tolerances. All comparisons show that larger tolerances reduce costs. However, no such study can be exact because the part design, material, hardness, processing, management practices, and a host of other conditions affect the outcome.

Accurate costs and processing determination can only be determined by a study of the part design and by developing a detailed estimate of cost wherein all factors are considered. However, the costs of similar parts and a family of parts can usually be charted after the first part has been studied and estimated.

Figure 4.12 shows typical processing for each of three diameters that have increasingly closer dimensional tolerances.

4.4.4 Tolerances—Surface Finish

The same statements made concerning producibility versus dimensional tolerances can be made for *surface finish tolerances* versus producibility. Figure 4.13 shows typical processing for each of three diameters that have increasingly closer surface finish tolerances.

Surface finish tolerances and dimensional tolerances must, of course, both be specified for a surface diameter or a flat surface. Also, they are interrelated in the following ways:

1. It is possible to have a part designed with a large diametral tolerance but with a small, or close, surface finish tolerance.
2. It would be very difficult to check, or inspect, a part designed with a very large (rough) surface finish tolerance but with a small, or close diametral tolerance. The reason for this is that the waviness and heights of the rough tool marks could well exceed the dimensional variation (see Fig. 4.14).

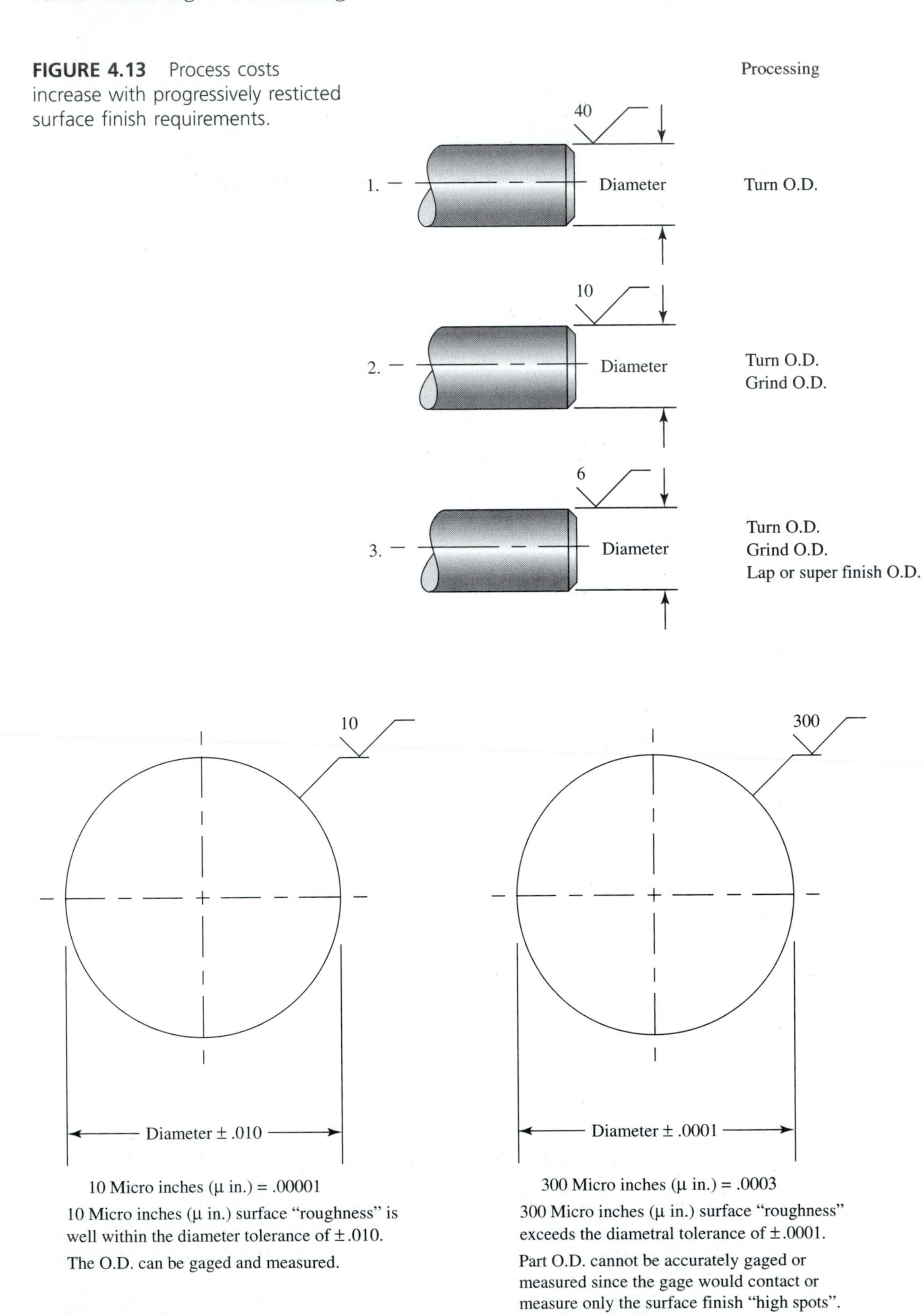

FIGURE 4.13 Process costs increase with progressively resticted surface finish requirements.

FIGURE 4.14 Part dimensioning must relate to surface finish specifications.

In summary, producibility will be furthered by specifying the maximum allowable dimensional and surface finish tolerances as determined by the part's function.

The product engineer should not reduce the amount of tolerance by any percentage "for insurance." Manufacturing Engineering must provide equipment that is capable of producing parts within specified tolerances; Quality Control must monitor for any variances.

4.4.5 Geometric Dimensioning and Tolerancing

> Geometric Dimensioning and Tolerancing (GD&T) saves money directly by providing for maximum producibility of the part through maximum production tolerances. It provides "bonus" or extra tolerances in many cases.
>
> In particular, it is a means of dimensioning and tolerancing a drawing with respect to the actual function or relationship of part features that can be most economically produced. "Function" and "relationship" are the key words.
>
> In general, it is a system of building blocks for good drawing practice and provides the means of stating necessary dimensional or tolerance requirements on the drawing not otherwise covered by implications or standard interpretation.
>
> It ensures that design dimensional and tolerance requirements, as they relate to actual function, are specifically stated and thus carried out.
>
> It adapts to, and assists, computerization techniques in design and manufacture.
>
> It ensures interchangeability of mating parts at assembly.
>
> GD&T reduces errors and misunderstandings. It is specific, convenient to use and is uniformly applicable to all designs.

Geometric dimensioning and tolerancing is now the required drafting standard for all U.S. defense work. It is also fast becoming the standard (ASME 714.5M, 1994, "Dimensioning and Tolerancing") for most major manufacturing companies.

GD&T is almost a universal engineers' language in that a part print utilizing this technique can be read and understood by an engineer of any nationality. For this reason, it has even more value for manufacturers who have worldwide production facilities.

4.5 MATERIAL

A material must be selected and specified by Product Engineering to meet the part's minimum functional requirements. Then, within the material group selected, the most producible alloy, form, and possibly heat treat method must be specified.

Some products require the ultimate in material physical properties, and no alternates are available or permissible. When this is true, processing must be planned to provide facilities capable and with capacity for producing parts to the part design.

When optional material alloys, form, or heat treating do exist, significant time, material, equipment, and potential scrap costs can be saved by studying the alternatives and their effects on producibility *before* releasing the final part design.

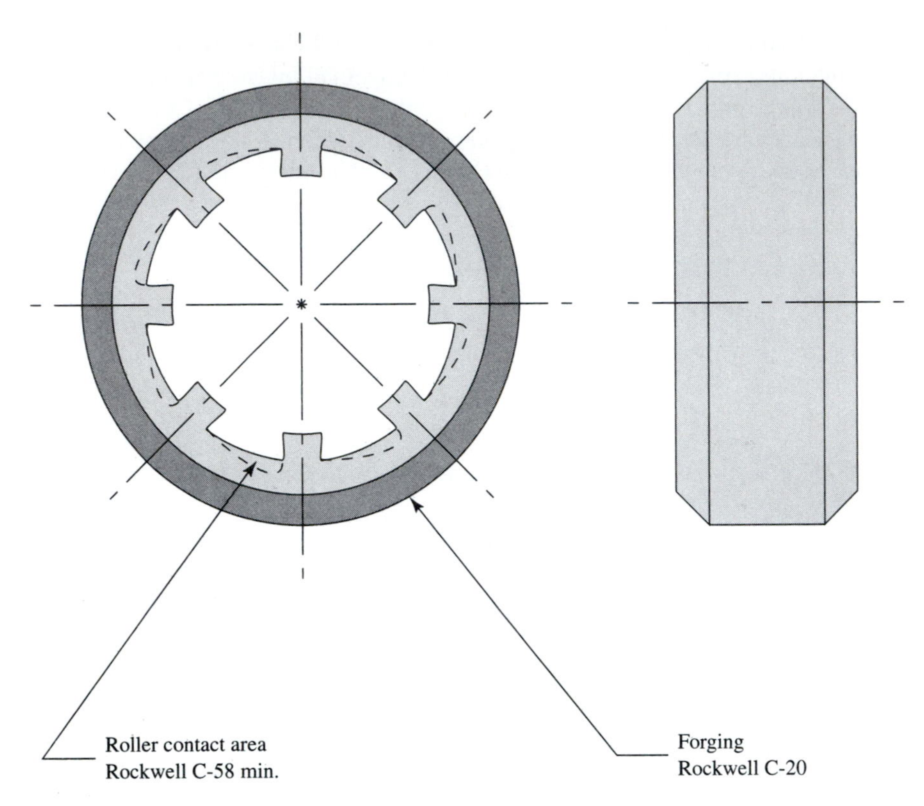

FIGURE 4.15 Overrunning Clutch Cam.

As an example: One automobile manufacturer designed an Over-Running Clutch Cam for a torque converter as shown in Figure 4.15. The part was to be machined from a forging having a hardness of Rockwell C-20 (HRC-20). The material specified was SAE-4340. After machining, the roller contact areas were to be hardened to 58 Rockwell C minimum (HRC-53MN).

The processing developed for this design was as follows:

Operation No.	Operation Description	Equipment
10	Receive and inspect forging Hardness to be Rockwell C-20	
20	Machine O.D. Machine other end face	Chucking Machine
30	Machine I.D. Machine other end face	Chucking Machine
40	Broach cam surfaces	Broaching Machine
50	Harden cam surfaces	Flame Hardener
60	Grind O.D.	Cylindrical Grinder
70	Grind both end faces	Disc Grinder
80	Grind, roll contact surfaces	Special Grinder

Problems were encountered almost at once, when extremely hard parts broke teeth in the costly form broaches. Meetings were held with the forging supplier, trial lots of many differing heat treats were run, 200% check of parts was initiated—but the breakage continued. The hardness was found to be in spots and not the entire part. The hardness checking brale missed finding the hard spots.

To keep the production line going, all incoming forgings were normalized and re–heat treated at added cost, then the part was redesigned. All options were studied, and the most producible version adopted. The material was changed from SAE 4340 forging to SAE 4340 seamless steel tubing and the process was changed accordingly:

Operation No.	Operation Description	Equipment
10	Receive and inspect tubing Hardness to be Rockwell C-20	
20	Machine blank complete and cut off	Auto Screw Machine
30	Broach cam surfaces	Broaching Machine
40	Harden cam surfaces	Induction Hardener

The balance of the operations remained the same.

Results

- Material cost was increased but uniform hardness was assured.
- One operation was eliminated, reducing labor, material-handling, and perishable tool costs.
- Floor space requirements were reduced.
- The total, net cost per piece was lowered and continued production was realized.
- Higher-quality, more-uniform parts were produced.

Hindsight is too often used to reprocess a part—at a penalty. The time to evaluate other alternatives is before the part is released to production. Too often, there is not enough time to do it right the first time, but time is found, somehow, to do it over.

4.5.1 Other Considerations: Castings, Forgings, Extrusions, and Stamping

Certain product designs require specific materials and properties to ensure a functionally acceptable product, but where any latitude exists in selecting material for a product design, the most producible material should be specified. For example, if a design requires the use of *castings*, the most producible *casting material* and *casting process* that satisfies the design must be determined and used.

Where an alternative process such as forging or extruding can provide satisfactory physical properties and prove to be more producible, the alternative process should be used. A few relevant considerations of castings, forgings, and extrusions follow.

Sand Castings

Definition. Molten metal is poured into a shaped cavity made in sand. Sand castings are considered to be the most flexible of casting methods and probably represent the majority of material form selections for all non-bar-stock/tubing products. They can be poured with extremely complex sections in almost limitless shapes and sizes using any commercially available metal.

The principal limitations of sand castings are that they cannot be poured with thin sections and almost all functional surfaces (locating, mating, etc.) must be machined to maintain dimensional tolerances and surface finishes.

Tolerances and Surface Finishes. Dimensional tolerances for aluminum, copper, and magnesium castings range up to .032 in. per 6-in. length for in-line or same-side dimensions. Dimensional tolerances for these metals across the parting line range up to double the same-side tolerances. Surface finishes are 250+μin. for nonferrous and 500+μin. for ferrous. Dimensional tolerances for irons and steels range up to 20% more than tolerances for aluminum, copper, and magnesium castings.

Applications.

- Gears, hubs, pulleys, covers
- Pump housings
- Machine bases, slides, and components
- Engine blocks
- More

Tooling Cost. Low.

Permanent Mold Castings

Definition. Molten metal is poured by gravity into a permanent mold which is a metal die composed of hinged and sliding sections. The permanent mold casting process can be used for relatively simple part designs of moderate size.

The molds are usually mounted on a revolving table, allowing the process to be run at a relatively high rate of production—controlled by the chill time required to solidify the castings before opening the die sections.

Savings result from less required machining due to the ability of the process to hold closer tolerances and smoother surface finishes. Material is also saved due to better control of casting sections. The process is limited to small to medium-sized parts, and only nonferrous metals can be cast.

Tolerances and Surface Finishes. Dimensional tolerances are generally .016 in. for the first inch plus .006 in. for each additional inch. Surface finishes generally range from 125 to 250 μin.

Applications. Castings are poured for many products using the permanent mold process, but the aluminum internal combustion engine piston casting accounts for the greatest production. Pistons of many types are cast, some with metal control struts cast in place.

Tooling Cost. Moderate.

Die Casting

Definition. Molten metal is forced under pressure into a closed metal die. The diecasting process produces parts with the closest dimensional tolerances and the smoothest finishes of all metal casting processes. Very complex shapes can be produced with very thin walls by the diecasting process. Also, inserts can be cast in place. Sizes of parts range from small electronic components to engine blocks with cast-iron sleeves cast in place.

One limitation of the process is that only nonferrous metals can be cast. Another limitation is the fact that almost all diecast parts have porosity, part of which can be eliminated by purging the die passages with oxygen after each cast. Also, an acceptable method of impregnating and filling porous areas is available as a secondary operation.

Tolerances and Surface Finishes. Dimensional tolerances as close as .005 in. per inch can be held for non-parting-line dimensions; .010 in. per inch is normal. Surface finishes normally range from 32 to 125 μin., and smoother surfaces can be achieved by special means.

Applications. Almost any nonferrous part can be produced by die casting provided the tooling cost can be justified, including

- Electronic parts
- Valves and valve bodies
- Appliances and components
- Transmission and transaxle cases
- Cylinder blocks with lines cast in place

Tooling Cost. High.

Forgings

Definition. Forging is the process of forming metal to a desired shape by applying impacts of pressure from hammers or presses and usually operating at elevated temperatures.

Rates of production and kinds of material must be determined for parts to be considered for production by one of the four forgings methods in use. A further discussion of forgings can be found in Chapter 21, "Cross-Section-Changing Processes."

Smith Forging. This forging method evolved from the blacksmith practice of hand hammering heated metal to shape. Smith forging does, indeed, still employ some hand hammering, but this is greatly augmented by the use of steam hammers, helve hammers, and pneumatic hammers. Presses are also used, as are simple, flat dies.

Smith forging is used to produce parts where requirements are too low to justify very much tooling expense. Some very large forgings are produced with this method.

Tolerances. Surprisingly close tolerances can be held and depend on the requirements and on the operator's skill.

Application.

- Railroad items
- Heavy truck I-beam axles
- Metal preforms to be finish machined

Tool Cost. Low.

Drop Forging Drop forgings are produced by impacts from hammers dropped by gravity or powered. Metal is formed between die halves—one mounted on the hammer, the other on the machine bed or anvil. Several blows or drops are usually required to produce a completed forging.

This is the most used of all forging processes and can be used for all metals except high-strength magnesium alloys. A trimming operation must also be planned to remove excess flash from the forging with a trimming die and press.

Tolerances. Wide tolerances.

Applications. These are too numerous to list but include gear blanks, connecting rods, rocker arms, farm implement parts, aircraft parts, railroad parts, and many others.

Production Rates. Very high.

Total Cost. First cost relatively low, but maintenance cost of dies high.

Press Forgings The press forging process is similar to drop forging with the exception that pressure is applied to the metal to complete a forging in one cycle rather than by repeated impacts. Upon completion, the forging normally must be forced out of the die by ejectors.

The process can be used for all metals but is mainly used for aluminum and magnesium. A trimming die is used to remove excess flash.

Tolerances and Surface Finish. Tolerances are complex and involve width, length, thickness, straightness, draft angles, mismatch, and die closure allowances. In general, the quality of a press forging can be thought of as approximating that of the permanent mold process.

Surface finishes of aluminum and magnesium forgings are normally within 250 μin. Surface finishes for steel and titanium are normally within 500 μin.

Applications. The press forging process is used for parts very similar to those produced by the drop forging process, but are of a higher quality. Also some larger parts can be produced.

Production Rate. High.

Tool Cost. Moderate.

Upsetting Originally, this process, sometimes called **heading**, was capable of upsetting only the ends of a length of steel wire and was used to form nail heads; present-day upsetters are capable of increasing the diameter or cross-sectional area at any location along the length of a part.

The process is accomplished by cutting off and transferring a length of steel wire of up to 1 in. in diameter progressively along and through succeeding steps of a multistage die until the part is completed—or fed into secondary operations for threading, cross-drilling, and so on.

Tolerances. Close tolerances can be held.

Applications. See Chapter 21.

Production Rates. Very high.

Tool Cost. Relatively high.

Extrusions The extrusion process consists of confining a billet of metal in a die, after which a ram applies pressure to the metal, forcing it through an opening in the die. The extruded metal emerging takes the shape of the die opening.

Many extrusions are performed cold, but some have heat applied to the billet before extruding. A more complete discussion and illustrations of the extrusion process can be found in Chapter 21.

Tolerances. Close tolerances can be held.

Applications. Shapes, shafts, pins, hubs.

Production Rates. Moderate.

Tool Cost. Relatively low.

Stamping It is beyond the scope of this book to describe and illustrate how cast or forged parts could be redesigned for more economical production as stampings. However, this is a viable procedure and many companies are giving increasing attention to this approach to improving productivity.

The many potential advantages include

- Reduced weight
- Reduced material cost
- Reduced labor costs
- Reduced tooling costs
- Increased flexibility of design

4.6 COMPUTER-AIDED DESIGN

Producibility can be greatly increased with the aid of a computer. Quality, physical, and functional elements can be optimized; manufacturing problems can be addressed. Total required engineering time can be reduced. The majority of all engineering and manufacturing companies now employ the computer-aided design (CAD) system to accomplish these ends. The use of CAD is so vital to engineering and manufacturing that all engineers involved in these areas would do well to study and become knowledgeable of the system and its capabilities.

4.7 REVIEW QUESTIONS AND PROBLEMS

1. Who should participate in the simultaneous engineering of a product and why?
2. Which areas of parts manufacturing may have costs reduced through the use of simultaneous engineering?
3. Develop a process for each design of the 4 drilled holes shown in Figure 4.16. Which design will cost the least to drill? Use scaled length from drawings for drilled hole depths.

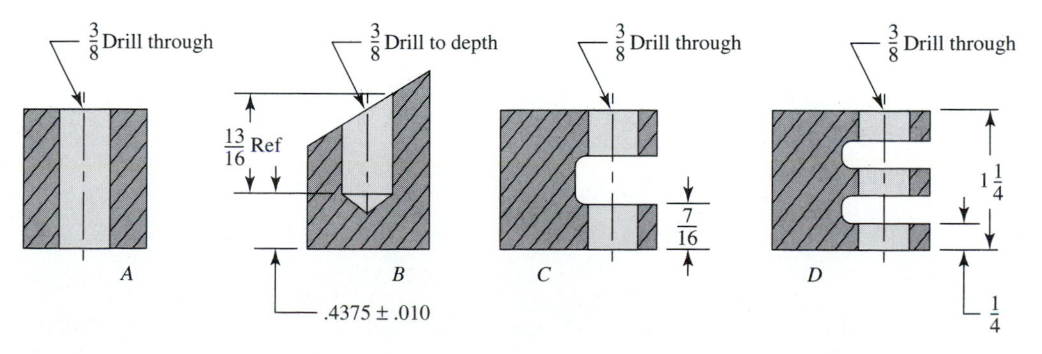

FIGURE 4.16 Parts requiring drilled holes in varying configurations.

4. What advantages pertain to the use of sand casting?
5. What are the disadvantages in the use of cast iron?
6. What limits the use of permanent molding in parts design?
7. All die castings have porosity to some degree caused by gases and air entrapment in the mold. Exactly what detrimental effects can such porosity cause— and what can be done about it?
8. Under what conditions are smith forgings used?
9. When is the drop forging process used in preference to press forging?
10. What are the advantages of using the press forging process?
11. The review of the extrusion process is better covered in Chapter 21 where the process is explained in more detail.

CHAPTER FIVE

PART DESIGN ANALYSIS

CHAPTER OVERVIEW

5.1 INTRODUCTION

In the previous chapter it was shown that Manufacturing Engineering can, and must, collaborate with Product Engineering in the development of a producible, functional product design at the outset.

This chapter, and subsequent chapters, will show how Manufacturing Engineering—having the responsibility for developing a manufacturing plan for the product—does so in a logical, almost methodical, manner. All manufacturing begins and ends with the part design. Manufacturing planning must begin with a study of the part design; manufacturing must ultimately produce a part that satisfies the part design.

5.2 VISUALIZING

The manufacturing engineer must clearly visualize the part. He must understand how each dimension, note, and specification applies, and to which surfaces and elements. Without a clear concept of exactly what the design intends and how the part physically looks, it is impossible to develop a manufacturing plan. To this end, if after thoroughly studying the part design, the manufacturing engineer still has difficulty in visualizing the part, he may find the following actions of value:

1. Talk to the product engineer.
2. Color, or pencil shade, the main surface and elements on a print of the part.
3. Remove all notes and dimensions from the drawing. They may distract from viewing the part outline, configuration, or shape.

4. Construct a cross-sectional view.
5. Project an auxiliary view.
6. Develop a 10–x view of the part or section of the part.
7. Obtain a sample part, if available, or construct one from clay, wood or other material.

The forced thinking and reasoning involved in accomplishing items 4 through 7 will usually resolve the problem.

The design will show, list or refer to, the part's official description (Bracket, Pulley, Shaft, etc.) along with the scale, material, engineering changes, notes, specifications, and standards. The design under study may show related parts and assemblies; it usually will give the material form.

Scale The scale should be carefully noted to gain a mental picture of the part's physical size. This will have a direct influence later, in considering the part's weight, material handling, fixturing, and processing. A caution is in order here—many part designs are stored in microfiche or in CAD system's memory, so the print *as supplied* may be reduced from the original. In many instances, a reference to "scale, full" can be misleading.

Engineering Changes Previous engineering changes should be reviewed to determine the part's original form and any problems. Also, a print of the part under study must be the latest issue—a later issue could have been released but not yet received.

Notes and Specifications All notes must be carefully read and analyzed. General notes are usually associated with fillets, radii, chamfers, and tolerances that are not otherwise specified. Specific notes call out conditions pertaining to specific features or functions of the part.

Standards Many design features of products have been reduced to, or developed into, national and international standards. These include

- Gearing
- Heat treating
- Materials
- Sheet metal thickness—gage
- Wire sizes

and many others.

References to SAE, ANSI, Steel Wire Gage, UNC, UNF and many others will be found on part prints either in the title block area or as they apply to some specific element. Each must be looked up to determine all information and its effect on the part's manufacture.

Corporate Standards In addition to the use of national and international standards, many manufacturing companies have found the need to develop their own standards to cover certain areas peculiar to their requirements. These standards will have that company's identification, can be anywhere on the part point, and will cover such areas as

- Chemical surface treatment
- Drafting practice
- Heat treatment
- Material
- Testing

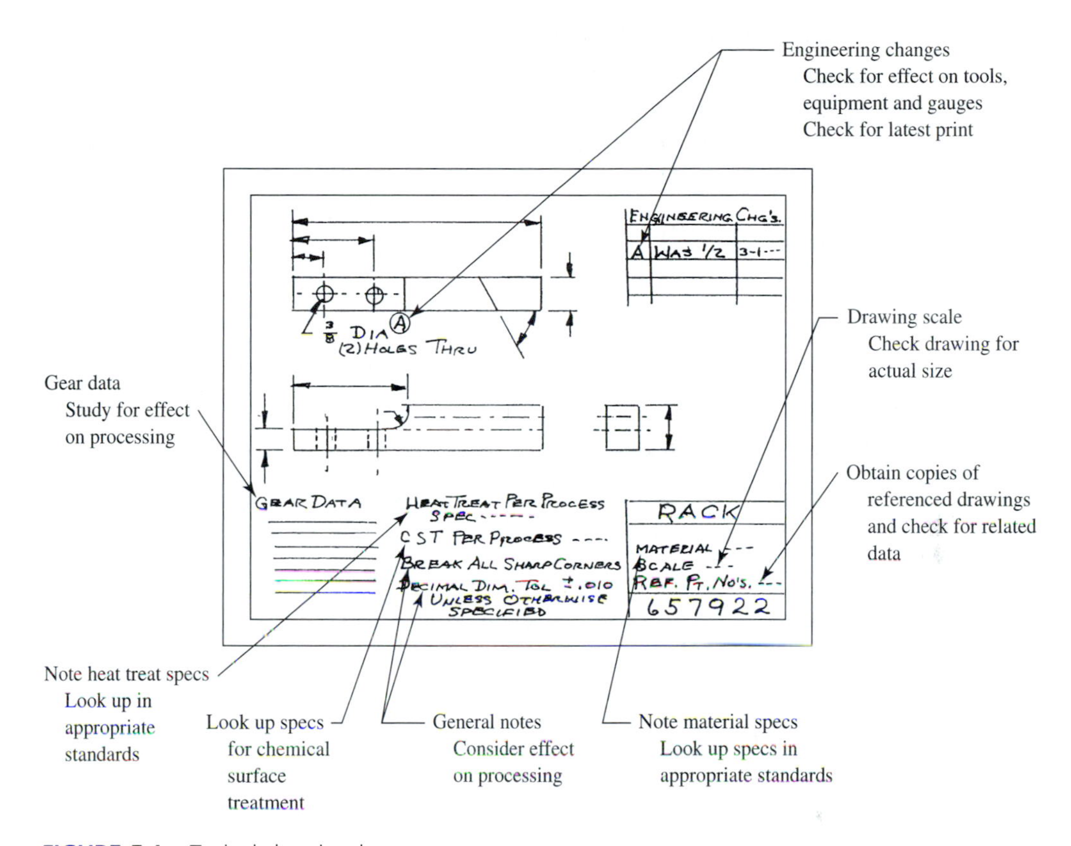

FIGURE 5.1 Typical drawing layout.

and many others. These standards, too, must be looked up and studied, and their effect on manufacture determined.

Related Parts and Assemblies All related parts and assemblies must be obtained and studied, not only for the data and dimensions included, but also for the *intent*.

Many part designs list related part numbers and also prior and subsequent assembly numbers. If these are not given, Product Engineering must be requested to provide this information. If a bill of material or an exploded view is available, it should, of course, be obtained, for these will greatly aid understanding the overall function of the part concerned. Figure 5.1 illustrates the various areas and items of information referred to.

5.3 MATERIAL FORM

The *form* in which the basic material is received will partially indicate the manufacturing process that will be required. Further thought and inquiry should be given to the *condition* the material will be in when received. The condition may very well require an initial preparatory process of washing, pickling, shot blasting, snag grinding, chemical surface treatment immediately following the changes receiving inspection operation.

At this time, some consideration and thought should be given to how the material will be introduced into the production line—racks, baskets, boxes, special containers, conveyers, chutes, or other.

Some materials must be run in "heat lots"; other materials must be sampled and approved metallurgically; most forgings must be approved for grain flow before being released to production for manufacture.

Provisions must be in place, or provided, to inspect and sample material received—including an area to impound and store until approved.

Quality Control is responsible for maintaining the level of quality specified for all incoming productive material. The "lab" that monitors and checks chemical composition, hardness, grain structure and other metallurgical properties is usually part of Product Engineering. It is well to know organizational responsibilities and the flow of material in the event of subsequent troubleshooting. Also, these areas may require new, additional, or different inspection facilities to inspect and control new, or different, materials required for a new product.

5.3.1 Product Design Elements

Geometry and Configuration Geometry, the relationship of points, lines, angles, surfaces, and solids, defines elements of a product design; configuration, the relationship of individual elements, is the product's final shape.

The design of simple products can consist of as few as two or three elements—a round pin, a shaft, a length of tubing, and a rectangular plate are examples. Bar stock is supplied by the

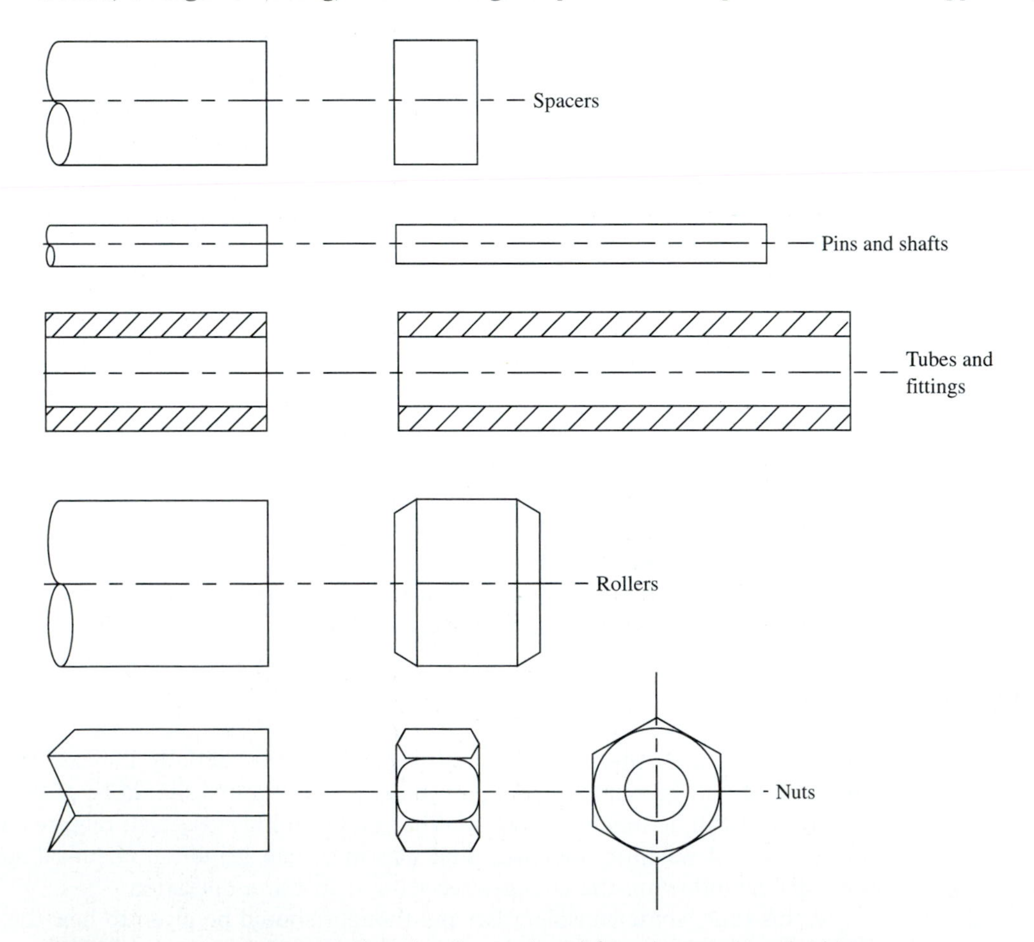

FIGURE 5.2 Simple parts designed to be produced from bar stock and tubing. Many sizes and variations of each are used in actual practice.

mill in standard sizes and shapes (round, hexagonal, square, tubing), and in many different materials, alloys, and finishes.

Product Engineering can and does design parts to take advantage of such standard shapes. Manufacturing Engineering must analyze the part design, recognize the intent, and process the part to realize the full extent of productivity available. Figure 5.2 illustrates parts designed to be produced from bar stock; only cutoff and deburr operations are required.

As product design becomes more complex, manufacturing costs increase with the manufacturing operations added to accommodate the added elements. In analyzing the more complex part, the manufacturing engineer must

1. Determine the total manufacturing process required.
2. Determine the leadoff or main operation.
3. Determine the required secondary operations.

Figure 5.3 illustrates parts requiring secondary operations.

The final shape, or configuration, of a complex part is still a relationship of basic geometrical elements. When no indication of a main or leadoff operation is indicated (such as machining from bar stock), the manufacturing engineer must study the part and develop a tentative, applicable manufacturing process. Figure 5.4 illustrates a tentative processing for a design where no leadoff operation is indicated.

Geometry of Form. Drawings must clearly define the geometric shape of a surface—flat, round, parallel with, perpendicular to, and so on, and also specify the amount of manufacturing tolerance allowable. Difficulty in these areas has been experienced due to implied conditions and also due to the inherent ambiguity of notes. All conditions cannot be shown dimensionally. Much of the misunderstanding and confusion in these areas has been eliminated by the development and adoption of Geometrical Dimensioning and Tolerancing (GD&T) Standard (the ANSI Y14.5M).

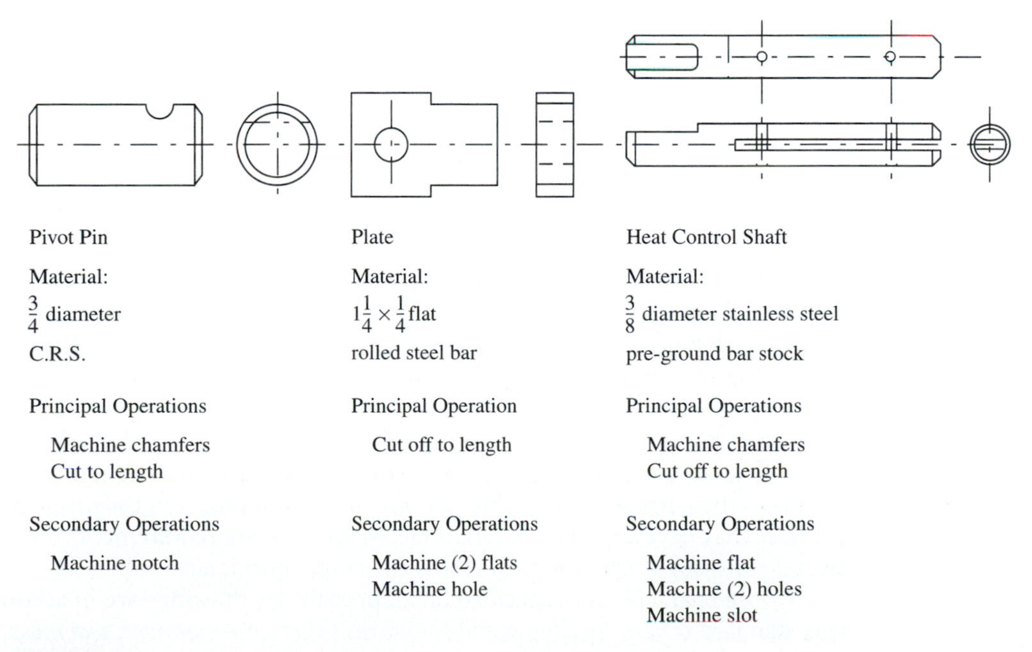

FIGURE 5.3 Secondary operations required for screw machine parts.

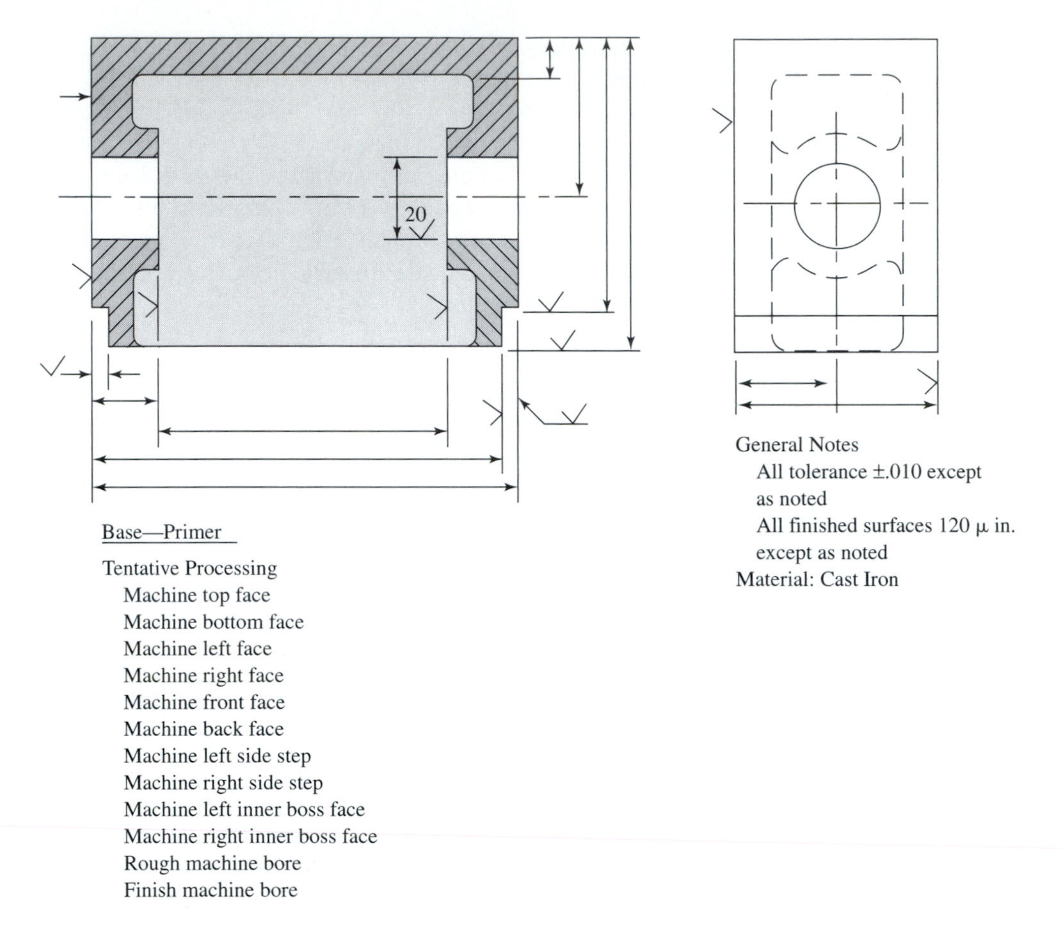

FIGURE 5.4 Part design with no apparent or obvious leadoff operation.

The section of this standard that illustrates and defines the geometry of form is shown in Figure 5.5. The manufacturing engineer must be able to read and interpret the entire GD&T standard because it is being used increasingly in all areas of product design. In fact, it is now required 100% for U.S. defense work.

In addition, when considering the GD&T symbols, definitions of *concentricity* and *eccentricity* may prove helpful:

- **Concentricity** is a condition where all diameters concerned revolve around a common center line.
- **Eccentricity** is a condition where the involved diameters revolve about center lines that are not common, but are displaced by the amount of eccentricity.

Surface Texture The manufacturing engineer must study and include a part design's surface texture requirements in his planning. A surface may have no designating symbol shown, which can mean the surface is acceptable "as cast" or as received from the originating supplier. Again, a surface may have a symbol designating a surface texture requirement so stringent as to require several subsequent operations to satisfy: machine, grind, lap.

Surface characteristics specified on all present-day drawings are in accord with ANSI B46.1. This standard covers specifying and measuring *surface roughness, waviness,* and *lay*.

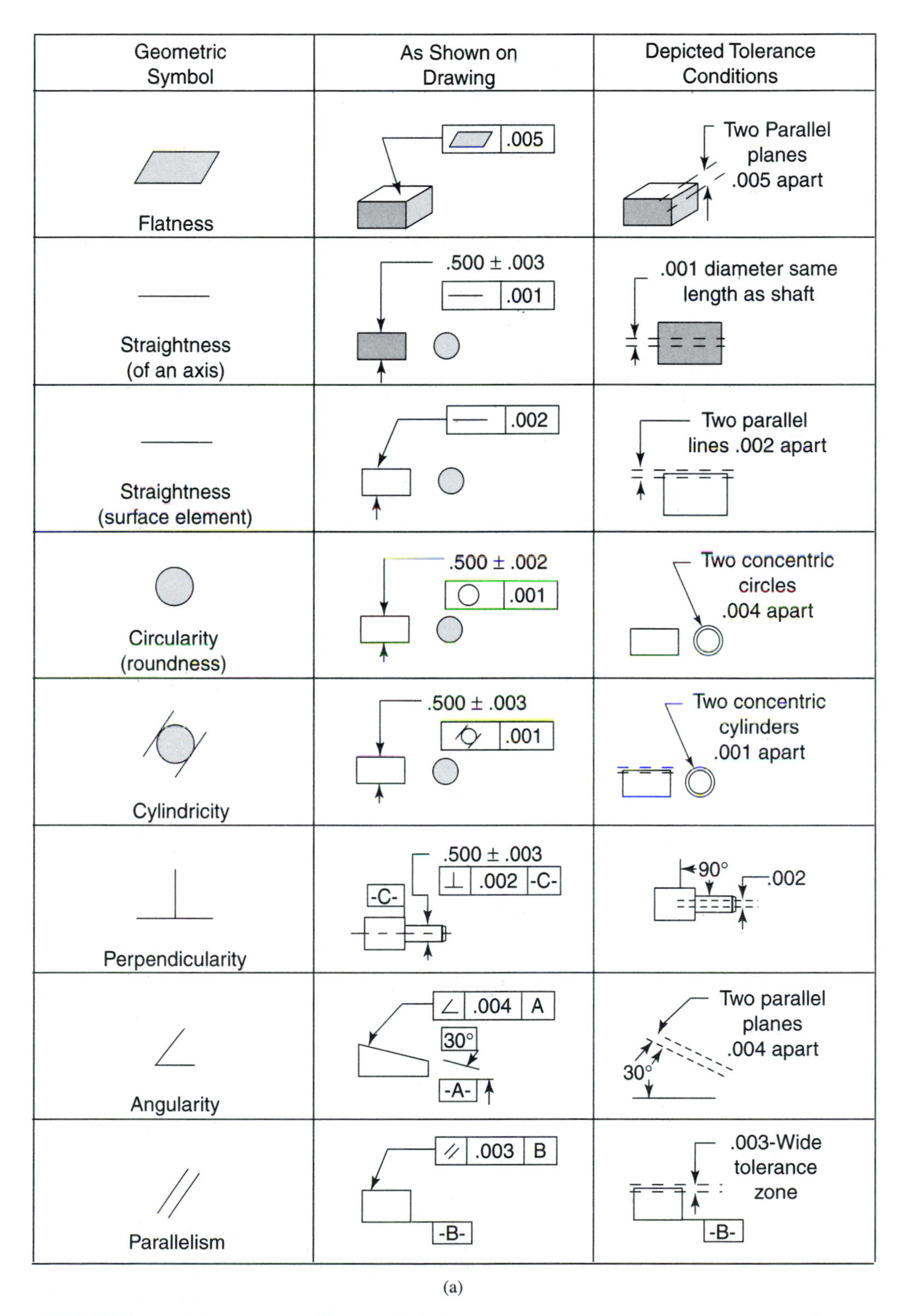

Geometric Symbol	As Shown on Drawing	Depicted Tolerance Conditions
Flatness	.005	Two Parallel planes .005 apart
Straightness (of an axis)	.500 ± .003 .001	.001 diameter same length as shaft
Straightness (surface element)	.002	Two parallel lines .002 apart
Circularity (roundness)	.500 ± .002 .001	Two concentric circles .004 apart
Cylindricity	.500 ± .003 .001	Two concentric cylinders .001 apart
Perpendicularity	.500 ± .003 .002 -C- -C-	90° .002
Angularity	.004 A 30° -A-	Two parallel planes .004 apart 30°
Parallelism	.003 B -B-	.003-Wide tolerance zone -B-

(a)

FIGURE 5.5 Geometric dimensioning and tolerancing symbols.

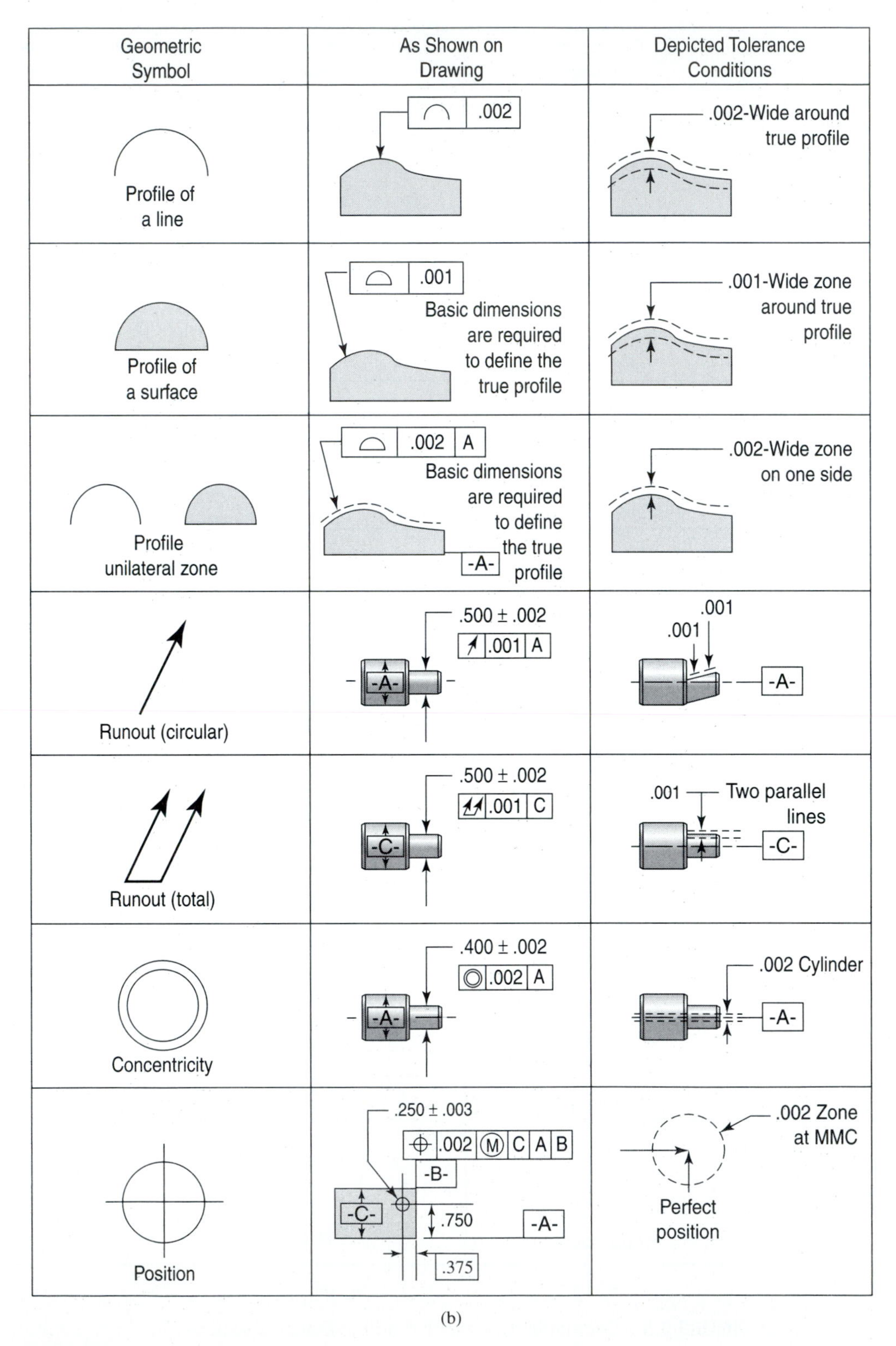

(b)

FIGURE 5.5 Continued

Symbol	Description
-A- (boxed)	Datum Symbol This symbol represents physical features or surfaces that must be used for location in machining or inspection.
∅	Diameter Symbol This symbol replaces the word "diameter".
C–1	Datum Target Symbol This is the symbol for a datum on irregular parts such as castings and forgings.
Ⓜ	MMC modifier Meaning maximum material condition applies
Ⓛ	LMC modifier Meaning least material condition applies
Ⓢ	RFS modifier Meaning regardless of feature size
Ⓟ	Projected Tolerance Zone Modifier Meaning that the stated tolerance zone is longer than the feature
0.500 (boxed)	Basic Dimension Basic dimensions are located by position tolerances only (except basic angles).
(0.500)	Reference Dimension For computation purposes only

(c)

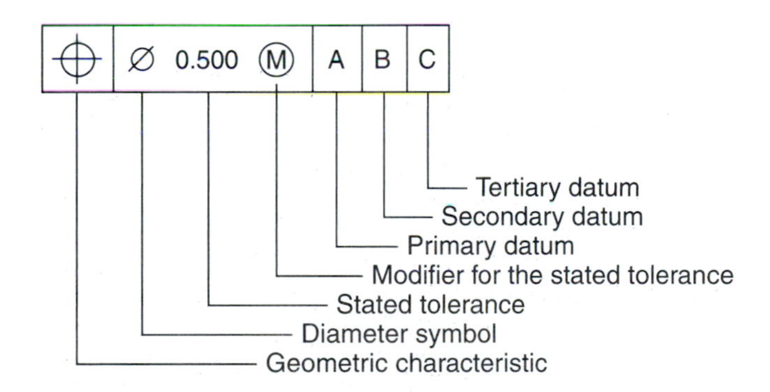

Tolerance Zones
All tolerance zones shown in feature control frame are total.

Feature
A feature is a physical portion of a part such as a hole, surface, or slot, etc.

(d)

FIGURE 5.5 Continued

Some older drawings, still in use in various areas, use symbols such as "f," "ff," "G" and "GG" for finish, fine finish, grind, and finish grind. Such symbols have no way of indicating the degree of required surface texture control. Companies using these symbols work out their own levels of quality acceptance. There is no correlation between the "G" and "f" notations and present-day use of ANSI B46.1.

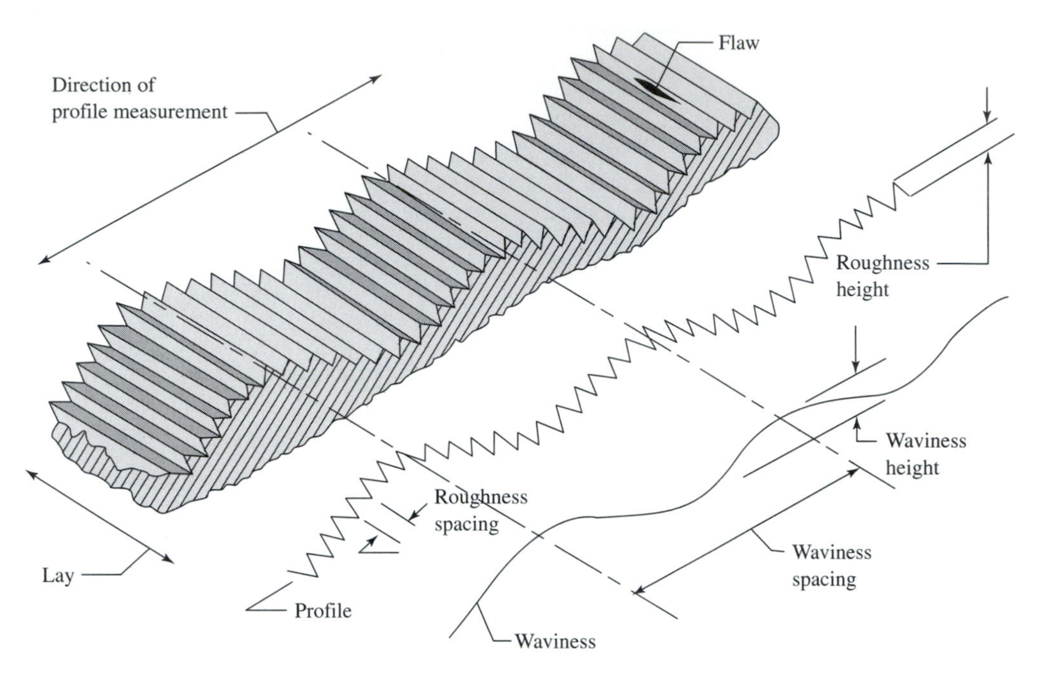

FIGURE 5.6 Surface texture elements.

Following and in Figure 5.6 is a condensed version of ANSI B46.1 sufficient to read and interpret part print surface nomenclature used in this text book. However, all manufacturing engineers should be thoroughly knowledgeable of the data included in this standard.

SURFACE TEXTURE STANDARD ANSI B46.1

Definitions

Flaws are unintentional irregularities which occur at one place or at relatively infrequent or widely varying intervals on the surface. Flaws include such defects as cracks, blow holes, inclusions, checks, ridges, and scratches.

Lay is the direction of the predominant surface pattern ordinarily determined by the production method used.

Microinch is the unit of measure for surface roughness. A microinch is one millionth of an inch (0.000001) and may be abbreviated 'μin.' 1 μin = 0.0254 μm.

Profile is the contour of the surface in a plane perpendicular to the surface, unless some other angle is specified.

Roughness means the fine irregularities of the surface texture, usually including irregularities that result from the inherent action of the production process.

Surface means the boundary of an object that separates that object from another object, substance, or space. In manufacturing, a surface is produced by any of the manufacturing processes.

Waviness is irregularity of the surface occurring at greater spacing than roughness. Waviness may result from such factors as machine or work deflections, vibration, chatter, and heat treatment. Roughness may be considered as superimposed on a wavy surface.

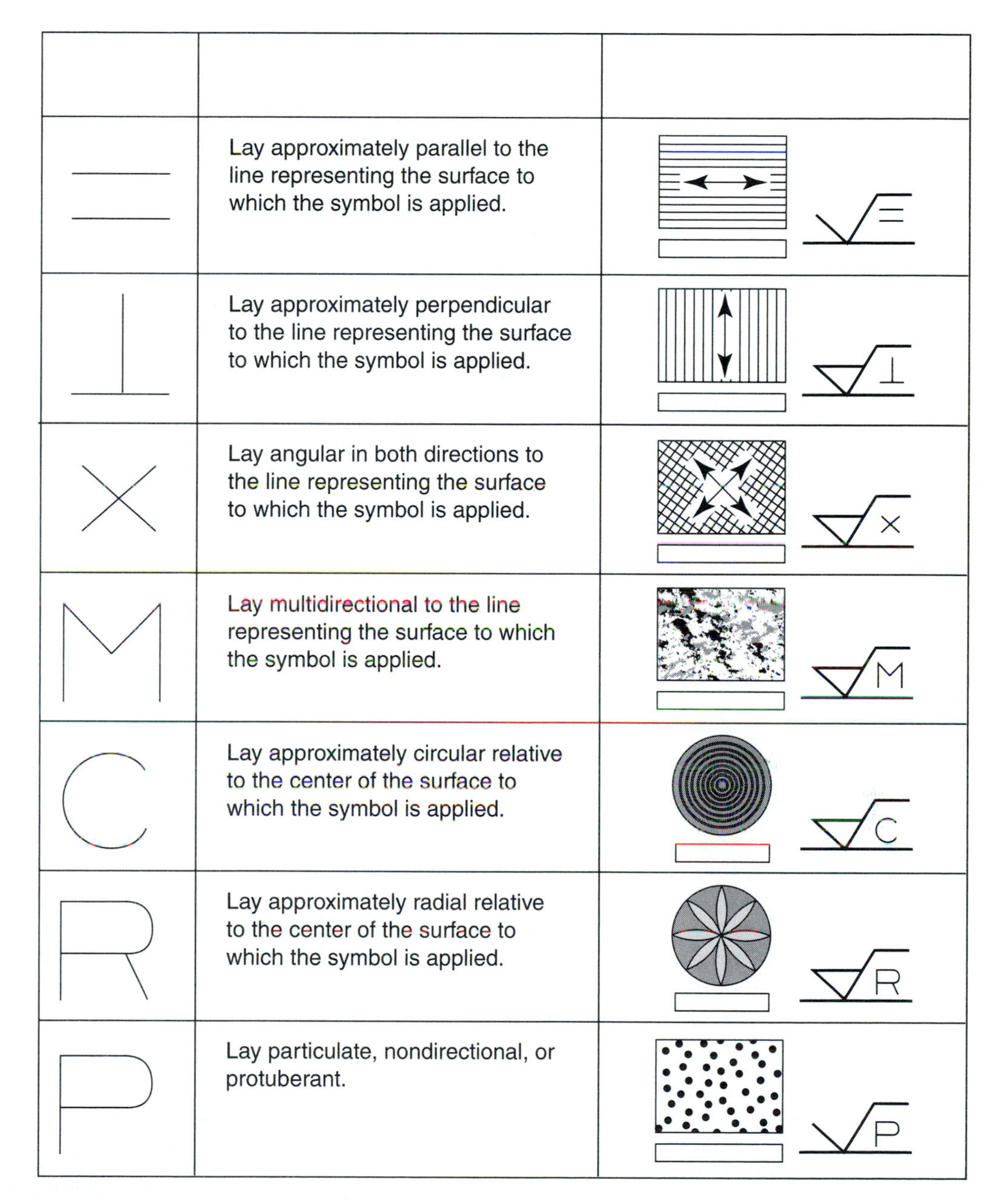

FIGURE 5.7 Surface lay symbols.

Lay. The direction of the predominant surface pattern, or lay, of surfaces resulting from many different manufacturing processes is illustrated in Figure 5.7 along with the direction in which a reading of surface roughness is to be taken.

The standard symbol used to designate surface roughness is shown in Figure 5.8(a). This symbol indicates that the surface may be produced by any method. The roughness specification is placed to the left of the long leg as shown in Figure 5.8(b). Other surface requirements may be specified as shown in other views of Figure 5.8.

(a)	Basic surface texture symbol. Surface may be produced by any method except when a bar or circle (view b or d) is specified.
63 (b)	Material removed by machining is required. The horizontal bar indicates that material removal by machining is required to produce the surface and that material must be provided for that purpose.
032 (c)	Material removal allowance. The number indicates the amount of stock to be removed by machining is inches or millimeters. Tolerances may be added to the basic value shown or in a general note.
(d)	Material removal prohibited. The circle in the V indicates that the surface must be produced by processes such as casting, forging, hot finishing, cold finishing, die casting, powder metallurgy, at injection molding without subsequent removal of material.
(e)	Surface texture symbol. To be used when any surface characteristics are specified above the horizontal line or to the right of the symbol (see Figure 5-9). Surface may be produced by any method except when the bar or circle (view b and d) is specified.

FIGURE 5.8 Surface texture symbols.

Figure 5.9 illustrates the placement of waviness, roughness, and lay specifications on the symbol along with the interpretation of each.

Applying Surface Texture Symbols. The point of the symbol is to be on a line representing the surface, an extension line of the surface, or a leader line directed to the surface. The symbol may be specified following a diameter dimension. The long leg and extension are to the right as the drawing is read. For parts requiring extensive and uniform surface roughness control, a general note may be used as shown in Figure 5.10.

Figure 5.11 illustrates a method of controlling surface texture values from one manufacturing operation to subsequent operations on the same surface. This method can be used to control critical or costly surfaces where assurance is required that the surface will clean up all tool marks 100% at each succeeding operation.

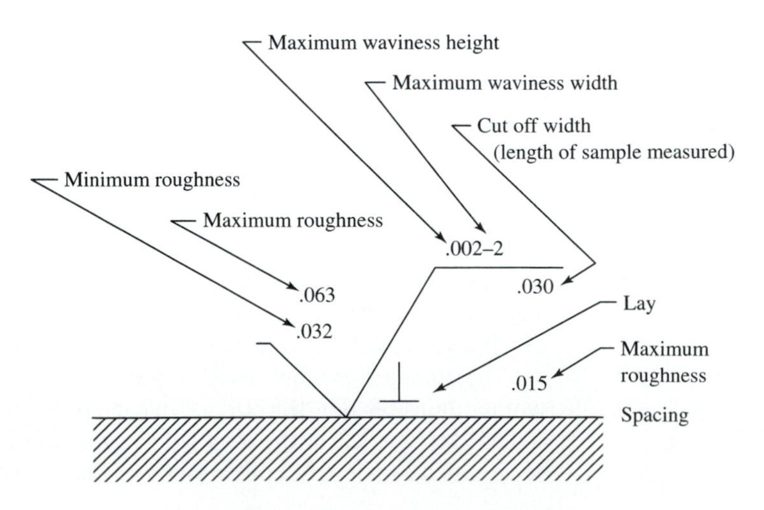

FIGURE 5.9 Placement of surface texture component symbols on design.

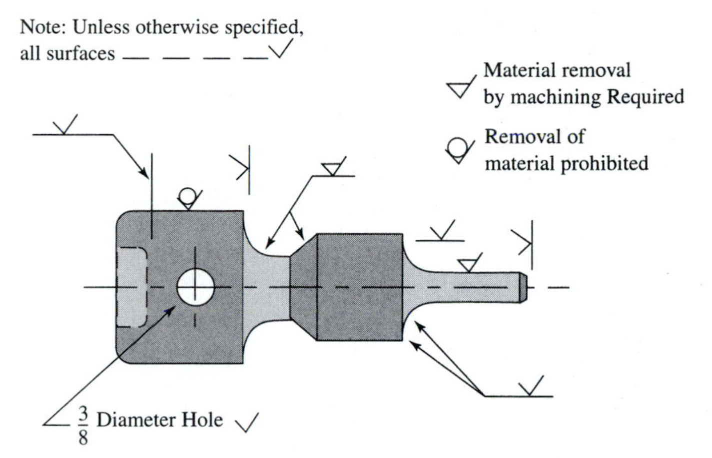

FIGURE 5.10 Application of surface texture symbols.

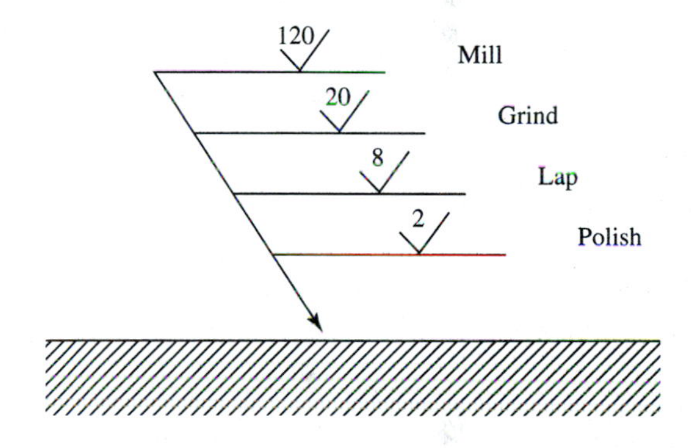

FIGURE 5.11 A method of specifying surface control for succeeding operations on the same surface.

Measuring Surface Roughness. Surface roughness is measured by one of three different methods: profile, area, and comparison.

Profile. The profile method of surface roughness measurement is the one in most general use. It consists of a measurement taken with a stylus of a single line, of a designated length, that is representative of the entire surface. Two types of stylus instruments are shown in Figure 5.12.

Area. Capacitance, optical, and pneumatic systems are employed to measure an area of a surface and, when compared to a known acceptable master, may be used to check a process. A profile check is normally also used in conjunction.

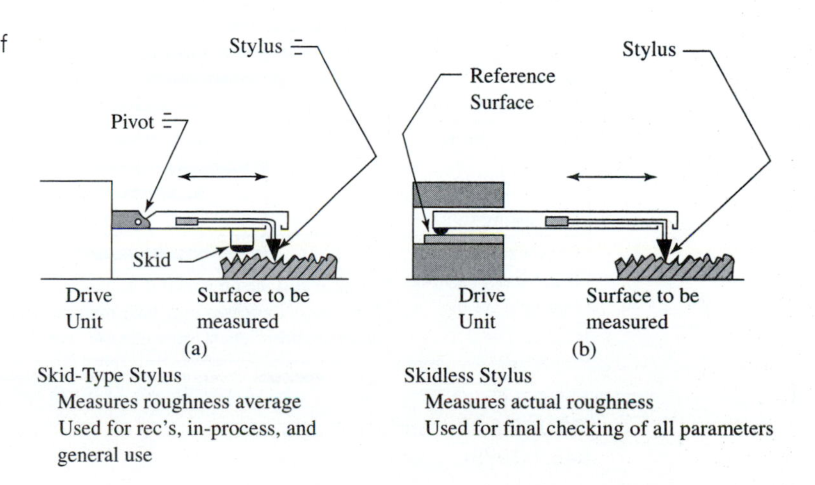

FIGURE 5.12 Types of roughness measuring stylus.

Comparison. Surfaces with a comparatively "rough" surface texture requirement (above 60 μin.) may be compared with a known sample such as an approved sample part or with commercial standard roughness specimens. The check is made by visual comparison and sometimes by tactual fingernail check.

Stereoscopic comparison microscopes are also available and usually used in the surface roughness range of from 20 μin. to 60 μin. or above.

Surface Roughness Produced by Common Production Methods. A chart of surface roughness produced by common manufacturing methods is shown in Figure 5.13.

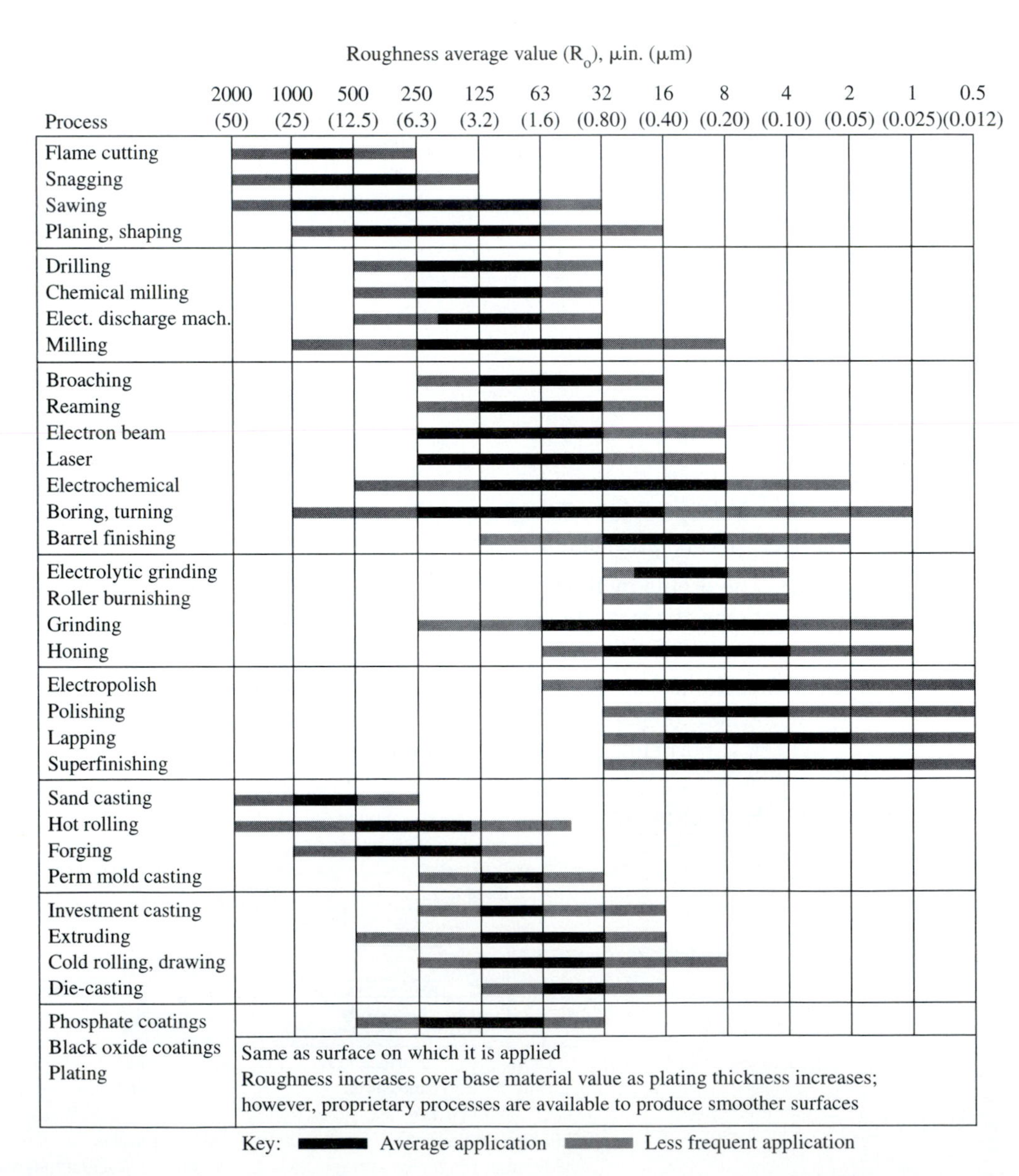

FIGURE 5.13 Surface roughness produced by common production methods. (Based on ANSI B46.1-1978).

5.4 DIMENSIONING

Rectangular Coordinates A part's size and shape are described with dimensions—augmented with notes defining surface characteristics such as round, flat, and so on. The rectangular coordinate system is generally used for dimensioning; it is implied and assumed that surfaces shown normal to each other are at 90° angles.

Product Engineering employs $A–B–C$ dimensioning for width, depth, and length. Machine tool equipment has $X–Y–Z$ axis of movement to translate design dimensions into machined surfaces as shown in Figure 5.14(a).

Nonrectangular parts, such as cams and wedges, may be dimensioned with the required angle, basic (no tolerance) and a set of coordinate dimensions with tolerances provided to establish a reference point for manufacturing and checking as shown in Figure 5.14(b) and (d).

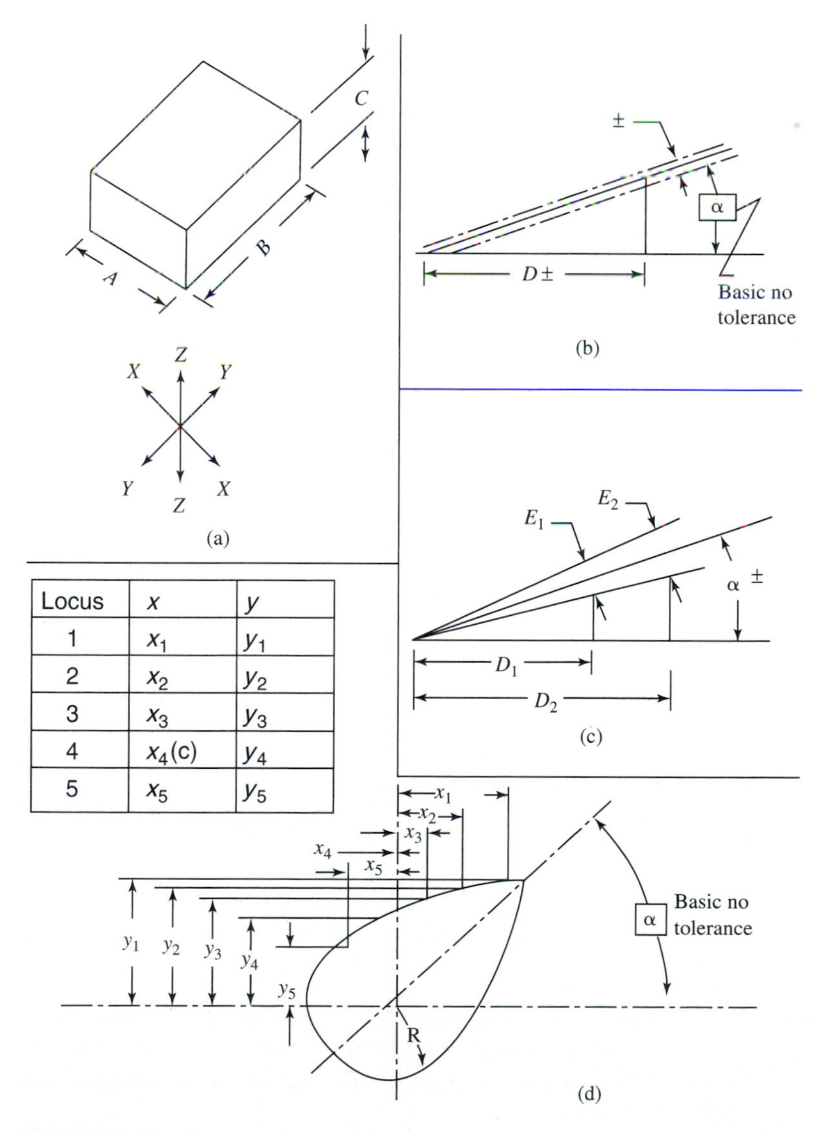

Locus	x	y
1	x_1	y_1
2	x_2	y_2
3	x_3	y_3
4	x_4(c)	y_4
5	x_5	y_5

FIGURE 5.14 Rectangular coordinates.

Manufacturing tolerances should not be applied to the angular specifications because the amount of dimensional error increases or decreases as the distance from the apex of the angle varies, as shown in Figure 5.14(c). Confusion and misinterpretation result when manufacturing attempts to interpret how much manufacturing tolerance is allowable at what distance from the apex. This is illustrated in Figure 5.14(c).

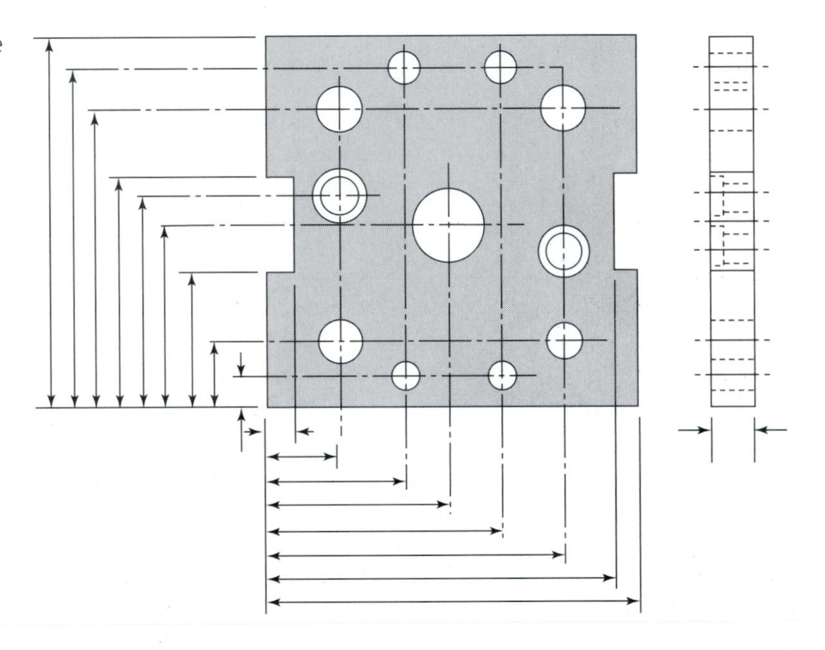

FIGURE 5.15 Baseline dimensioning.

Baseline Dimensioning This method of dimensioning, which originated in die making and other precision work, utilizes two finished edges at right angles as base of reference lines and measures all dimensions from these lines. The jig plate, Figure 5.15, is an example. The advantage of this method is that tolerance errors are not cumulative.

In the production of multiple, successive parts, the purpose of utilizing baseline dimensioning in the part's design is the same—reduction of cumulative tolerances. To manufacture parts so designed, the baseline should be *located* on the surfaces from which the part print is dimensioned.

Baselines. For purposes of manufacture, a baseline, to be used as a locating surface, can be defined as a line depicting an area or surface best qualified for locating the part while machining is performed. It must meet these requirements:

1. It must be an accessible area free of parting lines, flash, ridges, identification numbers, and so on.
2. It must be related to the dimensional system so as to render machining and holding manufacturing tolerances feasible when the part is located on that surface.
3. The part should be stable when so located, although supports may be used if needed.

The part shown in Figure 5.16(a) does not employ baseline dimensioning. It would be impossible to machine all of the stepped surfaces in one operation, from one location, and maintain the manufacturing tolerances. The reason is called "tolerance stacking" and is detailed further in Section 5.5, "Tolerances."

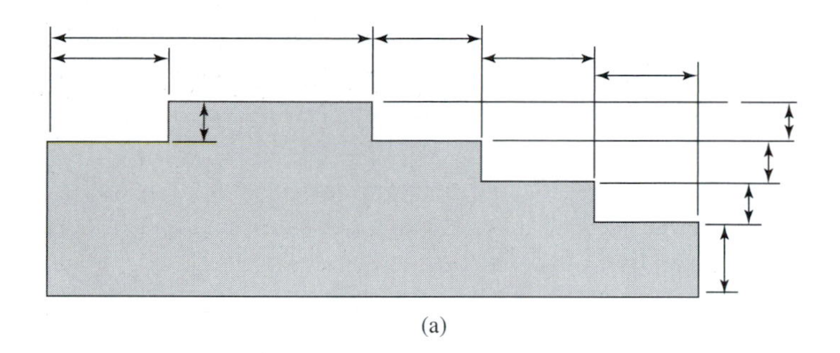

(a)

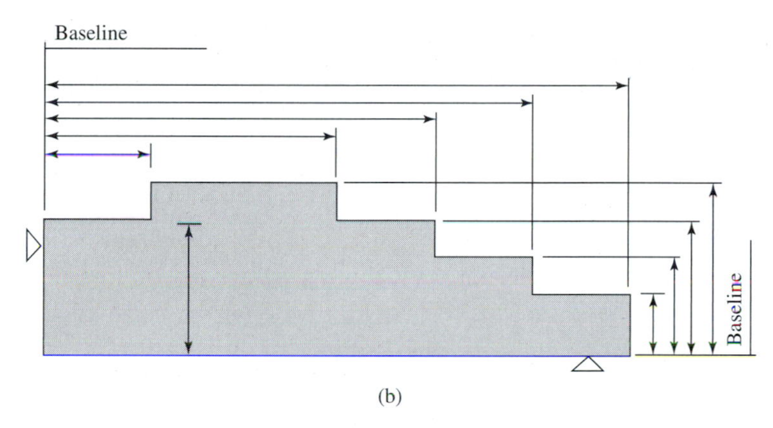

(b)

FIGURE 5.16 (a) Dimensions not acceptable for manufacturing, no common surface usable for locating. (b) Baseline dimensions acceptable for manufacturing, baselines from common locating surfaces.

The part shown in Figure 5.16(b) has been designed utilizing baseline dimensioning. This part, when located on the surfaces indicated, can have all stepped surfaces machined in one operation while holding all the specified manufacturing tolerances.

Direction of Process Dimensions Product Engineering need only define the distance from one product feature, or surface, to another; Manufacturing Engineering must additionally define directions. Process dimensions must specify a direction to identify the surface to be machined, ground or otherwise processed.

Process dimensions are always *from* the locator *to* the processed surface. This understanding is needed to clarify instructions given in manufacturing plan process sheets and is required for the development of a tolerance chart.

5.5 TOLERANCES

Definitions Manufacturing tolerance may be defined as the total permissible variation from the specified basic size of the part. Manufacturing tolerances are required to allow for errors and deviations inherent in one or more of the following contributing sources.

1. **Machine.** All machines have a level of quality capability that is at its highest (best conditions) when the machine is new. As the machine surfaces, components, and bearings wear, the capability drops, necessitating repair, rebuilding, and perhaps eventual replacement.

2. **Fixtures.** Fixtures locate and hold the part during machining. Deflection under load and/or wear on locating surfaces contributes to the need for manufacturing tolerances.
3. **Cutting Tools.** Manufacturing tolerance needs attributed to cutting tools include breakage, deflection, wear, and changes of size due to sharpening.
4. **Materials.** Machinability is affected by variations in alloy, hardness, and/or the amount of stock left to be removed. Machinability changes cause changes in part deflection and breakage, wear, and deflection of the cutting tool.
5. **Human Error.** Manufacturing tolerance is lost when tools are incorrectly set or adjusted, when speeds or feeds are incorrectly used, or when parts are incorrectly loaded in fixtures.

Definitions of Terms Commonly Used

1. **Nominal size** is a general classification used for general identification such as a standard $\frac{1}{2}$-in. bolt or $\frac{3}{4}$-in. bar stock. It carries no tolerance specifications.
2. **Basic size** is the exact theoretical size and carries no tolerance specification.
3. **Tolerance** is the total permissible variation from the specified basic size of the part and can be expressed in several different ways:
   - *Specific tolerances* are tolerances that are given for one particular dimension.
   - *General tolerances* are usually given in a note on the part design such as, "All 2-place decimal dimensions are ±.010 unless otherwise specified."
   - *Unilateral tolerances* specify all variations to be in one direction only as in Figure 5.17(a).
   - *Equal bilateral tolerances* specify that tolerances are to be equal either way (plus or minus) as shown in Figure 5.17(b).
   - *Unequal bilateral tolerances* specify that tolerances are to be in either direction (plus or minus) but unequally as shown in Figure 5.17(c).
4. **Limits** are the maximum allowable dimensions of a part as shown in Figure 5.17(d).
   - External dimensions of a part are expressed with the larger limit placed on top.
   - Internal dimensions of a part are expressed with the larger limit placed on the bottom.
   - An example of limit dimensioning use is shown in Figure 5.18.

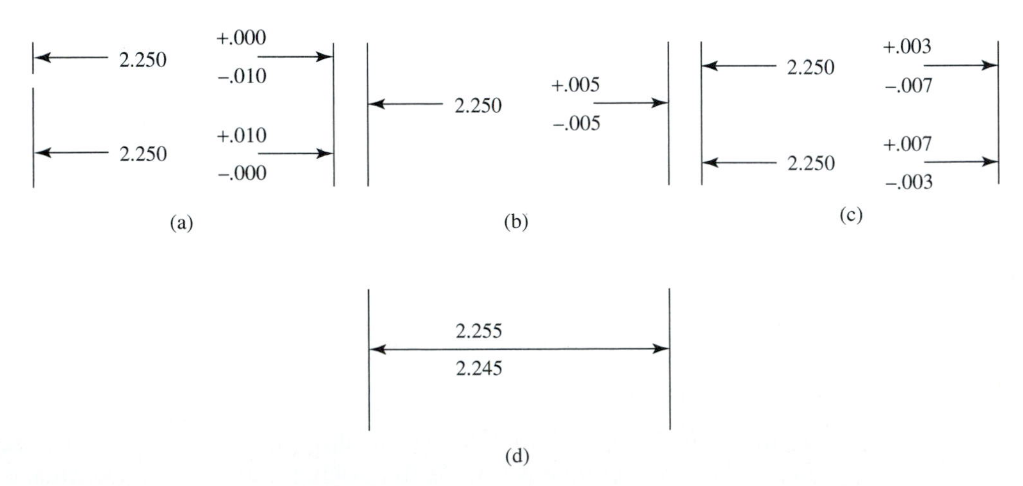

FIGURE 5.17 Examples of tolerance terms: (a) unilateral tolerance, (b) equal bilateral tolerance, (c) unequal bilateral tolerance, (d) limit dimensioning.

FIGURE 5.18 Example of limit dimensioning use.

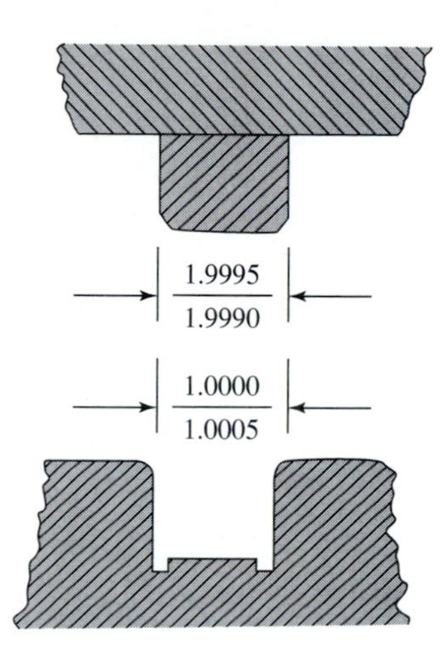

5. **Allowance** is the amount a part dimension is made to vary usually for clearance purposes and may be positive or negative.
   - A positive allowance provides *clearance* between a shaft and its mating part as shown in Figure 5.19(a).
   - A negative allowance results in an *interference* or press fit between a shaft and its mating part as shown in Figure 5.19(c).
   - A transitional fit has an allowance that positions it between the clearance and interference fits as shown in Figure 5.19(b).

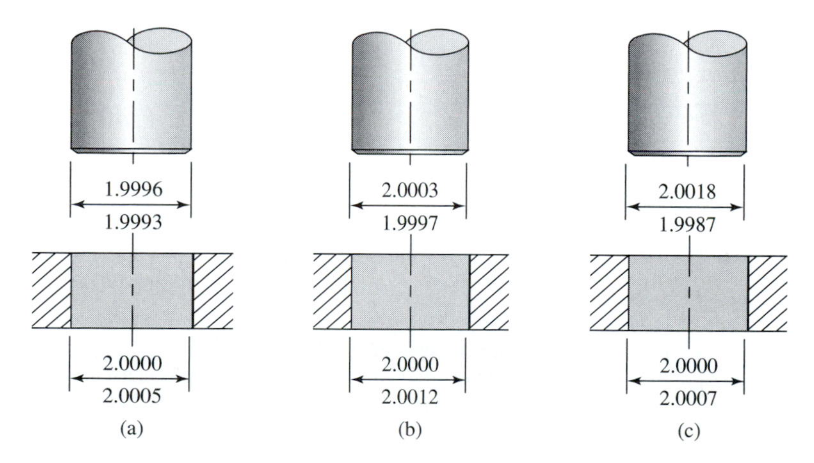

FIGURE 5.19 Allowances as applied to classes of fit: (a) clearance fit, (b) transitional fit, (c) interference fit.

Tolerance Stacks Tolerance stacks are the sum of tolerances of two or more dimensional tolerances. The stack may be generated by the addition of the overall length tolerance of two or more parts; it may be from the addition of two or more dimensions of the same part as shown in Figure 5.20(a) and (b).

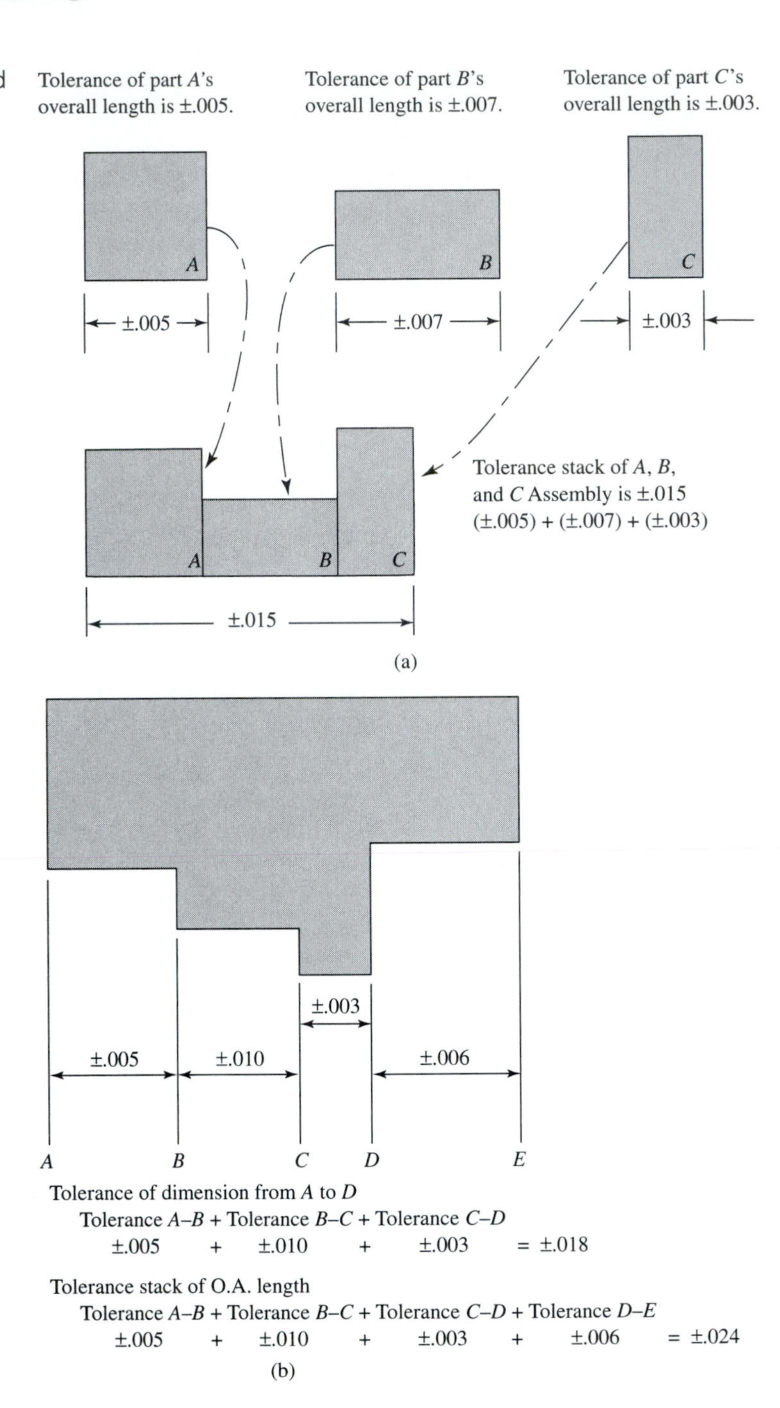

FIGURE 5.20 (a) Assembled parts tolerance stack. (b) Dimensional tolerance stacks.

Unacceptable tolerance stacks are those stacks that exceed some specified total. The countershaft shown in Figure 5.21(a) has a general note specifying all tolerances, not shown, to be within ±.010. The tolerance of x in the figure is ±.030, which exceeds the general note.

The same part redesigned, utilizing baseline dimensioning, will have an acceptable tolerance stack of ±.010 as shown in Figure 5.21(b), which satisfies the general note.

Unacceptable tolerance stacks can result from the part design (design tolerance stack) or from processing (process tolerance stacks) and some from a combination of both. The corrective actions for unacceptable tolerance stacks are

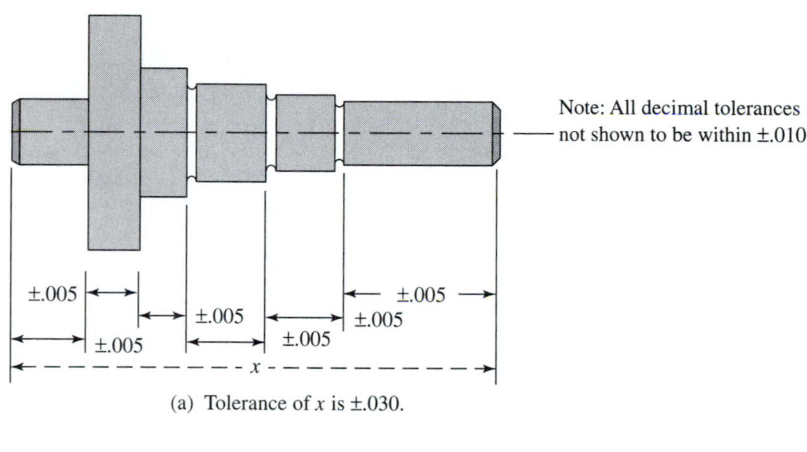

(a) Tolerance of x is ±.030.

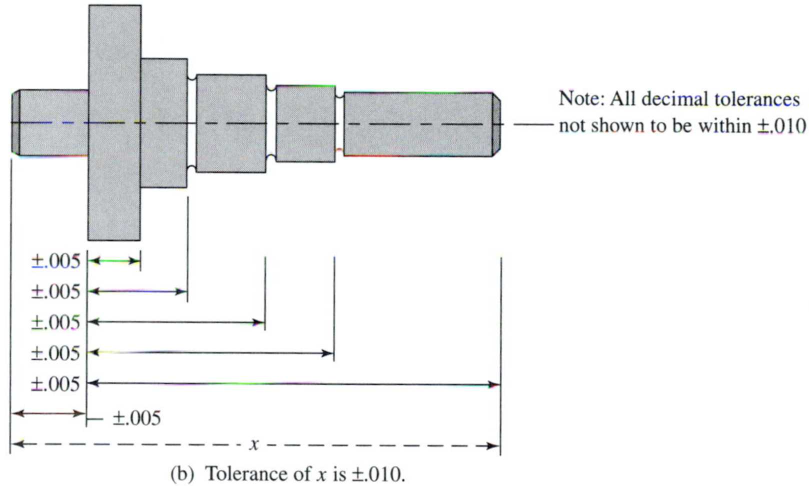

(b) Tolerance of x is ±.010.

FIGURE 5.21 (a) Countershaft design with unacceptable tolerance stack. (b) Countershaft design with acceptable tolerance stack.

1. **Redesign**
   - Open up unrealistically close tolerances.
   - Redimension: employ baseline dimensioning.
2. **Reprocess**
   - Make certain that locaters use the same baselines as part design and are consistent from operation to operation.
   - Make certain extremely restrictive tolerancing is actually required. (Too often, it provides only "insurance.")
   - Develop a tolerance chart to pinpoint where, and by how much, the problem exists. See Chapter 7, "Tolerance Charting."
3. Consider selective assembly where applicable.
4. Cut tolerances as required.

Cutting the tolerances is usually considered as a last resort because this action can result in adding qualifying operations, slowing speeds and feeds, increasing the amount of scrap and rework (when the equipment does not have the capability), and generally increasing the cost of production.

5.6 REVIEW QUESTIONS AND PROBLEMS

1. A vital first step in the manufacture of a part is the study of the part design by the manufacturing engineer to become completely familiar with the part, its properties, and its intent. Complete familiarity with the part design and intent is required before any meaningful attempt can be made at processing. Make a list of part properties that must be clearly understood and visualized.
2. What technique can be used to aid in visualizing extremely small parts?
3. All pertinent data required to produce a part may not be shown on the part design. Where else can this data be found?
4. What are some of the areas or part features that must be closely studied when analyzing parts related to other parts and assemblies?
5. List the major material forms.
6. Identify the geometrical tolerance symbols shown in Figure 5.22.

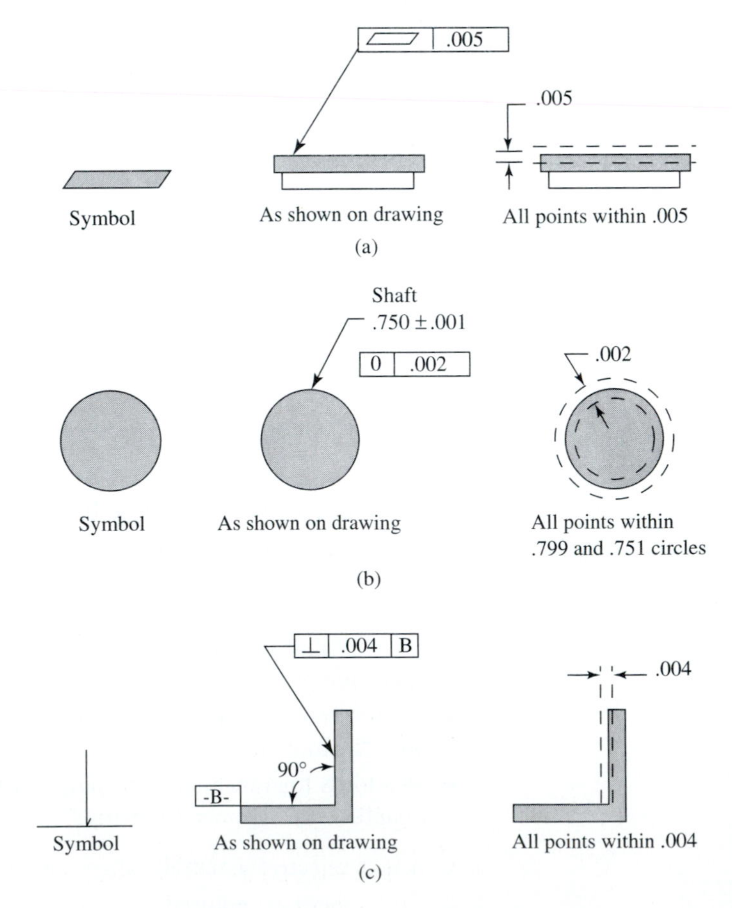

FIGURE 5.22 (a) Tolerancing symbols—flat surface. (b) Tolerancing symbols—shaft diameter. (c) Tolerancing symbols—perpendicularity.

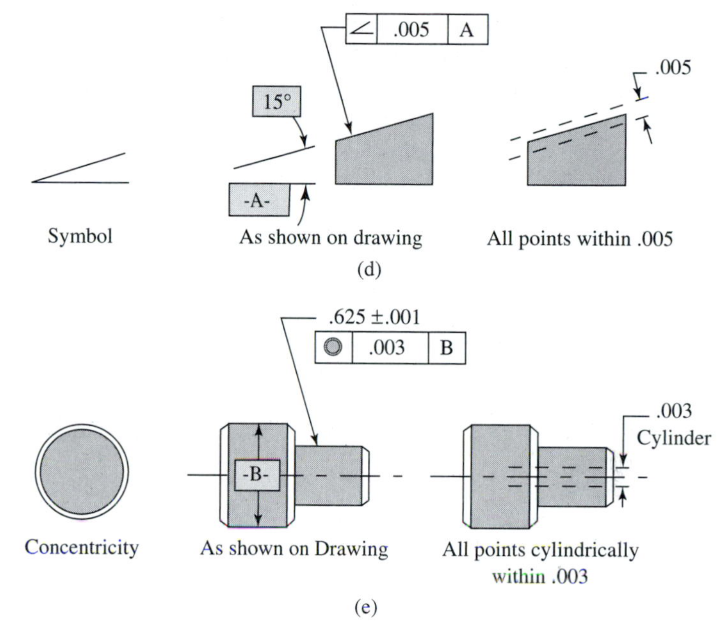

FIGURE 5.22 *Continued* (d) Tolerancing symbols—angularity. (e) Tolerancing symbols—concentricity.

7. Define and describe the three basic elements of surface finish created by machining and/or grinding processes.
8. Define the surface finish symbols shown in Figure 5.23.

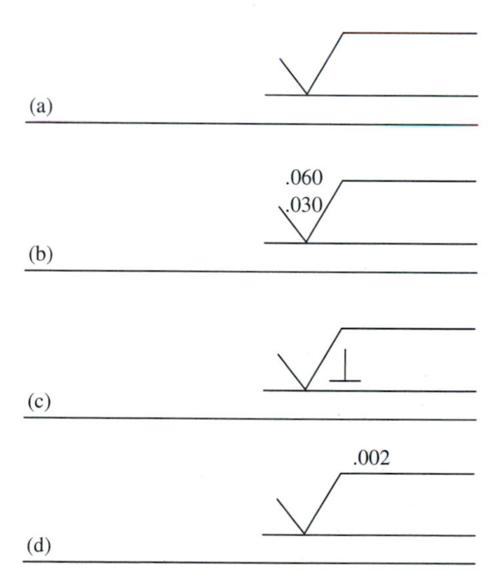

FIGURE 5.23 Surface finish symbols.

9. Refer to Figure 5.13:
   - Determine the full range of surface finishes that can be expected from a *drilling* operation. What is the range that can be expected from an average *drilling* operation application?
   - Determine the full range of surface finishes that can be expected from a *reaming* operation. What is the range that can be expected from an average *reaming* operation application?

- Determine the full range of surface finishes that can be expected from a *grinding* operation. What is the range that can be expected from an average *grinding* operation application?
- Determine the full range of surface finishes that can be expected from a *honing* operation. What is the range that can be expected from an average *honing* operation application?
- Determine the full range of surface finishes that can be expected from a *lapping* operation. What is the range that can be expected from an average *lapping* operation application?

10. Redimension parts shown in Figure 5.24 to a baseline dimensioning. Use surface A as the baseline.

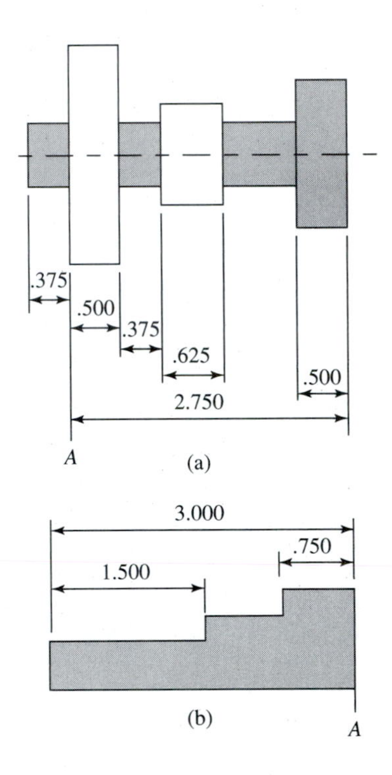

FIGURE 5.24 (a) Nonbaseline dimensioning of countershaft. (b) Nonbaseline dimensioning of stop block.

11. Calculate the dimensions and tolerance stacks for parts shown in Figure 5.25(a) and (b).

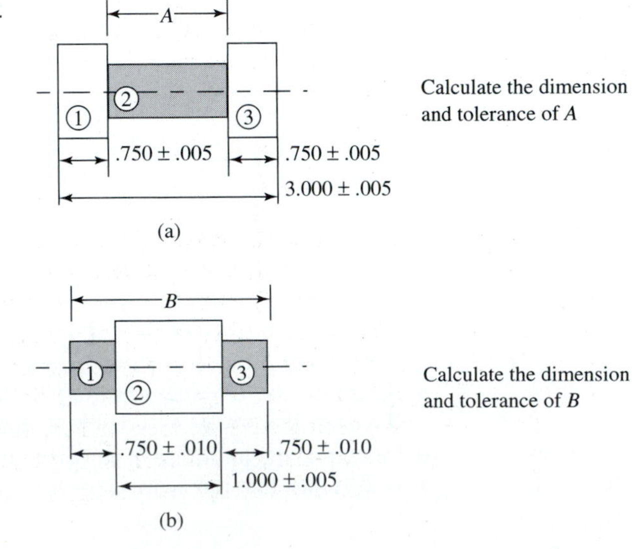

FIGURE 5.25 (a) Spool. (b) Roller.

- Calculate the amount of clearance and the tolerance stack of assembly shown in Figure 5.26.

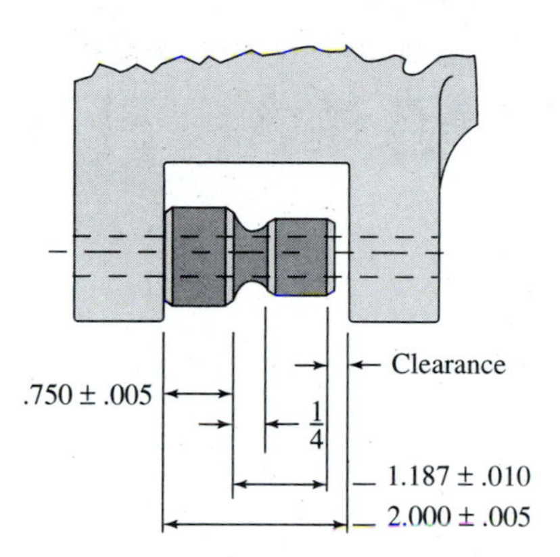

FIGURE 5.26 Assembly clearance.

CHAPTER SIX

MANUFACTURING PROCESSING

CHAPTER OVERVIEW

- Introduction
- Steps to Processing
  - Analyze Part Design
  - Summarize All Data
  - Itemize Basic Operations
  - Determine Locating Surfaces
  - Determine Feasible Methods
  - Combine Operations and Develop Sequence
  - Determine Manufacturing Dimensions and Tolerances Required to Define and Control Machine Surfaces and Provide Gaging
- Processing Examples
- Review Questions and Problems

6.1 INTRODUCTION

Manufacturing processing is at the heart of developing a manufacturing plan; central to manufacturing processing are the basics:

- Analyze the part design.
- Resolve the locating surfaces.
- Develop the processing.
- Develop manufacturing dimensions and tolerances (see Chapter 7, "Tolerance Charting").
- Design workpiece holders to position parts on designated locating surfaces (see Chapter 8).

These processing essentials apply whether the part has low production requirements and will probably be produced on a single-spindle, manually operated machine, or high enough production requirements to justify full automated, multistation transfer-type machines.

The manufacture of a part starts and ends with the part design. Processing must be designed to produce a part that satisfies the design; dimensional planning (tolerance charting) must be provided to assure that it does.

Following is a condensed example of processing that illustrates the preceding four basic rules. These rules, or steps, are amplified and detailed in the following section of this chapter. Figure 6.1 illustrates the four steps, A, B, C, and D, in developing the process for a part.

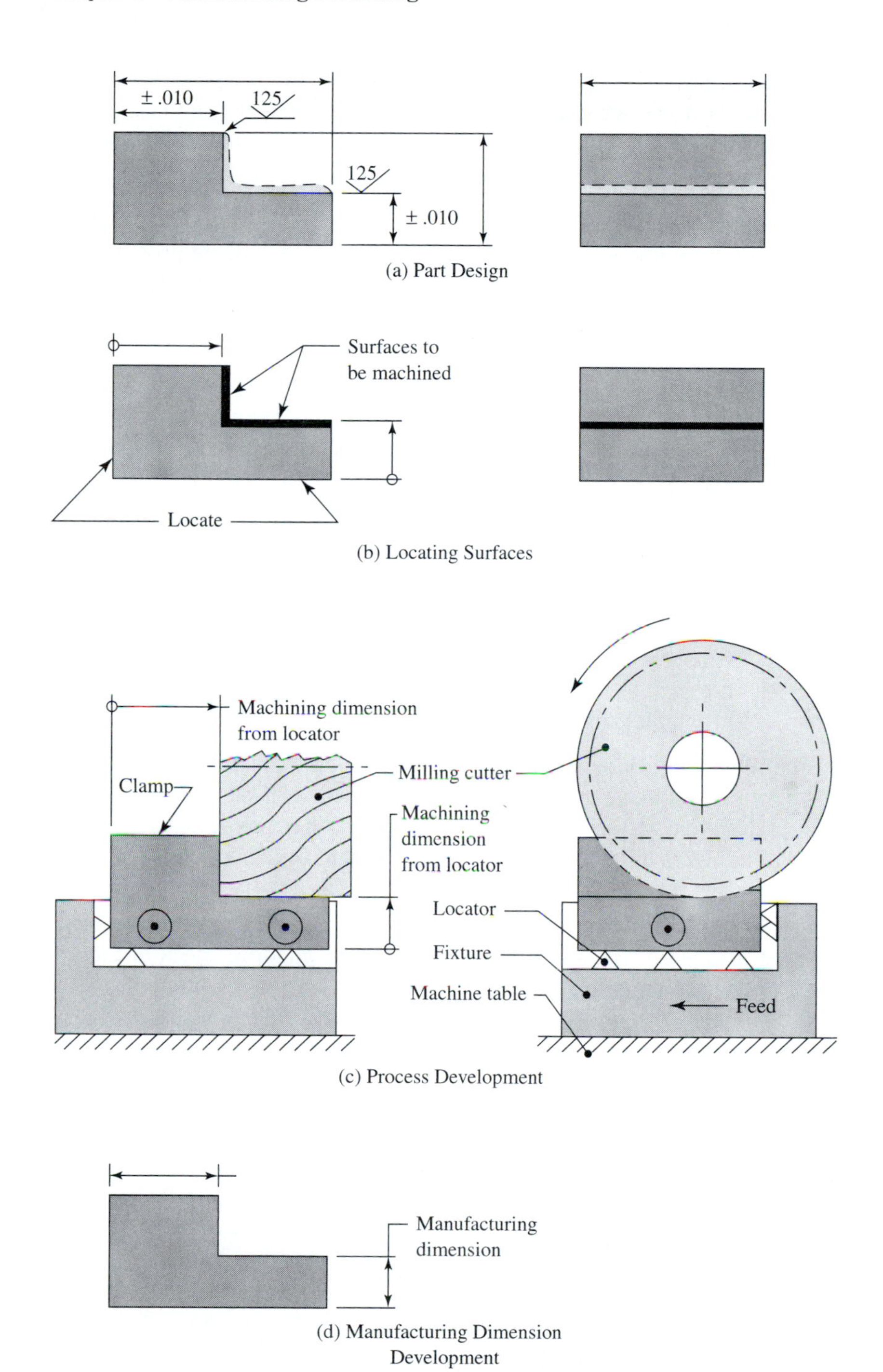

FIGURE 6.1 Process development steps.

A. **Part Design.** Analysis of the part design reveals that

- Two surfaces must be machined to 125 μin. surface finish.
- The machined surfaces are dimensioned from the left end face and the bottom face.
- Dimensional tolerances are ±.010.

B. **Locating Surfaces.** The machined surfaces are dimensioned directly from the left end face and the bottom face. Since the left end face and the bottom face are accessible and provide a large enough area to be stable, these areas are selected as locating surfaces.

C. **Process Development.** Production requirements for this part were determined to be 10,000 pieces total. This quantity would justify expenditures for only the most basic type of tooling.

A milling operation was selected as the manufacturing process; special tooling requirements would consist of a simple, manually operated milling fixture plus a standard milling cutter. Further, the surface finish requirements and dimensional tolerances are well within milling operation capabilities.

The view in *C*, Figure 6.1, shows the selected processing and the required machine and tool setup:

- The milling fixture has been designed with locators to contact the end and bottom face locating surfaces of the part.
- The job setter has set up the milling machine so that the fixture locators are in proper lateral and vertical relationship with the cutting tool.
- The part has been loaded in the fixture with the left end face and bottom face in contact with the fixture locators—and in this position, clamped.
- When the machine is cycled, the parts' surfaces will be machined to the set relationship and, as long as the tooling is properly maintained and the part properly loaded, acceptable parts will be produced.

D. **Manufacturing Dimension Development.** Manufacturing dimensions and tolerances must be developed to define and control the machined surfaces within the limits established in the part design. Gages must be provided to control the two dimensions machined, and they must be designed to check from the locating surfaces to the machined surfaces as they were milled and as the part design was dimensioned.

Not only must the gages provide an immediate check of the part just produced, they must also aid in the job setup and provide an ongoing check of tooling and loading conditions as production continues.

6.2 STEPS TO PROCESSING

Expanding from the condensed introduction example to a full step-by-step planning sequence, the following must be considered in developing a plan that will economically produce a part that satisfies the part design:

STEP 1: Analyze part design.
STEP 2: Summarize all data.
STEP 3: List basic operations.
STEP 4: Determine locating surfaces.
STEP 5: Determine feasible methods.
STEP 6: Combine operations and develop sequence.
STEP 7: Determine manufacturing dimensions and tolerances.

6.2.1 Analyze Part Design

Make a thorough analysis of the part design. Read, consider and apply the procedures discussed in Chapter 5, "Part Design Analysis."

6.2.2 Summarize All Data

Summarize Step 1 and gather all applicable facts and conditions effecting the part's manufacture including

- Total required production, rate per hour, and any quantities required for service
- Length of production run (life of product)
- Lead time (date of required production startup)
- Time study or standards data

6.2.3 Itemize Basic Operations

List all required manufacturing operations with no attempt made to combine operations at this stage, and no attempt made to identify precise machine types.

Use of the general term "machine," where possible, rather than specific manufacturing processing terms such as "drill," "turn," "mill," and so on will help to keep options open. When applying the following steps (where operations are combined and sequenced), many other feasible combinations will be suggested if limiting, preconceived decisions have not been made.

6.2.4 Determine Locating Surfaces

In addition to base line relationships and tolerances, many other considerations must go into the selection of locating surfaces.

Productivity versus Locating Surface Selected Consider the amount of machining that is attainable from the selected surfaces. If a part has more than one locating surface, as shown in Figure 6.2, surface A would be chosen in order to potentially accomplish the most machine work; more surfaces can be combined and machined from surface A than from surface B.

Locating surfaces must provide a stable rest area large enough to resist clamping and tool forces. It should be noted that the locating surfaces may require machining first to provide a flat, qualified surface to clamp the part on or against. The choice of locating surfaces will determine which operation will be the main or leadoff operation and which surfaces will then require secondary operations.

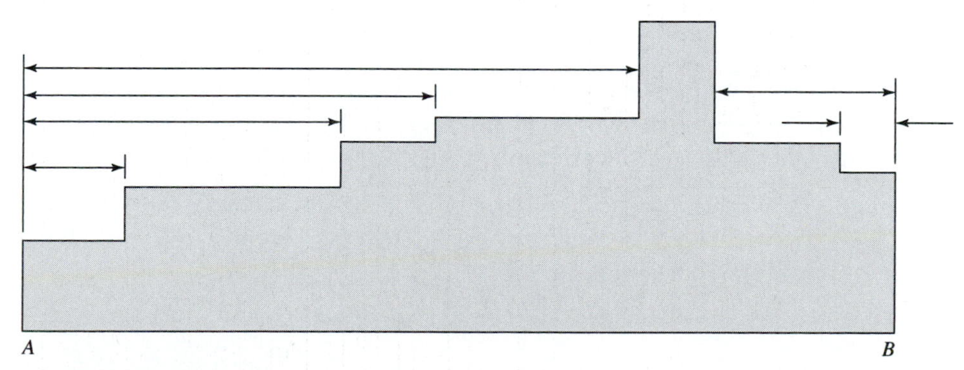

All surfaces machined to 125 μ in.
All decimal tolerances ± .010

FIGURE 6.2 Baselines relative to work accomplished.

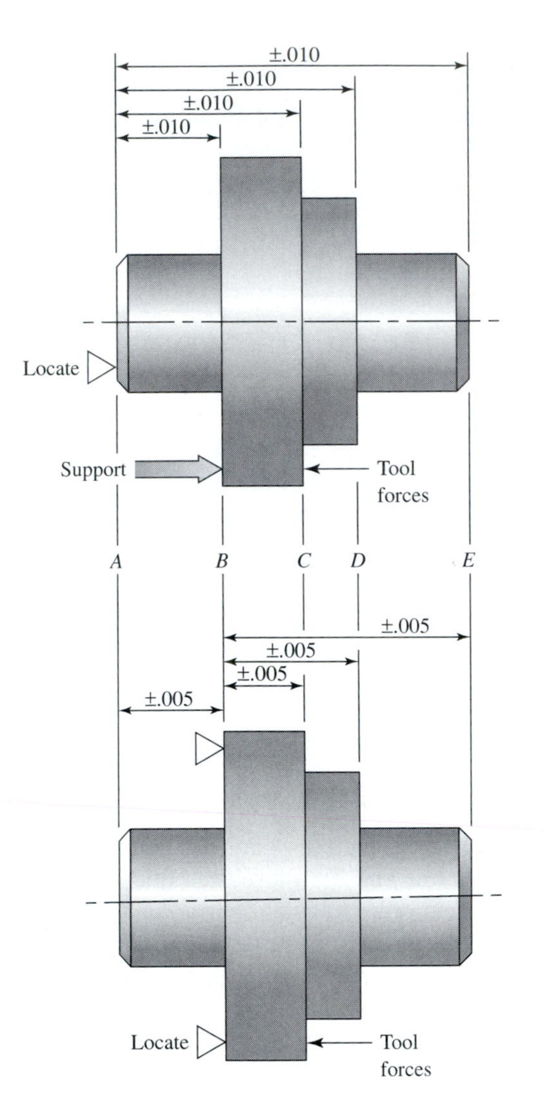

Method 1:
Part Located on Baseline

Surface *A* is baseline. Dimensionally, part should be located on surface *A*, but will require support for stability.

Full dimensional tolerance can be used—if supports can be successfully developed.

Method 2:
Locating Surface Transferred

Locating surface can be transferred to surface *B*, providing the piece part with stability.

Processing and manufacturing tolerances must be developed such that tolerance totals equal original part print tolerances:

$$AB + BE = \pm.010$$
$$AB + BD = \pm.010$$
$$AB + BC = \pm.010$$

FIGURE 6.3 Locating surface transferred.

Transfer of Locating Surfaces Parts might prove unstable when located on specific surfaces even though the surfaces are correctly dimensioned directly from the baseline. Supports* may sometimes be used, but it could prove desirable to locate on another surface. The locating surface may be transferred to one that is more stable—sometimes more accessible—provided

1. The locating surface transferred to is dimensionally related to the part baseline.
2. Dimensional tolerances are held closer, if need be, to assure that all part print final dimensions and tolerances are met.

Figure 6.3 illustrates the transfer of locating surfaces.

*Supports are further discussed in Chapter 8, "Workpiece Holding."

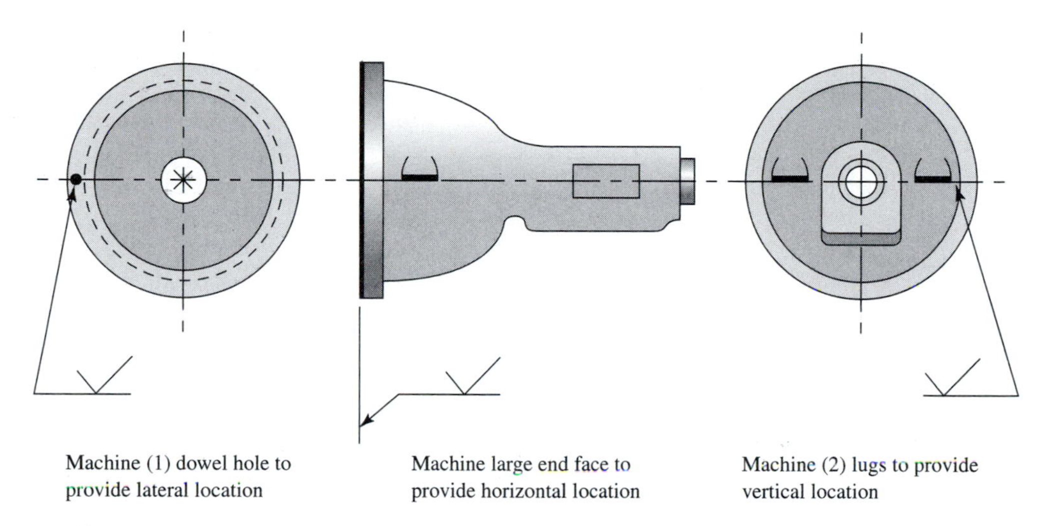

FIGURE 6.4 Providing locating surfaces for an automatic transmission case casting.

Locate on Provided Lugs or Bosses Certain parts may require a qualifying operation to machine an area, lugs, or bosses provided to act as locating surfaces only. These areas, lugs, or bosses do not enter into the function of the part and may be left, or they may be machined away at a later operation. An example of this type of location surface is shown in Figure 6.4.

Establishing locating surfaces and points is critical and must be done before planning process sequences.

6.2.5 Determine Feasible Methods

Each operation in the list of basic operations must be considered individually and an attempt made to determine the most feasible process for it. Areas that must be evaluated include

- Volumes
- Available equipment
- Dimensional tolerances required
- Surface finishes specified

Volumes Probably the first question a manufacturing engineer asks when starting to process a new part is, "How many pieces are we talking about?" The rate of production per hour is usually the deciding factor between a slower, manually controlled operation and a more highly specialized, higher-production piece of equipment. This is especially true for a new product where there is no existing, available equipment.

The total yearly production required plus the anticipated life of the product serve as fairly reliable guides on the comparative amounts of money that can be spent on a tool-up.

Available Equipment The use of available equipment must be considered when processing a part. The equipment may be surplus, on hand, or equipment that will become available for use in time for the contemplated part's manufacture.

Short lead times—the time available from product release until the start of production—often make the use of available equipment mandatory; new, more productive equipment may not be obtainable in time.

Variations from Basic Dimensions							
Diameter or Stock Size		to .250	.251 to .500	.501 to .750	.751 to 1.000	1.001 to 2.000	2.001 to 4.000
Reaming	Hand	±.0005	±.0005	±.0010	±.0010	±.0020	±.0030
	Machine	±.0010	±.0010	−.0015 +.0010	+.0010 −.0020	±.0020	±.0030
Turning			±.0010	±.0010	±.0010	±.0020	±.0030
Boring			±.0010	±.0010	±.0015	±.0020	±.0030
Automatic screw Machining	Internal		Same as in Drilling, Reaming or Boring				
	External forming	±.0015	±.0020	±.0020	±.0025	±.0025	±.0030
	External shaving	±.0010	±.0010	±.0010	±.0010	±.0015	±.0020
	Shoulder location, turning	±.0050	±.0050	±.0050	±.0050	±.0050	±.0050
	Shoulder location, forming	±.0015	±.0015	±.0015	±.0015	±.0015	±.0015
Milling (single cut)	Straddle milling	±.0020	±.0020	±.0020	±.0020	±.0020	±.0020
	Slotting (width)	±.0015	±.0015	±.0020	±.0020	±.0020	±.0025
	Face milling	±.0020	±.0020	±.0020	±.0020	±.0020	±.0020
	End milling (slot widths)	±.0020	±.0025	±.0025	±.0025		
	Hollow milling		±.0060	±.0080	±.0100		
Broaching	Internal	±.0005	±.0005	±.0005	±.0005	±.0010	±.0015
	Surface (thickness)		±.0010	±.0010	±.0010	±.0015	±.0015
Precision boring	Diameter	+.0005 −.0000	+.0005 −.0000	+.0005 −.0000	+.0005 −.0000	+.0005 −.0000	+.0010 −.0000
	Shoulder depth	±.0010	±.0010	±.0010	±.0010	±.0010	±.0010
Hobbing		±.0005	±.0010	±.0010	±.0010	±.0015	±.0020
Honing		+.0005 −.0000	+.0005 −.0000	+.0005 −.0000	+.0005 −.0000	+.0008 −.0000	+.0010 −.0000
Shaping (gear)		±.0005	±.0010	±.0010	±.0010	±.0015	±.0020
Burnishing		±.0005	±.0005	±.0005	±.0005	±.0008	±.0010
Grinding	Cylindrical (external)	+.0000 −.0005	+.0000 −.0005	+.0000 −.0005	+.0000 −.0005	+.0000 −.0005	+.0000 −.0005
	Cylindrical (internal)		+.0005 −.0000	+.0005 −.0000	+.0005 −.0000	+.0005 −.0000	+.0005 −.0000
	Centerless	+.0000 −.0005	+.0000 −.0005	+.0000 −.0005	+.0000 −.0005	+.0000 −.0005	+.0000 −.0005
	Surface (thickness)	+.0000 −.0020	+.0000 −.0020	+.0000 −.0030	+.0000 −.0030	+.0000 −.0040	+.0000 −.0050

FIGURE 6.5 Processes versus dimensional tolerances.

Also, when capital expenditures must be curtailed for economic reasons, the use of on-hand, available facilities is necessary. However, a study of costs should be made to determine the actual cost per piece when produced on available equipment compared to the cost when produced on new, more productive facilities. Of course, in some instances, there is a middle course where existing, available equipment can be updated to be more productive.

A good starting point for considering the selection of feasible processes is the use of a process versus dimensional tolerance chart as shown in Figure 6.5 and also a process versus surface finish chart as illustrated in Figure 6.6. Figure 6.7 will prove helpful when considering drilling operations' capabilities.

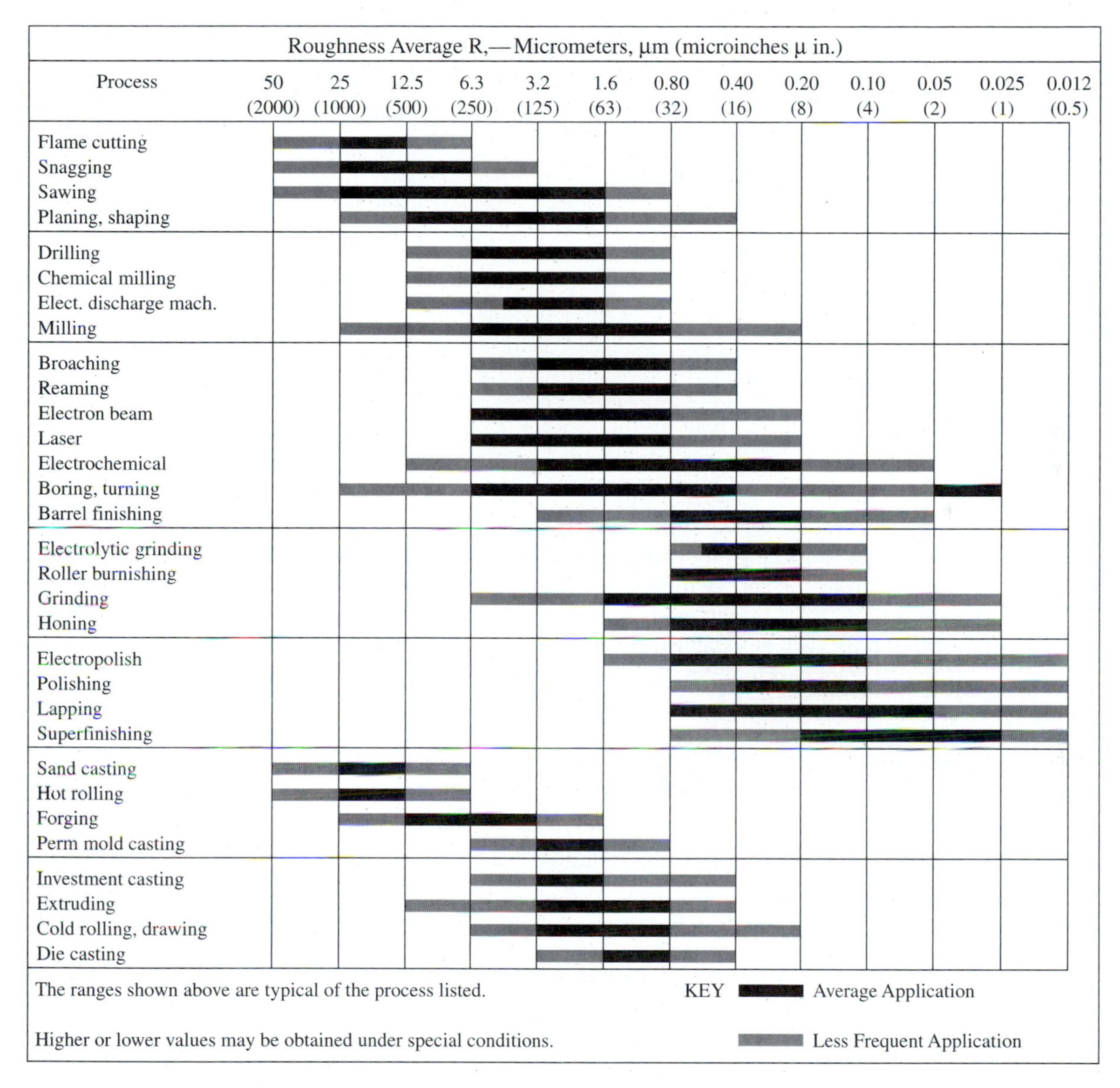

FIGURE 6.6 Processes versus surface finishes chart.

The use of such charts can prove helpful, especially in the early part of a manufacturing engineer's career. As an engineer's experience increases, he will become accustomed to, and acquainted with, the capabilities of the various types of manufacturing process equipment, but he will never know them all. He must learn to consult with those who are more knowledgeable and who specialize in the process in question—machine tool representatives, tool suppliers, his own supervisor, and peers.

When working with the process versus dimensional tolerance chart, it will be noted that larger parts generally require greater amounts of manufacturing tolerance than do smaller parts. Factors influencing the amounts required include cutting-tool speeds and feeds, cutting-tool design, part material type and hardness, and tool and machine deflections ± dimensional changes with temperature changes.

Tolerances on Drilled Holes			
Drill Size Range		Tolerance	
Smallest	Largest	Plus	Minus
.0135 (#80)	.042 (#58)	.003	.002
.043 (#57)	.093	.004	.002
.0935 (#42)	.156	.005	.002
.1562	.2656	.006	.002
.266 (H)	.4219	.007	.002
.4375	.6094	.008	.002
.625	.750	.009	.002
.7656	.8437	.009	.003
.8594	2.000	.010	.003

FIGURE 6.7 Drilled hole tolerances.

When the process versus surface finish chart, Figure 6.6, is used as an aid, the manufacturing engineer must be aware that the processes shown that produce the finer surface finishes are also the slowest. A part requiring a very fine surface finish, that is, low in microinches will normally be processed to remove as much material as possible by machining—then, the specified surface finish will be established with a following manufacturing process such as grinding.

The solid color bars in Figure 6.6 portray the usually expected finishes from the various manufacturing processes listed; lighter shaded bars show variances, and these can be expected from

- Differences between machines
- Differences between cutting-tool material and design
- Differences in cutting-tool speeds and feeds
- Differences of part material and hardness
- Differences of coolant

and so on.

It is better practice to start planning conservatively by selecting a process from the solid bar area of the chart and adding a finishing operation. Then, when more conditions are known, the plan can be modified.

6.2.6 Combine Operations and Develop Sequence

After investing the considerable time and thought required to analyze a part design, list basic operations—and, in fact, after accomplishing the study suggested in Steps 1 through 5, certain conditions will become apparent in regard to combining operations:

- Multiple holes in the same plane that can be combined for drilling in one operation
- Multiple surfaces in the same plane that can be machined by milling or broaching
- Multiple diameters, grooves, chamfers, and faces that can be machined in one turning operation
- Certain contours, diameters, steps, and grooves that can be machined simultaneously with a form tool on a turning operation
- Certain holes, not in the same plane as other holes, that can be machined at the same time by the addition of an auxiliary drilling unit to a drilling operation

While considering the foregoing, much of the mandatory sequencing of operations will become obvious:

- Drilling must precede tapping.
- Machining should precede hardening.
- A qualifying operation may be required to prepare a locating surface.
- When two or more different surfaces or holes are interrelated dimensionally, they must be processed so as to hold the relationship.
- Rough grinding can sometimes precede heat treating, but finish grinding is almost invariably accomplished after hardening to provide the specified surface finish and to remove heat treat scale and distortion.

Combining operations is generally thought of as advantageous, and this is generally so; however, the manufacturing engineer must be aware that some drawbacks exist and must consider possible disadvantages before committing to combined processes and operations.

Advantages of Combined Operations

- Improved accuracy, especially for dimensions held within the cutting-tool configuration
- Reduced labor costs where operations can be omitted
- Possibly fewer machines required at less cost
- Less material-handling labor and facilities required where operations can be omitted
- Fewer machines to set up—lower setup costs
- Possibly less floor space required

Disadvantages of Combined Operations

- Higher tool costs for complex tools—both purchasing and sharpening costs
- Difficulty of maintaining dimensions and tolerances where more than one baseline exists
- Compromising or averaging cutting speeds for multidiameter parts rather than providing optimum speeds for each diameter

Summing Up: Recording and Formalizing Data The use of a summary sheet such as that shown in Figure 6.8 is needed to gather data, and preserve a record of the process planning made up to this point.

A first tentative processing including any decided upon combining of operations should be listed in the Operation Description column. An operation number, in a 10–20–30 series, should be entered in the appropriate column to help define the sequence. Machine or manufacturing processes are to be entered in their respective locations in the column provided.

Of utmost importance is the inclusion of a description of the locating points, or surfaces, decided upon for each operation. Such descriptions need not fully describe the part, its dimensions, tolerance, and so on, but must include enough to make clear the intent with a definition such as

- Locate on O. D. and against left end face.
- Locate in I. D. and against right end face.
- Locate on left end face and bottom face.
- Locate between centers.
- Locate on large end face and in two, bored dowel holes.

and so on. *Note*: The use of summary sheets is further illustrated in the process examples that follow.

Op. No.	MACHINE OR PROCESS	OPERATION DESCRIPTION	LOCATING SURFACE

FIGURE 6.8 Manufacturing processes summary sheet.

6.2.7 Determine Manufacturing Dimensions and Tolerances Required to Define and Control Machine Surfaces and Provide Gaging

Gages designed and made to check the dimensions and tolerances shown on the part design are known as finish, or final dimension gages.

Other gages, known as process or work gages, check dimensions of unfinished parts in process of manufacture. These gages may check calculated dimensions, dimensions reflecting machining or grinding stock still to be removed, or dimensions held to closer tolerances than part print for purposes of tolerance control.

Process, or work, gages are never used as final inspection gages; only gages designed to check part design dimensions to their full specified tolerances assure holding the required level of outgoing quality while providing all of the producibility available from the part design specifications.

6.3 PROCESSING EXAMPLES

Example 1: Bracket–Door Roller
Low Production Requirements
Bar Stock Material

Example 2: Bushing Block
Moderate Production Requirements
Steel Casting

Example 3: Oil Pump Housing
High Production Requirements
Aluminum Die Casting

Example 4: Adapter
High Production Requirements
Bar Stock Material

PROCESSING EXAMPLE 1

Bracket–Door Roller (Fig. 6.9)

PRODUCTION REQUIREMENTS

600 pieces total

Life of Product

One-time requirement

STEP 1: Analyze part design.
STEP 2: Summary.

From the part design analysis a summary of the manufacturing requirement highlights can be made:

- Material is a length of hot-rolled steel, SAE 1020.
- No heat treatment is specified.

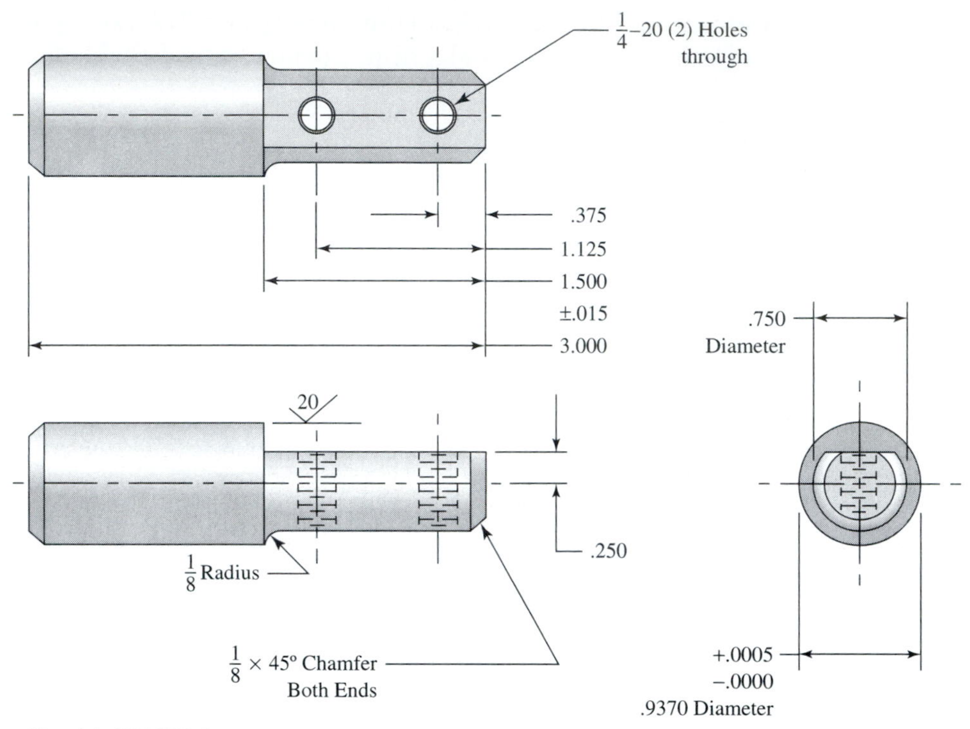

Material: 1020 H.R.S.
All surface finishes 120 μin. unless otherwise specified
All dimensional tolerances +.005 unless otherwise specified

FIGURE 6.9 Bracket–Door Roller.

- The closest dimensional tolerance is

$$\begin{matrix}+.0005\\-.0000\end{matrix}$$

  as it applies to the .937 diameter.
- All other dimensional tolerances are ±.005.
- The most stringent surface finish is 20 μin. for the .937 diameter.
- All other surface finishes are 120 μin.
- In configuration, the part is machined from a length of 1 inch-diameter hot-rolled bar stock with two diameters, two chamfers, two threaded holes, and a flat surface.
- Production requirements were found to be 600 pieces total.

STEP 3: Itemize basic operations.

The separate basic operations required to produce the part are

1. Machine part to length from round bar.
2. Machine flat.
3. Machine .937 diameter.
4. Machine .750 diameter.
5. Machine $\frac{1}{8} \times 45°$ chamfer one end.

6. Machine $\frac{1}{8} \times 45°$ chamfer other end.
7. Machine one hole.
8. Machine other hole.
9. Thread one hole.
10. Thread other hole.
11. Countersink one hole.
12. Countersink other hole.

STEP 4: Determine locating surfaces.

The right end face of the part is the baseline for all lateral dimensions; therefore, the right end face of this part is the correct locating surface for all operations establishing these dimensions.

Two diameters are concentric about the part's centerline, and two threaded holes are located on the centerline. Since centers are not provided for this part, the outside diameter must be used as the locating surface to establish the centerline.

Additionally, the threaded holes are normal to the flat surface; the flat surface must be used as a locating surface to establish this relationship while machining the holes.

STEP 5: Determine feasible methods.

STEP 6: Combine operations and develop sequence.

Steps 5 and 6 are separate considerations, but they must be considered together when making operation decisions and planning their sequence. The following illustrates this approach.

General. All dimensional tolerances, except for the .937 diameter, are within the capabilities of basic machine operations. This can be confirmed with the process versus dimensional tolerance chart, Figure 6.5.

Also, surface finish requirements are within basic machine-tool capabilities, except for the .937 diameter, and this can be verified with the process versus surface finish chart, Figure 6.6.

The dimensional tolerance and surface finish requirements of the .937 diameter will require a finish grinding operation.

Summary Sheet. From this point on, a summary sheet must be used to record the first tentative processing. This, the starting point, will then be modified with operations added, subtracted and altered, and the sequence changed until the optimum processing is developed—one that satisfies the print economically. Entering into the tentative processing are judgments such as the following.

Further Processing Conclusions.

1. Standard, single-spindle, manually operated machines will be used due to the low production of 600 pieces total.

2. With the general processing decided on, as in Part 1 above, the first operation will be to cut the part to length from bar stock, as shown in Figure 6.10(a). With higher production requirements, an automatic screw machine would include cutoff as part of this operation; however, the much higher cost of tooling the automatic screw machine cannot be justified.

3. It is better practice to machine both the .950 and .750 diameters before machining the flat surface in order to avoid an interrupted cut. Since no centers were provided, the part must be held (chucked) on the .950 diameter end when machining the .750 diameter, as shown in Figure 6.10(b). Conversely, it must be chucked on the .750 diameter end when machining the

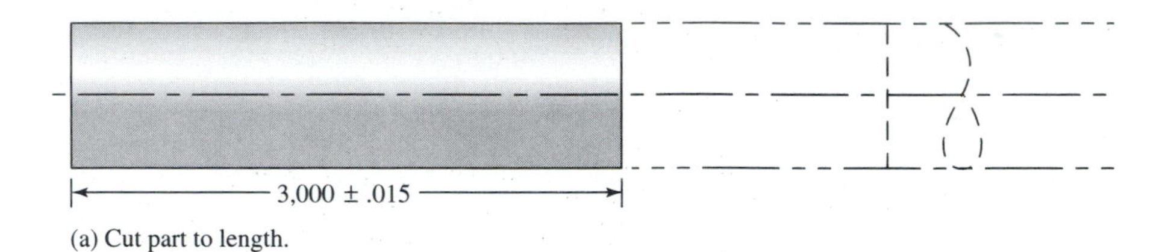

(a) Cut part to length.

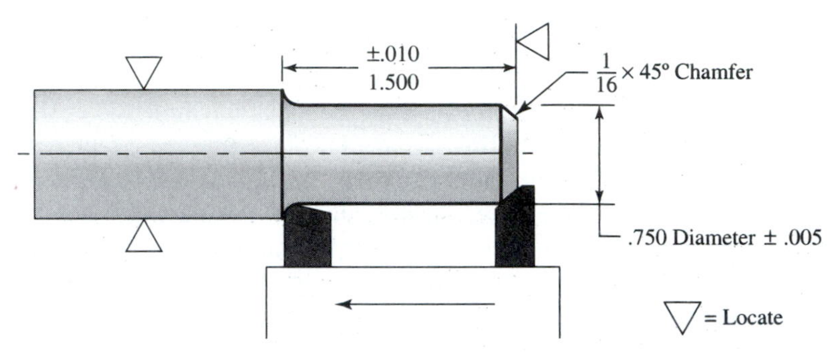

(b) Machine .750 diameter and chamfer (1) end.

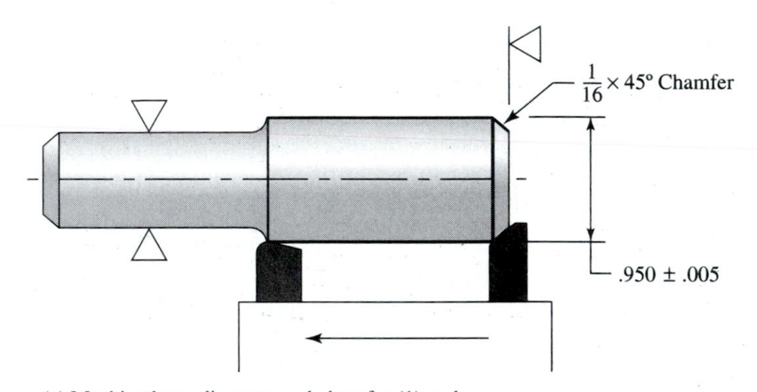

(c) Machine large diameter and chamfer (1) end.

FIGURE 6.10 Bracket–Door Roller Processing.

.950 diameter as shown in Figure 6.10(c). The chamfers can be combined with the diameter machining as shown in Figure 6.10(b) and (c).

4. The flat surface must be machined next to permit drilling the 2 holes. Milling would be an appropriate process because its productivity is good and the tooling costs are less than broaching. The part must be located from the right end face and on its centerline. Figure 6.10(d) illustrates this operation.

5. Drilling the 2 holes must be done in a fixture that locates the part from its right end face, on its centerline, and radially against the flat surface. The drill will be located with a bushing plate. Figure 6.10(e) illustrates this operation.

The holes must, of course, be drilled before tapping; they should be countersunk before tapping to allow better entry of the tap and to avoid burring of the thread if countersinking were done last.

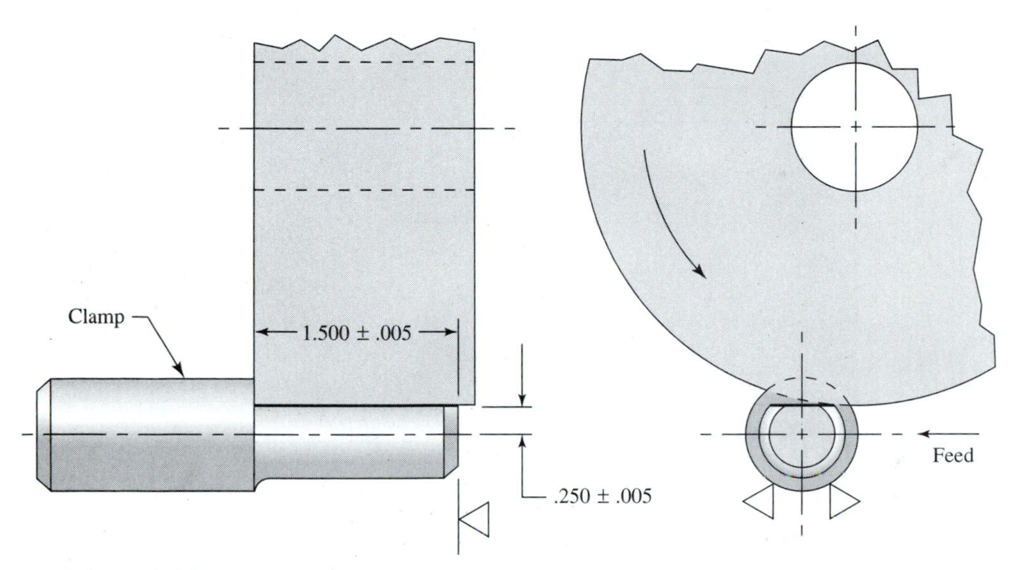

(d) Machine flat surface.

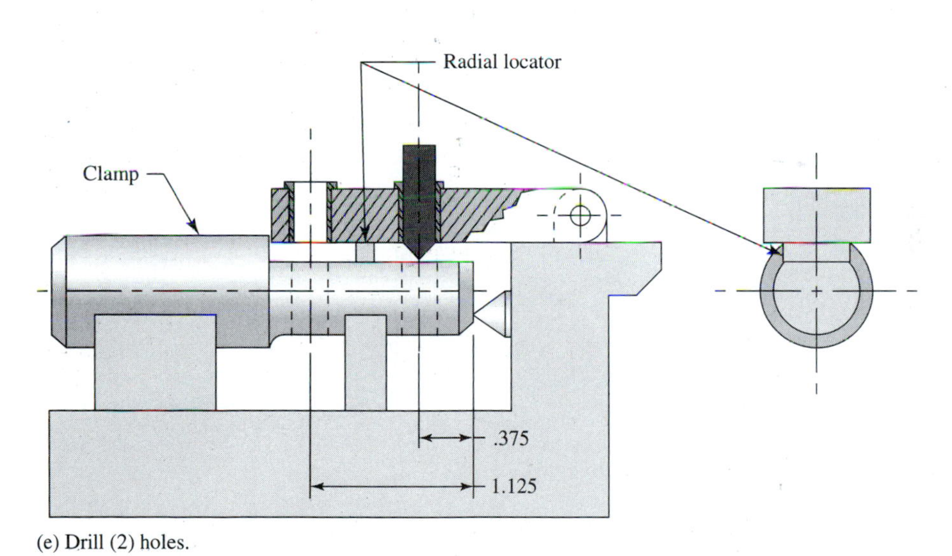

(e) Drill (2) holes.

FIGURE 6.10 Continued

The same fixture that is used to drill the holes can be used to hold the part while countersinking and tapping. The bushing plate will be designed to swing away as shown in Figure 6.10(f).

If the shop has an adjustable two-spindle drill head available, drilling the 2 holes can be combined—if not, they can be drilled, one at a time, on the same fixture.

6. The .937 diameter can now be finished by grinding, which will establish both the diametral tolerance and the surface finish. An appropriate machine would be a centerless grinder using in feed. A chucking grinder can also be used, chucking on the .750 diameter and against the right end face as shown in Figure 6.10(g).

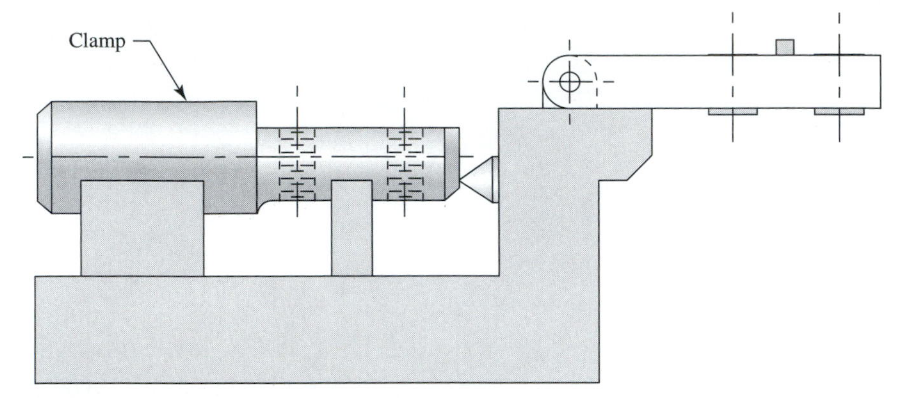

(f) Countersink and tap (2) holes.

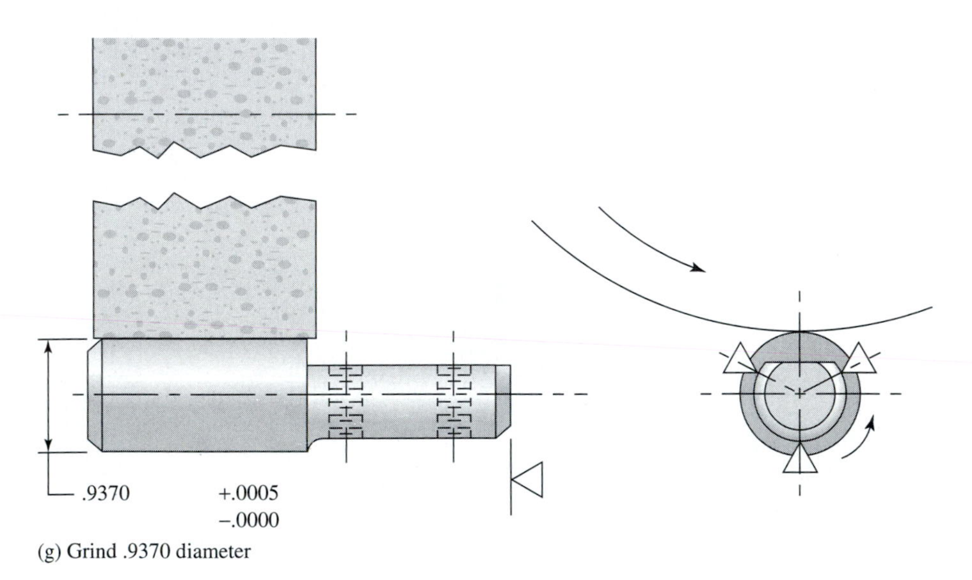

(g) Grind .9370 diameter

FIGURE 6.10 Continued

The foregoing has been added to the summary sheet Figure 6.11.

STEP 7: Determine manufacturing dimensions and provide gaging.

Gages are required to check each of the part dimensions, and, for this part, the gages will be the same at final inspection as those used for the manufacturing operation, except

- A work, or process, gage will be required at operation 30 to check the rough machined .937 diameter.
- A gage to check the finished ground .937 diameter to its specified tolerance will be required at operation 110 and also at final inspection.
- Location of the 2 holes will be checked in layout inspection to save the cost of a fixture gage.

Op. No.	MACHINE OR PROCESS	OPERATION DESCRIPTION	LOCATING SURFACE
10	Band Saw	Cut part to length	Right end face
20	Lathe	Turn .750 diameter	Outside diameter
		Turn $\frac{1}{8} \times 45°$ chamfer	Right end face
30	Lathe	Turn .937 diameter	.250 Diameter
		Turn $\frac{1}{8} \times 45°$ chamfer	Large end face
40	Mill	Mill flat face	.750 Diameter
			Small end face
50	Drill Press	Drill (1) hole	.750 Diameter and .937 Diameter
			Small end face
			Flat surface
60	Drill Press	Drill other hole	.750 Diameter and .937 Diameter
			Small end face
			Flat surface
70	Drill Press	Countersink (1) hole	.750 Diameter and .937 Diameter
			Small end face
			Flat surface
80	Drill Press	Countersink other hole	.750 Diameter and .937 Diameter
			Small end face
			Flat surface
90	Drill Press	Tap (1) hole	.750 Diameter and .937 Diameter
			Small end face
			Flat surface
100	Drill Press	Tap other hole	.750 Diameter and .937 Diameter
			Small end face
			Flat surface
110	Chucking Grinder	Grind .937 diameter	.750 Diameter
			Small end face
120		Inspect	

FIGURE 6.11 Manufacturing processes summary sheet—bracket–door roller.

PROCESSING EXAMPLE 2

Bushing Block (Fig. 6.12)

PRODUCTION REQUIREMENTS

12,000 pieces annually

Life of Product

Five years (est.)

STEP 1: Analyze part design.
STEP 2: Summary.

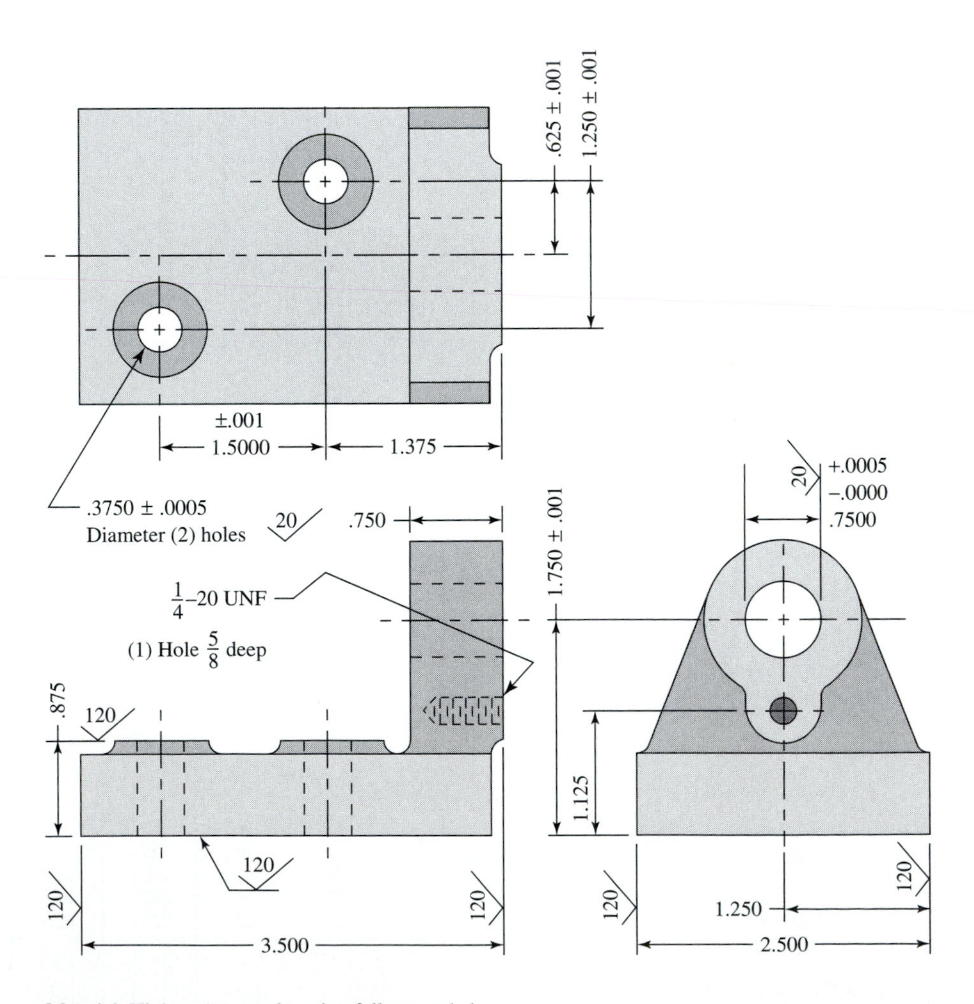

Material: High-carbon steel casting fully annealed
All dimensional tolerances +.005 unless otherwise specified
Scale full

FIGURE 6.12 Bushing block.

From the part design analysis a summary of the manufacturing requirement highlights can be made:

- Material is a steel casting.
- A heat treatment is specified to normalize the casting and relieve stresses. This operation will also improve machinability.
- The closest specified dimensional tolerance relates to the holes.
- The two .375 diameter holes have a diametral tolerance of ±.0005.
- The .750 diameter bore has a diametral tolerance of

$$\begin{matrix} +.0005 \\ -.0000 \end{matrix}$$

- Center-to-center hole locations have tolerances of ±.001, including the relationship between the two .375 diameter holes and the .750 diameter bore.
- All other dimensions have tolerance of ±.005.
- The most stringent surface finishes are 20 μin. specified for the .750 diameter bore and the two .375 diameter holes.
- All other surface finishes are 120 μin.
- In configuration, the part is an L-shaped casting with a bore and threaded hole in the upright leg; two mounting holes are provided in the base. Holes and bore all have cast bosses, which require machining.
- The bottom face, 2 sides, left end face, and right end face are machined.
- Product requirements are 12,000 pieces annually including service parts. Life of the product (a machine-tool part) is estimated at 5 years.

STEP 3: Itemize basic operations.

1. Machine bottom face.
2. Machine rear face.
3. Machine front face.
4. Machine left end face.
5. Machine right end face.
6. Machine .750 diameter bore.
7. Machine $\frac{1}{4}$–20 threaded hole.
8. Machine $\frac{1}{4}$–20 thread.
9. Machine $\frac{1}{4}$–20 threaded hole countersink.
10. Machine one .375 diameter hole.
11. Machine other .375 diameter hole.
12. Machine one .375 diameter hole boss.
13. Machine other .375 diameter hole boss.

STEP 4: Determine locating surfaces.

The Bushing Block design has 3 baselines. For clarity in referring to manufacturing processes and the locating surfaces related to each, the following nomenclature is used:

- *X*-axis: Dimensions left to right
- *Y*-axis: Dimensions front to back
- *Z*-axis: Dimensions bottom to top

***X*-Axis**

- The right end face of the part is the baseline for all lateral dimensions; the right end face is dimensioned from the left face of the upright leg.
- The left face of the upright leg is the locating surface for machining the right face, after which it becomes the locating surface for all dimensions in this axis.

***Y*-Axis**

- The back face is the baseline for dimensions to the front face and centerline of the $\frac{1}{4}$–20 thread hole, two .375 diameter holes, and .750 diameter bore. After machining, it is the locating surface for dimensions in this axis.
- While machining, the part must be so located as to center it; that is, machining stock must be divided as shown in Figure 6.13(b).
- A dimensional interrelationship also exists between the two .375 diameter holes centerline and the .750 diameter bore centerline, which can be held by locating in the two .375 diameter holes.

***Z*-Axis**

- The bottom face is the baseline for dimensioning to the top of the two .375 diameter hole bosses, centerline of the .750 diameter bore, and $\frac{1}{4}$–20 threaded hole. After machining from the top faces of two .375 diameter hole bosses, the bottom face is the locating surface for dimensions in this axis.

STEP 5: Determine feasible methods.

STEP 6: Combine operations and develop sequence.

General. Dimensional tolerances and surface finish requirements are not all within the capabilities of basic machine operations: Hole location dimensions in both x- and y-axes for the two .375 diameter holes, their centerline relationship with the .750 diameter bore, and surface finish specifications will require precision boring operation. All other dimensions and specifications are within basic machine capabilities.

Summary Sheet. A summary sheet should be started at this point to record all judgments and processing decisions as they develop. From the preceding steps and considering production requirements, dimensional tolerances, and surface finish specifications, it can be concluded that development of processing can continue along the following lines:

1. Flat surfaces can be milled.
2. All holes can be drilled.
3. The .750 diameter bore and two .375 diameter holes require a finish precision boring operation.

Further Processing Considerations.

1. Generally, the best processing practice is to start with the largest surface area to be machined and continue with each succeeding operation—machining progressively smaller surfaces.

2. Machining the bottom face will establish the largest surface as a stable surface from which to start processing. A milling operation would be suitable for this first operation which is illustrated in Figure 6.13(a).

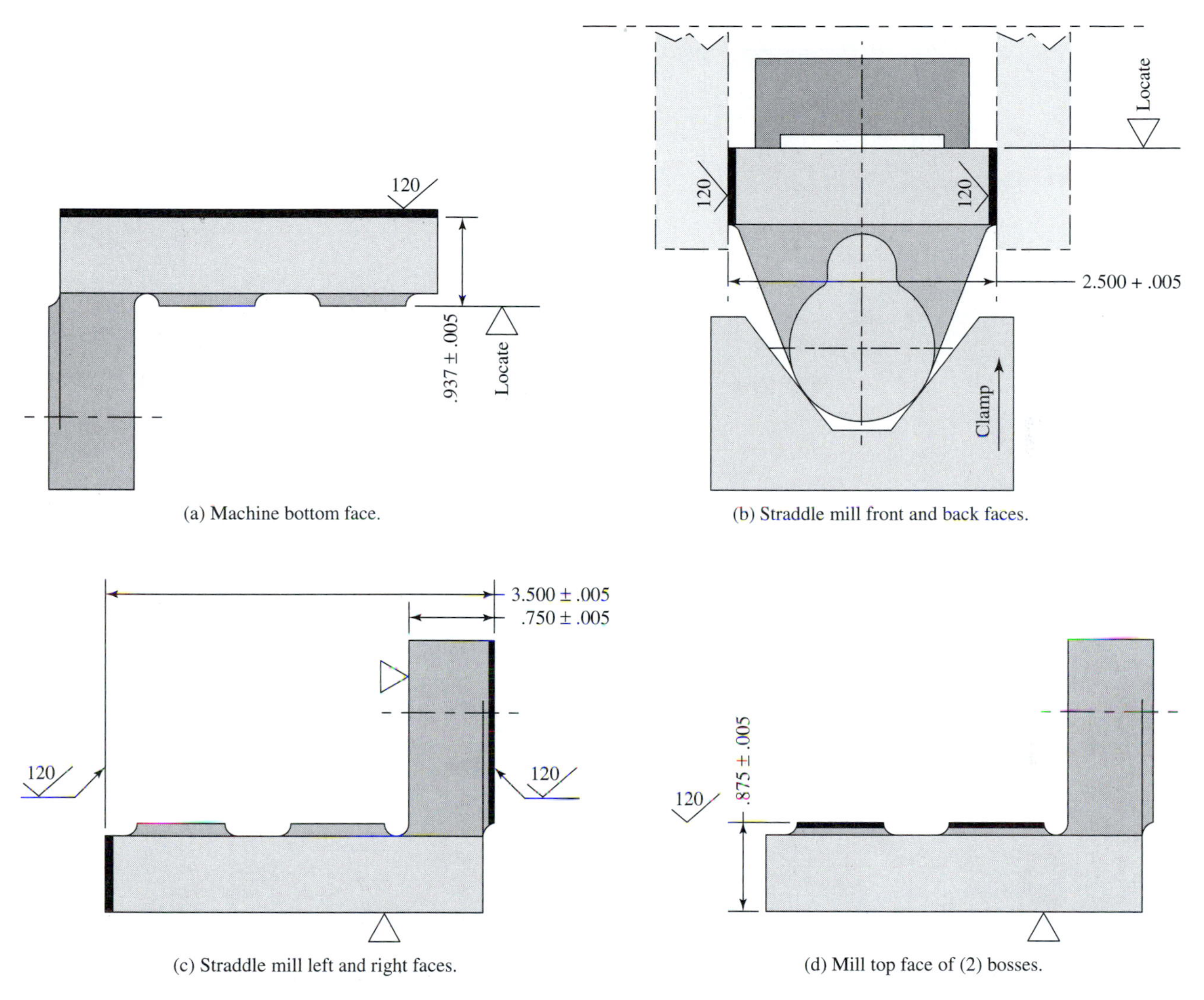

FIGURE 6.13 Bushing block processing.

3. Next in order, by size, is machining the front and back faces. Since these surfaces are parallel, milling them can be combined into a straddle milling operation as shown in Figure 6.13(b).

4. Machining the right end face must be done with the part located on the bottom face, against the left side of the upright leg and squared against one side face as shown in Figure 6.13(c).

Since the left end face of the part is parallel with the right end face, machining these surfaces can be combined in a straddle milling operation as shown in Figure 6.13(c).

5. The top faces of the two .375 diameter hole bosses may be spot faced, following drilling, or they can be milled in a single operation. Milling, being more productive, will be specified here.

The part must be located on the bottom surface—all other locators need only be approximate since there are no *X*- or *Y*-axis dimensions to hold as shown in Figure 6.13(d).

6. Since the two .375 diameter holes can serve as locating points to hold the interrelationship between them and the .750 diameter bore centerline, they are drilled next leaving stock for

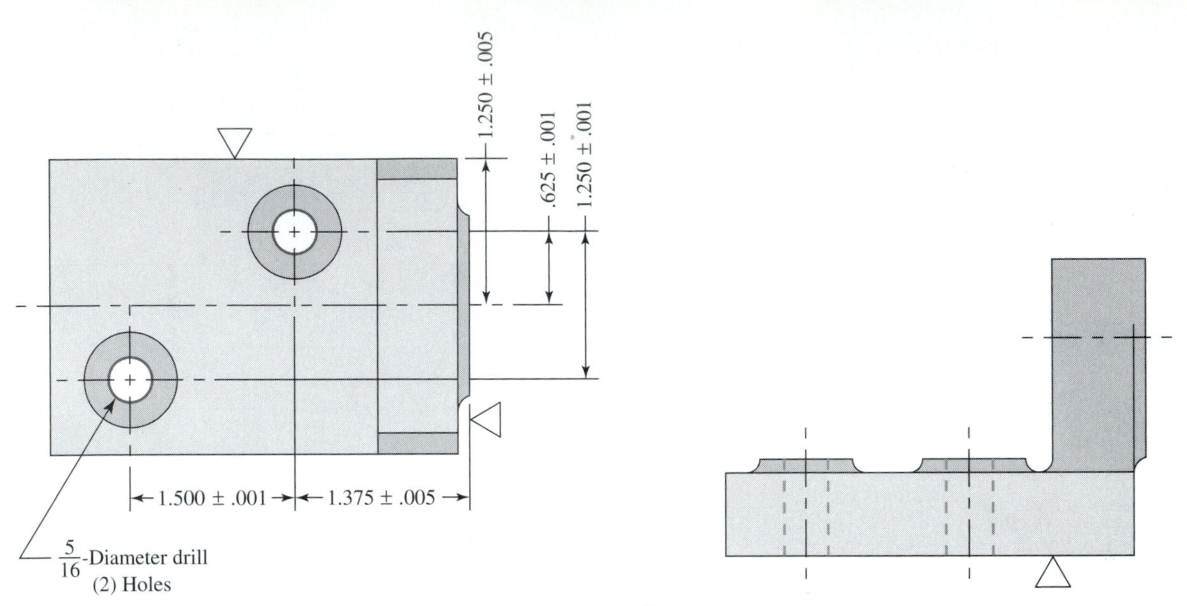

(e) Drill (2) .375 diameter holes to $\frac{5}{16}$ diameter.

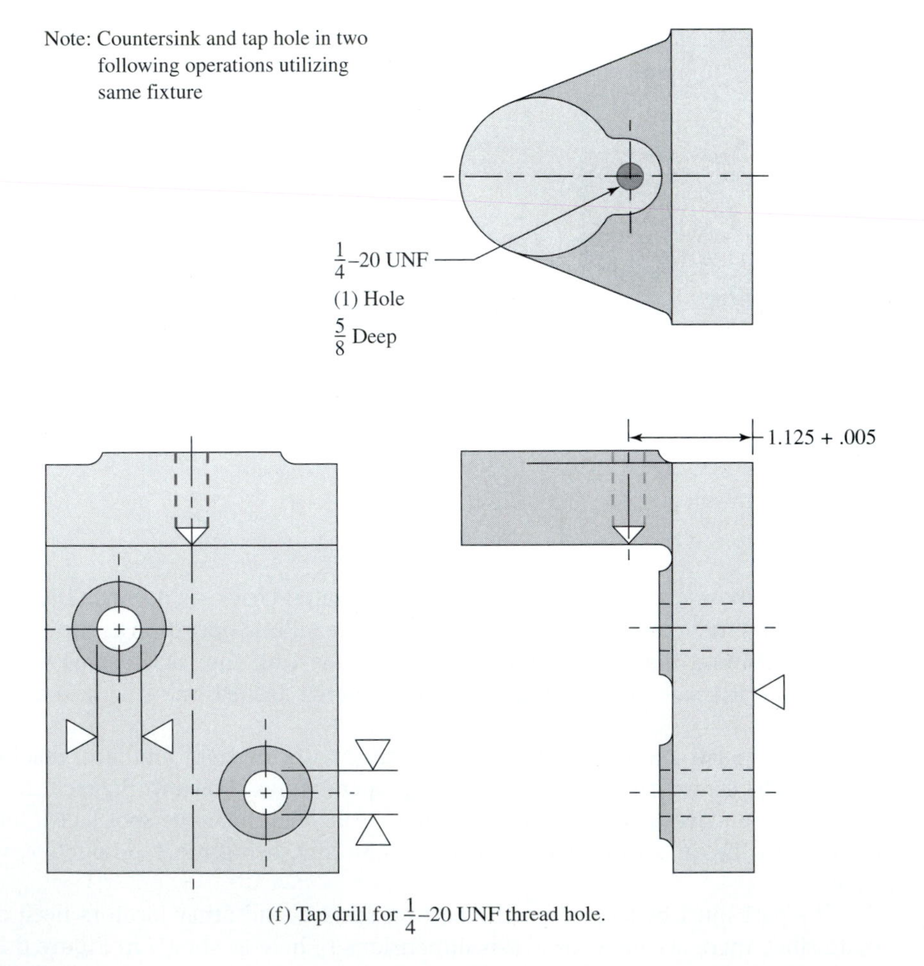

(f) Tap drill for $\frac{1}{4}$–20 UNF thread hole.

FIGURE 6.13 Continued

a final precision boring operation. Location is from the bottom surface, rear face, and right face as shown in Figure 6.13(e).

7. Drilling, countersinking, and tapping the $\frac{1}{4}$–20 threaded hole can be done on a drill press, but the operations cannot be combined.

Location will be on the part's bottom surface and in the two previously drilled .375 diameter holes as shown in Figure 6.13(f).

8. The .750 diameter bore is not cored and must be drilled from the solid, but the hole location dimensional tolerances and surface finish requirements are beyond the capability of a drilling operation; therefore, the hole must also have a finishing operation. A precision boring process will produce both the required dimensional tolerance and the surface finish.

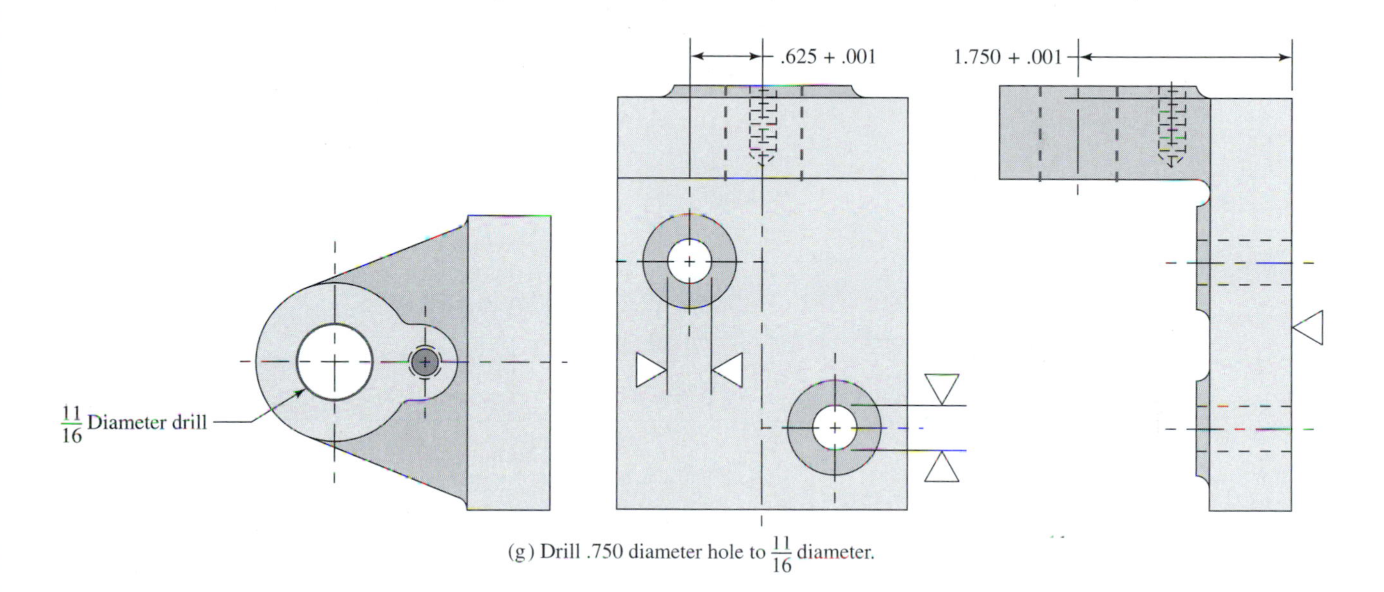

(g) Drill .750 diameter hole to $\frac{11}{16}$ diameter.

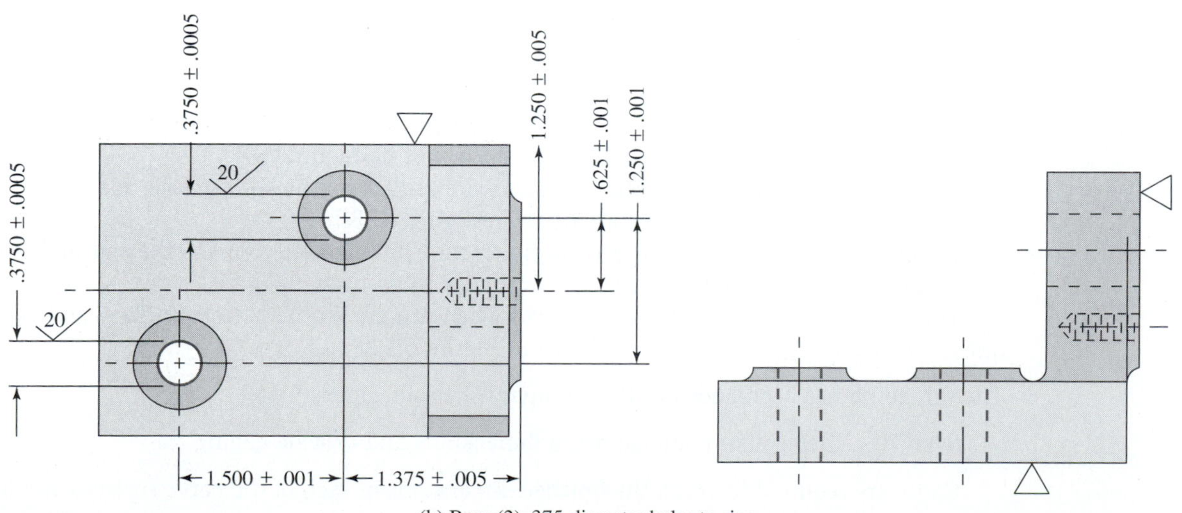

(h) Bore (2) .375-diameter holes to size.

FIGURE 6.13 Continued

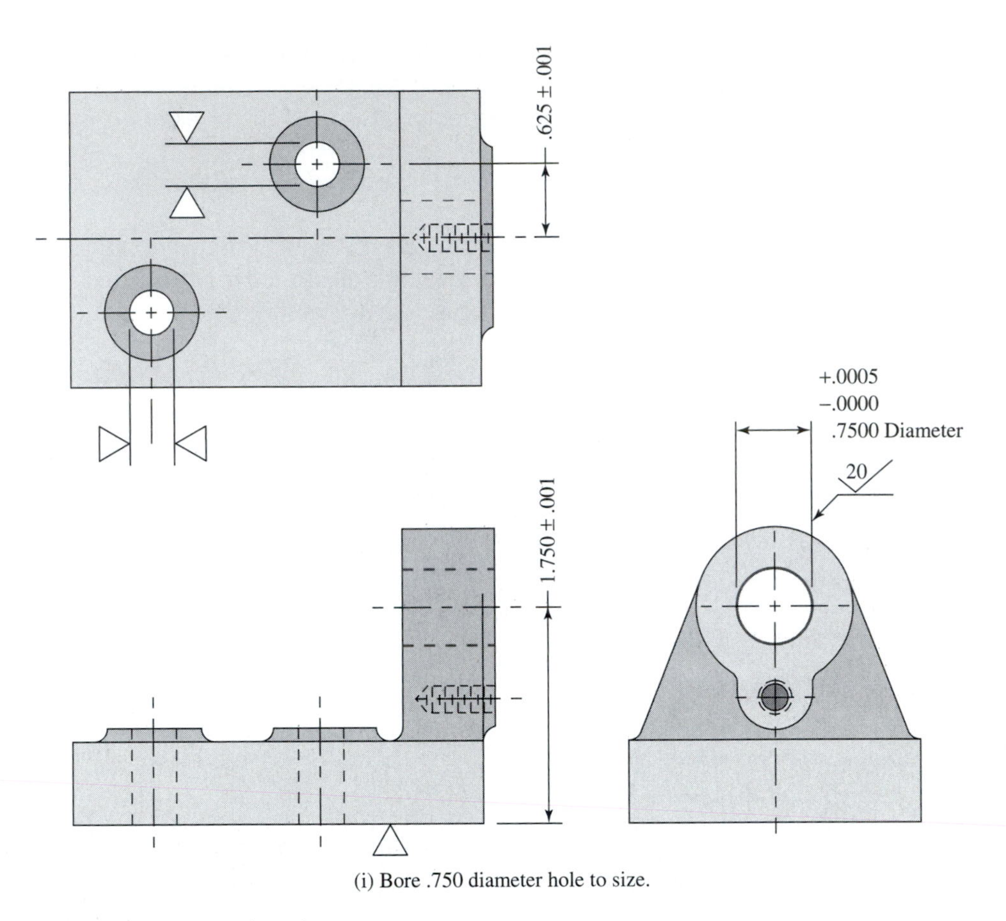

(i) Bore .750 diameter hole to size.

FIGURE 6.13 Continued

Location on the bottom face and in the two previously drilled .375 diameter holes to the specified location dimension and also the specified interrelationship as shown in Figure 6.13(g). Precision boring is done as late in the process sequence as possible to minimize damage from handling.

9. The two .375 diameter holes must be finished to size, specified surface finish, and hole location dimensions. A precision boring operation will be required.

Location must be as it was when the holes were drilled: on the bottom face, against the right end face, and against the rear face as shown in Figure 6.13(h).

10. An operation must be provided to finish the .750 diameter bore to the size and surface finish specified. Precision boring has the capability to do both.

The part must be located again, as it was when drilled, on the bottom surface, and in the 2 *finish bored* .375 diameter holes as shown in Figure 6.13(i).

The foregoing has been added to the summary sheet in Figure 6.14.

STEP 7: Determine manufacturing dimensions and provide gaging.

Gages are required to check the finished dimensions of each of the part's surfaces and holes. All gages will be the same at final inspections as those used at the manufacturing operation except:

- A work, or process, plug gage will be required to check the rough-drilled .750 bore.

Op. No.	MACHINE OR PROCESS	OPERATION DESCRIPTION	LOCATING SURFACE
10	Heat Treat Equipment	Normalize casting	
20	Mill	Machine bottom face	Top of .375-diameter hole bosses
			Inside face of extended leg
			Back face
30	Mill	Straddle mill front and back faces	Bottom face
			Inside face of extended leg
			Centered (in y-axis) on extended leg
			See Figure 13B
40	Mill	Straddle mill right (boss) face and left end face	Bottom face
			Inside face of extended leg
			Back face
50	Mill	Machine top faces of (2) .375 diameter hole bosses	Bottom face
			Right end face
			Back face
60	Drill Press	Drill (2) .375 diameter holes to rough size	Bottom face
			Back face
			Right end face
70	Drill Press	Drill $\frac{1}{4}$ –20 hole	Bottom face
			Back face
			Right end face
80	Drill Press	Countersink $\frac{1}{4}$ –20 hole	Bottom face
			Back face
			Right end face
90	Drill Press	Tap $\frac{1}{4}$ –20 thread	Bottom face
			Back face
			Right end face
100	Drill Press	Drill (1) .750 diameter hole to rough size	Bottom face
			(2) drilled .375 diameter holes
110	Double-End Precision Boring Machine	Bore (2) .375 diameter holes to finish size	Bottom face
			Back face
			Right end face
120	Single-End Precision Boring Machine	Bore (1) .750 diameter hole to finish size	Bottom face
			(2) finish bored .375 diameter holes
130		Inspect	

FIGURE 6.14 Manufacturing processes summary sheet—bushing block.

- A work, or process, plug gage will be required to check the rough drilled two .375 diameter holes.
- A work, or process, fixture gage will be required to check the location and relationship of the .750 and .375 diameter holes. This gage is shown in Figure 6.15.

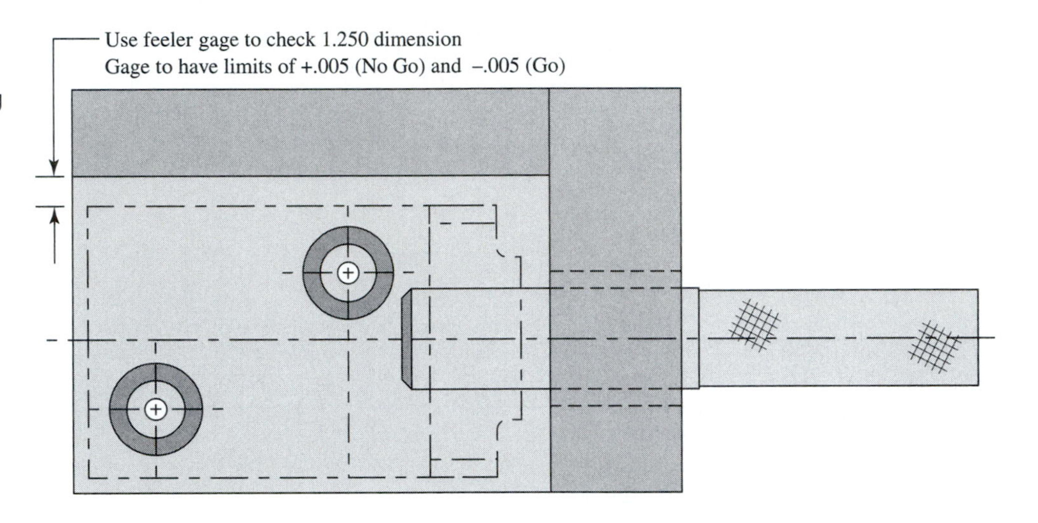

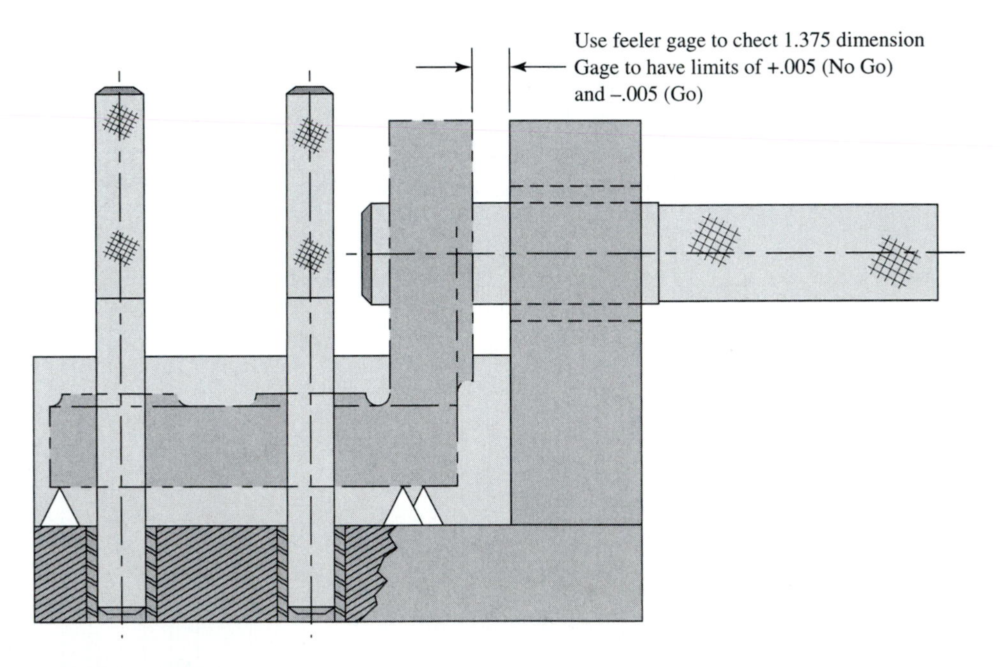

FIGURE 6.15 Gage to check hole relationships and center lines to locating surfaces dimensions.

PROCESSING EXAMPLE 3

Oil Pump Housing (Fig. 6.16)

PRODUCTION REQUIREMENTS

400,000 annually
100 pieces per hour
16 hours per day (2 shifts)

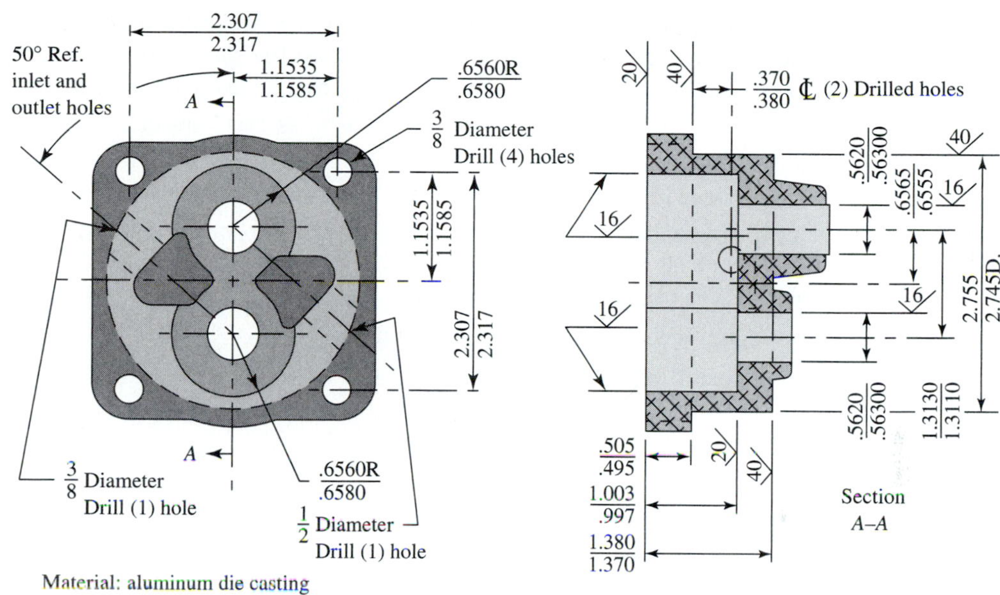

FIGURE 6.16 Oil Pump Housing.

Life of Product

Three years (est.)

STEP 1: Analyze part design

STEP 2: Summary.

From the part design analysis, a summary of the manufacturing requirements highlights can be made:

- Material is an aluminum die casting.
- No heat treatment is specified.
- Closest dimensional tolerances are

  Gear shaft centerlines, ±.001

  Depths of gear pockets, ±.002

  Hole locations, ±.005
- Closest diametral tolerances are

  Gear shaft bores, ±.0005

  Gear pocket (radius), ±.001
- Closest surface finish requirements:

  Gear pocket radii, 16 μin.

  Gear shaft bores, 16 μin.

  Large end face, 20 μin.

  Gear pocket bottom face, 20 μin.

- In configuration, the part is the housing of a geared-type oil pump. Two pockets are provided for the gears; two bores are provided on the same centerline for the gear shafts. The pocket areas are cored, but the shaft bores and all drilled holes are not.
- Dimensions to diecast surfaces that are left "as cast" are deleted to clarify the machining dimensions.
- All surfaces to be machined have .062 per surface of machining stock.
- All finished machined castings are to be pressure tested at 100 psi for porosity for potential leakage.
- The annual production requirements are 400,000 pieces total to be run at a rate of 100 pieces per hour, 16 hours per day on a 2-shift basis.
- The life of the product was estimated as 3 years.

STEP 3: Itemize basic operations.

1. Machine 2.750 diameter.
2. Machine left end face.
3. Machine back of flange to .500 dimension.
4. Machine right end face to 1.375 dimension.
5. Machine 1 gear pocket to .657 radius × 1.000 depth.
6. Machine other gear pocket to .657 radius × 1.000 depth.
7. Machine four $\frac{3}{8}$ diameter holes.
8. Machine one $\frac{1}{2}$ diameter hole.
9. Machine one $\frac{3}{8}$ diameter hole.
10. Machine one .5625 diameter bore.
11. Machine other .5625 diameter bore.
12. Remove all burrs.
13. Pressure test.

STEP 4: Determine locating surfaces.

The left end face is the baseline for dimensions to the back of the flange, bottom faces of the gear pockets, and right end face. After machining, the left end face will be the locating surface for machining these surfaces.

Centerlines of the inlet and outlet holes ($\frac{3}{8}$ and $\frac{1}{2}$ diameters) are dimensioned from the back face of the flange, which will be the locating surface for machining these holes. These 2 holes also have an angular relationship with the gear pocket centerlines that must be held.

The four $\frac{3}{8}$ diameter holes have coordinate dimensioning and are related to the part's flange profile. The locating surfaces will be the 2.750 diameter, back of the flange face, and, radially, some portion of the flange profile.

Locating surfaces for the gear pockets will be the left end face and the 2.750 diameter, and, radially in relation to this flange profile, one of the four $\frac{3}{8}$ diameter holes may be used for the radial location. Also, each pocket centerline must be offset .656 from the part's true centerline, and this must be provided in the fixture design. Location surfaces for the shaft bores will be the same as those for the gear pockets.

STEP 5: Determine feasible methods.

STEP 6: Combine operations, develop sequence.

General. Dimensional location tolerances and drilled hole size tolerances of the four $\frac{3}{8}$ diameter holes and the $\frac{3}{8}$ and $\frac{1}{2}$ diameter inlet and outlet holes are all within drilling operation capabilities.

Lateral and diametral tolerances and surface finish requirements for the left end face, flange back face, 2.750 diameter, and 1.375 face are in the range of precision turning and boring on special equipment.

Machining the gear pockets to .657 radius, 1.000 depth, and specified surface finish will require precision boring on special equipment utilizing a specially designed offset fixture. Machining the shaft bores after the drilling operation will require precision boring equipment utilizing a specially designed offset fixture.

Summary Sheet. A summary sheet should be started at this point to record all judgments and processing decisions as they develop. From the preceding steps, considering the high production requirements, dimensional tolerances, and surface finish specifications, it can be concluded that development of processing should continue along the following lines:

1. Facing and turning operations must be performed on special, precision, high-production equipment.
2. Drilling the four $\frac{3}{8}$ diameter mounting holes and the $\frac{1}{2}$ and $\frac{3}{8}$ diameter inlet and outlet holes can be performed on high-production drilling equipment—perhaps dial-type rotary-table machines.
3. Boring the gear pocket radii and the shaft holes must be done on special, precision, high-production boring machines equipped with specially designed offset fixtures.

Further Processing Considerations.

1. It is extremely important to maintain the maximum possible wall thickness of the casting between the 2.750 outside diameter and the gear pocket radii to minimize leakage when porosity is uncovered during machining. Processing must be planned accordingly.
2. Machining the left end face while the part is chucked on the 2.750 (cast) diameter and on the back face of the flange will provide a locating end surface square with the part's main centerline as shown in Figure 6.17(a).
3. Machining the 2.750 diameter while locating in the (cast) gear pockets and against the left end face will equalize and maintain wall thickness to the maximum possible. Machining the .500 and 1.375 lateral dimensions can be combined and machined with the 2.750 diameter as shown in Figure 6.17(b).
4. The radial relationship of the four $\frac{3}{8}$ diameter holes, the gear pockets, and the inlet and outlet holes can be established by
   - Drilling the four $\frac{3}{8}$ diameter holes while located on the 2.750 diameter, flange back face, and radially on the flange profile as shown in Figure 6.17(c).
   - One of the $\frac{3}{8}$ diameter holes can then be used at subsequent operations as the radial locator.
5. Drilling the $\frac{1}{2}$ diameter inlet hole and $\frac{3}{8}$ diameter outlet holes require the part to be located on the 2.750 diameter, against the flange back face and radially with a locating pin in one of the four $\frac{3}{8}$ diameter holes as shown in Figure 6.17(d).
6. Drilling the gear shaft holes next will relieve the center of the gear pocket bottom faces—enhancing the starting surface condition for the gear pocket bottom faces, in turn enhancing the starting surface condition for the gear pocket precision bore and face operation. As shown in Figure 6.17(f).

Location surfaces for drilling the 2 gear shaft holes are the 2.750 diameter, against the left end face and radially in one of the four $\frac{3}{8}$ diameter holes as shown in Figure 6.17(e). When so located, the .656 offset dimension is provided for in the fixture design.

7. Boring the 2 shaft holes requires the same locating surfaces as shown in Figure 6.17(f).

8. Boring the gear pockets to .657 radii and generating the bottom face to the 1.000 dimension requires the same locating surfaces as shown in Figure 6.17(g).

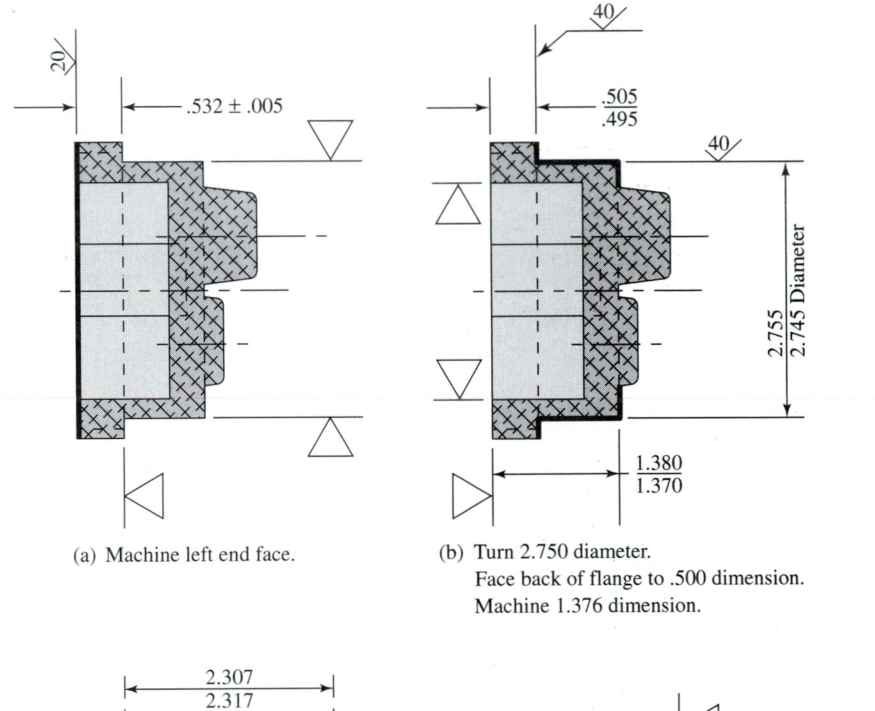

(a) Machine left end face.

(b) Turn 2.750 diameter.
Face back of flange to .500 dimension.
Machine 1.376 dimension.

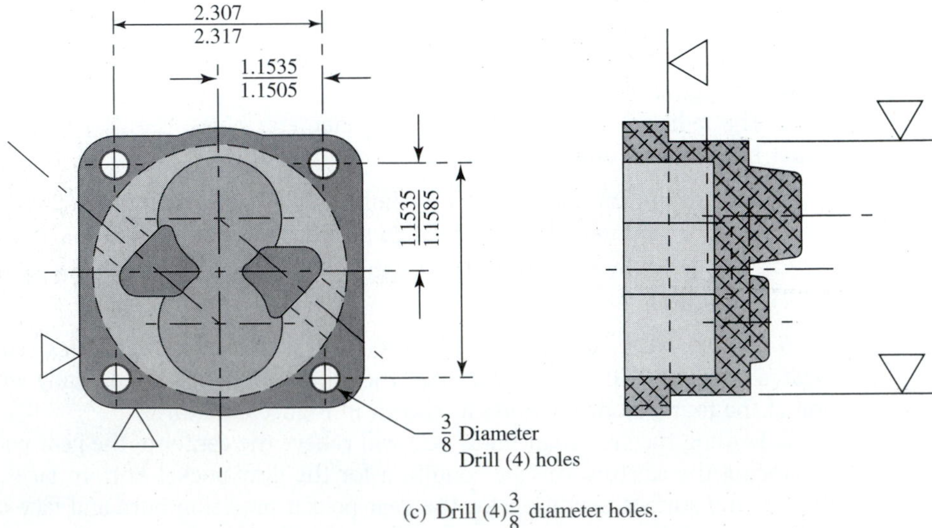

(c) Drill (4) $\frac{3}{8}$ diameter holes.

FIGURE 6.17 Oil pump housing processing.

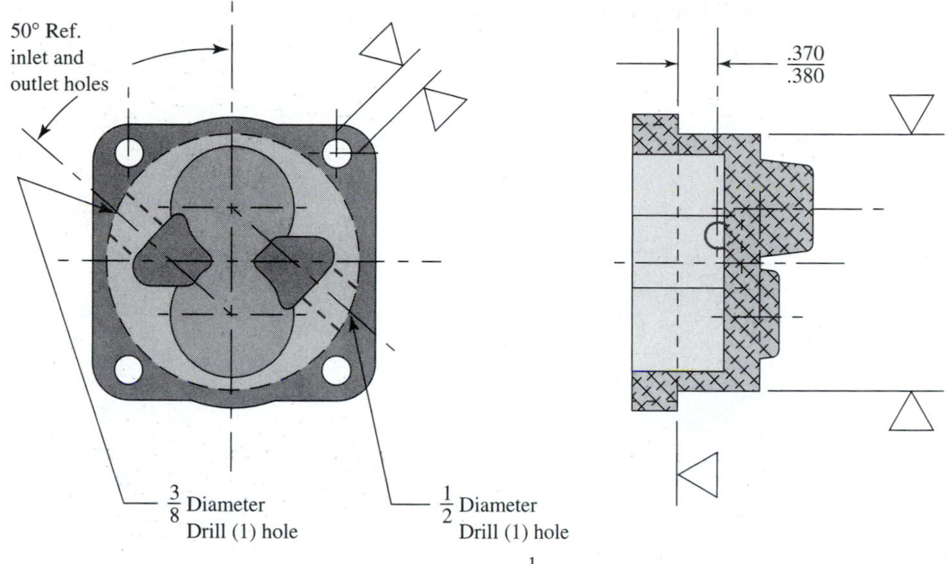

(d) Drill (1) $\frac{1}{2}$-diameter hole.
Drill (1) $\frac{3}{8}$-diameter hole.

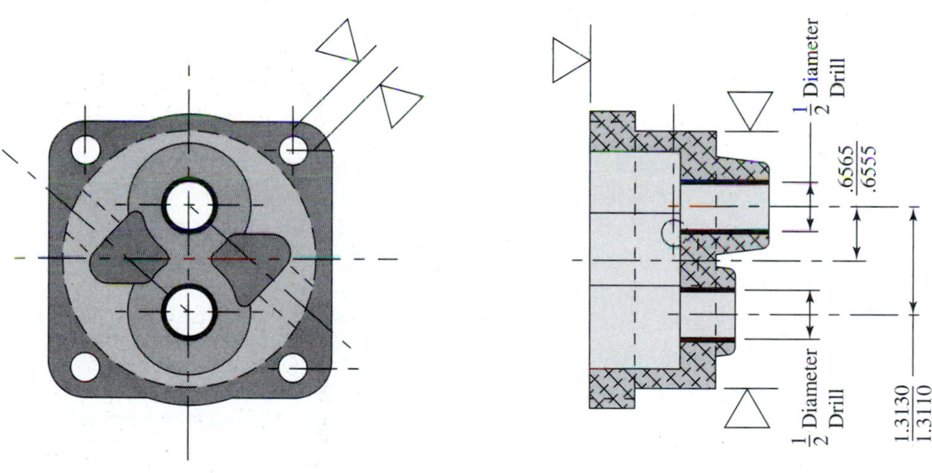

(e) Drill (2) gear shaft holes.

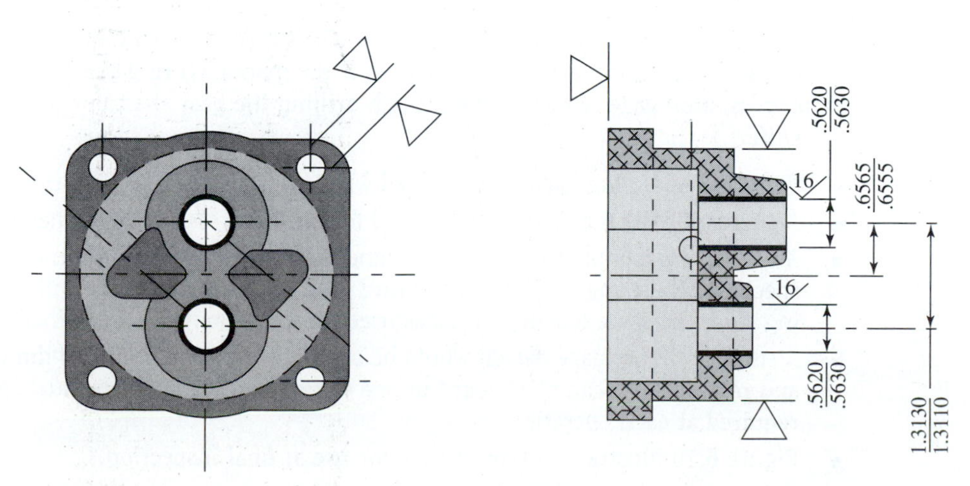

(f) Bore (2) gear shaft holes.

FIGURE 6.17 Continued

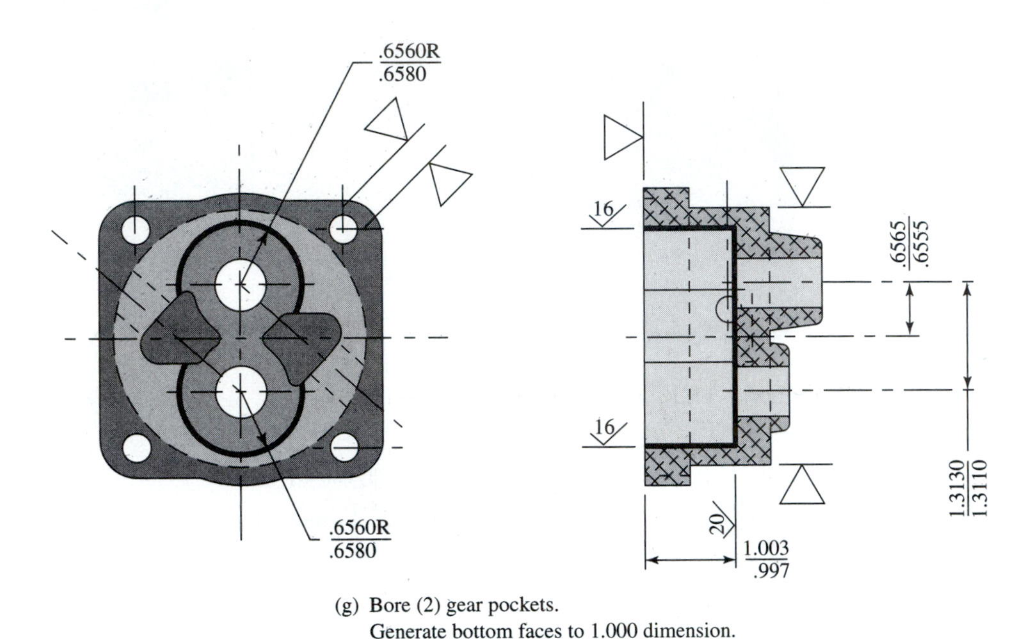

(g) Bore (2) gear pockets.
Generate bottom faces to 1.000 dimension.

FIGURE 6.17 Continued

9. A part design note specifies removing all burrs and breaking all sharp edges. An aluminum die casting is relatively soft, and precise. Highly finished surfaces now exist, so care must be taken to avoid damage at this operation. One acceptable way to deburr and break edges on parts such as this is to process the part in a media, or slurry, type of deburring equipment.

10. Pressure decay detection can quickly check for leakage, but this type of equipment does not identify the location of the leak. To identify where a leak is, the part must be pressurized in a fixture, submerged in water, and observed for air bubbles from the source of the leak.

Some porous parts can be repaired acceptably with an impregnation operation if the area and amount of leakage can be identified. The foregoing has been added to the summary sheet in Figure 6.18.

STEP 7: Determine manufacturing dimensions and provide gaging.

Gages are required to check each of the part's machined surfaces and holes. All gages will be the same at final inspection as those used at the manufacturing operation except for the work, or process, plug gage required when rough drilling the gear shaft bores.

Other:

- Standard AGD plug gages can be used when drilling all holes.
- A standard AGD snap gage can be used for checking the 2.750 diameter.
- All other gages must be specially designed to suit the part print dimensions and tolerances specified. These gages will be of a fixture gage design; that is, they will receive the part locating it on the same locating surfaces used when the part was machined.
- A suitable fixture gage design would be one based on air gaging. All dimensions, tolerances, and relationships can be checked in one or two such gages. Individual specific gages will be required at each operation.
- Figure 6.19 illustrates a typical gage for use at final inspection.

Op. No.	MACHINE OR PROCESS	OPERATION DESCRIPTION	LOCATING SURFACE
10	Precision Turn and	Machine left end face	2.750 cast diameter
	Facing Equipment		Back face of flange
20	Precision Turn and	Turn 2.750 diameter	(2) cored gear pockets
	Facing Equipment	Face back of flange to .500 diameter	Left end face
		Machine 1.375 diameter	
30	4-Station Rotate	Drill (4) $\frac{3}{8}$ diameter holes	2.750 diameter
	Table Drilling Machine		Back face of flange
			Flange profile
40	3-Station Rotate	Drill (1) $\frac{1}{2}$ diameter hole	2.750 diameter
	Table Drilling Machine	Drill (1) $\frac{3}{8}$ diameter hole	Back face of flange
			Flange profile
50	3-Station Rotate	Drill (2) gear shaft holes	2.750 diameter
	Table Drilling Machine		Left end face
			(1) $\frac{3}{8}$ diameter hole
60	2-Spindle Precision	Bore (2) gear shaft holes	2.750 diameter
	Boring Machine		Left end face
			(1) $\frac{3}{8}$ diameter hole
70	2-Spindle Precision	Bore (2) gear pockets	2.750 diameter
	Boring Machine	Generate bottom faces to 1.000 diameter	Left end face
			(1) $\frac{3}{8}$ diameter hole
80	Media Type Deburring	Deburr and break sharp edges	
	Equipment		
90		Pressure test	
100		Inspect	

FIGURE 6.18 Manufacturing processes summary sheet—oil pump housing.

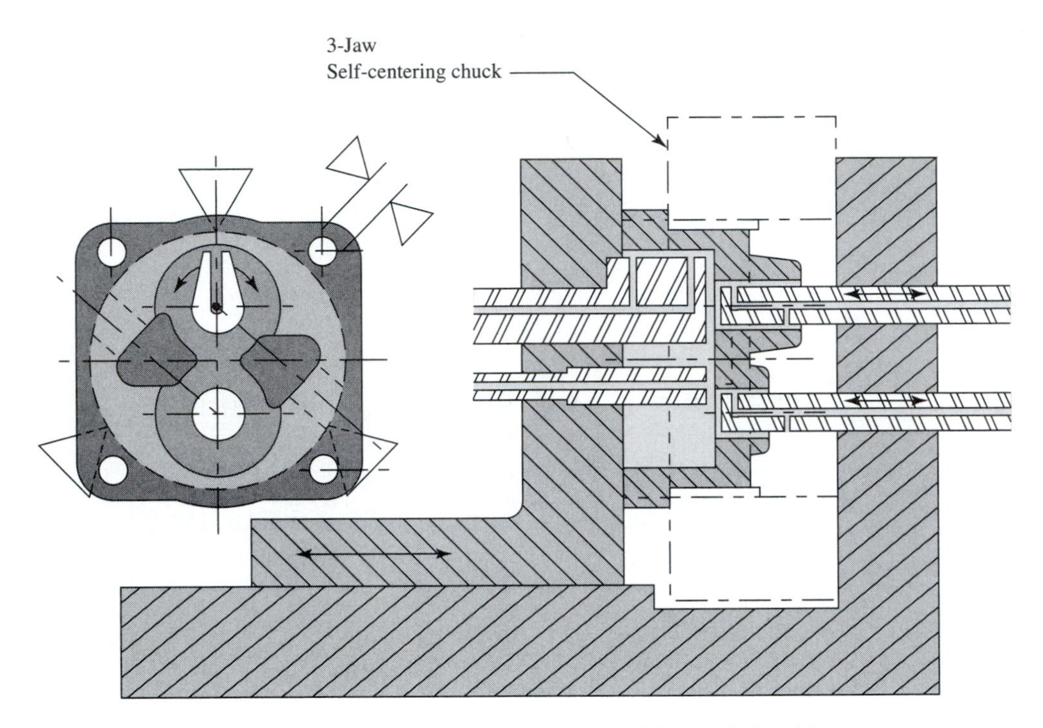

FIGURE 6.19 Pneumatic fixture gage to check bores and bore relationships.

PROCESSING EXAMPLE 4

Adapter (Fig. 6.20)

PRODUCTION REQUIREMENTS

125,000 pieces annually
May be run in 1 lot

Life of Product

Four years (est.)

STEP 1: Analyze part design.
STEP 2: Summary.

From the part design analysis, a summary of the manufacturing requirements' highlights can be made:

- Material is SAE 1020 hot-rolled bar stock.
- A heat treatment of carburizing and hardening to a depth such as to leave .015 minimum of case after grinding is specified.
- Closest dimensional tolerances are ±.005.
- Closest diametral tolerances are ±.0005. Additionally, each diameter of .0005 tolerance has a specified runout tolerance with its adjacent face of .002, and with each other of .001 total indicator reading.

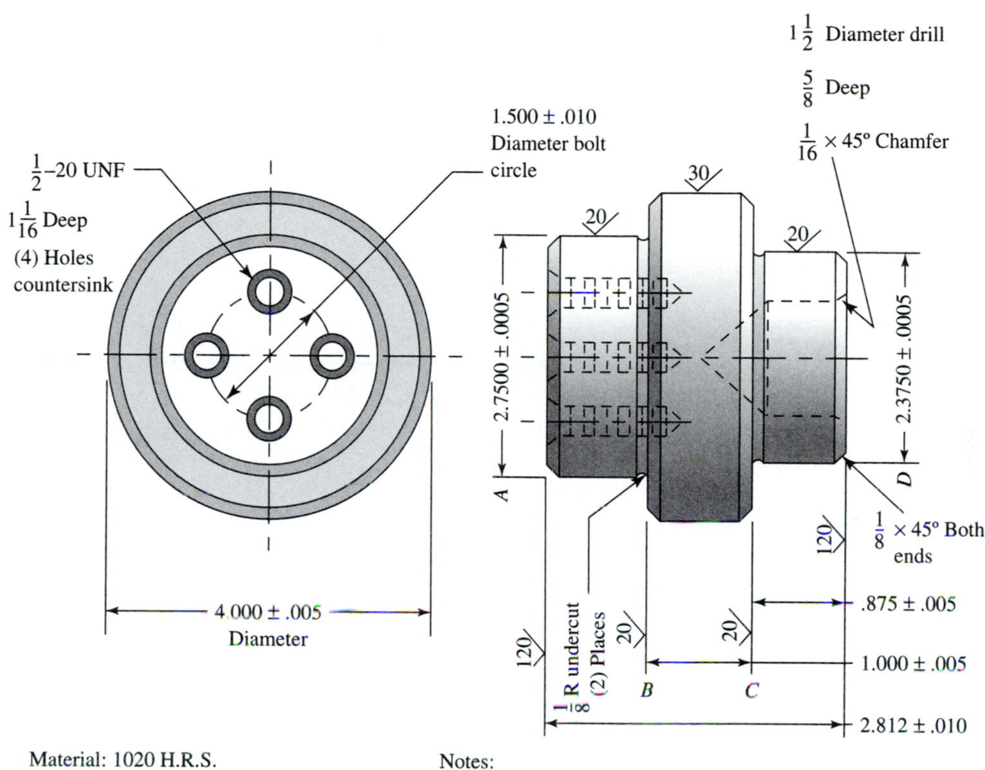

FIGURE 6.20 Adapter.

- Closest surface finish requirements are the 2.375 and 2.750 diameters' finish of 20 μin.
- Adjacent faces also have a surface finish requirement of 20 μin.

In configuration, the part is a 3-diameter round adapter with 4 mounting holes. Of military origin, the adapter's design employed geometric dimensioning and tolerancing. To avoid the required translation, at this stage, coordinate dimensioning and notes have been substituted.

- An annual production requirement of 125,000 pieces may be run in one lot, requiring only one setup annually.
- The life of the product has been estimated at 4 years but could run longer.

STEP 3: Itemize basic operations.

1. Machine 2.375 diameter.
2. Machine 4.000 diameter.
3. Machine 2.750 diameter.
4. Machine chamfers and undercuts.
5. Cut off.
6. Drill four $\frac{1}{2}$–20 UNF holes
7. Countersink four $\frac{1}{2}$–20 UNF holes

8. Tap four $\frac{1}{2}$–20 UNF holes
9. Heat treat.
10. Grind 2.375 diameter and adjacent face.
11. Grind 2.750 diameter and adjacent face.

STEP 4: Determine Locating Surfaces.

Since the part has been designed to be produced from bar stock, and considering the relatively high production requirements, an automatic screw machine is the logical choice for manufacture.

Automatic screw machines utilize a stop to locate the bar stock on its end face, making the right end face of the part its locating surface. Accordingly, the part has been designed so that its right end face is the baseline for lateral dimensions.

After each part is cut off, the bar is fed to the machine stop, making the end face the locating surface for all succeeding parts. The centerline of the 3 diameters is controlled by locating and chucking the bar on its outside diameter. Locating surfaces for machining the four $\frac{1}{2}$–20 UNF holes are the left end face for depth and the parts outside diameter for centrality of the bolt circle.

Since the part does not have centers, grinding the 2.375 diameter must be done while located on the 2.750 diameter and adjacent face; grinding the 2.750 diameter must be done while located on the 2.375 diameter and adjacent face.

STEP 5: Determine feasible methods.

STEP 6: Combine operations.

General. Lateral tolerances and tolerance of the 4.000 diameter are within the capabilities of an automatic screw machine. Dimensional hole locations and depths of the four $\frac{1}{2}$–20 UNF holes are all within standard drilling practice capabilities. Surface finish requirements, diametral tolerances, and runout relationships of the 2.735 and 2.750 diameters will require a finish grinding operation on each diameter following heat treating.

Summary Sheet. A summary sheet should be started at this point to record all judgment and processing decisions as they develop. From the preceding steps, considering the high production requirements, dimensional tolerances, and surface finish specifications, it can be concluded that processing development can continue along the following lines:

1. Machine the part blank complete, on an automatic screw machine, less the four $\frac{1}{2}$–20 UNF holes and also leaving finishing stock on the 2.735 and 2.750 diameters and adjacent faces for final grinding as per the tool layout in Figure 6.21.

2. Drill, countersink, and tap the four $\frac{1}{2}$–20 UNF holes on a 4-station indexing, rotary-table drilling machine equipped with 4-spindle drill heads as diagramed in Figure 6.22.

3. Heat treat to produce a hardness depth as much deeper as the amount of grinding stock that will be removed; that is, if .006 is left per surface for grinding, the hardness depth must be .021 minimum at the heat treat operation to maintain the specified .015 minimum after grinding (.021 − .006 = .015).

4. Utilize chucking-type grinders to grind the 2.735 and 2.750 diameters to hold surface finish, diametral tolerances, and runout specifications. The part must be located (chucked) on the 2.750 diameter and against the adjacent face when grinding the 2.375 and its adjacent face, and the reverse is then required: The part must be located (chucked) on the 2.735 diameter and its adjacent face when grinding the 2.750 diameter and its adjacent face as shown in Figures 6.23 and 6.24.

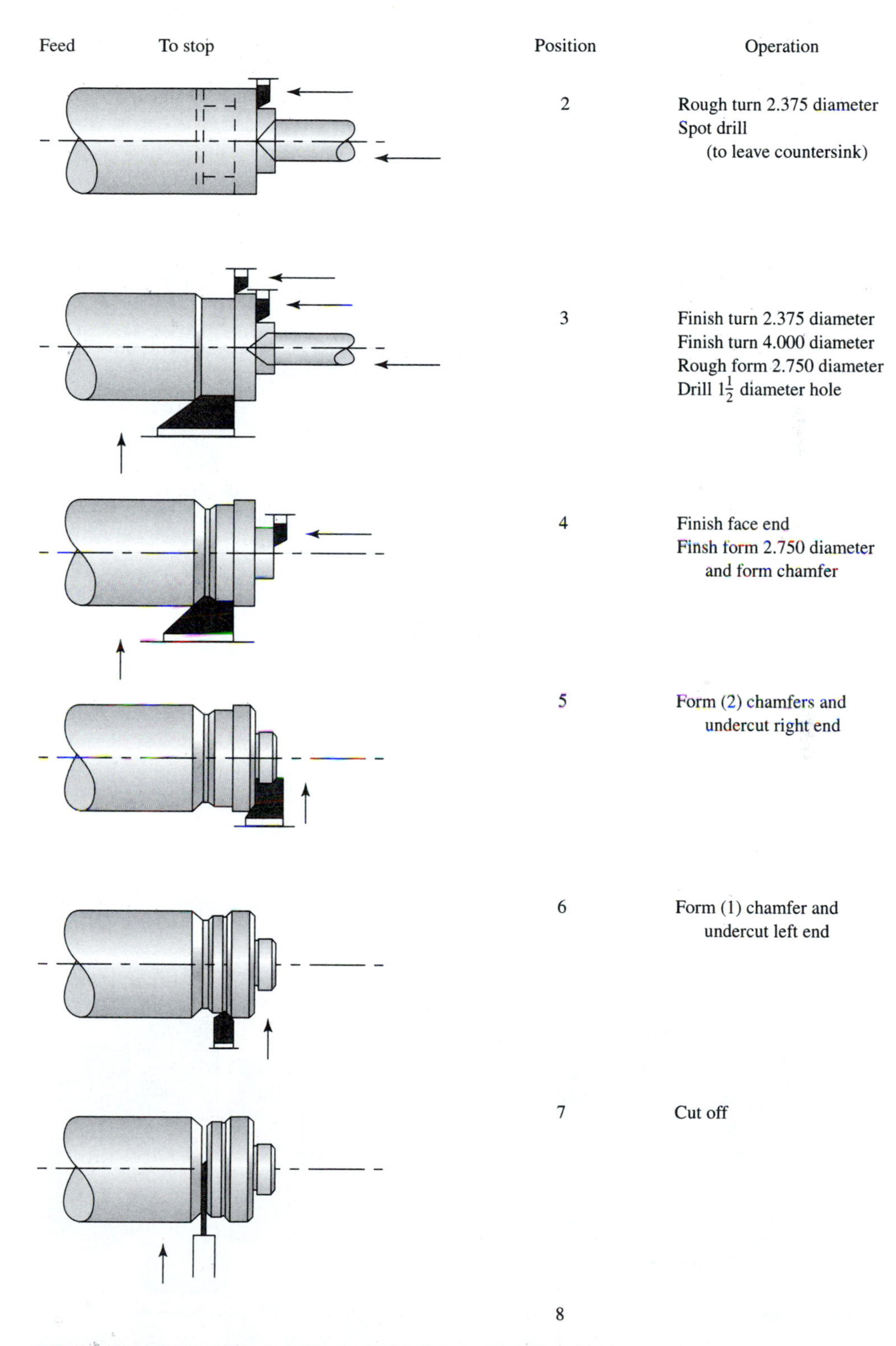

FIGURE 6.21 Tool layout–Adapter (8-Station Automatic Screw).

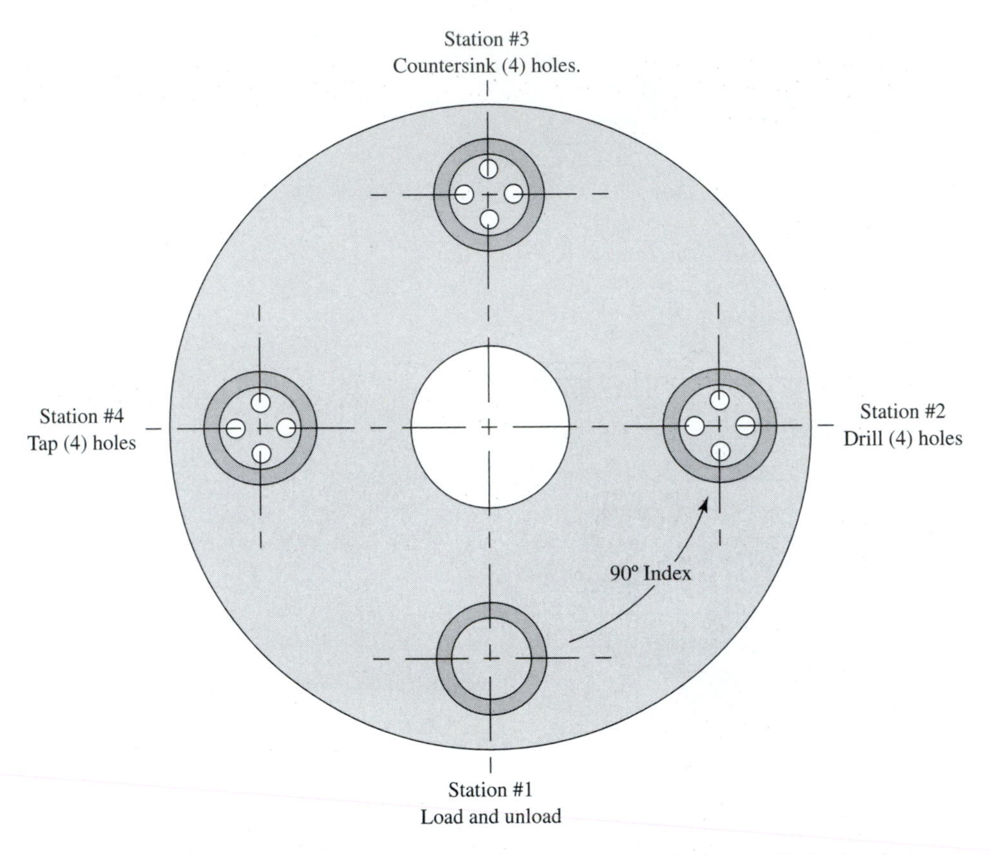

FIGURE 6.22 Drill, countersink, and tap (4) holes in adapter hub face (4-Station Index Drilling Machine).

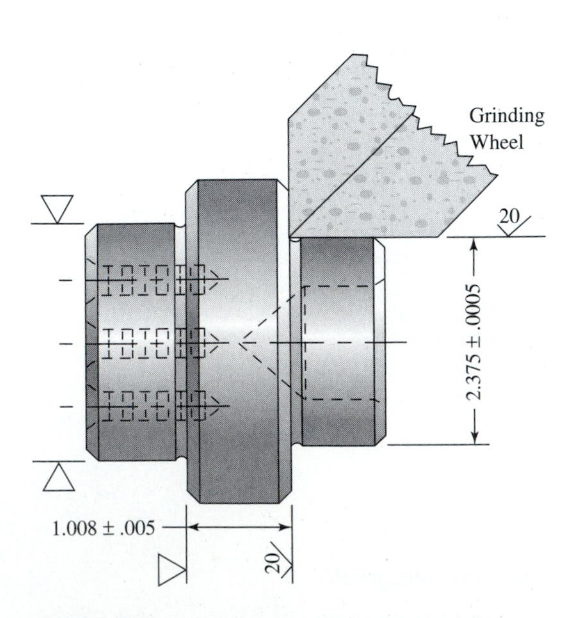

FIGURE 6.23 Grind 2.375 diameter and flange face (chucking grinder).

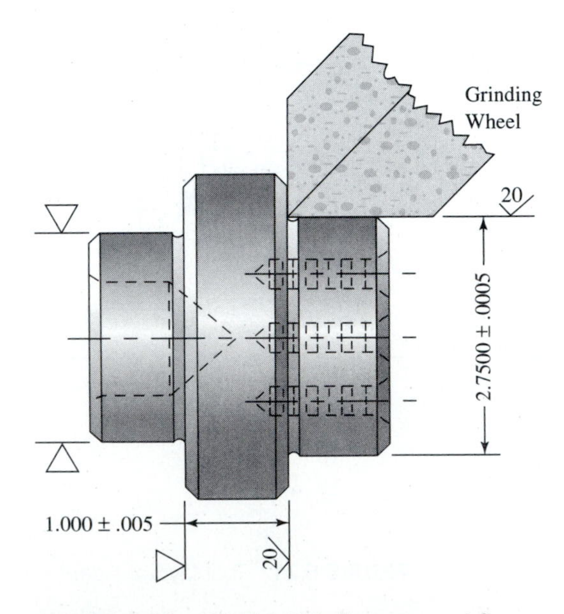

FIGURE 6.24 Grind 2.750 diameter and flange face (chucking grinder).

Further Processing Considerations.

1. A part processed on an automatic screw machine must have the work to be done balanced between the different stations and slides. The length of time to produce a part is the same as its slowest cut. Not only must the length of time for each segment of the operation be the same and at a minimum for productivity, but also for maximum tool life.

Figure 6.21, a typical tool layout, shows an even breakdown of work. Each element of the operation must be balanced by actual calculations.

2. The net number of pieces that can be machined from a bar of a given length is a multiple of the part's length, plus the width of the cutoff tool, plus machining stock left on the end face for a facing cut.

The narrowest feasible width of cutoff blade plus the minimum required amount of facing stock must be specified for the operation—then monitored during production to realize the maximum yield of parts in the bar length.

The foregoing judgments and decisions have been added to the summary sheet in Figure 6.25.

STEP 7: Determine manufacturing dimensions and provide gaging.

Gages are required to check each of the part's lateral dimensions, 3 diameters, four $\frac{1}{2}$–20 UNF holes, and their location. Also, gaging must be provided to check the diametral and adjacent face runouts as specified.

All gages can be the same at final inspection as those used at the manufacturing operation except for the work, or process, gages required to check the diameters' turned sizes and also adjacent faces before grinding.

- Standard AGD snap gages can be used to check all 3 diameters' turned sizes and the overall length of the part.
- Indicator, electronic, or air gages will be required to check the 2.375 and 2.750 diameters due to the close tolerance ($\pm$.005).
- Specially designed gages will be required to check the .875 and 1.000 lengths.
- Standard thread plug gages can be used to check the $\frac{1}{2}$–20 UNF holes, but a special fixture gage will be required to check their location.
- A specially designed receiving gage will be required to check face and diameter runouts. One design that can be used is a diaphragm chuck mounted on a "zero-lash" (sleeve-bearing) spindle. Dial indicators can be used to check runouts of faces and diameters as shown in Figure 6.26. A gage for each diameter will be required.

Op. No.	MACHINE OR PROCESS	OPERATION DESCRIPTION	LOCATING SURFACE
10	6-Spindle Auto	Machine Adapter blank	Bar stock O.D.
	Screw Machine		Right end face
20	4-Station	Drill (4) $\frac{1}{2}$–20 holes	4.000 Diameter
	Drilling Machine	Countersink (4) $\frac{1}{2}$–20 holes	2.750 Diameter end face
		Tap (4) $\frac{1}{2}$–20 holes	
30	Heat Treat Equipment	Heat Treat	
40	Chucking Grinder	Grind 2.375 diameter	2.750 Diameter and
			Adjacent face
50	Chucking Grinder	Grind 2.750 diameter	2.375 Diameter and
			Adjacent face
60		Inspect	

FIGURE 6.25 Manufacturing processes summary sheet—adapter.

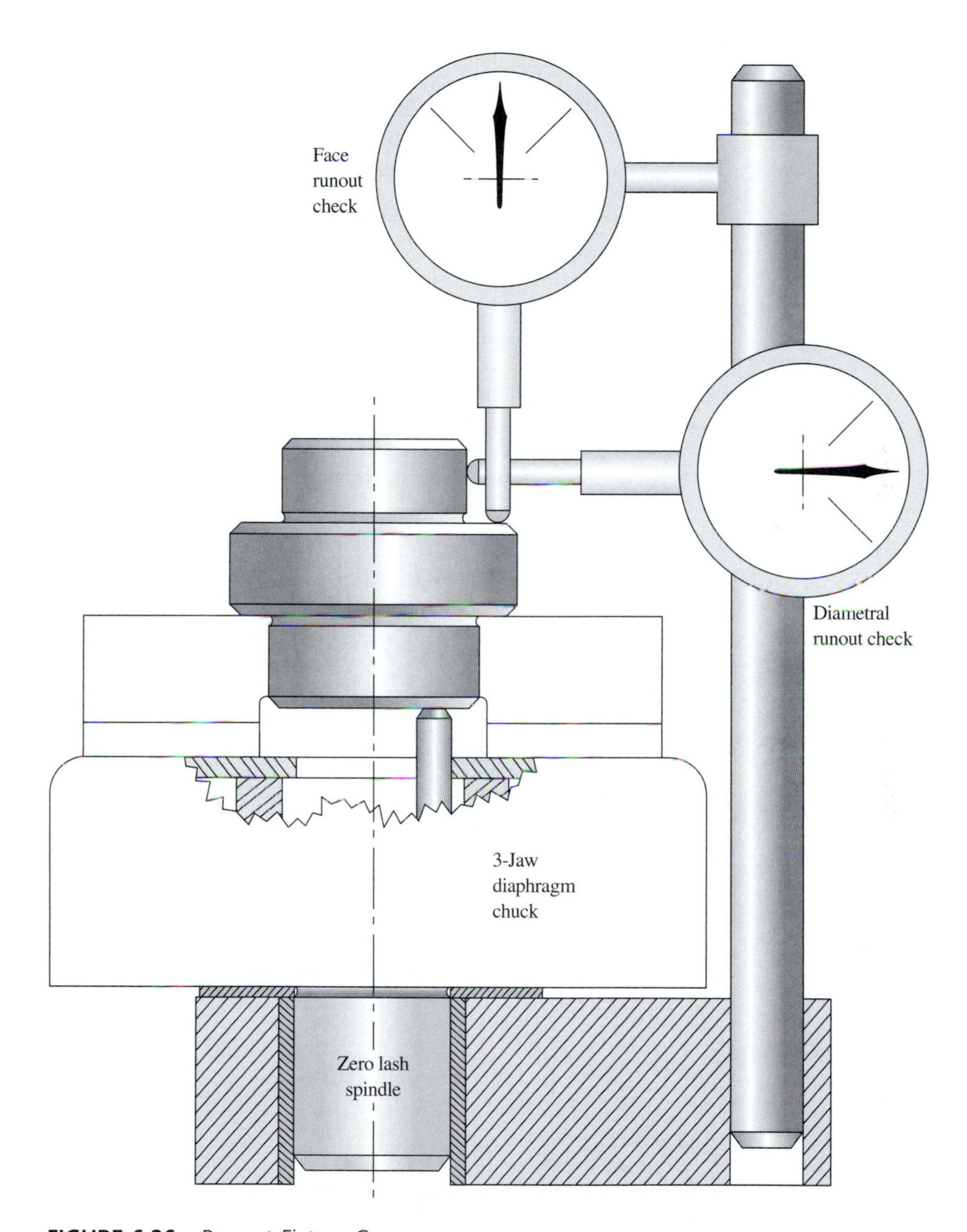

FIGURE 6.26 Runout Fixture Gage.

6.4 REVIEW QUESTIONS AND PROBLEMS

1. Where does planning for processing begin?
2. What must first be established before any process sequence can be developed?
3. When selecting locating surfaces, what criteria or part design properties must be looked for?
4. What dictates process requirements and sequence of operations?

5. From the standpoint of a product engineer, design 4 parts each requiring one of the following processes:

Part *A*	Drilling	Show holes to be drilled.
Part *B*	Turning	Show diameters to be turned.
Part *C*	Machining	Show flat surfaces to be machined.
Part *D*	Grinding	Show surfaces to be ground.

   Complete each design to the extent required to show dimensions, tolerances, surface finish requirements and notes required to perform the process specified.
   Use good practice in establishing a baseline for dimensions. Also note the locating surfaces or points you, as the product engineer, would expect the manufacturing engineer to use.
6. Where, or when, should unnecessary process operations be eliminated?
7. (a) State the *advantages* of combining process operations.
   (b) State the *disadvantages* of combining process operations.
8. State, generally, what process operations and part design features have potential for combining operations.
9. Develop a tentative process sequence for the Bearing Hub Roller, Figure 6.27.

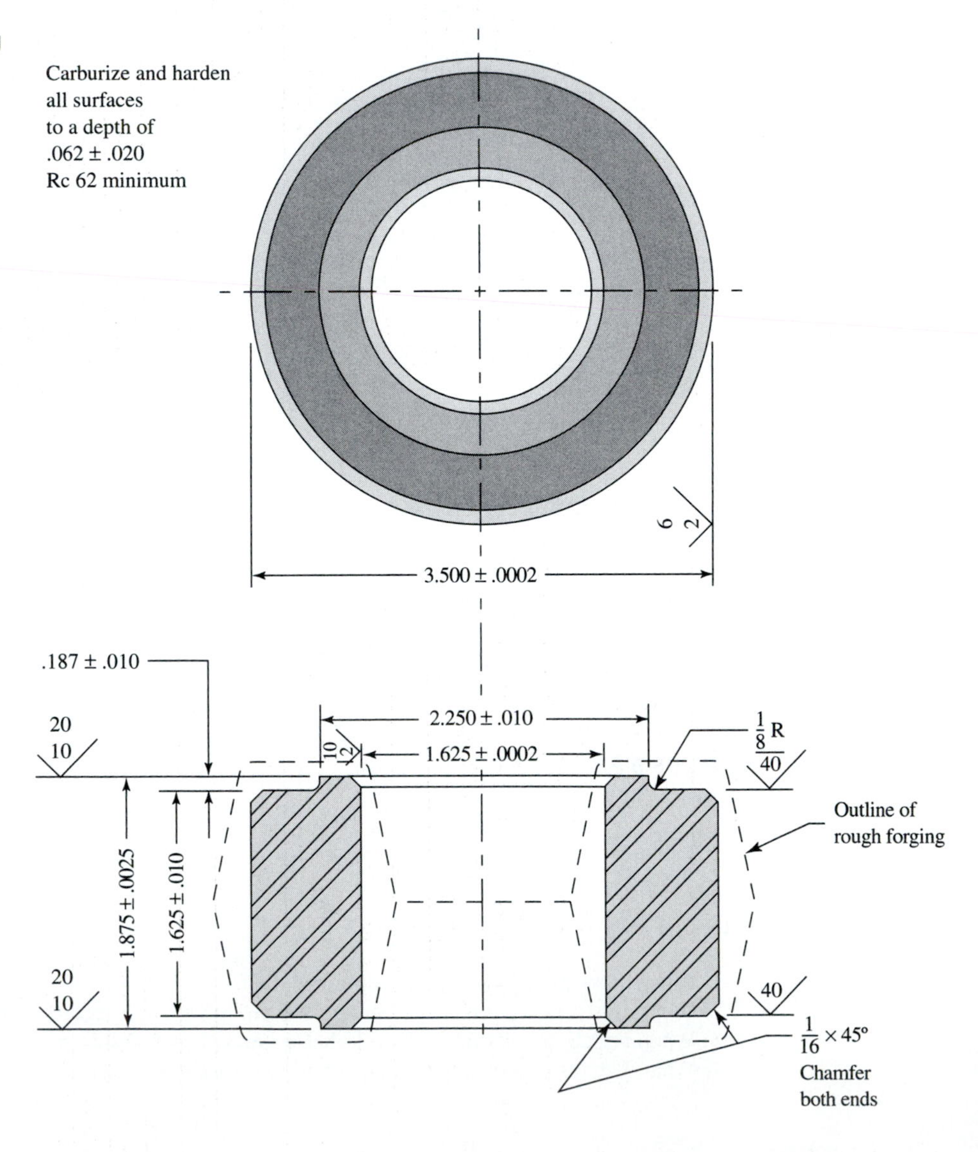

FIGURE 6.27 Bearing Hub Roller.

Establish
- Analysis of part design
- Processes required
- Locating points or surfaces to be used
- Tentative sequence

Production requirements
- 60 pieces/hour
- 1350 pieces/day
- 270,000 pieces/year

10. Develop a tentative process sequence for a Torsion Bar Front Suspension Strut, Figures 6.28 and 6.29.

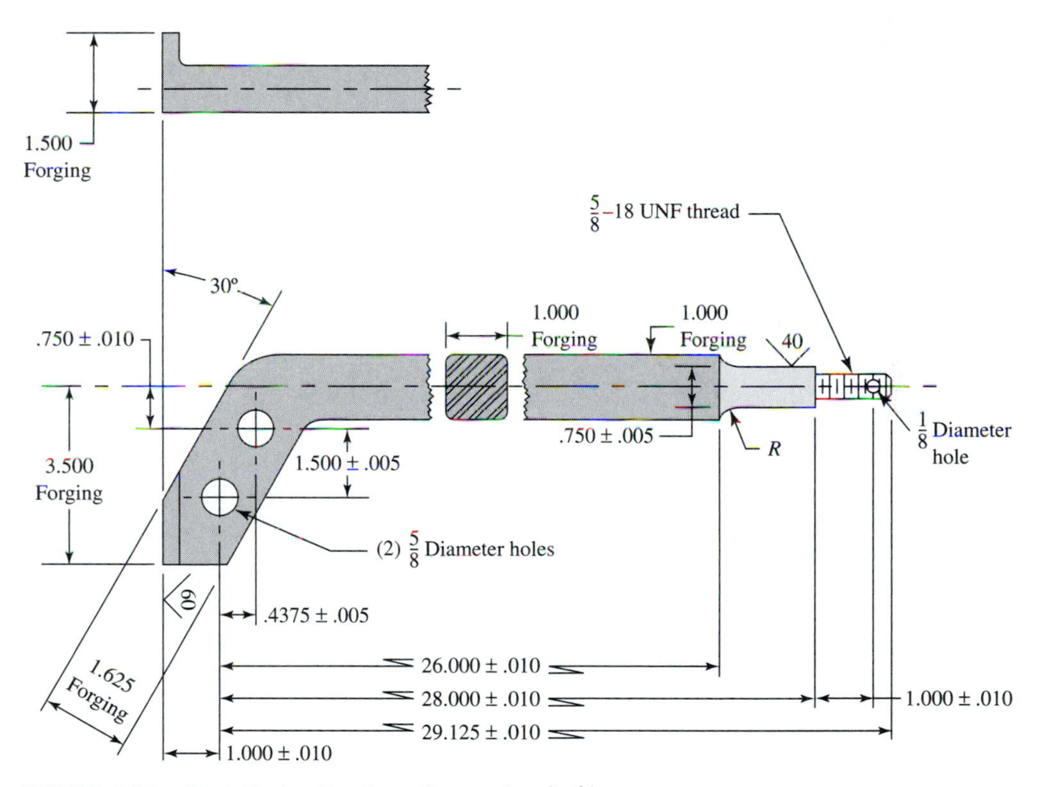

FIGURE 6.28 Strut-Torsion-Bar Front Suspension (Left).

Establish
- Analysis of part design
- Processes required
- Locating points or surfaces to be used
- Tentative sequence

Production Requirements
- 133 pairs/hour
- 3000 pairs/day
- 600,000 pairs/year

FIGURE 6.29 Strut-Torsion-Bar Front Suspension (Right).

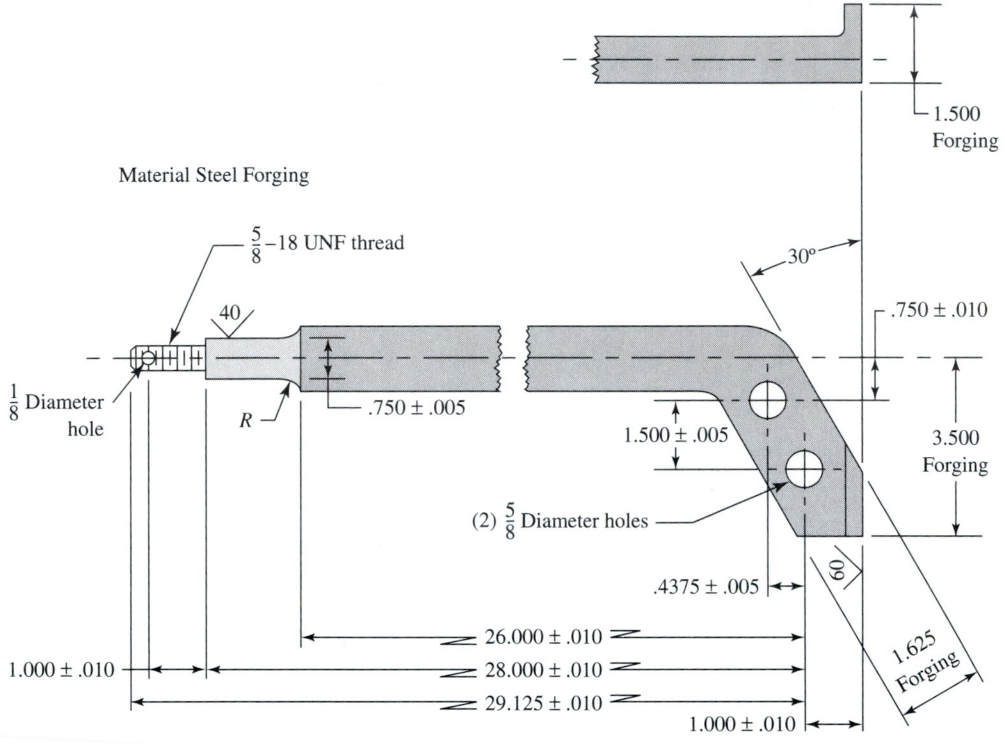

11. Develop a tentative process sequence for each of the shaft designs shown in Figure 6.30.

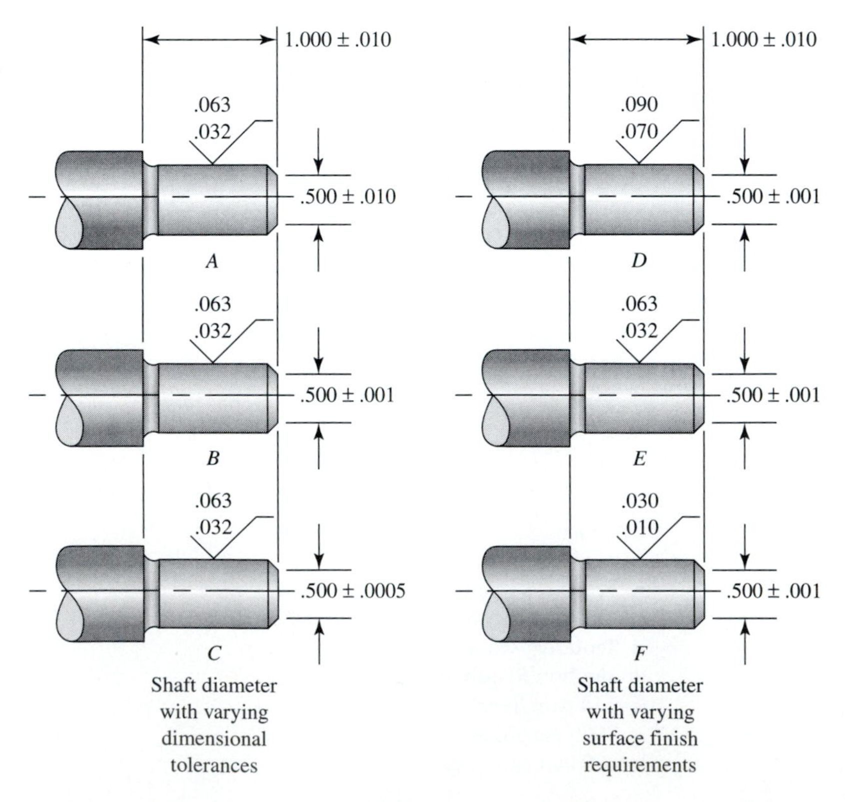

FIGURE 6.30 Shafts with varying dimensional tolerances and surface finish requirements.

Establish
- Analysis of parts design
- Processes required
- Locating points or surfaces to be used
- Tentative operations sequence

Production Requirements
- 100 parts/hour
- 2250 parts/day
- 450,000 parts/year

12. Develop a tentative process sequence to drill the hole shown in Figure 6.31.

FIGURE 6.31 Lifter.

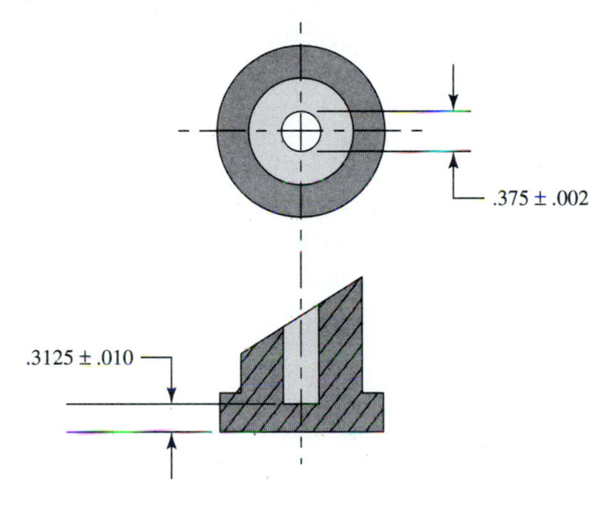

- What operations are required in addition to the drilling operation?
- Production requirements are 1000 pieces total.

CHAPTER SEVEN

TOLERANCE CHARTING

CHAPTER OVERVIEW

- Introduction
- Definitions, Symbols, and Tolerance Forms
- Tolerance Chart Construction
- Tolerance Chart Examples
- Selective Assembly
- Computer-Aided Tolerance Charting
- Review Questions and Problems

7.1 INTRODUCTION

Tolerance charting is a system that seems destined to be rediscovered by each succeeding generation of manufacturing engineers. Some form of dimensional tolerance control must have been used by Eli Whitney to produce the first 10 truly interchangeable U.S. military rifles in 1800. Tolerance charting was used in the late 1920s by the Wright Aeronautical Corporation to manufacture their J-series aircraft engines—one of which powered Lindbergh's plane to Paris in 1927. Wright Aeronautical developed the system further and used it in the manufacture of all parts produced during World War II, as did many other manufacturers of the time.

The use of tolerance charting has increased and continues to increase. The wonder is that so many manufacturers, having so much to gain by its use, are unfamiliar with the system and its many cost and quality advantages.

This text covers the fundamental concepts, uses, and basic applications of tolerance charting, but a manufacturing engineer should augment his education with further study to master all variations and realize all of the benefits of the system.

A toolmaker does not need a tolerance chart to manufacture 1 part. He can "zero-out" his tool settings; that is, he can set up his machine such that it will machine the first dimension from its reference point to the mean of the specified dimension and tolerance of the part print. Then he can *reset* to the reference point, or surface of all subsequent dimensions, and machine those dimensions to their specified mean dimensions and tolerances. In so doing, he will produce a part to the part print and still utilize the full amount of allowable tolerance. Numerical Control (NC) can work very much the same way if the part design is dimensioned accordingly.

Machines do not presently have the capability of zeroing out to successive locating surfaces (as a toolmaker can)—nor, perhaps, would it be desirable from the standpoint of complexity and time added to each operation.

The continuous, progressive production of a part (having more than one locating surface) requires that each operation be set up to accommodate machining the surface(s) peculiar and related to each locating surface.

As parts progress along the production line and are successively located for each operation, tolerances between surfaces can and do add—sometimes beyond the total allowable amount—if no system is employed to control them.

A tolerance chart's main functions are to exactly control machining dimensions, tolerance accumulations, and amounts of stock removed from each involved surface. An actual, or proposed, sequence of operations is a prerequisite for developing a tolerance chart. Also, it aids greatly, when processing a new part, to develop the chart concurrently.

Most tolerance charts are developed for one view, or elevation, of a part, but related views can be developed for as many elevations as required. Additional views are developed simultaneously and, preferably, on the same sheet of paper utilizing the same chart form.

Diameters are generally not charted unless stock removal from multiple, succeeding operations becomes very complex. Tolerances do not accumulate on surfaces (diameters) machined from a common center line.

What a Tolerance Chart Is A tolerance chart is a graphical display of manufacturing dimensions, tolerances, and amounts of stock to be removed at each operation of a manufacturing sequence; the chart exactly delineates the part's dimensions and tolerances at each stage of manufacture.

What a Tolerance Chart Does A properly developed and checked tolerance chart will *guarantee* that the processing, dimensioning, and tolerancing proposed for a new part will produce the part to the part design dimensions and tolerances.

A tolerance chart will detect an unfeasible part design that cannot be made to the part design at all. A tolerance chart will detect faulty processing, dimensioning, and tolerancing of existing parts, reducing scrap and enhancing quality in so doing.

Specifically, a tolerance chart aids in

Developing

- ✓ The most feasible manufacturing process and sequence (combined operations, minimum labor, assured capacity, and capability)
- ✓ More exact machining dimensions
- ✓ Maximum allowable working-dimension tolerances
- ✓ Required amounts of stock removal

Determining

- ✓ Proper locating surface for each operation
- ✓ Specific gaging requirements

Providing

- ✓ Means to check proposed processing versus machine capability
- ✓ Means to compare optimal processes
- ✓ Means to determine raw material or blank sizes
- ✓ Means to facilitate and incorporate later dimensional and processing changes

7.2 DEFINITIONS, SYMBOLS, AND TOLERANCE FORMS

Certain expressions and terms are associated with the development of tolerance charts. A familiarity and understanding of these terms is essential for study, use, and communications relative to successfully utilizing tolerance charting. The following definitions are illustrated in Figure 7.1.

1. **Part Design Dimension.** A part design dimension is a dimension relating to a surface or location of a part and usually includes the allowable deviation or tolerance.

2. **Working Dimension.** A working dimension is a processing dimension and its tolerance used in a manufacturing operation. It is sometimes known as a machining dimension and is the distance between the locator and the involved surface.

3. **Dimensional Direction.** Dimensional direction is the direction a machining, or working, dimension is from the locator. It is expressed as *from* the locator *to* the surface.

4. **Resultant Dimension.** A resultant dimension, sometimes called a balance dimension, is the difference between two dimensions; it may be the difference between two working dimensions or between a working dimension and another resultant dimension. An *intermediate* resultant dimension indicates further machining stock is to be removed from one (or both) of the affected surfaces.

5. **Stock Removal.** Stock removal, as the term implies, is the amount of material to be removed from a surface. There must be enough stock, in the case of machining, for the cutting

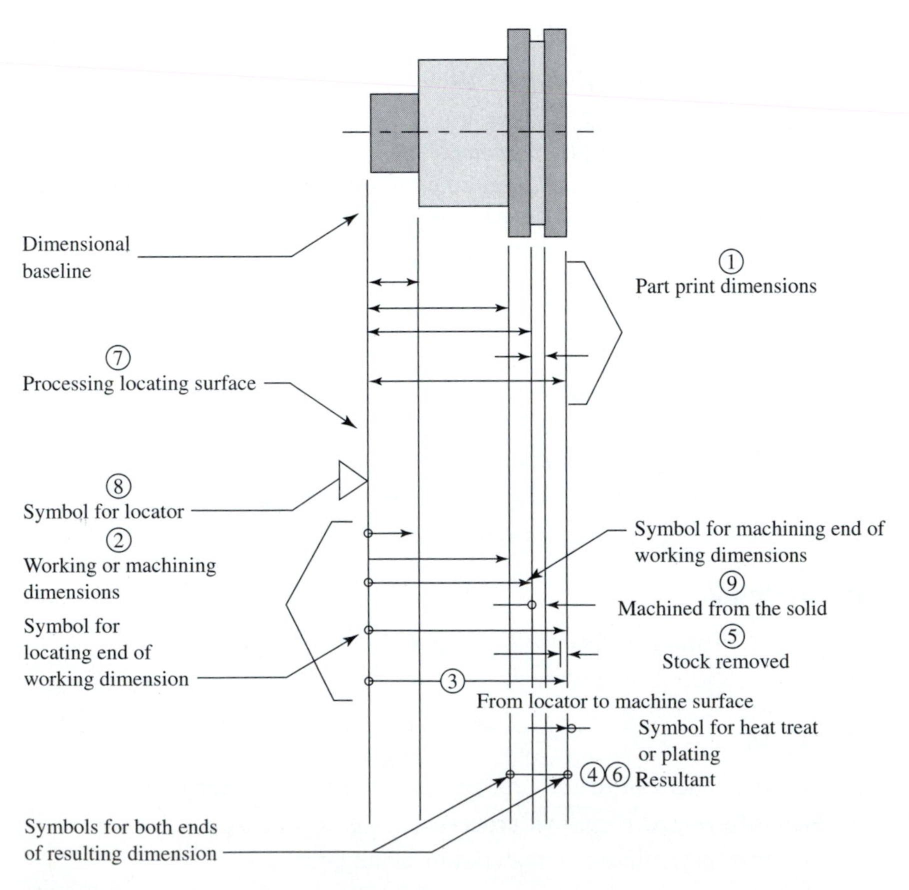

FIGURE 7.1 Tolerance chart symbols.

edge of the tool to "get under the surface" and not have a dulling, burnishing action. Stock removal for a grinding operation, however, must be kept to the minimum that will clean up because time for stock removal, by most grinding processes, is directly proportional to the amount of material to be removed. When the proper grinding wheel, speeds, feeds, and coolants are used, burnishing is not a problem. In any case, enough stock must be provided to assure cleanup; that is, to remove the entire surface resulting from the preceding operation, including surface variations and tool marks. A new, true surface must be established to and within the newly specified dimensions and tolerances.

6. **Actual Dimension.** An actual dimension is a final, existing dimension, established at the conclusion of a part's processing. It may be a dimension machined in relation to a locator, or it may be a calculated resultant. It is a dimension that is compared to a part design dimension at the conclusion of developing a tolerance chart and must, of course, be the same as (or of closer tolerances than) the part print dimensions.

7. **Locating Surface.** A locating surface is the surface upon which a part rests while a manufacturing process operation is performed. It is the surface from which working dimensions originate.

8. **Locator.** A locator is the element of a work fixture upon which the locating surface of a workpiece rests. The element may be in the form of rest buttons or a full surface. It is indicated as a triangle on the tolerance chart.

9. **"Solid" or "From the Solid."** This expression is commonly used when a groove, hole, or slot is machined in virgin material where no previous operation has been performed.

10. **Clean Up.** A surface is said to **clean up** when the entire surface left from a preceding operation is removed by a succeeding machining and grinding operation. Enough stock must be provided at each operation to assure this condition regardless of surface variations caused by dimensional tolerances.

Tolerance Forms, Conversion, and Addition

For certain specific reasons, some part designs are dimensioned with tolerances expressed as *unilateral,*

$$2.000^{+.005}_{-.000}$$

unequal bilateral,

$$2.000^{+.005}_{-.010}$$

or in *limits,*

$$\frac{1.997}{2.000}$$

However, as will be seen in examples following, the development of a tolerance chart requires that dimensions be controlled by *equal, bilateral tolerances* such as 2.000 ± .005.

Mean dimensions with equal, plus, or minus tolerances (equal bilateral) can be determined arithmetically for any unilateral or unequal bilateral dimension, but a quicker way is to use a tolerance conversion chart such as the one shown in Figure 7.2.

Adding and Subtracting Dimensions and Tolerances

✓ The mean of *dimensions* (working, machining, actual, or part design) is *added* or *subtracted,* as required, to determine resultant dimensions or stack removals.

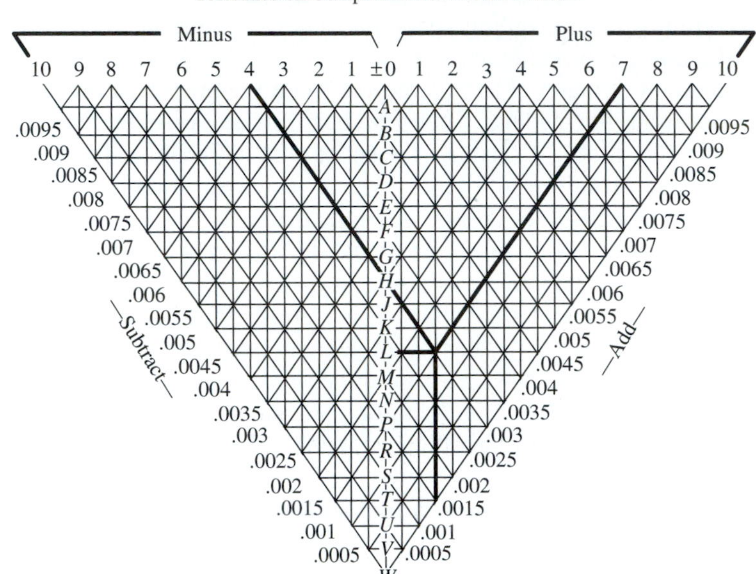

Key letter	A	B	C	D	E	F	G	H	J	K
Plus-and-minus tolerance	.0005	.001	.0015	.002	.0025	.003	.0035	.004	.0045	.005
Key letter	L	M	N	P	R	S	T	U	V	W
Plus-and-minus tolerance	.0055	.006	.0065	.007	.0075	.008	.0085	.009	.0095	.010

Problem:

Convert the dimensions $5.000^{+.007}_{-.004}$ to a dimension with equal bilateral tolerances.

Solution:

From the top of the chart, select the two given tolerances, –.004 at the left and +.007 at the right.

From these two points, follow the diagonal lines to a point where they intersect within the chart as shown.

To the left of the intersection, read the key letter *L* from the vertical center scale.

From the intersection, follow the vertical line downward as shown. The reading is .0015. This figure is added to the original basic dimension. The new basic dimension now becomes 5.000 + .0015 = 5.0015.

The key letter *L* indicates an equal bilateral tolerance of .0055. The new dimension is 5.0015 ± .0055

FIGURE 7.2 Tolerance conversion chart.

✓ Equal, bilateral *tolerances* are *always added* to determine actual or total tolerances.
✓ Figure 7.3 illustrates the additions and subtractions of dimensions and tolerances.

7.3 TOLERANCE CHART CONSTRUCTION

The specific steps required to develop a tolerance chart are

1. **Processing.** Study the part's processing—or, in the case of a new part, develop the processing. The data required from either source must include

- Operation sequence
- Description of each operation

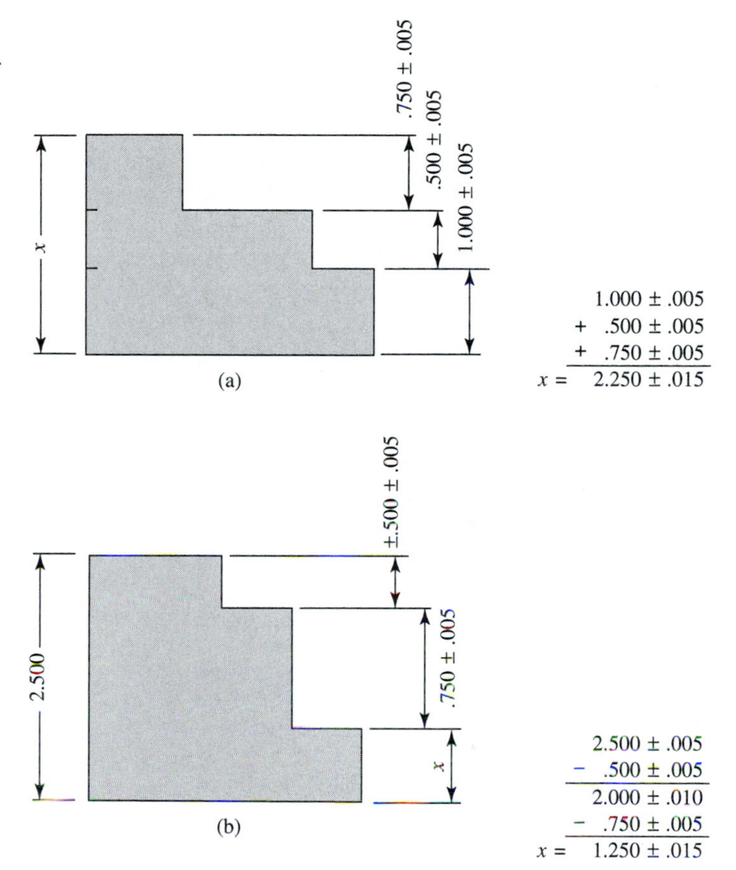

FIGURE 7.3 (a) Adding dimensions and tolerances. (b) Subtracting dimensions and tolerances.

- Locating surfaces for each operation
- Dimensional data—proposed dimensions and tolerances for each operation
- Machine selected for each operation
- Types of tooling—form tool, single point, straddle milling cutters, and so on

2. **Strip Layout.** Illustrate the part processing with a sketch (strip layout) showing, for each operation, the following:

- Piece part
- Machining dimensions
- Working tolerances
- Locating surfaces
- Surfaces machined

3. **Tolerance Chart Form.** Draw the chart form with the part centered at the top and columns as shown in the examples. It is best, for arithmetical reasons and subsequent checking, to develop the chart completely on one sheet of paper. Attempting to follow a dimensional sequence from one page to another can lead to confusion and errors. The scale and profile of the part drawing should be exaggerated enough to clarify all surfaces that are involved.

4. **Show Strip Layout Data on Tolerance Chart Form.** Transfer all data from the strip layout to the tolerance chart form using the appropriate lines and columns as shown in the examples. Identify surfaces to be machined, locating surfaces, and resultant dimensions with symbols as shown in Figure 7.1.

5. **Calculate Resultant Dimensions.** For clarity and later use, calculate and show resultant dimensions on the chart as they develop.

6. **Calculate Stock Removal.** Calculate and show stock removal in the appropriate column as it occurs.

7. **Utilize a Lines Involved Column.** Identify the origins of dimensions used in Steps 4, 5, and 6 in the Lines Involved column for later use in checking and also in comparing actual dimensions with part design dimensions. It will become a road map of where dimensions came from.

8. **Indicate Conclusion of Tolerance Charting.** For clarity, draw a heavy black line to separate the termination of tolerance charting from the subsequent comparison of the actual developed dimensions and specified part design dimensions and tolerances.

9. **Add Part Design Dimensions and Tolerances.** Show specified part design dimensions and tolerances in appropriate columns as shown in the examples.

10. **Determine Actual Dimensions and Tolerances and Compare with Part Design Specifications.** Determine final, actual dimensions and tolerances for each surface from the charted data and compare with specified part design dimensions and tolerances. Show these in the appropriate columns as shown in the examples.

11. **Make Any Required Changes or Corrections.** Retain a copy of the tolerance chart as now developed for a base. Develop a new chart to show any required changes in processing, dimensioning, or tolerancing.

7.4 TOLERANCE CHART EXAMPLES

Example 1 Pivot
Material: Bar stock
Processing: Turning
Production: Low

Example 2 Over Running Clutch Hub
Material: Steel forging
Processing: Chuckers
Heat Treat
Grinders
Production: High

EXAMPLE 1

Develop and check a tolerance chart for a Pivot, Figure 7.4.

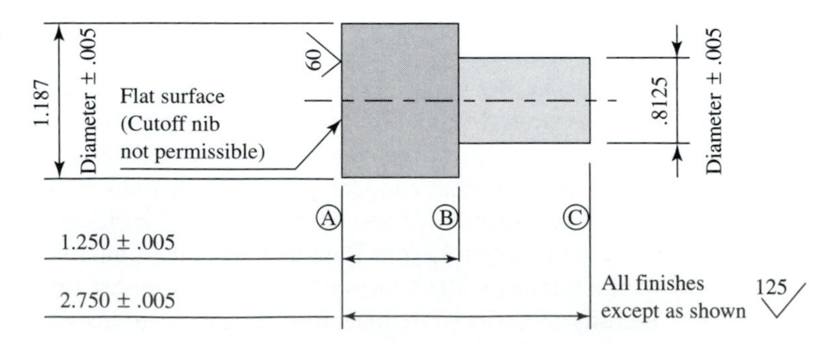

FIGURE 7.4 Pivot.

FIGURE 7.5 Processing of Pivot.

Op. No.	Equipment	Operation Description
20	Turret Lathe	Feed bar stock to stop Collet on O.D. Turn (2) diameters Cut off to length
30	Lathe	Chuck on small hub diameter Locate against shoulder Face large end face

Part Design Analysis. The part in this example, Figure 7.4, is a Pivot, machined all over from steel bar stock. No heat treatment or grinding operations are specified. Machining tolerances are suitable for lathes, turret lathes and automatic screw machines. The large end face is to be machined flat with no protuberances or cutoff nib permitted.

1. **Processing** (Fig. 7.5). The part is completed in two machining operations. A turret lathe is used at operation 20 to turn 2 diameters and cut the part off to length. In this operation, bar stock is fed to a lateral locator and the cutting tools are set in relation to the locator.

A lathe is employed for operation 30 to machine the large end face flat removing the cutoff nib. The right surface, or shoulder, of the large diameter is used as the locating surface and the facing tool is set in relation to it.

2. **Strip Layout** (Fig. 7.6). There are no rules or order to the form of the strip layout as long as it clearly and completely portrays the information given in the part's processing. Where feasible, it is helpful to use the same scale and form that will be used in the tolerance chart form.

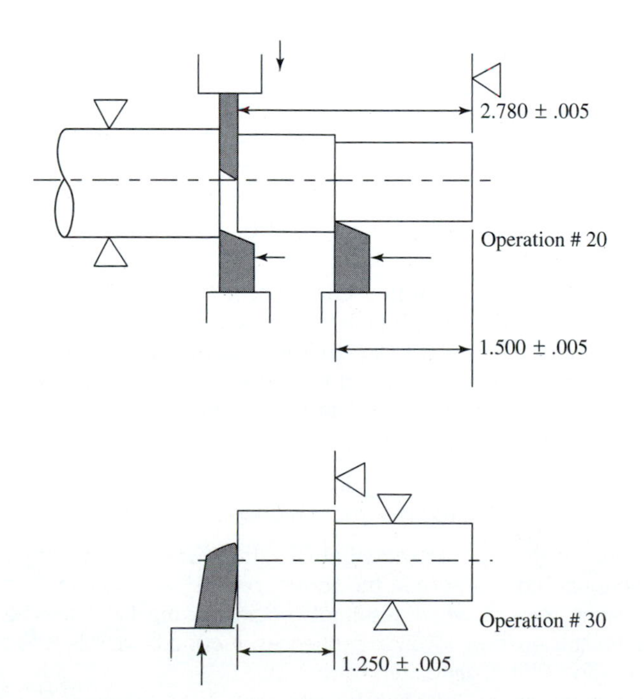

FIGURE 7.6 Strip layout illustrating processing of Pivot.

3. **Tolerance Chart Form** (Fig. 7.7). The format, columns, and data shown are generally accepted as standard form. The part is shown and dimensioned, centered at the top with lettered lines drawn from the various surfaces. Lettering of the lines is for purposes of clarity and identification when referring to a surface in processing and in the tool and operation sheets.

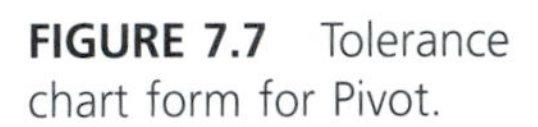

FIGURE 7.7 Tolerance chart form for Pivot.

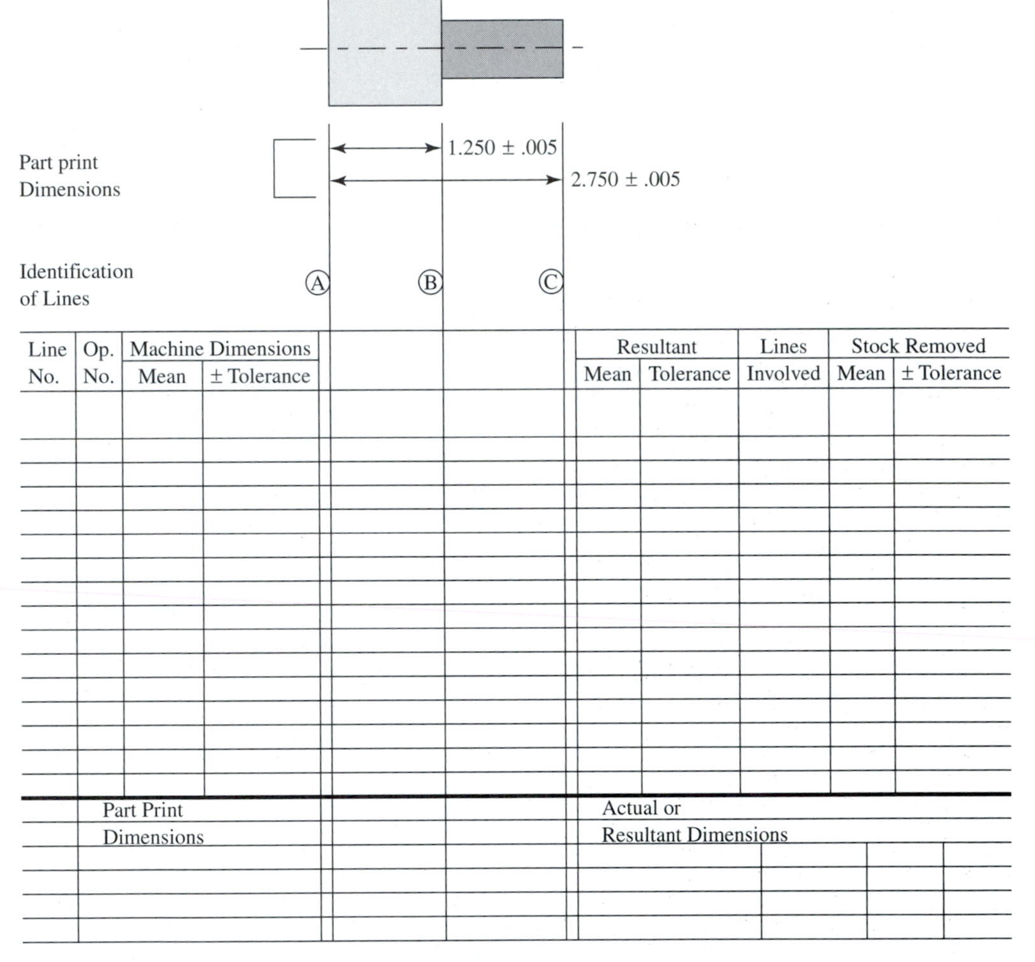

4. **Show Strip Layout Data on Tolerance Chart Form** (Fig. 7.8). Start charting the part by drawing dimensional lines for operation 20 from the locating surfaces to the involved surfaces as shown at ①. Identify the locating end of the lines with small circles and the machining ends with arrow heads. Add operation number, machining dimensions, and tolerances in the appropriate columns.

5. **Calculate Resultant Dimensions.** Calculate the resultant length of the large diameter:

$$\begin{array}{r} 2.780 \pm .005 \text{ (line 2)} \\ \underline{-1.500 + .005 \text{ (line 3)}} \\ 1.280 \pm .010 \text{ (resultant)} \end{array}$$

Draw the resultant line on the chart as shown at ② with small circles identifying each end. Show the resultant mean dimension and tolerance in the correct column. Add ① and ② in the lines involved for future reference. Keeping track of where resultants originate simplifies later checking of the chart and is especially helpful for tracing long, involved sequences. A good basic rule is to calculate and chart a resultant dimension as soon as it is generated.

Draw the machining dimension on the chart for operation 30 as shown at ③. Identify the locating end with a small circle and the machining end with an arrowhead. Add operation number, machine mean dimension, and machine dimension tolerance data in the appropriate columns.

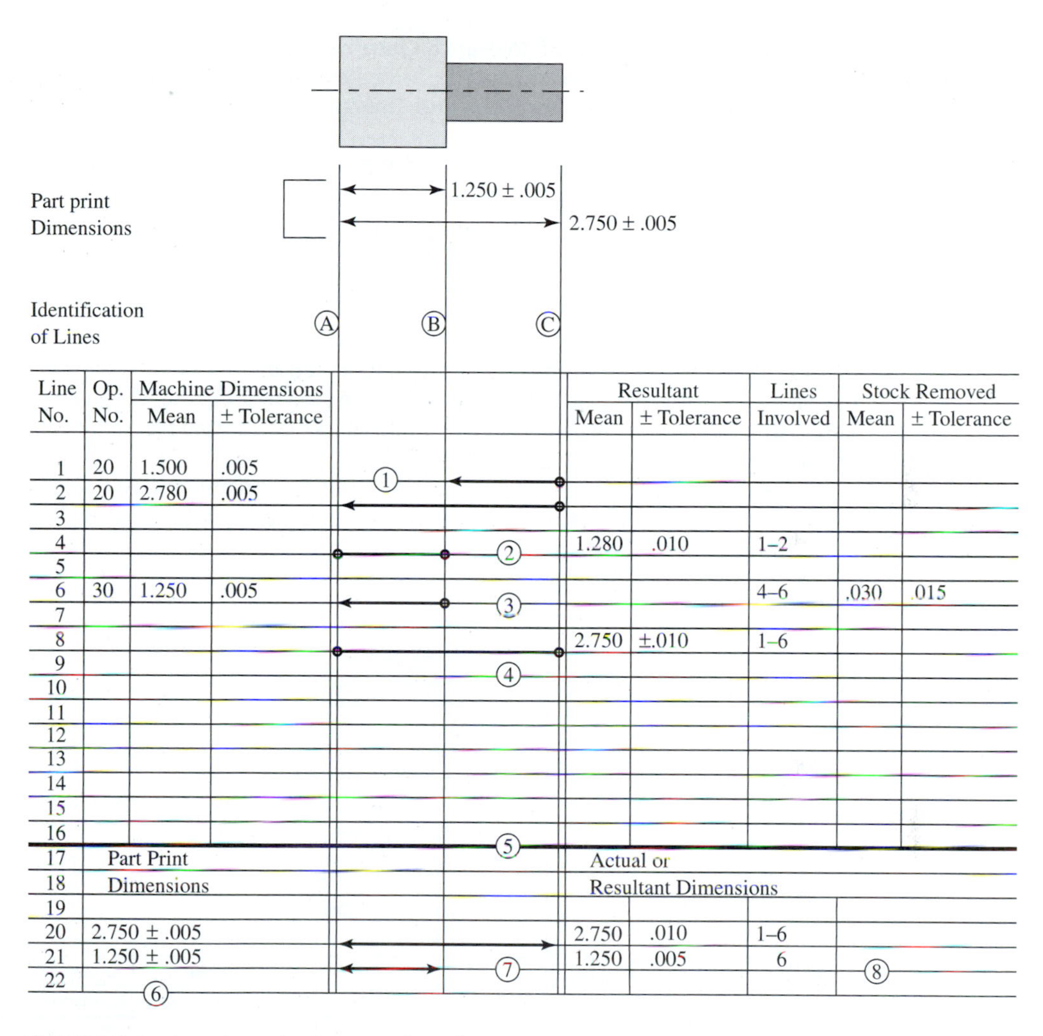

FIGURE 7.8 Developed tolerance chart for Pivot.

6. **Calculate Stock Removal.** Calculate the stock removed at operation 30.

$$
\begin{array}{r}
1.280 \pm .010 \text{ (line 4)} \\
\underline{-1.250 + .005 \text{ (line 6)}} \\
.030 \pm .015 \text{ (stock removed)}
\end{array}
$$

This is acceptable. Variation of the surface to be machined is within the total amount of stock to be removed and cleanup of the surface is assured. Had the variation of the surface to be machined varied by an amount greater than the total amount of stock available (.030 ± .040 for example) the surfaces of some parts would not clean up. Show stock removed data in the correct columns and show 4–6 as the lines involved.

Calculate the resultant overall length.

$$
\begin{array}{r}
1.500 \pm .005 \text{ (line 1)} \\
\underline{+1.250 + .005 \text{ (line 6)}} \\
2.750 \pm .010 \text{ (resultant)}
\end{array}
$$

Add the line and data as shown at ④.

7. **Indicate Conclusion of Tolerance Charting.** Draw a heavy black line as shown at ⑤ to indicate the conclusion of tolerance charting and beginning of checking and verification.

8. **Add Part Design Dimensions and Tolerances.** Add part design dimensions and tolerances as shown at ⑥.

9. **Determine Actual Dimensions and Tolerances: Compare with Part Design Specifications**

- The .250 ± .005 part design dimension was machined to 1.250 ± .005 at operation 30 (line 6). This dimension and tolerance satisfies the part design.
- The 2.750 ± .005 part design dimension was developed partially in operation 20 and completed in operation 30. The resultant is shown at ④. The tolerance of ±.010 does *not* satisfy the part design.
- Dimensions and tolerances are added to the chart as shown at ⑦ and ⑧.

10. **Make Any Required Changes and Corrections.** Changes must be made to processing, machining dimensions, or tolerances such that the part design is satisfied while retaining a producible part with the maximum allowable working tolerances. In this example, reduction of working tolerances has been employed as the most feasible solution.

To reduce the tolerance of the 2.750 ± .010 resultant dimension to 2.750 ± .005, shown at ⑨ of Figure 7.9, tolerances of machining dimensions must be reduced to ±.0025. These changes and results are shown in Figure 7.9.

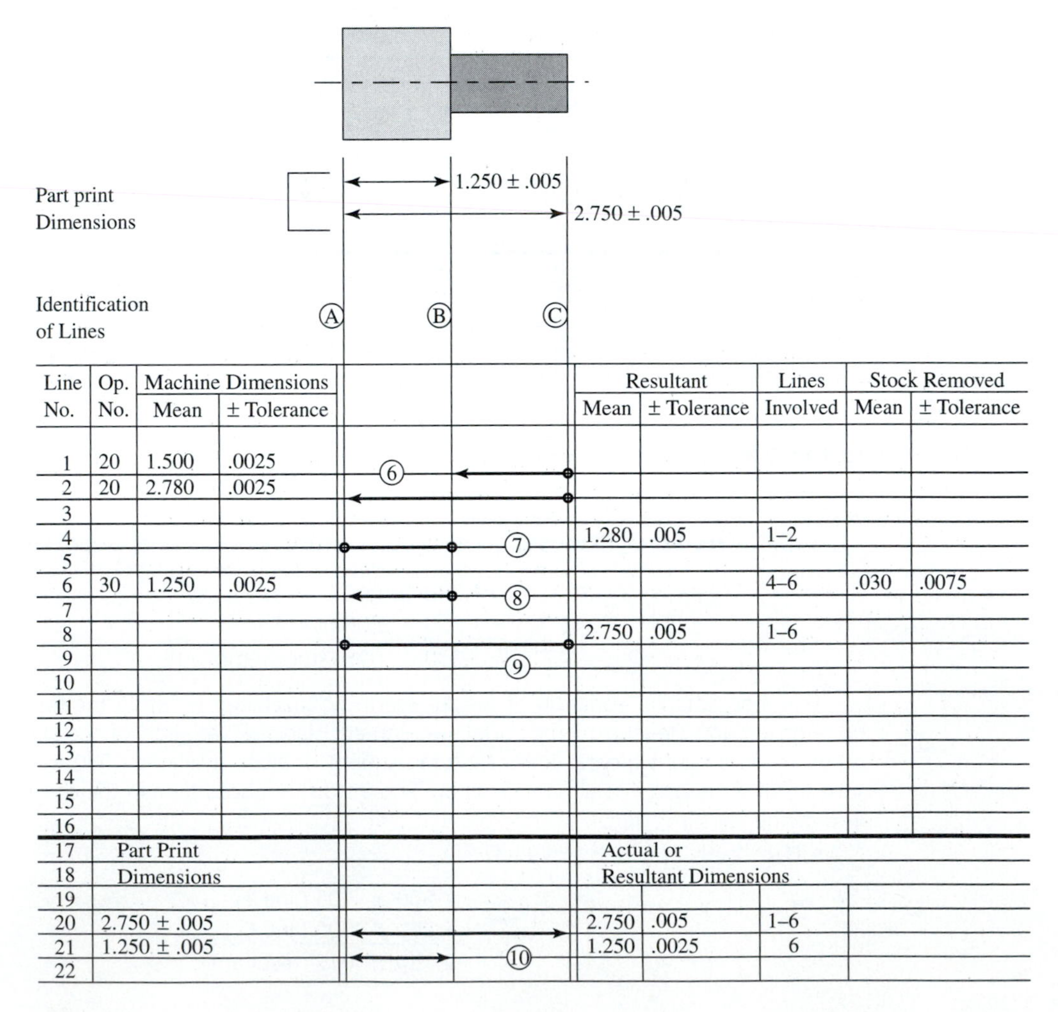

Line No.	Op. No.	Machine Dimensions Mean	Machine Dimensions ± Tolerance	Resultant Mean	Resultant ± Tolerance	Lines Involved	Stock Removed Mean	Stock Removed ± Tolerance
1	20	1.500	.0025					
2	20	2.780	.0025					
3								
4				1.280	.005	1–2		
5								
6	30	1.250	.0025			4–6	.030	.0075
7								
8				2.750	.005	1–6		
9								
10								
11								
12								
13								
14								
15								
16								
17	Part Print			Actual or				
18	Dimensions			Resultant Dimensions				
19								
20	2.750 ± .005			2.750	.005	1–6		
21	1.250 ± .005			1.250	.0025	6		
22								

FIGURE 7.9 Corrected tolerance chart for Pivot.

Changes

⑥ Change the machining tolerance of the 1.500 dimension to ±.0025.
⑦ Change the tolerance stack of the 1.280 resultant to ±.005.

$$
\begin{array}{rl}
2.780 \pm .0025 & \text{(line 2)} \\
-1.500 + .0025 & \text{(line 1)} \\
\hline
1.280 \pm .005 & \text{(new resultant)}
\end{array}
$$

⑧ Change the machining tolerance of the 1.250 dimension to ±.0025. Change stock removal tolerance to ±.010.

$$
\begin{array}{rl}
1.280 \pm .005 & \text{(line 4)} \\
-1.250 \pm .0025 & \text{(line 6)} \\
\hline
.030 \pm .0075 & \text{(new stock removal tolerance)}
\end{array}
$$

⑨ Change the overall length resultant tolerance to ±.005.

$$
\begin{array}{rl}
1.500 \pm .0025 & \text{(line 1)} \\
1.250 \pm .0025 & \text{(line 6)} \\
\hline
2.750 \pm .005 & \text{(new resultant tolerance)}
\end{array}
$$

⑩ Change the accumulated tolerance of the 2.750 dimension to ±.005 (line 8). Change the tolerance of the 1.250 dimension to ±.0025 (line 1). The actual dimensions and tolerance of the part as produced now satisfy the part design.

EXAMPLE 2

Develop and check a tolerance chart for an Over-Running Clutch Hub, Figure 7.10.

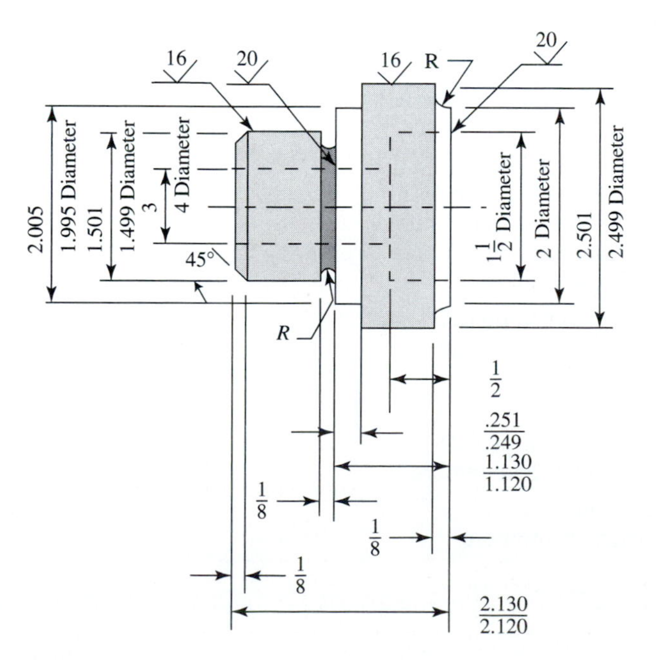

Notes
Carburize and harden all exterior surfaces .060 minimum depth
All finishes except as noted 125
Tolerances of fractional dimensions .015
Remove all burrs and break sharp corners .032 maximum

Part
Over Running Clutch Hub
Material
1020 SAE Steel Forging
Scale
Full

FIGURE 7.10 Over-Running Clutch Hub.

Part Design Analysis. The part in this example (Fig. 7.10) is a clutch hub machined on all surfaces from a steel forging. The part requires heat treating and subsequent grinding operations. A minimum depth of hardness must be maintained after finish grinding. Tolerances for surfaces left "as machined" are suitable for lathes, turret lathes, and automatic screw machines or chucking machines. Surfaces with specified finishes of 20 μin. or less, should be ground. All corners must be broken and all burrs removed.

1. **Processing** (Fig. 7.11). The part is completed in two machining operations, a heat treat, three separate grinding operations, and a deburr.

- The production rate per hour, although not stated for this example, is sufficient to justify the use of automatic lathes.

Op. No.	Equipment	Operation Description
20	4–Spindle Chucking Machine	Station #1
		Load and unload
		Chuck on large diameter
		Locate against rough forging
		large end face surface (J) (See Fig. 7.14)
		Station #2
		Rough turn small hub O.D.
		Rough turn 2.000 diameter
		Face end
		Station #3
		Finish form 2.000 diameter and undercut
		Form 1/8 x 45° chamfer
		Drill 3/4 diameter hole half way
		Station #4
		Finish turn small hub O.D.
		Drill 3/4 diameter hole balance of way through
30	4–Spindle Chucking Machine	Station #1
		Load and unload
		Chuck on small hub O.D.
		Locate against small hub
		end face surface (B) (See Fig. 7.14)
		Station #2
		Rough turn large O.D.
		Form 2 diameter hub
		Face end
20		Station #3
		Finish turn large O.D.
		Rough drill counterbore
		Station #4
		Finish counterbore
40		Inspect

FIGURE 7.11 Tentative processing of Over-Running Clutch Hub.

Op. No.	Equipment	Operation Description
50	Heat Treat Equipment	Heat Treat Carburize and harden exterior surfaces to .066 minimum depth of case
60	Cylindrical Chucking Grinder	Chuck on large O.D Locate against large hub end face Grind small hub O.D.
70	Cylindrical Chucking Grinder	Chuck on small hub O.D Locate against shoulder surface (E) (See Fig. 7.14) Grind large O.D.
80	Cylindrical Chucking Grinder	Chuck on small hub O.D. Locate against shoulder surface (E) Grind large hub end face
90	Deburr Equipment	Deburr and break sharp edges
100		Inspect

FIGURE 7.11 Continued

- The depth of case in the heat treat operation must be as much deeper than specified as the stock removed in the subsequent grinding operations.
- For stability in the first operations (getting the part out of the rough), the rough forging is chucked on the larger diameter and located on the large end face. Once the small diameter and end face are machined, they will form a stable chucking and locating surface for subsequent operations.
- The closest, or most limiting, lateral tolerance is the .249–.251 dimension width of the 1.995–2.005 diameter shoulder. This dimension will be produced by a form tool where the manufacturing tolerances (of the tool) is in tenths of a thousandth of an inch. The tool dimension will be reproduced in the part. See the strip layout.

2. **Strip Layout** (Fig. 7.12 on page 174). Processing, with the exceptions of heat treat and deburring, for this example, is shown visually in the strip layout. This step is vital to visualizing, regulating, and seeing the process development. Processing a part should not be attempted without some form of this visual aid.

3. **Tolerance Chart Form** (Fig. 7.13 on page 175). The chart form has been drawn, as it was in Example 1, with the part centered at the top. Part design dimensions are included for easier reference when developing working dimensions and tolerances. Columns, operational lines, and identification data have been included as before.

4. **Transfer Strip Layout Data to T/C Form** (Fig. 7.14 on page 176). Data from the strip layout has been added to the tolerance chart form, including all dimensions, tolerances, locating surfaces, and so on in the appropriate columns. Note the word "solid" has been used where a feature is machined from "solid" material.

5. **Calculate Resultants.** The resultant dimensions have been calculated as they occurred and the operational line numbers have been listed in the Lines Involved column.

This applies to the resultants shown on operational lines 7, 13, 14, 15, 16, 17, and 23.

6. **Calculate Stock Removed.** Stock removed has been calculated and shown in the appropriate columns of operational lines 3, 9, and 21, and the originating dimensions for the calculations are noted in the Lines Involved column.

7. **Indicate the Conclusion of Tolerance Charting.** A heavy black line has been added to the tolerance chart form following line 23.

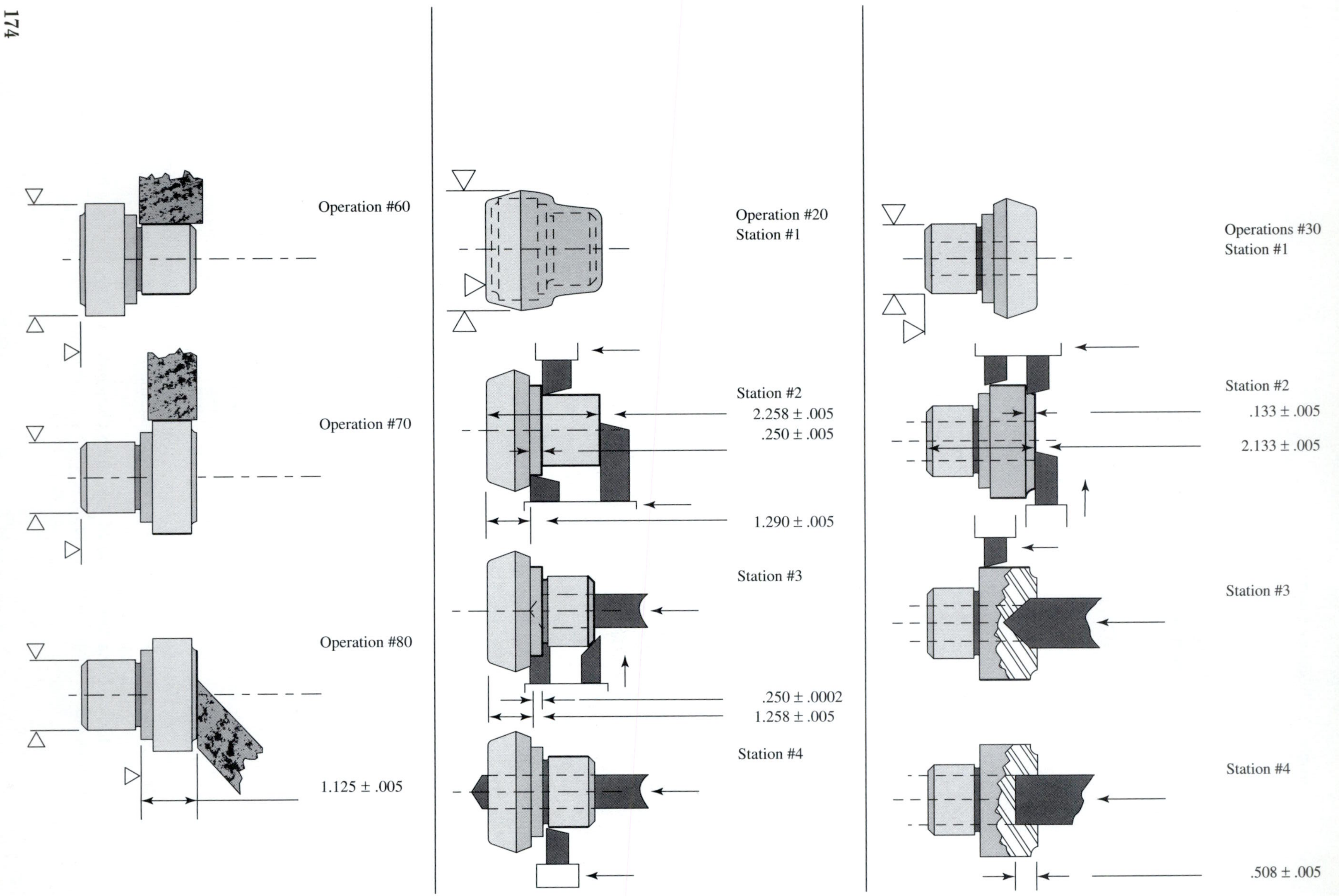

FIGURE 7.12 Strip layout illustrating processing of Over-Running Clutch Hub.

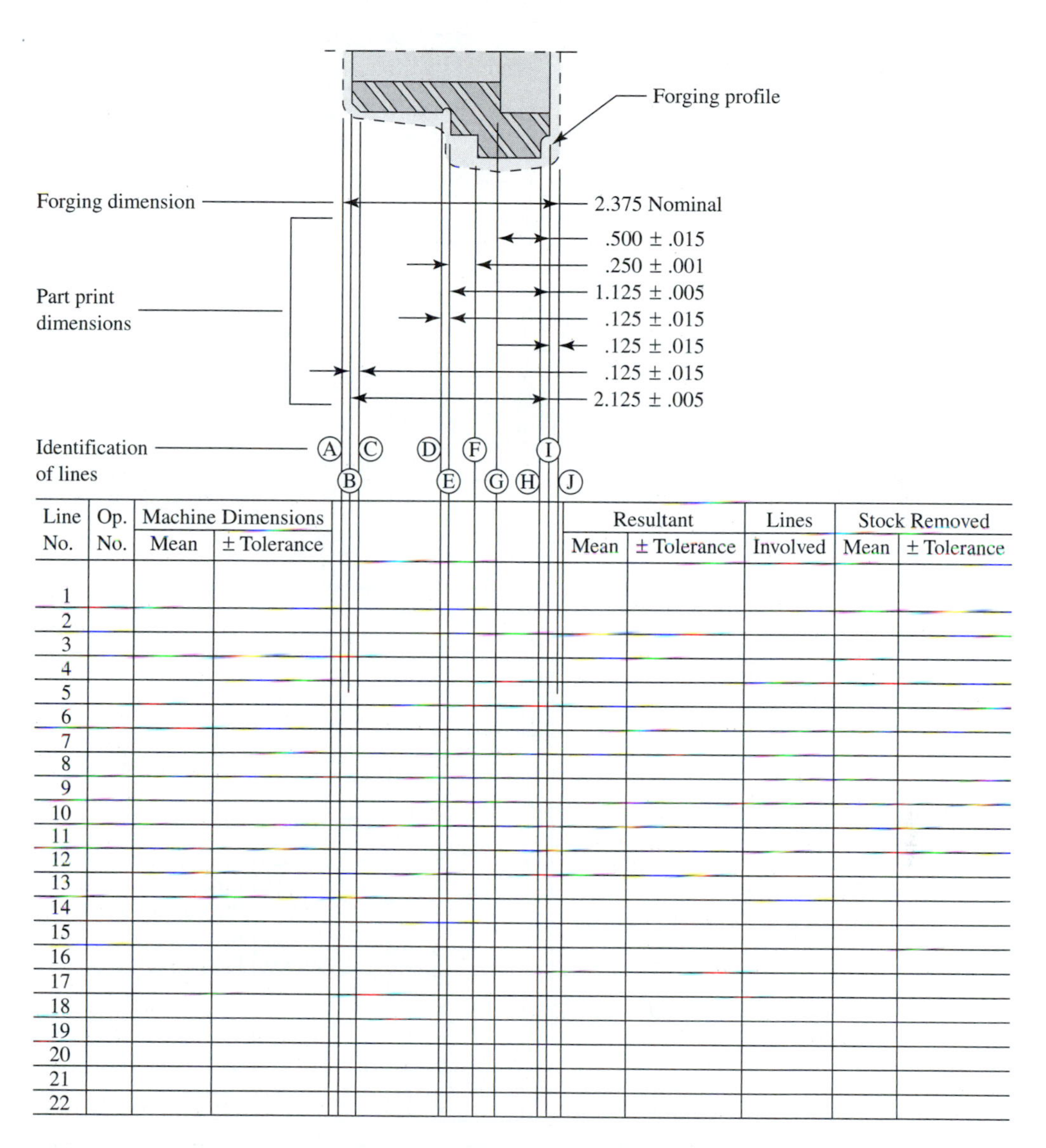

FIGURE 7.13 Tolerance chart form for Over-Running Clutch Hub.

(Continued from page 173.)

8. **Add Part Design Dimensions and Tolerances.** This data has been added to the tolerance chart form following line 23.

9. **Determine Actual Dimensions and Tolerances; Compare with Part Design Specifications.** After calculating the dimensions and tolerances as processed, it is found that certain part print specifications have not been satisfied:

(a) Stock removal at operation 80 is unsatisfactory.

(b) Tolerances of the following actual dimensions do not meet part design specified dimensions:

- .500 ± .015 depth of bore is actually .500 ± .025.

$$\begin{array}{r} 1.125 \pm .005\ (21) \\ \underline{-.625 + .020\ (16)} \\ .500 \pm .025\ (27) \end{array}$$

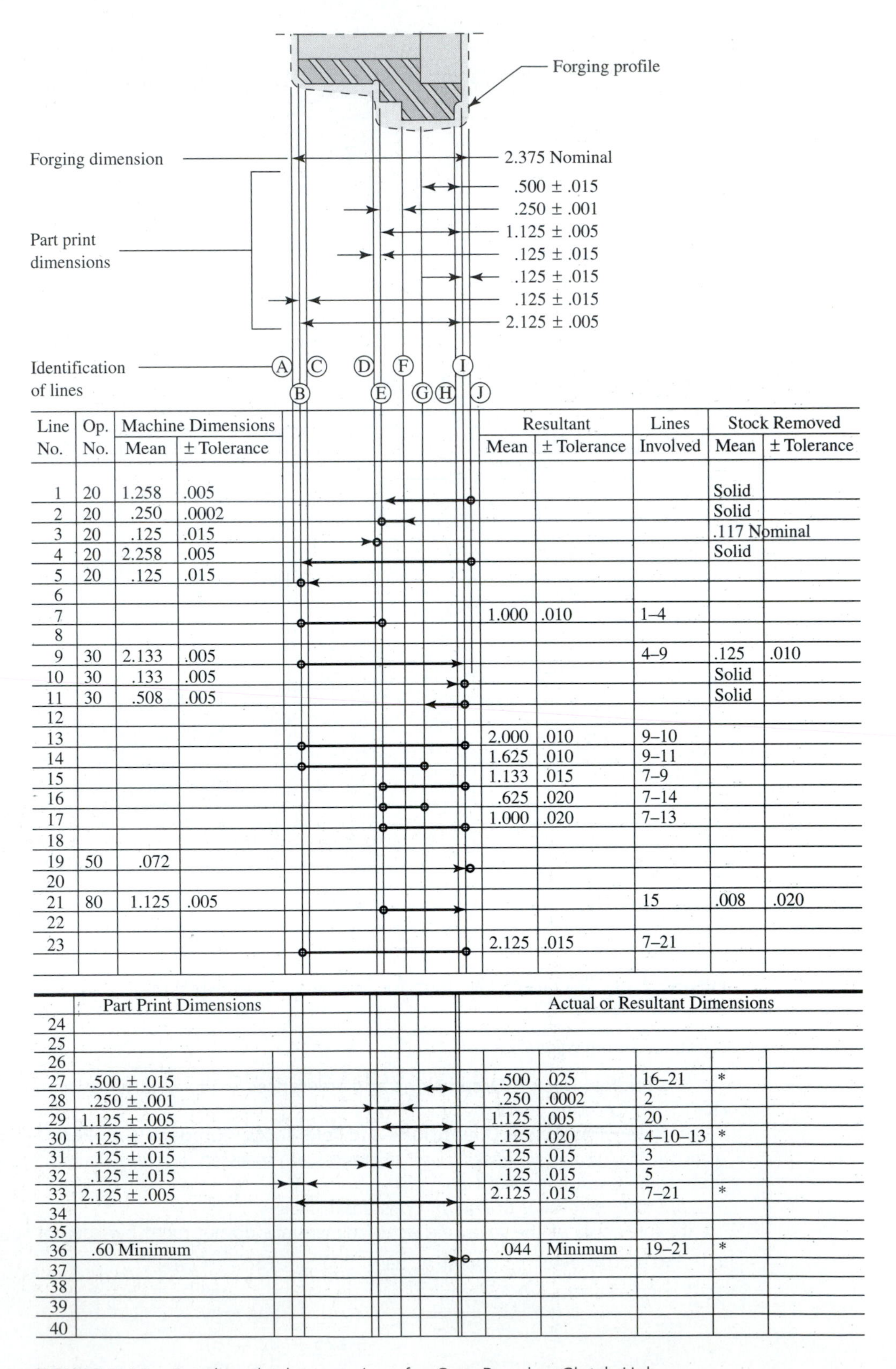

Line No.	Op. No.	Machine Dimensions Mean	Machine Dimensions ± Tolerance	Resultant Mean	Resultant ± Tolerance	Lines Involved	Stock Removed Mean	Stock Removed ± Tolerance
1	20	1.258	.005				Solid	
2	20	.250	.0002				Solid	
3	20	.125	.015				.117 Nominal	
4	20	2.258	.005				Solid	
5	20	.125	.015					
6								
7				1.000	.010	1–4		
8								
9	30	2.133	.005			4–9	.125	.010
10	30	.133	.005				Solid	
11	30	.508	.005				Solid	
12								
13				2.000	.010	9–10		
14				1.625	.010	9–11		
15				1.133	.015	7–9		
16				.625	.020	7–14		
17				1.000	.020	7–13		
18								
19	50	.072						
20								
21	80	1.125	.005			15	.008	.020
22								
23				2.125	.015	7–21		

Line No.	Part Print Dimensions	Actual or Resultant Dimensions			
24					
25					
26					
27	.500 ± .015	.500	.025	16–21	*
28	.250 ± .001	.250	.0002	2	
29	1.125 ± .005	1.125	.005	20	
30	.125 ± .015	.125	.020	4–10–13	*
31	.125 ± .015	.125	.015	3	
32	.125 ± .015	.125	.015	5	
33	2.125 ± .005	2.125	.015	7–21	*
34					
35					
36	.60 Minimum	.044	Minimum	19–21	*
37					
38					
39					
40					

FIGURE 7.14 Developed tolerance chart for Over-Running Clutch Hub.

(Continued from page 175.)

- 2.125 ± .005 overall length is actually 2.125 ± 0.15.

$$\begin{array}{r} 1.000 \ \pm .010\ (7) \\ +1.125 + .005\ (21) \\ \hline 2.125 \ \pm .015\ (33) \end{array}$$

(c) .060 minimum depth of case specification has not been met.

$$\begin{array}{l} .072 \text{ Depth of case } (18) \\ -.028 \text{ Maximum stock removal } (20) \\ \hline .044\ (35) \end{array}$$

3. **Make Required Changes and Corrections** (Fig. 7.15 on page 178).

- Corrections can easily be made to the amount of stock removed and to the depth of hardening by recalculating the dimensions involved. This is shown.
- Corrections to the instances of excessive tolerance stacking can be made by

  Redimensioning of product
  Reprocessing
  Reducing working tolerances

Redimensioning of Product. From the greatest producibility or least cost of manufacture standpoint, the part must be dimensioned the way it will be produced; that is, the surfaces must be dimensioned from the manufacturing locating points. Assuming, for this example, the part's function precludes this change, reprocessing or the reduction of working tolerances must be considered.

Reprocessing. Excessive tolerance stacking can be remedied by reprocessing. For example, the .500 ± .015 depth of bore can be established by leaving stock in the bottom of the bore during the machining operation. Then, while the part is located against the right end face, the .500 ± .015 dimension can be completed to depth in a secondary operation, which can utilize the full allowable tolerance.

However, adding to the cost of labor, additional equipment and tooling, and floor space for the added operations is the fact the part is hardened at this stage. The added operation would now necessarily be a comparatively slower and more costly grinding operation. Similar consideration will apply and must be made to the other excessive tolerance stacks.

Reduce Working Tolerances. Reducing working tolerances is a common way of reducing tolerance stacks. It is used where the equipment is capable or where not too great a penalty is imposed on producibility in the forms of slower feeds, rework of borderline parts, scrap, and so on.

Dimensional tolerance requirements have been developed that will produce this part to the part print specifications. These are illustrated in Figure 7.15.

In practice, at this point, Manufacturing Engineering will review machine capabilities and also determine manufacturing costs by the processes and reprocessing considered. A decision can then be made directing the use of either reprocessing or reducing manufacturing tolerances.

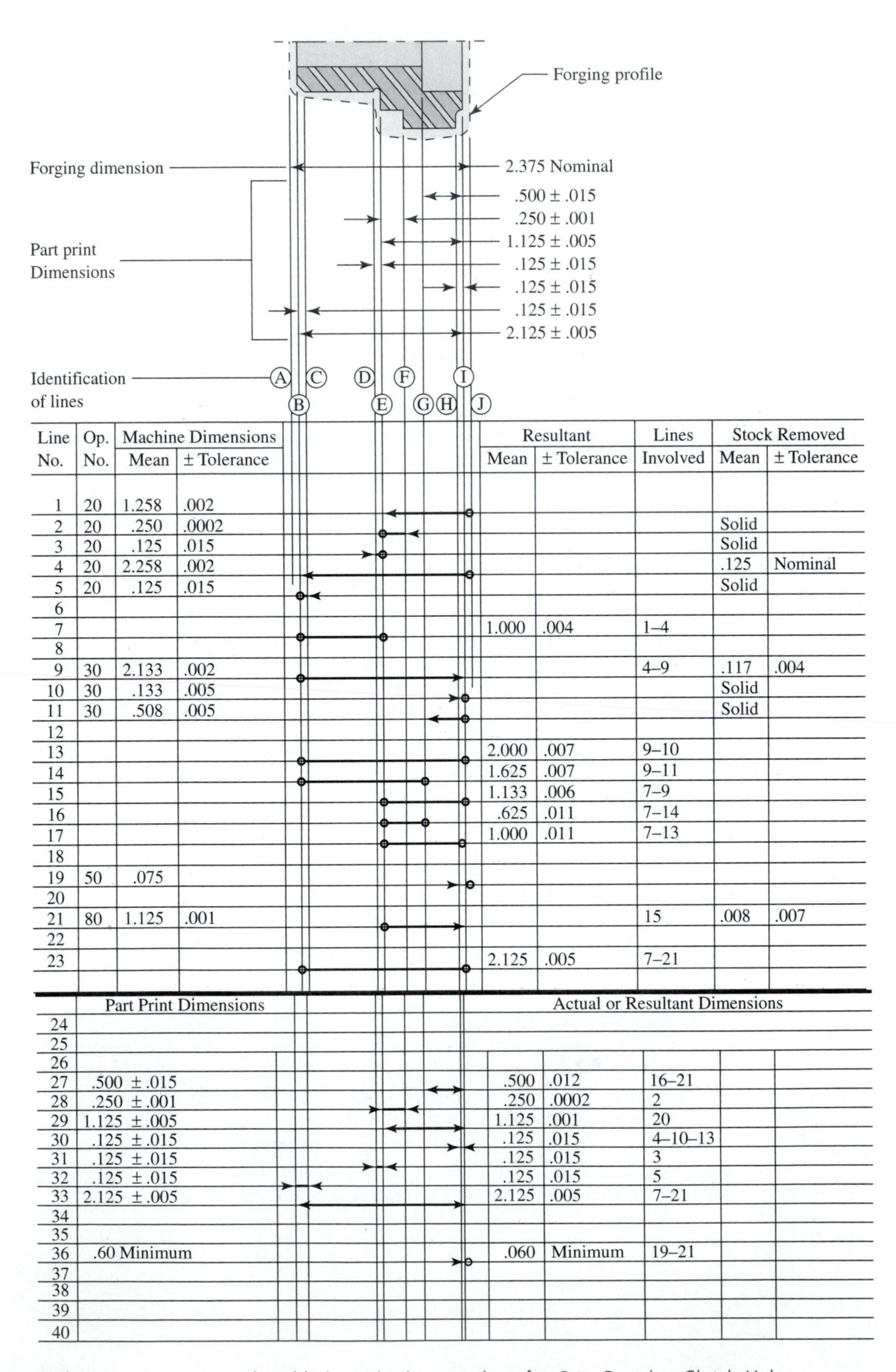

Line No.	Op. No.	Machine Dimensions Mean	Machine Dimensions ± Tolerance	Resultant Mean	Resultant ± Tolerance	Lines Involved	Stock Removed Mean	Stock Removed ± Tolerance
1	20	1.258	.002					
2	20	.250	.0002				Solid	
3	20	.125	.015				Solid	
4	20	2.258	.002				.125	Nominal
5	20	.125	.015				Solid	
6								
7				1.000	.004	1–4		
8								
9	30	2.133	.002			4–9	.117	.004
10	30	.133	.005				Solid	
11	30	.508	.005				Solid	
12								
13				2.000	.007	9–10		
14				1.625	.007	9–11		
15				1.133	.006	7–9		
16				.625	.011	7–14		
17				1.000	.011	7–13		
18								
19	50	.075						
20								
21	80	1.125	.001			15	.008	.007
22								
23				2.125	.005	7–21		

Line No.	Part Print Dimensions	Actual or Resultant Dimensions Mean	± Tolerance	Lines Involved
24				
25				
26				
27	.500 ± .015	.500	.012	16–21
28	.250 ± .001	.250	.0002	2
29	1.125 ± .005	1.125	.001	20
30	.125 ± .015	.125	.015	4–10–13
31	.125 ± .015	.125	.015	3
32	.125 ± .015	.125	.015	5
33	2.125 ± .005	2.125	.005	7–21
34				
35				
36	.60 Minimum	.060	Minimum	19–21
37				
38				
39				
40				

FIGURE 7.15 Corrected and balanced tolerance chart for Over-Running Clutch Hub.

- Stock Removal

Originally Developed Dimensions (Fig. 7.14)	Corrected Dimensions (Fig. 7.15)
2.258 ± .005 (4)	2.258 ± .002 (4)
− 1.258 + .005 (1)	− 1.258 + .002 (1)
1.000 ± .010 (7)	1.000 ± .004 (7)
2.133 ± .005 (9)	2.133 ± .002 (9)
− 1.000 + .010 (7)	− 1.000 + .004 (7)
1.133 ± .015 (15)	1.133 ± .006 (15)
− 1.125 + .005 (21)	− 1.125 + .001 (21)
.008 ± .020 (21)	.008 ± .007 (21)
Insufficient stock to clean up at one extreme of tolerance	Sufficient stock to clean up at either extreme of tolerance

- .500 ± .015 Dimension

Originally Developed Dimensions (Fig. 7.14)	Corrected Dimensions (Fig. 7.15)
2.133 ± .005 (9)	2.133 ± .002 (9)
− .508 + .005 (11)	− .508 + .055 (11)
1.625 ± .010 (14)	1.625 ± .007 (14)
2.258 ± .005 (4)	2.258 ± .002 (4)
− 1.258 + .005 (1)	− 1.258 + .002 (1)
1.000 ± .010 (7)	1.000 ± .004 (7)
1.625 ± .010 (14)	1.625 ± .007 (14)
− 1.000 + .010 (7)	− 1.000 + .004 (7)
.625 ± .020 (16)	.625 ± .011 (16)
1.125 ± .005 (21)	1.125 ± .001 (21)
− .625 + .020 (16)	− .625 + .011 (16)
.500 ± .025 (27)	.500 ± .012 (27)

- 1.125 ± .005 Dimension

Originally Developed	Corrected
1.125 ± .005 (29)	1.125 ± .001
Acceptable as is	Held closer than part design due to excessive tolerance stack of associated dimensions

- 125 ± .015 Width of Shoulder Dimension

Originally Developed	Corrected
2.258 ± .005 (4)	2.258 ± .002 (4)
− 1.258 + .005 (1)	− 1.258 + .002 (1)
1.000 ± .010 (7)	1.000 ± .004 (7)
2.133 ± .005 (9)	2.133 ± .002 (9)
− .133 + .005 (10)	− .133 + .005 (10)
2.000 ± .010 (13)	2.000 ± .007 (13)
2.000 ± .010 (13)	2.000 ± .007 (13)
− 1.000 + .010 (7)	− 1.000 + .004 (7)
1.000 ± .020 (17)	1.000 ± .011 (17)
1.125 ± .005 (21)	1.125 ± .001 (21)
− 1.000 + .020 (17)	− 1.000 + .011 (17)
.125 ± .025 (30)	.125 ± .015 (30)

- 2.125 ± .005. Overall Length

Originally Developed	Corrected
2.258 ± .005 (4)	2.258 ± .002 (4)
− 1.258 + .005 (1)	− 1.258 + .002 (1)
1.000 ± .010 (7)	1.000 ± .004 (7)
+ 1.125 + .005 (21)	+ 1.125 + .001 (21)
2.125 ± .015 (23)	2.125 ± .005 (33)

- .060 Minimum Depth of Case

Originally Developed	Corrected
.072 (19)	.075 (19)
− .028 Maximum stock removal (21)	− .015 Maximum stock removal (20)
.54 (36)	.060 (36)

7.5 SELECTIVE ASSEMBLY

The functional performance of some products requires extremely close fits between parts—a fit so restrictive as to require impractically limiting manufacturing tolerances of component parts. Instead of conventional machining dimensions and tolerances being used to manufacture such parts, the technique of selective assembly is employed:

1. Component parts are manufactured to a feasible manufacturing tolerance.
2. Finished component parts are inspected and sorted into groups or classes of specific sizes.
3. Parts are selected from appropriate groups, or classes of sizes, to provide the required fit.

Examples of the use of selective assembly follow.

Hydraulic Controls Shifter valves, operated by hydraulic pressure, are designed to cover or uncover ports in a hydraulic circuit, which in turn actuate servos or clutches. The valves must operate smoothly in their bores but with no leakage past them. An extremely close fit and a very high degree of surface finish are required.

It was found in performance tests for one particular shifter valve that the clearance fit of the valve in its bore had to be within .0002–.0006. This amount of clearance dictated tolerances of ±.0001 for the shifter valve outside diameter and the bore inside diameter.

However, the capability of machines for either part was ±.0003. The problem was resolved by manufacturing the parts to ±.0003 tolerances and utilizing selective assembly to provide the required fits as shown in Figure 7.16 on page 181.

Pistons and Cylinder Bores Automotive pistons are assembled into the bores of engine blocks utilizing the selective assembly concept. Individual bores are measured, and a piston of the correct diameter is selected and assembled into it. It is possible for an engine to have a different diameter of bore and piston for each of its bores, although not probable due to the extreme accuracy of machining processes.

Sophisticated equipment, automated to a very high degree, is used to measure, mark, select, and assemble the great quantity of pistons and bores involved in a high-production engine line.

Pump Components Many automobile and truck automatic transmissions use “Gerotor” pumps to provide the hydraulic pressure required to operate the various units and components.

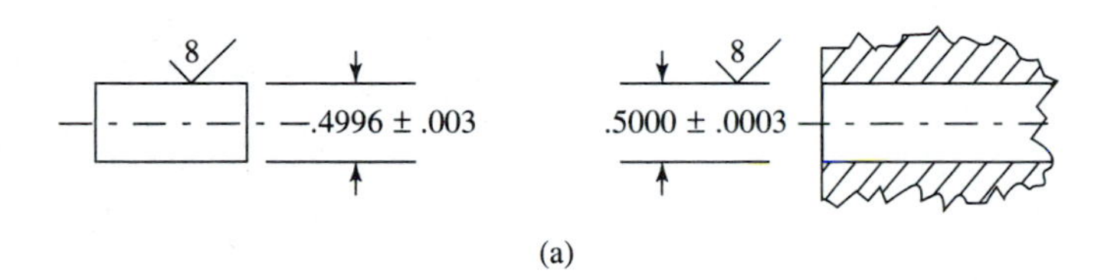

(a)

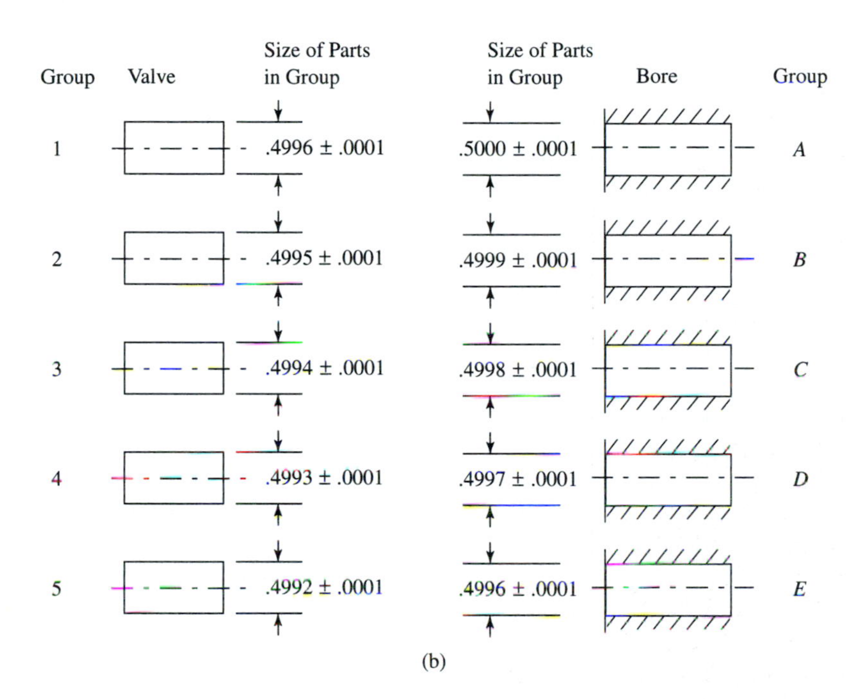

(b)

Valves from lettered group are assembled into bores from respective numbered group.

Example:

4 .4993 .0001 .4997 .0001 D

.4997 .0001 Bore (group *D*)
.4993 .0001 Valve (group 4)
.0004 .0002 Clearance range

.0004
−.0002
.0002 Minimum clearance

.0004
+.0002
.0006 Maximum clearance

(c)

FIGURE 7.16 Selective Assembly. (a) Capability range of manufacturing process. (b) Parts as sorted and grouped. Valves selected from: lettered group are assembled into bores from respective-numbered group. (c) Resulting operating clearance fits.

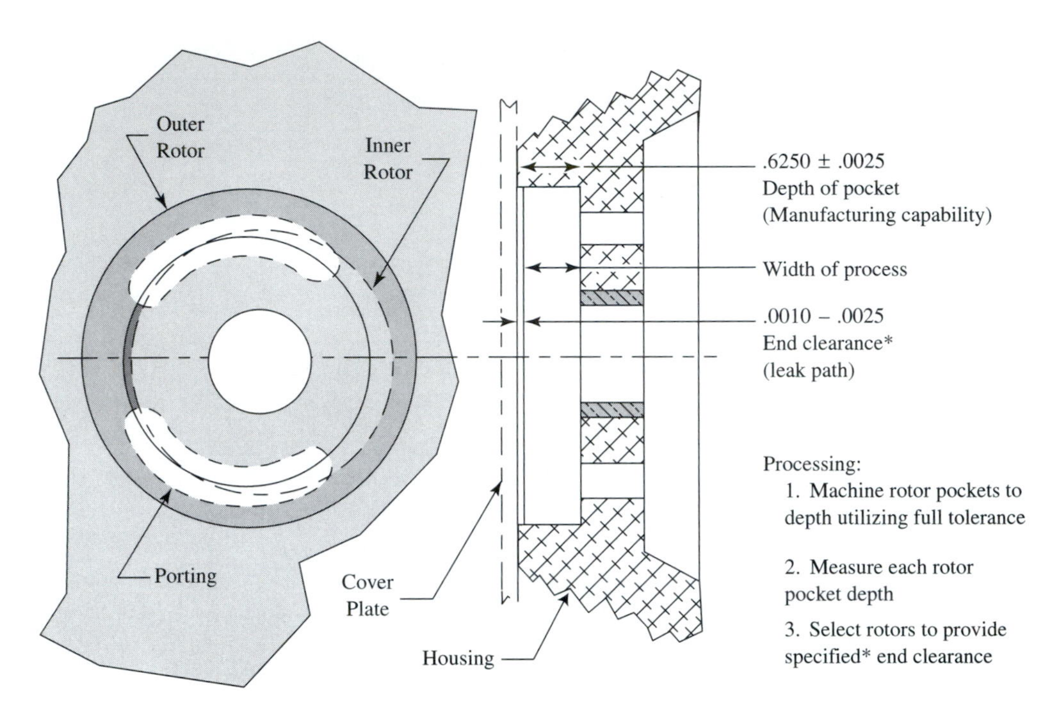

FIGURE 7.17 Oil pump assembly rotor pockets machined using full tolerance; rotors are selected for width to provide optimum end play when assembled.

For these pumps to develop maximum rated capcity, leak paths between rotor ends and the cover plate must be kept to a minimum.

Utilizing the selective assembly concept, rotor pocket depths are machined to a feasible manufacturing tolerance. The pocket depths are then measured, and rotors with suitable widths are selected and assembled to provide the required operating clearance as shown in Figure 7.17.

Provide Specified End Play A slightly different application of the selective assembly process is to provide specified end play regardless of the tolerance stacks of assembled parts. This process sequence is as follows:

1. Manufacture component parts to full allowable tolerances.
2. Subassemble parts using randomly selected parts.
3. Assemble the resulting subassembly into place in a randomly selected housing.
4. Measure the resulting gap between subassembly and housing.
5. Select and assemble a suitable spacer to provide specified end play.

This technique is illustrated in Figure 7.18 on page 183.

End Play in a Transmission Figure 7.19 on page 184 illustrates the application of the selected spacer technique in 3 different areas of a transaxle. Each of the decks of gears and components requires end play of very closely controlled tolerances. The actual amount of end play tolerance is less than 10% of the possible total tolerance stack of each assembled deck.

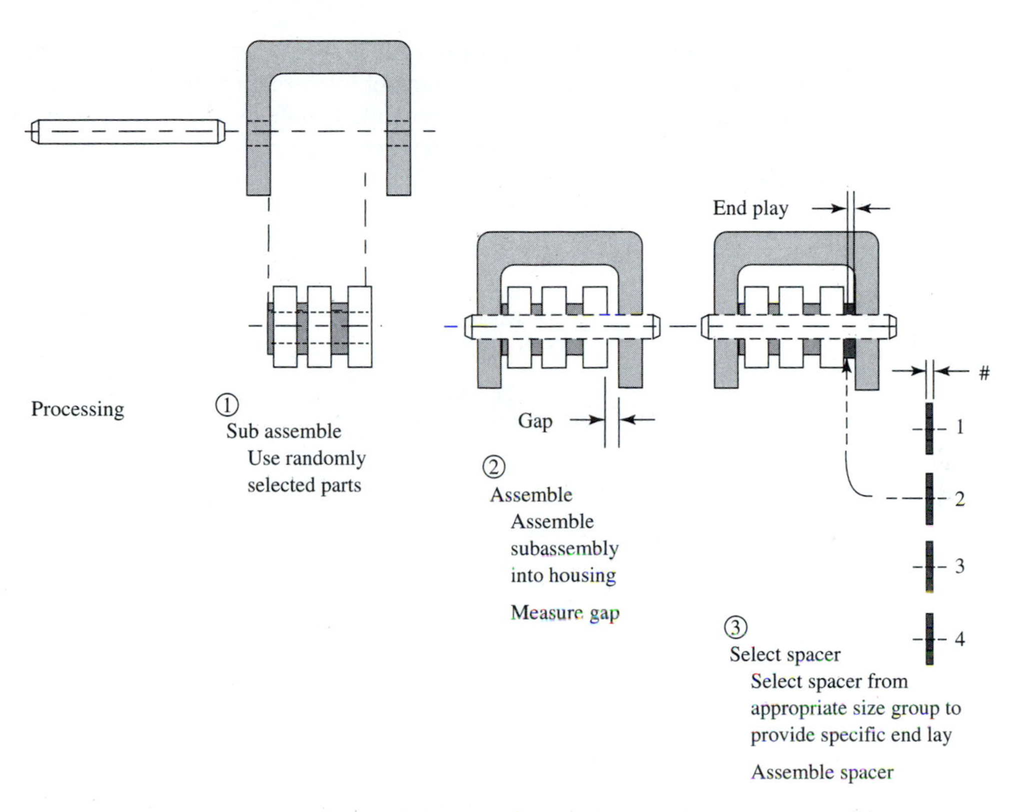

FIGURE 7.18 Provide specified end play by selecting Spacer of appropriate width.

7.6 COMPUTER-AIDED TOLERANCE CHARTING

The development of a tolerance chart requires a person with experience in processing, the ability to apply creative thinking, and time. Computers can aid in developing a chart in three very significant ways:

1. Reducing the time required for calculations
2. Determining optimum manufacturing tolerances
3. Developing tolerance charts for a family of similar parts

Reducing the Time Required for Calculations The time required to develop a tolerance chart depends on the complexity of the part. Some very complex parts require as many as 200 to 300 hours to chart, including the tentative process and strip layout. A computer can be used to significantly reduce the portion of time required for the dimensional and tolerance calculations. Computers can also save time by keeping track of resultants, stock removals, and tolerance accumulations.

Determining Optimum Manufacturing Tolerances A major advantage in utilizing a computer's aid is to determine optimum allowable manufacturing tolerances for each affected surface in a part's processing. This is done by calculating and checking many combinations of manufacturing tolerances, that is, trying tolerances that range from the minimum to the maximum for each surface and the effect on all other surfaces. The final combination must, of course, be the one that provides the greatest manufacturing tolerance for each surface while still producing a part that satisfies the part print.

Too often under the constraints of time, the first set of feasible manufacturing dimensions and tolerances developed are adopted and put into production. The computer's speed can be used to correct this by providing many incrementally differing iterations of the originally developed chart. In so doing, manufacturing tolerances are fine-tuned to the maximum allowable. Very often major increases in producibility result.

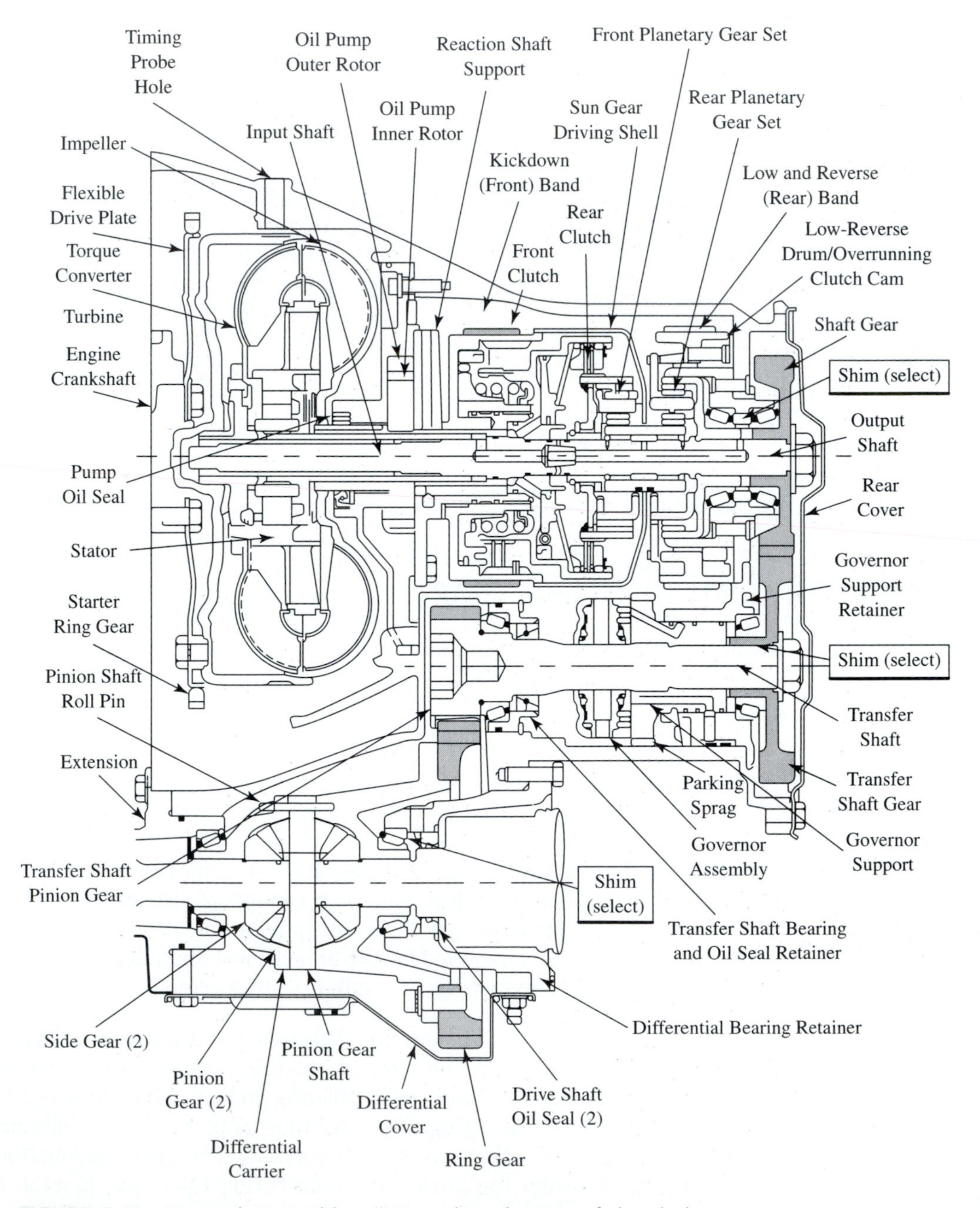

FIGURE 7.19 Transaxle assembly utilizing selected spacers for end play.

Develop Tolerance Charts for a Family of Similar Parts Although a computer cannot be utilized to aid in developing a tolerance chart until the part is processed, a computer can be programmed to fully develop tolerance charts of similar parts based on the first part charted. The family of parts must have similar processing.

Many attempts have been made to develop a tolerance chart solely by computer. When top management of one large automobile division realized the value of tolerance charting, they issued a directive to the effect that all their current and future designs of machined parts were to be tolerance charted. To accelerate the task, the parts were to be charted by computer.

Manufacturing Engineering attempted to comply but failed; only a few engineers had experience in tolerance charting, and none had hands-on experience with computers. The company brought in a contract shop to provide the computer service, but this too failed due to the computer programmers' lack of manufacturing experience and ability to process. The manufacturing engineers could not program the computer; computer programmers could not process.

The company then, logically, set up teams of manufacturing engineers and computer operators to work together in developing the charts. Also, the company initiated training programs for their manufacturing engineers in both tolerance charting and computers. All new hires of manufacturing engineers must now have some knowledge of both.

It is beyond the scope of this book to attempt instructions in programming a computer to aid in developing tolerance charts. However, the function is so vital that anyone involved in tolerance chart development would do well to take an appropriate study course in computers. Computer courses and their application are now required in many of our leading engineering schools.

7.7 REVIEW QUESTIONS AND PROBLEMS

1. Define a tolerance chart.
2. What are its main functions?
3. What steps are required to create a tolerance chart?
4. Define the following terms:

Locating surface	Stock removal
Working dimension	"From the solid"
Resultant dimension	Machining dimension

5. What are the rules for adding and subtracting tolerances?
6. Convert the following dimensions and tolerances to dimensions with equal bilateral tolerances:

$$4.500^{+.005}_{-.010} \qquad 1.250^{+.010}_{-.000}$$

$$3.250^{+.008}_{-.004} \qquad 2.375^{+.006}_{-.003}$$

$$\frac{2.012}{1.996} \qquad \frac{4.250}{4.244}$$

7. Refer to the tolerance chart in Figure 7.8.
   - Substitute machining dimension tolerances of ±.002 in lines 1, 2, and 6.
   - Determine the actual or resultant tolerance for lines 4, 8, 20 and 21.
   - Does this satisfy the part design dimension?
   - If not, is the tolerance "too tight" or "too loose?"
   - What action should be taken?
   - Calculate stock removed for line 6.
8. Develop a tolerance chart for Figure 7.20.

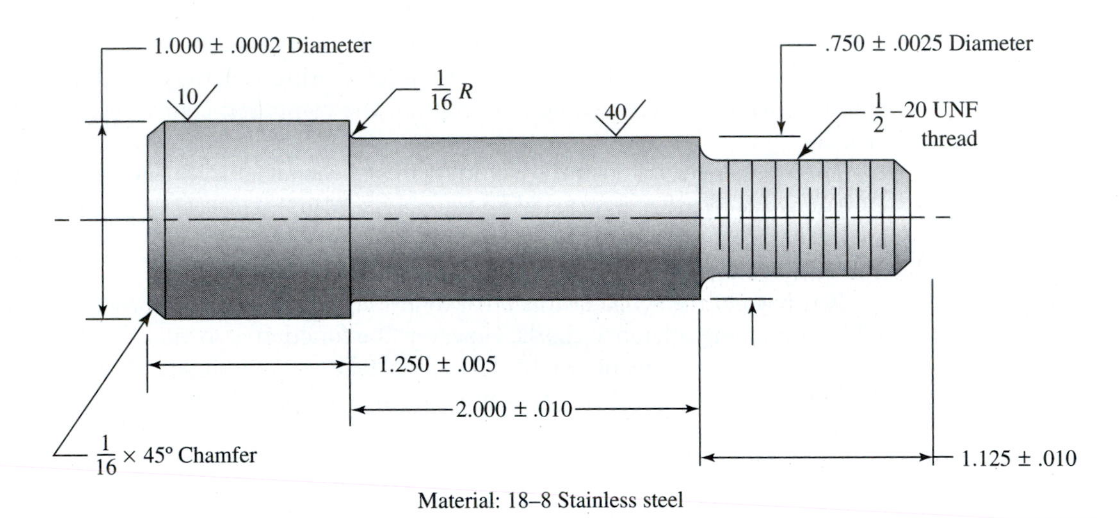

FIGURE 7.20 Pump Idler Shaft.

9. Develop a tolerance chart for Figure 7.21.

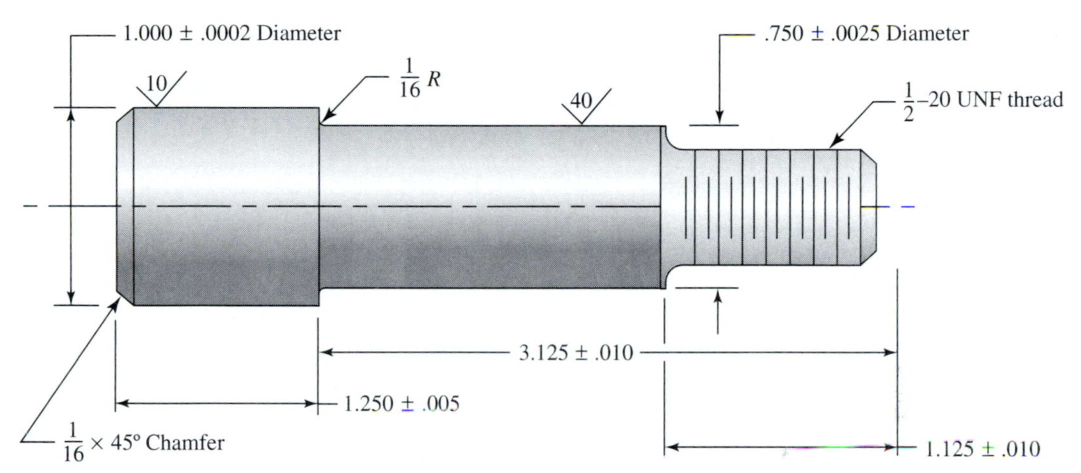

FIGURE 7.21 Pump Idler Shaft.

10. Compare and discuss the chart developed for Figure 7.20 with the chart developed for Figure 7.21.

CHAPTER EIGHT

WORKPIECE HOLDING

CHAPTER OVERVIEW

- Introduction
- Terms and Definitions
- Theory
- Workpiece Holder Design
- Practice
- Review Questions and Problems

8.1 INTRODUCTION

This chapter describes how unstable workpieces—parts free to move in any direction—are brought into a condition of stability; theory is augmented with line drawings illustrating the control of rectangular and cylindrically shaped parts. Forces acting on a workpiece during a machining process are also described, together with the methods used to control them.

The workpiece holder design must provide for locating a workpiece on the locating surfaces determined by processing and tolerance charting, and must maintain constancy of part position regardless of variables of tool and clamping pressures, dirt, chips, or workpiece variations. A workpiece holder must be maintained at design standard levels through regularly scheduled maintenance.

8.2 TERMS AND DEFINITIONS

Certain words and terms are associated with workholding and the design of workpiece holders. To help clarify intent and meanings, definitions follow:

1. **Arbors and Mandrels:** An arbor, or mandrel, is a spindle upon which a workpiece is mounted for purposes of machining or grinding. It may run between centers and also may be chucked. The workpiece may be retained by friction resulting from expanding the arbor sleeve, or from a press or taper fit between the arbor and the workpiece. The workpiece may also be driven by splines, gear teeth, or keys.

2. **Chuck:** A mechanical device attached to the spindle of a machine to hold a workpiece by means of pressure from jaws or jaw inserts. It may be operated by air, hydraulically, electrically, or manually.

3. **Centerline Control:** Maintaining centerline relationship to specifications by means of an arbor collet, chuck, or V-block, or other devices.

4. **Clamp:** A mechanical detail of a workholder that contacts a workpiece to press it against locaters. It can be actuated manually or with power.

5. **Clamping Force, or Holding Force:** The amount of pressure or force developed by a clamp.

6. **Collet:** A type of chuck, used for bar stock, round, and tubular material, that is operated by the closing action of a tapered ring.

7. **Deflection:** Bending or displacement of a tool, spindle, or workpiece from its true position. It can be caused by clamping force or tool pressure.

8. **Equilibrium:** The state of balance between opposing forces.

9. **Fixtures:** Workpiece holders that locate and clamp the part in position but do not have features to guide the cutting tool. They may, however, have set up blocks or other means of aligning with the machine and with the cutting tools.

10. **Foolproofing:** A workholder designed with a pin, block, or other device, such that a workpiece cannot be loaded incorrectly, is said to be foolproofed.

11. **Jigs:** Workpiece holders that locate and clamp the part in position and have features to guide the cutting tool through a specific path, thus controlling the relationship between workpiece locating surface and machined surface or hole.

12. **Locater:** The surface of the workholder upon which the workpiece rests. It may be a pin or a surface. Symbols for locaters used in a workpiece holder diagram are shown in Figure 8.1.

13. **Positioned:** A workpiece properly loaded, chucked, or placed in a workholder is positioned.

14. **Resultant Clamping:** A single clamp, or clamping force, that is used to push a piece part against all three planes of location in place of three separate clamps.

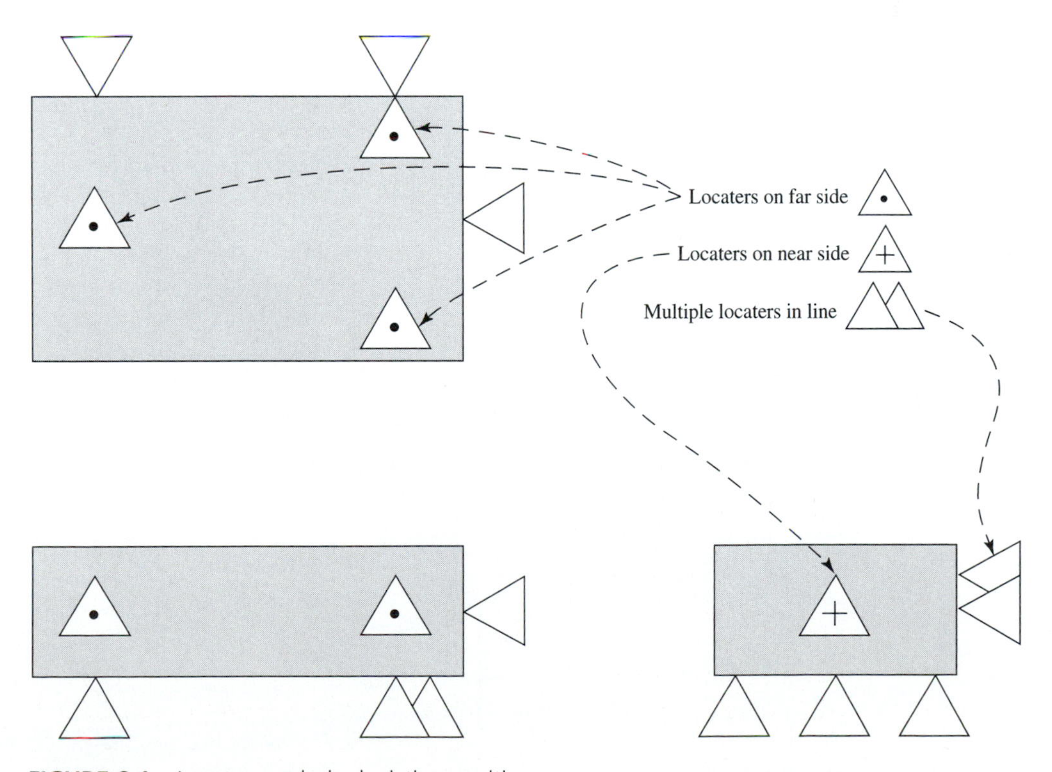

FIGURE 8.1 Locater symbols depicting position.

15. **Support:** A fixture component that resists deflection of an area of the workpiece. It may be fixed or adjustable. The adjustable support may be manually or automatically adjusted.

16. **Tool Force:** The amount of force applied to the workpiece by the tool or grinding wheel.

17. **Vise:** A workholding device consisting of two jaws and a means of bringing them together to grip the workpiece. This may be screw, cam, or air operated.

18. **Workpiece:** A partially completed part, or a part in process.

19. **Workpiece Control:** The degree of accuracy within which the workpiece is held to specifications, dimensions, and tolerances regardless of material, machine, or tool variances. Total workpiece control is composed of dimensional, geometric, and mechanical controls.

20. **Workholder:** The term "workholder" is used interchangeably with the terms "workpiece holder" or "fixture."

8.3 THEORY

Concept of Movement of a Free Body If a workpiece were in space, it would be free to move either way along its three axes and could revolve either way about each axis for a total of 12 separate movements, as shown in Figure 8.2. The proper placement of locaters will establish equilibrium and part location in all axes, as will be shown.

Concept of Equilibrium The object, or workpiece, illustrated in Figure 8.3(a) will move in the direction of the force applied if not arrested. A locater positioned opposite the force will arrest motion and bring the part into equilibrium in one plane or along one axis.

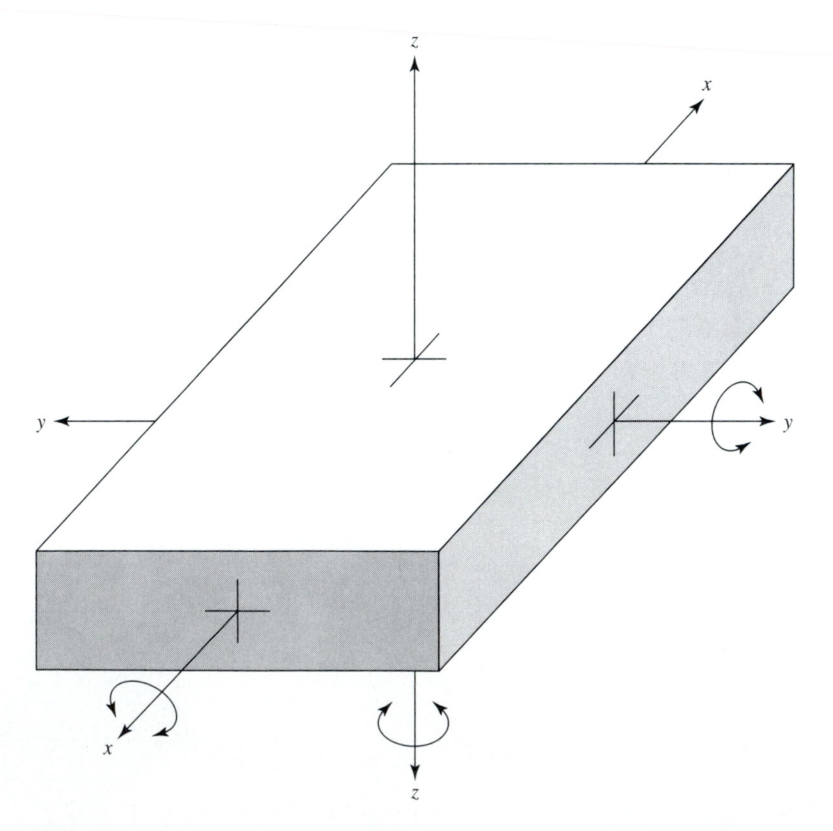

FIGURE 8.2 Rectangular workpiece free to move—(12) movements possible.

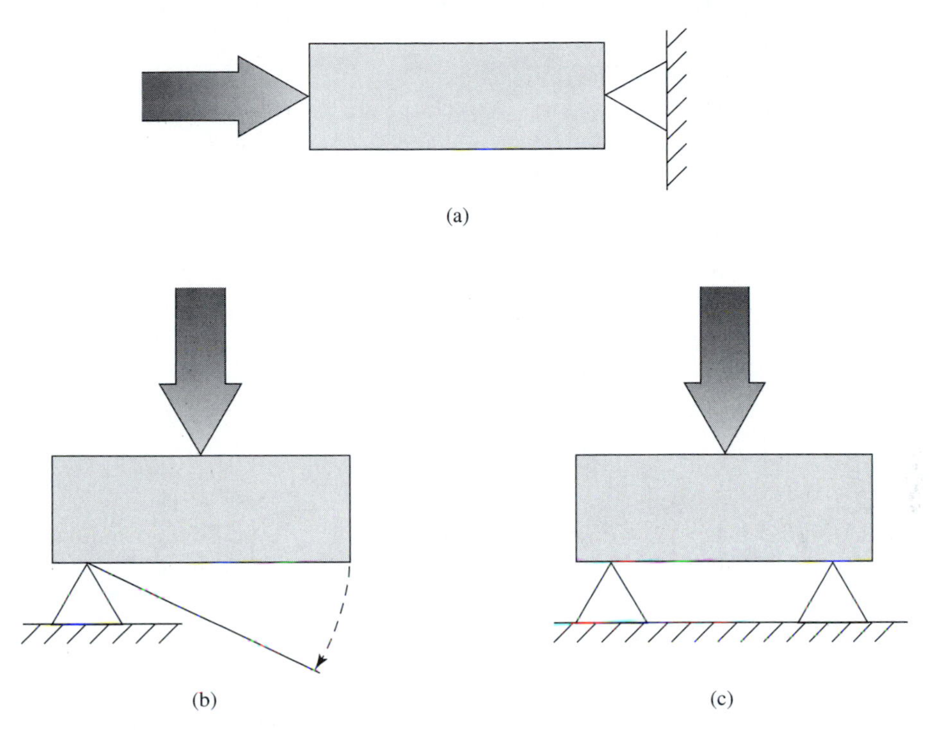

FIGURE 8.3 (a) Locater resisting directly applied force. (b) Workpiece revolving from force applied off-center. (c) Rotation stopped by application of second locater.

The object, or workpiece, illustrated in Figure 8.3(b) will rotate about a single locater in response to a force applied off-center. Positioning a second locater, as shown in Figure 8.3(c), will stop rotation and bring the part into equilibrium in one plane or axis.

Rectangular Workpiece Equilibrium Equilibrium and workpiece control are established for a rectangular part when it is restrained in its three axes as shown in Figures 8.4 through 8.7. The workpiece can be restricted in one plane by placing it on three locaters as shown in Figure 8.4. Three locaters are used because three will always make contact, even on a rough surface. Four, or more, locaters can cause the part to "rock." However, in practice, finished surfaces may have more than three locaters; the locater, indeed, will be a full surface when used on a flat machine locating surface.

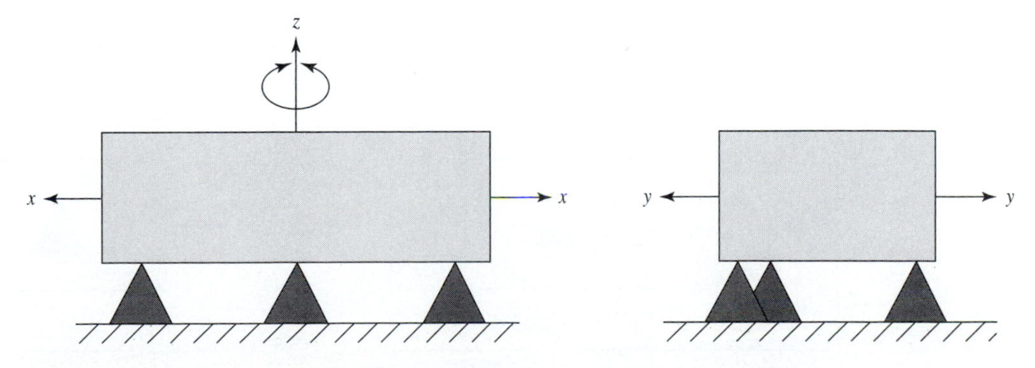

FIGURE 8.4 Rectangular workpiece positioned on (3) locaters. Workplace restricted in (1) plane; (5) Movements restricted; (7) Movements possible.

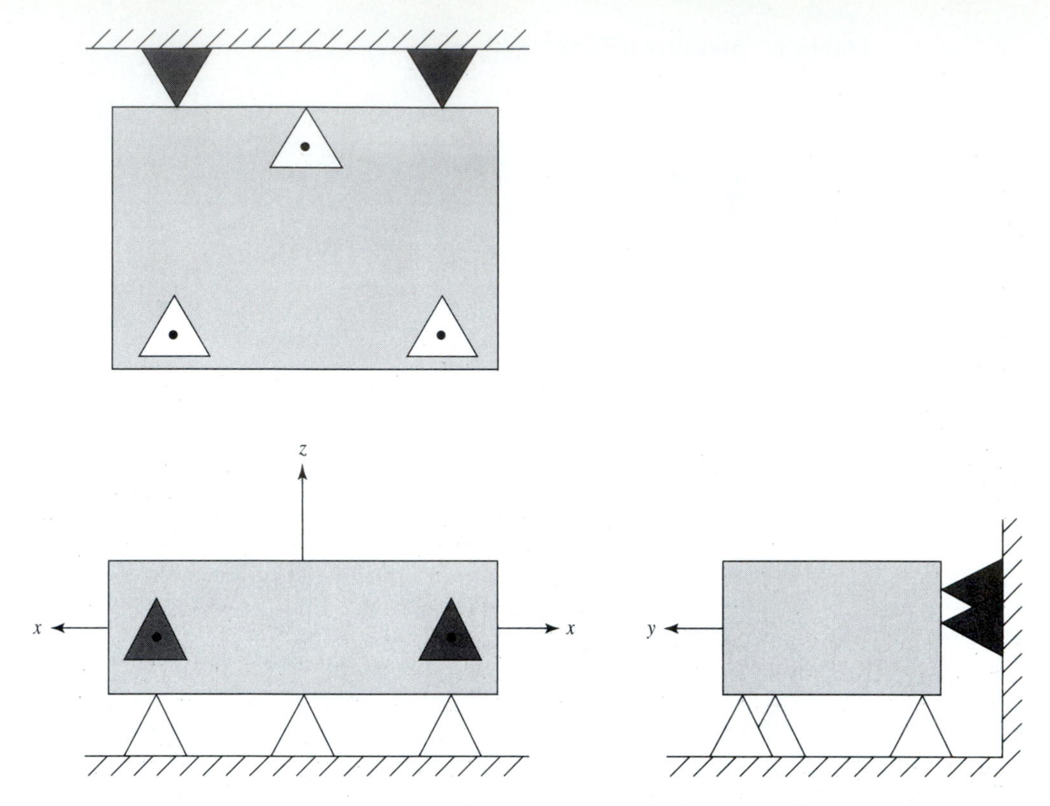

FIGURE 8.5 Rectangular workpiece positioned on (2) additional locaters. Workplace restricted in (2) planes; (8) Movements restricted; (4) Movements possible.

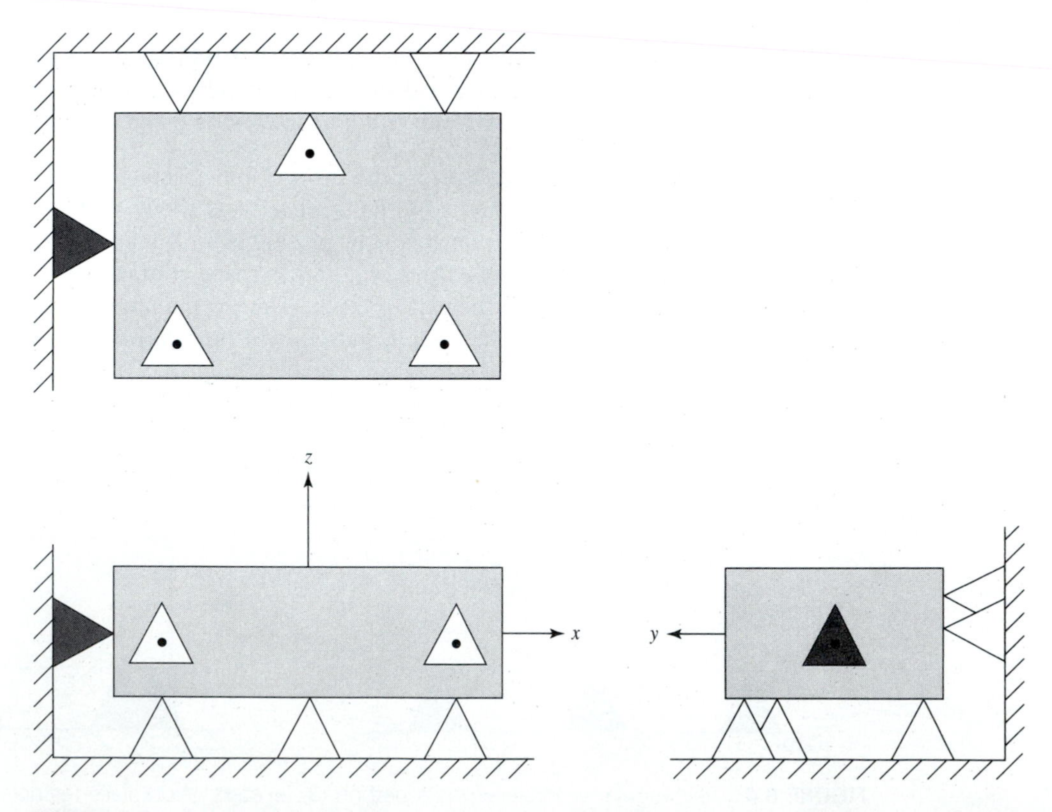

FIGURE 8.6 Rectangular workpiece positioned on (1) additional locater. Workplace restricted in (3) planes; (9) Movements restricted; (3) Movements possible.

The workpiece can be restricted additionally in a second plane with two additional locaters as shown in Figure 8.5. Three locaters are usually not used here because, in practice, the two workpiece surfaces can be out of square so that only two of the additional locaters would make contact.

The workpiece can be restricted additionally in the third plane with one additional locater as shown in Figure 8.6. More than one additional locater would be redundant because of possible out-of-square conditions in two planes.

Three, Two, One Location System A workpiece positioned on six locaters as shown in Figure 8.6 utilizes what is known as the 3–2–1 locating system. The remaining three movements can be restrained, or brought into equilibrium with a clamp or clamps, as shown in Figure 8.7.

Cylindrical Workpiece Equilibrium If free in space, a cylindrical part could move in 12 different ways or motions as illustrated in Figure 8.8. Equilibrium and workpiece control are established for a cylindrical part when it is restrained in its three axes as shown in Figures 8.9 through 8.12.

The workpiece can be restricted in one plane by placing it on two locaters as shown in Figure 8.9. The workpiece can also be restricted additionally in a second plane with two additional locaters as shown in Figure 8.10. The workpiece can be restricted additionally in the third plane with one additional locater as shown in Figure 8.11.

A workpiece positioned on five locaters as shown can have three remaining motions restrained and brought into equilibrium with a clamp or clamps, as shown in Figure 8.12. Note that cylindrical parts require one less locater than rectangular parts to obtain complete equilibrium.

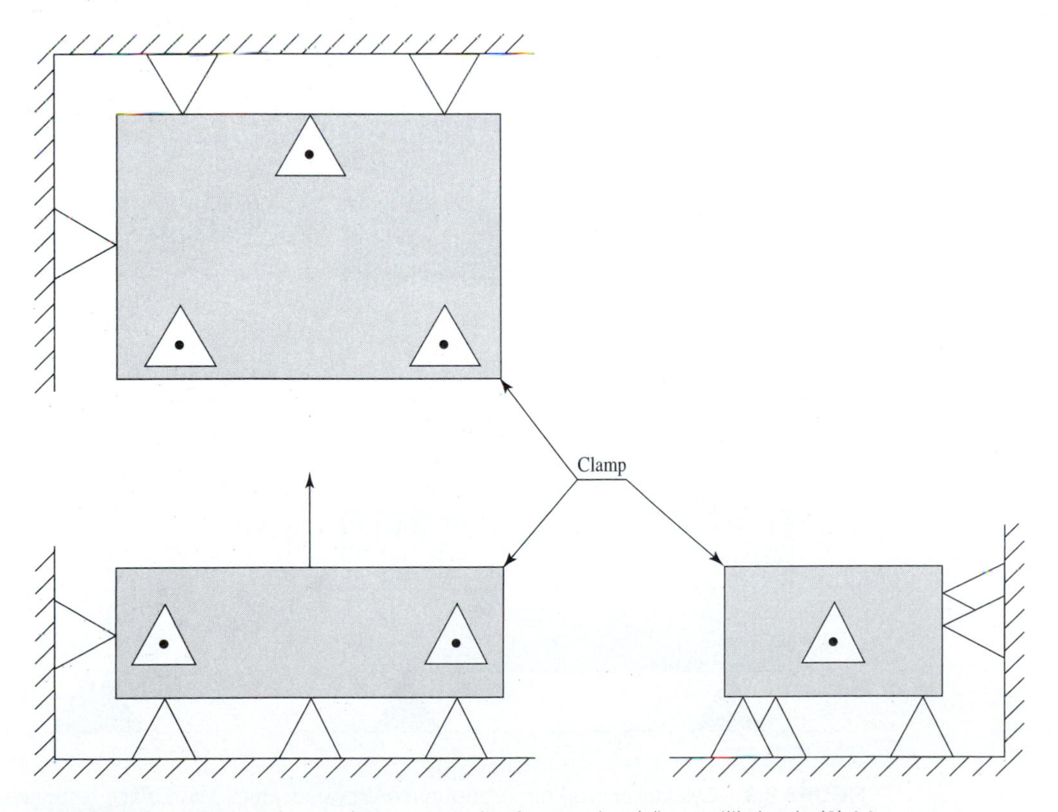

FIGURE 8.7 Rectangular workpiece completely restrained (in equilibrium). (9) Movements restricted with locators; (3) Movements restricted with clamps.

FIGURE 8.8 Cylindrical workpiece free to move. (12) Movements possible.

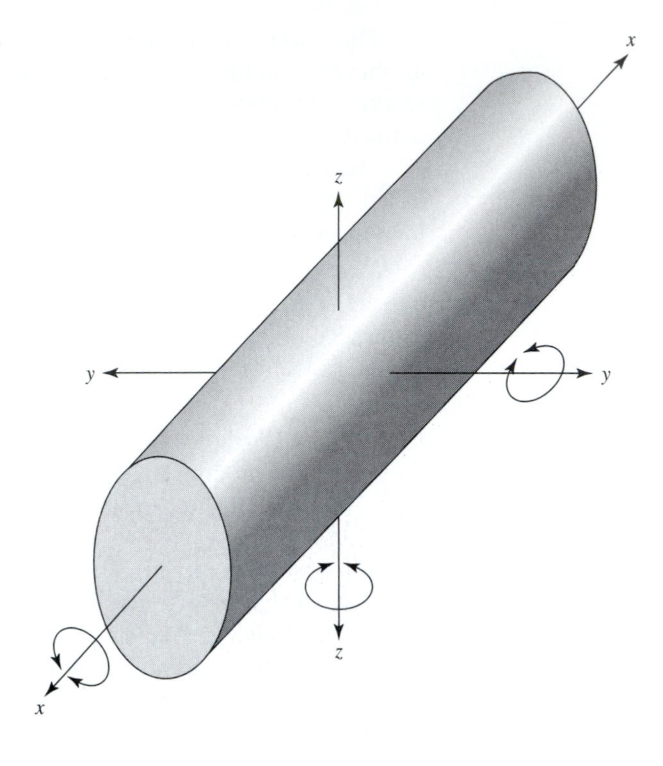

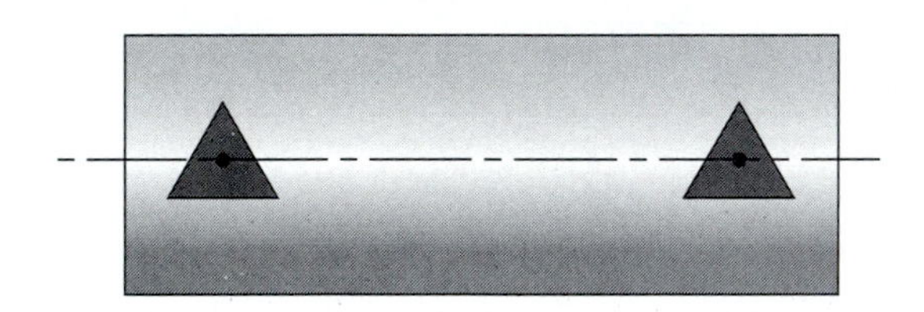

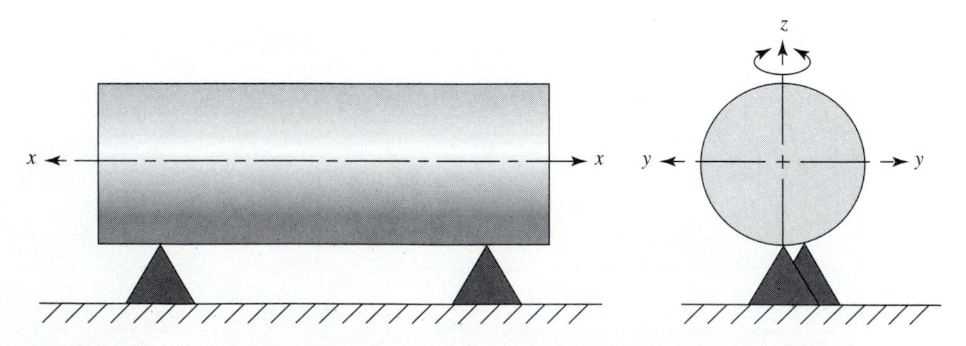

FIGURE 8.9 Cylindrical workpiece positioned on (2) locaters. Workplace restricted in (1) plane; (5) Movements restricted; (7) Movements possible.

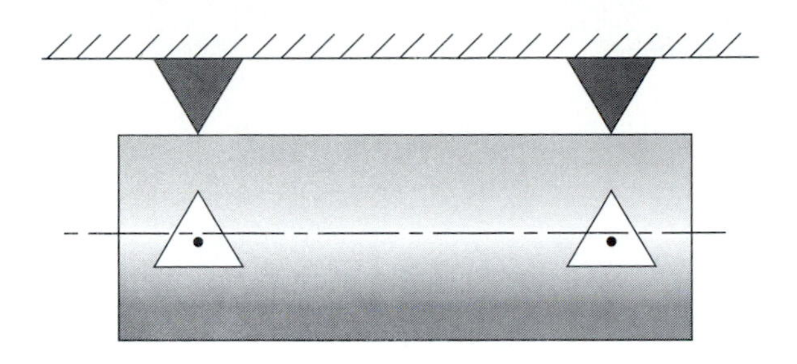

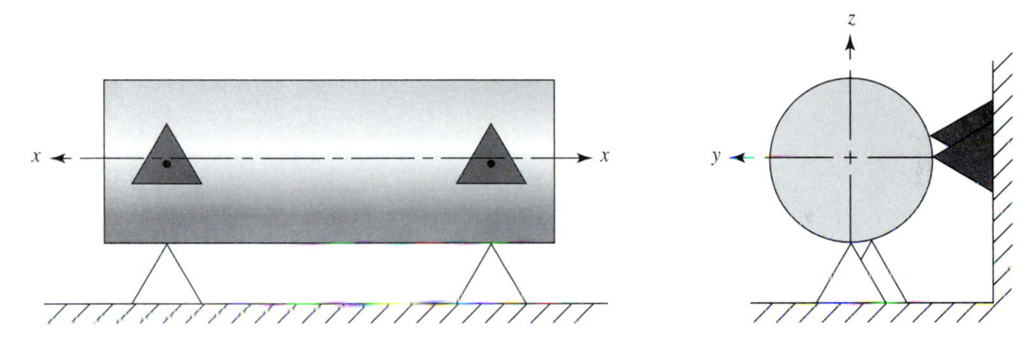

FIGURE 8.10 Cylindrical workpiece positioned on (2) additional locaters. Workplace restricted in (2) planes; (8) Movements restricted; (4) Movements possible.

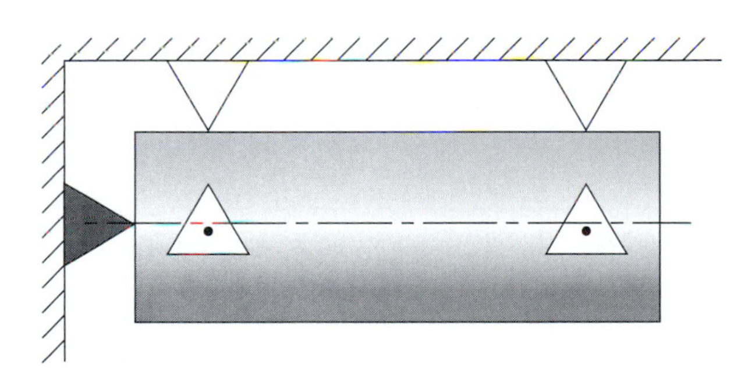

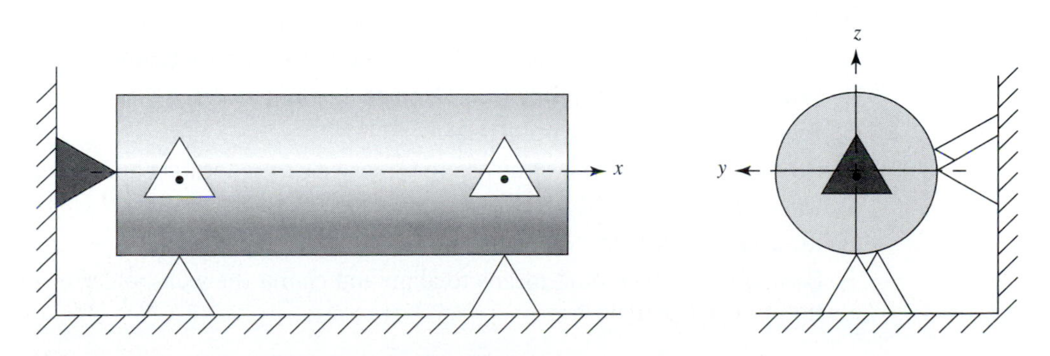

FIGURE 8.11 Cylindrical workpiece positioned on (1) additional locater. Workplace restricted in (3) planes; (9) Movements restricted; (3) Movements possible.

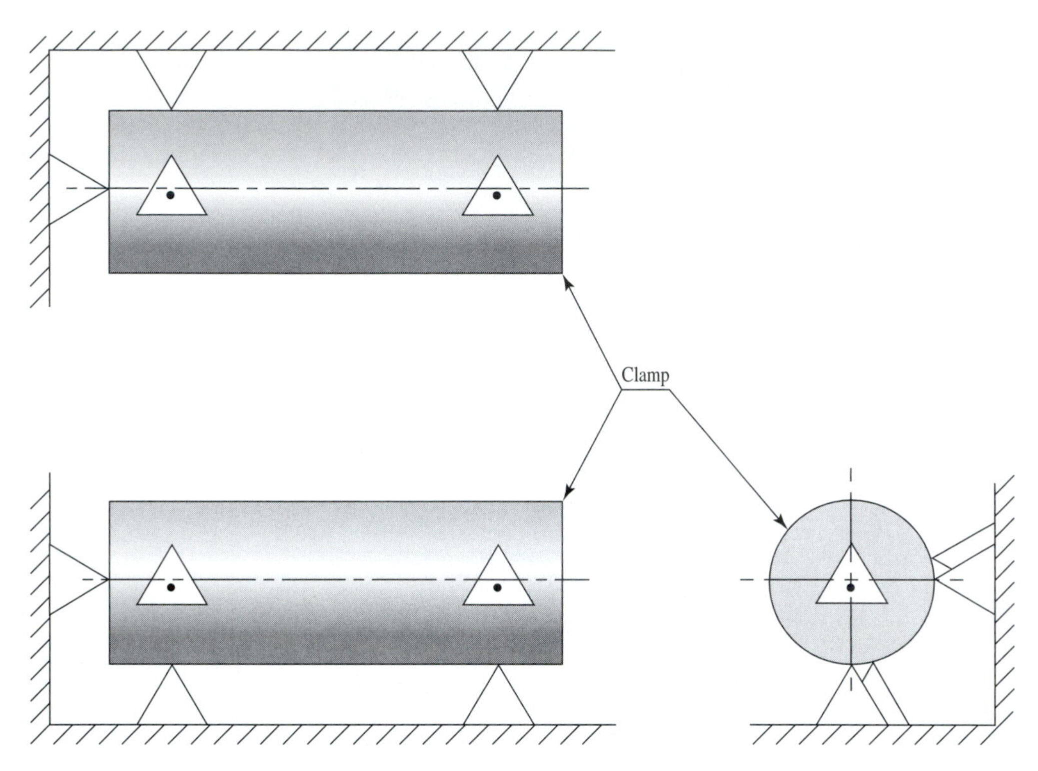

FIGURE 8.12 Cylindrical workpiece completely restricted (in equilibrium). (9) Movements restricted with locators; (3) Movements restricted with clamps.

8.4 WORKPIECE HOLDER DESIGN

The successful design of a workpiece holder employs both practical and theoretical considerations.

Practical Considerations Practical considerations are empirical, resulting from years of manufacturing use and experience. Workholders for some specialized processes reflect specialized development, but most designs and applications require and include the following:

1. Ease of loading and unloading the workpiece
2. Avoiding locating on parting lines, flash, or other irregularities of the workpiece
3. Ease of clamping (if manual) and accessibility to clamp actuators
4. Protection of the workpiece from damage during process
5. Provisions to replace worn details
6. Dirt and chip exclusion, or means to remove clearing locaters and clamps for next operation
7. Safety guards
8. Rugged construction
9. Ease of setup, including means to align and clamp the workholder in the machine in relation to cutting tools

Theoretical Considerations Theoretical concepts are perhaps best covered with a discussion of what have come to be known as dimensional, geometrical, and mechanical controls.

1. ***Dimensional Control***: Dimensional control governs workpiece holding design features that affect producing parts to specified dimensions and tolerances.
2. ***Geometrical Control***: Geometrical control governs workpiece holding design features that affect workpiece positioning and stability when located in the workpiece holder.
3. ***Mechanical Control***: Mechanical control governs workpiece holding design features that retain, resist clamping and tool forces on, and support the workpiece.

Dimensional Control. Dimensional control is best effected by locating the workpiece on the locating surface from which the part's dimensions are based.

As illustrated in Figures 8.13(a) and 8.13(b), parts are located on the surfaces from which the part design's dimensions originate. Location for the milling operation, Figure 8.13(a), is on the outside flange surface. The locating surface selected for the turning operation, Figure 8.13(b), is the inner flange face. Both surfaces, in addition to originating the part print dimensions, are proportionately large in relation to part lengths. The parts will be stable when so located.

FIGURE 8.13

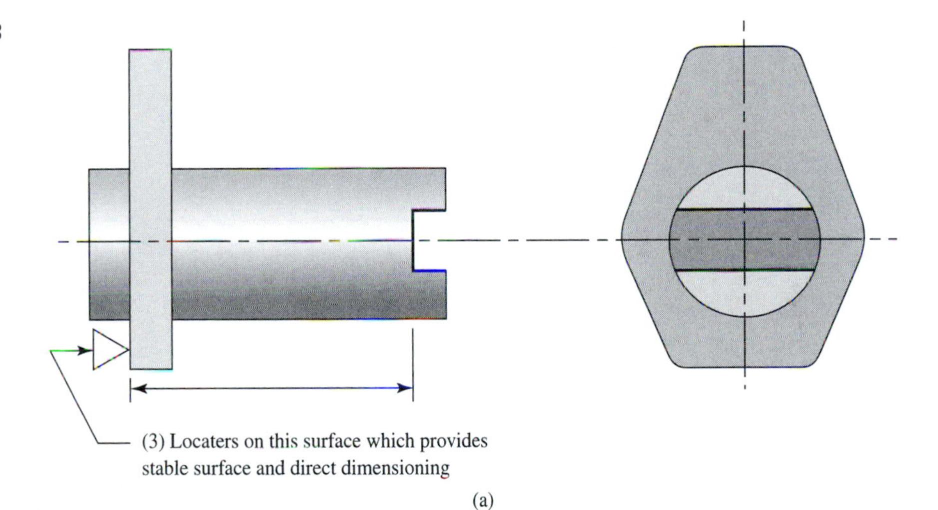

(a)

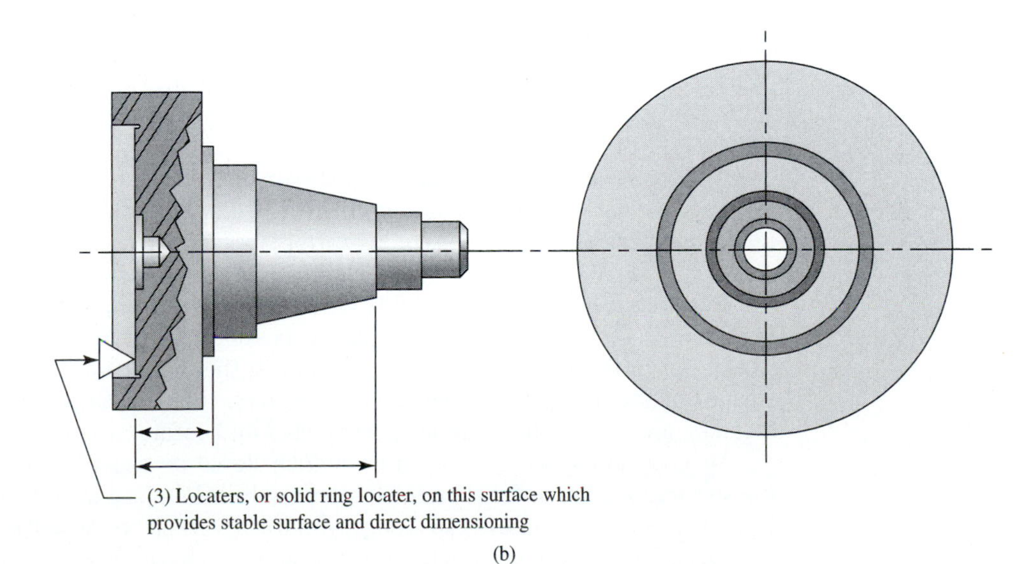

(b)

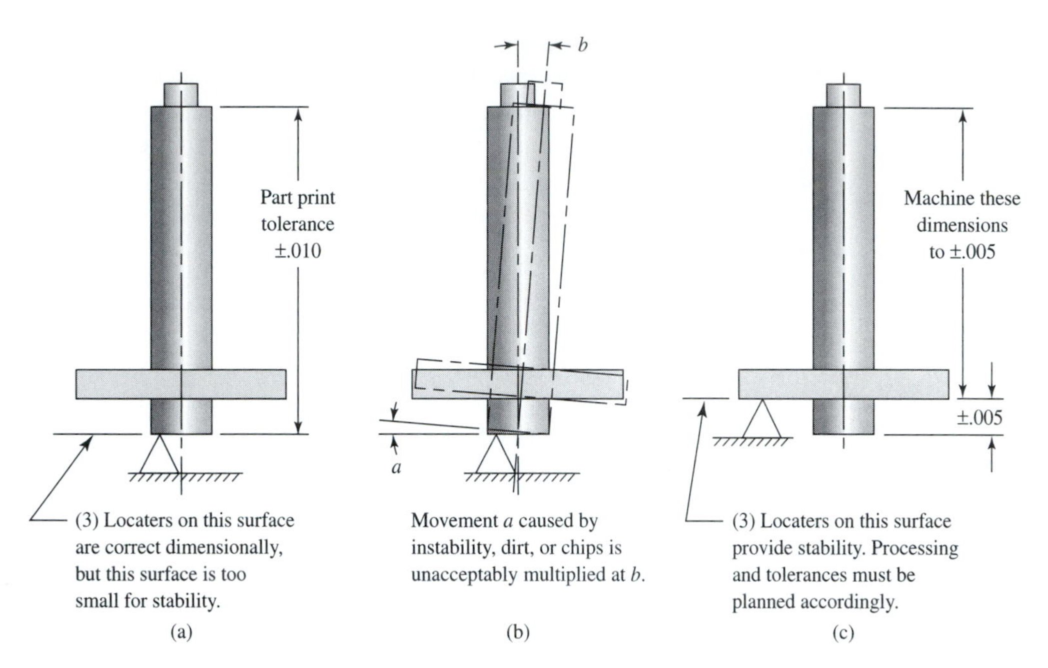

FIGURE 8.14

When it is not feasible to locate on the dimensional originating surface, for reasons of workpiece instability, unsuitable clamping areas, or inability to compensate for tool forces, the workpiece must be located on an alternate surface as shown in Figure 8.14(a), (b), and (c).

Requests for part design dimensional changes should be made, but, if not permissible, an alternate locating surface must be used with manufacturing dimensions and tolerances altered accordingly.

It is not feasible to locate the part shown in Figure 8.14(a) on the small end face that originates the part design dimension. This surface is too small proportionate to the part's length, so any movement of the part on the locating surface (caused by chips, dirt, or unbalance) will compound the error unacceptably as shown in Figure 8.14(b). An alternate locating surface must be selected—in this case, the lower flange face. Use of this location will require the lower end face to be subsequently machined in a secondary operation. Machining dimensions and operations must be established for both surfaces that will satisfy the part design dimension as shown in Figure 8.14(c).

It becomes apparent that the interrelationship of all parameters—diametrical, geometrical, and mechanical—must be considered when planning and designing a workpiece holder. The ultimate criterion is that dimensional control takes precedence. The part design must be satisfied.

Geometrical Control. Stability is best accomplished by locating a workpiece on its largest surface with locaters spaced as far apart as possible. Additional locaters applied to position the workpiece in the remaining planes accomplish the 3–2–1 location system, where the largest surface of the workpiece is placed on 3 locaters, the next largest surface is contacted with 2 locaters, and the smallest surface is contacted by 1 locater as shown in Figure 8.15.

Also, the amount of positional or angular deviation caused by locater wear, a rough locating surface, or chips and dirt on the locater can be minimized with the use of more widely spaced locaters as shown in Figure 8.16. Long, cylindrical workpieces, as well as rectangular ones, should be placed horizontally on widely spaced locaters as shown in Figure 8.17.

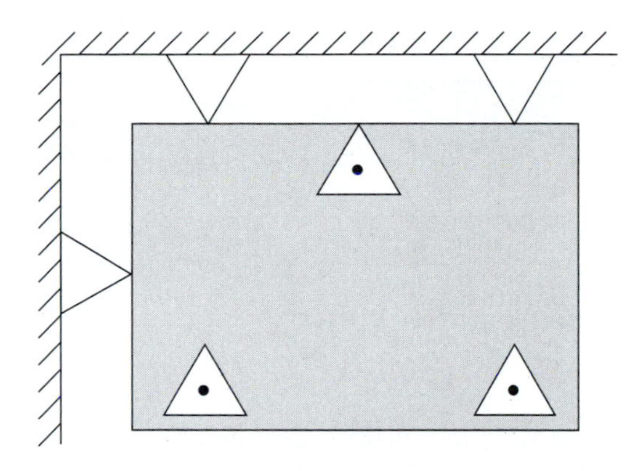

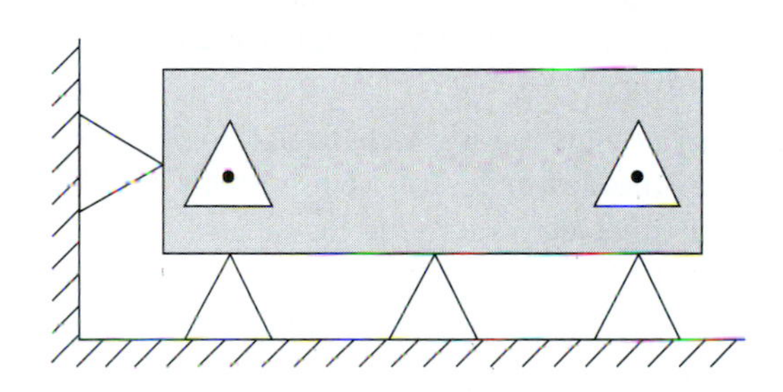

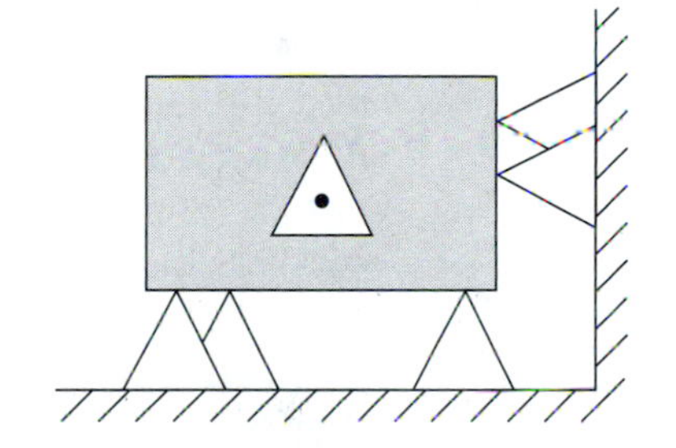

FIGURE 8.15 3-2-1 Locating system.

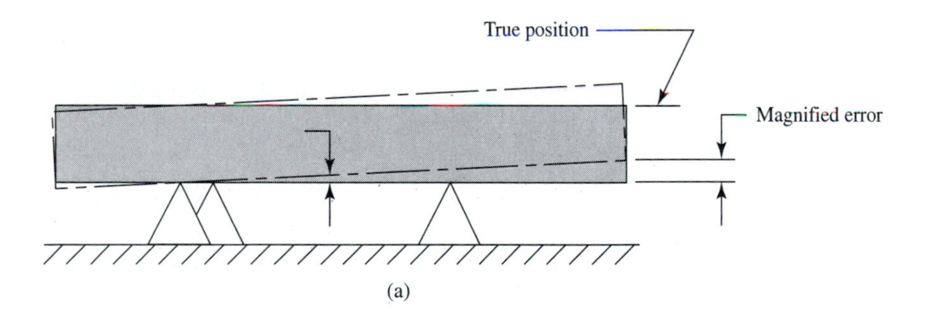

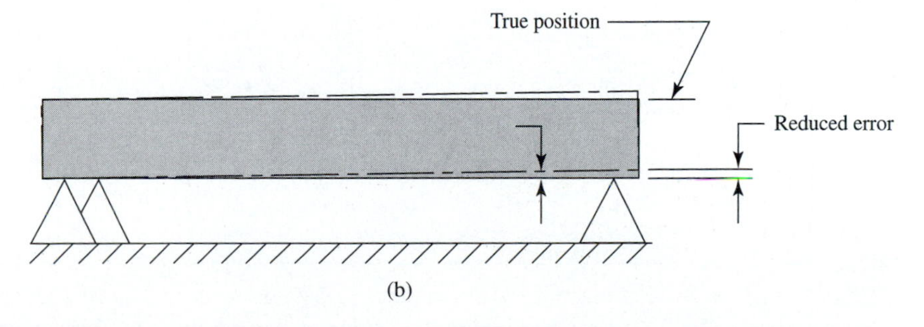

FIGURE 8.16 Incorrectly spaced locaters magnify errors caused by locater wear, dirt, and chips. (a) Errors magnified from too closely spaced locaters. (b) Errors reduced with correctly spaced locaters.

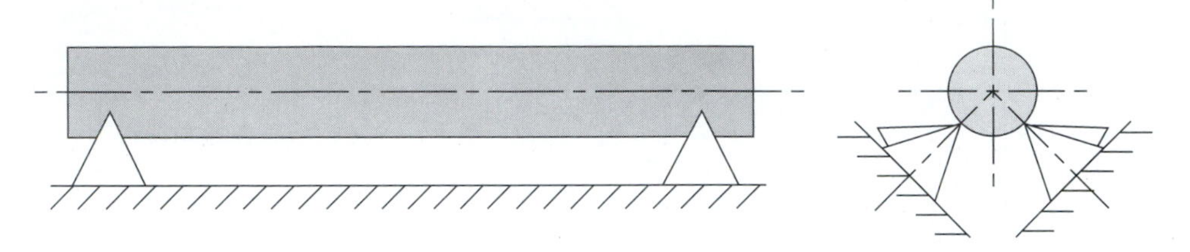

FIGURE 8.17 Long cylindrical workpiece correctly located.

The configuration of equipment used to process a workpiece affects workpiece position and locating surfaces; a machine with a vertical spindle may require the workpiece to be positioned to suit, as shown in Figures 8.18 and 8.19. In such instances, temporary retainers such as an approximate locater should be employed to hold the workpiece in position until clamps are applied to press and retain the part against its locaters. Relatively short cylindrical parts should be located horizontally on the end surface as shown in Figure 8.20.

Radial Control. Certain part designs require radial relationships to be held between surfaces, holes, keyways, splines, and other features. This requirement must be provided for in the workpiece holder. Figure 8.21 illustrates use of a radial locater.

Centerline Control. The element of geometric control known as centerline control is provided by locating a cylindrical part in relationship to its centerline such that dimensions and tolerances related to it can be held. Four conditions of centerline control must be considered:

1. When machined surfaces of cylindrical parts are dimensioned from the *outside diameter* and the part is located on the *outside diameter*, no loss of manufacturing tolerance occurs, as shown in Figure 8.22(a).
2. When machined surfaces of cylindrical parts are dimensioned from the *centerline* and the part must be located on the *outside diameter*, loss of some manufacturing tolerance occurs due to the diametral variance, as illustrated in Figure 8.22(b).

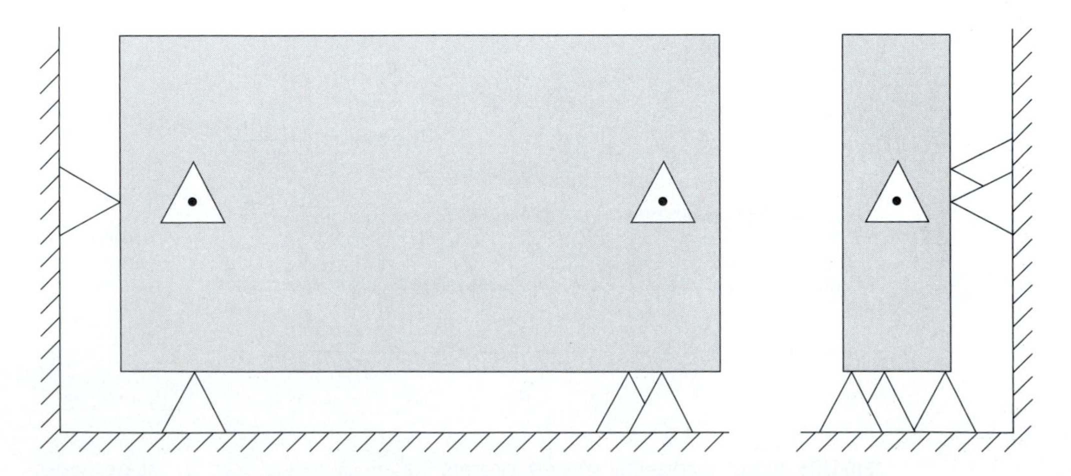

FIGURE 8.18 Narrow, rectangular workpiece located to satisfy a specific condition.

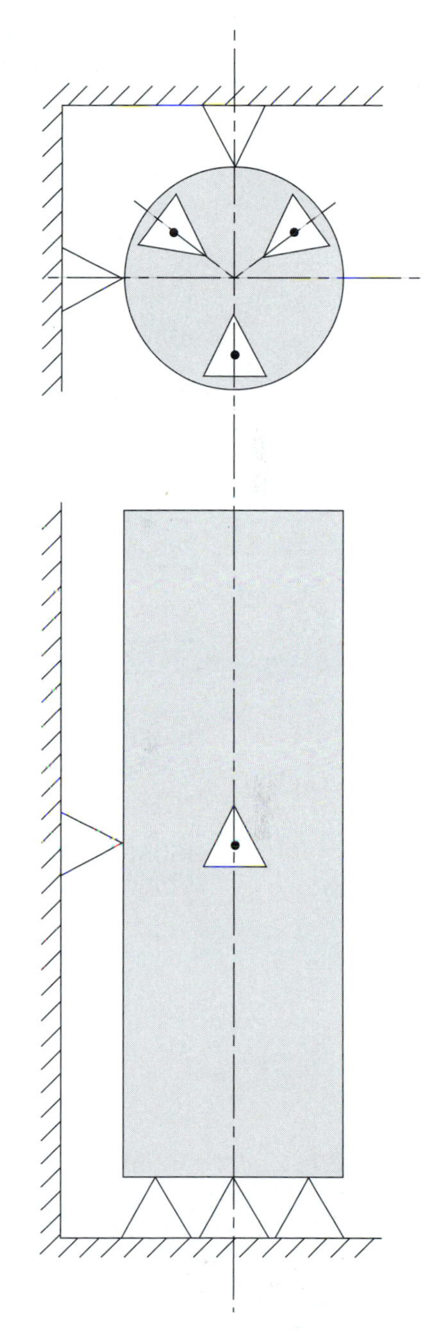

FIGURE 8.19 Long, cylindrical workpiece located to satisfy a special condition.

3. When machined surfaces of cylindrical parts are dimensioned from the *centerline* and the part is located in *V-blocks* on its *outside diameter*, some loss of manufacturing tolerances occurs—depending on the angle of contact with the locaters as shown in Figure 8.23.

4. When cylindrical part surfaces are dimensioned from the *centerlines* and the part can be *centralized* (located on its centerline) by means of equalizing devices such as centers, collets, chucks, or expanding arbors, no loss of manufacturing tolerance occurs.

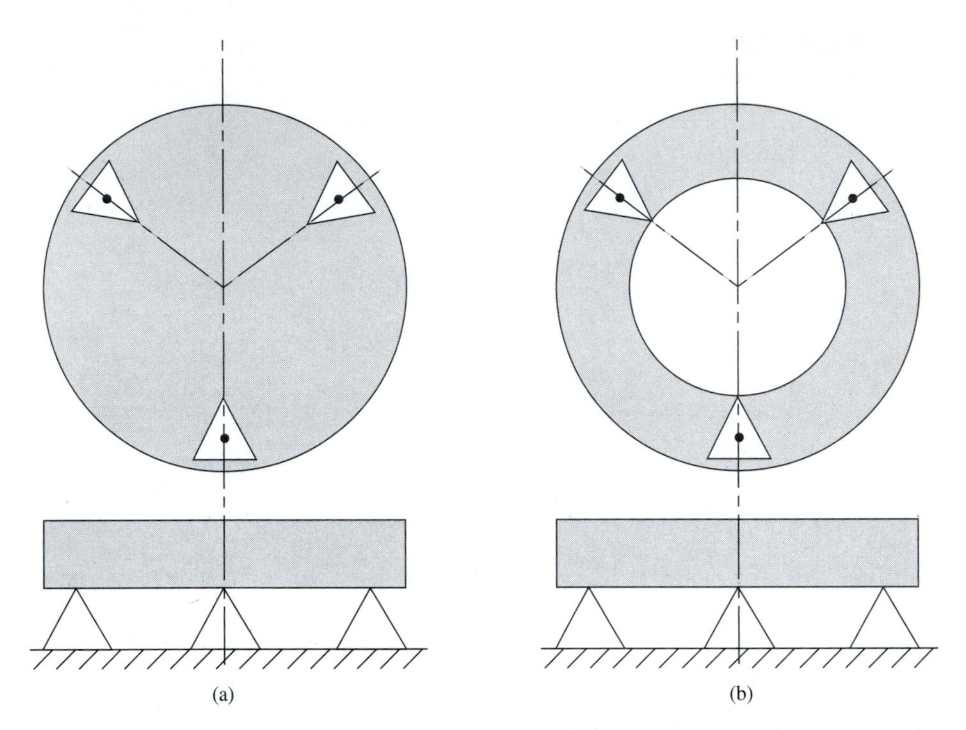

FIGURE 8.20 Short, cylindrically shaped workpieces are located on their end surfaces. (a) Short cylindrical workpiece. (b) Short tubular workpiece.

FIGURE 8.21 Radial control of angle about centerline.

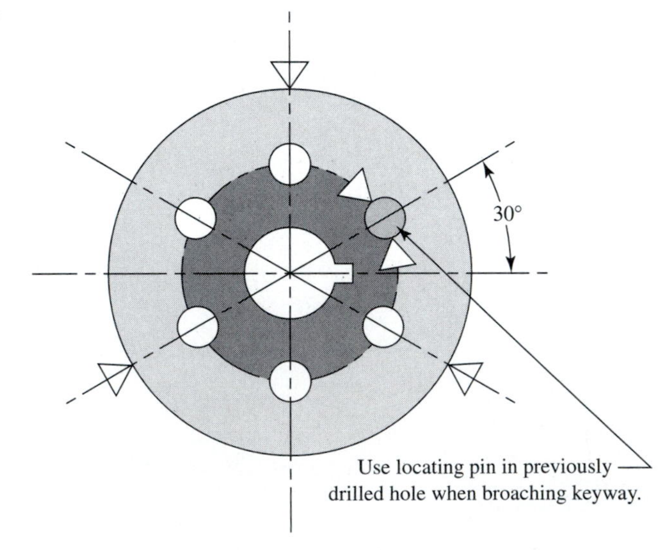

Figure 8.24 illustrates the correct placement of locaters for a cylindrical part, without centers, to maintain full manufacturing tolerance for a dimension from the part's centerline to a surface. The locaters, as shown, eliminate any vertical movement of the part caused by outside diameter variations and transfer it to horizontal movement. Horizontal movement does not affect the manufacturing tolerance of the dimension.

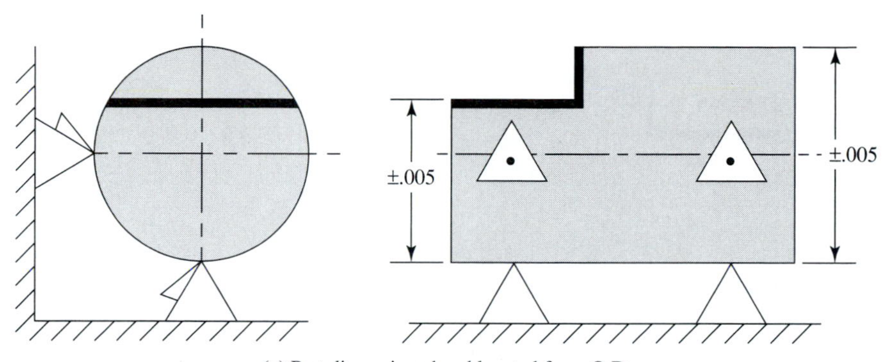

(a) Part dimensioned and located from O.D.
No tolerance stack
No loss of manufacturing tolerance

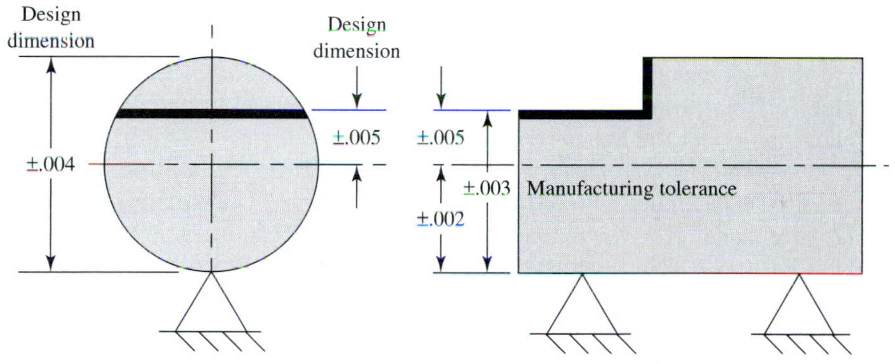

(b) Part dimensioned from center line but located on O.D.
Manufacturing tolerance reduced due to O.D. variation.

FIGURE 8.22 Methods of locating cylindrically shaped parts to reduce or eliminate tolerance stacking.

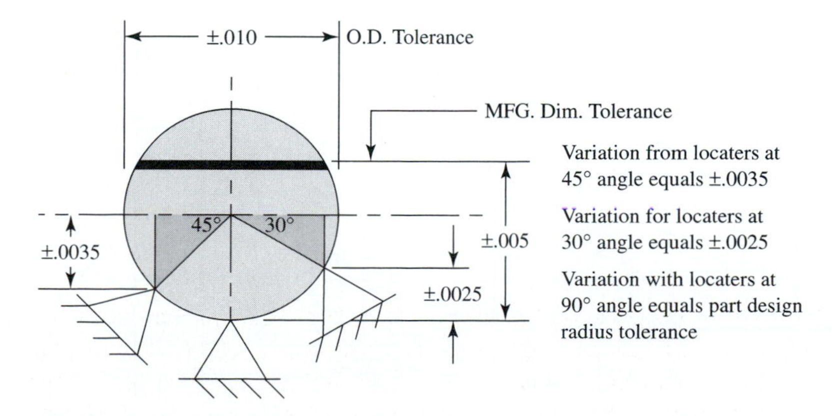

FIGURE 8.23 Variation (centerline shift) varies with angle of locater contact.

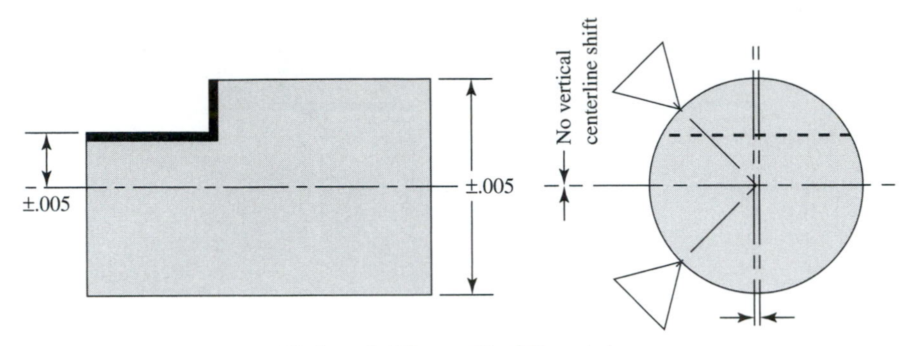

FIGURE 8.24 Method of controlling centerline shift.

Figure 8.25 illustrates the correct placement of locaters for a cylindrical part, without centers, to drill a cross hole centrally with no lateral shift caused by outside diameter variation. The locaters, as shown, eliminate horizontal movement and transfer it to vertical movement, which does not effect the hole's central position.

Workpieces with bores, or inside diameters large enough to accommodate, may be located as shown in Figure 8.26. When so located, however, the part will be off centerline in response to any variation of the diameter. A single movable clamp cannot compensate for I.D. variation.

Bar stock, tubing material, and individual cylindrical parts may be established centrally on their centerlines with self-centering chucks and collets that locate an outside diameter as shown in Figure 8.27(a) and (b).

Workpieces with suitable bores or inside diameters may be located with expanding arbors as shown in Figure 8.28(a). Workpieces are controlled centrally on arbors with no shift or movement due to part diameter variation. Self-centering chucks may also be used as shown in Figure 8.28(b).

Geometric Specifications. A part's design may specify squareness, run out, parallelism, and/or other requirements in addition to dimensioning and tolerancing. The workpiece holder must provide for all such requirements.

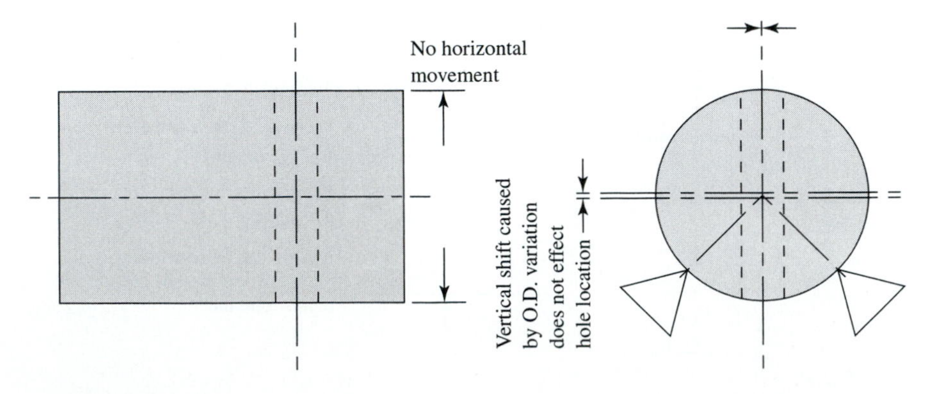

FIGURE 8.25 Part correctly located on O.D. to control location of drilled hole on centerline.

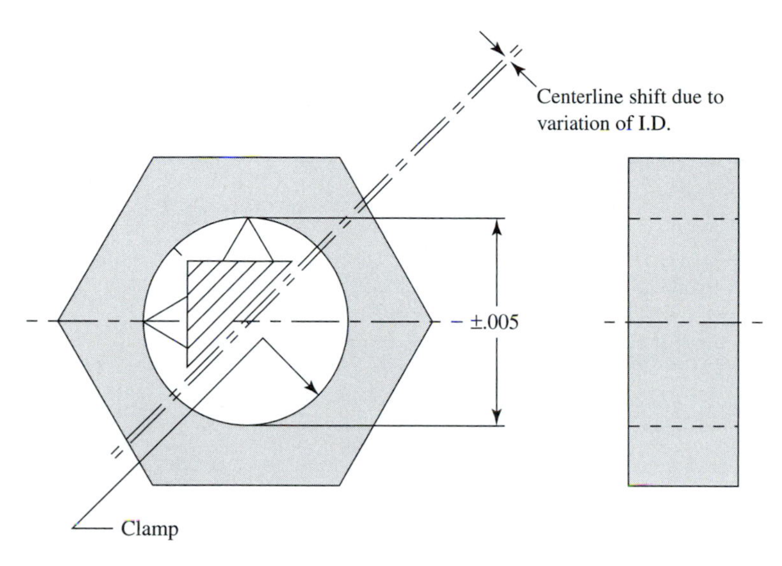

FIGURE 8.26 Part location suitable for facing cut or other operation not requiring centerline control.

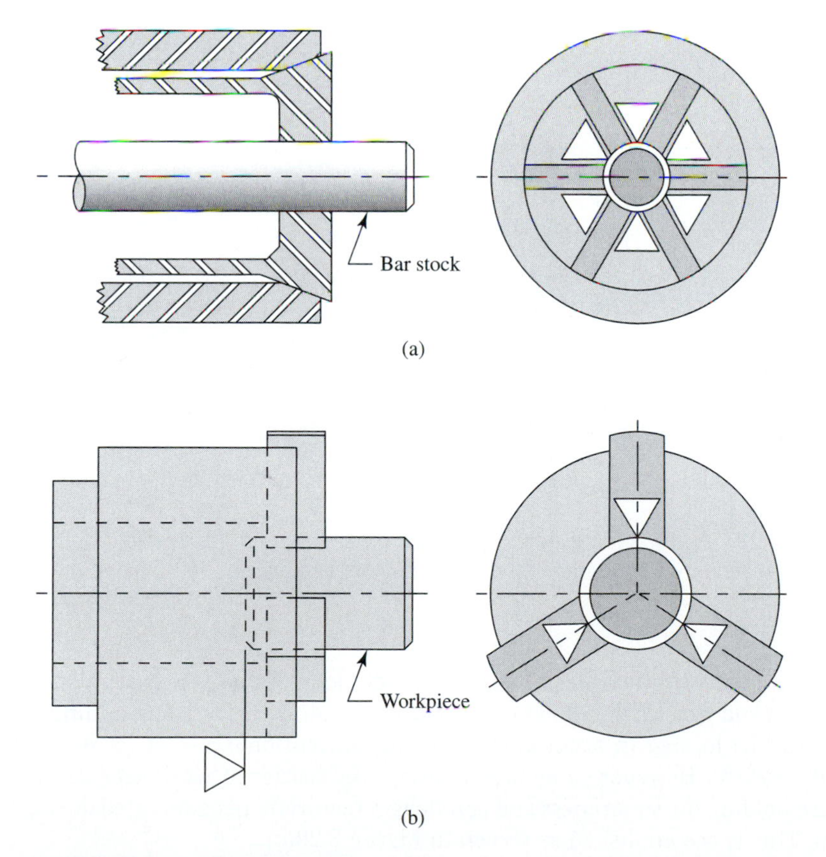

FIGURE 8.27 (a) Self-Centering Collet permits bar stock to be fed to desired length and clamped on O.D. (b) Self-Centering Chuck positions part laterally and on centerline.

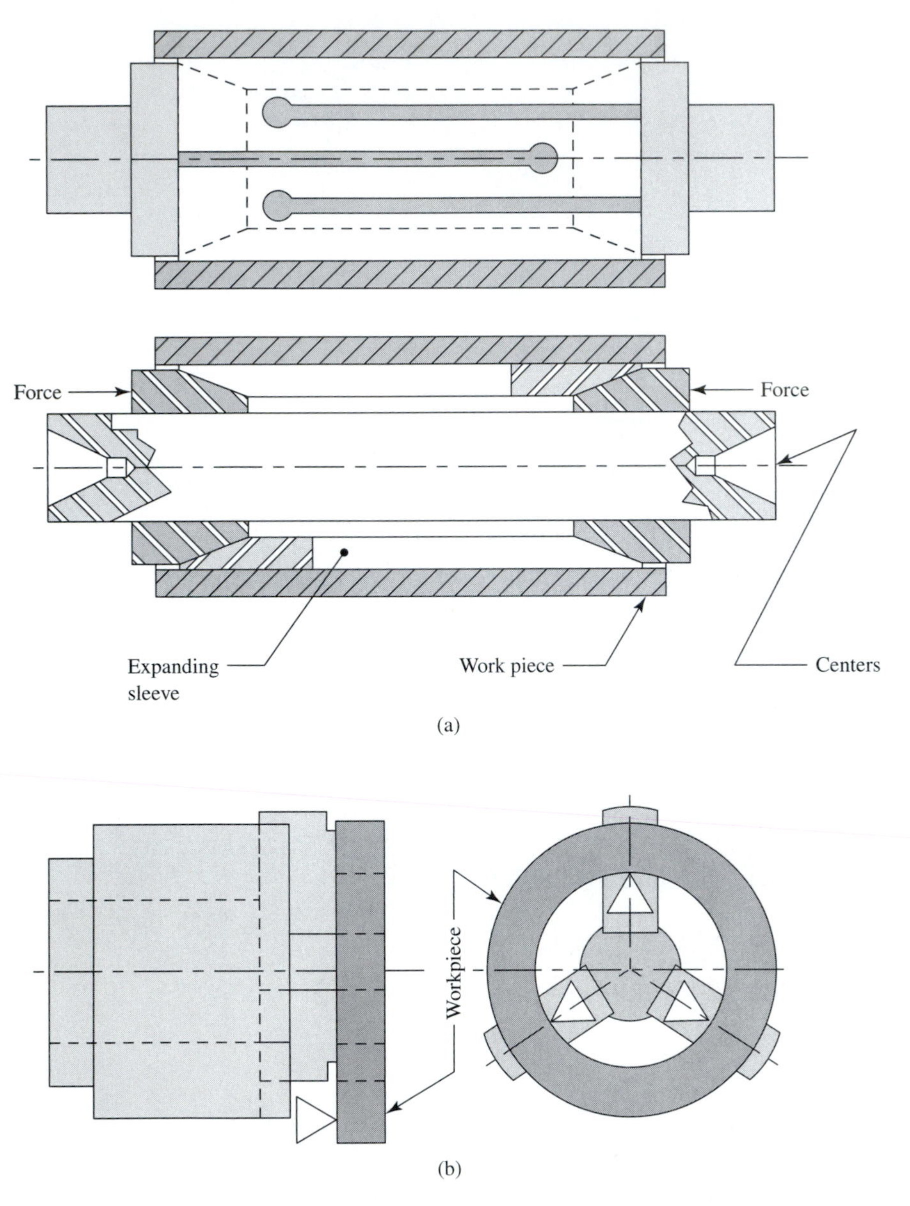

FIGURE 8.28 (a) Expanding Arbor grips and centers part on centerline. (b) Self-Centering Chuck centers and holds part on centerline.

Figure 8.29(a) illustrates a hardened steel block with a slot that must be square and parallel to one end face within .0005 total indicator reading. During the milling operation, the workpiece can be located in accordance with the conventional 3–2–1 locating system as shown in Figure 8.29(b). However, after heat treating, the hardened workpiece must be located so as to ensure holding the squareness and parallelism requirements generated during the grinding operation. This is accomplished as shown in Figure 8.29(c).

The workpiece is located on the left end face ("squared up") with 3 locaters. Clamping force is used to hold the part in contact. The 2 locaters are applied to the largest surface to control the height dimension, and 1 locater controls the workpiece in the third plane.

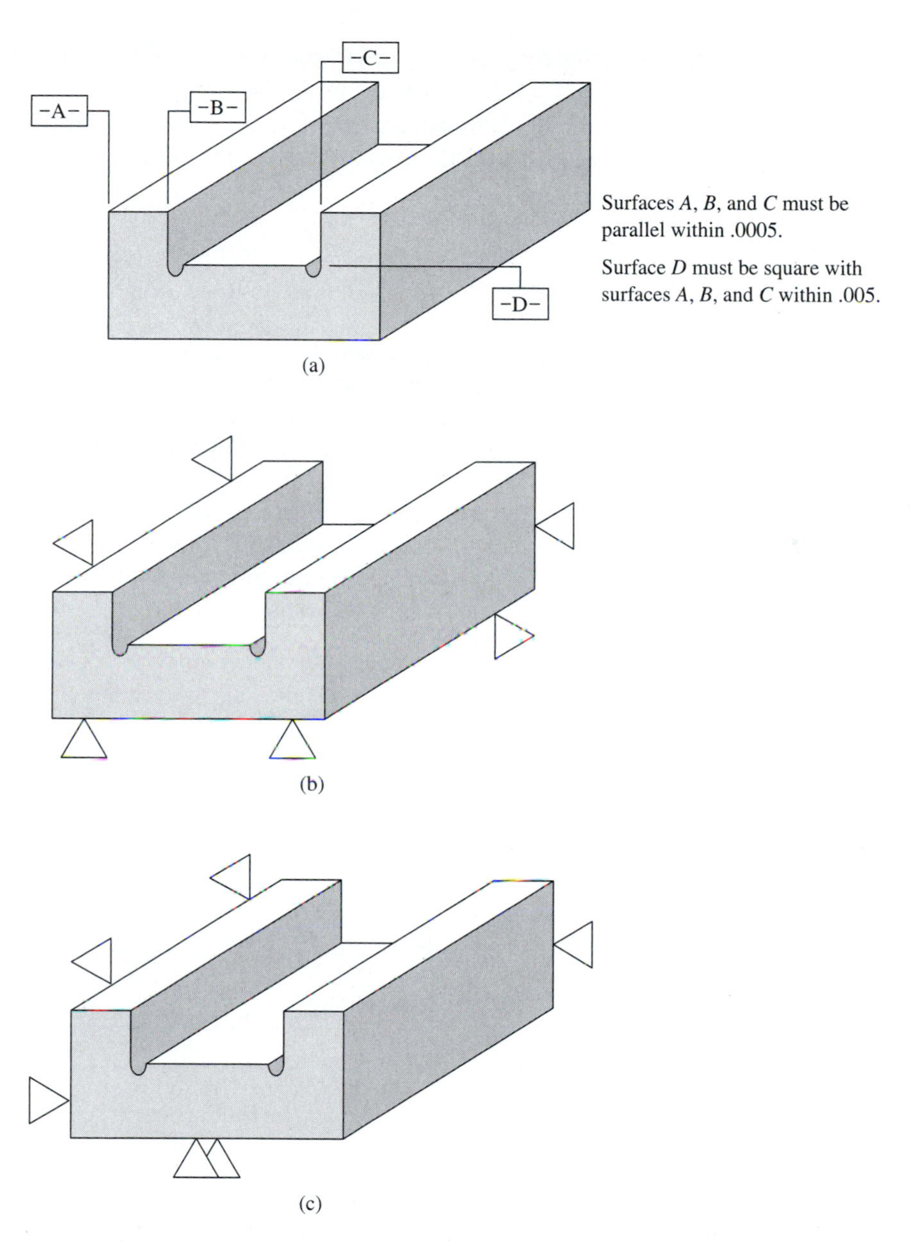

FIGURE 8.29 (a) Block design. (b) 3-2-1 Locating system used while machining slot. (c) 3-2-1 Locating system used while grinding slot and establishing squareness and paralellelism requirements.

Mechanical Control. The retention and support of a workpiece positioned on predetermined locating surfaces is known as mechanical control. Retention is achieved with the use of clamps to press and hold the workpiece in positive contact with the locaters; support, when required, is provided by devices known as supports to prevent workpiece deflection. Both function to resist tool forces and control deflection.

- Clamps must, in addition to properly retaining the workpiece, be capable of resisting any tool forces applied to them.

- Supports must be properly located to resist workpiece deflections caused by tool or clamping forces, or workpiece deflection caused by its own weight.
- Some parts' designs require approximate or alternate locaters for temporary retention of the workpiece (until it is clamped). Others will remain in the desired location due to their inherent stability.

Forces and Deflection. The workpiece shown in Figure 8.30(a) and (b) illustrates the forces involved in two different methods of milling a surface. The difference is the rotation of the milling cutter.

Conventional milling, shown in Figure 8.30(a), is performed with the cutting teeth cutting "up"—starting with zero thickness of chip and increasing to maximum thickness at breakout. Tool forces are to the left, ahead of the cutter, and up. The cutter attempts to lift the part with a torque or turning force. This force must be restrained by the clamp. Deflection of the part is minimal.

Climb milling, shown in Figure 8.30(b), is performed with the cutting teeth cutting "down," starting with maximum thickness of chip and decreasing to zero thickness. Tool forces are to the right, behind the cutter, and down. The cutter attempts to climb onto the workpiece, forcing it against the locaters with a torque or turning force. Deflection of the workpiece can be high, and very rugged workpiece holders are required.

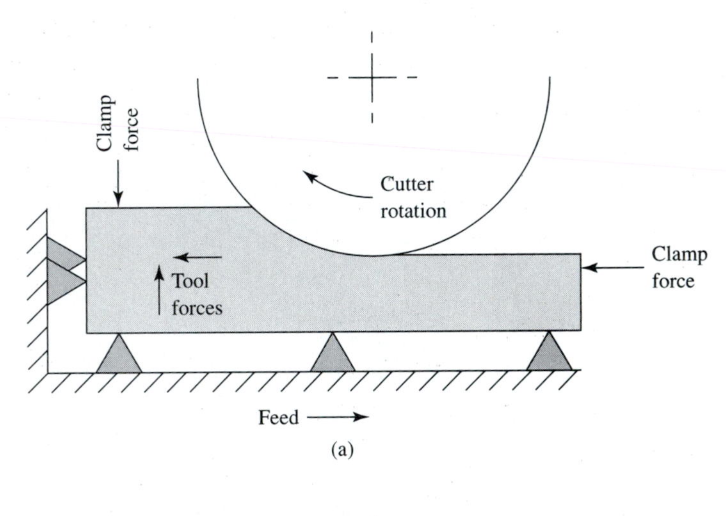

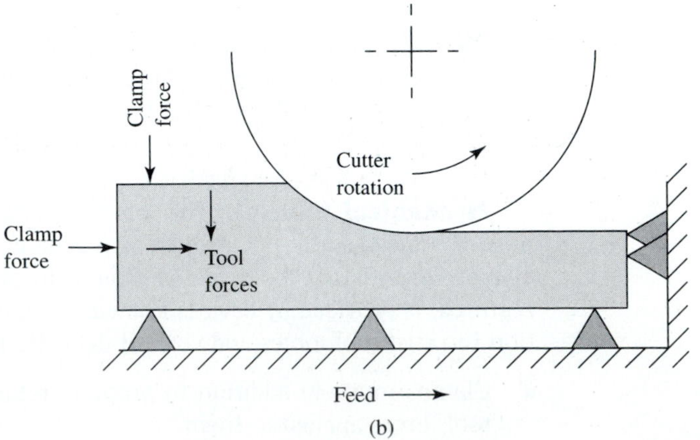

FIGURE 8.30
(a) Conventional horizontal milling. (b) Climb horizontal milling.

Clamps. Clamps should be placed directly opposite locaters, sometimes must be placed between locaters, but must always be positioned so their action presses and retains the workpiece against all locaters regardless of the amounts of tool forces. Illustrated, and noted, in Figure 8.31(a) through (c) are several considerations that must be made concerning proper positioning or location of clamps.

Incorrect clamping can result in workpiece instability, and improperly controlled clamping pressures can result in workpiece distortion. As the shape or configuration of a workpiece becomes more complex, clamping cannot always be provided exactly according to theory. Clamps must sometimes be located on the only areas accessible, or in locations such as to keep them out of the path of cutting tools. Instability and distortions may exist, requiring additional measures to control. Supports, discussed a bit further on, may be required.

Clamping should be kept as simple as possible, but the basic purpose of clamping must not be compromised: Parts must be pressed to and maintained against locaters regardless of part and tool pressure variations. In many instances, two or more clamps may be used as shown in Figure 8.31(d). In other instances, several clamps may be combined as shown in Figure 8.31(e).

Machining dimensions and tolerances can be effected by part deflection, caused in turn by tool and/or clamp forces. Generally, the resulting inaccuracy is the amount of deflection and "spring back" the part experiences from the forces. This inaccuracy may or may not be acceptable, and a countermeasure may or may not be required to correct the condition shown in Figure 8.32(a).

When workpieces are deflected beyond the yield point, the part is distorted and will not spring back to its original configuration. Other, previously machined, surfaces and dimensions are affected, as is shown in Figure 8.32(b). This condition is unacceptable, and preventive measures must be taken.

The amount of distortion of a workpiece can be minimized by the proper placement of locaters on the locating surface. If the locating surface has been previously machined, a full workpiece holder surface can be provided and no deflection will occur in that plane.

A rough workpiece locating surface requiring three locaters can distort under pressure between contacts, as shown in Figure 8.33(a). Repositioning the locaters may be possible, bringing one locater into the correct position to counter tool pressure as shown in Figure 8.33(b).

Since a workpiece can move and deflect in any plane in response to tool and/or clamp pressures, all forces applied to a part must be analyzed and provided for. Figure 8.33(c) illustrates a workpiece that is subjected to a tool force in the X- and Y-axes and is also subjected to a torque, or rotational force, in the Z-axis. The workpiece holder locaters must be capable of resisting all forces applied to them; the clamps must be capable of restraining the workpiece to the locaters and also of countering the torque of the tool.

Supports. Deflection and location problems of some workpieces require the help of "supports" to effect control. Supports are mechanical devices that are used to prevent movement or deflection of a workpiece from forces applied. Supports may be fixed or adjustable. Figure 8.34(a) illustrates a fixed support that is used to limit the amount of total deflection where some deflection is permissible. The design of this type of support is a simple hardened block that, to function, needs only to be kept clear of chips.

Figure 8.34(b) illustrates an adjustable support that is used to limit all deflection. It may be hydraulically or mechanically operated, and in some applications, it is automatically applied. In use, it is unlocked and retracted behind the plane of the locaters. After the workpiece is loaded and clamped, the support is brought into contact with the workpiece surface—then locked to resist forces and deflection. *The adjustable support must be unlocked and retracted before loading the next workpiece.* Failure to do so will make it difficult, or impossible, to load the next workpiece and can cause tool damage and part scrap. For this reason, in addition to saving time, many applications of adjustable supports are automatically controlled.

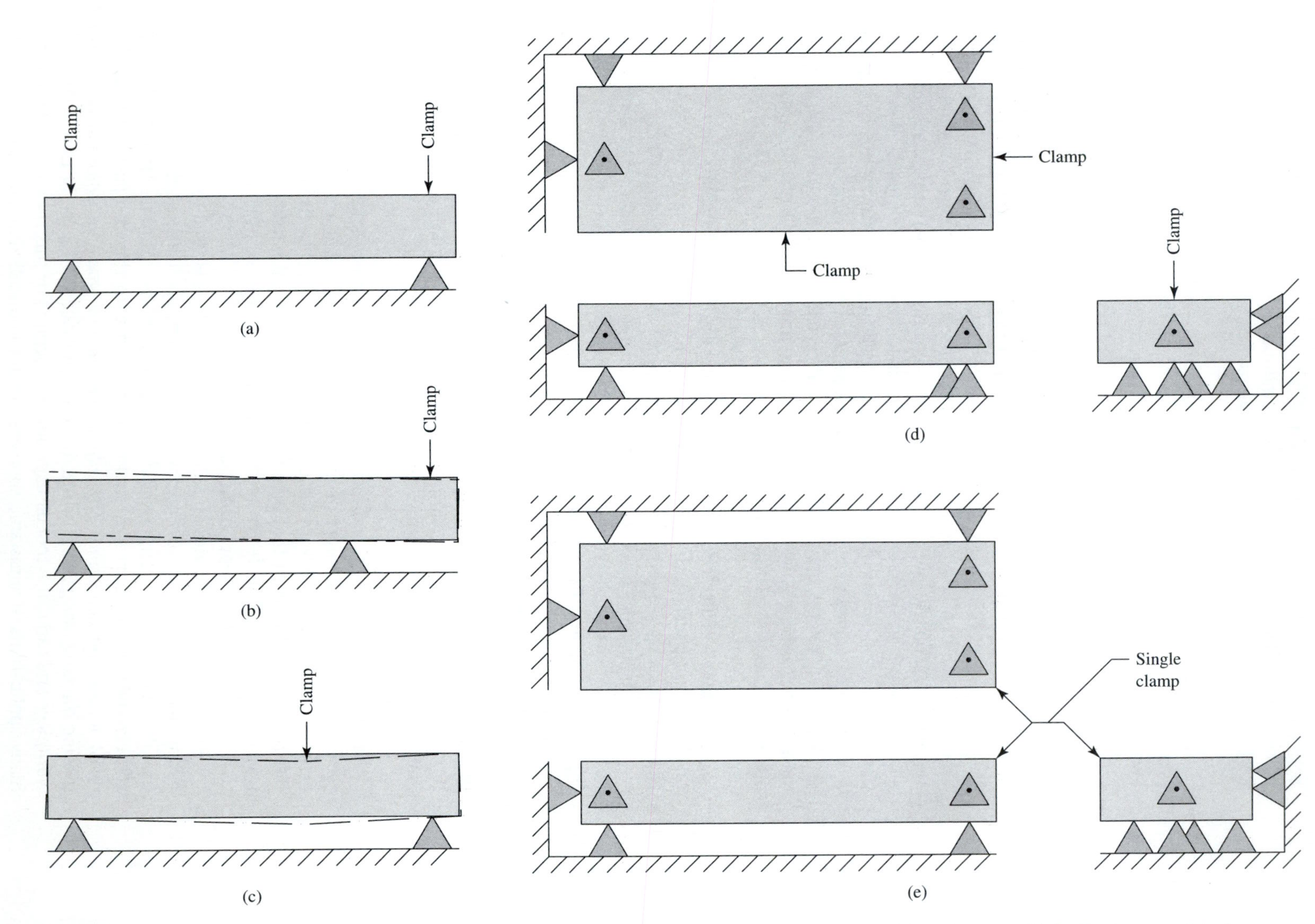

FIGURE 8.31 (a) Clamps correctly positioned opposite locaters. (b) Instability from incorrectly positioned clamp. (c) Deflection possible from incorrectly positioned clamp. (d) Individual clamps applied in each axis. (e) Single, composite clamp.

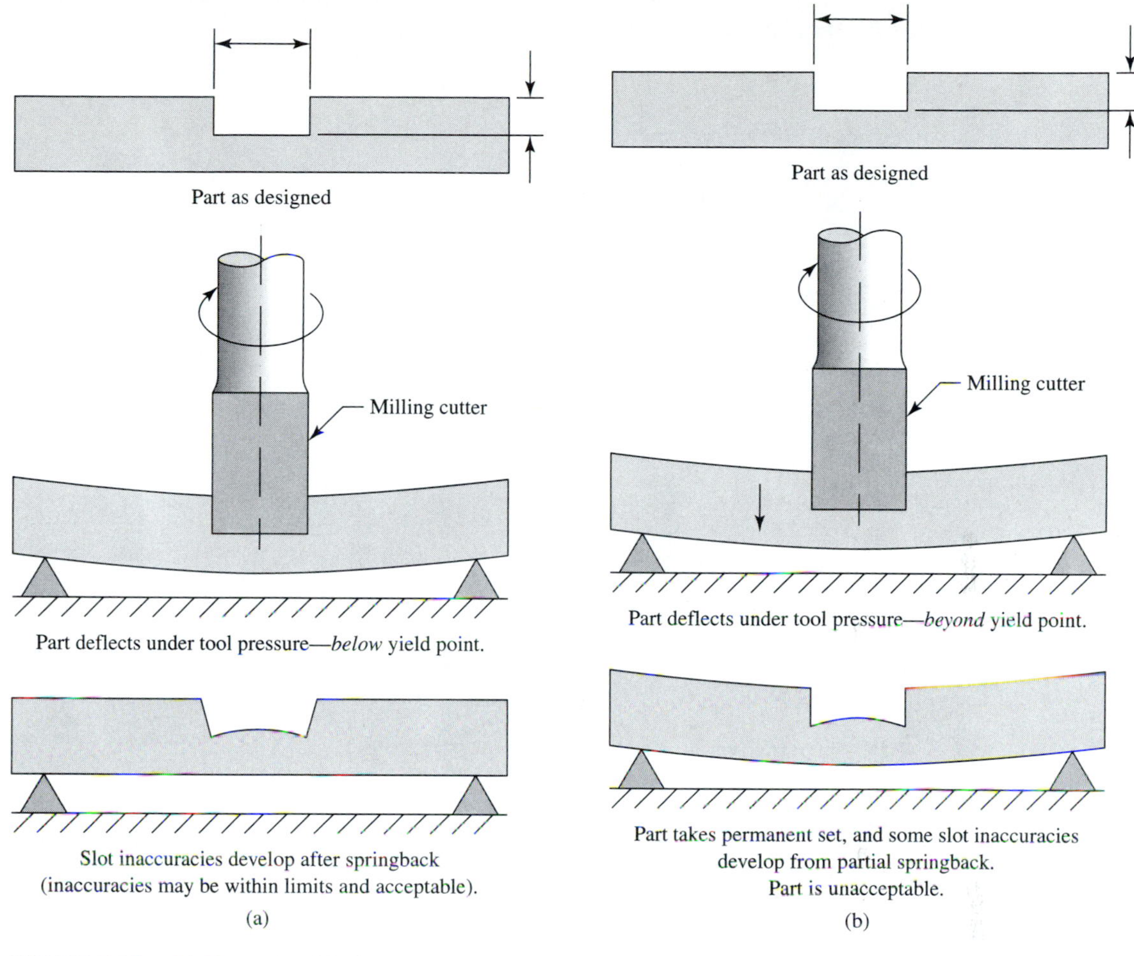

FIGURE 8.32 (a) Sheet #1. (b) Sheet #2.

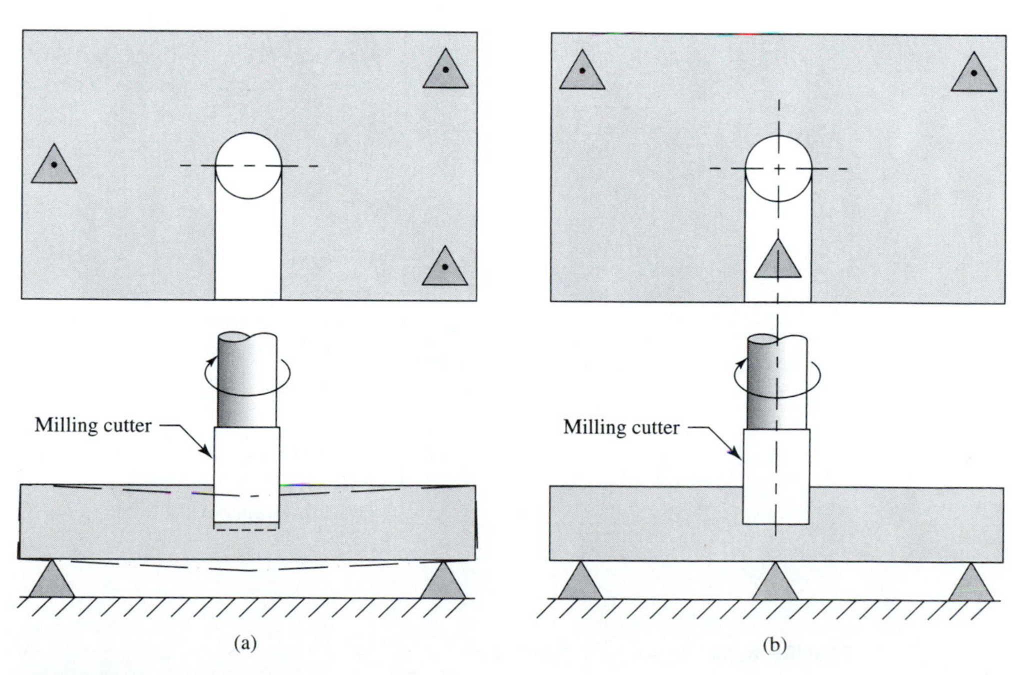

FIGURE 8.33 (a) Part deflection resulting from tool pressure. (b) Locaters repositioned to resist tool pressure.

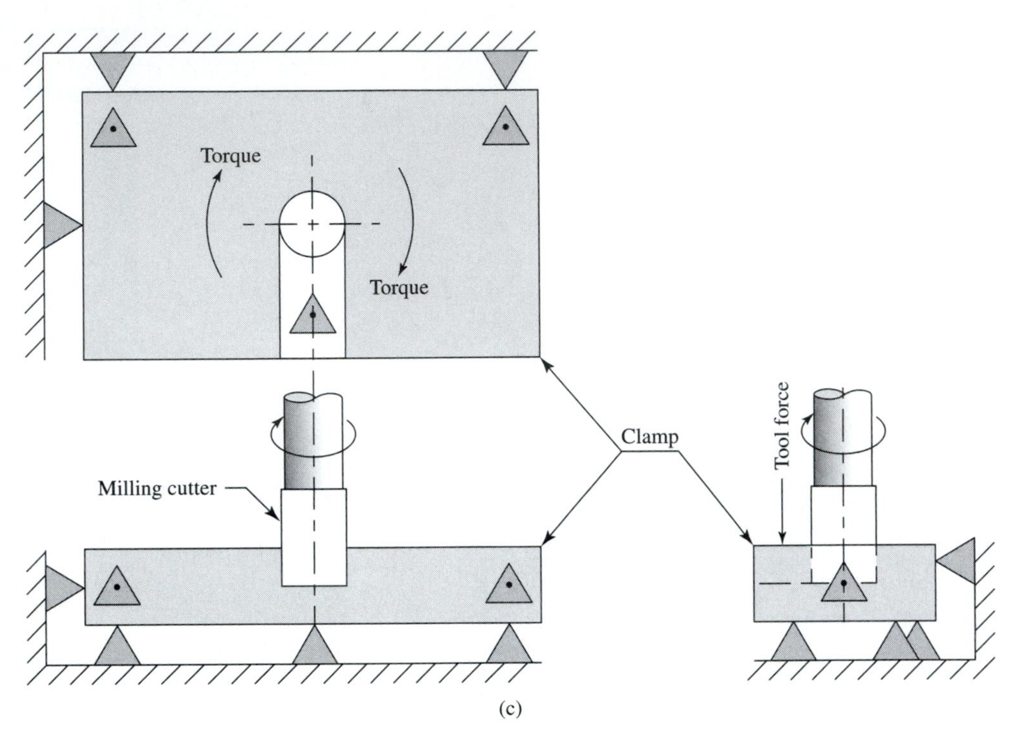

FIGURE 8.33 Continued (c) Proper positioning of locaters and clamps restrains part resisting tool force and torque during machine in milling fixture.

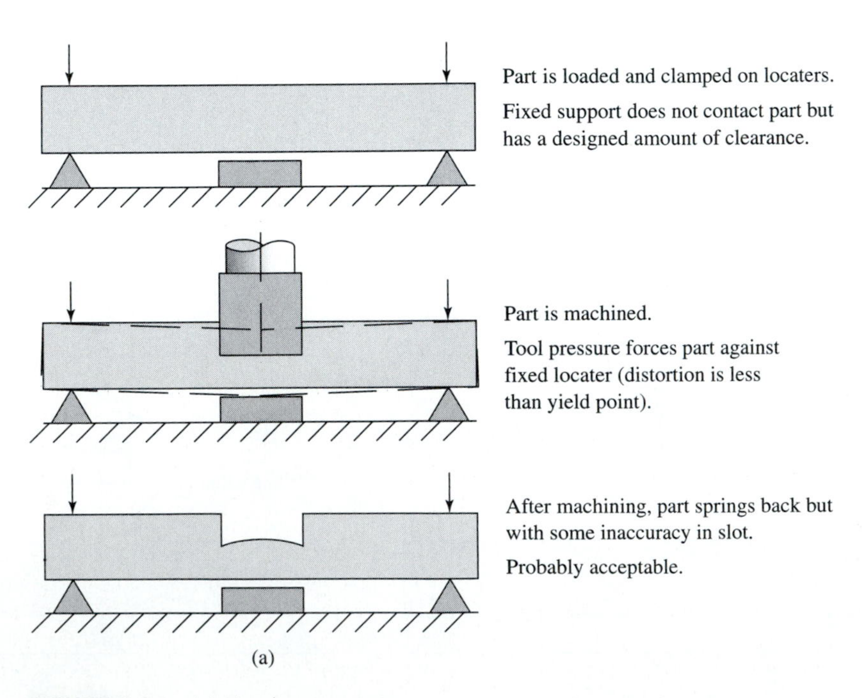

FIGURE 8.34 (a) Fixed support use.

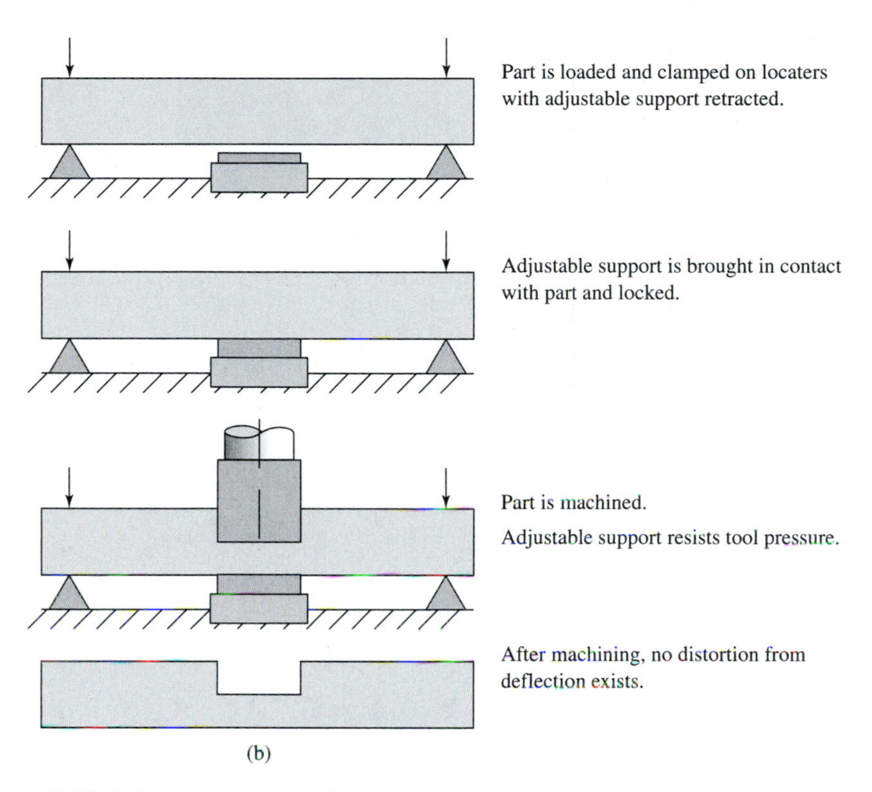

FIGURE 8.34 Continued (b) Adjustable support use.

EXAMPLE

Workholder with Adjustable Support. The Center Link, Figure 8.35, will require the use of a workholder employing an adjustable support during machining of the second surface, establishing a critical face-to-face dimension. Processing of the part and use of the adjustable support follows.

Operation 20—Machine Large-Diameter Boss Outer Face. The outer face of the large-diameter boss is machined first. The rough forging is located with the back (inner) face of the large-diameter boss resting on two locaters and the small-diameter boss resting on one locater as shown in Figure 8.36. This arrangement tends to equalize machining stock on each boss.

The part is clamped between a fixed and a movable jaw, both of which have a "V" configuration for aligning the part over the locaters. The jaws contact only the upper draft angle of the forging—pressing the part down onto the locaters.

Operation 30—Machine Small-Diameter Boss Outer Face. The outer face of the small-diameter boss is machined from the finish machined large-diameter boss outer face. The part is located and clamped on the large-diameter boss outer face as shown in Figure 8.37. The part is then clamped between a fixed and a movable jaw similar to the clamping arrangement used for operation 20. The locked adjustable support will resist tool pressure when the small-diameter boss outer face is machined.

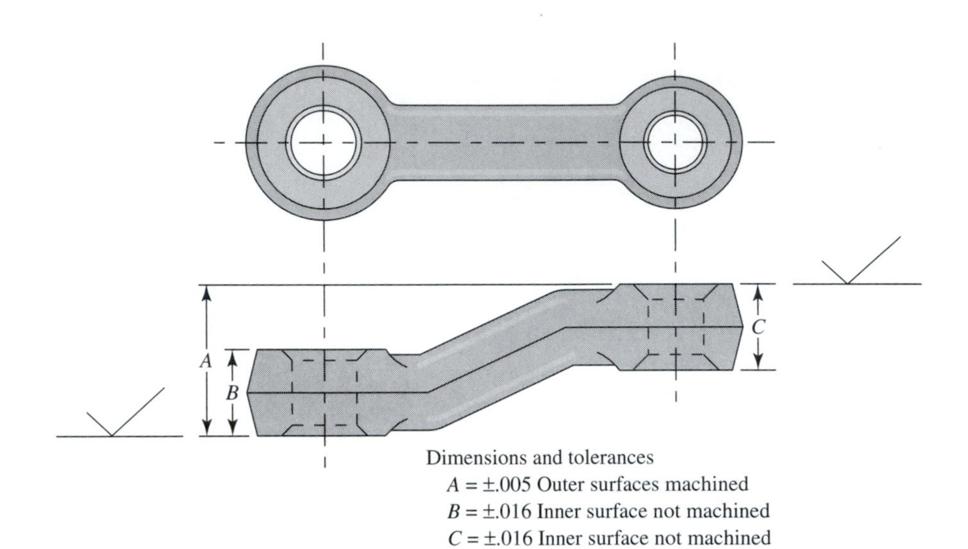

FIGURE 8.35 Center Link.

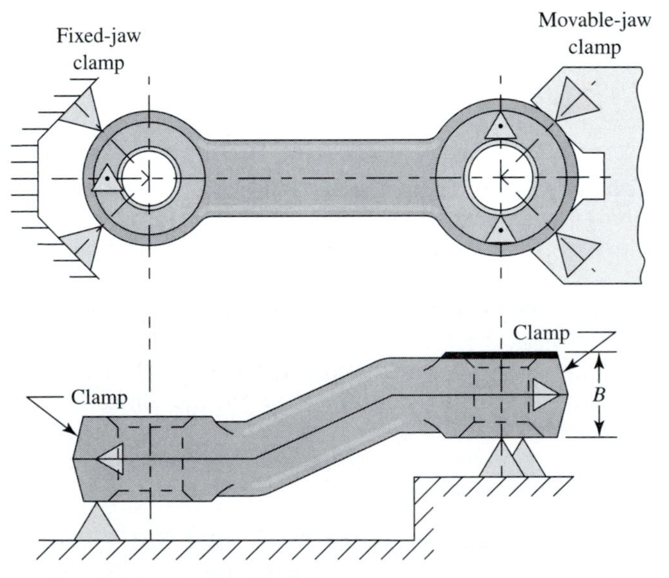

FIGURE 8.36 Operation #20.

FIGURE 8.37 Operation #30.

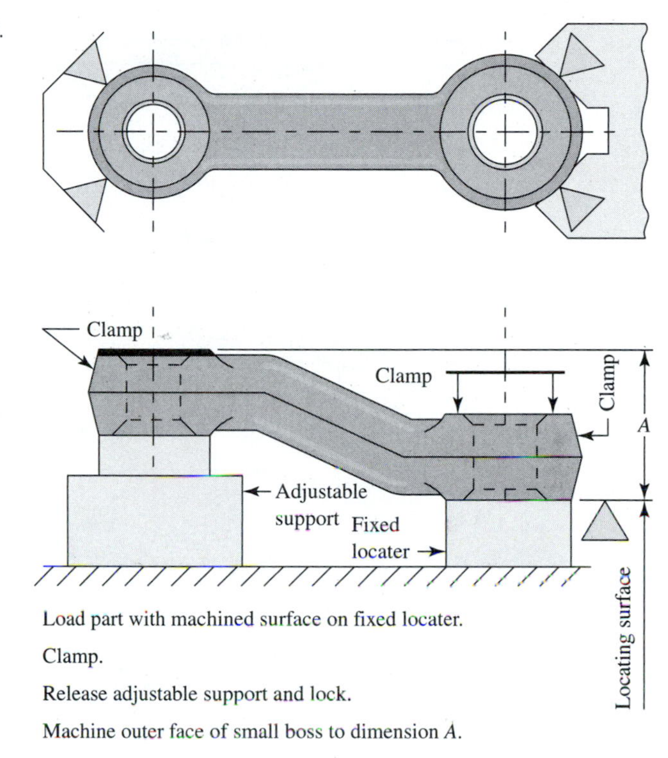

Approximate Locaters. Approximate locaters are used to facilitate loading suitable workpieces into a workpiece holder. Where these locaters can be used, they aid in positioning and retaining a part approximately, until clamping action presses and retains it against the locaters. Figure 8.38 illustrates a workpiece positioned generally with approximate locaters, one of which also foolproofs so the part cannot be incorrectly loaded.

Alternate Locaters. Alternate locaters are duplicate locaters used to position a part from either of two acceptable surfaces. The hole may be drilled from either set of locaters and the tolerance held for the part illustrated in Figures 8.39 and 8.40. Alternate locaters ease and speed part loading much as approximate locaters do; the part print must be analyzed to determine which, if either, can be used.

Figure 8.41 illustrates a cylindrical part that is located from alternate locaters—any two of the three total. The part centerline can shift .002–.006 from its true position. This is acceptable because the centerline of the three drilled holes has a total allowable tolerance of ±.005. The drilling process will be permitted a manufacturing tolerance of ±.003, which is adequate for this process.

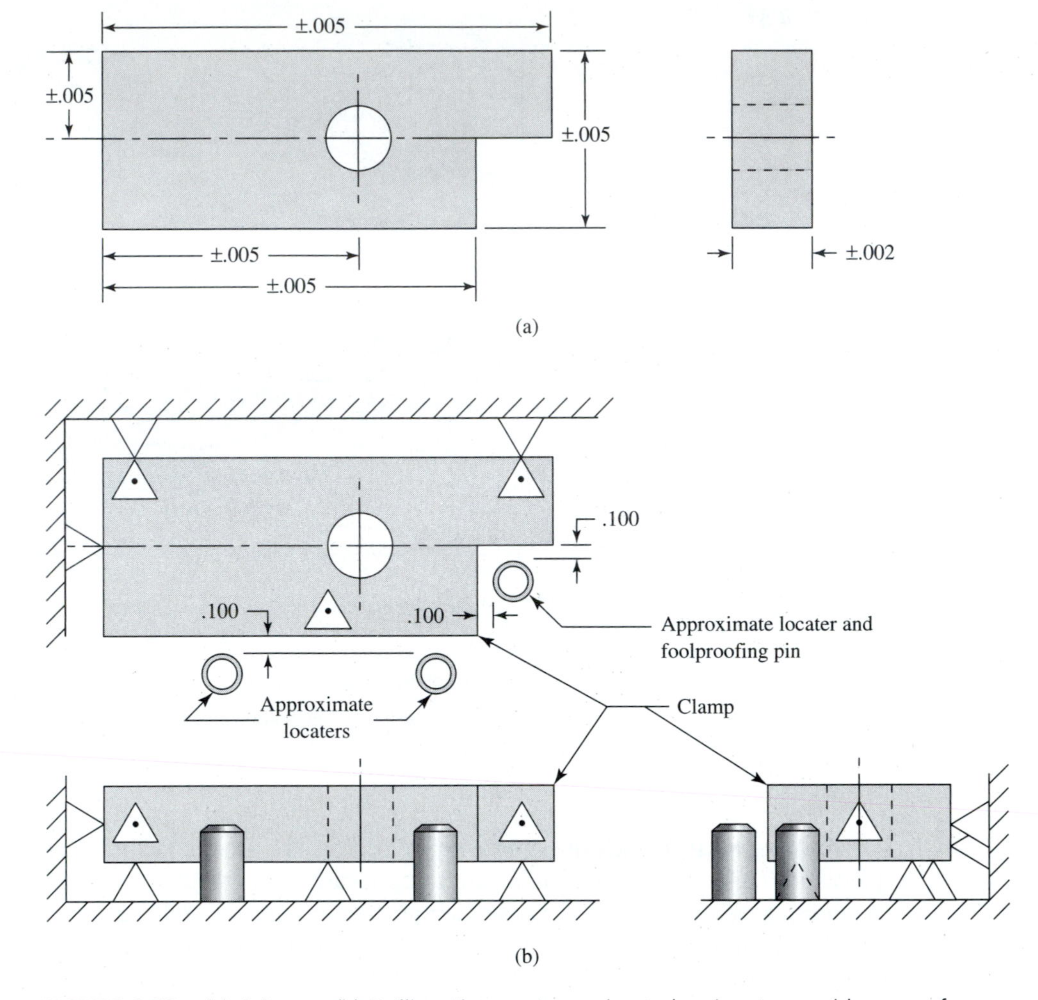

FIGURE 8.38 (a) Adapter. (b) Drilling Fixture. Approximate locaters to provide ease of loading part into fixture. Combination approximate locater and foolproofing pin assures proper orientation of part in fixture.

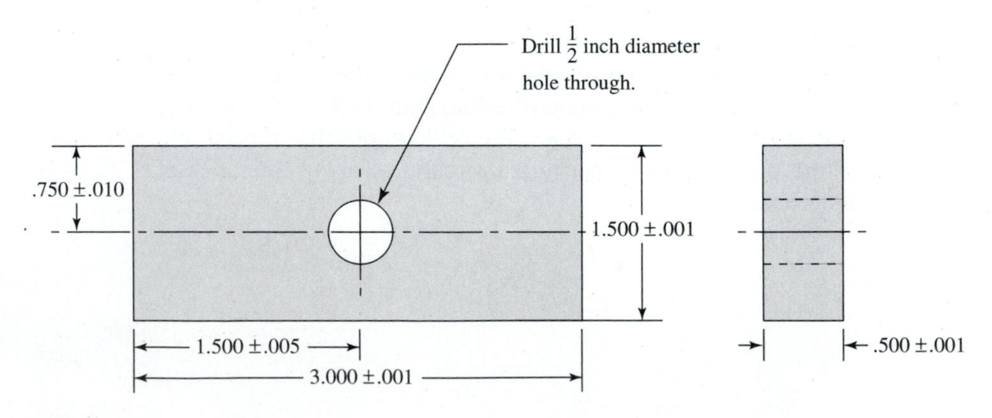

FIGURE 8.39 Spacer.

FIGURE 8.40 Drilling Fixture with alternate locaters designed for a rectangularly shaped part.

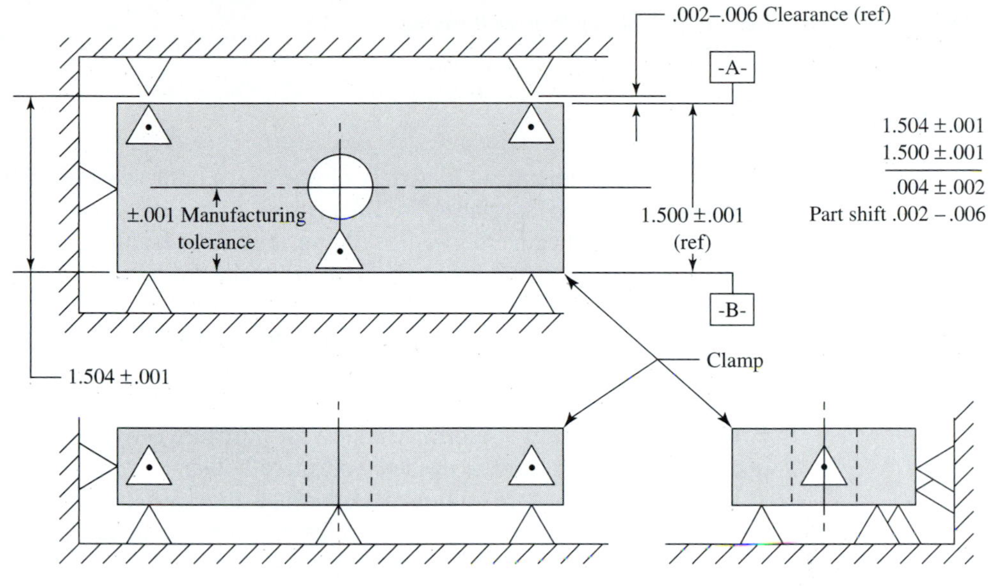

Location of part on either of (2) qualified surfaces produces an acceptable part.

FIGURE 8.41 Drilling Fixture with alternate locaters designed for a cylindrically shaped part.

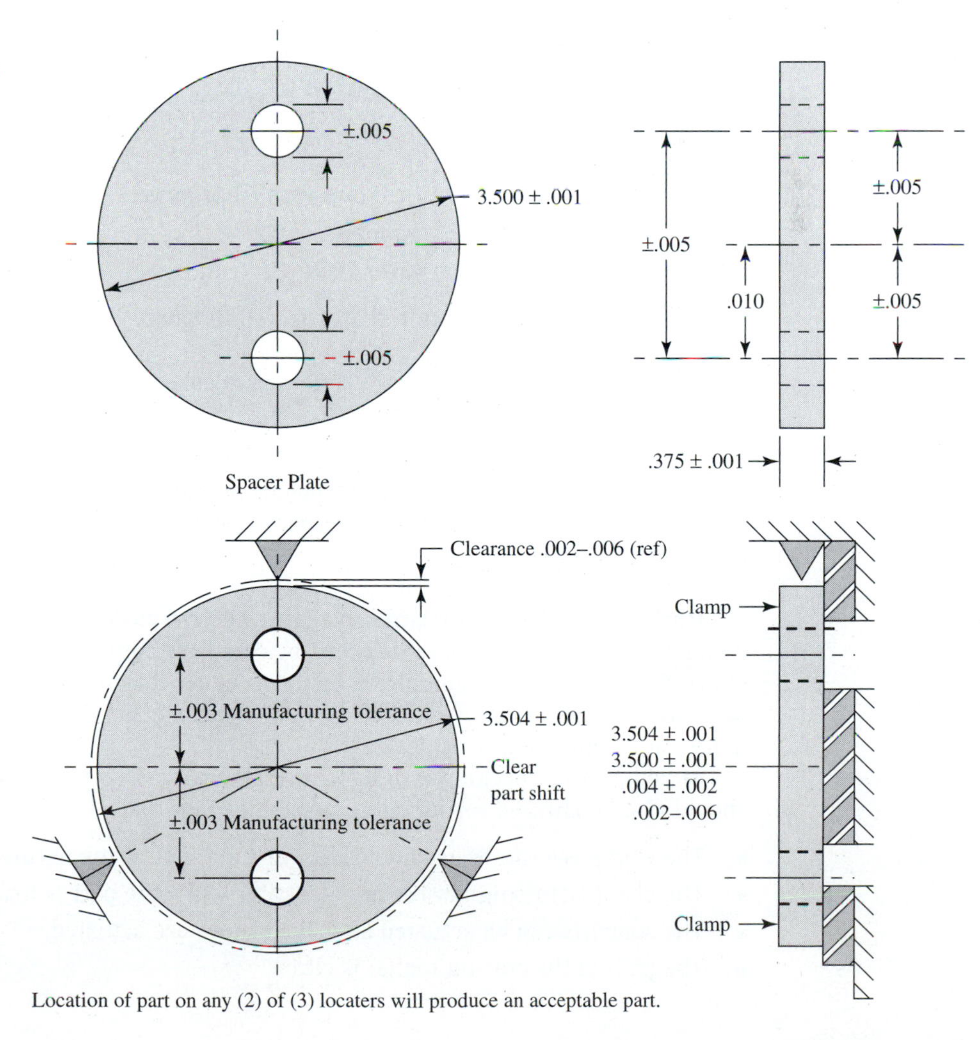

Location of part on any (2) of (3) locaters will produce an acceptable part.

8.5 PRACTICE

Most of the large quantity and variety of workpiece holders required by industry are supplied by manufacturers who specialize in their production. Many are standardized units produced, stocked, and supplied from inventory by various manufacturers. Others are made to order from special designs. Some workpiece holders are made "in-house" when economic or time requirements dictate.

Whatever the source, a successful workpiece holder will be obtained when a procedure requiring careful analysis of requirements and control of designing and making is followed.

Procedure to Obtain A formal request for purchase is written, usually by the requiring manufacturing engineer, for the needed workpiece holder.

Requests for standard tooling are forwarded directly to purchasing; requests requiring design and make are sent to appropriate sources. Usually, the design is completed, then sent out for competitive bids for making or building.

Design and Make

Determining the Requirements. All factors affecting the operation of a specific workpiece holder must be determined by the designer before starting work on a design. This data can be obtained from the tool and operation sheets and will include:

- Machine description
- Operation description
- Locating surfaces
- Clamping requirements (including manual or power)
- Any part loading devices required
- Any tool setting or setup gages required

Other pertinent information that it is well to have:

- Production rate per hour
- Pieces produced per cycle
- Coolant requirements

Many tool and engineering sheets include picture sheets, which help clarify. Also, where any item is not clear, it is the designer's responsibility to ask questions—of his supervisor or of the manufacturing engineer.

Design. The layout for large and intricate workpiece holders is usually done by a senior designer, and components are designed by detailers. Some designs, depending on the amount of detailing required, are completed by the designer making the layout. It is prudent, however, to always have all design work checked by a separate checker. Both conventional and CAD techniques are employed in tool design.

In addition to checking a design for mathematical correctness and materials specified, it should also be checked for function, making certain that

- The workpiece can be readily loaded into, and out of the fixture.
- The clamp actuating devices are accessible when the part is loaded.
- The supports can be actuated after the clamps are actuated.
- The path of the cutting tool(s) is clear.

- The locating surfaces can be cleared of chips before loading a successive workpiece by means of coolant flow, air blast, or manually.
- Provisions for coolant, if required, have been made.
- Wear surfaces may be maintained by means of replaceable details.
- Provisions have been made to integrate with any loading or unloading devices used.
- Set up blocks, or other provisions, are made to align the fixture on the machine and in relationship with the cutting tools.
- Provisions are made for clamping the fixtures to the machine (keys, slots, lugs, etc.).

Make. Checked and approved copies of the workpiece holder design are sent out for competitive bids. The successful bidder must supply the unit exactly to the design and within the time specified. Upon delivery the fixture is inspected; any discrepancies must be corrected by the supplier, and when the fixture is finally approved, payment is made.

Standard Tooling Available

Complete Workpiece Holders. Many standard items of workpiece holders are available "off the shelf." A manufacturing engineer should become thoroughly familiar with what is available to provide his company not only of lower-cost tooling, but also savings in lead time to obtain the tooling.

Some of the standard workpiece holders available from stock include

- Chucks to suit any machine tool spindle application are available in 2-, 3-, or 4-jaw configurations and may be manually or power operated.
- Gear chucks, which locate a gear workpiece on its pitch diameter are available from stock.
- Diaphragm chucks, which chuck with an interference fit and require pressure to unchuck, can be supplied with only the diameter of the jaws requiring finishing.
- Arbors and mandrels with mechanically expanded split sleeves, and also hydraulically expanded types, are available from stock with flanged mounting or centered ends.
- Drill jigs and fixtures are available in many sizes and types, most of which require only a locater and drill bushings to complete.
- Collets are available with mechanical or hydraulic actuation and mounting ends to suit applications.

Several of these units are illustrated in appropriate chapters of Part IV.

Finished Workpiece Details. Many standard fixture details are available complete and ready for use and inclusion in special designs. These include locaters, clamps, supports, jacks, angle blocks, detents, knobs, and others, some of which are also illustrated in Part IV.

Workpiece Holder Setup and Maintenance. The manufacturing engineer should be familiar with the way his company sets up and maintains tooling. This knowledge not only can help in developing more effective tooling, but will be required in the event of troubleshooting a problem.

- Setup. The responsibility for setting up tooling is usually that of the manufacturing department (as is its storage when not in use). The actual setup is accomplished by a job setter. In the case of a fixture or workpiece holder, the job setter places the fixture on the machine, aligns it, clamps it in place, and adjusts it in relation to the cutting tools such that specified operational dimensions and tolerances are generated.

Major subsequent adjustments and/or replacement of cutting tools are normally performed by the job setter. Minor tool adjustments may be made by the operator.

- **Storage.** When not in use, tooling should be properly identified and stored in a designated area not subject to damage, rust, or corrosion. Protective coatings of rust and corrosion inhibiting materials may be required, which—if used—must be removed at the time of next use.
- **Maintenance.** Workpiece holders should be regularly inspected for wear and damage of locating areas, rest surfaces, drill bushings, and any other component critical to the fixture's ability to produce acceptable workpieces. All such substandard items must be reconditioned or replaced as soon as they are identified. It is less than cost effective to wait and allow the workpiece scrap rate of an operation dictate the need for fixture repairs.

  Tool maintenance is usually the responsibility of the toolroom, aided in some manufacturing companies, by an area or function known as "tool trouble."

8.6 REVIEW QUESTIONS AND PROBLEMS

1. Define a workpiece holder.
2. Make a drawing illustrating the concept of a rectangular workpiece free to move in any direction in space. Add descriptive notes.
3. Make a drawing illustrating the requirements of bringing a rectangular workpiece into equilibrium. Add explanatory notes.
4. Make a drawing illustrating the requirements of bringing a cylindrical workpiece into equilibrium. Use explanatory notes.
5. Define the elements of workpiece control that must be considered when designing a workpiece holder. Discuss their application.
6. What is a support? How many types of supports are used, and what is the purpose of each? Show the use of each in a drawing.
7. Why are locaters symbolized in some workholder locater applications?
8. - What is an alternate locater and how is it used?
   - What is an approximate locater and how is it used?
9. Discuss centerline control, including the difficulties of application and the principles involved.
10. Give some examples of workholding devices that utilize friction rather than locaters and clamps.

PART III

PROCESS AND MACHINE SELECTION

INTRODUCTION

Machines and equipment, additional to those required for the specific manufacturing processes, are required to provide a complete and viable manufacturing facility. These are discussed in Chapter 9 and include

- **Ancillary Systems:** systems used by and with the basic production machines
- **Auxiliary Systems:** systems that perform work on a part but do not significantly change the part
- **Support Systems:** systems that provide a needed service during the manufacture of a part

All are discussed and illustrated in Chapter 9.

Machines are purchased for many reasons in addition to tooling up for a new product. Manufacturing Engineering, knowing the reason for a machine purchase, must select facilities that best answer the need. In addition, the characteristics, costs, capabilities, and design features of each machine considered must be known and compared. Significant machine properties are discussed and illustrated in Chapter 10, and a method of comparison is illustrated.

Special machines range from near-standard drilling, milling, turning types to fully automatic self-loaders and include special-purpose types. Machining systems are available that, in effect, combine many special machines either in a common base or as separate machine units to which workpieces are delivered by such means as a special type of vehicle or by some form of conveyor.

Much of what is termed "automation" is now designed and built into a machine at the time of the machine's manufacture. Some automation can be, and is, added, especially in the area of automatic loading and unloading of parts. In Chapter 11, special machine types and automation are discussed and illustrated.

Group technology is a plan to advantageously manufacture groups or families of parts that have some similarities of design or process. The concept is illustrated and compared with conventional methods of manufacture. The different types of machine layouts are discussed and illustrated in Chapter 12.

CHAPTER NINE

ANCILLARY, AUXILIARY, AND SUPPORT SYSTEMS

CHAPTER OVERVIEW

9.1 INTRODUCTION

This chapter is to provide a study and discussion of operations, systems, and components that are required in addition to production equipment to render a manufacturing plant complete and operational.

These elements may be adjuncts to the production equipment or a completely separate entity supplying different required services. All must be studied for need and capacity, and all must be checked for availability by the manufacturing engineer.

Ancillary systems are those systems that are required by and used with, or in conjunction with, a production machine during manufacture, and include

- Cutting fluids
- Electrical power
- Compressed air
- Water
- Gas

Auxiliary systems are those systems that perform work known as direct labor on a part during manufacture, but do not significantly change the part dimensionally. Included are

- Deburring
- Cleaning
- Heat treating
- Finishing

Support systems are those systems that do not perform any work on a part, are known as indirect labor, and provide a function or service required in the manufacture of a part. Included are

- Quality assurance
- Material handling
- Maintenance

Figure 9.1 illustrates the flow of parts and assemblies through basic manufacturing, auxiliary, and support operations.

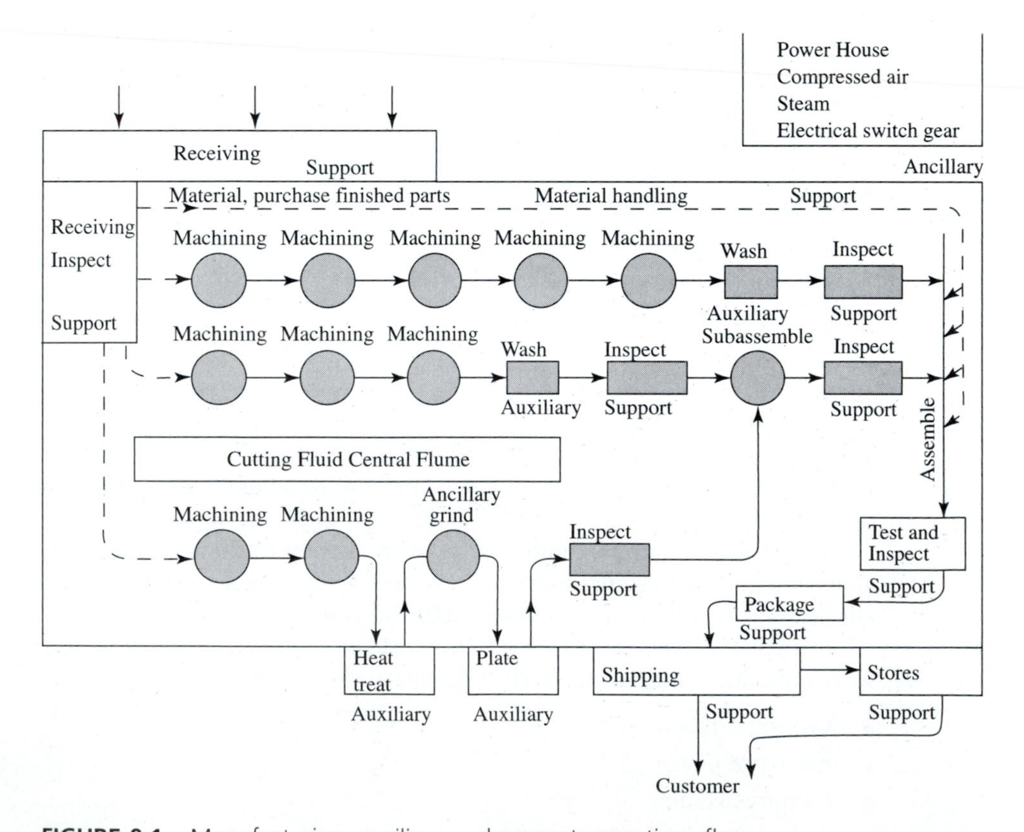

FIGURE 9.1 Manufacturing, auxiliary, and support operations flow.

9.2 ANCILLARY SYSTEMS

9.2.1 Cutting Fluids

Manufacturing engineers must have a good basic understanding of coolants and their properties to be able to make the proper selection for each application. The following material is provided to aid in the selection and application of coolants and also to provide sources for additional on-the-job help.

Machining Requirements Cutting fluids, or coolants, are used primarily to remove the heat generated as a chip is formed and to reduce friction and subsequent heat caused by chips sliding across tool surfaces. Additionally, coolants must lubricate machine surfaces, ways, and slides, and also wash chips into machine pans and chip conveyors.

Some coolants are formulated to prevent corrosion, or rust, on newly machined surfaces—straight mineral oils have this quality inherently.

Germicides are added to control bacteria, fungi, and yeasts, and to prolong coolant life, and not to control dermatitis. Dermatitis has been prone to result from reactions between the skin and the cutting fluid itself, including mechanical impurities such as fines; dermatitis is not caused by bacteria.

Types Countless formulations and concentrations of coolant are developed from four basic types:

- Water-soluble or emulsified oils
- Chemical or synthetic fluids
- Straight mineral oils
- Gaseous fluids

Processes that generate considerable heat from high metal removal rates such as turning, milling, and broaching usually employ the superior cooling qualities of a water-soluble or emulsifiable oil formulation. Chemical or synthetic fluids are also used but are more difficult to dispose of.

Straight mineral oils are generally used on lighter-duty applications where heat removal is not a prime problem. Mineral spirits or kerosene are used successfully for honing and bearingizing.

Gaseous fluids such as compressed air have been used in some applications; inert gases such as argon, helium, and nitrogen have been used where oxidation of the chip is a problem—but at a cost penalty compared to air or CO_2.

Selection Factors Many varied, even diverse, considerations must be accommodated when selecting a coolant:

- Material type and alloy
- Material hardness
- Material machinability
- Material removal rates

- Surface finish specified for part
- Tools—type or process
- Tools—material
- Machine type
- Machine arrangement
- Coolant applications method
- Corrosion protection
- Biocides

Also, *process* mix must be considered: Where a machine is restricted to a specific process, such as turning, and where the cutting fluid is restricted to that machine, an optimum coolant can be selected and used. Many machines, however, include a mix of processes such as turning, drilling, boring, and reaming, as well as utilizing coolant from a central system that serves other machines with their processes and materials. Where this is the situation, a cutting fluid must be selected to satisfy the most critical process. The coolant so selected will exceed the quality required by many processes and will be more costly, but there is no alternative.

Aids in Selection A manufacturing engineer must be familiar with the types, properties, and costs of coolants but need not, probably cannot, be an ultimate authority on the subject. As in so many other areas affecting manufacturing engineering, good basic knowledge of the phase under consideration, coupled with the knowledge of how and where to obtain expert, professional help, is what gets the job done satisfactorily.

Machine-tool builders and suppliers will provide recommendations and choices of a coolant. Producers and suppliers of the many metalworking fluids will study, recommend, and supply a cutting fluid to satisfactorily meet a need. Among the many specialists in this field one company, Master Chemical, in addition to aiding with specific selections, regularly offers educational seminars to qualified people. Indeed, it has been said that this company has educated much of its competitors' sales forces through these seminars! Aid is also available in many reference books on the subject.

Application For cutting fluids to satisfactorily perform their functions of cooling, lubricating, and chip removal, it is basic that the fluid must be sufficiently and efficiently applied to the cutting tool and workpiece. Various methods are employed to accomplish this, depending on the process and on the part configuration. Whichever method is used, the goal must be to supply coolant to the tool and workpiece at a constant rate and in copious quantities. Intermittent flow and inadequate volume result in shortened tool life, cracked tools, and poor surface finish of the workpiece. Methods used to apply cutting fluids are manual, flooding, high pressure, and misting. Gaseous fluids are used in special cases to cool the tool only.

Manual application is limited to use in the toolroom, prototype work, very short runs, and maintenance. Cutting fluids, pastes, and solid lubricants are applied with a brush or oil can, or by pouring over the tool and workpiece. Foams and pressure dispensers are also used. This method is not for use in the continuous repetitive machining of parts.

Flooding is the most commonly used method of supplying coolant to the tools and workpiece. Coolant is pumped to a nozzle, or nozzles, which direct the flow to the desired area. Coolant returns to the sump by gravity, carrying chips to the chip pan or conveyor. Special fan-shaped nozzles, or rings, may be required, depending on the cutting-tool arrangement and part shape.

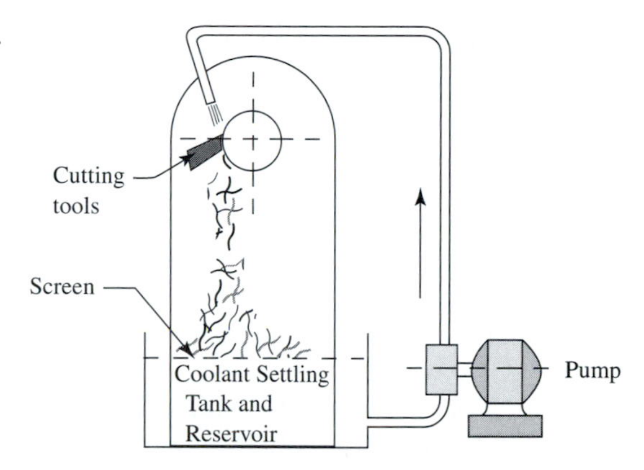

FIGURE 9.2 Basic coolant system.

High pressure is used to force coolant to the cutting edges of deep-hole drills and gun drills through fluid passages formed by grooves, holes, or tubing running the length of the drill. Chips are forced out along the grooves or flutes of the drill.

Misting is the dispersion of cutting fluid by means of air pressure, which then carries the fine drops to the cutting-tool edges. Some special applications require misting, but this method has serious disadvantages: The machine operators and nearby personnel inhale the mist, and a potential fire hazard exists if a combustible cutting fluid is used. Mist collectors must be used but are never entirely satisfactory.

Individual and Central Systems Coolant systems vary in complexity from a simple pump, screen, and sump arrangement for a simple machine to a central system capable of supplying multiple machines with adequate quantities of consistently controlled cutting fluid.

Figure 9.2 illustrates a basic individual system wherein coolant is pumped to the cutting tools and workpiece, from where it returns to the machine base by gravity, washing the chips into the chip pan. Screening in the bottom of the pan catches the chips and allows the coolant to flow through to the sump. Most of the fines that get through the screening settle out in the sump. Chips are periodically raked out manually.

Figure 9.3 illustrates a somewhat more sophisticated individual system, designed to provide cutting fluid to the cutting tools and workpiece, to wash chips into the machine base, from

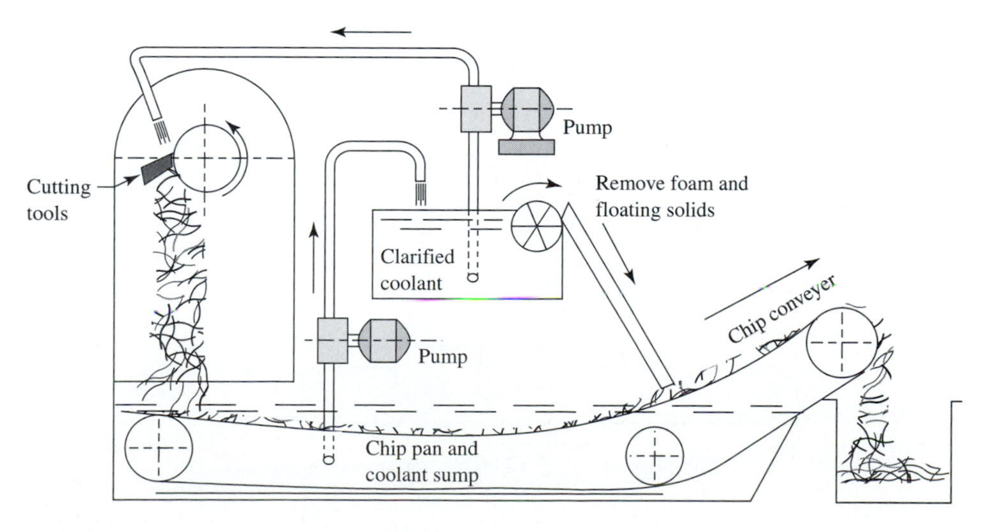

FIGURE 9.3 Improved coolant system.

where they are conveyed mechanically, and to remove foam and floating fines from the coolant before it is again pumped to the cutting tools. Some fines will eventually settle out in the sump, requiring manual cleaning. In this system, the cutting fluid would be manually added and would have been manually premixed as required.

Figure 9.4 illustrates a central coolant system designed with a common central flume around which many individual machine tools are arranged. The system is sized to have capacity adequate for all machines planned to be serviced by it. This particular system was designed to maintain and use the cutting fluid indefinitely, and many such systems have done just that for several years without dumping and refilling.

In operation, the system draws used and contaminated coolant from the common flume, or sump, and pumps it through the various clarification elements into a header tank that is connected with piping to the various cutting-tool zones. After use in the various cutting zones, the

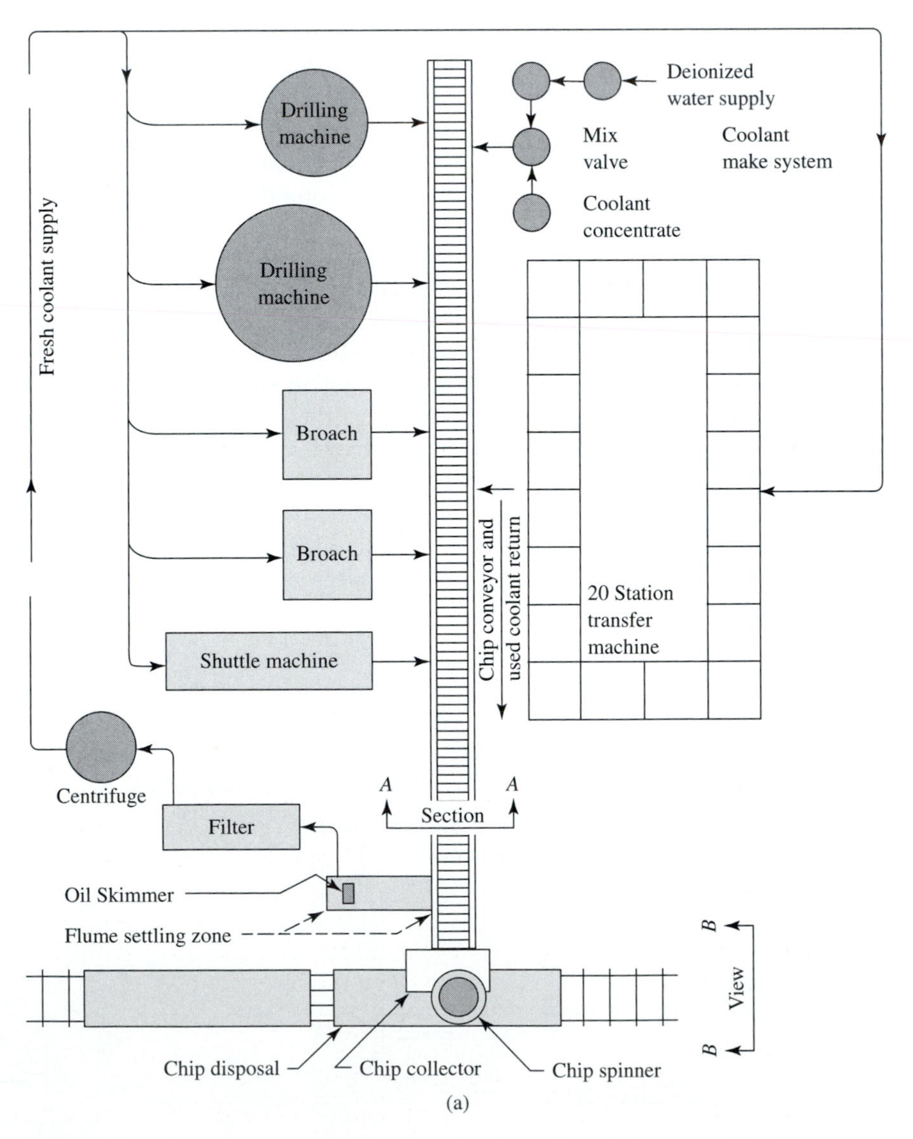

FIGURE 9.4 (a) Central coolant system designed for extended coolant life.

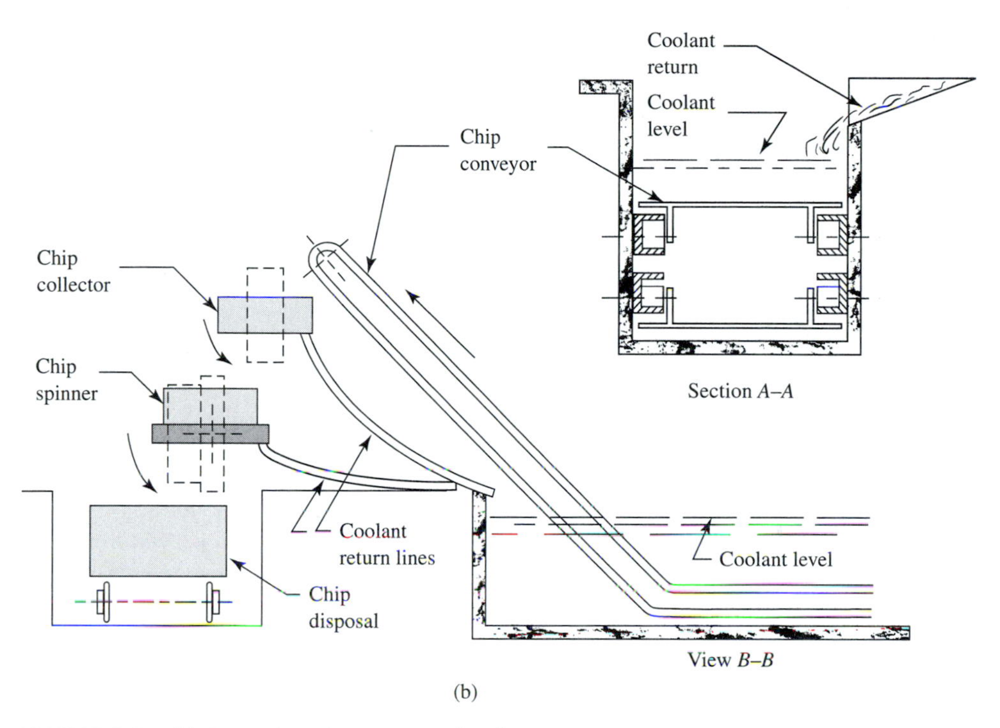

FIGURE 9.4 (b) Central coolant system details.

coolant returns by gravity through short return flumes, carrying chips with it into the common flume or sump. Components of the systems are designed to continuously maintain the cutting fluid within defined limits of mix, pH factor, contaminants, bacteria, and total quantity.

Flumes and sumps are trenches cut into the floor of the manufacturing plant and are of poured concrete construction. The main, or central, flume acts as the system's sump.

Coolant return flumes are short, fairly steeply pitched (10°–15°) and are usually lined with stainless steel sheet metal for smoothness. The pitch and low friction of these flumes aid in washing chips into the main flume.

Pumps used in coolant central systems are subjected to abnormally high rates of wear and jamming caused by chips, fines, abrasives, and contaminants mixed in the used coolant. One pump developed for this purpose, known as the "trash pump," has been used very successfully in many installations.

A chip conveyor is installed along the bottom of the main, or central, flume to remove any chips that settle out and deliver them to a low-speed centrifuge, where they are spun dry. The spun-off coolant is returned to the central flume, and the dried chips are deposited in a hopper. Dry chips command a higher price when sold, and environmental problems caused by coolant dripping along roads and rail beds as wet chips are shipped are eliminated.

Oil skimmers are used to remove free-floating tramp oil from the surface of cutting fluid. Tramp oil, which has its origin in hydraulic system leaks and machine lubricants, is a major source of problems associated with water-emulsifiable cutting and grinding fluids.

Tramp oil is

- A major contributor to coolant-destroying bacteria
- A major contributor to the formation of undesirable residues on machines, tools, and workpieces

- A major contributor to smoke and oil mist in shop atmospheres
- A major inhibitor to good material removal rates and production of good finish and size control with good tool and grinding-wheel life

Two types of oil skimmers are available: wheel types and belt types. The principles of operation are similar. They are mounted above the surface level of the cutting fluid such that the belt, or wheel, revolves through the fluid surface as shown in Figure 9.5(a) and (b). Free-floating oil adheres to the belt, or wheel, is carried to a stripper, where it is separated, and collected in a tramp oil reservoir.

Belt-type skimmers are able to accommodate a greater range of fluid level fluctuations. Both types are more effective when located in a "quiet" section of the sump where tramp oil has time to separate from the roiling and float to the surface.

Oil skimmers should be sequenced and mounted ahead of filters because tramp oil has a decidedly detrimental effect on many types of disposable filter media, tending to seal the filter so that coolant does not flow through. Pressure sensors cause fresh medium to index until flow is again established, resulting in considerable waste of medium.

Filters of several types are in use to remove chips and contaminants from cutting fluid. All types pass coolant through some form of filtering medium by means of gravity, vacuum, or pressure. Filtering media may be permanent or disposable.

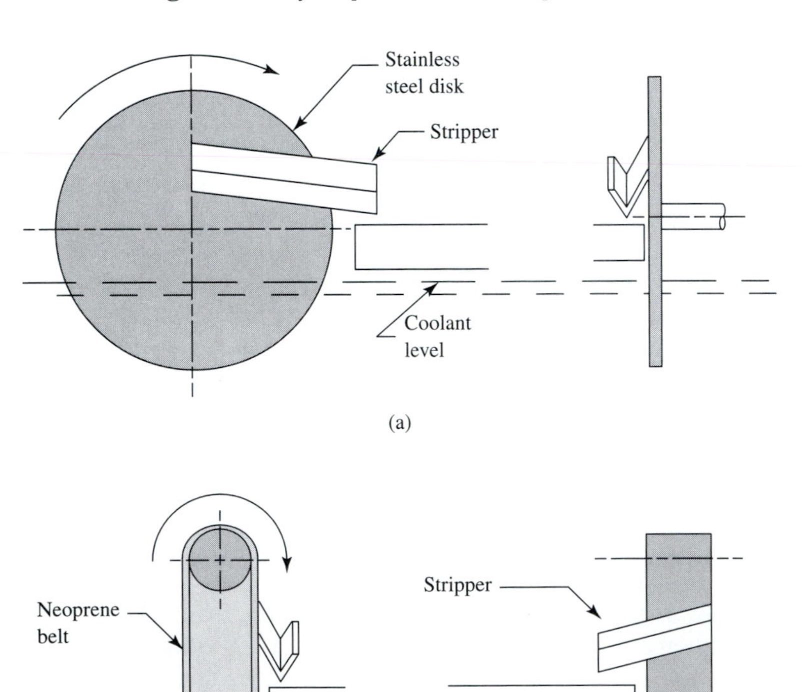

FIGURE 9.5 (a) Tramp oil skimming wheel. (b) Tramp oil skimmers.

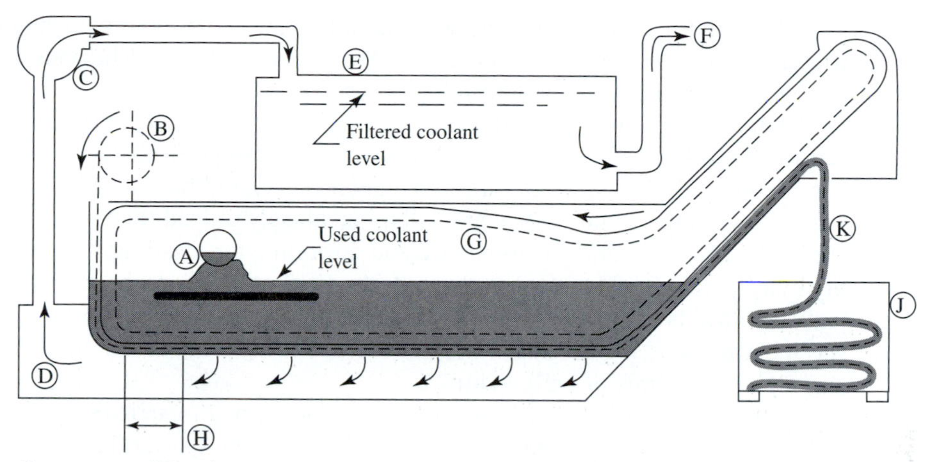

Component and Function

A. Used coolant inlet
B. Filter cloth in roll form
C. Pump—creates partial vacuum
D. Coolant drawn through filter media by vacuum
E. Filtered coolant reservoir
F. Filtered coolant returned to cutting tools
G. Open mesh conveyor belt to carry filter media
H. Length of index of filter media (determined by trial)
J. Used media container
K. Used media

FIGURE 9.6 Vacuum coolant filter.

Permanent media take the form of metal screens, meshes, and braided material, and may be back-washed, scraped, cleaned, and reused.

Disposable media are supplied in paper, cloth, fibers, and combinations. Many filters have a prefilter screening stage to eliminate rags, floating debris, and any large object that could damage the equipment or filter media.

Figure 9.6 illustrates a filter designed to operate on the vacuum principle. Components of this filter are a large used coolant reservoir with a vacuum chamber at the bottom, covered by a flight conveyor and disposable filter medium, a clean coolant reservoir, pump, sensors, and all necessary piping and electrical items.

The filtration cycle consists of the following:

- Used coolant is pumped into the used-coolant reservoir.
- Used coolant is pulled into the vacuum chamber by the pump, passing through the disposable medium in so doing.
- Chips, fines, and dirt particles are trapped by the filter medium and accumulate into a filter cake, which further filters incoming used coolant.
- As the filter cake increases in thickness, vacuum increases to a predetermined level, which initiates an index of the flight conveyor and filter media.
- A length of disposable medium and built-up filter cake is indexed into a hopper for disposal, and a fresh length is brought in by the flight conveyor.
- Clean coolant is pumped to a clean coolant reservoir and from there back to the cutting tools or, as in the installation, to the high-speed centrifuges for further clarification.

Filters and filter media can be selected to remove contaminants as fine as 1/2 micron and will also remove bacteria plus some desirable coolant additives. The additives must, of course, be replaced if removal is unavoidable.

High-speed disk bowl centrifuges are becoming indispensable in the operation of an efficient central coolant system. They can effectively remove tramp oil, bacteria, and fines from coolant, but are normally used downstream from a filter that would have a larger capacity to remove the bulk of the solids. Some centrifuges will automatically unload their accumulation of solids and should be used to reduce maintenance and downtime for cleaning. Centrifuges can be used to final clean and clarify coolant, following filtration, or they can be used alone where the contaminant load is within their capacity. Centrifuges cannot remove particles that have the same specific gravity as the cutting fluid, but the specific gravity of the fluid can be changed by a slightly different mix or with additives.

A **coolant makeup device** is required to replace coolant carried off on parts and chips, and lost to evaporation. Losses must be made up of the same type and mix of cutting fluid used in the system and must be made up in the quantity required to maintain the system at its designed operating level.

Manual mixing and addition of cutting fluid should not be relied on; mixing valves and proportioning displacement cylinders coupled with fluid-level sensors will reliably control mix and level. Indirect labor costs will also be reduced in so doing.

High-quality water has been proven to be the single most important factor affecting cutting fluid performance. Since water composes 90 to 99% of emulsifiable coolants, large quantities of high-quality, low-cost water are required for central coolant systems.

Most water in the United States is "hard"; that is, it contains calcium and/or magnesium. Hard water used in formulating emulsified cutting fluids develops sticky, gummy residue and also tends to "crack," or separate, the emulsion. Coolant flumes and sumps act to concentrate and increase hardness levels by evaporation: Water evaporates from the systems but minerals in solution remain.

Four methods of water conditioning or improving are used—with varying degrees of success:

- **Water softening** is a process whereby calcium and magnesium ions are replaced with sodium ions. The "hardness" is corrected, but the resulting sodium ions tend to corrode. This system finds considerable use, due partially to the low cost of the regenerating agent—salt or sodium chloride.
- **Deionizing**, which exchanges calcium and magnesium with hydrogen and also replaces sulfates, chlorides, and carbonates with hydrochloric acid, is more costly than sodium chloride, but the resulting water far outperforms softened water in cutting fluid usage. Also, deionized water can greatly extend the sump life of water-miscible fluids. Deionized water is the water of choice for use in emulsifiable cutting fluids.
- **Reverse osmosis** produces relatively pure water by a process of forcing water through a semipermeable membrane utilizing very high pressure. Water passes through, but impurities are retained. A purity of about 90% can be attained, but membranes are unpredictable and unreliable. Also, large quantities of water are lost in the process.
- **Distillation** produces water of 100% purity by evaporating and recondensing the water, leaving all impurities behind. Use of this process is limited due to the relatively high cost of capital equipment and energy to operate.

It was earlier suggested that the manufacturing engineer consult with a cutting fluid supplier when selecting a coolant for use in a process or processes. It is further suggested that when a central coolant is required as part of a manufacturing plan, the services of a cutting-fluid supplier who additionally supplies coolant system components be contracted for. Also, when purchasing, sourcing of both the coolant and coolant systems to a single source eliminates split responsibility.

Environmental Considerations Cutting fluids can be a major contributor to environmental pollution:

- Leaks from plumbing, pumps, storage and supply tanks and flumes permit coolant to find its way into water supply systems and navigable streams.
- Wet chips drip onto roadways and bridges, and into ditches, as they are transported to reclamation centers, and, again, the coolant finds its way into streams and water sources.
- Improper handling of effluent from parts washing machine cleanouts results in noxious materials discharged into city drains and sanitary sewers.
- Tramp oil in coolant causes fumes to generate when it comes into contact with hot cutting tools. Fans and blowers used to clear fumes from the worker's area exhaust them into the atmosphere.

Such environmental pollution can be eliminated by corrective action. Corrections include the following:

- Eliminate leaks.
- Wash and spin-dry chips before shipment. In addition to eliminating pollution due to this cause, appreciable amounts of coolant can be reclaimed and returned for further use.
- Monitor discharges from parts washing machines when cleaned out. Have all sludge and foreign matter transported to appropriate landfills or reclamation centers.
- Remove tramp oil with oil skimmers, filters, or centrifuges.
- Employ collectors to remove fumes and oil mist from the air before exhausting it into the atmosphere.

Also, consideration should be given to operations that can be machined without coolant. For example, many cast-iron parts are successfully machined "dry."

Coolant must be kept clean and free of trash; not only bacteria growth fostered by unclean coolant is a health hazard to workers who must come in contact with it, bacteria can foster the coolant "cracking," or de-emulsifying, or going out of solution. Further, the organic oils and elements putrefy, resulting not only in an unusable product, but one with an unbearable stench.

The only correction for this condition is to pump out the spoiled coolant into holding tanks and send it to reclamation for possible salvage of the elements, plus, all affected flumes, pumps, filters, piping, and so on must be washed out and recharged with fresh coolant—a big job in the case of a central flume, and an expensive one.

Some coolant suppliers have developed coolant cleaning facilities that run continuously and are so effective they have kept coolant in continuous use for several years without the need to pump, clean, and recharge. This equipment should be considered and used where it can be justified.

Maintenance A properly designed, installed, and managed coolant system can be a highly effective ongoing cost-saving activity—provided it is continuously and rigidly maintained.

Manufacturing engineering should develop a preventive maintenance system to maintain all systems, components in top operating condition and should see to it that a schedule of laboratory checks is set up to monitor the coolant condition.

Faithfully carried out, these two activities will extend coolant sump pump life indefinitely, and savings and improvements will be realized in

- Reduced labor and material costs due to reduced or eliminated coolant pump outs and refills
- Reduced tool costs due to extended cutting-tool life
- Improved quality due to better control of surface finishes, reduced corrosion, and reduced residual deposits on part surfaces
- Enhanced environmental conditions due to reduction of foul odors, reduction of smoke in atmosphere, and cleaner machines and plant surfaces.

9.2.2 Utilities

Electrical Considerations Total electrical requirements must be determined and compared to total available capacity when a manufacturing plan requires significant increases of devices, such as motors and heaters, and processes that consume electrical energy.

It would be prudent for Manufacturing Engineering to survey the entire manufacturing plant's electrical system, including

- Capacity of primary switch gear
- Transformers
- Capacity and requirements of all plant lines and areas

Action must be taken to add capacity and/or electrical facilities where needed and before needed.

Most machine tools require a 440-volt, 60-cycle electrical supply. Some use 220 volts; a few operate on 110 volts. Voltages are supplied by means of step-down transformers in increments of 480, 208, and 120 volts. High-frequency electrical energy and direct current requirements are generally handled by special means such as a motor–generator set placed in the production line adjacent to the source of need. An example of such a need would be an induction-hardening, heat treating process.

Compressed Air The total volume of compressed air required must be determined and compared with the total capacity available. The addition of only one major piece of manufacturing equipment can cause an extremely large increase in compressed air requirements for such uses as air cylinders used for clamping or movement; air used to eject parts from fixtures or dies; air used to blow off moisture, dust, or chips; and air gages and pneumatic controls.

It may prove cost effective to study and make an effort to reduce compressed air usage by means of substitutes such as mechanical movement devices, hydraulic clamping cylinders, mechanical cams, and electrical components—especially where air use versus air capacity is borderline.

When a large amount of compressed-air capacity must be added, a rotary air compressor might be considered due to its inherent large capacity but with the least amount of required plant floor space. The extremely shrill sound of a rotary compressor is usually controlled with sound insulation or shielding.

Means must be provided to remove all moisture from the compressed air; whatever water vapor was in the air before compression condenses into water, which is, of course, detrimental to piping, valves, equipment, and tooling.

These systems are in general use for the removal of moisture from compressed air:

- Salt or desiccant
- Mechanical—traps, drains
- Refrigeration

Gas Gas availability must be determined and checked against the amount required. Gas is used in large quantities in manufacturing areas of

- Heat treating
- Comfort heating
- Steam generation
- Drying processes
- Incineration

Usually, like electrical energy, gas is brought into a manufacturing plant area known as the "power house," where a considerable amount is used to fire boilers to create steam. Some gas is distributed to the plant for production process uses. Some boilers operate on coal or oil with natural gas as a backup.

Whatever the distribution plan is, it is vital to determine gas availability and *future* availability, and also the utility company's policies: Many manufacturing plants have been purposely located and built in areas that have ample quantities of natural gas available. Much of the U.S. automotive catalytic converter core and pellet production was sourced to be manufactured in Oklahoma for this reason.

Water Water availability must be determined and compared with the proposed total requirements for this element. Manufacturing plant water uses include

- Steam
- Fire systems
- Human consumption—potable
- Sanitary systems
- Coolant systems

Many manufacturing plants have two water systems due to the unavailability or high cost of a sufficient quantity of high-quality water—making water recycling a necessity in either case.

One system is a process water system for use in manufacturing processes; the other is a potable water supply for human consumption. Rigid controls, valves, and antisiphon devices keep them separate.

Whether the manufacturing plan is for a new plant, or for a revised, existing facility, *all utilities* must be considered and provided for well in advance of the final project approval. Failure to do so can result in embarrassment, emergency measures, and even a complete failure of the plan.

9.3 AUXILIARY SYSTEMS

Auxiliary operations must be included in the sequence when a part is originally processed. All such operations must satisfy part print specifications; most require some type of process equipment or systems. Some auxiliary processes, such as manual deburring, are labor intensive; others require only loading and unloading into fully automated equipment.

Each operation must be analyzed and planned with the same aggressive attitude and attention to detail given a major machining operation. Some considerations to be given selection of equipment for various auxiliary operation follow.

9.3.1 Deburring and Polishing

Burrs must be removed from parts and sharp edges broken in auxiliary operations when it is not feasible, or possible, to accomplish in a major machining operation. *Manual deburring* is normally performed at a bench location with special hand tools, picks, and files when efforts to include the work in the machining operation fail. A manual operation may have to be resorted to in order to get the job going, but efforts to eliminate the manual work must continue. An example of this involves the deburring of a small vent hole in the bore of an automotive master cylinder as illustrated in Figure 9.7.

The cups and piston, normally in position as shown in Figure 9.7(a), allow brake fluid to flow from the reservoir into the cylinder bore. As brakes are applied, piston and cup move forward, successively trapping fluid in the cylinder and then pressurizing fluid in the cylinder and complete brake system as shown in Figure 9.7(b). Hydraulic brake fluid under pressure forces the neoprene cup to bulge into the vent hole as shown in Figure 9.7(c) and enlargement.

If the edge of the vent hole has a burr, or is sharp, successive applications of the brake will cut a groove in the cup, causing a leak and eventual master cylinder failure. Efforts made to deburr the edge of the hole included a special scraping tool, dental burrs, and vapor blasting. The problem was eventually resolved with a wedge-operated punch that coined the edge into a radius as shown in Figure 9.8. This tool was included in the last station of the machine that drilled the vent hole. Most mechanized deburring processes also upgrade (polish) all part surfaces.

Tumbling is a process whereby parts and abrasive media are loaded into a hexagon-shaped barrel, which rotates about its central axis. The resulting tumbling action causes parts and media to rub in a mass movement that removes burrs and sharp edges, and polishes surfaces. The medium consists of plastic and ceramic preforms (triangles, squares, and rectangles of appropriate sizes), polishing compounds, and water.

Tumbling is a general-purpose process used to remove burrs and break sharp edges of parts' exterior surfaces. Some difficulty is experienced with media shapes lodging and jamming in parts' holes, slots, and interior surfaces. Medium shapes, abrasives, loads, and tumbling time are found by trial.

Vibratory finishing is, in actuality, an efficient, mechanized form of tumbling. A mix of parts and medium is developed as in tumbling, but relative motion between the two is provided by vibrating the container or tub, rather than revolving the mass in a barrel.

Parts can be loaded, processed, and unloaded—separated from the medium—automatically. The process is suitable for almost the same range of parts as would be processed in a tumbling barrel.

Slurry-type deburring machines suspend parts in a bowl of abrasive paste, or slurry, after which the parts are rotated and the bowl revolved to obtain the relative motion required to abrade burrs, break sharp edges, and polish surfaces as shown in Figure 9.9.

Parts selected to use this process must be configured to permit chucking, clamping, or mounting on an arbor. The slurry process can produce a very fine, controlled finish on surfaces and controlled radii on corners.

Abrasive jets direct a blast of air and abrasive against a burr, edge, or surface. Sand blasting is an example. This process can be used for low production, or on small parts in a cabinet, or it can be automated for high production. Jets can be used on internal surfaces and holes if such are accessible.

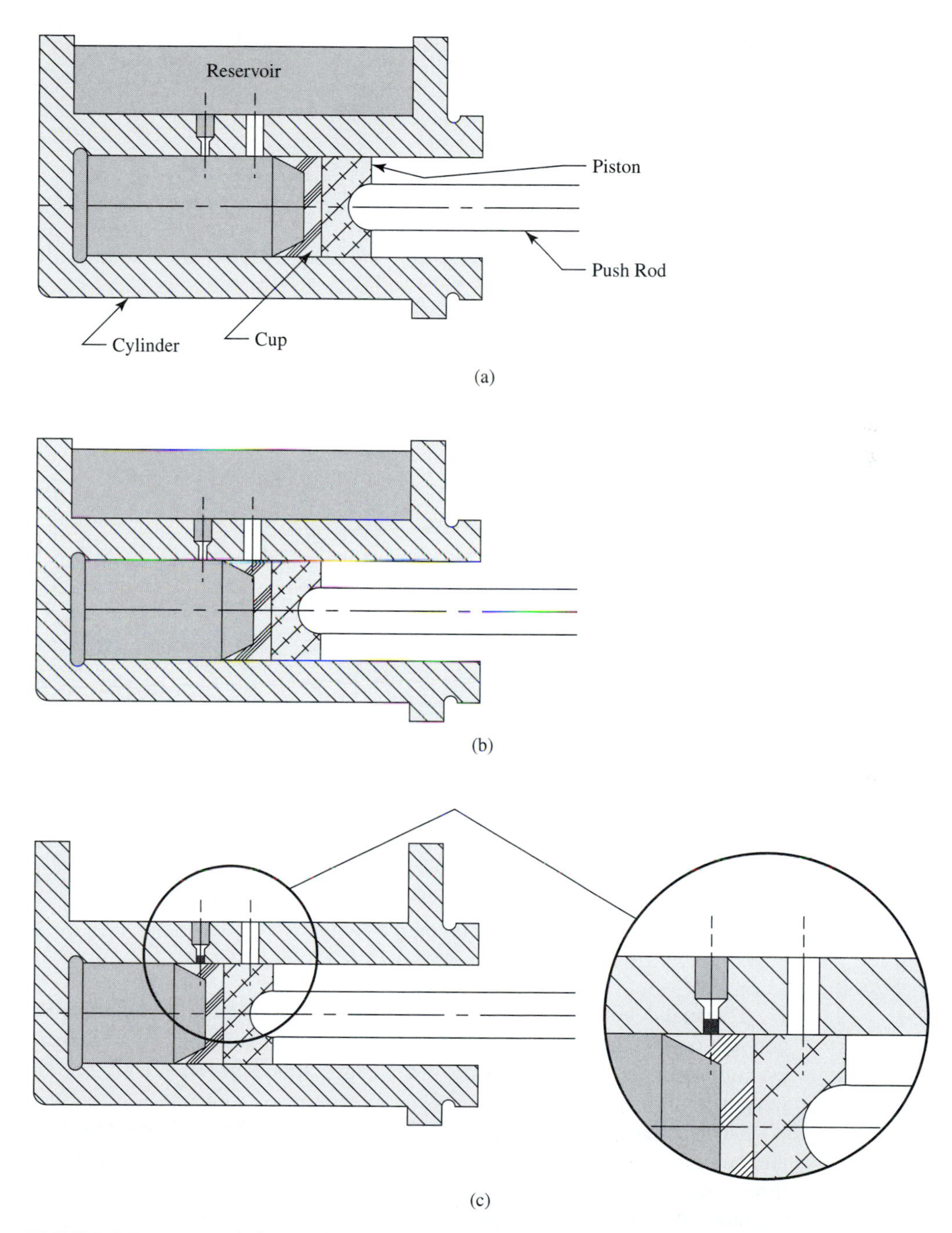

FIGURE 9.7 Burr and sharp edge of hole cause failure of automotive master cylinder. (a) Master cylinder—filling. (b) Master cylinder—venting. (c) Master cylinder under pressure. Neoprene cup is forced into vent hole by hydraulic pressure. Sharp corner of hole cuts cup and causes leak path.

Water jets or hydro jets are devices similar to abrasive jets except that only water under extremely high pressure is used to cut away burrs and sharp edges.

Most equipment utilizing this principle clamps the part in a fixture and programs the desired burrs and edges to move past the jet. This process leaves no residue and requires no subsequent washing operation.

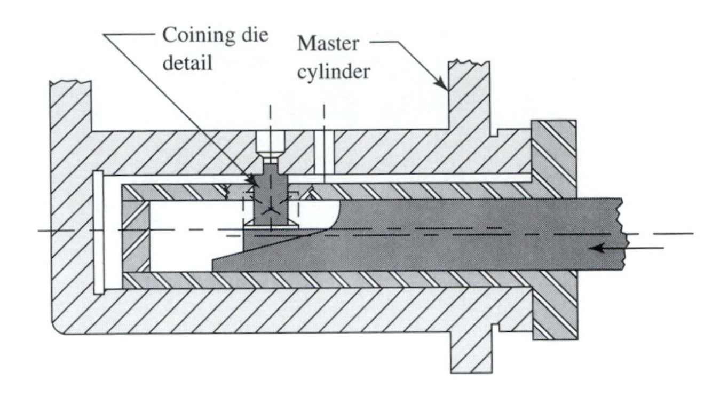

FIGURE 9.8 Burr and sharp edge removed with tool to form radius on sharp edge of vent hole.

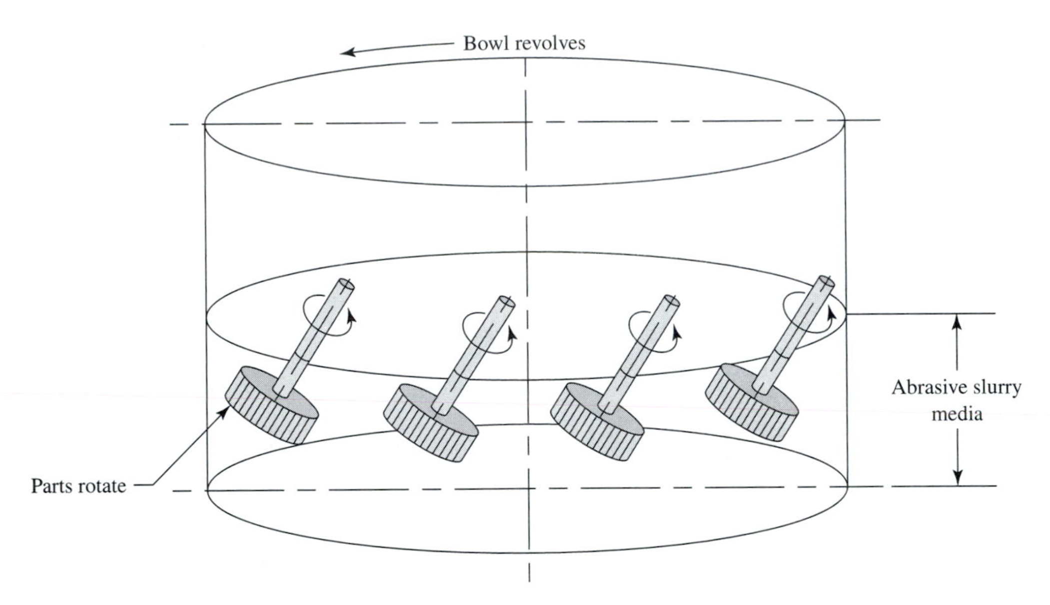

FIGURE 9.9 Wet slurry–type deburring machine.

Thermal deburring is accomplished by placing a part in a sealed chamber into which hydrogen and oxygen are introduced and ignited. The extremely high temperatures that result vaporize and oxidize small burrs or fine, sharp edges (fins) without affecting other surfaces. Explosive noise is a problem.

Electrochemical, or electro-polishing is a process wherein burrs and sharp edges are removed and surfaces polished by making the workpiece the anode of an electroplating process. The surfaces, burrs, and sharp edges are, in effect, then "deplated." A very fine surface finish can be obtained, but this is very dependent on the surface roughness of the part as it comes to the process.

Wire brushes and buffing wheels are used to deburr and polish a wide range of parts with an equally wide range of surface requirements. The parts may be handheld or automated. Surfaces may be polished and brought down into the very low microinch readings by the proper selection of buffing wheels and abrasives.

One technique used very successfully to remove grinding burrs from centerless ground parts is shown in Figure 9.10. A centerless grinder is set up in tandem with the finish grind centerless grinder to remove the fine sharp edge and break the corners. This grinder has a wire brush mounted in place of a grinding wheel.

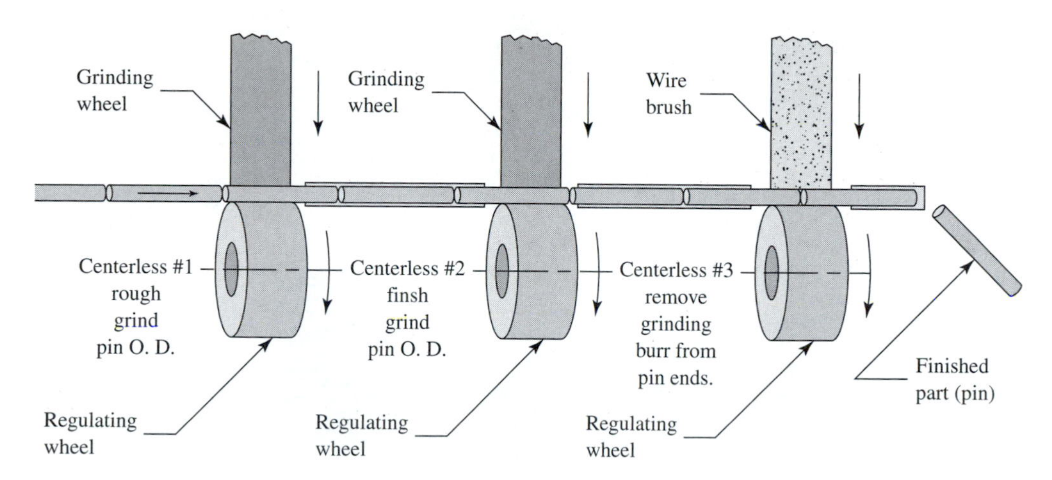

FIGURE 9.10 Centerless wire brushing to automatically remove grinding burrs from O.D. of pin ends.

9.3.2 Cleaning

Parts are cleaned on auxiliary equipment to remove oil, chips, dirt, or residue and prepare them for succeeding operations such as accurately locating in work-holding fixtures, heat treating, and/or finish applications. The reason for cleaning often dictates the type of cleaning equipment required.

A degreaser, shown in Figure 9.11, does an excellent job of cleaning when removing oil from a part's surface is the only requirement. If chips are present, perhaps adhering to the part by means of an oily, sticky film, a washing machine with a mechanically activated washing action will be required to remove both.

Degreasing systems must be properly installed and vented to control the heavier-than-air fumes, which can create a potentially hazardous (suffocating) situation.

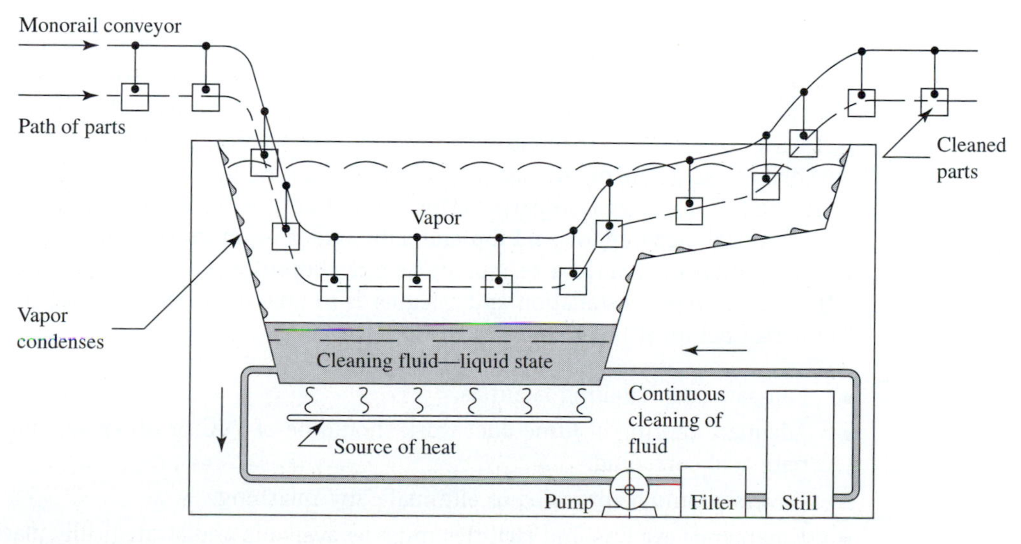

FIGURE 9.11 Degreaser-type cleaner for oily parts.

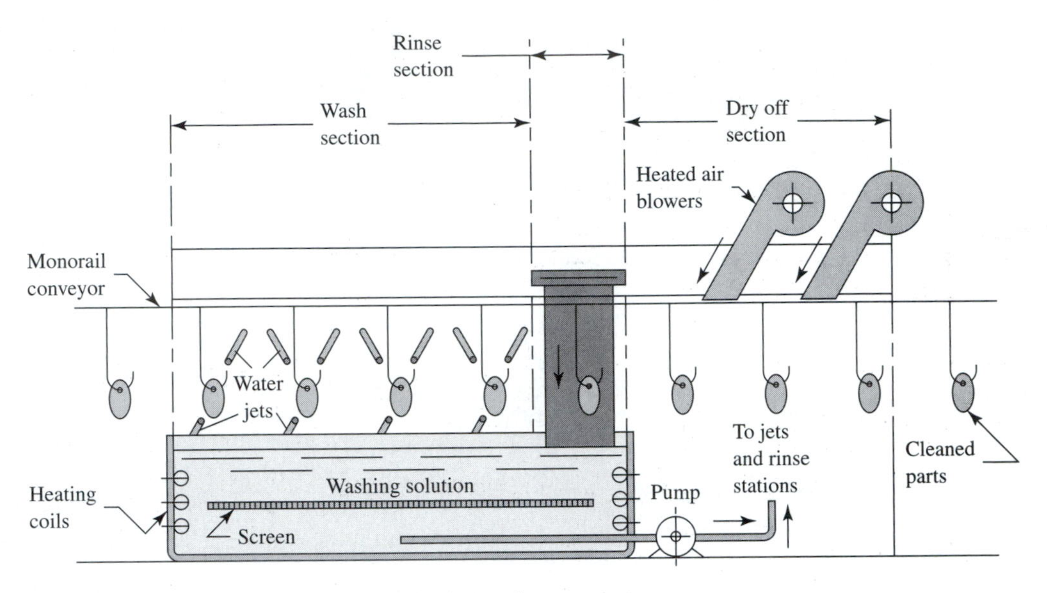

FIGURE 9.12 Washing machine with drying feature.

Fresh air must be supplied for machine operators and interlocks provided to prevent personnel entering unsafe areas. Some types of degreasing fluids such as trichloroethylene are no longer used in open degreasers due to toxicity.

The *washing machine* shown in Figure 9.12 is typical of most general-purpose washing machines used to clean parts. These machines pump hot or cold water and cleaning solutions through nozzles arranged to impinge on the part's surfaces. Many washing machines include a drying station utilizing heat and forced air.

Usually, parts are carried through the washer hanging on some type of monorail conveyor hook. Some water and cleaning solution is lost to carry-off, and provisions must be made for makeup.

Like all manufacturing equipment, washers must receive regularly scheduled maintenance to remain effective: Washers are self-contained, and all oil, chips, and dirt washed off parts remain, progressively degrading cleaning action to an unacceptable level. The washer must then be pumped out and cleaned, screens changed, and clean water and solution added.

The quality level of machined parts is so dependent on part cleanliness that many manufacturing companies have instituted formal "Parts Cleaning Committees" with responsibility and authority to schedule maintenance of all plant cleaning facilities. The committees are guided by "dirt checks" of production parts performed regularly in clean rooms by lab technicians.

Chemical cleaning, or acid dip and subsequent water rinsing may be performed as a separate operation to remove a residue or as one station in a sequence such as electroplating. In either case, proper installation and safeguards to guarantee safe working conditions for operating personnel must be provided:

- Fail-safe acid-handling facilities.
- Adequate venting of fume duct work should be of PVC or other suitable material to eliminate acid corroding.
- Covers should be provided to eliminate any splashing.
- Emergency eye washing facilities must be available and strategically placed.

9.3.3 Heat Treating

Appropriate and adequate facilities must be available, or provided new, to process parts to required metallurgical specifications and standards. Actual times and required machine capacities are determined by Industrial Engineering working with the plant metallurgist. Manufacturing Engineering, guided by the developed times and capacities, must provide additional and/or new equipment where required.

Heat Treat Process

Furnaces

- Stress relieve or normalize.
- Carburize low-carbon steel parts to prepare for subsequent surface hardening.
- Through-harden alloy and high-carbon steel parts.
- Surface-harden carburized steel parts.
- Reheat and temper.

This type of equipment is normally located in a separate area of a plant due to the uncomfortably high heat generated and also due to its potential for fires.

Induction-Hardening Equipment

- Surface hardening bearing areas of previously heat treated parts is used for high-carbon steel parts that have been heat treated to increase toughness of core by hardening and drawing back to a lower hardness. An example of part and process is shown in Figures 9.13 and 9.14. Induction equipment, including a high-frequency alternating power source, is normally located in-line of a production line.

Flame Hardening

- Surface-harden steel parts with flame from oxyacetylene gas, then quench with spray.
- Flame may traverse the part, or the part may rotate to provide relative motion.
- Equipment may be located in-line or in separated heat treat department. In-line, the crackling and snapping sounds of gas ignition may prove unsettling to nearby workers.
- The process is normally used for larger parts, ring gears, and the like.

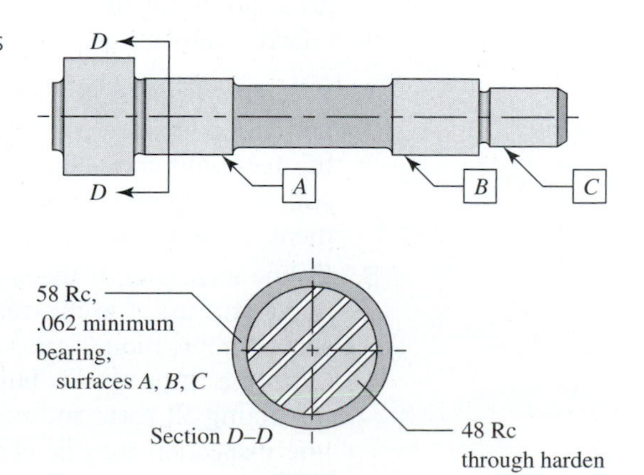

FIGURE 9.13 Part with two levels of hardness specified.

Operation No.	Operation	Equipment
10	Through harden:	
	Heat	Furnace
	Quench	Quench tank
20	Draw back to 48 Rc:	
	Reheat to required temperature	Furnace
	Quench	Quench tank
30	Surface harden bearing diameters to 58 Rc,	
	.062 minute depth	Induction hardener
	Reheat to required temperature	Quench ring
	Quench	

FIGURE 9.14 Heat treat processing required for part in Figure 9.13.

9.3.4 Surface Treatment

The field of surface finishes is extremely specialized and varied, and includes painting, plating, and chemical surface treating. Coverage is beyond the scope of any one textbook, yet a manufacturing engineer must know enough about the field to be able to plan and provide adequate facilities when so specified. A manufacturing engineer would do well to study, do general reading on the subject, and, when confronted with a specific problem, consult with specialists—in particular, vendors of the specific type of equipment required for assistance. In that way, capacities, space required, and estimated costs can be obtained for the manufacturing plan.

9.4 SUPPORT SYSTEMS

9.4.1 Quality Control

Quality control's function of monitoring quality levels of all raw material, purchased finished parts, and manufactured parts requires dedicated gages, inspection equipment, material-handling facilities, work areas, and storage space.

A manufacturing plan for a new plant must provide for all such requirements, and a plan for a change or addition to an existing facility must provide for any addition and/or changes required.

Areas of the quality control function requiring consideration of facilities for inclusion in a manufacturing plan include the following:

A. **Receiving inspection** checks and controls quality levels of all incoming castings, forgings, purchased finished parts, tooling, and raw material. A well-laid-out area in proximity with the receiving dock—including such material-handling devices as hoists, trucks, recording scales, and cranes—must be provided. Also, all pertinent gages, general measuring equipment, surface plates, and hand tools must be included.

B. **In-line inspection** is the system that has been used by most U.S. manufacturers in the past. Some still use it exclusively, some now employ *statistical process control*, and others use some combination of the two.

   In-line inspection is built on the philosophy that a quality level will be maintained by inspecting all parts and removing from the system any parts not within specifications. In-line inspection may be accomplished by roving inspectors or in work stations located at strategic points in a line and at the end of a line (final inspection). Workstations must be properly and functionally designed, with special attention being given to providing good lighting. All checking fixtures, particular gages, and material-handling requirements must also be provided.

C. **Statistical process control** depends on the philosophy (here greatly simplified) of rendering all processes capable and maintaining them at a specified quality level by strategically sampling and inspecting parts from each operation, lot, run, heat, or shift of resulting production.

   Corrective action of tool adjustments, replacement, machine maintenance, or other is immediately taken to correct any discrepancies before further production is permitted.

   Gaging and measuring devices are required, and areas must be provided where a lot, or a bank, may be held while parts are sampled. Additional areas are often required to impound and hold rejected lots while sorting and repairs are made to parts and to the faulty process.

D. **Testing** a product usually means putting it through its functional cycle while monitoring, observing, and comparing the results with the required performance parameters.

   Such equipment may be simple devices or, for complex products, highly specialized facilities located in dedicated, enclosed areas and requiring specially trained machine operators.

   Whatever the requirements, they must be provided for in the manufacturing plan. Two examples of testing equipment developed for complex products follow.

Testing–Transmissions. Figure 9.15 illustrates a computer-controlled automatic-transmission-testing machine. In use, the tester manually loads a transmission into the testing machine from the transporting conveyor by means of an air hoist.

When the transmission is properly located, the tester pushes a control button to initiate the test cycle. The computer control then

- Clamps the transmission in location
- Connects hydraulic lines, drive sleeves, and mechanical (shift) controls
- Drives the transmission through a test cycle for
  - ✓ Shift points
  - ✓ Pressures

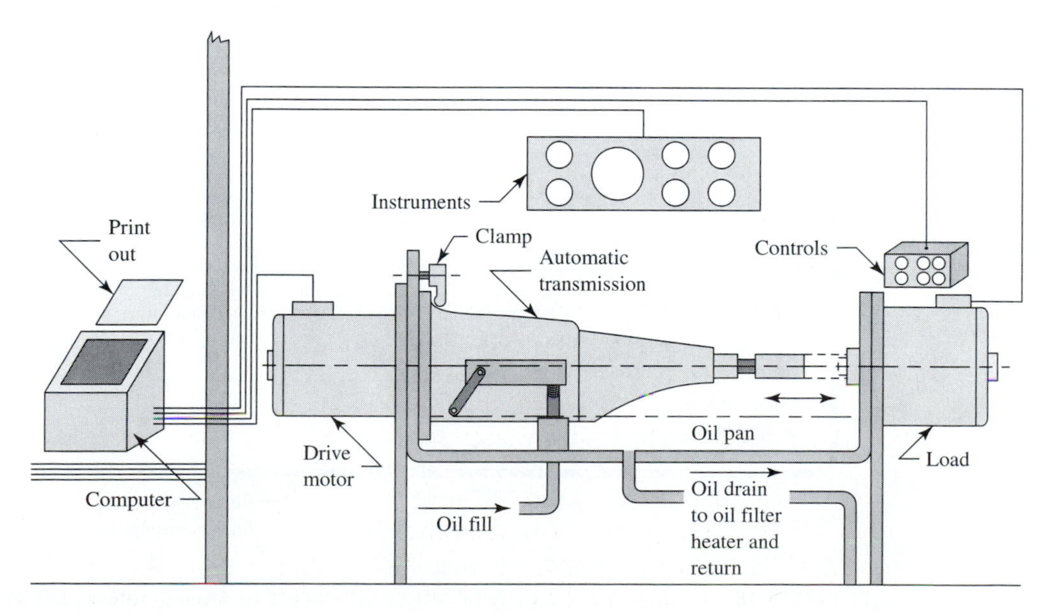

FIGURE 9.15 Manual load/unload—computer controlled automatic transmission test stand.

✓ Noise
✓ Reverse and Park

- Records and marks OK transmissions
- Rejects and supplies a printout of specific defects for faulty transmissions
- Unclamps and disconnects

The tester removes the transmission, those marked OK are forwarded to shipping, and the rejects are routed to repair and retest.

Engine Testing. Figure 9.16 illustrates a test facility designed to test, repair, retest, and ship V-6 automobile engines. Included are fully automated, computer-controlled test stands, conveyors and switch gear, repair equipment, and repair area test stands.

Programming provides the following:

1. Convey the engine from the final assembly to the test area on a special test/repair pallet that has magnetic tape coding capacity.
2. Transport the engine to the first available test stand wait station.
3. Position the engine in the wait station.
4. Move the previously tested engine from the test stand to the conveyor, and move the engine now in the wait station into the test stand.
5. Automatically clamp the engine in position and automatically connect water, oil, and electrical services.

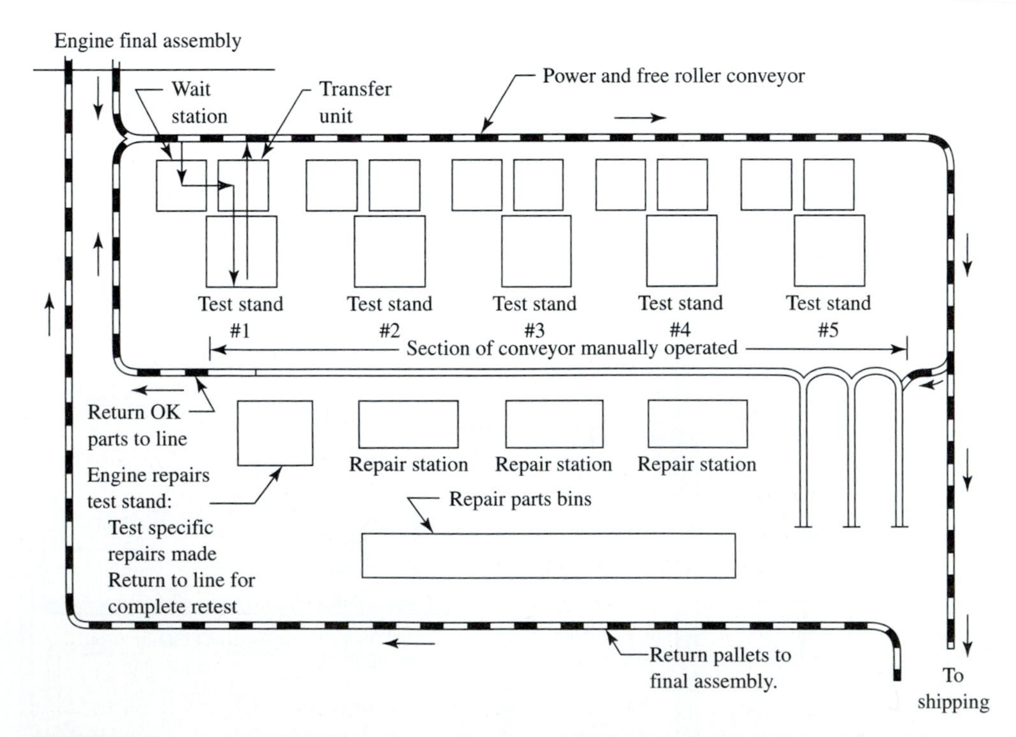

FIGURE 9.16 Computerized facility to test, make necessary repairs, retest, and ship V-6 engines.

6. Test the engine for function and any defect in six groups:
   ✓ Mechanical
   ✓ Leaks
   ✓ Cooling
   ✓ Ignition
   ✓ Carburation
   ✓ Oil pressure
7. Code the engine pallet magnetic tape with test results and the engine forwarding destination:
   A. OK engine
      - Route to shipping
   B. Defective engine
      - Route to repairs
      - Specific defect
8. Unclamp the engine and disconnect the services
9. Move the pallet onto the conveyor
10. - Automatically transport OK engines to shipping
    - Automatically transport defective engines to repair

Defective engines are repaired, tested in repair area test equipment, and returned for final retest.

Note: Magnetic tape records the number of times each engine is returned for repairs, the reason, and the repairman. Production supervision can monitor the tapes and take corrective action, especially in instances when an engine has been sent back repeatedly for repair and retest of the same item.

9.4.2 Production Control

Production Control is responsible for all production material: receipts, movement, shipping, and storage. Four areas of this activity—receiving, material handling, shipping, and by-products—require space and facilities appropriate to their functioning, and the requirements for each must be included in the manufacturing plan. Some considerations of requirements for each of the four Production Control areas of responsibility follow.

Receiving A smooth, uninterrupted flow of incoming material is mandatory to sustain efficient production schedules—especially with the trend to reduced inventories and just-in-time planning. Receiving docks and unloading facilities must be planned to enhance flow and reduce unloading time. Some desirable features are illustrated in Figure 9.17.

1. Adequate truck positions and dock plates must be provided to handle the number of trucks anticipated.
2. Railroad docks must be adequate to spot the number of cars expected to unload simultaneously.
3. Parking areas for dropped-off semitrailers must be provided.
4. Driveways must be planned to promote the easy flow of trucks to and away from docks and trailer drop-off parking areas.
5. Safe, fast-acting, strategically located cranes and hoists must be provided.
6. Adequate storage areas for incoming materials must be planned in relationship with the dock and point of use.

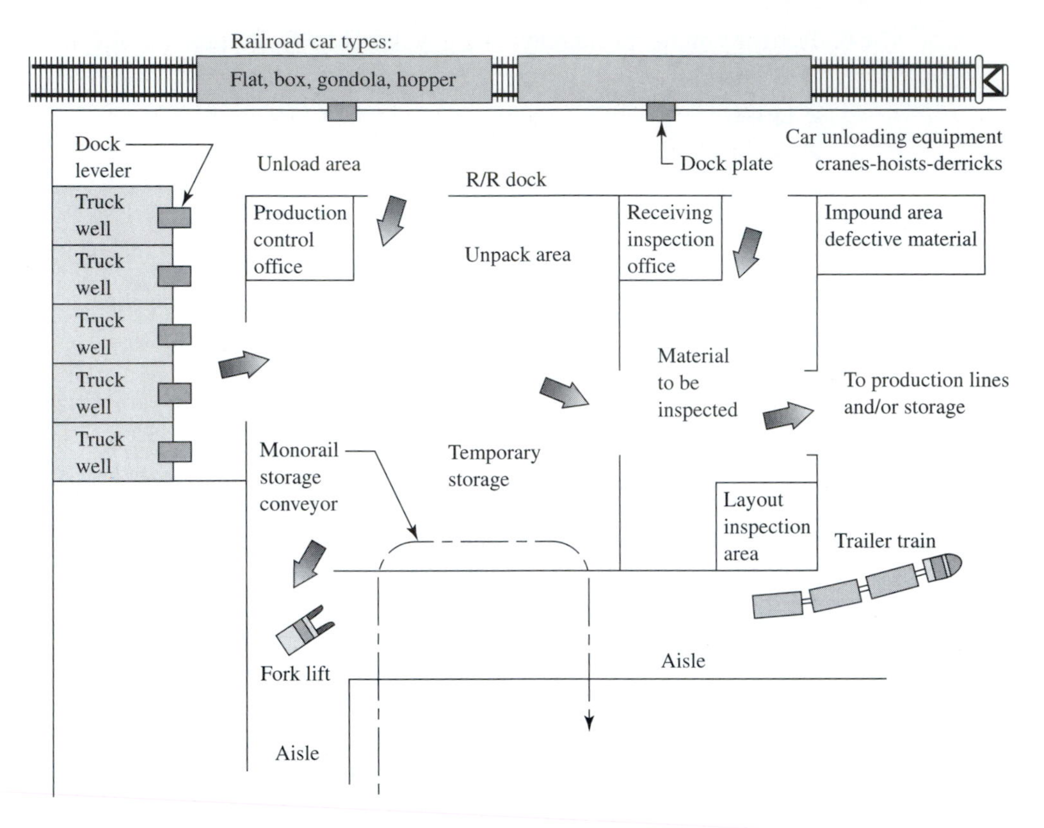

FIGURE 9.17 Receiving dock area relationships and functions.

Material Handling Material-handling facilities and systems must be planned to have adequate capacity and flexibility at all planned (and indeed, unplanned) production schedules. Elements of the systems and some of their requirements include the following:

1. **Parts containers**, which range in size and form from large steel tubs to small perforated baskets, must be selected to suit the parts involved, quantities required, and method of transport and use at destination. Containers may consist of original cardboard cartons supplied by vendors and transported intact to point of use.

Arrangements may be made with material suppliers to use permanent, returnable containers, which can be transported to the production lines as received, then returned for refill. This arrangement is usually the most cost effective in that it obviates continued purchase, opening, and disposal of cardboard containers.

Container sizes and loaded weights must be considered in relation to their mode of transport (manual, truck, conveyor) and disposition at point of use (dump, place on rack, etc.).

A container washing machine should be included in the plans to maintain clean and oil-free parts and to enhance good plant housekeeping.

2. **Vehicle** types and capacities must be selected to suit parts and material configuration, weight, and quantities.

- *Hand trucks* are best used for small loads moved short distances and to reposition skids and containers within a localized area.

- *Platform trucks* are used to move heavy loads and material that has been loaded on raised skids, but *forklift trucks* can be so used and also will handle low shipping pallets. Attachments are available for forklift trucks to adapt them to better move cartons, barrels, bales, steel carts, and so on.
- *Specialized trucks* are available to handle rods, pipes, steel coils, lumber, hoists, machinery moving, and so on.
- *Trailers* are used to transport containers, large parts, and raw material. Towed by a tractor, trailer trains of any desirable length can be formed. Single trailers and trailer trains can be automated: Designed to follow a cable in the floor, self-powered, driverless trailers can be programmed for any desired route including stops and starts for loading and unloading.

3. **Conveyors** transport the major portion of raw material, workpieces, and finished product in most manufacturing plants. The diversity of product size, shape, weight, and quantities has caused conveyor suppliers to develop a great many types of conveyors from which to select. Conveyor specialists and the intended eventual supplier should be contacted early in the development of the plant layout and manufacturing plan. Some basic conveyor types and components follow.

Conventional Conveyors

1. **Monorail conveyors** are perhaps the single most used type of conveyor installed. Their advantages are relatively low installation costs, low operating and maintenance costs, flexibility, and the saving of floor space because the entire conveyor is hung.

The monorail's single track can be bent to follow any contour and grade, including installations that change floors.

Monorails can be arranged to carry a very great variety of parts, assemblies, cartons, and containers by means of fixtures, hooks, and trays hung from the conveyor chain.

Since chain, rail, drives, and trolleys are standardized, they may be reused in other conveyor installations or changes, and monorail may easily be lengthened or altered (see Fig. 9.18).

2. **Roller conveyors** may be gravity or powered; that is, this conveyor may be inclined so that objects to be carried move along by gravity (or may be assisted with a manual push), or it may be powered to move the load.

Roller conveyors find wide use in transporting cardboard cartons, boxes, and containers. A variation of the conveyor is the wheel conveyor, which is suitable for lighter loads and may be used in corners (see Fig. 9.19).

3. **Belt conveyors** are powered endless belts used for much the same purposes as roller conveyors (see Fig. 9.19).

Special Conveyors

1. **Power and free conveyors** are a combination of powered conveyor and unpowered, interconnected with track switches, elevating and lowering devices, and repair areas and controls. The controls may be manual or automatic, and the system may be computer controlled.

The concept of power and free can be applied to either monorail or roller conveyors. The conveyor incorporated in the automated engine test illustrated in Figure 9.15 is an example of power and free as applied to roller conveyors.

2. **Floor conveyors** are designed to operate with their top surfaces at floor level, eliminating the need to lift or lower loads. In use, parts and/or containers of parts are placed on, or pulled off, the conveyor at desired locations, sometimes by use of a hook.

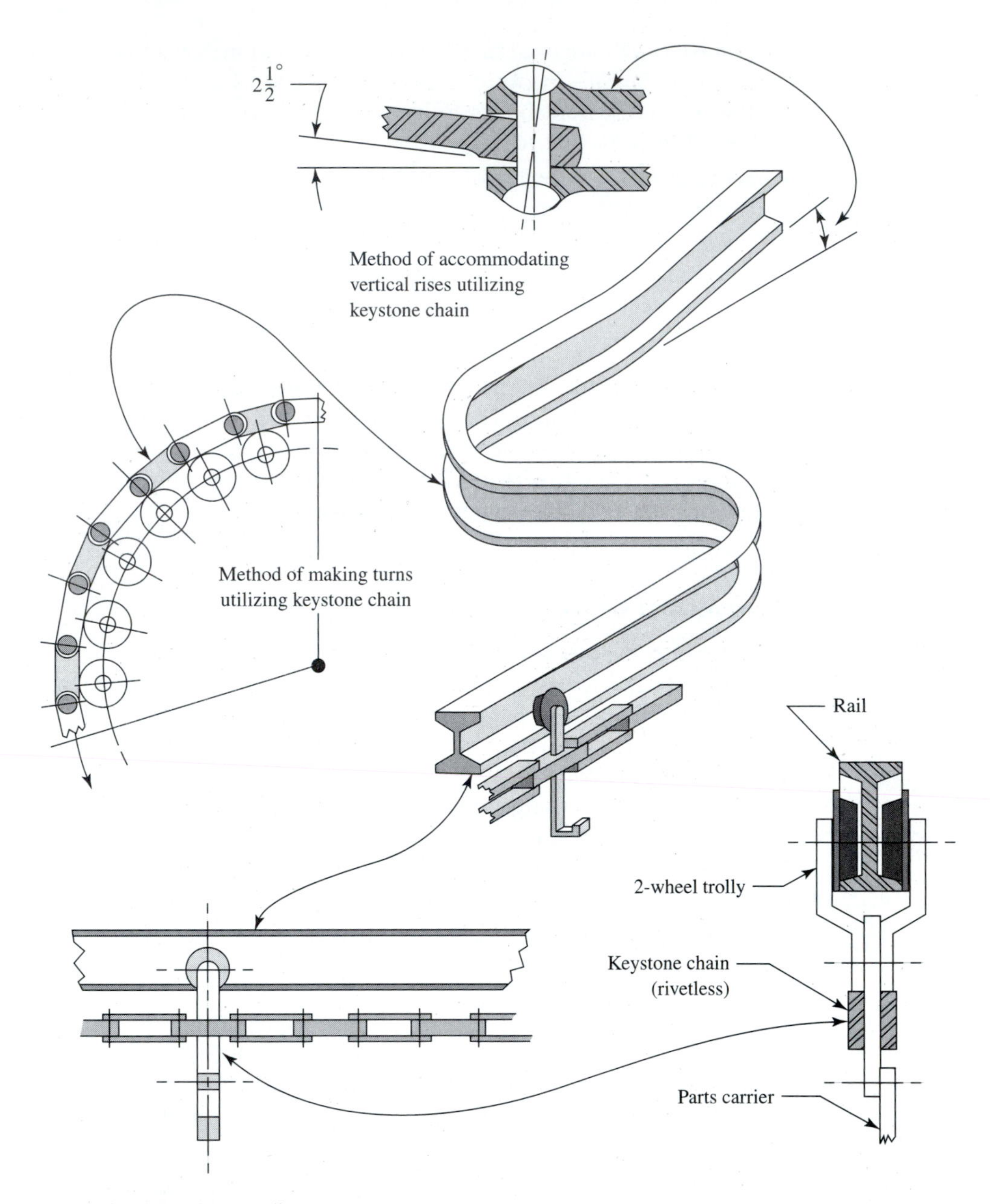

FIGURE 9.18 Monorail conveyor.

This conveyor is frequently used for such heavy loads as containers of parts as found in cold healing operations. An example is shown in Figure 9.20.

3. **Storage conveyors** can be of any type, but the most adaptable are roller and monorail conveyors. Monorails can be routed to take material to, and store it in, remote areas that are less frequently used. Some monorails have been designed to store material in serpentine fashion near the ceilings of plants.

Storage conveyors accumulate tremendously heavy loads; their supporting structure must be designed accordingly, including any plant structure involved. In one instance, a railroad was used as a storage "conveyor"—a plant that manufactured automobile fuel tanks shipped the

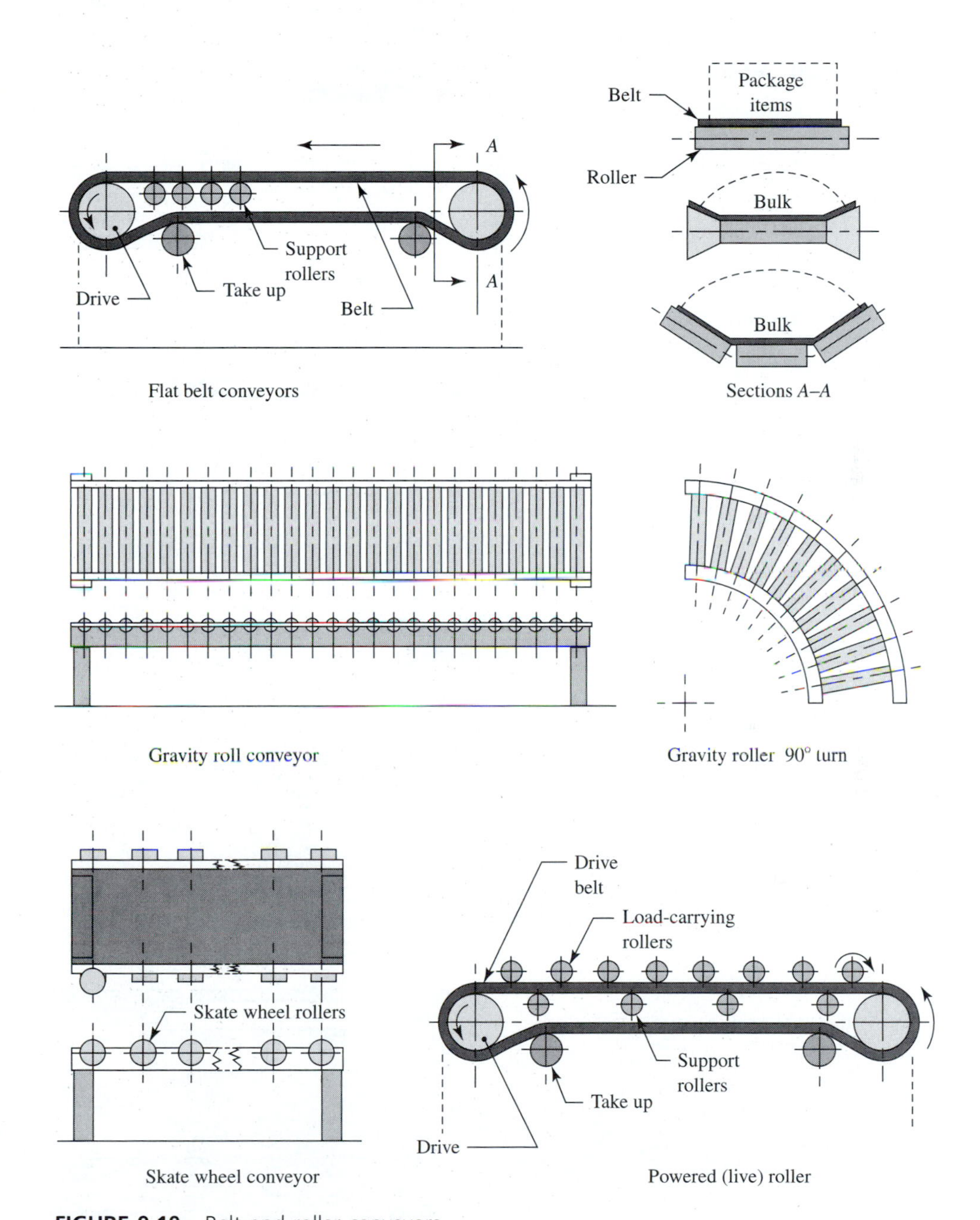

FIGURE 9.19 Belt and roller conveyors.

tanks to a car assembly plant located in the same city. Since the tanks occupied a large amount of space, railroad cars were chosen as the mode of transportation. After determining the length of time required "on the road," the necessary number of railroad cars were loaded and shipped daily—the railroad thus becoming an en route storage facility.

4. **In-line banks** must be provided and maintained at certain operations. A new progressive manufacturing line has the capacity of each operation so balanced that it operates on the normal flow of parts from one operation to the next. However, some machines must be down for periods of time for normal tool changes, scheduled maintenance, and so on.

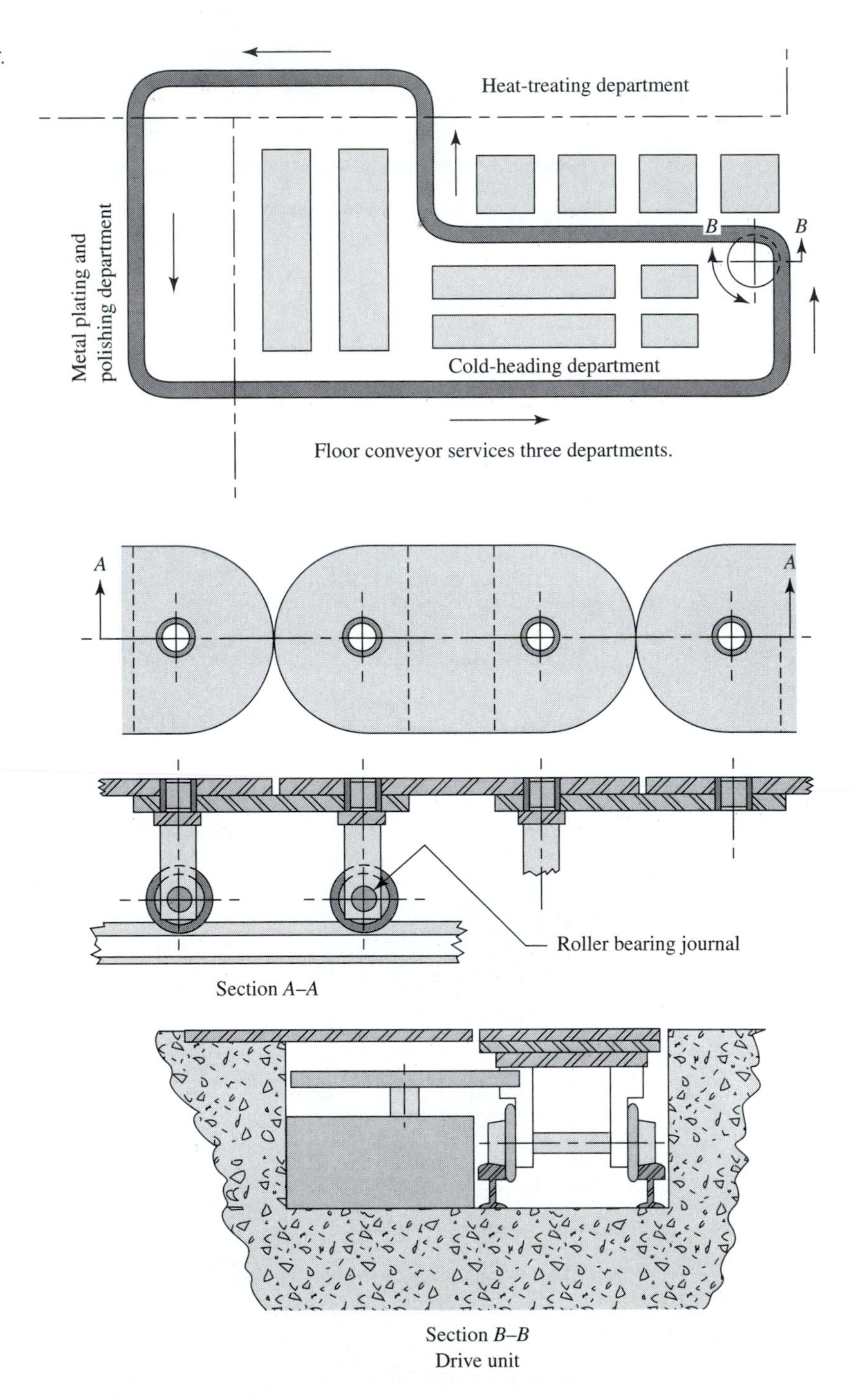

FIGURE 9.20 Floor conveyor.

During down time an in-line bank of parts must be supplied to keep the balance of the line producing. After the down machine is again operational, the bank is restored, either by the machine's excess capacity or by operating the machine overtime.

Appropriate parts storage facilities must be planned and provided for all in-line banks. A simple container may suffice for the parts required for some operations, whereas others, such as pinion gear blanks, require a special storage facility such as silos or roller spirals.

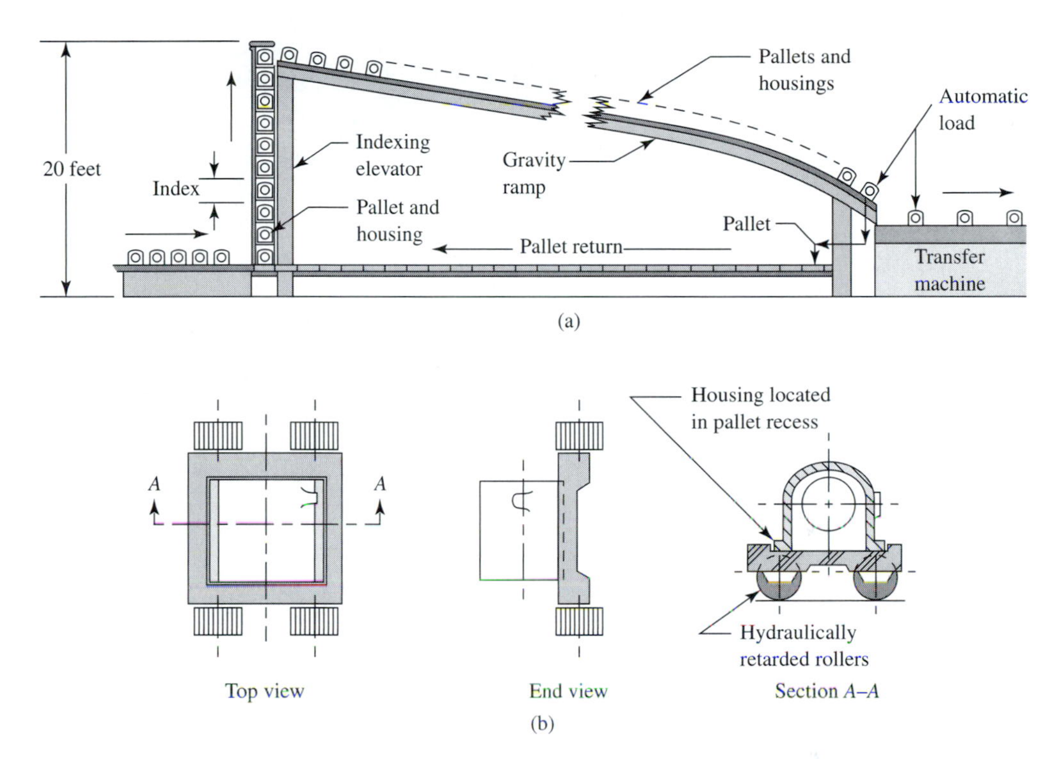

FIGURE 9.21 Storage ramp for semifinished machined housings. (a) In-line parts bank for transfer machine. (b) Pallet details.

In another instance, a bank of housings had to be maintained before and after a transfer machine. The parts, which were large castings, did not lend themselves to stacking due to the damage that would occur to several machined surfaces.

The problem was resolved by designing and building unique storage ramps as shown in Figure 9.21. The housings were clamped to a pallet, which was elevated to the top of the ramp. Then, as needed, they descended the ramp to the point of use controlled by a built-in hydraulic retarder.

Shipping The finished product must be shipped to the customer regularly and promptly to satisfy established shipping orders and schedules.

Some manufacturing plants have common shipping and receiving docks and handling facilities; other plants have separate docks as shown in Figure 9.22. For either arrangement, total facilities must be adequate for both shipping and receiving activities.

Packaging, when required, is logically performed by Production Control in an area adjacent to shipping. Packaging facilities must be provided for in this plan.

By-Products Chips, scrap, bar ends, and other by-products of manufacturing processes are usually handled by Production Control, including picking up, transporting, storage, and subsequent shipping to scrap dealers.

Higher selling prices of by-products can be realized by proper handling and preparation of such material. Actions that enhance by-product selling prices include these:

- Segregate steel, aluminum, cast iron, and other materials.
- Segregate same-kind material by sizes of bar ends, scrap, chips, and stamping offal.
- Compress stringy material such as long shavings and stamping offal in a baler.

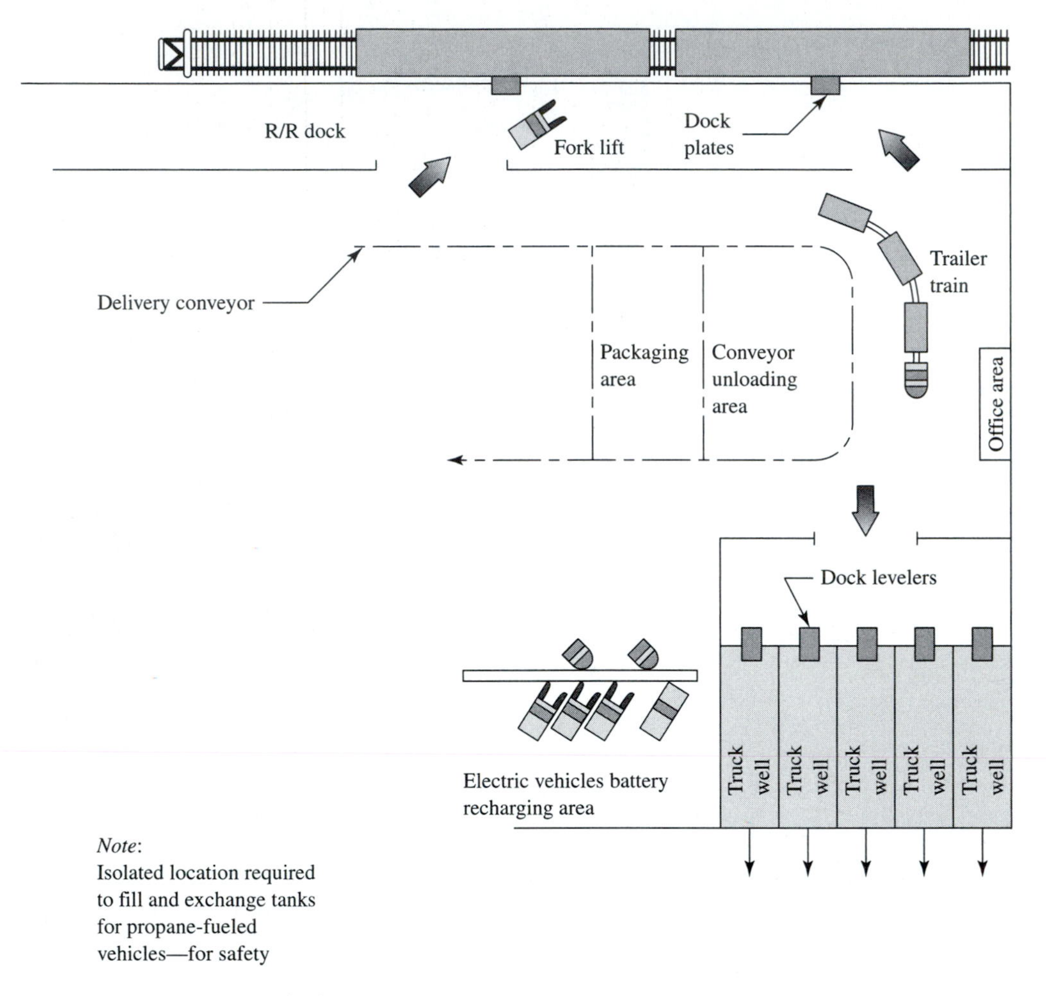

FIGURE 9.22 Shipping facilities.

- Separate oil from water in used coolant.
- Filter lubricating oils and sell—or use up as hydraulic oil or fuel.
- Make certain that machine chips are wrung oil-free in the production line's chip wringer.

An analysis of by-products (or potential by-products) should be made to determine space and facilities required and provided for in manufacturing plan. Facilities required may include bins, tanks, pumps, chip compressors (balers), and vehicles.

9.4.3 Maintenance

The Maintenance Department has the responsibility of maintaining all manufacturing plant equipment, tooling, gaging, production control facilities, building, plumbing, electrical—in fact, all facilities, sometimes with the exception of telephones, computers, and the like.

Maintenance must be staffed with capable, knowledgeable supervision, skilled tradesmen, and other help. Maintenance machinery and equipment must be maintained in an excellent state of repair, capable of any process required of it. Some special work is "farmed out," and sources for that must be available.

The manufacturing plan must contain a review of maintenance equipment—both quantities and types—to determine if it is adequate for a new program or parts. As an example, an automobile manufacturer released a new model power steering pump for production that had a special elliptically shaped rotor. The special production machines on order had a relatively long delivery time scheduled.

Manufacturing Engineering and Maintenance management determined that, because the toolroom made prototype parts for Product Engineering, a cam lathe must be obtained for the rotor. Delivery of this machine was from stock and, upon receipt, the machine was used in the production of many prototype power steering pumps for testing. The machine was later used in the maintenance of the production equipment.

9.5 REVIEW QUESTIONS AND PROBLEMS

1. Describe ancillary systems.
2. Name the major systems.
3. What properties must cutting fluids have, and what functions must they perform?
4. What fluid is superior for heat removal?
5. What sources of aid are available to the manufacturing engineer in selecting a coolant for a specific machine or process?
6. Describe briefly the different systems used to apply cutting fluids to cutting tools and workpieces.
7. One prime requirement in the use and management of cutting fluids is that they must be maintained at a high level of cleanliness and consistency. What actions must be taken and what facilities used to maintain cutting fluids at the proper levels of quality?
8. Which utilities must be checked for adequate capacity and which one is the most critical?
9. Describe auxiliary systems.
10. Name and describe four different types of auxiliary systems.
11. What is a support system?
12. Name and describe the major support systems.

CHAPTER TEN

MACHINE SELECTION

CHAPTER OVERVIEW

- Introduction
- Reason for Purchase
  - New Product
  - Additional Capacity
  - Engineering Changes
  - Replace Obsolete Equipment
  - Cost Reduction
- Required Machine Features and Characteristics
  - Quotation Must Comply with Request
  - Capabilities
  - Machine Design Features
  - Environmental
- Other Machine Purchasing Considerations
  - Machine Delivery Time
  - Floor Space
  - Ancillary and Support Systems
  - Special Maintenance Required
  - Standard—Special—Retool
  - Turnaround Space and Time
- Cost Comparisons
- Final Selection and Acquisition
- Review Questions and Problems

10.1 INTRODUCTION

The correct selection and acquisition of new machines and equipment is essential for the successful launch and manufacture of a new product and will enhance production rates, quality levels, and cost-reduction goals.

Incorrect machine selection will penalize these areas for the life of the machine. Comparisons of each area influencing the machine's purpose must be made. The machine ranking highest in each comparison must be determined; the machine ranking highest in the most areas usually is purchased. Some overriding detriment such as an unacceptable delivery date can shift the selection to the next highest ranking machine.

Most manufacturing companies use a formal, defined procedure when purchasing new equipment. All areas involved in machine acquisition must follow the procedure. The manufacturing engineer must be thoroughly familiar with all aspects of this because he or she will be involved in the original request for quotation, analysis of subsequent bids, final selection, follow-up, and installation. The procedure for obtaining used equipment or retooling machines on hand is usually the same as for purchasing new equipment. The risk—in terms of time, money, lost production, or lower quality resulting from an incorrect selection—exists whether the machines are new or used.

10.2 REASONS FOR PURCHASE

When the company requires new machines, the manufacturing engineer, knowing the primary reason(s) for acquisition, selects suitable suppliers and initiates requests for quotations.

Machines are purchased for many reasons—in some instances to answer a single need such as increasing production capacity, but often to also meet additional needs such as holding closer tolerances or incorporating an engineering change. To make a completely satisfactory selection of supplies and the machine required, the manufacturing engineer must resolve all needs and specify a machine whose capabilities and characteristics will answer all requirements. Also, as competitive potential machines and suppliers are selected to bid, all costs must be considered and compared.

The primary reasons for purchasing machines are

1. New product
2. Additional equipment for more capacity
3. Engineering change
4. Replacement
5. Cost reduction

10.2.1 New Product

A new product that is released for production, requiring all new machines, provides the greatest opportunity to obtain equipment that is state-of-the-art with respect to productivity, accuracy, and longevity.

A newly released product may consist of a single part, several parts, an assembly, or a complete component such as an engine or transmission. The release may involve machined parts, subassemblies, final assembly, and testing. A manufacturing engineer's objective remains the same, regardless of the complexity of the product or number of parts involved: to specify and obtain by a specified date, the equipment best suited to produce the desired quality parts economically in the required quantities.

Machine first-costs, which cover either new or retooling costs, and machine operating costs are the two major economic considerations governing machine selection. These costs vary with the machine type selected; that is, whether standard, single-purpose, special, or retooled facilities are planned. Affecting this choice are product complexities and rates of production. For example, a round shaft with low production requirements would almost certainly be produced on a simple lathe, whereas a housing that contains many machined surfaces, holes, and bores, and has high hourly rate requirements would probably be a candidate for a transfer machine.

Some products will be produced on available equipment or equipment that will become available. Such equipment may require overhauling and reconditioning at the time of retooling. Costs for this must be included in cost comparisons.

When newly released parts are similar to previously produced parts (with certain dimensional differences) and when the new parts are to run on the same equipment (as presently run parts), special attention must be paid to tooling design for ease of machine setup and changeover to minimize downtime between runs.

An early decision must be made concerning the use and choice of machine types (standard, single-purpose, special, or retooled) because the Plant Layout and Industrial Engineering departments need this information so their work can be conducted concurrently.

The complexity of the product can alter a manufacturing engineer's assignments but not his responsibilities. More engineers will share the workload of a complex product. Some form of overall control of the program must be maintained.

10.2.2 Additional Capacity

Increased sales and production requirements can, indeed, require additional *facilities*, but studies should first be made to determine if some other action could increase production *capacity* in a less costly, more prudent manner.

Actions to Consider for Additional Capacity

1. Eliminate bottlenecks in the present line to increase the production rate of the entire line.
2. Work the present facility's overtime to produce the needed added requirements.
3. Retool the present equipment to increase production rates, including changing tool grades and/or coolant.
4. Substitute material with one of a higher machinability rating to permit higher rates of production on present equipment.

When studies prove additional equipment must be purchased, one of three general approaches should be investigated:

1. Purchase the required amount of additional equipment to be the same type as that being presently used. Advantages are that operators and Maintenance are familiar with the facilities, and spare parts probably are on hand. Disadvantages could be the loss of potential labor savings and possible lower quality level as compared to more modern equipment.

2. Purchase additional equipment to be state-of-the-art. Advantages are possible labor savings and an increase in quality level. Disadvantages might include the learning curve for the new equipment plus potential operating problems in running two different machines at two different rates.

3. Purchase a new state-of-the-art machine with enough capacity to obviate the need to run the older equipment at all. Advantages are potential labor savings, increased quality, and machine reliability. Disadvantages are higher first-costs. This option can be considered as a replacement as well as increased capacity.

10.2.3 Engineering Changes

All engineering changes are studied and costs determined before being released for production. Some engineering changes are mandatory due to product failure, legal requirements, and/or some other unforseen occurrence, but costs and any resulting effects of the change must always be determined.

Whatever the cause of a change may be, a manufacturing engineer's attitude must always be to take advantage of changing conditions to obtain any other possible benefits while satisfying the required change. Processing, plant layout, material handling and any other service or area that relates to the product at the point of change should all be reviewed for possible improvements.

Some engineering changes are so extensive, or add so much more processing, that the only feasible answer is to purchase the required additional machines. Other engineering changes can be incorporated into existing equipment. Only a thorough study of the change and of the equipment presently used will determine which approach is the most economical and feasible. Following are two examples of engineering changes that were successfully included in existing equipment in a relatively low-cost manner.

EXAMPLE 1: TRANSMISSION CASE

An automobile manufacturer requested a limited quantity of regular production automatic transmissions to be modified for a motor home application. This application required a different valve body, which, in turn, required the drilling of 4 additional holes for oil passages.

1. The production cases were machined on a multistation transfer machine that had no idle stations or any working stations that could accommodate the added tooling required.
2. Altering or "stretching" (inserting an added station) was impractical: The machine base, coolant flume, chip conveyor, and electrical controls were all involved. Costs to change were prohibitive—and there was no space available to lengthen the machine.
3. The required special cases were provided by manufacturing the cases complete on the regular production machine—then transporting them to another area of the plant where a newly purchased tape-controlled drilling machine provided the additional oil holes.
4. The new drill was not 100% loaded, so the excess capacity was made available to the maintenance department to produce certain replacement parts at a lower cost than previously.

EXAMPLE 2: SUSPENSION LINK

A manufacturer of automotive suspension components produced a link, illustrated in Figure 10.1, on a 5-station column-type indexing drilling machine, with one completed part produced with each index of the machine. Total requirements were produced in two 8-hour shifts. Figure 10.2 illustrates the processing.

An additional link was released for production as shown in Figure 10.3. This part was identical to the link in Figure 10.1, except for a different center distance between the two bores. Production requirements for the new part would require 4 hours of machining time per day. The proposed additional production was studied and all available options for producing were determined.

Option 1. Purchase an additional machine to produce the newly released part.

Option 2. Design and build new fixtures, drill heads, and bushing plates. Changeover fixtures, heads, and plates between production runs for each part on the present machine.

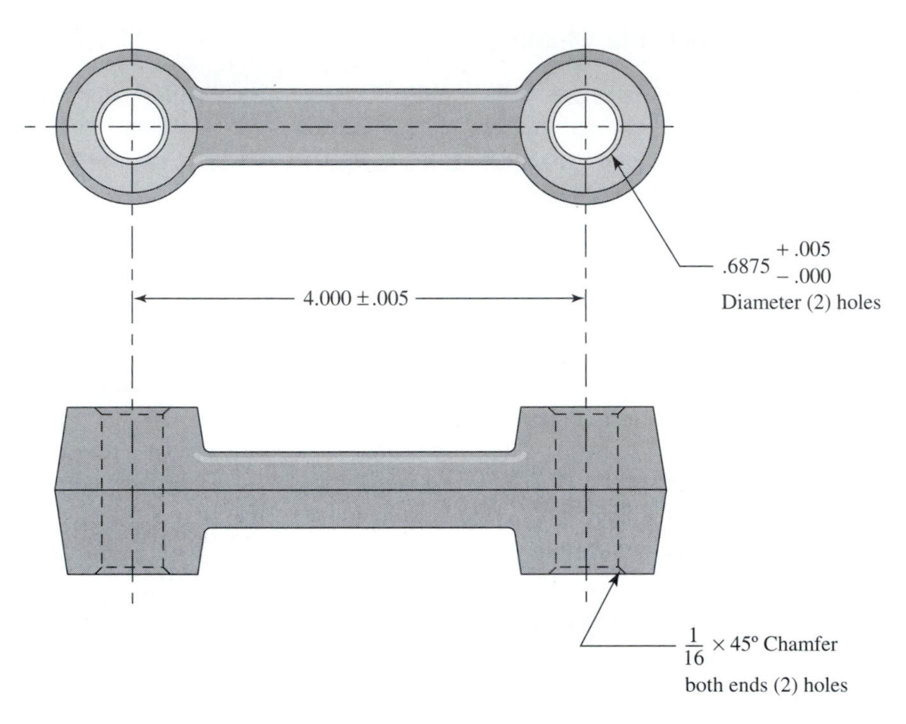

FIGURE 10.1 Link part #1.

Option 3

- Provide replacement details for the present fixtures, permitting either part to be loaded.
- Design and build new drill heads with an additional spindle that would permit either part to be machined, depending only on which part was loaded into the fixture.
- Design and build new bushing plates with an additional bushing permitting bores in either part to be machined.
- As illustrated in Figure 10.4, tooling will accept either part without changeover at job setting.

Option 3 was adopted. No loss of production from downtime for changeover was incurred. Savings in capital expenditures, as well as floor space, were realized in comparison with purchasing newer equipment. The design for the new drill heads also provided the opportunity to include an improved cutting-fluid system for added tool life.

10.2.4 Replace Obsolete Equipment

Most progressive manufacturing companies have a replacement and funding procedure established to regularly replace and upgrade equipment as it ages, becomes worn out, or is otherwise obsolescent, and also to effect cost reductions. Each replacement requires a formal study, proposal and justification.

The *advantages* accruing with the purchase of new facilities must be considered along with *unsatisfactory* features of existing equipment. Potential advantages and improvements of new machines are discussed in more detail in the following sections of this chapter; undesirable traits of present equipment can include the following:

Oper. No.	Operation	Machine
10	Receive and inspect forging	Bench
20	Straddle mill boss faces	Mill
30	Spot drill to chamfer one end, drill, ream (2) .6875 diameter holes	5-Station drill machine
40	$\frac{1}{16} \times 45°$ Chamfer one end—(2) holes	Drill
50	Wash	Washer
60	Inspect	Bench

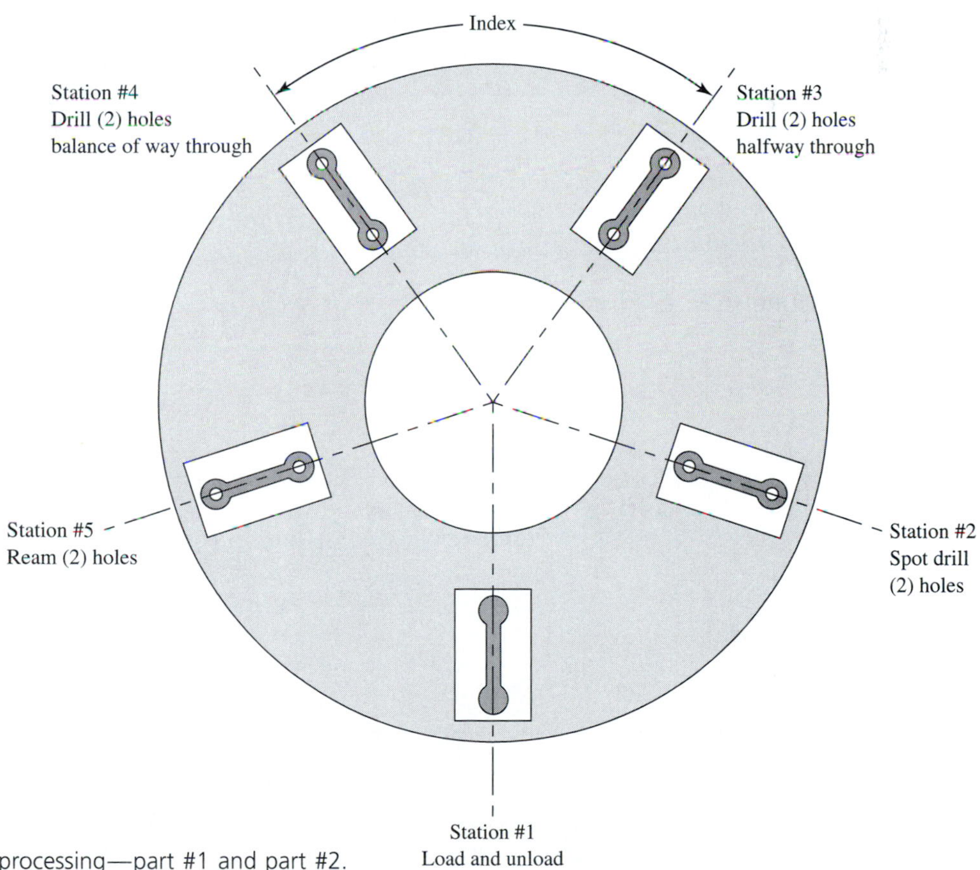

FIGURE 10.2 Parts processing—part #1 and part #2.

10.2.4 Replace Obsolete Equipment (Continued)

A. *Losses from worn out and unreliable equipment*

- Lost production and inventory
- Scrap
- Rework of borderline material
- Excessive maintenance
- Overtime penalty—required to make up lost production

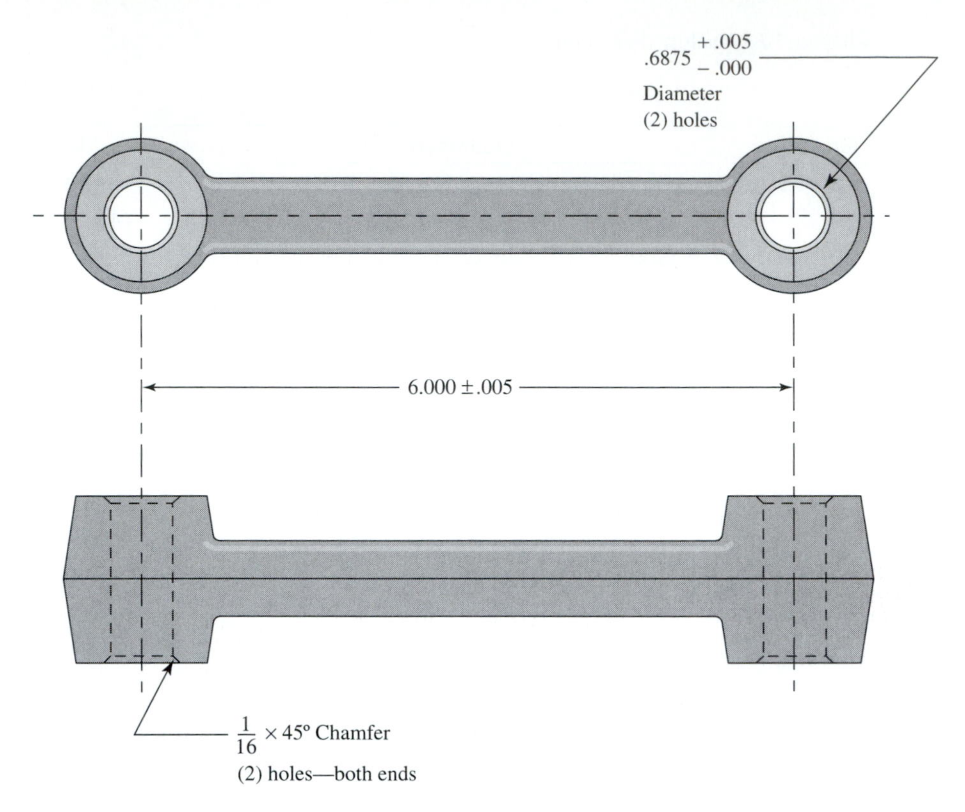

FIGURE 10.3 Link part #2.

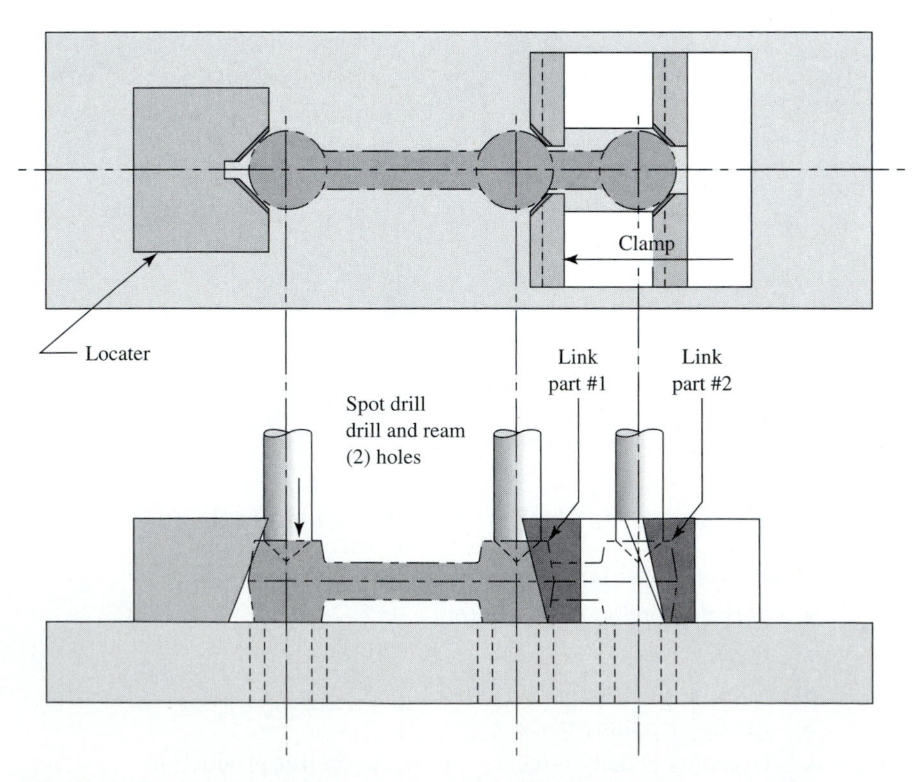

FIGURE 10.4 Common holding fixture for part #1 and part #2.

10.2.4 Replace Obsolete Equipment (Continued)

B. *Operational costs too high*

- Direct labor costs excessive
- Indirect labor costs (job setting, maintenance) too high
- Utilities and other costs too high

C. *Quality level not acceptable*

A machine must be replaced when, after all feasible repairing, rebuilding, and adjusting, it proves incapable of maintaining the required quality level. There is no alternative.

10.2.5 Cost Reduction

It has been said the manufacturing engineer must hurry to process his assigned parts in order to have time to complete cost-reduction studies before new model parts are assigned to him to process. Cost reductions are vital to all companies' competitive positions and profits. Most have a backlog of proposed, unfinished cost-reduction and machine-replacement studies waiting for available manufacturing engineering time to complete. Many cost-reduction proposals result in, or depend on, replacing present equipment to effect the following:

1. **Eliminate operations** through reprocessing. The ultimate cost reduction is realized in the complete elimination of an operation.

No machine purchase is necessary, labor cost is eliminated, total maintenance costs are reduced, floor space is gained, and all at a cost, perhaps, of a change in material handling.

Many cost reductions in this category are a result of a specification's change, or a dimensional tolerance change. Product Engineering must be in accord with such changes and are, indeed, the originators of many such changes.

2. **Reprocess to combine operations.** Altering tooling and/or the cycle of existing equipment is sufficient to effect many cost-reduction suggestions involving combining operations such as

- Substituting subland drills to handle 2 concentric diameters instead of 2 different drills.
- Altering tool blocks to accept additional turning or facing tools to "break up the cut" and shorten the cycle time. Some cost-reduction efforts will require new, or possibly rebuilt, machines and tooling to *automatically* transfer, load, unload, or dimensionally check the part. Other, simpler cost reductions may be implemented with the replacement of the present equipment with a like machine of higher production and efficiency.

3. **Reduce material costs,** scrap, and rework costs by replacing inefficient, worn machines with more capable equipment. Some present equipment can, of course, be reconditioned, rebearinged, scraped, or whatever else is required to render it more capable. A more accurate, more capable machine is often able to save material by operating with a smaller-diameter bar stock, with less facing material, or with a narrower cutoff blade. Also, such a machine can often eliminate subsequent qualifying operations and the accompanying required material. Scrap resulting from operator error or incorrect setup must be corrected by appropriate measures of training and/or discipline. Machine replacement is not a cure-all.

4. **Reduce labor costs** or produce more with the same amount of labor. Almost all production employees of the larger manufacturing concerns are members of a union who have contracts with the manufacturing companies for a stipulated period of time with stated agreements and conditions. One of the conditions covers "labor standards," or how much labor each employee classification will perform. This is a firm agreement allowing no unilateral speedup of a line by management, or slowdown by employees.

However, present-day unions do recognize that their company's management must stay competitive if they in turn, are, to retain their jobs. To this end, most labor contracts include an agreement that labor standards may be altered upon mutual agreement that a "technological improvement" has been made. Technological improvements are physical improvements such as new machines, tooling, work stations, plant rearrangement, or other actual physical changes that make a job easier or more productive.

Industrial Engineering, which is responsible for developing and maintaining labor standards, can, for instance, make requests for equipment changes that will permit the agreed-upon labor standard to be changed to a more competitive figure. The labor standard changes must, of course, justify the machine costs.

Labor costs can be reduced by

- Automating present equipment with the addition of automatic loading and unloading devices
- Replacing present equipment with new, more productive equipment.

Parts production can be increased with the same labor by

- Retooling present equipment to take heavier cuts, producing parts in multiples per cycle, and changing tool grades and/or coolants.
- Replace with new, more productive equipment.

10.3 REQUIRED MACHINE FEATURES AND CHARACTERISTICS

When a new machine is to be acquired, the relative merits of each machine considered must be compared, including function, machine design, performance, and cost. Figure 10.7 illustrates a "spreadsheet," or form, that is commonly used to make comparisons easier to see and evaluate.

10.3.1 Quotation Must Comply with Request

The quotation for any machine under consideration must answer and provide for exactly what the formal request specifies. This does not preclude a supplier from proposing improvements. One acceptable way to handle this is for the supplier to submit a quotation as requested, then submit an additional, alternative quotation that includes suggestions for improvements. All quotations are treated as highly confidential, and this must also apply to any alternative quotations. Quotations cost a considerable amount in terms of time, money, and engineering, and any supplier who goes an extra mile in supplying additional recommendations must be assured that his ideas will not be given to his competitors to see how they would price them. Divulging a vendor's idea, plan, or suggestion is not only unethical, it will stifle any future special efforts of help from that source, and it will badly damage the reputation of the associated manufacturing engineer.

10.3.2 Capabilities

A request for quotations for all manufacturing equipment specifies the production rates and quality levels required. The resulting quotation from each bidding machine builder must state both the proposed machine's production and quality capabilities.

A. Production rates requested include, in addition to the basic *production requirements,* an additional quantity to cover losses to production known as allowances, which include

✓ Tool setup and tool change time
✓ Setup scrap
✓ Normal maintenance

- ✓ Bar stock reloading (screw machines)
- ✓ Cooling or chill time (die castings)
- ✓ Operator
  - Personal
  - Lunch
  - Fatigue
  - Complexity

The term "net practical" is often used to define the actual production attained after deducting all recognized allowances.

Scrap parts will further reduce the net practical. Some companies allow for a percentage of scrap; others choose not to recognize scrap as an allowance, but, either way, scrap reduces the net total, and, acceptable replacement parts must be produced within the machine's total production capability or on overtime.

The successful bidder will be required to prove the machine's total production capability at tryout.

B. The specified *machine capability* is normally much more stringent than producing a part to the part print dimensions, tolerances, and specifications. Machine components such as spindles and bearings wear; runout and other inaccuracies resulting from such wear are reflected in the machined part. A new machine must therefore be more accurate or "capable" than the part design tolerances to allow for machine wear in use.

This capability must be proven during machine tryout. A specific quantity of workpieces will be supplied to the machine builder, the parts will be machined, and the resulting machined parts will be inspected and checked for compliance with the requirements. Any discrepancies must, of course, be corrected by the machine builder before the machine is shipped.

The tryout material or workpieces are supplied by the machine purchaser and must be of the same material, hardness, dimensions, and finish that will be used in production.

10.3.3 Machine Design Features

The statement that "form follows function" is certainly true in the design of a piece of production equipment. A modern machine capable of high production rates of very accurate parts exhibits massive cast bases, large-diameter spindles, adequate power, and sophisticated controls as compared to the lighter-construction, smaller-diameter spindles and bearings—and perhaps manual controls—of older equipment that is only capable of much lower production rates and accuracies.

In considering the machine to purchase, the machine builder's selection of components to produce the required output and accuracies must be examined. Adequate components will assure continued rated quality output with less downtime for maintenance.

Many types of machines, particularly transfer machines, benefit from a basic design system known as modular construction, wherein units, slides, heads, tables, and other components are standardized. These units are more economically manufactured in quantity in appropriate sizes. They can be stockpiled for eventual use and can thus reduce delivery time at time of ordering.

Weight A machine's weight is a good indication of the machine's ruggedness and resistance to distortion, and also of its ability to absorb vibrations set up by the cutting-tool action, thereby contributing significantly to accuracy and smoothness of cut. Many older, more lightly constructed machines now in use are aided with pigs of lead placed in strategic locations to resist vibration.

Bases of new machines are more desirable as castings than as lighter-weight weldments. Slides and tables are almost universally castings. Spindles should be of a generous size, not only to absorb vibration and resist deflection but also to provide reserve capacity for potential future increased loads from product or tooling changes.

Horsepower Motors driving machining spindles should be sized larger than just adequate for insurance against possible surges from overloads and potentially increased future demands due to product, tooling, or material changes. The cost penalty for the larger motors and the energy to run them is in reality a very low-cost insurance factor.

Bearings All contemporary manufacturing equipment is designed with adequate ball or roller bearings in all critical areas. A comparison of these is not usually made, but the lubricating systems that serve all wear points and bearings of potential machines should be compared. The lubricating system should be automatic; that is, it should operate continuously while the machine is in operation. Most good systems include a pump, a sump, lines, metering valves or orifices, and a replaceable filter element. One excellent filter has two replaceable elements or cartridges, one of which is on-line, and also a spare. When the on-line element becomes loaded with dirt and metal particles, increased line back-pressure causes a changeover to the spare clean cartridge and activates a signal light to alert maintenance to the need for filter element replacement.

Controls Control panels for machine tools are generally solid state, having no moving parts, vacuum tubes, or relays. Limit switches and any other electrical control components exposed to oils or coolants must, of course, be sealed and protected. The electrical circuit, or system, design must be compatible with the intended machine operating method.

- A flexible system must be provided to aid in setup and changeover when many different parts are to be run on the same machine.
- A fixed, perhaps computer-controlled, circuit should be provided for high-production, repetitive use.

A diagnostic system wired into the control panel and used for troubleshooting is a valuable timesaving feature that can greatly aid in keeping a machine in operation.

Safety All machines must be designed in accordance with the Occupational Safety and Health Act (OSHA), and older existing equipment must be brought up to these requirements. All such equipment must be maintained at the safe prescribed levels or shut down until corrections have been made.

Manufacturing engineers must be aware of the OSHA intent and study its applications to the machines under consideration for purchase. One excellent source of information concerning the application of OSHA requirements to manufacturing equipment can be found in Volume V of *The Tool and Manufacturing Engineers Handbook*. Some major areas of concern include

1. Nip or pinch points
2. Guards
   - Moving machine parts
   - Cutting tools and grinding wheels
   - Chips and coolant
3. Electrical

4. Startup signals
5. Emergency stop provisions
6. Lock-out system to disable a machine when workers must be in or exposed to the machine, moving parts, and so on.

10.3.4 Environmental

Proper machine selection requires that consideration be given each machine's effect on the environment. A machine can have an effect in each, or all, of three different areas:

1. **Noise.** Allowable sound levels are specified in OSHA standards as shown in Figures 10.5 and 10.6. Machines must be designed to conform to the allowable limits. Sources of noise must be eliminated or controlled by means of isolation, sound-deadening materials, mufflers, and acoustical screens. Where it is not feasible or possible to control sound within allowable limits, operator hearing protection must be provided.

FIGURE 10.5 OSHA noise exposure limits.

Hours of Exposure	Sound Level, dB (A)
8	90
6	92
4	95
3	97
2	100
$1\frac{1}{2}$	102
1	105
$\frac{1}{2}$	110
$\frac{1}{4}$ or less	115

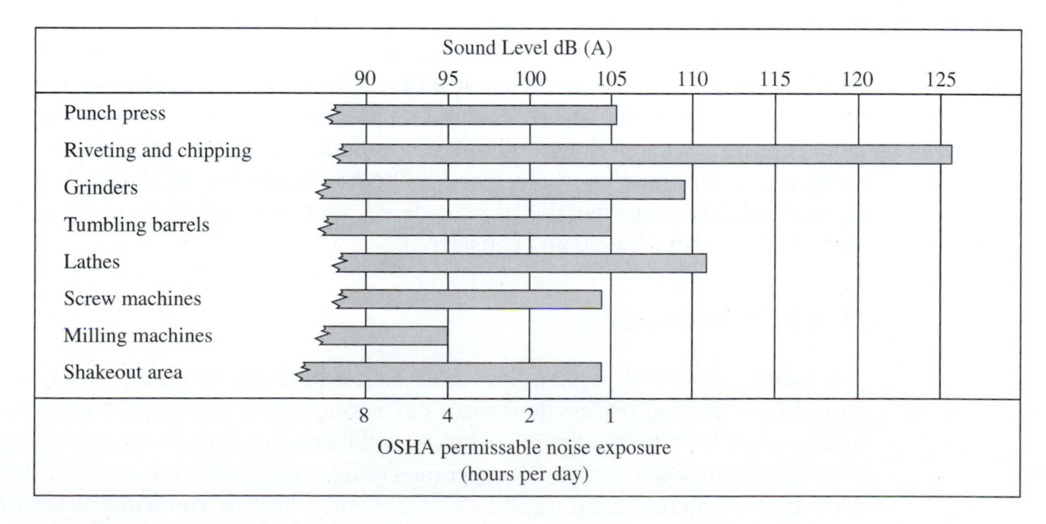

FIGURE 10.6 Typical sound levels at machining operator's position.

Major sources of noise in a machine are hydraulic pumps, fans, motors, gearing, and cutting tools. Many manufacturing companies employ an engineer whose expertise is sound to deal with sound problems exclusively; other concerns employ outside acoustical consultants to aid in the control of noise.

2. **Water Pollution.** One common type of water pollution is oil from a leaking machine hydraulic system that finds its way into a storm sewer and then to a navigable stream or a drinking water supply. Any special treatment of this problem in a machine-tool builder's quotation should be given special attention, with feasible solutions given credit on the comparison sheet.

A machine—or a process such as painting, plating, and heat treating—can pollute a water system when process by-products are incorrectly disposed of or discharged. Consideration must be given early-on to the correct method of handling such by-products, because this may require providing ancillary facilities, which may require plant floor space and certainly will have delivery time requirements.

3. **Air Pollution.** Machine tools can pollute air when they generate fumes, smoke, or mist, which is then exhausted into the atmosphere. Many processes—including heat treating, casting, painting, and cleaning—must be provided with filters, scrubbers, or other devices to control exhausted by-products.

Some types of machine tools are lubricated with an oil mist system. The oil mist is forced through the unit to the various surfaces, bearings, and wear points. Exhausted mist must be collected and recirculated. Oil mist that escapes into the atmosphere not only pollutes the air, but some will also condense, coating the factory roof and exhaust stack with oil, which then becomes a fire hazard.

10.4 OTHER MACHINE PURCHASING CONSIDERATIONS

Many significant points to be considered when purchasing new manufacturing equipment do not fall neatly under uniform headings. The following material is not in order of relative importance: All areas are vital. Some items may not apply to all machine purchases; others may need to be added. Any areas affecting the total cost and total requirements of each machine must be identified and the costs determined.

10.4.1 Machine Delivery Time

Delivery time must be specific in a quotation. Terms such as "*x* weeks after receipt of part prints and all information" are unacceptable and lead to misunderstandings and problems. A proposal must specify the actual date of delivery or a specific number of working days, weeks, or months from receipt of the purchase order. Any questions concerning part print specifications, dimensions, and so on must be resolved before the purchase order is placed. Additionally, when quotations' delivery promise dates are compared, each vendor's past record and reliability in this area must be remembered and considered.

10.4.2 Floor Space

Since each square foot of plant floor space affects profits (new construction, maintenance, taxes), a machine that requires less floor space has a competitive advantage. Each quoted machine must be compared for total required floor space, and its physical layout must be compared to the user's plant layout to make certain the machine configuration suits the shape of the plant floor space provided. Each competing machine's *height* must also be compared with the intended space. Any machine that requires less plant rearrangement or structural changes has a cost advantage.

10.4.3 Ancillary and Support Systems

Certain activities and systems provide services required to support productive equipment. Examples are cutter grinding, coolant supply, material handling, and so on.

Each potential machine selected must be analyzed for its peculiar requirements and compared with its competition. It may well be that all machines under consideration will require the same services and support. At any rate, in addition to resolving what new services will be required for the potential new equipment, *existing* ancillary facilities must be reviewed for availability and capacity.

If additional ancillary equipment is required, costs must be determined, selections from competitors made, and facilities obtained concurrently with the manufacturing equipment. The lead times for ancillary equipment can be as long, in some types, as that of the manufacturing facilities it serves. Required ancillary and support systems are discussed in more detail in Chapter 9.

10.4.4 Special Maintenance Required

Some thought should be given to the degree of machine sophistication that can be maintained in the user plant. Not only must the plant's Maintenance Department be evaluated, so must the local outside services available. A manufacturing plant located in a manufacturing area such as Detroit or Cleveland has little trouble hiring maintenance men with the required expertise—or, in an emergency, calling in outside help from locally available sources. A manufacturing plant located in an outlying agricultural area may not have such aid immediately available.

Certain proposal features of a new machine could pose the likelihood of calling in the machine-tool builder's service personnel to keep the machine operational. Any such proposed unit or feature of a machine under consideration should be questioned for its actual worth and for possible substitutions with more easily maintained items. The distance a machine-tool builder's service personnel must travel, or more to the point, the *time* it takes them to arrive when needed, should also be considered.

Continuous training of maintenance personnel is an ongoing normal practice, but advanced training in hydraulics, electronics, and computers is now a necessity. Many user plants send maintenance personnel, and sometimes production help, to the machine builder's plant during the building and tryout phases to become familiar with the equipment in advance of actual use.

10.4.5 Standard—Special—Retool

One more decision must be made early: whether to produce the released product or parts on general-purpose, standard machines, on specially designed, special-purpose equipment, or on existing facilities retooled to suit the product.

Whichever approach is chosen, all vendors must be chosen and instructed to quote the same way. Apples must be compared with apples. Some advantages and disadvantages of each type of tool-up follow.

General-Purpose, Standard Machines

Advantages

- Purchase costs are probably lower than those of special-purpose machines even though more of them may be required to produce the same quantity as a single special-purpose machine.
- Delivery times are almost always shorter—sometimes from inventory.

- They are more flexible:
  1. When one machine of a group is down, others can still produce.
  2. Setup and debugging is easier and quicker.
  3. They are easier and quicker to retool for other products or to include engineering changes.
  4. They have lower obsolescence costs—machines can be retooled or sold if no longer required.

Disadvantages

- Direct labor costs are probably higher than for a special-purpose machine.
- Indirect labor costs for material handling, and so on are probably higher.
- More in-process banks and inventory may be required.
- They may require more floor space.

Special-Purpose Machines

Advantages

- Incorporate latest technological advances
- Higher rate of production
- Lower direct labor cost
- Lower indirect labor cost
- Potential material savings

Disadvantages

- Usually highest purchase cost
- Longer delivery time
- Less flexible:
  1. The machine is completely down when any component is inoperative.
  2. Setup and debugging are more intricate and take more time.
  3. The machine is more difficult and costly to retool for other parts or to absorb an engineering change.
  4. The machine is more likely to be obsolete, with little resale value at end of product run.

Retool Available Equipment

Advantages

- Lowest first-cost—for tooling only
- Shortest delivery time—for tooling only
- Shortest learning curve—operators and maintenance familiar with present equipment

Disadvantages

- Probably lacks some desired technological features that more up-to-date equipment would have.
- Has been used—may require time and cost for rebuilding.
- Possibly has less quality capability.

10.4.6 Turnaround Space and Time

At times, arrangements must be made to supply a continuing flow of production parts for use while presently operated equipment is retooled, replaced, or rearranged. Usually this can be provided for by building a bank of parts ahead of time to cover the turnaround time required.

When a sufficient bank of parts cannot be supplied and the present equipment must be relocated to make room for replacement facilities, a place must be provided where the older equipment can be operated until the newer machinery is operational. This is known as turnaround space.

10.5 COST COMPARISONS

Many books have been published with formulas developed to aid in cost justification of machine replacements and in cost-effective selection of the machine replacement. Other user plant departments besides Manufacturing, notably Industrial Engineering and Finance, aid in determining present actual machine costs.

However, the Manufacturing Engineering Department must originate, coordinate, and total the cost estimates. Records of labor costs, material costs, scrap, maintenance, taxes, and so on are made available to Manufacturing Engineering for their use.

New Machine Purchasing Cost Inclusions

1. Purchasing costs include all design and make costs associated with the basic machine and associated tooling such as fixtures, drill head and plates, tool blocks, and, usually, up to three sets of perishable cutting tools. Perishable tools are expense items, and additional replacements are not capitalized.

2. Shipping costs include all labor, material, fees, and common carrier costs associated with moving the machine from the builder's plant to the user's plant—usually to the receiving dock.

3. Installation costs include labor, material, and all other costs associated with placing a machine in location on its pads, or footings, and include all electrical, pipe fitting, millwright, or other such work to ready the machine for start-up. In most manufacturing companies, all costs of purchasing, shipping, and installation are capitalized and amortized.

Leasing of new facilities is a viable alternative to purchasing; it conserves the user plant's capital funds but does not permit depreciation or cost amortization until the final residual payment is made. If this approach is selected, the user plant's Finance Department obtains all required financial data from the machine builder to develop costs associated with leasing.

Machine Operating Costs

- Comparisons developed for machines quoted to supply equipment required for newly released parts must, of course, identify the best buy among all those quoted.
- Comparisons developed to compare the costs of the best buy of all machines quoted and the operating costs of an existing machine to be replaced from the primary documents needed to justify replacement.
- Included in such comparisons are

  1. **Labor Costs**
     - It is usually considered sufficient to compare the number of machine operators per shift per machine for purposes of comparing direct labor.
     - Similarly, the difference in indirect labor should be determined in the areas of job setter setup times, tool-changing times, and any indirect costs associated with material handling.

2. Maintenance Costs
   - It is common practice to assume that maintenance costs will be approximately the same for each machine quoted, provided each is of the same level of quality in regard to general construction, material, bearings, and so on.
   - An unusual feature of any potential purchase, whether tending to increase or decrease expected maintenance costs, should be noted.
3. Utilities
   - Utility costs, like maintenance costs, are presumed to be the same in comparing similar proposed equipment.
   - Should any piece of equipment, machine, or process require notably more or less electrical energy, gas, water, or compressed air than its competition, that should also be noted.
4. Tool Costs. Planned tooling for each candidate machine should be studied for features affecting costs and also for reasons to justify more costly tools. For example, carbide, ceramic, and diamond tooling *should* be used where they will provide longer tool life before changing; multiple tools *should* be employed where multiple cuts will increase production rates. Some design and cost features include the following:
   - Standard catalog item tool bits, cutoff blades, indexable inserts, and toolholders are less costly than special individually designed cutting tools and holders.
   - Cutting-tool materials should be specified to provide the longest practical tool life.
   - An effort should be made to balance the tool life of each individual tool; that is, if all tools could have the same life, the entire set could be changed with only one interruption, or shutdown, of the machine.
   - The number of duplicate tools per setup should be examined and reasons for any multiple tools noted.
   - Circular or flat-form tools are more costly but may produce a surface in less time than generating that same surface with a single-point tool.
   - The method of replacing and resetting cutting tools should be looked at. The use of keyed, quick-change tool blocks in which the cutting tools have been preset with setup gages is one method used to minimize machine downtime for tool changes.

Direct Material Costs Some machines require less (direct) material to produce a part than others. That is, they can manufacture a part with less offal in the form of chips, cutoff loss, bar ends, or stock left for subsequent operations.

A machine sufficiently rigid and smooth in operation may eliminate a roughing cut or may require less finishing stock to be left for a subsequent operation. This capability could result in producing parts form forgings or castings having less material to be machined off, or from bar stock having a smaller outside diameter.

Significant bar stock material savings can be realized from the use of narrower cutoff blade widths and from less required stock left for end facing. Any machine able to realize such savings will have a competitive advantage for both the seller and the users.

Productive Rates To be considered for purchase, a machine must be capable of producing the minimum quantity of parts to satisfy production requirements. Any additional parts produced in the stipulated time may reduce the cost per piece. This is not entirely true, however, in the use of a fully integrated, continuous line, where each machine feeds the next in a progressive cycle. The excess capacity of a machine in such a line is usable perhaps only for the important function of supplying a bank of parts to cover downtime during maintenance and tool changing.

Excess capacity of a machine operating independently of a continuous line can reduce the cost per piece. The machine, after supplying its quota of parts, can be set up to run other parts, or—less desirable—can be idled and the labor reassigned. An idle machine is still "on the books," still occupies valuable floor space, and still must be made to pay its way in profit return.

Manufacturing time per piece equates to cost per piece. Some areas of productive time loss chargeable to the machine that must be considered when comparing machines are

- **Wait time**—the time during each cycle when a machine is nonproductive, waiting for part loading or unloading, tool approach time, or any other delay extraneous to actual machining or index time
- **Index time**—the time required for each machine, in each cycle, to index the part, or tools, one station

10.6 FINAL SELECTION AND ACQUISITION

Selecting Vendors Narrowing the field of potential suppliers to those best able to supply the requirements is a crucial step. All suppliers should be of about the same quality level and reputation. If all bids were to come in with the same promised delivery date, production rate, and other comparable features (not too likely), the lowest-cost machine would probably be awarded the purchase order. Manufacturing Engineering should be in accord with the entire list of bidders and able to live with anyone selected.

The final list of acceptable suppliers is sent, along with forms known as Requests for Quotations (RFQs) to the purchasing department. The list, based on Manufacturing Engineering's knowledge of requirements and the supplier's past performance, can be supplemented by Purchasing. However, care must be taken that any additional bidders be of the same quality level and ability as those on the base list. Needless dilemmas have been created when an additional bid has been requested by Purchasing for "competitive reasons" from a supplier who will build a less costly machine but cannot meet the machine quality and capability levels required. The dilemma, and needless work and lost time, is created when Purchasing wishes to buy the lowest-cost machine, forcing Manufacturing Engineering to accept a potentially inferior machine or justify the reasons for the more costly selections. An early understanding between Manufacturing Engineering and Purchasing of goals and ground rules is certainly the proper way to proceed.

Bidding Package A package of all the information that will be required by each company requested to supply a quotation is prepared and sent to purchasing along with the RFQs. This package includes

- All product specifications, production rates, and specified machine capabilities
- Part prints to the latest engineering changes and any other related prints or drawings
- All completion, tryout, shipping, installation, and required machine startup dates
- Any data pertaining to tryout material and approval
- Any required user plant design standards and any required electrical and hydraulic standards

Handling of Quotations Each company has a procedure for handling quotations, which, for ethical and legal reasons, must be adhered to. Usually, Purchasing is designated as the department to officially request quotations, receive them from the bidder, hold them unopened until the stated closing date, and open, record, and distribute copies to those involved for analyzing and final decision.

Some companies have tried having unpriced copies of quotations supplied to Manufacturing Engineering earlier than the closing date to provide more time for machine feature comparison. Problems developed when early bidders objected, believing their proposals might be made public and of benefit to their competitors who turned in their quotations later.

At any rate, Manufacturing Engineering and Purchasing must go beyond just abiding with the established purchasing procedures. A manufacturing engineer or a purchasing agent, especially one beginning a career, must be extremely conscientious in this matter of ethics and confidentiality of knowledge and data. The reward for uncompromising integrity is earned confidence. There is no greater business asset for engineers or purchasing agents than the unquestioning confidence and trust of those they deal with. A slip, whether inadvertent or intentional, will do more damage to a reputation than can be overcome in years—if ever.

Spreadsheet Some system must be used to record and display each feature of all considered machines for easy comparison. One method is to use a form developed for this purpose and known as a spreadsheet. Figure 10.7 illustrates a spreadsheet designed to compare machine tools. As shown, the different machines quoted are listed in the left-hand column, with the various comparable features entered in their appropriate columns.

After all machines and features are recorded, it becomes relatively easy to identify which machine or machines have more mass, more horsepower, higher rates of production, and so on. The data highlighted by the spreadsheet, coupled with a satisfactory delivery date and purchase price, helps narrow the final selection.

Similar spreadsheets are used to aid in the selection of sheet metal, forging, plastic, foundry, and other production equipment by designing the form to compare significant features peculiar to those kinds of facilities.

Final Order Placement When a decision is reached resolving which machine to purchase, Manufacturing Engineering issues a requisition to purchase specifying the machine(s), equipment, and tooling required, along with the necessary accounting data. Purchasing issues a formal purchase order to the selected vendor authorizing design and construction of the equipment as per quotations.

All dates of completion, tryout, shipping, and installation should be confirmed in this document. Terms of payment must also be defined to clarify the method, progressive payments or payment in full after successful runoff and proof of capability.

Any further change or changes that may be required by the purchaser after formal issuing of this purchase order will require a Purchase Order Change (POC). Such changes must be understood and agreed to by the supplier. POCs usually add cost to the machine purchases.

Alternately, any change requested by the supplier is usually resisted due to the additional costs. There are exceptions. POCs are sometimes approved for a mutually desirable change or where there is a proven misunderstanding or hardship.

Follow-Up The manufacturing engineer must act as coordinator and perform the follow-up function for any new machine from its purchase through building and tryout to launch and debugging. Some of the areas of concern follow:

1. **Visits to Vendor Plants.** A schedule should be set up of proposed visits to the machine builder's plant. Visits should be scheduled as follows:

- An immediate visit to ascertain that all prints, information, and other data is available and on hand, and that all questions have been resolved
- A visit to review drawings and designs at the completion of the engineering phase
- Progressive visits, as required, to review machine building
- A final visit at the time of machine tryout period to approve for shipping

Vendor	Machine Weight	H/P	Hyd System	Lubrication System	Bearings	Controls		No. of Stations	Automatic Clamping	Net Production per hour
						Electrical	Coolant			

FIGURE 10.7 Typical quotation comparison spreadsheet.

2. **Tryout Material**

- The required quantity of tryout material must be determined and arrangements made to provide it in ample time for vendor's needs.
- Production control should be checked to be sure production material will be on hand when the machine is installed and tried out preparatory to actual production.

3. **Shipping.** Shipping arrangements should be checked, even though the machine's builder is normally responsible for this. Some large equipment may require the use of special trailers or railroad lowboys, and some special routing may be required to avoid tunnels, and so on.

Maintenance in the receiving plant should be briefed on when and how the new equipment will arrive. Aisles may need to be cleared or objects moved in preparation for the new machine's arrival.

4. **Installation.** Installation is normally provided by the machine builder with preparatory work such as foundations, pits, flumes, and any required plant rearrangement provided by the machine buyer.

At the time of purchasing, it is well to have stipulated who is responsible for the shipping. Also, if riggers, millwrights, or other outside help is required to move equipment from the truck or railroad car to the proper area, responsibility for obtaining such help must also be spelled out.

If aisles must be closed and traffic temporarily disrupted, production must be notified ahead of time so that other routings can be provided. Manufacturing Engineering must act as coordinator and follow up for all areas to assure a smooth, successful machine installation.

5. **Quality Control and Inspection Needs.** All special gages and inspection equipment needed to check parts produced on the new machine during tryout at the machine builder's plant must be made available in a timely way. These gages will, of course, also be required during the tryout and startup of the equipment after installation in the buyer's plant.

6. **Ancillary Requirements.** The status of all ancillary equipment such as coolant filters, pumps, material-handling facilities, special or additional cutter-grinding machines, and any other provisions that are on order for use with the new facilities must be checked. Some of these items may require a separate, formal follow-up schedule of their own.

7. **Plant Engineering.** Items usually known as plant engineering items must be checked to assure timely availability. These items can include

- Excavations, roadways, parking areas, pits, foundations, and so on
- Additions or changes to public utilities, water, electric, gas, and drains
- Additional compressed air
- Changes to heat, light, and ventilation

8. **Production.** Plant production management must be briefed on changes that will affect manufacturing. In particular, the projected needs for setup personnel and machine operators during machine tryout, and also the possible need for their training.

9. **Vendor's Vendors.** Sometimes overlooked in checking and follow-up are the purchased items required by the machine-tool builder to incorporate into the basic machine. Included are such standard units and items as motors, pumps, hydraulic systems, valves, switches, bearings, and perhaps complete control panels.

Machine builders have their own follow-up departments, and it is advantageous for the manufacturing engineer involved to become acquainted with the personnel in that area.

A double-check can prevent a slip that can have a disastrous effect on the machine delivery date. As an example, one automotive manufacturer placed an order for a line of machines to produce tubing from flat steel strips for a new model shock absorber. Routine vendor-submitted follow-up reports showed this machine build line on schedule. However, when the machine builder set up the line in his plant for tryout, it was discovered that one unit, a special press, was missing. Panic, commotion, accusations, and fixing of the blame could not produce the press in time. The machine had simply not been ordered. The tubing line was late in delivery, and the product introduction date was delayed. A double-check would have uncovered the omission; the line would have gone into production as planned, and, incidentally, two jobs might have been saved.

10.7 REVIEW QUESTIONS AND PROBLEMS

1. What one factor above all others must a manufacturing engineer keep in mind when selecting and acquiring an additional machine?
2. What other action should be taken before a decision is made to acquire an additional machine?
3. What is meant by "machine capability"?
4. What are the main advantages of purchasing a special-purpose machine over purchasing a standard machine?
5. What are the disadvantages?
6. What are the advantages of purchasing a new special-purpose machine over rebuilding and retooling an older same-make machine?
7. What is the ultimate cost reduction?
8. Why is weight a good indicator of a better machine?
9. What environmental areas can machines affect? What are some ways their detrimental effects can be controlled?
10. What is meant by "turnaround space"?
11. What is direct material?
12. What one personal characteristic above all others must a manufacturing engineer possess and demonstrate, especially when working with vendors in selecting machines?
13. What related areas must also be studied when planning for a new machine?

CHAPTER ELEVEN

MULTIFUNCTION MACHINES, MACHINING SYSTEMS, SPECIAL-PURPOSE MACHINES, AND AUTOMATION

CHAPTER OVERVIEW

11.1 INTRODUCTION

Several of the standard machines described in previous chapters of this text are capable of performing multiple operations:

- A drilling machine is capable of drilling, reaming, countersinking, tapping, and so on.
- A lathe is capable of turning, facing, drilling, threading, and so on.

However, when increased production rates require the purchase of additional machining capacity, it is almost always more economical and feasible to purchase multifunction machines capable of quick changes and simultaneous machining, and possessing automatic cycle controls.

Descriptions and applications of machines described in this chapter range from equipment considered "standard" to machines that are custom designed and built to suit a specific application and that are considered "special purpose." Machining systems described include machining centers and flexible manufacturing systems.

- Machining Centers are numerical-control (NC) or computer numerical-control (CNC) machines having automatic tool-changing capabilities. These machines are often used for low-volume complex parts and also for large workpieces that are not easily transported.
- Flexible manufacturing systems employ one, or more, machining centers, usually along with other equipment, to produce medium-volume workpieces. A workpiece-handling system is required, and the entire arrangement is typically controlled by a central computer.

"Automation" is a term coined in the 1950s and applied to devices that automatically loaded and unloaded stamped parts in die equipment and tooling. The term is now accepted to cover the automatic handling of any workpiece and includes the functions of

- Loading and unloading
- Processing
- Measuring resulting workpiece size
- Adjusting a machine to maintain workpiece size
- Repeating the cycle

Some elements of automation are now built into machines considered standard machines. Other elements must be designed, made, and added—or designed and built into a special-purpose machine as it is constructed.

The selection of most machines for specific applications follows a common practice:

The manufacturing engineer:

1. Determines the process or processes required for each operation of a specific part
2. Makes a general selection of a suitable machine for each operation
    - Ascertains machine capacity
    - Ascertains machine process capability
    - Ascertains machine availability
3. Makes, or has made, a tool layout showing
    - Workpiece
    - Process
    - Tooling required

The layout is an invaluable aid when writing the process operation sheets and when designing and making or ordering the tooling. More important, it will confirm the manufacturing engineer's selection of a machine—or it could uncover a problem and cause a different machine or model to be used.

11.2 MULTIFUNCTION MACHINES

Machines selected for medium to high levels of production vary from those used, for instance, in a toolroom to perform the same function in two vital ways:

1. Production machines are designed to run continuously; that is, they are fitted with controls and mechanisms that cause the machine to repeat its cycle automatically, and they usually have some form of mechanized loading and unloading devices.

2. Operations are performed simultaneously, not consecutively, whenever possible. A lathe may turn, face, groove, and cut off a workpiece all at the same time, and a special drilling machine may spot face, countersink, and tap a part within the time it takes to drill as illustrated in Figure 11.1.

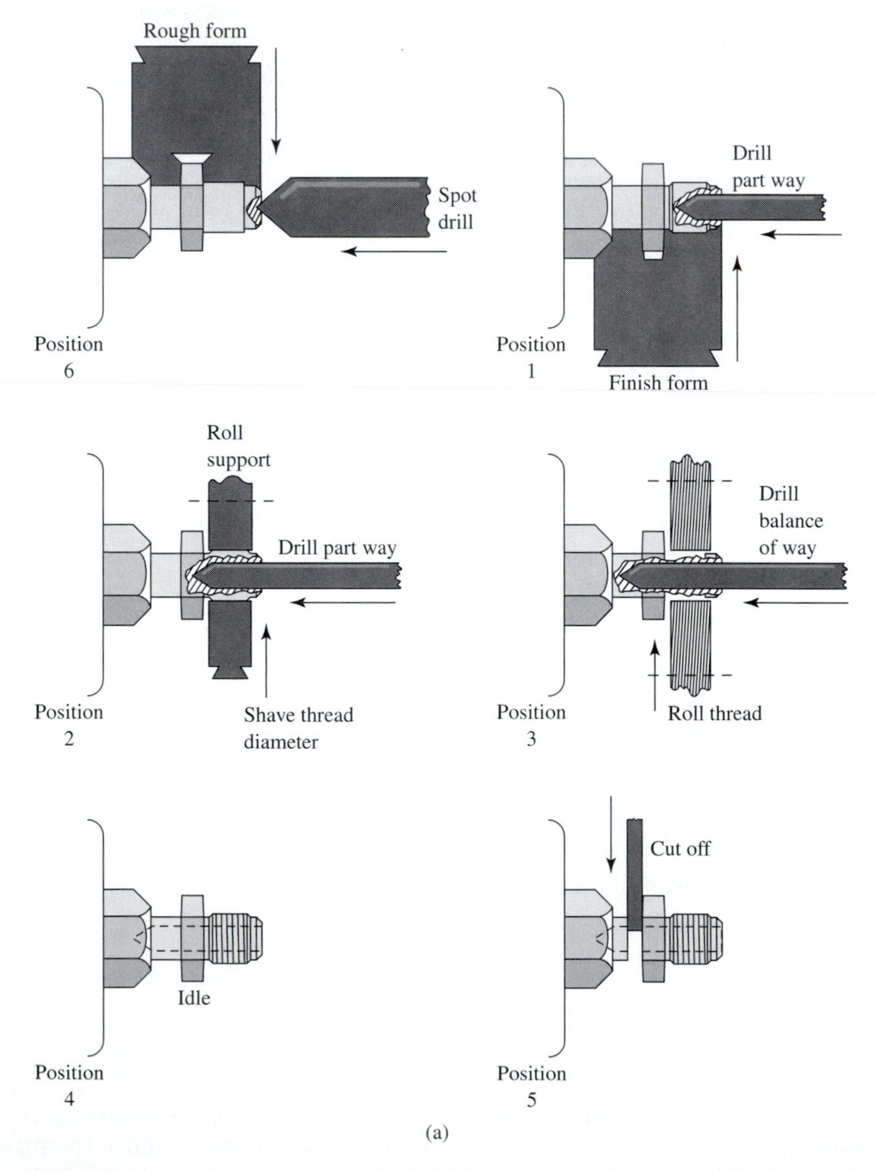

FIGURE 11.1 (a) Part processed on 6-spindle automatic bar machine to provide equal-length machining times for each position.

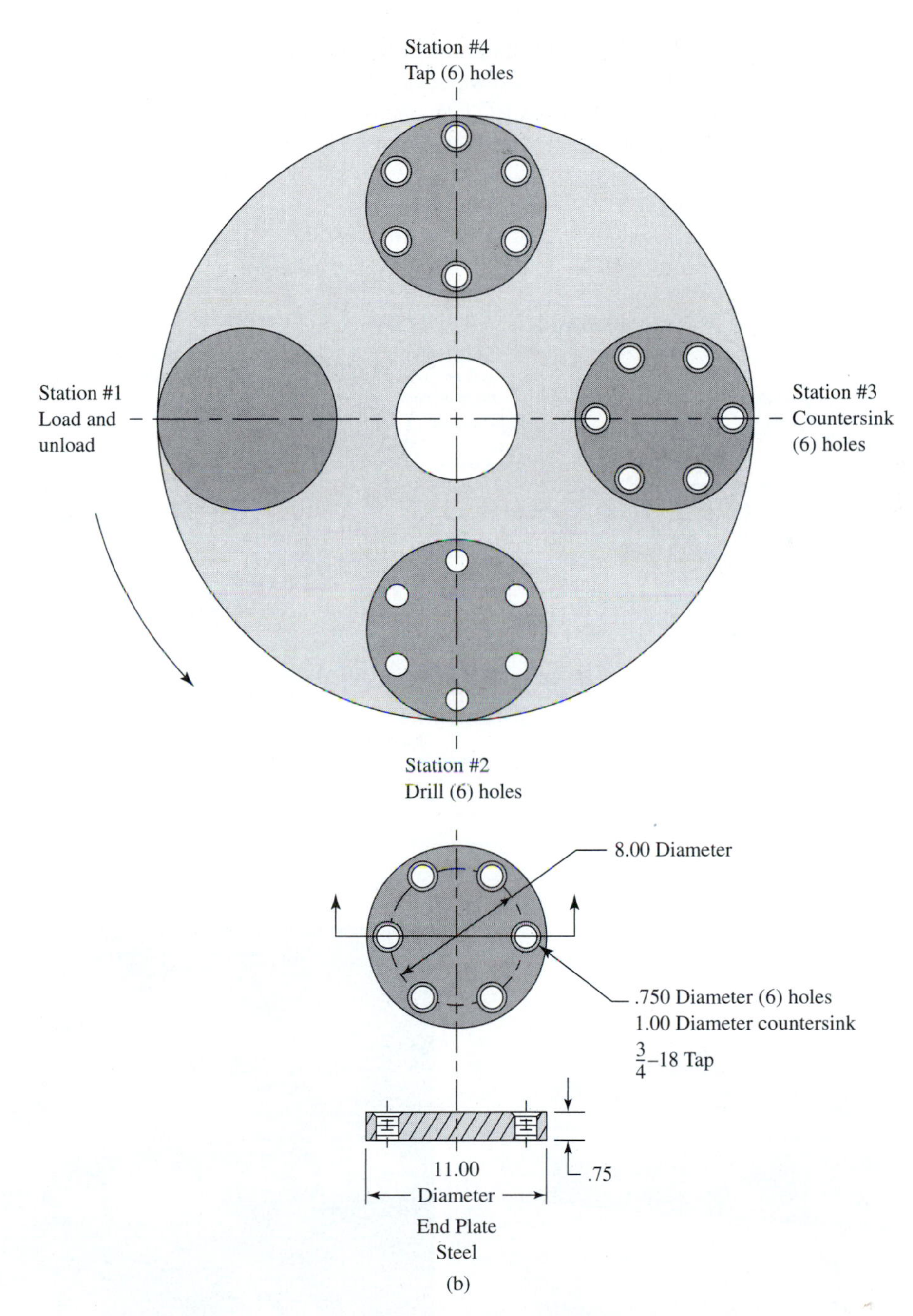

FIGURE 11.1 Continued (b) Part processed on 4-station indexing drilling machine. Countersinking and tapping are accomplished within drilling time.

11.2.1 Single-Spindle Automatic Lathes

Machine Design The majority of single-spindle automatic lathes are designed to machine workpieces that are located between two centers. Some, however, hold the workpiece in a chuck, collet, or specially designed fixture. Most have horizontal spindles.

As illustrated in Figure 11.2(a and b) conventional single-spindle automatic lathes have six major components: base, bed, and ways; headstock; tailstock; work spindle; front tool slide; rear tool slide. A special 2 × 2 axis CNC lathe is illustrated in Figure 11.2(c). Typical parts produced by 2 × 2 axis CNC lathes are shown in Figure 11.2(e) to (i).

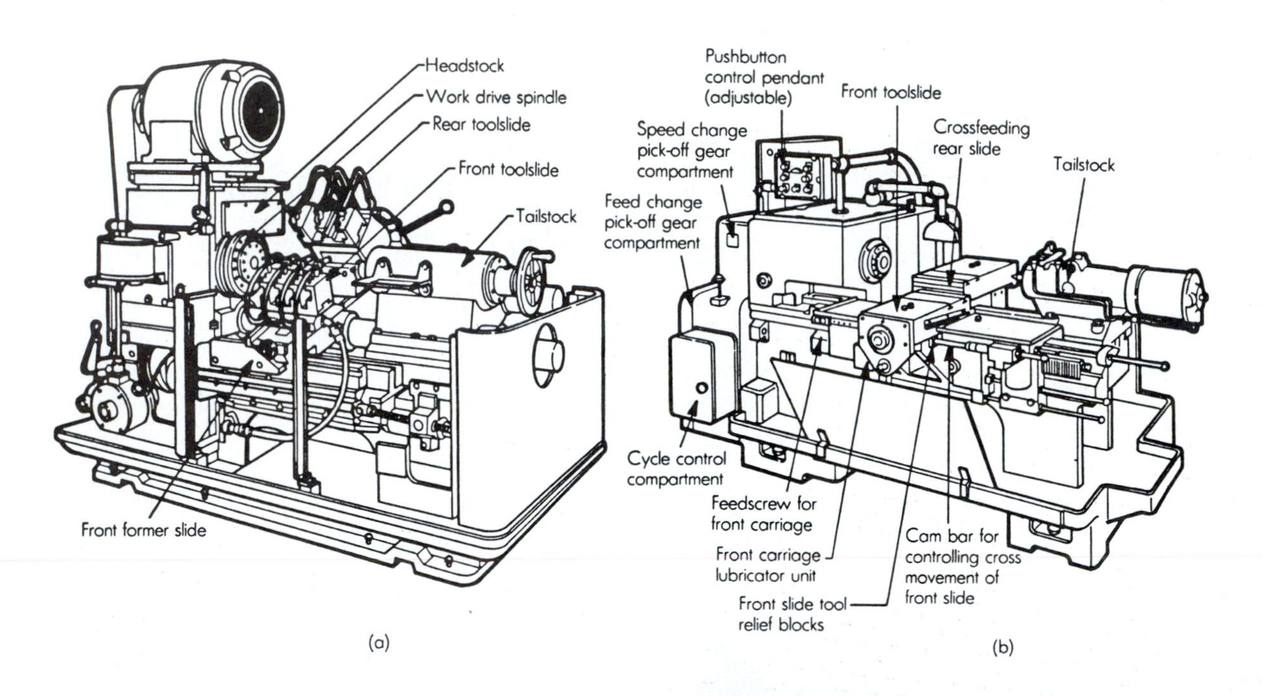

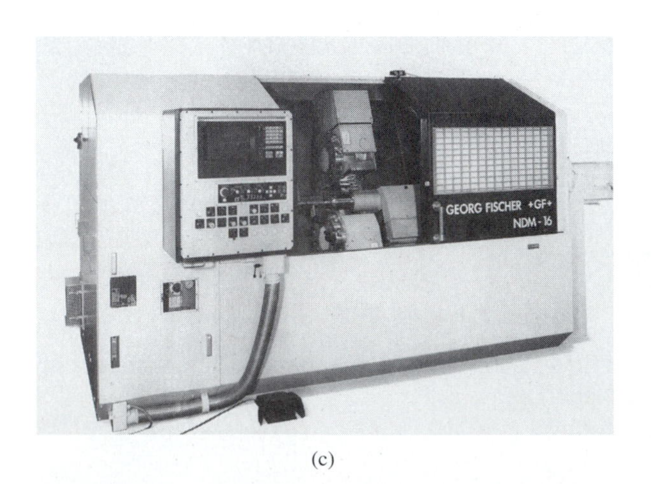

(c)

(d)

FIGURE 11.2 (a and b) Conventional single-spindle automatic lathes. (c) A 2 × 2 axis CNC lathe feautures two independent tool carriers with their own contouring control system, enabling a significant improvement in metal removal rates through a reduction in non-cutting "parasitic" time. Simultaneous processing of workpieces using both tool carriers is possible in some applications. (d) The structural characteristics of this CNC lathe are decisive for a reliable manufacturing process, even more in case of simultaneous turning on lathes with 2 × 2 axis systems.

FIGURE 11.2 Continued (e) Front view of the 2 × 2 axis CNC lathe. (f) An aircraft component machined by CNC lathe. Note the work rest support at the outboard end of the part. (g) A compressor component machined by CNC lathe. (h) A component for a railway application machined by CNC lathe. (i) A cylindrical component for a sewing machine processed by CNC lathe. (a and b, Courtesy National Acme Co., Div. DeVlieg–Bullard Inc., c–i, Courtesy George Fischer FMS Turning Technology Ltd., Schaffhausen, Germany and Farmington, Michigan)

(e)

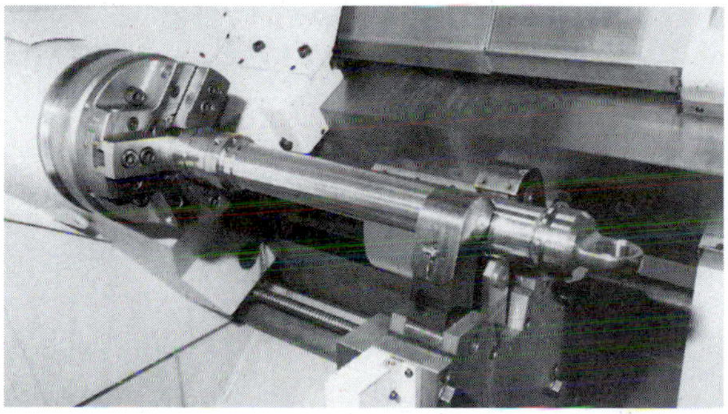

(f)

(g)

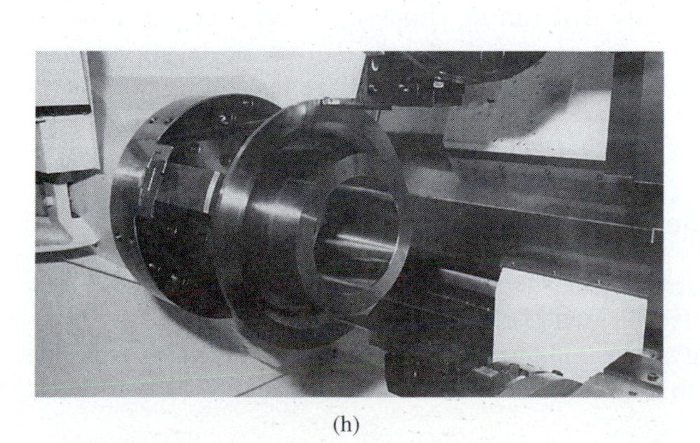

(h)

(i)

The feed rates of the tool slides are controlled by cams, hydraulics, or lead screws. Spindle speeds are changed to suit workpiece diameter/material requirements by means of change gears in the headstock.

Tooling and Attachments

Workpiece Holders. Any of the several available workpiece holders that are suitable for the particular application may be used, including chucks, faceplate drives, collets, and specially designed fixtures. Chucks, where used, should be power operated to avoid the time lost to manually actuate chucks.

Toolholders. Toolholders are normally designed with slots to locate and clamps to hold individual cutting tools in their required locations. The assembled toolholders are, in turn, keyed and clamped in a specific location on the front and rear tool slides.

It is good practice to provide spare toolholders wherein a set of sharpened tools can be preset and clamped, ready to exchange for a set of dull tools. Setup time can also be saved by having spare toolholders preset with the tools required for the next part to be run as shown in Figure 11.3.

Attachments

- Steadyrests are available in 2-roll and 3-roll types to support long, slender workpieces for resisting tool pressures. They can be applied and retracted automatically.
- Double-end and center-drive attachments are available to accommodate unusual machining requirements.
- Automatic parts loaders and unloaders must be designed and made to suit the operation. Many loaders are quite simple, because round workpieces can be rolled into position, many times by gravity alone, as shown in Figure 11.4.

Safety Requirements Appropriate guards and splash shields must be provided for operators' and bystanders' safety. Operators must also be provided with safety glasses and any required protective hearing devices and headgear.

Applications Axle and transmission shafts, gear blanks, pump drives, and pinions are all particularly well suited for machining on single-spindle automatic lathes. In fact, almost any machinable metal part falling within its size capacity that can be chucked, fixtured, or run between centers is a potential candidate for this machine. Single-spindle automatic lathes perform turning, facing, chamfering, grooving, and forming operations, and are usually used for parts with moderate production rates.

11.2.2 Conventional Single-Spindle Automatic Screw Machines

Automatic screw machines are the present-day developments of earlier machines whose only function was the production of screws. Modern machines not only retain thread-cutting capabilities but also are capable of performing all turning operations. These machines produce a wide range of parts from bar stock fed through a hollow work spindle. Some machines are arranged to produce parts from coil stock.

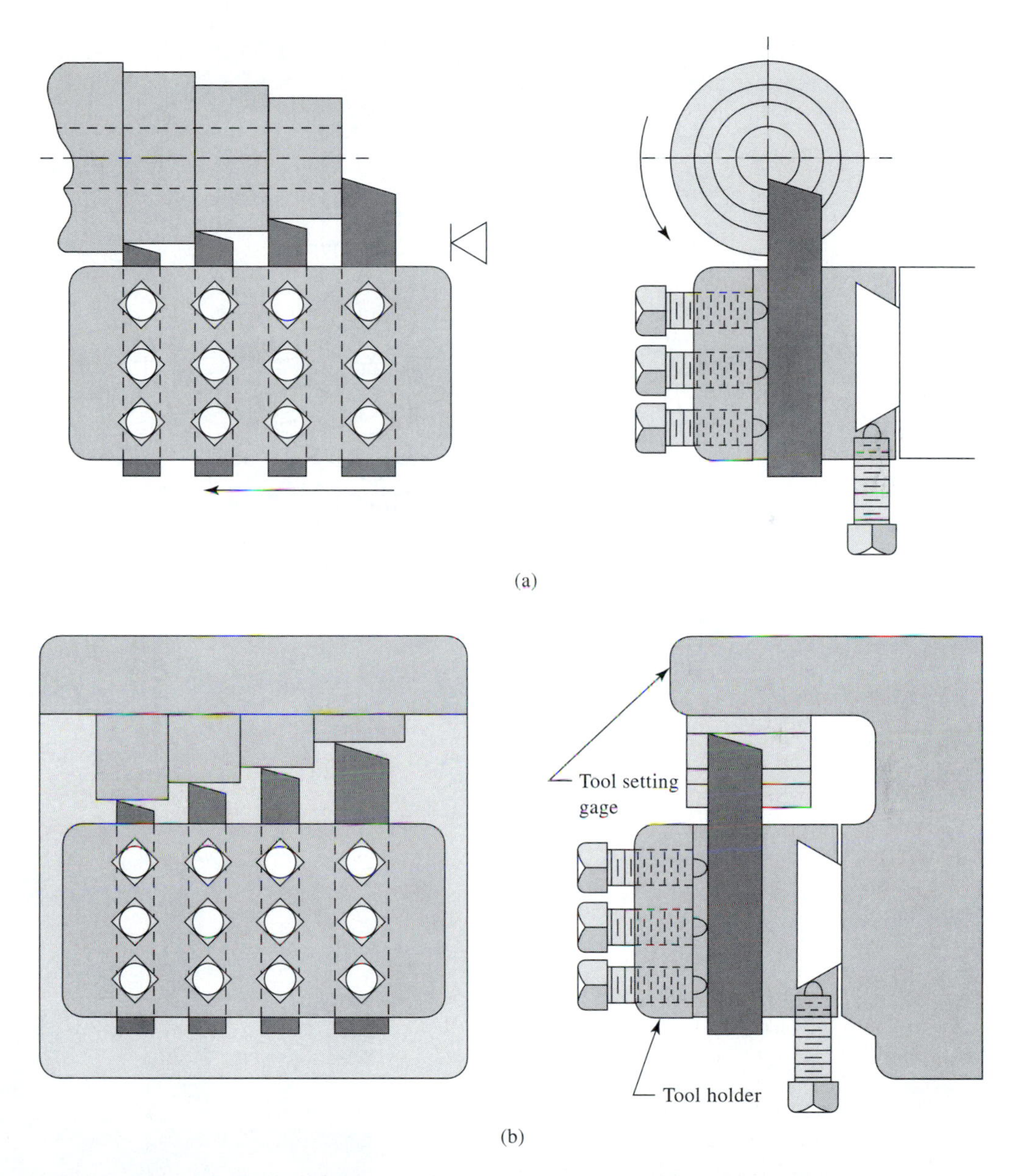

FIGURE 11.3 (a) Quick-change multiple tool holder—preset. (b) Tool setting gage.

Machine Design Single-spindle automatic screw machines have horizontal hollow spindles aligned with stock feeding tubes. Most are cam controlled but camless versions, sometimes NC or CNC controlled, are more flexible and quickly set up, making them more suitable for shorter production runs. Machines are available in several sizes and have six major components: base, headstock, hollow work spindle, front tool slide, rear tool slide, and turret, as shown in Figure 11.5.

The feed rates and motion of tool slides are controlled by cams or hydraulics. Spindle speeds are changed to suit workpiece diameter/material by means of change gears in the machine base.

Bar stock is fed automatically to a swing stop, or a turret stop, after each part is completed and cut off. The collet is automatically released during stock advances.

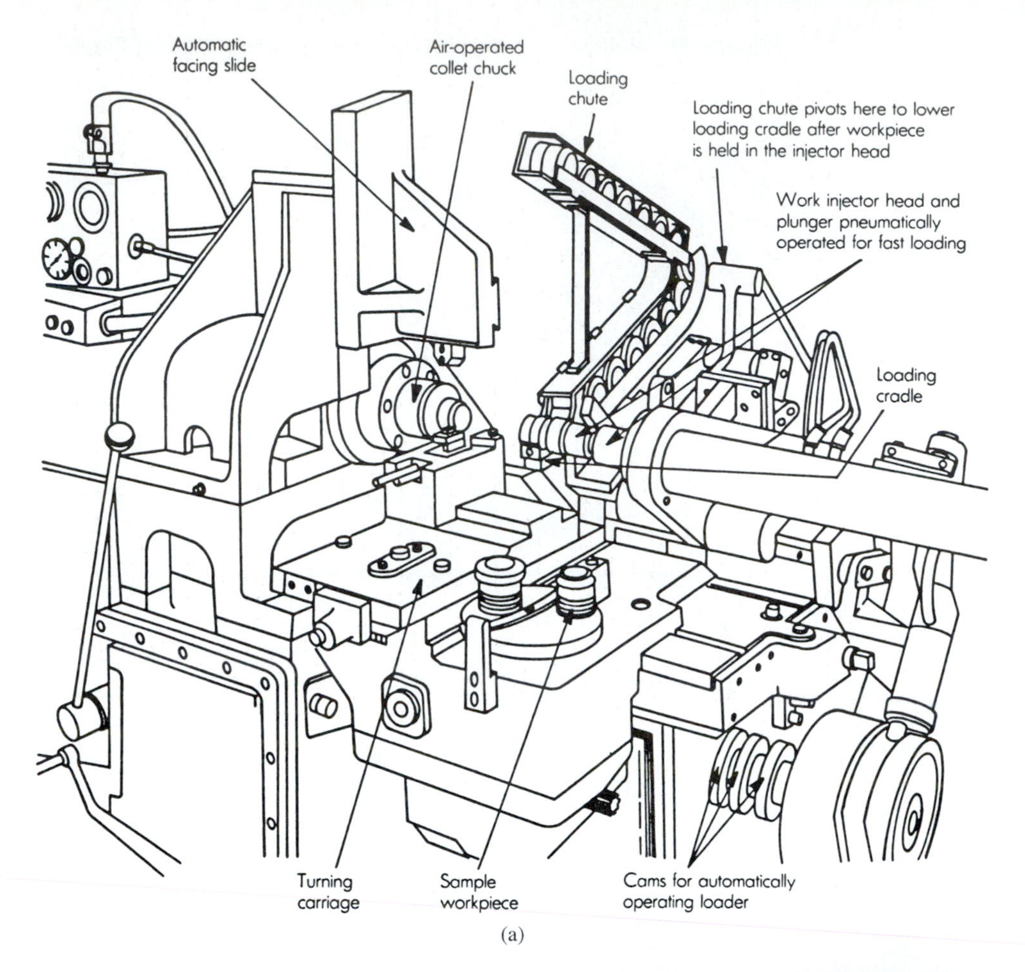

(a)

(b)

FIGURE 11.4 (a) Gravity/pneumatic automatic workpiece loader for roller. (Courtesy National Acme Co., Div. DeVlieg–Bullard Inc.) (b) Mechanized automatic loader for pinion incorporated on CNC lathe. (Courtesy George Fischer FMS Turning Technology Ltd, Schaffhausen, Germany and Farmington, Michigan)

FIGURE 11.5 Conventional single-spindle automatic screw machine. (Courtesy National Acme Co., Div. DeVlieg–Bullard Inc.)

Tooling and Attachments

Collets. Round, square, hex, and other standard-shape collets are available in sizes to suit commercial bar stock sizes. Specials are also made to suit.

Tools and Toolholders. Many special tools and toolholders are designed and made for certain applications, but a significant savings of time and money can be realized by the use of the standard tools and holders available. A large selection of standard tools are available from stock, some of which are illustrated in Figure 11.6.

Attachments. Standard attachments are available for such operations as screw head slotting, milling and right-angle drilling, deburring, and bushing assembly.

Safety Requirements Appropriate guards and splash shields must be provided for operators' and bystanders' safety. Operators must be provided with safety glasses as well as any required hearing protection and headgear devices.

Applications Single-spindle automatic screw machines are used to produce an extremely wide range of small parts including shafts, pins, knobs, screws, bolts, and so on from any machinable metal. Flats and slots can be milled and cross holes drilled. It is normal for one operator to operate several machines, the number depending on the frequency required for reloading bar stock and adjusting or changing tools.

Cut-Off Blades/Tool Bits & Bolts/Adjusting Plates

6 Spindle Machines:

Cut-Off Blades

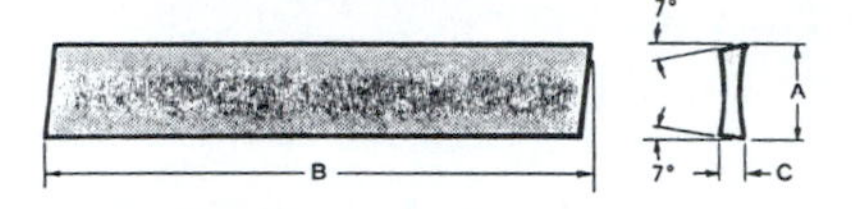

Cut-Off Blades for Blade Type Cut-Off Toolholders

Blade Part Number	Dimensions in Inches			Blade Part Number	Dimensions in Inches		
	A	B	C		A	B	C
AZ-50	.484/.483	4⅝	.040	02958-13	.484/.483	5¾	.062
02958-2	.484/.483	5¾	.093	02958-14	.484/.483	5¾	.109
02958-5	.484/.483	5¾	.125	—	—	—	—

NOTE: For former part and code numbers, see page 24.

Tool Bits – Standard Tool Bits for Turners (All Surfaces ground)

NOTE: For former part and code numbers, see page 24.

For Plain & Reversible Knee Turners, Pg. 15
Ordering Part Number AZ-18

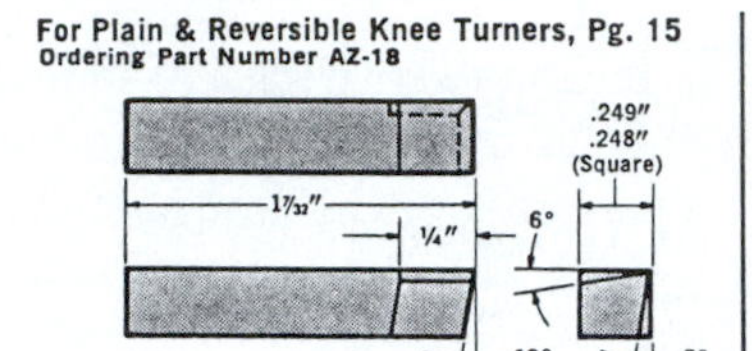

For Angular Knee Turners, Pg. 15
Ordering Part Number AZ-19

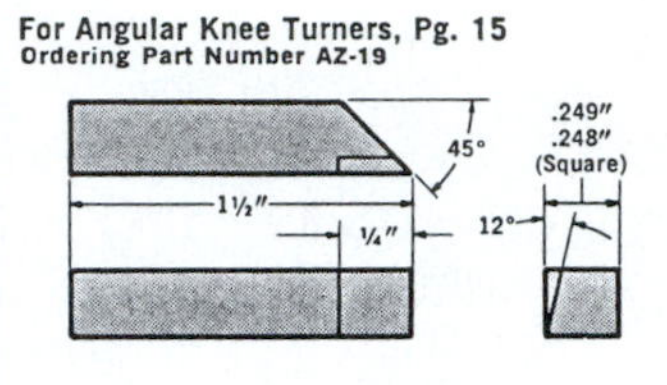

For Roll Turners, Pg. 16
Ordering Part Number AZ-17

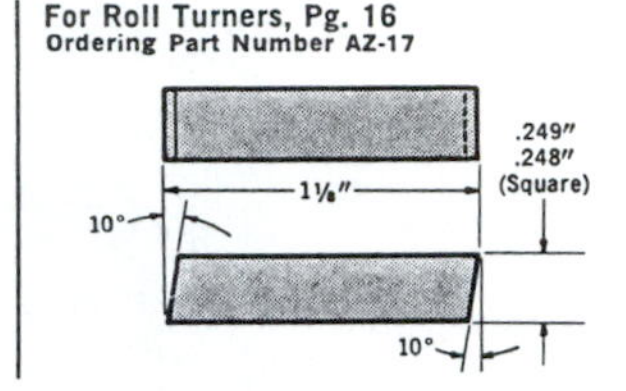

Tool Bolts

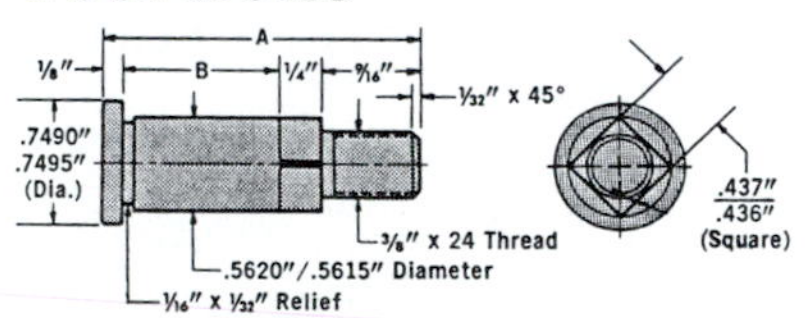

Tool Bolts for Circular Form Toolholders, Pgs. 6 & 7

Tool Bolt Part Number①	Dimensions in Inches			Tool Bolt Part Number①	Dimensions in Inches		
	A	B	Tool Width		A	B	Tool Width
956100	1⅞	15⁄16	¼	956104	2⅜	17⁄16	¾
956101	2	11⁄16	⅜	956105	2½	19⁄16	⅞
956102	2⅛	13⁄16	½	956106	2⅝	111⁄16	1
956103	2¼	15⁄16	⅝	956107	2¾	113⁄16	1⅛

NOTE: For former part and code numbers, see page 24.
① Pin and screw furnished with adjusting plate.

Circular Tool Adjusting Plates

For Form Toolholders, Pgs. 6 & 7 — **Ordering Part Number AZ-198130**
For Cut-Off Toolholders, Pgs. 3 & 5 — **Ordering Part Number AZ-19500** | **Ordering Part Number AZ-19494** | **Ordering Part Number AZ-19497**

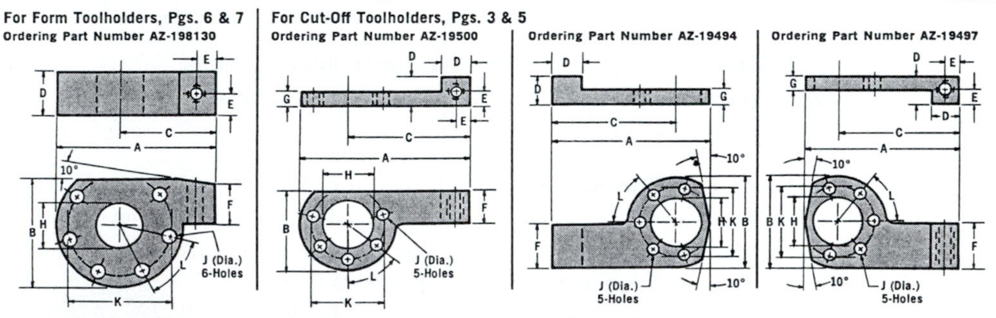

Circular Tool Adjusting Plates for Circular Form and Circular Cut-Off Toolholders

Circular Toolholder Type	Ordering Part Number	Dimensions in Inches										
		A	B	C	D	E	F	G	H(Diameter)	J(Diameter)	K(Diameter)	L
Form	AZ-198130①	1⅞	15⁄16	1⅛	½	¼	½	—	.5625/.5630	.143/.144	1.250/1.251	51° 26′
Cut-off	AZ-19500①	115⁄16	29⁄32	113⁄32	5⁄16	5⁄32	⅜	.156/.155	.5625/.5630	.143/.144	.8125	51° 26′
Cut-off	AZ-19494③	1¾	11⁄16	1⅜	5⁄16	—	½	.156/.155	.5625/.5630	.143/.144	.8125	51° 26′
Cut-off	AZ-19497①	1¾	11⁄16	1⅜	5⁄16	5⁄32	½	.156/.155	.5625/.5635	.143/.144	.8125	51° 26′

FIGURE 11.6 Standard tools for conventional automatic screw machines. (Courtesy National Acme Co., Div. DeVlieg–Bullard Inc.)

11.2.3 Multiple-Spindle Automatic Bar and Chucking Machines

Conventional multiple-spindle automatic bar and chucking machines have two major advantages over single-spindle automatics—both of which reduce the time required to produce a part:

1. The multiple-spindle machine performs work on each of its working stations concurrently; it is also possible to complete a different operation on a part at each position within the same time.

2. The maximum time required to complete one piece is the time required for the longest cut, plus index time, and in certain instances the longest cut can be broken up into increments. For example, a drilled hole that is the longest cut of a certain part may be completed in three or more positions.

Part sizes and complexity of design can be accommodated equally well on multispindle or single-spindle machines. Shorter changeover time favors single-spindle machines for short production runs, but the shorter machining time per piece of the multispindle machine makes it more economical for long runs.

Machine Design

Classifications. Multiple-spindle automatic machines are generally identified and classified by four features:

1. Capacity
2. Number of spindles
3. Method of chucking
4. Method of indexing

Examples of terminology are

- Two-inch, four-spindle bar machine with single index
- Six-inch, six-spindle bar machine with single index
- Eight-inch, eight-spindle chucking machine with double index

1. **Capacity.** Capacities of bar-type machines range from 1 in. to 8 in., with 8 in. the present maximum diameter of stock that can be fed through the spindles. The capacities of machines used as chuckers are limited only by the distance between spindles.

2. **Number of Spindles.** The great majority of multiple-spindle machines have an even number of spindles, such as four, six, or eight, but a machine is manufactured with five spindles. Discussion in this text refers to the even-numbered spindle machines. Machines with a greater number of spindles can accommodate more complex parts and, in many instances, can produce a part in less time by "breaking up the cuts" into shorter time increments.

3. **Method of Chucking.** Bar-type machines employ collets, pushers, and feed tubes to handle bar stock. Chucking machines may use three-jaw, four-jaw, or specially designed work-holding fixtures to hold forgings, castings, or tubular parts previously cut off.

4. **Method of Indexing.** Both bar and chucking types of multiple-spindle machines are capable of single or double indexing of the spindle carrier. Most use single indexing to produce one part per cycle. Double indexing may be used to machine both ends of a rough workpiece in a chucking machine. A synopsis of the processing follows and is illustrated in Figure 11.7(c):

- A rough forging Figure 11.7(a) is chucked on one end in load position A–1.
- The outer end is machined in one cycle.
- The forging is reversed and rechucked on the finished end in load position B–2.
- The second end (rough) is machined in one cycle, and then the part is disposed of.

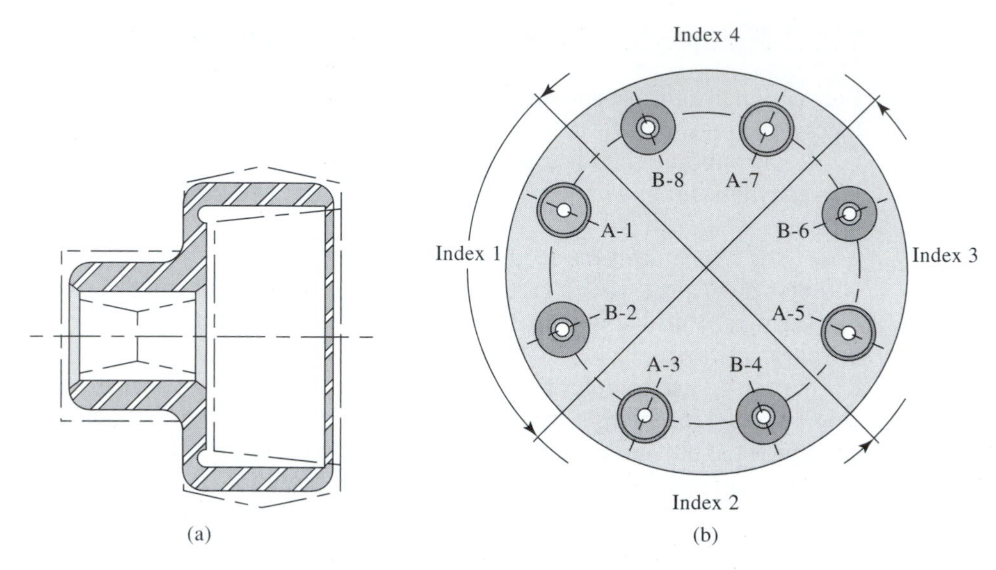

FIGURE 11.7 (a) Hub—steel. (b) Spindle layout, 8-spindle double-index chucking machine.

Two loading stations with their respective machining positions are required as shown in Figure 11.7(b). One finish machined part is produced every two complete machine cycles. Multiple-spindle bar machines may be double indexed to produce two parts per cycle in some applications.

Machine Construction. The general construction and main operating features of six-spindle bar and chucking machines are shown in Figure 11.8.

Spindle Carrier Assembly. The spindles revolve in the spindle carrier assembly, which indexes and is locked in each position in the headstock by means of a locking pin.

Main (End-Working) Tool Slide. The main tool slide is given longitudinal or horizontal movement to and away from the spindle carrier by means of a cam. End-working toolholders for end-working tools such as drills, boring bars, counter bars, and reamers are mounted on the tool slide by means of T-slots and bolts.

Cross Slides. The upper, intermediate, and lower cross slides (six in all) are mounted on the headstock. They are given inward, radial feeding by means of cams. Toolholders for slide tools such as forming, shaving, grooving, and cutoff are mounted on the slides by means of T-slots and bolts. Figure 11.9 illustrates the relative positions and paths of a six-spindle machine's cross slides.

Stock Feeds. Stock is fed to a stop in one position of a single-indexed machine, or two positions of a double-indexed machine. Pushers are used and collets are synchronized to be opened before and closed after feeding.

Standard Tooling and Toolholders. Standard toolholders are available for most end-working tools, including drills, boring bars, counter bars, turning tools, and reamers. Similarly, standard toolholders are available for cross-slide tooling, including form tools, shaving tools, grooving tools, and cutoff blades. A shaving tool and toolholder are illustrated in Figure 11.10.

Index	Position	Spindle	Operation	Process Picture
1	*A*	1	Load part into *A* position large end out Chuck on small diameter Locate against large end back face	
1	*B*	2	Unload finished part from *B* position Reload part from *A* position to *B* position small end out Chuck in I.D. Locate against inner face	
2	*A*	3	Rough turn O.D. Rough bore large diameter and form recess Rough face inner face and end face Form chambers	
2	*B*	4	Rough turn small diameter Rough bore I.D. Rough face end face Form chambers	
3	*A*	5	Finish turn O.D. Semi-finish bore large diameter Finish face end face and semifinish face inner face	
3	*B*	6	Finish turn small diameter Semifinish bore I.D. Finish face end face	
4	*A*	7	Finish bore large diameter Finish face inner face	
4	*B*	8	Finish ream I.D.	

(c)

FIGURE 11.7 Continued (c) Operational sequence for machining steel hub on 8-spindle double-index chucking machine.

Standard carbide inserts of square, round, or triangular shapes are used where possible due to the cost advantage. Appropriate grades of carbide must be selected for each application, and some form of monitoring systems should be developed to ascertain that the specified grades of tooling specified are, in fact, in use.

Special Attachments. The versatility of multiple-spindle bar and chucking machines is further enhanced with the use of special attachments. They make feasible performing certain

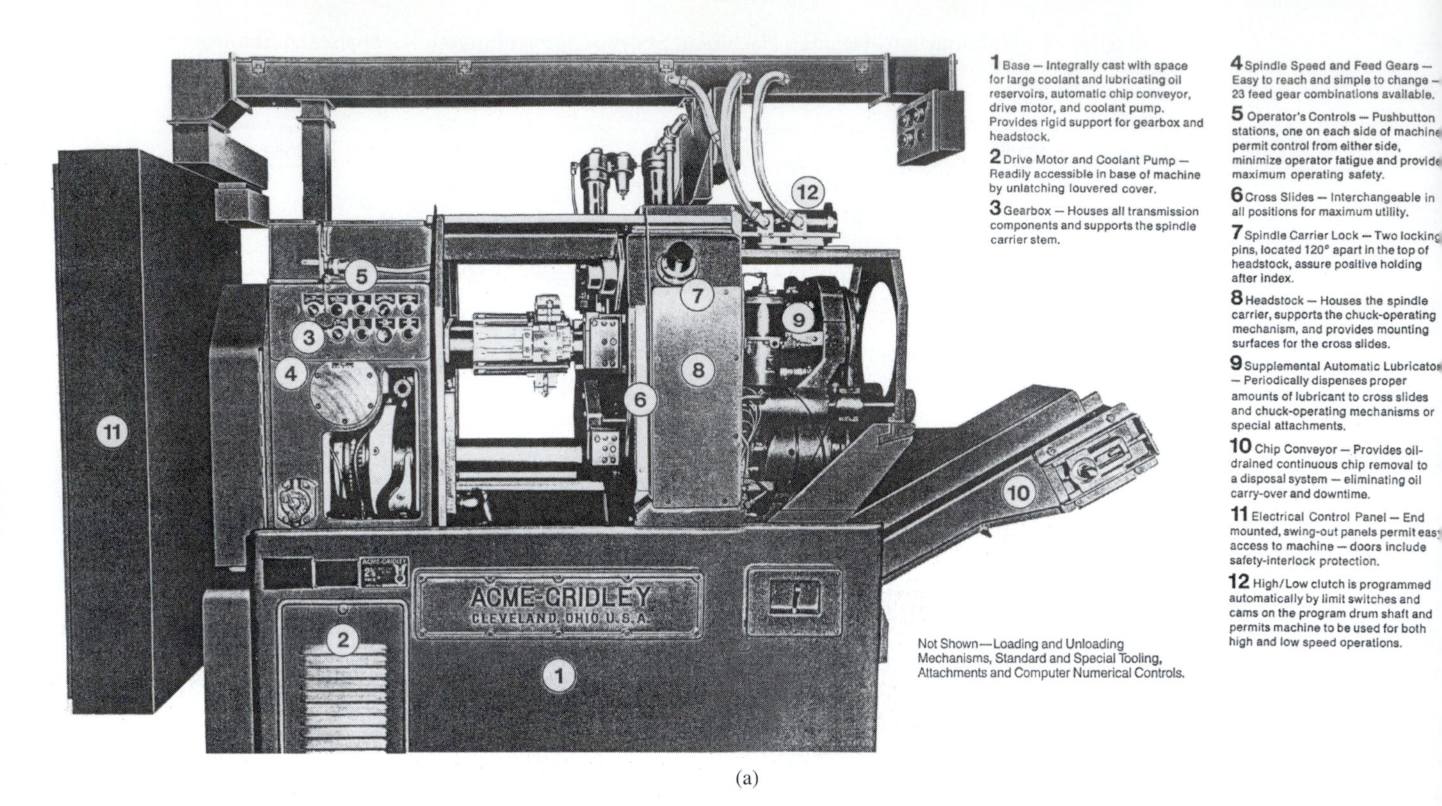

(a)

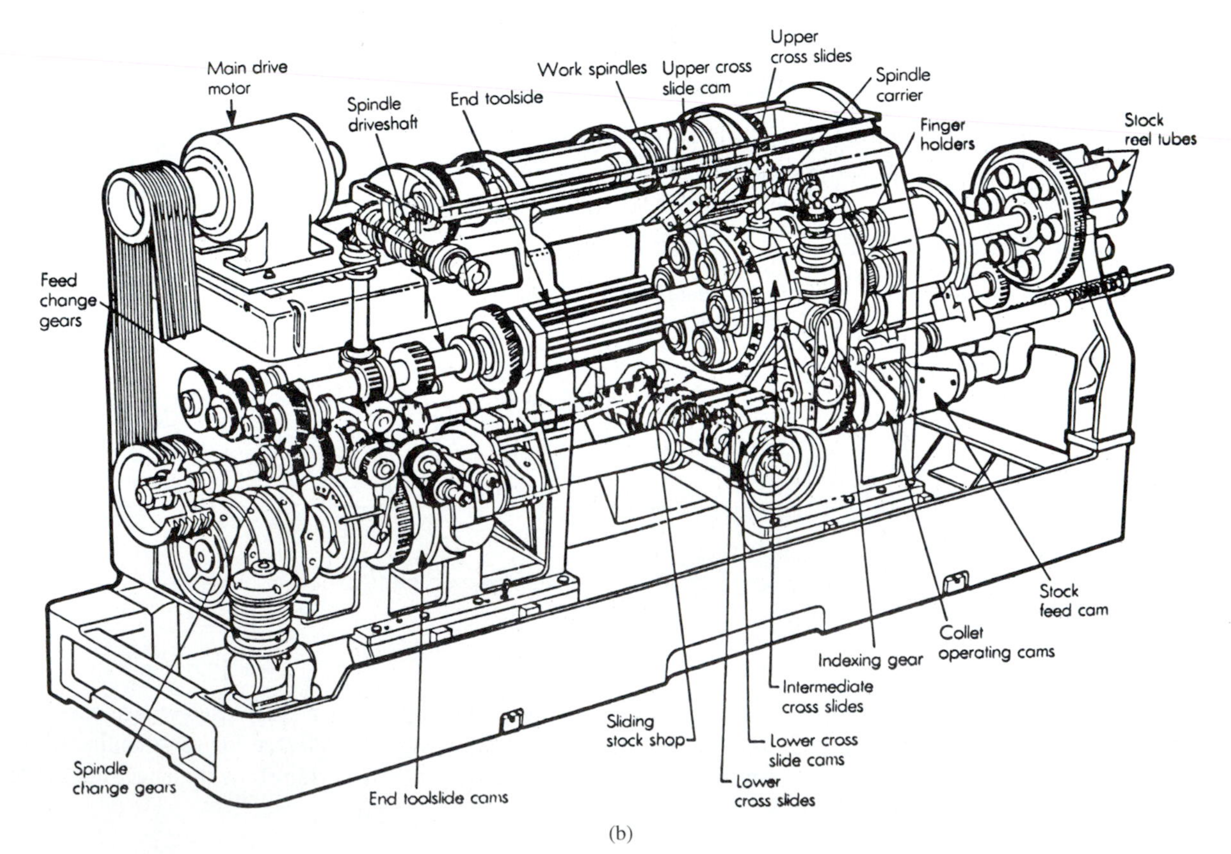

(b)

FIGURE 11.8 (a) Six-spindle automatic chucking machine. (b) Six-spindle automatic bar machine. (Courtesy National Acme Co., Div. DeVlieg–Bullard Inc.)

FIGURE 11.9 Cross slides, end slide, and spindle arrangements for multiple-spindle machines. (Courtesy National Acme Co., Div. DeVlieg–Bullard Inc.)

FIGURE 11.10 Shaving tool and holder. (Courtesy National Acme Co., Div. DeVlieg–Bullard Inc.)

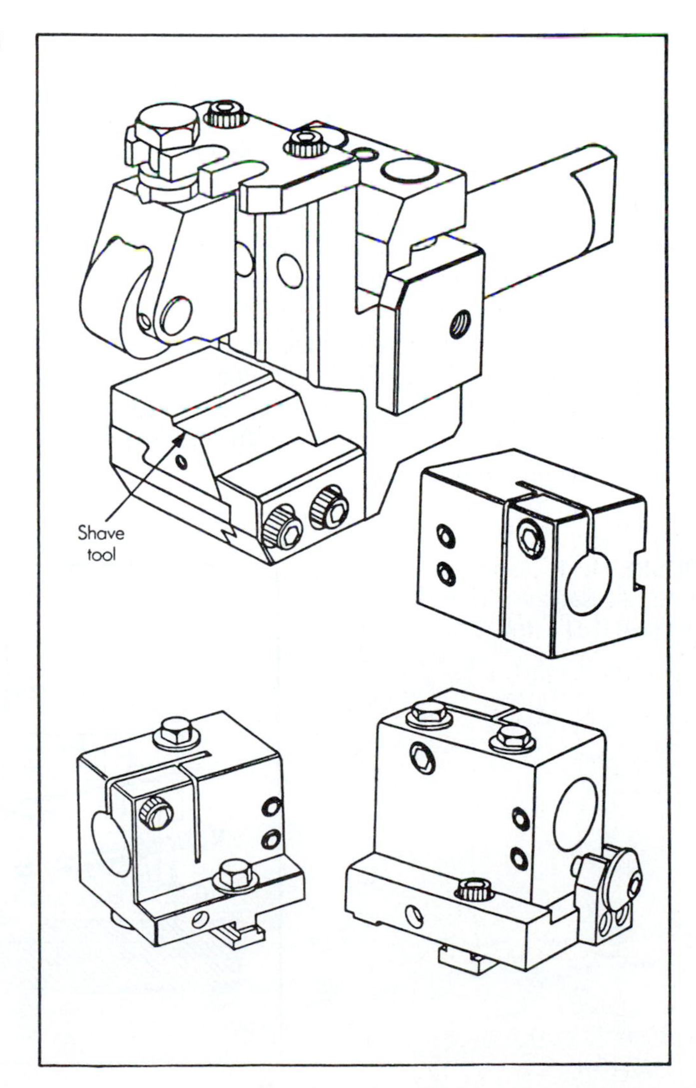

operations that are beyond the scope and motions of a basic machine. Although these attachments are denoted "special," they are available, usually from stock. They include

1. High-speed drilling attachments, for use on workpieces necessarily processed at RPM too slow for efficient drilling
2. Cross-drilling and cross-milling attachments for use with spindle stopping mechanisms
3. Threading attachments for the main tool slide for use with dies, taps, and thread rollers
4. Thread-rolling attachments for cross slides, which allow work to be done on the collet end and close to shoulders
5. Pickoff attachments, which allow secondary operations to be completed on multiple-spindle machine following part cutoff
6. Recessing attachments, which are usually for use with the main tool slide at and provide motion at right angles to the motion of the tool slide, useful in generating chamfers or undercuts in inside diameters

One form of recessing attachment is illustrated in Figure 11.11.

Tool Layouts—Processing A tool layout depicting the proposed processing and setup of a multiple-spindle automatic machine and, indeed, most machines, must be made in advance of actual setup to study tool costs, machining time, and, possibly, material costs and scrap. Variations of the setup, each one capable of producing a quality part, should be made and compared.

The development of tool layouts, especially for complex parts, must be done by a tool engineer or manufacturing engineer possessing the required skills and experience.

As in other areas of manufacturing engineering, there is no substitute for experience. Usually a younger, dedicated engineer is teamed with a senior engineer who possesses the required experience, skill, and attitude to understudy and "learn by doing."

One other source for obtaining optimum part processing and layouts is to purchase it with the machine; often, machines are purchased "tooled," with all tooling designed, made, and tried out. Purchasing the entire package from the original equipment manufacturer almost invariably provides optimum processing.

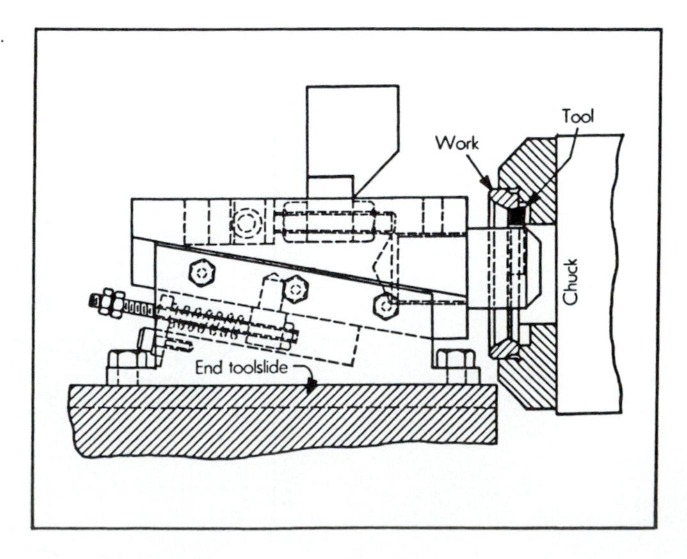

FIGURE 11.11 Recessing Attachment. (Courtesy National Acme Co., Div. DeVlieg–Bullard Inc.)

Safety Appropriate guards and splash shields must be provided for operators' and bystanders' safety. Operators must be provided with safety glasses as well as any required hearing protection and headgear devices.

11.2.4 Multiple-Spindle Vertical Automatic Chucking Machines

Multiple-spindle vertical automatic chucking machines are manufactured by several machine-tool builders in several sizes and models ranging from 4 to 8 spindles. One maker supplies a 16-spindle machine that is, in reality, double spindles for each position of an 8-spindle machine.

These machines use less floor space than an equivalent horizontal model and are more flexible in application. They do not, however, accept bar stock. Some other advantages are that they are convenient to load, operate, and adjust or change tooling.

Machine Design and Construction The machine illustrated in Figure 11.12 has three major components: base and center column, carrier and work spindles, and machining heads. The machine is designed to permit each spindle to operate independently, having independent speeds and feeds. In effect, the machine illustrated can operate as seven individual machines all loaded and unloaded at a common station.

Machines with dual spindles and multiple-tool machining heads are available, permitting duplicate setups, or first and second chucking work to be performed (both ends). Double indexing is available and is used with dual-spindle setups.

FIGURE 11.12 Multiple-spindle vertical automatic chucking machine. (Courtesy National Acme Co., Div. DeVlieg–Bullard Inc.)

Tooling and Attachments Several types of multiple-tool machining heads are available, including vertical motion, swivel units, and universal (vertical and horizontal) motion.

Chucks are power operated, and unloading aids (lifters and ejectors) are available for heavy work. Attachments available are

1. Drilling and tapping heads, both single- and multispindle
2. Boring heads
3. Turning heads
4. Facing heads
5. NC and CNC heads

Controls can be applied to individual spindles. Automatic loading and unloading devices that can be modified to suit many applications are available.

Applications These machines tend to be used for the larger and heavier workpieces. Many castings, forgings, and subassemblies are particularly well suited for multispindle machines due to ease of loading and unloading.

Tool Layout—Processing A typical application and processing of a conventional six-spindle bar machine are illustrated in Figure 11.13.

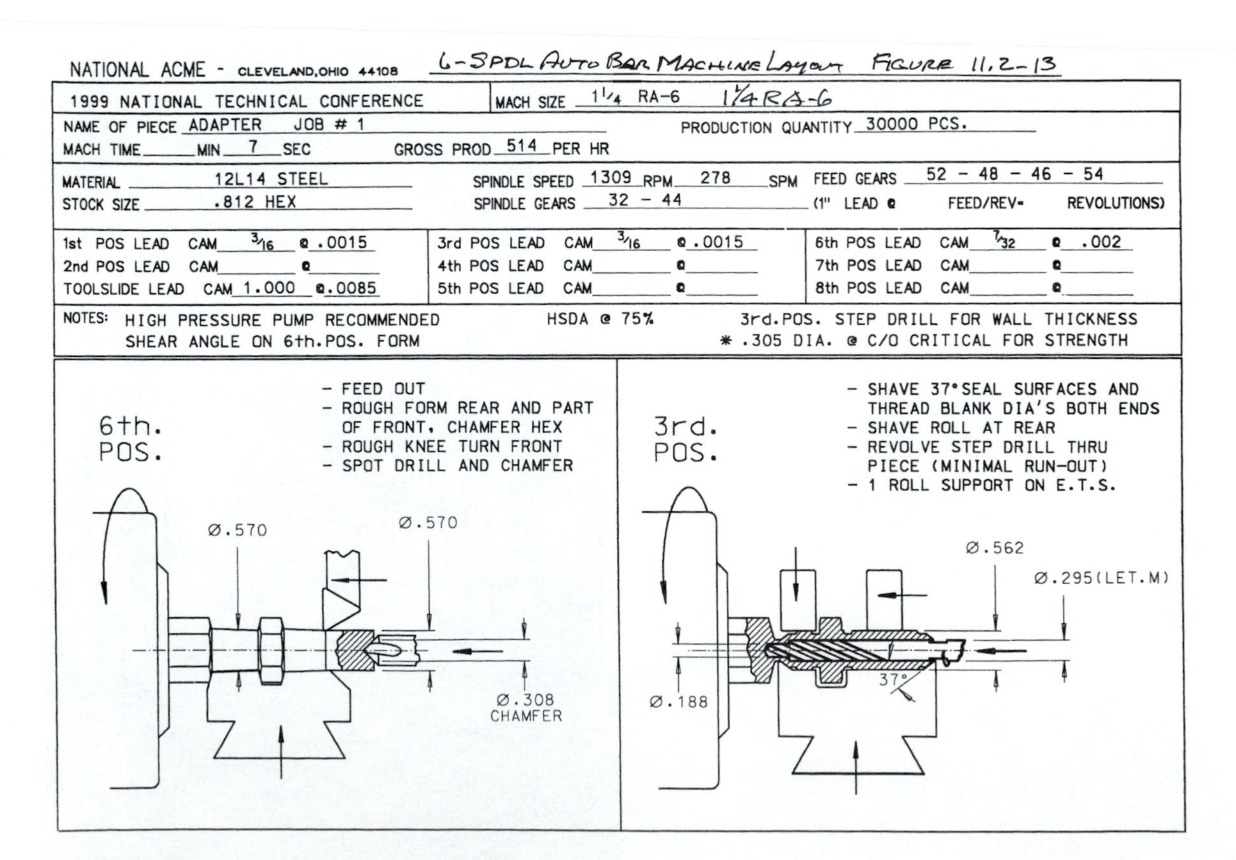
NATIONAL ACME - CLEVELAND,OHIO 44108 6-SPDL AUTO BAR MACHINE LAYOUT FIGURE 11.2-13

1999 NATIONAL TECHNICAL CONFERENCE | MACH SIZE 1 1/4 RA-6 1 1/4 RA-6

NAME OF PIECE ADAPTER JOB # 1 PRODUCTION QUANTITY 30000 PCS.

MACH TIME ___ MIN 7 SEC GROSS PROD 514 PER HR

MATERIAL 12L14 STEEL SPINDLE SPEED 1309 RPM 278 SPM FEED GEARS 52 - 48 - 46 - 54

STOCK SIZE .812 HEX SPINDLE GEARS 32 - 44 (1" LEAD @ FEED/REV= REVOLUTIONS)

1st POS LEAD CAM 3/16 @ .0015 | 3rd POS LEAD CAM 3/16 @ .0015 | 6th POS LEAD CAM 7/32 @ .002

2nd POS LEAD CAM ___ @ ___ | 4th POS LEAD CAM ___ @ ___ | 7th POS LEAD CAM ___ @ ___

TOOLSLIDE LEAD CAM 1.000 @.0085 | 5th POS LEAD CAM ___ @ ___ | 8th POS LEAD CAM ___ @ ___

NOTES: HIGH PRESSURE PUMP RECOMMENDED HSDA @ 75% 3rd.POS. STEP DRILL FOR WALL THICKNESS

SHEAR ANGLE ON 6th.POS. FORM * .305 DIA. @ C/O CRITICAL FOR STRENGTH

FIGURE 11.13 6-Spindle auto bar machine layout. (Courtesy National Acme Co., Div. DeVlieg–Bullard Inc.)

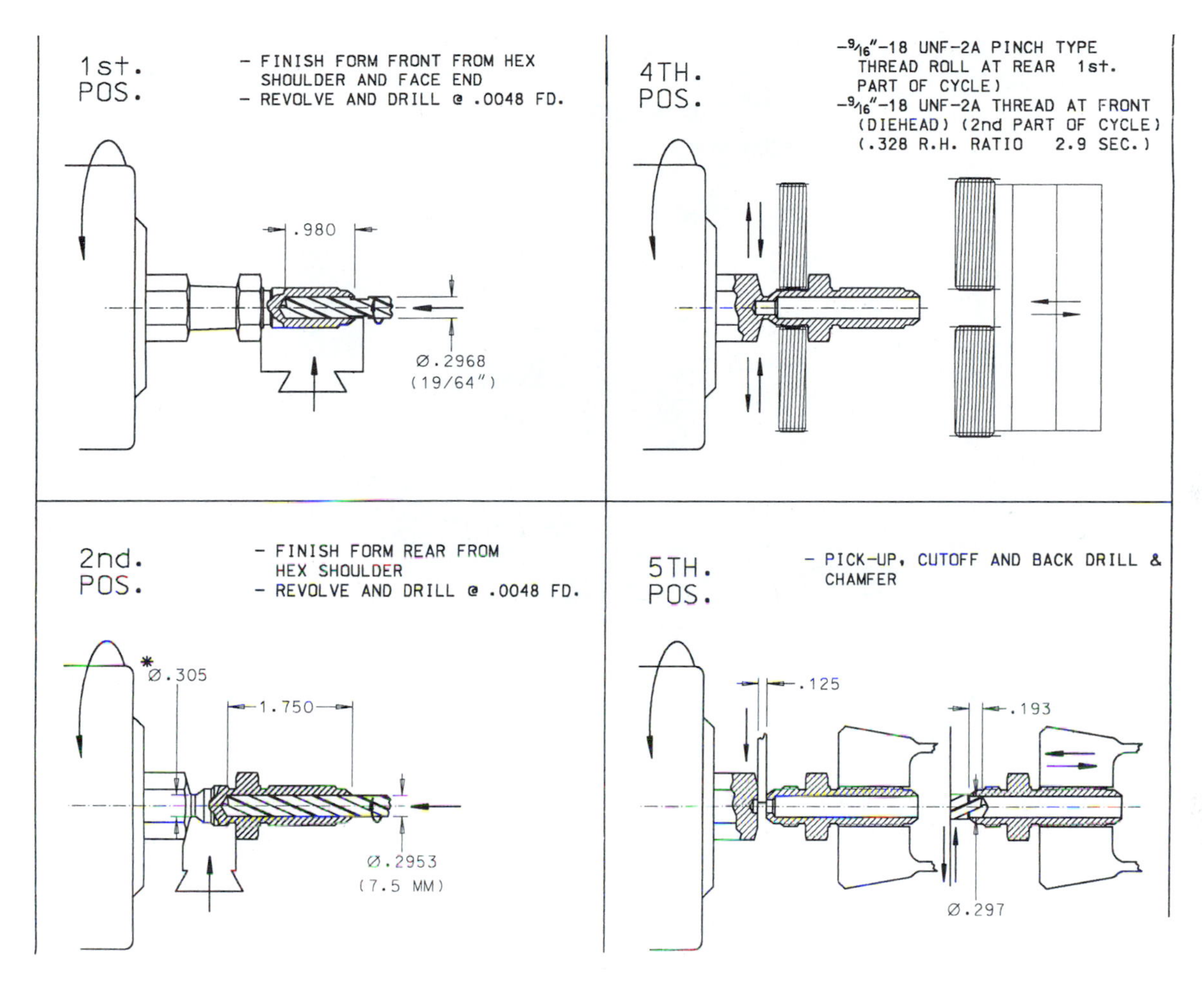

FIGURE 11.13 Continued

Safety Appropriate guards and splash shields must be provided for operators' and bystanders' safety. Operators must be provided with safety glasses as well as any required hearing protection and headgear devices.

11.3 MACHINING SYSTEMS

Machining systems have evolved into two general types:

- **Machining centers** are single, self-contained programmable machines.
- **Flexible manufacturing systems** include one or more machining centers along with other programmable production machines such as lathes, special-process machines, and assembly and test equipment.

11.3.1 Machining Centers

Machining centers can be defined as multifunctional self-contained machines that have automatic tool-changing capabilities, rotating tools, and NC or CNC controls. They are capable of

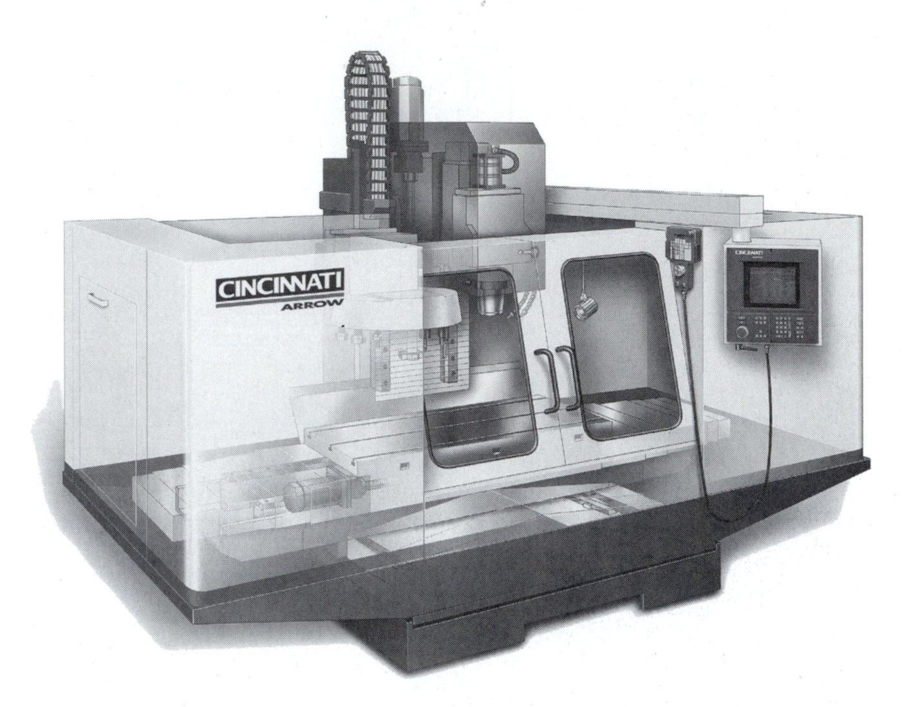

FIGURE 11.14 Vertical-spindle machining center. (Courtesy Cincinnati Machine, a UNOVA Co., Cincinnati, Ohio)

drilling, reaming, boring, tapping, milling, counterboring, and spotfacing operations. They usually include a system of automatic tool changing, and some special machines have automatic head changers.

Machine Design and Construction

Two major types of machining centers are the vertical- and horizontal-spindle models. Some machines are designed with one vertical and one horizontal spindle; others have multiple spindles in various arrangements. Many factors determine the type used, including size, shape, and tolerances of the part to be machined: complexity of part design: and families of similar parts to be produced.

Vertical-spindle machines provide a very accessible table for setup and loading of parts. They are used primarily for plate types of workpieces. Figure 11.14 illustrates a vertical-spindle machine.

Horizontal-spindle machines are more flexible and are much more in use. Many models are available in this type. Two models of horizontal spindle machines are illustrated in Figure 11.15.

Spindles. Spindles, which are the heart of dimension and tolerance capabilities as well as of maintaining production, are available with many desirable features:

- Water-cooled spindles, bearings, and carriers to control thermal expansion
- Special thrust bearings to absorb heavy boring and milling operations
- Roller bearings with automatic variable preloading for operating at different speeds with optimum preload

Power—Speed—Feeds—Accuracies. Machining centers are available powered with motors ranging from 2 to 75 horsepower and spindle speeds ranging from 0 to 4000 RPM. A spindle speed of 6000 RPM is generally considered as maximum but some special machines can operate up to 60,000 RPM.

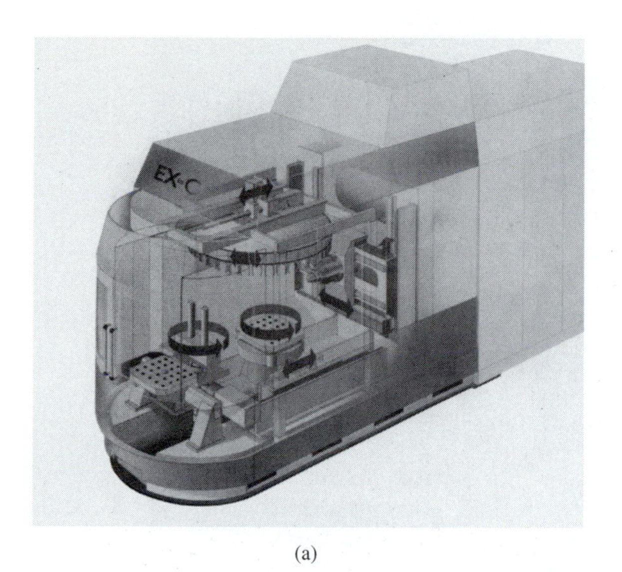

(a)

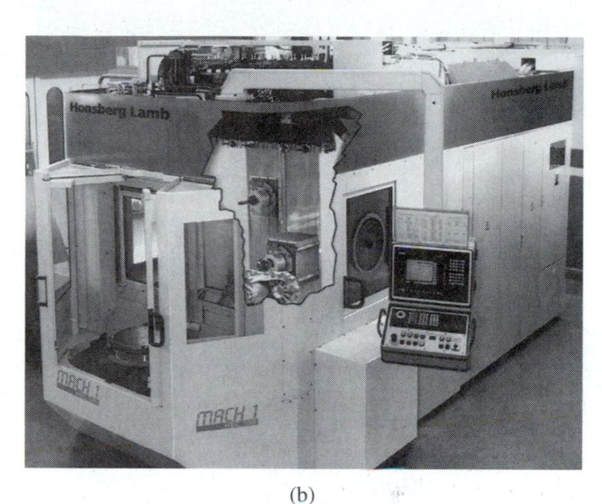

(b)

FIGURE 11.15 Horizontal-spindle machining centers. (a) EX-CELL-O high-speed, single-horizontal-spindle machining center for general applications. (b) Honsberg Lamb, 2-spindle, horizontal, high-speed cutting machining center, preferred for workpieces of aluminum and magnesium alloys. (a, Courtesy EX-CELL-O Machine Tools, Inc., Sterling Heights, Michigan; b, Courtesy Lamb Technicon Machining Systems, a UNOVA Company)

Feeds of up to 300 in. per minute are provided by preloaded precision ball screws that can maintain tolerances of ±.001 for positioning and ±.0005 for repeatability.

Tool Changers Tool changers are available in many different designs, including chains, dials, and drums, but all include idle tool storage and a device to interchange stored tools with a tool in the machine spindle. Several types of tool changers are illustrated in Figure 11.16. Tool storage capacities range from 3 to over 100, depending mainly on the size of the tools.

Tooling Standard cutting tools, drills, and reamers are generally usable but may require slight modifications. Toolholders are made to several different standards depending on the make of the machine. This can cause problems of cost and inventory if more than one make of machine is employed.

Controls Machine centers are controlled by NC or CNC with most employing built-in solid-state CNC units. Some systems are capable of automatically adjusting spindle speed rates to suit changing workpiece or tool size conditions.

Fixture and Workpiece Holders To optimize productivity, workpiece-holding fixtures must be designed to permit a maximum number of operations to be performed in one setup. Also, some machine centers are constructed to allow the use of multiple fixtures simultaneously; others are designed to shuttle additional fixtures into position as work is completed.

Other methods used to enhance productivity include

- Using fixtures capable of holding multiple parts
- Using duplicate fixtures, which allows one part to be machined while another is unloaded and reloaded

(a)

(b)

(c)

(d)

FIGURE 11.16 Tool changers. (a) A rack style system provides storage for a large quantity of tool holders. This tool management system is palletized to permit automatic storage and retrieval for servicing a flexible, automated machining system. (b) A rotary style tool changer is incorporated on this horizontal machining center. In this application, gripper devices are not required since the tools are changed directly with the spindle. The forty-position magazine uses a short axial movement to remove the tool from the spindle shaft and then installs the next tool in the same way. This provides a fast chip-to-chip time of 4.5 seconds. (c) Tool changer for 2-spindle horizontal machine. (d) Chain-type tool changer. (a, b, d courtesy EX-CELL-O Machine Tools, Inc., Sterling Heights, Michigan; c, Courtesy Lamb Technicon Machining Systems)

- Use of A and B positions in one fixture when two settings are required to complete a workpiece
- Use of machines with fixture pallet changing and shifting capabilities

Examples of these kinds of machines are illustrated in Figure 11.17.

Applications Originally, machining centers were used for small lots of many different parts and sizes requiring several operations, but present use includes medium runs of such parts. Some long runs of single-design parts have proven economical and are in increasing use. Predominant users are small to medium-sized companies and job shops.

Advantages Many advantages accrue to the use of machining centers, but chief among them are

- Higher productivity resulting from optimum cutting speeds and feeds for each operation
- Ease of setup and changeover, which reduces job setting and machine downtime

(b)

(a)

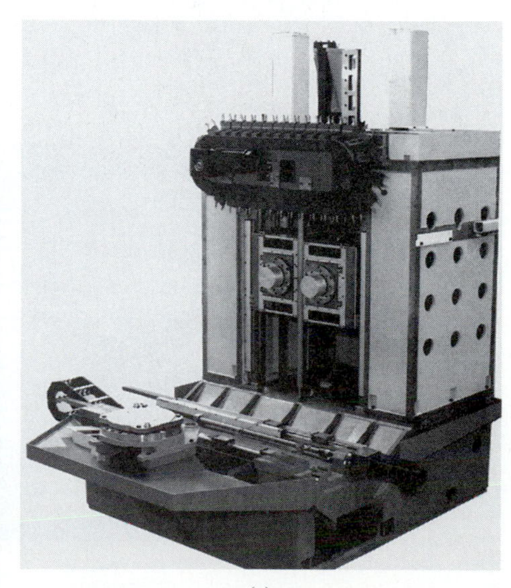

(c)

FIGURE 11.17 Pallet changing machines. (a) A pallet-type transfer machine handles ten different types of automotive throttle bodies in five families at 900 parts per hour, and requires less than one hour for tool changeover. The pallet-type system combines CNC-like flexibility with very short, 12-second cycle times and extreme precision on critical part tolerances. (b) Palletized parts are transferred among CNC work centers and selected special dedicated machining stations in this flexible system for producing automotive cylinder blocks. (c) Twin horizontal-spindle, machining center with pallet changer. (a, b Courtesy EX-CELL-O Machine Tools, Inc., Sterling Heights, Michigan; c, Courtesy Lamb Technicon Machining Systems, a UNOVA Company)

- Simple work-holding fixtures and use of standard tools, which result in lower tooling costs and shorter load time to produce the tooling
- Performing many operations per setup, which results in less machine downtime for job setting.

11.3.2 Flexible Manufacturing Systems

Flexible manufacturing systems can be described as machining systems that are composed of three major elements:

- Work stations
- Material handling
- Controls

Each element is designed and constructed to suit the intended application, or applications, and therefore differs from other flexible manufacturing systems; there is no standard. Figure 11.18

FIGURE 11.18 Flexible machining system arrangements. (a) A flexible machining system arrangement incorporating horizontal spindle machining centers is used for processing families of electric motor housings. (b) An integrated loading system is fixed directly to the CNC turning machine's bed, above the headstock, to permit automatic loading and unloading of parts.

(a)

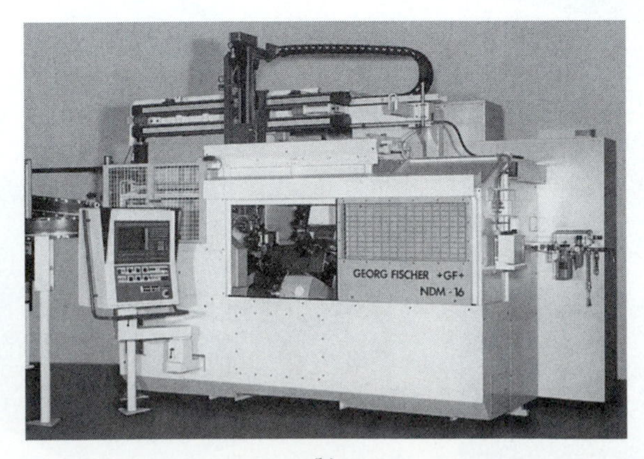

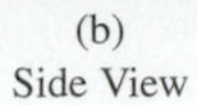

(b)
Side View

(b)
End View

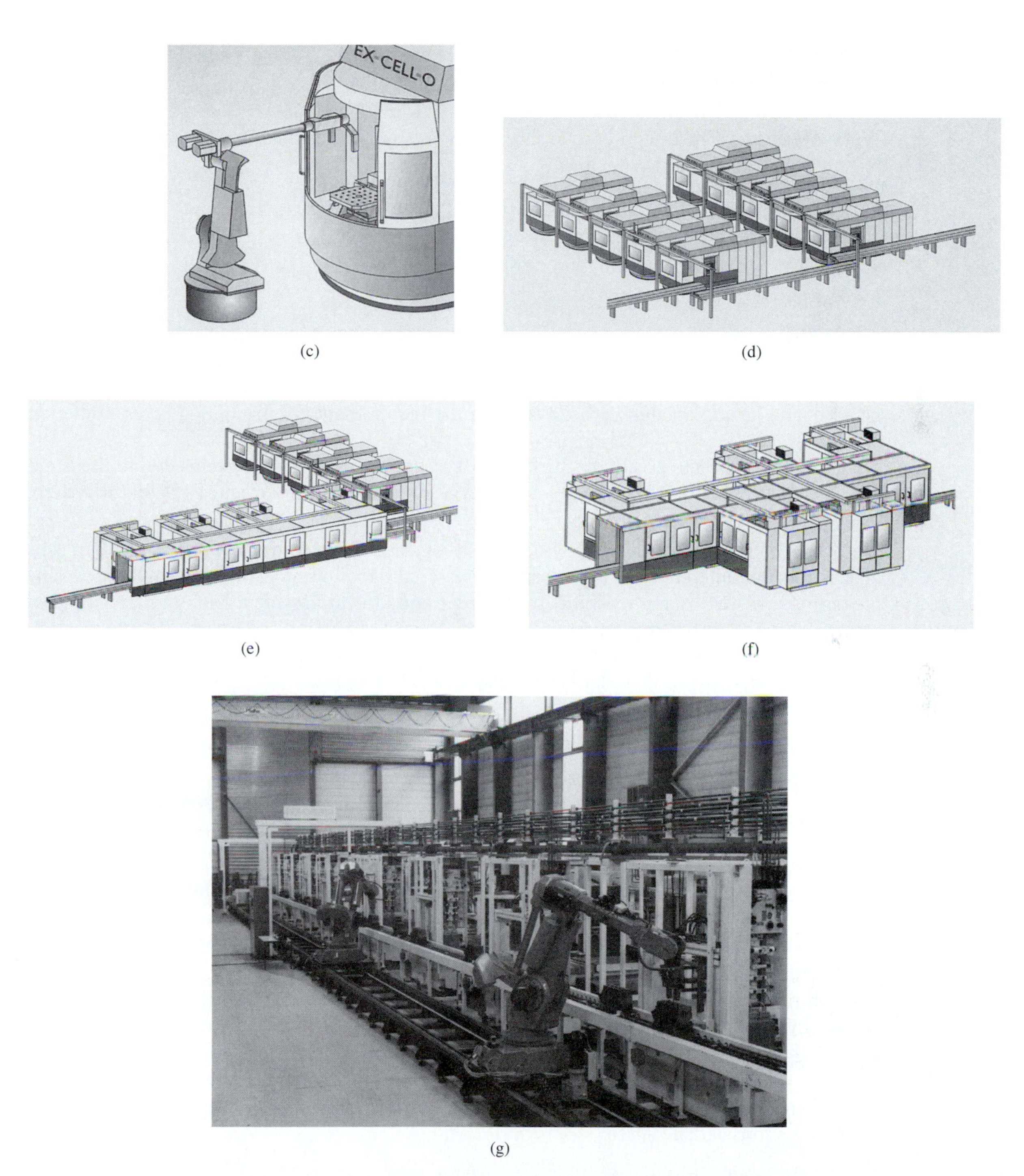

(c) (d) (e) (f) (g)

FIGURE 11.18 Continued (c) This schematic depicts a robot used for loading and unloading the work-switcher on the CNC machining center. (d) A machining system based on high-speed machining centers incorporating gantry-style loaders interfaced with the material handling system. (e) A "hybrid" combination of high-speed machining centers and transfer machine technology. (f) This production machining system incorporates transfer machines with flexible high-speed workstations. (a, c, d–f Courtesy EX-CELL-O Machine Tools, Inc. Sterling Heights, Michigan; b, Courtesy George Fischer FMS Turning Technology Ltd., Schaffhausen, Germany and Farmington, Michigan) (g) Flexible machining system arrangement with robot load and unload. (Courtesy Honsberg Lamb, a UNOVA Company)

illustrates a few of the great many possible flexible machining system arrangements. The system is capable of performing all machining processes.

Work Stations The design of a flexible machining system installation is based on the use of machining centers and may include

- Head changers
- Washing facilities for parts and fixtures
- Other programmable production equipment
- Measuring facilities

Fixtures used in flexible machining systems must be accurate and flexible; that is, they must be capable of precisely positioning and holding any of the family of parts they are designed for, and they must easily and quickly accept any workpiece presented. Similarly, tool storage must be adequate, having enough capacity to serve the requirements of machining several different parts.

Fixtures contribute greatly to the high cost of flexible machining systems due to the accuracy and ruggedness required and also to the large quantity of fixtures required to fill the system.

Material Handling Parts are moved from storage and between machine elements by means of one of several different types of systems. The material-handling system selected must be capable of routing any part to any machine in any order and also to provide a bank of parts ahead of each machine to realize maximum productivity. Material-handling systems for flexible machining systems are usually complex and costly. Parts are normally loaded and unloaded manually.

The various types of material-handling systems used include

- **Automated Guided Vehicles.** Workpieces are transported by carts guided electronically by buried cables in the floor or by rails on the floor. This is the most flexible system.
- **Towline System.** Carts are towed by chains buried beneath the floor along a path of slots cut in the floor. Some flexibility is provided by switches, disconnects, and reconnects.
- **Roller Conveyor System.** Fixture pallets are moved by powered, floor-mounted roller conveyors.
- **Overhead Conveyor System Including Monorails.** Fixtures suspended from and overhead conveyor move workpieces along from station to station. Some flexibility is provided by power and free sections where operators may be used to reroute specific workpieces.
- **Air-Film Conveyors.** Fixture pallets are floated and guided on a film of air along a smooth pathway. Power is provided by linear motors. Surprisingly heavy workpieces are moved by this system.
- **Cranes and Robots.** Stacker cranes and programmable robots are used to move workpieces from station to station. Certain parts, designs, and processing suit this method, but this system is not the most flexible. Several cranes or robots are usually required, which in turn require considerable space.

Control Systems The computer controls of flexible manufacturing systems have three levels of function:

- **Master Control.** The master control monitors and controls the entire system, including routing workpieces to appropriate machines, scheduling work, and monitoring machine functions. It also monitors and records tool breakage and wear.

- **Direct Numerical Control.** A DNC computer distributes appropriate programs to individual CNC machines and supervises and monitors their operation.
- **Element Control.** The third and lowest level of control is computer control of the machining cycles of individual machines.

Applications Flexible manufacturing systems are designed for use with many similar workpieces. They are economically effective on medium-sized lots (100–10,000), although exceptions cause them to be run continuously on only one or two different workpieces due to customer demand. Where this is true, the costly built-in flexibility of the system is lost.

Flexible manufacturing systems can be fed with random parts from a family of parts; the system is designed to identify each workpiece and forward it appropriately.

The productivity of flexible manufacturing systems can be enhanced by the use of group technology. Typically, the parts machined on this system are large and heavy, of various shapes, and require many machining operations. Examples are machine bases, tractor and truck transmission and differential cases, combine gear cases, hoist details, and locomotive components. The predominant users of flexible manufacturing systems are the larger manufacturing companies.

11.4 SPECIAL-PURPOSE MACHINES

Many special-purpose machines are basically standard machines to which accessories and elements of automation have been added. Most of the accessories are off-the-shelf items and include

- Index tables
- Multiple-spindle heads
- Automatic feed units

Some elements of automation, too, are standard, but others must be designed and made to suit the part requirements. These include

- Loading and unloading devices
- Devices to automate machine motions

Figure 11.19 illustrates a standard drill press equipped to function as a special-purpose machine.

When selecting a machine for a specific process, a manufacturing engineer should make a comparison of costs between standard and special-purpose machines versus production rates. Where built-up standard machines can be used, they usually have the advantages of lower cost and earlier availability.

11.4.1 Double-End Machines

A double-end machine is designed with a machining unit at either end of the base, or bed, and a workpiece-holding fixture located between them on a table. In use, both ends of the workpiece can be machined simultaneously as shown in Figures 11.20(a) and (b).

A variation in the use of a double-end machine is illustrated in Figure 11.21. In this example, a double-end machine is augmented by a third spindle mounted at right angles—resulting in a "three-way" machine.

Applications Operations such as drilling, milling, facing, and boring of small to medium-sized workpieces are routinely performed on double-end machines. Also, center drive units are available if a part must be rotated during machining.

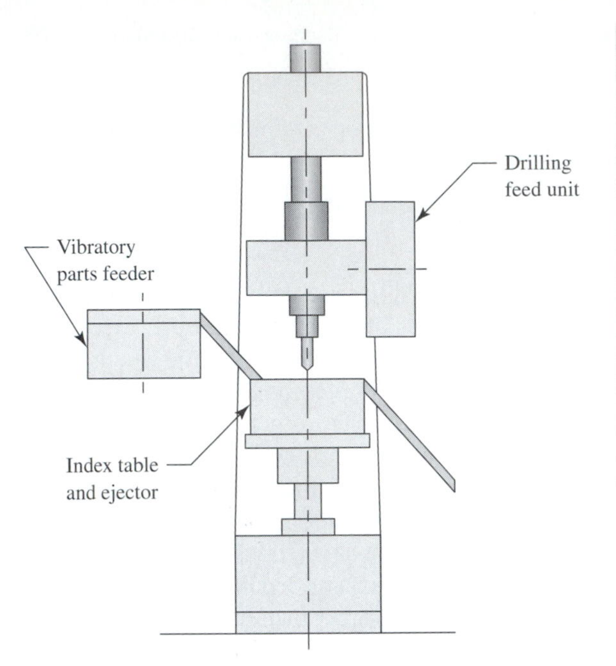

FIGURE 11.19 Standard drilling machine made fully automatic with addition of index and feed units.

(a)

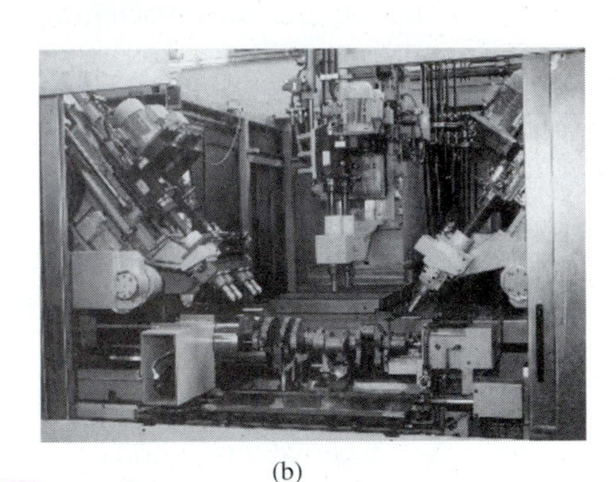

(b)

FIGURE 11.20 Double-end machines. (a) Double-end machine. (Courtesy EX-CELL-O Machine Tools, Inc.) (b) Double-end machine with angular mounted heads. (Courtesy Lamb Technicon Machining Systems, a UNOVA Company)

FIGURE 11.21 "Three-way" machine. (Courtesy EX-CELL-O Machine Tools, Inc., Sterling Heights, Michigan)

Flexibility Flexibility can be enhanced by providing such features as

- Interchangeable fixtures
- Interchangeable heads or machining units
- Speed changers
- Quick-change cutting tools
- Automation of cycle
- Automation of parts loading and unloading

11.4.2 Trunnion Machines

Trunnion machines can be thought of as double-end machines that have several workpiece holding fixtures mounted on an indexable vertical table located between the machining units. A trunnion machine is illustrated in Figure 11.22. This machine performs the following processes:

Station #1	Unload and load.
Station #2	Drill 10 holes.
Station #3	Countersink 7 holes.
Station #4	Counterbore 2 holes.
Station #5	Coredrill 1 hole.
Station #6	Ream 2 holes.
Station #7	Chamfer 3 holes.
Station #8	Tap 6 holes.
	Repeat cycle.

All machining operations are performed simultaneously. A finished part is produced with each index of the trunnion. Some trunnion machines are produced in single-end designs to accommodate specific part designs.

Applications Medium-sized parts with reasonably liberal tolerances are suitable for processing on trunnion machines. Larger parts with closer tolerances are usually processed on shuttle or transfer machines. Almost all processes can be performed on this machine.

Flexibility

- Interchangeable heads
- Interchangeable fixtures
- Variable speeds and feeds
- Automation applied to unload and reload parts

11.4.3 Dial (Rotary) Machines

Dial (rotary) machines are primarily drilling machines and consist of

- A horizontal index table mounted on a machine base and capable of indexing and carrying part fixtures and workpieces from one station to the next.

(a)

(b)

FIGURE 11.22 8-Position trunnion machine. (a) General view. Drill 10 holes, countersink 7 holes, counterbore 2 holes, coredrill 1 hole, ream 2 holes, chamfer 3 holes, tap 6 holes in crankshaft at the rate of 60 parts per hr at 80% efficiency. (b) View of trunnion with crankshaft in place for machining. (Courtesy Lamb Technicon Machining Systems, a UNOVA Company)

- Horizontal, vertical, and angular machining units mounted around the periphery of a machine base at locations corresponding to the index position. Multiple heads are in common use with the machining units.

Dial (rotary) machines are shown in Figure 11.23. One or more workpieces may be loaded in each fixture depending on the part design. Rotary fixtures can be designed to change part position, permitting machining of different work surfaces.

Dial machines are available in many sizes (diameters of work table) and number of stations. They are capable of drilling, reaming, countersinking, tapping, and spot facing.

Flexibility

- Interchangeable heads
- Interchangeable fixtures
- Variable speeds and feeds
- Automation applied to unload and reload parts

11.4.4 Center-Column Machines

Center-column machines, also a drilling machine, are similar in layout to dial-type machines except a horizontal index table rotates around a central column to which vertical machining slide units may be mounted.

Horizontal, vertical and/or angular machining units are mounted around the periphery of the machine base as with dial-type machines. Also, as with dial-type machines, parts are loaded and unloaded at one station and all machining operations are performed simultaneously, producing one or more finished parts per index. Center-column machines are shown in Figure 11.24. Center-column machines are capable of all drilling operations, plus they can drill additional holes from the center column.

Flexibility

- Interchangeable heads
- Interchangeable fixtures
- Variable speeds and feeds
- Automation applied to unload and reload parts

Center-column machines are available in many sizes (diameters of work table) and numbers of stations.

11.4.5 Shuttle-Type Machines

A shuttle machine, also known as an in-line indexing machine, is shown in Figure 11.25. This machine can perhaps best be described by describing its operation:

- A workpiece is loaded into a pallet fixture, which is then indexed in a straight line to two or more machining stations.
- Appropriate machining heads positioned at each station perform machining operations.
- After the pallet fixture and part have been indexed to, and machined at, all stations, they are returned to the starting position for unloading the finished part and reloading a new workpiece.

(a)

(b)

(c)

FIGURE 11.23 Rotary dial index machines. (a) Front view. (b) Rear view. (c) Plan view. (a, b, Courtesy Lamb Technicon Machining Systems, a UNOVA Company; c, Courtesy EX-CELL-O Machine Tools, Inc.)

(a)

(b)

(c)

FIGURE 11.24 Center-column machines. (a) Center-column machine for producing automotive transaxle component. (b) General view of 8-station center-column machine. (c) Dial machine with center column, plan view. (a, Courtesy Lamb Technicon Machining Systems, a UNOVA Company; b, Courtesy Kingsbury Michigan Machine and Engineering; c, Courtesy EX-CELL-O Machine Tools, Inc., Sterling Heights, Michigan)

(a)

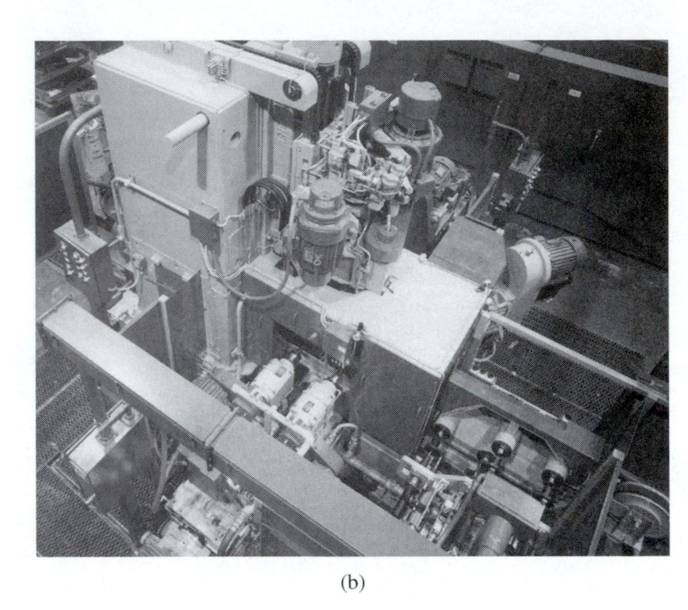

(b)

FIGURE 11.25 Shuttle-type machine. (Courtesy Lamb Technicon Machining Systems, a UNOVA Company)

- If this machine were provided with the capability of carrying pallet fixtures around a loop, returning them to the load and reload position without reversing, it would become a transfer machine. Transfer machines are described separately.

Applications Shuttle-type machines are well suited for low- to medium-production runs of heavy and odd-shaped parts. They are capable of most processes, including drilling, milling, and threading.

Flexibility

- Interchangeable heads
- Interchangeable fixtures
- Variable speeds and feeds

11.4.6 Transfer Machines

A transfer machine consists of work stations arranged in the required sequence to perform predetermined (processed) operations on a workpiece. Workpieces are usually located and clamped in a pallet-type fixture, which is moved from one station to the next by a powered, automatic transfer device. The pallet is precisely located at each work station by shot pins, bushings, wedges, and clamps. The design of some workpieces is such that they can be transferred by themselves from station to station without the use of a fixture.

Most transfer machines include a station where the workpiece and pallet fixture are washed and cleaned of chips following machining. Machine controls can become complex and costly because they control all movements and sequences; sense movements and positions, tool wear, and breakage; and—in increasing numbers of instances—measure part dimensions.

Transfer machines are built in two major types:

- Rotary transfer machines
- In-line transfer machines

Rotary Transfer Machines Rotary transfer machines resemble dial-type machines with certain basic differences:

- Rotary transfer machines do not have an index table but, rather, are provided with ways to guide fixture type pallets and also pallet locating devices at each station or position.
- The central area of the machine provides a base for mounting machining units to machine the inner surfaces of workpieces.
- Workpieces are located and clamped in pallet fixtures, which are moved, or indexed, in a circular path by an automatic transfer device. The pallets are precisely located at each station, usually by means of shot pins and bushings.

Rotary transfer machines are illustrated in Figure 11.26. Programmable controls are generally used. Since pallets are indexed back to their starting positions for unloading and reloading, no pallet return conveyor is required. Applications are usually small to medium-sized workpieces requiring fewer operations or stations than are usually provided in in-line transfer machines.

In-Line Transfer Machines Some in-line transfer machines are arranged to have the workpiece loaded and transferred through the required operations at stations that are arranged in a straight line along one span of the machine. Upon completion of the processes, the part is unloaded and the pallet is returned to the first station for reloading.

Other in-line transfer machines are arranged to have work performed on two or more spans of the machine. This arrangement also permits machining the ends, or all four sides of a part. In this design, the part is also returned to the original loading station for unloading. Parts and pallet-washing facilities are normally provided to remove chips. In-line transfer machines are illustrated in Figure 11.27. The design of certain workpieces is such they can be transferred to and clamped at each station directly without the use of fixtures or pallets.

Transfer mechanisms are powered by hydraulic power or by some form of electromechanical arrangement. Certain mechanical units employ harmonic drives and/or cycloidal motion to provide smoother acceleration and deceleration movements.

Most transfer machines utilize synchronous transfers, whereby all pallets are transferred at the same time. Work must be completed within a specific time at each station. Production is controlled by the time required for the longest operation.

However, where it is not feasible to transfer parts at all stations at the same time due to great differences of machining times, nonsynchronous transfer systems are available and are used to increase productivity.

Machining Units Manufacturers of transfer machines provide a complete range of machining units in their standard sizes along with a corresponding selection of columns, bases, and adapters for mounting.

Machining units are powered with electric motors of suitable sizes. Feeds are powered hydraulically, pneumatically, or electromechanically, depending on the application, although mechanical actuation is favored due to its constant feed and lower noise level. Servo-driven and

(a)

(b)

FIGURE 11.26 Rotary-transfer machines. (a) Rotary-transfer machines provide a highly reliable, compact system for medium- to high-volume parts production. (Courtesy EX-CELL-O Machine Tools, Inc.) (b) 7–Station rotary-transfer machine, plan view.

(c)

FIGURE 11.26 Continued (c) 7–Station rotary-transfer machine controls. (b, c Courtesy Michigan Machine and Engineering, a UNOVA Company)

hydraulic feed units are shown in Figure 11.28. Machining units are mounted on feed units as shown in Figure 11.29.

Tooling Preset, quick-change tooling must be used to generate optimum producibility, and tool-changing schedules must be developed and adhered to to maintain optimum producibility.

It is generally good practice to change all tools at the same time when changing any tool; repeated stops for changing groups of tools is usually more costly in terms of lost production time than changing some tool groups too soon (before they are dull).

Coolant Requirements Transfer machines require copious amounts of properly maintained coolant to optimize cutting-tool life, provide lubricity to various moving machine parts, and wash chips from machined surfaces and pallets.

Generally, these machines are installed with a flume system, which is a system of trenches adjacent to the machine and located below the floor. Coolant drains into the flume at the various machining stations, carrying machining chips with it. After use, the coolant is pumped through a system of centrifuges and filters to remove fines; additional coolant is provided to make up for coolant carried off with chips and parts.

Chips settle out of the coolant onto chip conveyors located at the bottom of the flumes, from where they are carried to a chip wringer centrifuge for removal of coolant. The spun-off coolant is returned to the flume, and dried chips are shipped to scrap and salvage companies.

Removal of coolant from chips is not only economically sound, but is required for environmental protection reasons; coolant dripping from wet chips as they are transported causes serious problems along railroad beds and highways. Figure 11.30 illustrates a transfer machine and flume system installation.

(a)

(b)

FIGURE 11.27 In-line transfer machines. (a) In-line transfer machine. (b) In-line transfer machine. Note ducting for fume and vapors control. (a, b Courtesy Lamb Technicon Machining Systems)

(c)

(d)

FIGURE 11.27 Continued (c, d) One in-line transfer machine from a system for producing automotive cylinder blocks in a high-volume manufacturing system. (c, d Courtesy EX-CELL-O Machine Tools Inc. Sterling Heights, Michigan)

FIGURE 11.28 Feed units. (a) Servo-driven feed unit on a transfer machine. Use of a belt from the servo motor to the feed unit eliminates gears and chains, permitting quieter operation and ease of maintenance while eliminating lubrication requirements. (a, Courtesy EX-CELL-O Machine Tools, Inc. Sterling Heights, Michigan) (b) Feed unit for hydraulic drive. (b, Courtesy Lamb Technicon Machining Systems, a UNOVA Company)

(a)

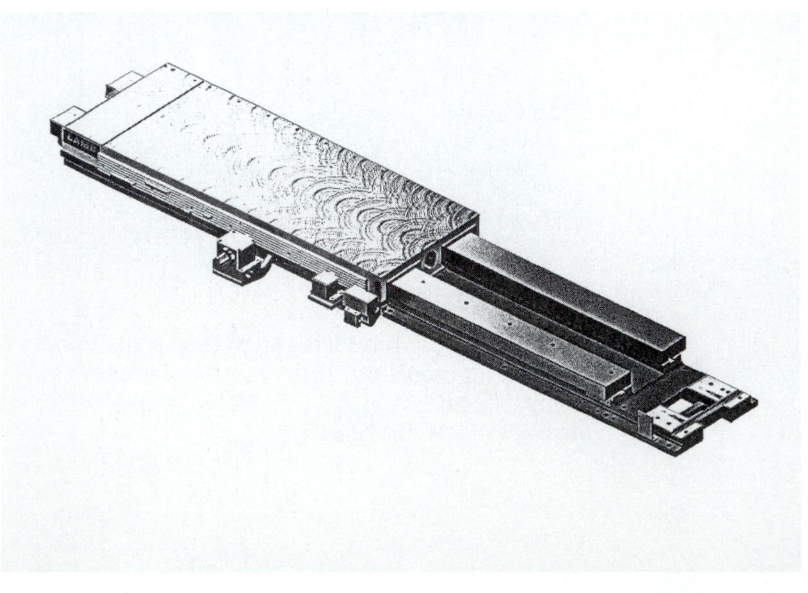

(b)

FIGURE 11.29 Machining unit from a special-purpose machine for medium to high production volumes. In this application, the tappet bores in cylinder blocks are machined locating from the crank bores. (Courtesy EX-CELL-O Machine Tools, Inc., Sterling Heights, Michigan)

FIGURE 11.30 In-line transfer machine and flume system. Flume system installed in the floor beneath this in-line transfer machine provides coolant management. The floor is fitted with "dry floor" guarding to help maintain a clean, safe working environment. (Courtesy EX-CELL-O Machine Tools, Inc., Sterling Heights, Michigan)

Controls Transfer machines, among the most complex production machines in use, require equally complex controls to operate efficiently. Functions must be monitored to ensure that all motions are made in proper sequence, that all motions are completed, and that each motion is electrically interlocked to prevent damage or wrecks caused by interfering or colliding machine or tool elements.

Included in motions and features that must be monitored and controlled are

- Correctness of part loading and unloading
- Correctness of pallet location and clamping at each station
- Completeness of each machining operation
- Completeness of each inspection operation
- Completeness of each assembly operation
- Correct and adequate coolant flow
- Lubrication and filter system operation

Tool Breakage The central control panel of most presently operated transfer machines uses electromechanical relays coupled with machine-mounted pressure switches and solenoid-operated valves. Many presently designed and most future transfer machine designs use—or will use—computers and microprocessors for controls.

All transfer machines require a means to rapidly identify improperly operating functions to minimize machine downtime. Some machines display the electrical circuit diagram, which includes pilot lights and also contacts for test probes to aid in locating problems quickly. Computer controls have superior diagnostic capabilities and are much quicker in locating problems.

Applications All machining operations, can be performed on a transfer machine as well as some assembly work such as pressing in bushings. Work requiring heating is also feasible as in the use of induction-heating coils. Gaging and measuring is performed increasingly with the increasing accuracies and capabilities available in the newer machines.

Rotary transfer machines are usually used for medium-sized, simpler designed parts with medium production runs. Many, tooled with quick-change tooling, are used to machine families of similar parts.

In-line transfer machines are ideal for large quantities of complex parts such as automotive cylinder blocks, transmission cases, cylinder heads, and brake and suspension components.

Many features to enhance flexibility can be designed and built into both rotary and in-line transfer machines. These include

- Quick-change heads
- Revolving machinery units
- Quick-change cutting tools
- Quick-change fixtures
- Idle stations, or provisions for the addition of added machining stations to provide for anticipated future engineering changes.

Processing of transfer machines is almost always a joint effort of the manufacturing engineer and the machine-tool builder's engineering staff.

11.5 AUTOMATION

11.5.1 General Concept

The term "automation" has had many definitions in the past, having originated in the 1950s. It seems to have been applied to the automatic loading and unloading of stampings and also to automatically handling sheet metal and strip stock.

Present-day usage and acceptance of the term is pretty well summed up in Webster's dictionary definition:

1. The technique of making an apparatus, a process, or a system operate automatically
2. The state of being operated automatically
3. Automatically controlled operation of an apparatus, process or system by mechanical or electronic devices that take the place of human organs of observation, effort and decision

The design of "standard machines" has matured significantly in accuracy, production capacity and lower maintenance cost, and also includes functions that would have been known as automation in the 1950s. Some of these functions include

- NC and CNC controls
- Variable speeds and feeds
- Preloaders and part ejectors

11.5.2 Loaders and Unloaders

Devices known as automatic loaders and unloaders further automate many present-day machines, rendering them capable of complete automatic operation from a supply of parts available to the loader.

Batch or Bulk Parts

Batch or bulk quantities of parts can be supplied by devices known as hoppers, magazines, and storage silos. These devices are used with symmetrical, usually round, workpieces. Nuts, bolts, and washers are fed to the point of use and automatically assembled. Figure 11.31 shows the use of one such application.

Individual Parts

Pick and Place Manipulators. Specially designed, single-purpose units are used to load and unload individual, symmetrical workpieces for certain suitable workpieces. Common applications are various grinding operations and also spline-rolling operations. These devices can be designed to handle multiple parts. Pick and place applications are illustrated in Figure 11.32.

Pick and place units are usually pneumatically or hydraulically powered mechanical linkages that have some versatility of motion such as amount of swing, number of stop positions, and amount of twist motion. They are not programmable. Where they can be used, they are economical to purchase and maintain.

Programmable Robots. Programmable robots are defined by The Robot Institute of America as follows:

> A robot is a reprogrammable, multifunctional manipulator designed to move material, parts, tools and special devices through variable programmed motions for the performance of various tasks.

FIGURE 11.31 Automatic assembly device. The hoppers shown in the foreground contain bolts, which are fed to the assembly station where differential carriers and caps are automatically removed from pallets, assembled, and torqued. (Courtesy EX-CELL-O Machine Tools, Inc., Sterling Heights, Michigan)

To satisfy the definition, a robot must meet the following requirements:

1. The unit must have multiaxis capabilities to move parts or tools to a specific location in space.
2. The unit must be capable of orienting a part or tool in any attitude.
3. The unit must have a control and memory system capable of driving the manipulator repeatedly through the desired motions.
4. The unit's control system must have the capability of signaling and interacting with related devices and operations.
5. The unit must have logical and easy-to-use programming and reprogramming capabilities.

Robot geometric configurations are shown in Figure 11.33. All motions are not required for all applications.

Applications Programmable robots are finding increasing acceptance and use both as machine-tool builders and equipment users become more familiar with their capabilities, advantages, and installation parameters. Principal uses include

- Tool changers
- Loading and unloading parts for a single machine
- Transferring parts from one machine to another
- Changing a part's position in one machine during an operation
- Welding operations, especially those requiring repeated welds such as respotting
- Hazardous or environmentally hostile operations or locations

FIGURE 11.32 Pick and place applications. (a) Pick and place manipulators. (b) A gantry-style load/unload device is used to transfer differential carriers from the machining system into an automatic gaging station. The gantry loader provides secure transport for large, heavy parts. (a, Courtesy World Std.); (b, Courtesy EX-CELL-O Machine Tools, Inc., Sterling Heights, Michigan)

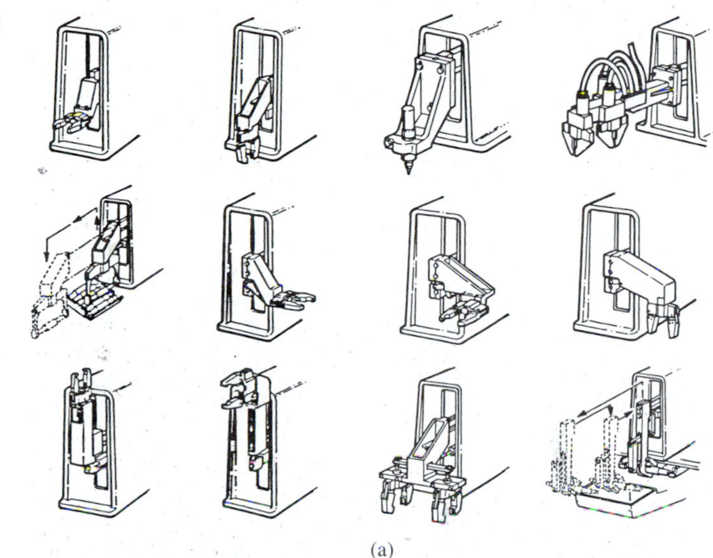

(a)

(b)

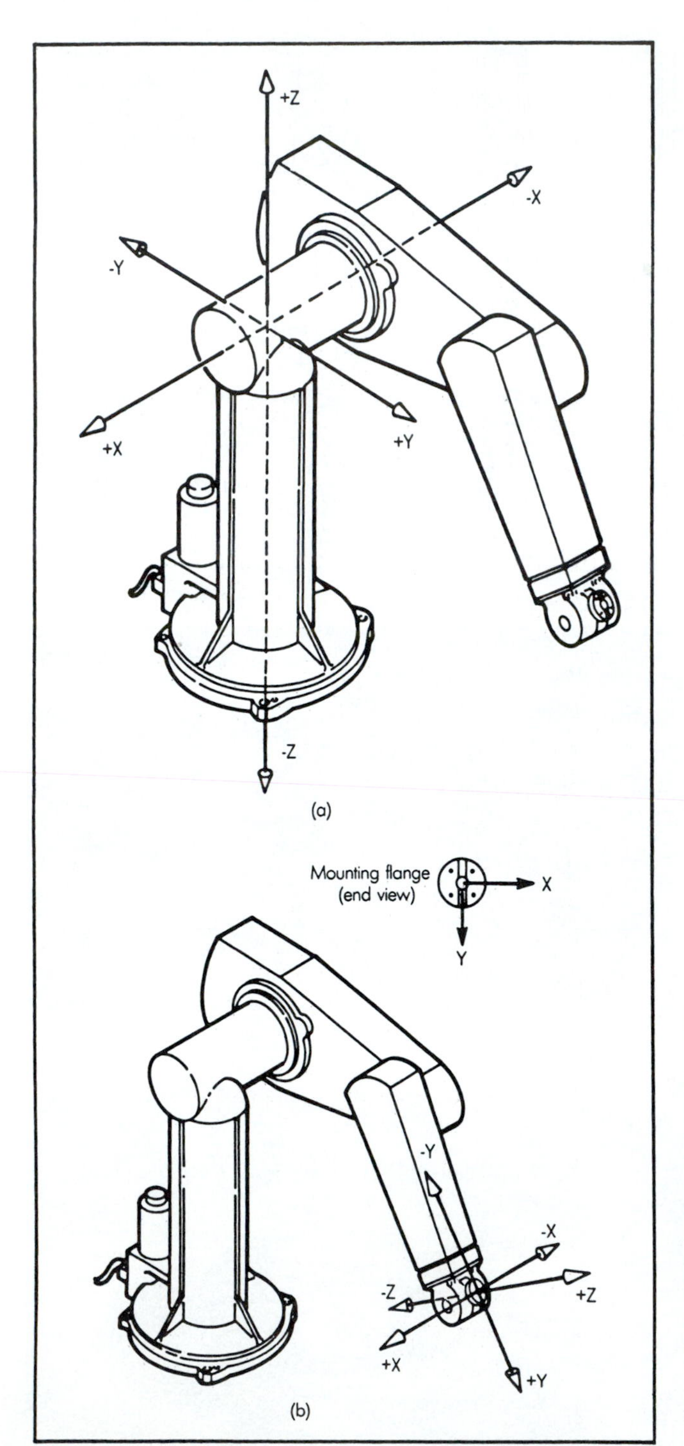

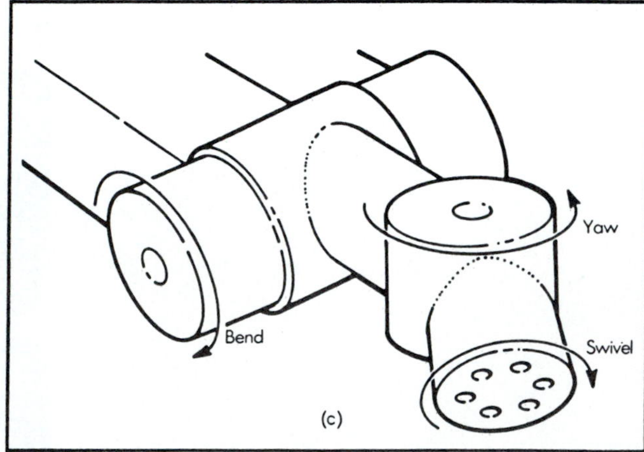

FIGURE 11.33 Programmable robots. (a) World coordinate system. (b) Tool coordinate system. (c) Wrist articulation of an industrial robot. (Courtesy of Society of Manufacturing Engineers)

11.5.3 Machine Controls

NC and CNC controls "automate" a machine's operating cycle as compared to operator control of similar manually operated machines. Also, a machine's function and movements may be monitored for completeness of cycle, stoppages, and breakdowns.

Productivity, part quality, and scrap reduction are enhanced by postprocess monitoring. Parts can be gaged and checked for completion of the operation: Tools can be checked for breakage and wear with sensors that have the capability of stopping equipment, and further production of faulty parts, as soon as a problem is discovered.

Worn tools and missing tools not only can cause scrap and rework but—as in a transfer machine—a missing bit from a broken drill can cause broken taps, with other potential damage. A portion of a broken tool that remains in a workpiece and is subsequently transferred to another station has the potential to cause severe damage and possibly a wreck. Many different types of sensors have been developed for application to the varying sensing needs:

Sensor Type	Application
Probe	Check for tool presence (touch)
	Check for hole presence
	Check for hole depth
	Check for hole size
Laser	Check for broken tools
	Check for broken edge of tool
	Check for missing hole in some part designs
Acoustic emissions	Check for tool breakage
Cameras	Check for tool wear
	Check for tool breakage and missing tools
Vibration sensors	Check for broken and missing tools
Home position measures	Check for dull or broken tools—not too efficient in terms of time lag involved and damage done before defective or missing tools are detected
Torque measure	Check for dull taps causing undersize tapped holes

Most sensors are costly to include in a machine control system, but an analysis of scrap and rework costs of high-production runs almost always justifies the costs.

11.6 REVIEW QUESTIONS AND PROBLEMS

1. Why should a tool layout be made for machines selected for processing?
2. What should the layout show?
3. What other uses does the layout have?
4. What is the main purpose of spare toolholders?
5. What processes are single-spindle automatic lathes capable of?
6. Describe parts best suited for production on a single-spindle automatic lathe.
7. What is the main reason to select a multiple-spindle automatic chucking machine over a single-spindle machine?
8. Describe, generally, the kind of part that would be most suitable to run on a machining center.
9. What are the parameters of parts most suitable for a flexible manufacturing system?
10. What is a double-end machine?

11. In what ways are trunnion machines similar to double-end machines?
12. Dial and rotary machines are drilling machines. In what ways are they similar, and how are they different?
13. Can a shuttle machine be considered a transfer machine, and can it do the same processes?
14. How does a rotary transfer machine differ from an in-line transfer machine?
15. Why use a rotary rather than an in-line transfer machine?
16. Describe the operation of an in-line transfer machine.
17. Which manufacturing companies are most likely to purchase in-line transfer machines?
18. What does the term "automation" mean and what functions does it cover?

CHAPTER TWELVE

GROUP TECHNOLOGY

CHAPTER OVERVIEW

12.1 DEFINITION

Group technology is a plan of manufacture for groups or families of parts that have similar design and/or manufacturing sequences.

12.2 INTRODUCTION

More total parts are produced in small lots than are manufactured by all existing in-line, mass-producing methods. Most of these parts are manufactured by job shops and by special job-shop-type departments set up within larger companies' plants.

Parts are produced for a wide range of users including automotive, aircraft, electronic, cameras, hardware, tooling, and more. Orders for parts are obtained by competitive bid. Total allowable costs do not permit time for formal process planning, and most required delivery dates force immediate machining action.

Commonly, job-shop processing consists of a single routing sheet listing only the manufacturing operations to be performed; it is left to the shop personnel to set up and complete. Not infrequently, scheduling takes the form of "notes on the back of an envelope."

Variations of manufacturing the same and similar parts in subsequent runs result from someone's inability to recall processing setup details of previous runs. Routing, too, may be done by different individuals—all of which leads to inconsistencies in cost accounting and misunderstandings with the customers.

Machine layouts in job shops are typically by process area: lathe area, drilling machine area, grinders' area, and so on. Parts are shuttled from area to area as required.

Problems inherent in the conventional job shop method of production include

- Cost accounting variations previously mentioned
- Inefficient loading and utilizing of machine capacity—there are never enough machines of some particular type while others are idle. This condition is exacerbated by keeping employees productively employed; no waiting for machines or idle time can be afforded.
- Extra parts and inventory build up, and material shortages develop from controlling labor rather than parts flow.
- Some variations of processing require more material.

To their credit, our job shops do an amazingly good job using the conventional system. They accomplish it through the dedication, good judgment and experience of their management and employees. The illustrations shown in this chapter are typical of present-day applications of group technology.

12.3 ESTABLISHING AND/OR CONVERTING TO GROUP TECHNOLOGY

Group technology is a major improvement in processing and manufacturing short runs of the same or similar parts. An increasing number of conventional job shops are converting to it. However, before any conversion of processing, scheduling, or machine rearrangement can begin, certain basic requirements must be resolved:

- Parts must be produced to meet commitments during the time of conversion and changeover. It may be possible to schedule production of enough parts to create a bank from which to ship during the downtime period—or it may prove more satisfactory to temporarily resource parts to other suppliers.
- Turnaround space must be available—or made available—in which to start arranging machines in groups or cells.
- Planning, scheduling, machine layouts, and training must all be completed.
- Additional supplies, machines, and any additional temporary help (such as consultants, machinery movers, etc.) must be available and contracted for.

Establishing or converting to the group technology concept requires the completion of two steps:

- Grouping parts
- Arranging machines

A third step, coding, refines and greatly improves the selection of parts for grouping. These actions are discussed in the following sections.

12.4 PARTS GROUPS OR FAMILIES

Parts groups, or families, must be defined; this can be done in one of three different methods:

1. Design similarity grouping
2. Similar production methods grouping
3. Coding

12.4.1 Design Similarity Grouping

Drawings or part prints of all candidate parts can be visually inspected and manually separated into groups based on design similarities such as those illustrated in Figure 12.1. This is the quickest method to establish parts families of existing parts. The parts thus sorted will generally be found to have similar processing and manufacturing features. These groups will realize savings of tool setup and changeover times. However, with this method, dissimilar parts manufactured by the same processing as parts in the similar designs group will not be included.

FIGURE 12.1 Parts of similar designs. Potentials for family groups.

1. Pins—push rods

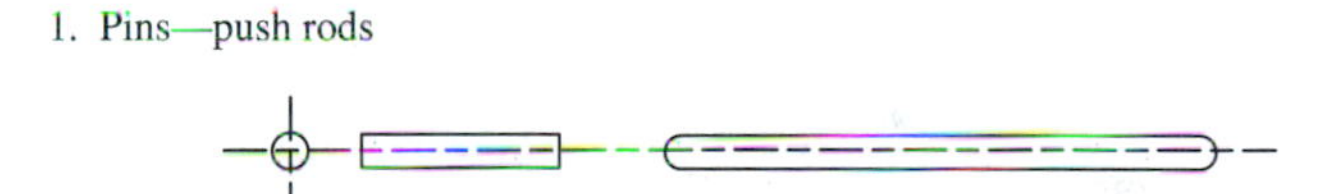

2. Shafts

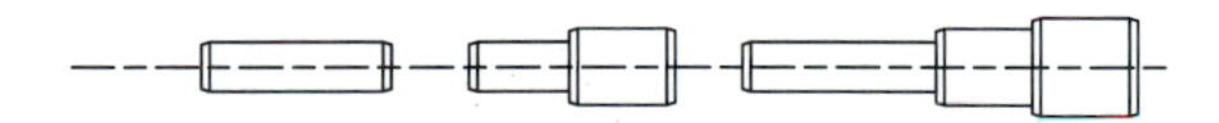

3. Hub-shaped parts

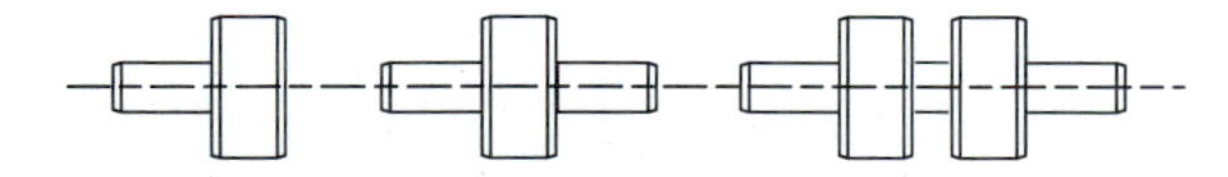

4. Valves

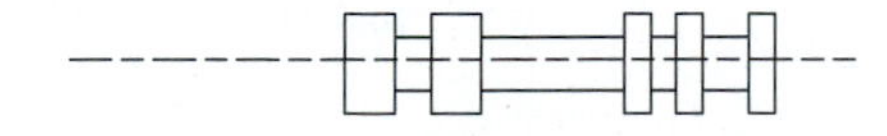

5. Gear blanks

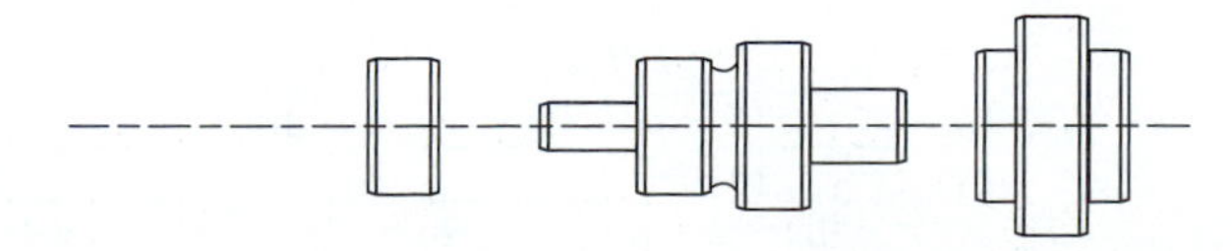

6. Hydraulic fittings

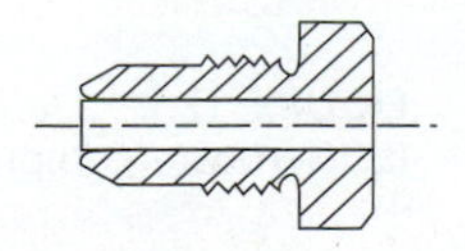

12.4.2 Similar Production Methods Grouping

Parts can be separated into groups by comparing process sheets and routings. A relatively small quantity of parts can be handled manually, but for a large manufacturing plant with hundreds or thousands of different part numbers involved, the task of grouping must be organized and accomplished in manageable steps. Typically, these steps are

1. Deciding on parts to be included in grouping
2. Obtaining specific process routing sheets along with part prints
3. Sorting parts into logical groups manually where feasible
4. Sorting large numbers of parts into logical groups with computer aid, which includes
   a. Assigning a numerical key to each unique routing of manufacturing process as shown in Figure 12.2(a), for example,

Determine Existing Unique Process Routings	Assign Key Number
Turn–Drill–Mill	10
Turn–Mill–Drill	20
Mill–Drill–Drill	30
Etc.	

(a)

Etc.

Part # xxxxx
Mill–Drill–Drill
#30

Part # xxxx
Turn–Mill–Drill
#20

Part # xxx
Turn–Drill–Mill
#10

(b)

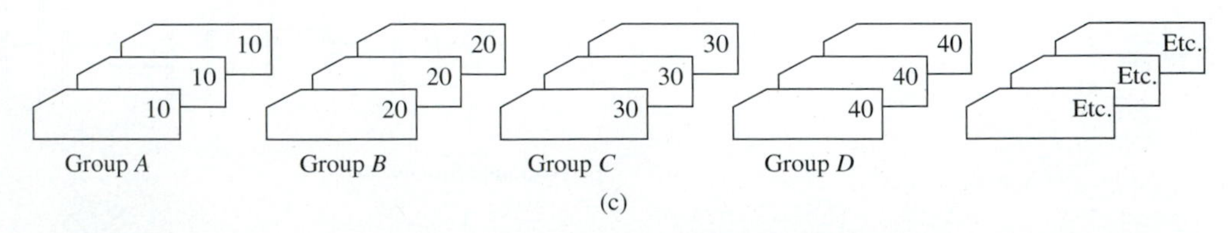

(c)

FIGURE 12.2 Determine parts of similar proccessing. (a) Assign numerical key to processing. (b) Generate computer cards with key numbers. (c) Sort cards into alphabetical groups.

Manufacturing Processes Routing	Key Number
Turn—Drill—Mill	10
Turn—Mill—Drill	20
Mill—Drill—Drill	30

b. Entering the key number on individual cards as shown in Figure 12.2(b)

c. Sorting cards, by key number, into groups and identifying each group alphabetically as shown in Figure 12.2(c).

5. Determining machine loading with the aid of a matrix chart, plotting groups versus machines as shown in Figure 12.3. (*Note*: More than one machine for each process may be required in a cell for capacity requirements.)
6. Analyzing the patterns generated in Figures 12.2 and 12.3, and developing cellular machine layouts. Units that do not fit into a group must be accommodated in some other manner—perhaps with a mini process arrangement.

The system of grouping by production methods is thorough, but is only as good as the material analyzed; it has the disadvantage of including different processing of the same part if such differences exist.

Machine Identification Number	Alphabetical Groups					
	A	*B*	*C*	*D*	*E*	Etc.
100	X				X	
101	X	X	X		X	
102		X		X		
103	X	X	X		X	
104	X		X			
105	X			X		
106			X		X	
Etc.						

FIGURE 12.3 Determine machine loading.

12.4.3 Coding

Coding of a part is the defining of the part's design and manufacturing requirements by assigning digits to each of the pertinent elements. Parts so coded may be stored in memory systems and recalled to aid in scheduling future orders and to help form groups or families of similar parts. A code system may be developed by the using company or it may be developed and custom made to order by specialists.

All parts manufactured by a company may be coded utilizing the combined design/manufacturing process system and a computer to select compatible groups yielding the greatest returns during manufacture.

Advantages of Coding

- New parts can be added at any time, and especially as new designs are released for manufacturing.
- Systems can be checked and purged of duplicate parts and processing.
- Systems can be checked and purged of variations of processing the same parts.

Several code systems exist and are in use to varying degrees as well as the custom-designed codes mentioned earlier. A description of how each works and is used is beyond the scope of this book, but they are identified for informational purposes.

A manufacturing engineer whose duties become involved in the development of group technology would do well to obtain one or more of the detailed periodicals and books dedicated to coding.

- **The Opitz Parts Classification System.** This system, developed by Professor Opitz of the University of Aachen, Germany, consists of 9 digits used to convey both part design and processing data. The system can be expanded to 13 digits.
- **The CODE System of Classification.** This system uses base 16 rather than the common base 10 numbering system to convey both part design and processing data. Other information such as tolerances, plating, and material may also be conveyed.
- **The MICLASS System of Classification.** This system uses 12 digits to classify parts design and processing data. Although this system is normally done by computer, special training in computer language is not required.

12.5 MACHINE LAYOUTS

The machine layout, or order in which machines are arranged, also known as the plant layout, is designed to suit the production method used to produce parts. There are three basic types, with variations to suit unusual problems:

1. Product layout (in-line)
2. Process layout (functional or departmental)
3. Group layout (cellular)

12.5.1 Product Layout

This layout, also termed in-line and progressive, is used for high production. Material is supplied to the first operation of a line and is moved sequentially through all required operations until a part is finished and leaves the end of the line. An individual line is dedicated to produce a part or an assembly; many dedicated lines may be required for the production of a complete product.

The product line illustrated in Figure 12.4 is balanced; that is, each operation has the same capacity to enable a constant, even flow of parts from one operation to the next. Features and characteristics of a product line layout include:

1. High investment cost of machines and equipment
2. Little or no flexibility of changeover to run other parts, but also little or no labor costs to changeover and re–setup
3. Excellent control of production and inventory
4. Material control inherent

FIGURE 12.4 Product layout. Machines are arranged in-line to produce one part with sequential operations.

Operation sequence:

Operation 10	Lathe
20	Mill
30	Drill
40	Drill
50	Grinder

Parts flow with machines arranged in product layout

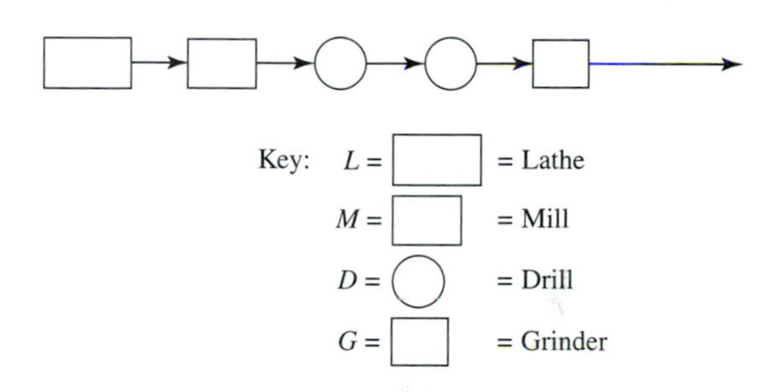

5. Low skill requirements to operate and low incremental labor cost (per operation)
6. Easy supervision
7. Production dependency on having all equipment in operation—one down machine can stop production of the entire line.

12.5.2 Process Layout

This layout, also known as functional and departmental, is used to produce parts in small- to moderate-sized lots. Machines are arranged by process; that is, lathes are in one common area, drilling machines are in another, and so on.

Material is shuttled from one process area to another to suit the required process sequence of the part. After running one requirement of parts of a specific design, tooling is changed on the machines required to run the next (different design) lot of parts. Conventionally laid out job shops use this arrangement, and it is illustrated in Figure 12.5.

Features and characteristics of a process or functional line layout include

1. Lower investment in machinery (than product line)
2. Line very flexible but at a penalty of setup and changeover costs
3. Control of production and inventory difficult
4. Moving parts from department to department incurs added costs
5. High worker skills required, at higher costs
6. Supervision more difficult
7. One down machine not usually a deterrent to other production

12.5.3 Group Layout

Group technology requires laying machines out in cellular form, or groups, as shown in Figure 12.6. Machines in each group are arranged to suit the process sequence of a family of parts. Each group of parts families requires a separate cell; cells are normally arranged adjacent to each other for material-handling, maintenance, and supervisory purposes. Machines within a group are usually arranged in-line or in a progressive order. Features and characteristics of a group layout include:

1. The investment in machines is lower than in a product layout, but more than in a process layout. Each cell requires a complete complement of machines, and there is some idle capacity.
2. Flexibility is provided by the ability to route each of a family of parts through a related cell with little or no changeover.
3. The smoother flow of material improves control of production and inventory over that of a process layout.
4. Some manual movement of parts is required to and from cells but at less labor cost than for a process layout.
5. Higher labor skills are required, with higher costs, but efficiency and no, or little, setup costs more than offset this. Each worker must be able to run several different types of machines.
6. A higher level of supervision is required.
7. One down machine can hold up production in a cell, but all cells are not affected.

Other

- Reorganization and training are big problems, but worker productivity and satisfaction improve greatly.
- Some parts may require redesign to permit inclusion in a parts family.
- Some parts may not fit into a cellular layout at all and require manufacture by other means.

Operation sequence:
Operation 10 Lathe
20 Mill
30 Drill
40 Drill
50 Grinder

Parts flow with machines arranged in process layout

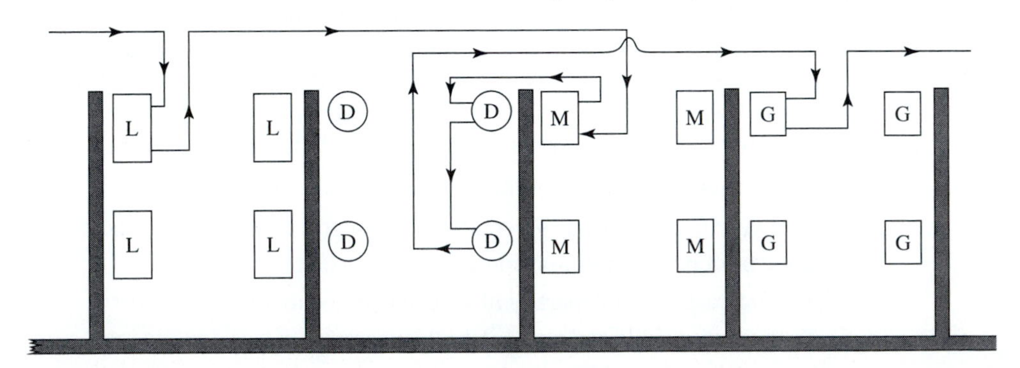

Key: L = [ ] = Lathe
M = [ ] = Mill
D = ( ) = Drill
G = [ ] = Grinder

FIGURE 12.5 Process layout. Machines are arranged by process areas or departments. Parts are shuttled from process to process as required.

Operation sequence:

Operation 10	Lathe
20	Mill
30	Drill
40	Drill
50	Grinder

Parts flow with machines arranged in group layout:

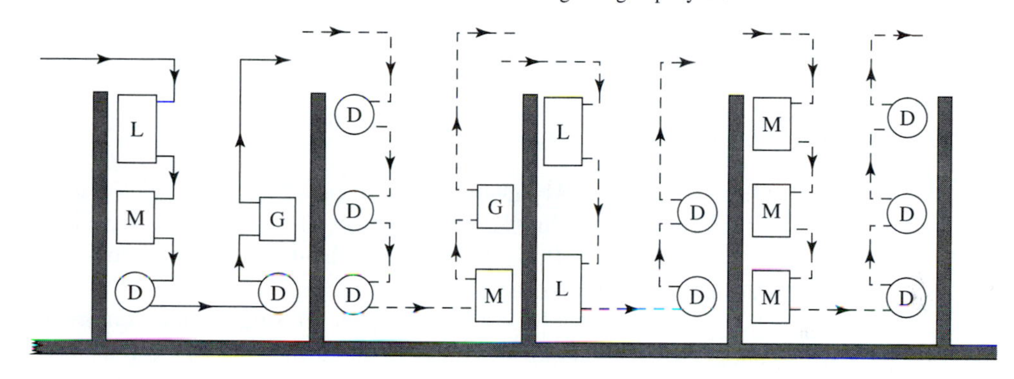

Key: $L =$ ☐ = Lathe
$M =$ ☐ = Mill
$D =$ ◯ = Drill
$G =$ ☐ = Grinder

FIGURE 12.6 Group layout. Machines are arranged in cells or groups. Each cell layout includes machines arranged in-line to produce a group of related parts.

12.6 TOOLING

The tooling used with any of the three major machine layouts has certain qualities in common:

- Sturdy and accurate workholders
- Quick-change toolholders
- Long-lived perishable cutting tools
- Perishable tools with the most economical costs

In addition, tooling for each of the three types of machine layouts has properties required by and peculiar to the particular machine layout where it is used.

12.6.1 For Product Layout (In-Line or Processive)

Tooling for product or in-line applications is designed for maximum production of a specific part in the shortest (cycle) time. It must also satisfy the sequential processing of the part. Because the line is not intended for changeover to other parts, the chucks, fixture, and tool blocks are custom designed for a part. Perishable tooling is, of course, designed for quick changes and replacement due to dulling and breakage.

The processing may consist of special machines and equipment such as broaches, slotters, trunnion machines, and so on that are specially designed and built to produce a certain part and cannot be changed over to produce another. Also, feed cams and gearing of machines in this line are calculated for optimum production and tool life, and are not quickly changed over.

Many of the material-handling devices are special for a part and integral with the line, making it difficult or impractical to change over. For these reasons and because a line may not include a specific process required for another part, changeover of tooling and equipment in a product line is not feasible.

12.6.2 For Process Layout

Tooling for use in a process layout is, where possible, designed for general use on standard machines and equipment. Chucks, workholders, and tool blocks may have some interchangeable details designed in, or the entire unit may be designed for change over. Perishable tooling is always designed for quick changes.

Speeds and feeds probably will be averaged, although some equipment—such as lathes—may be changed easily and quickly. Material handling is usually quite basic, consisting of slides, chutes, and so on, and depending to a large degree on labor for shuttling material and parts.

The conventional job shop is arranged with a process layout and is capable of producing any part it has the equipment (process) and the capacity to handle.

12.6.3 For Group Layout

Tooling designed and used in a group layout is the same as that used in a process layout. Since machines are arranged in-line within each group or cell, material handling can be more sophisticated, with accompanying labor savings.

Other Parts design, especially for a family of parts, should have common groove widths, undercuts, radii, and chamfers specified where possible to simplify tool design and tool changing. Where such features differ, requests should be made to Product Engineering for changes.

12.7 REVIEW QUESTIONS AND PROBLEMS

The group method of machining parts that are similar in design or process is reviewed with questions that also cover conversion requirements from process layout to group layout.

1. What is the basic philosophy of group technology?
2. Define a parts group or family.
3. What is the main difference between a process layout and a group layout?
4. What are the advantages of a group layout over a process layout?
5. Discuss what must be done to get ready for a conversion to group layout from process layout.
6. Define a code.
   a. State what a code does.
   b. What are the advantages of using a code with group technology over manual methods of developing a family or group of similar parts?
7. What are some features that perishable tooling that is used to machine a family of similar parts should have?

PART IV

SINGLE-POINT MACHINING

INTRODUCTION

"Machining" is a term that covers many manufacturing processes designed to remove unwanted material, usually in the form of chips, from a workpiece. Machining is used to convert castings, forgings, or preformed blocks of metal into desired shapes with size and finish specified to meet design requirements. Almost every manufactured product has components that require machining, making this group of processes one of the most important of the basic manufacturing processes, especially because of the value added to the final product.

The machine tool is the starting point of any metal-cutting operation. The type of cut to be taken, the speed and feed to be used, and the accuracy obtained all depend on the machine available for the job. All machine tools have certain common characteristics. All provide a way to support the cutting tool, which actually removes the excess stock. They also provide a means for locating and holding the workpiece, are equipped to position the tool accurately relative to the workpiece, and provide a means to produce powered motion between the cutting tool and the part being machined.

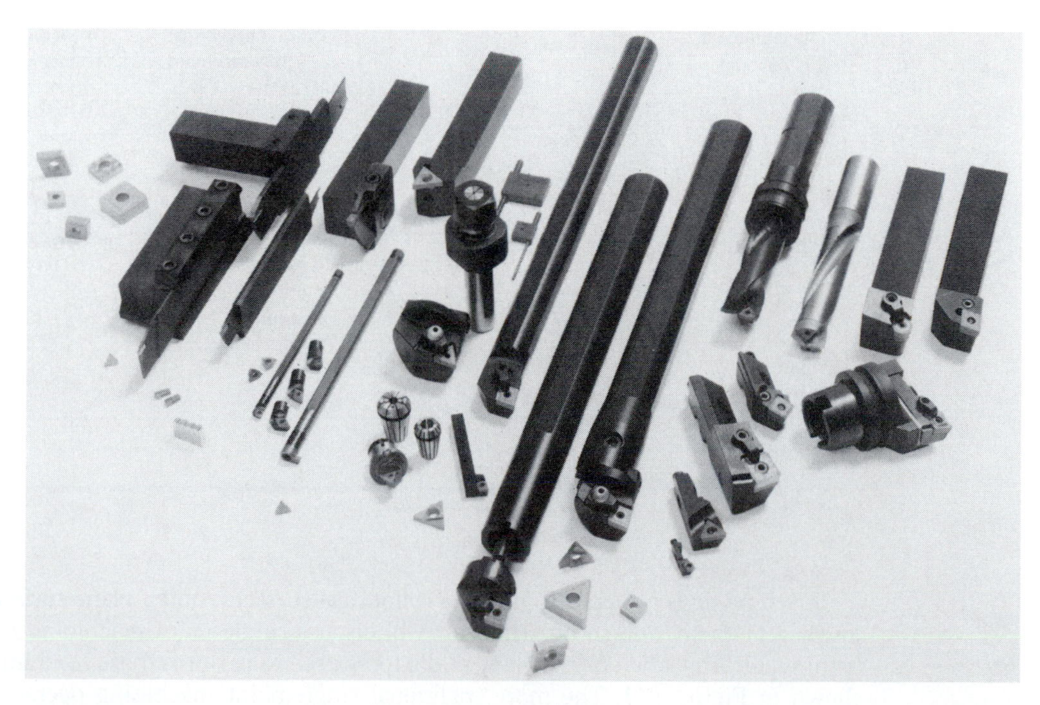

FIGURE IV.1 Various tools used to perform single-point machining operations such as turning, drilling, and boring. (Courtesy Valenite Inc.)

FIGURE IV.2 Single-point machining operations diagrams with machining characteristics and types of machines used.

Operation	Diagram	Type of Machines	Characteristics
Turning	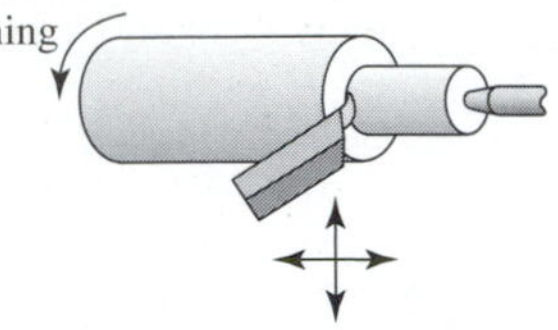	Lathe, vertical boring mill and machining centers	Work rotates, tool moves for feed.
Shaping	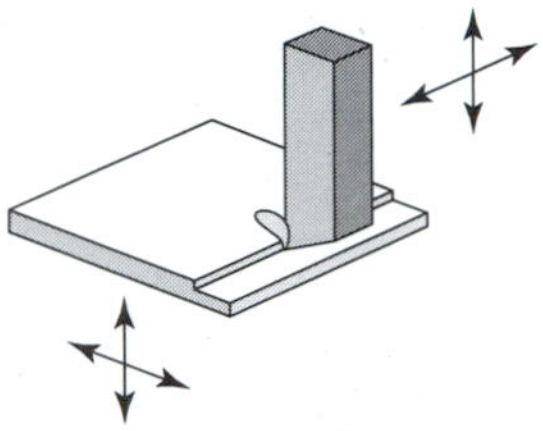	Horizontal and vertical shapers	Work is held stationary and tool reciprocates. Work can move in two axes. Toolhead can be moved up or down.
Planing	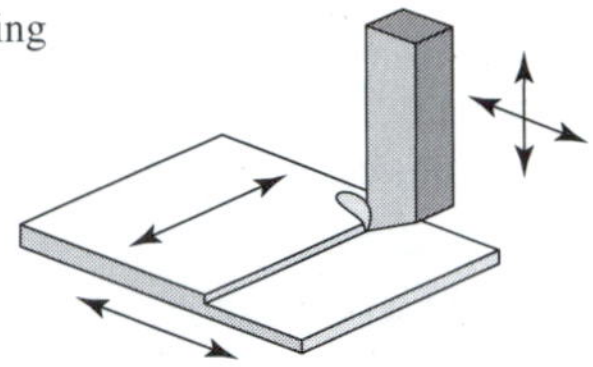	Planer	Work reciprocates while tool is stationary. Tool can be moved up, down, or crosswise. Worktable cannot be moved.
Drilling and reaming	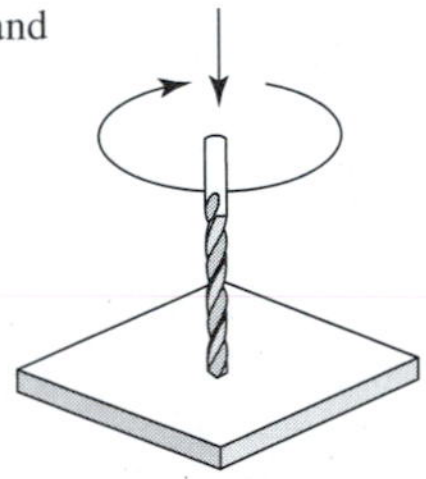	Drill presses, vertical milling machine and machining centers	Drill or reamer rotates while work is stationary.
Drilling and reaming	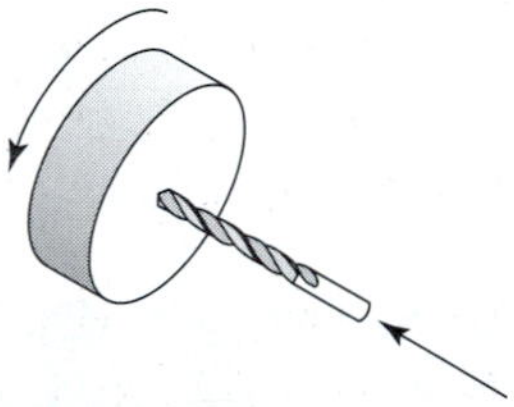	Engine lathes, turret lathes, automatic screw machines and machining centers	Work turns while drill or reamer is stationary.
Boring	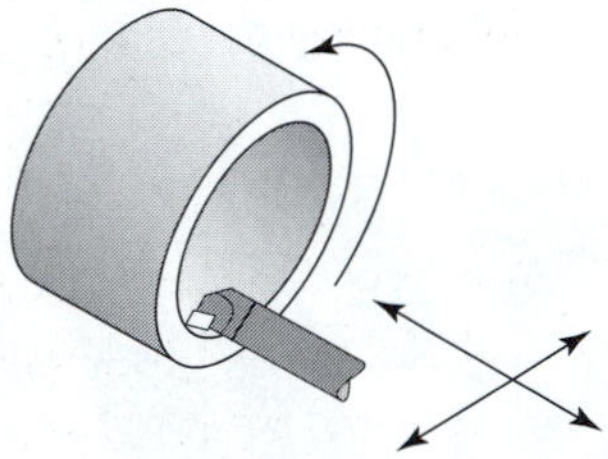	Engine lathes, horizontal and vertical turret lathes, vertical boring mills and machining centers.	Work rotates, tool moves for feed on internal surfaces.

Machine tools are built to produce cylindrical surfaces, holes, plane surfaces, irregular contours, gear teeth, and so on. This part will discuss single-point machining operations such as turning, drilling, and boring. Many of the tools used to perform these machining operations are shown in Figure IV.1. The more traditional single-point machining operations are shown in Figure IV.2 and are described throughout Part IV.

CHAPTER THIRTEEN

TURNING OPERATIONS

CHAPTER OVERVIEW

13.1 INTRODUCTION

Turning is a metal-cutting process used for the generation of cylindrical surfaces. Normally the workpiece is rotated on a **spindle** and the tool is fed into it radially, axially, or both ways simultaneously, to give the required surface. The term "turning," in the general sense, refers to the generation of any cylindrical surface with a single-point tool. More specifically it is often applied just to the generation of external cylindrical surfaces oriented primarily parallel to the workpiece **axis**. The generation of surfaces oriented primarily perpendicular to the workpiece axis is called facing. In turning, the direction of the feeding motion is predominantly axial with respect to the machine spindle. In facing, a radial feed is dominant. Tapered and contoured surfaces require both modes of tool feed at the same time, often referred to as profiling. Turning facing and profiling operations are shown in Figure 13.1.

The cutting characteristics of most turning applications are similar. For a given surface only one cutting tool is used. This tool must overhang its holder to some extent to enable the holder to clear the rotating workpiece. Once the cut starts, the tool and the workpiece are usually in contact until the surface is completely generated. During this time the cutting speed and cut

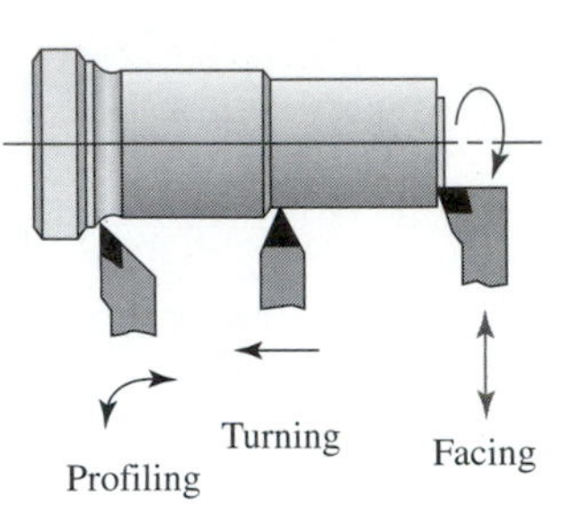

FIGURE 13.1 Diagram of the most common lathe operations: facing, turning, and profiling.

dimensions will be constant when a cylindrical surface is being turned. In the case of facing operations the cutting speed is proportional to the work diameter, the speed decreasing as the center of the piece is approached. Sometimes a spindle-speed-changing mechanism is provided to increase the rotating speed of the workpiece as the tool moves to the center of the part.

In general, turning is characterized by steady conditions of metal cutting. Except at the beginning and end of the cut, the forces on the cutting tool and the tool tip temperature are essentially constant. For the special case of facing, the varying cutting speed will affect the tool tip temperature. Higher temperatures will be encountered at the larger diameters on the workpiece. However, because cutting speed has only a small effect on cutting forces, the forces acting on a facing tool may be expected to remain almost constant during the cut.

13.2 RELATED TURNING OPERATIONS

As shown in Figure 13.2(a) through (f), a variety of other machining operations can be performed on a lathe in addition to turning and facing. Single-point tools are used in most operations performed on a lathe. Short descriptions of six additional lathe operations are given below:

Chamfering: The tool is used to cut an angle on the corner of a cylinder.

Parting: The tool is fed radially into rotating work at a specific location along its length to cut off the end of a part.

Threading: A pointed tool is fed linearly across the outside or inside surface of rotating parts to produce external or internal threads.

Boring: Enlarging a hole made by a previous process. A single-point tool is fed linearly and parallel to the axis of rotation.

Drilling: Producing a hole by feeding the drill into the rotating work along its axis. Drilling can be followed by reaming or boring to improve accuracy and surface finish.

Knurling: Metal-forming operation used to produce a regular crosshatched pattern in work surfaces.

Chamfering and face grooving operations are shown in Figure 13.3(a) and (b) respectively.

13.3 TURNING TOOL HOLDERS

Mechanical toolholders and the ANSI identification system for turning toolholders and indexable inserts were introduced in Chapter 2. A more detailed discussion of toolholder styles and their application will be presented here.

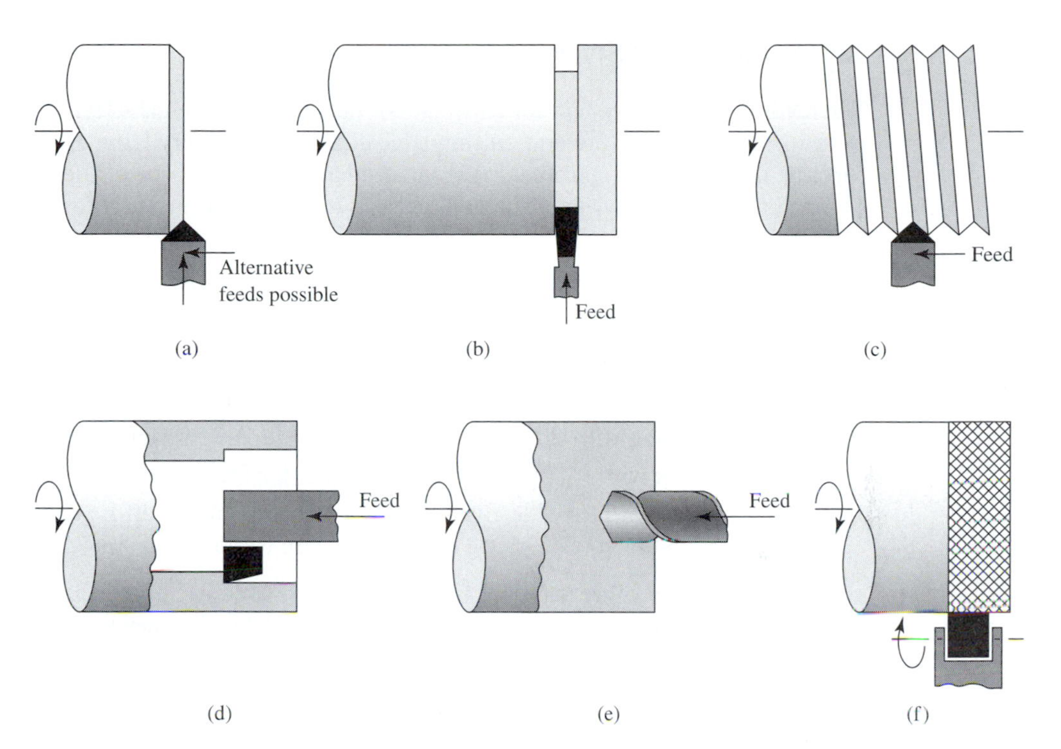

FIGURE 13.2 Related turning operations: (a) chamfering, (b) parting, (c) threading, (d) boring, (e) drilling, (f) knurling.

FIGURE 13.3 Chamfering (a) and face grooving (b) operations, typically performed on a lathe or a machining center. (Courtesy Valenite Inc.)

13.3.1 Toolholder Styles

The ANSI numbering system for turning tool holders has assigned letters to specific geometries in terms of lead angle and end cutting-edge angle. The primary lathe machining operations of turning, facing, grooving, threading, and cutoff are covered by one of the seven basic tool styles outlined by the ANSI system. The designations for the seven primary tool styles are A, B, C, D, E, F, and G.

A: Straight shank with 0° side cutting-edge angle, for turning operations.
B: Straight shank with 15° side cutting-edge angle, for turning operations.
C: Straight shank with 0° end cutting-edge angle, for cutoff and grooving operations.
D: Straight shank with 45° side cutting-edge angle, for turning operations.
E: Straight shank with 30° side cutting-edge angle, for threading operations.
F: Offset shank with 0° end cutting-edge angle, for facing operations.
G: Offset shank with 0° side cutting-edge angle; this tool is an A-style tool with additional clearance built in for turning operations close to the lathe chuck.

There are many other styles of turning tools available in addition to those shown here, as detailed by the ANSI numbering system (see Figure 2.35). The seven basic tools are shown in operation in Figure 13.4.

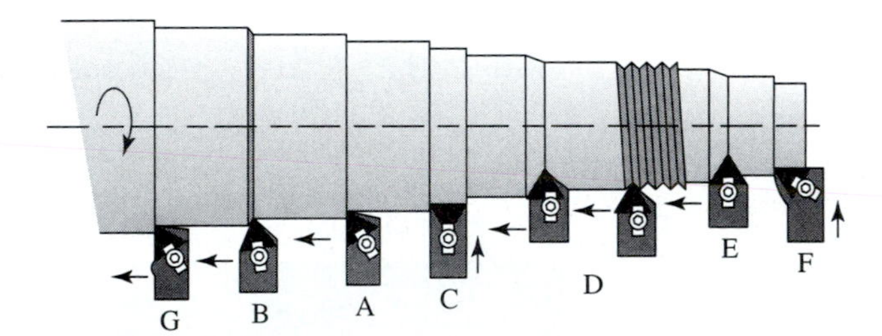

FIGURE 13.4 The primary lathe machining operations of turning, facing, grooving, threading and cut-off are performed with one of seven basic toolholder styles.

Right- and Left-Hand Toolholders The toolholder styles discussed here and shown in Figure 13.4 represent a fraction of those standard styles available from most indexable-cutting-tool manufacturers. ANSI standard turning tools can be purchased in either right- or left-hand styles. The problem of identifying a right-hand tool from a left-hand tool can be resolved by remembering that when you hold the shank of a right-hand tool as shown in Figure 13.5 (insert facing upward), it will cut from right to left. The left-hand tool, when held by the shank as shown in Figure 13.5 (insert facing upward), will cut from left to right.

13.3.2 Turning Insert Shapes

Indexable turning inserts are manufactured in a variety of shapes, sizes, and thicknesses, with straight holes, with countersunk holes, without holes, with chipbreakers on one side, with chipbreakers on two sides, or without chipbreakers. The selection of the appropriate turning toolholder geometry, accompanied by the correct insert shape and chipbreaker geometry, will ultimately have a significant impact on the productivity and tool life of a specific turning operation.

Insert strength is one important factor in selecting the correct geometry for a workpiece material or hardness range. Triangle inserts are the most popular shape of inserts primarily because of their wide application range. A triangular insert can be utilized in any of the seven basic turning holders mentioned earlier. Diamond-shaped inserts are used for profile turning

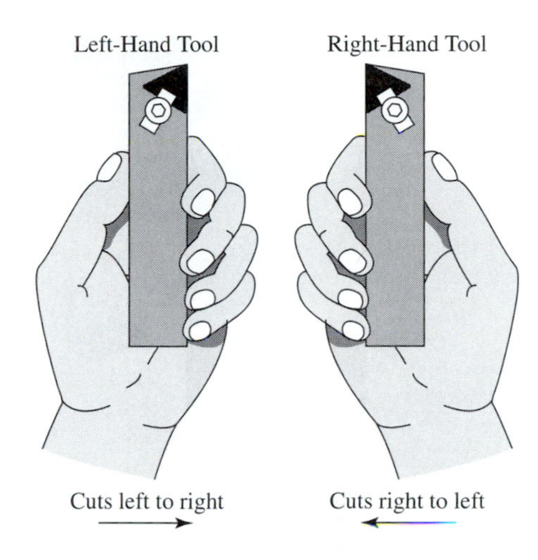

FIGURE 13.5 Identification method for right- and left-hand turning toolholders.

operations; squares are often used on lead angle tools. The general rule for rating an insert's strength based on its shape is, "the larger the included angle on the insert corner, the greater the insert strength."

The following list describes the different insert shapes from strongest to weakest. The relationship between insert shapes and insert strength was shown in Chapter 2 (see Fig. 2.28).

Insert Letter Designation	Insert Description	Insert Included Angle (°)
R	Round	N/A
O	Octagon	135
H	Hexagon	120
P	Pentagon	108
S	Square	90
C	Diamond	80
T	Triangle	60
D	Diamond	55
V	Diamond	35

Six common turning tool holders are shown in Figure 13.6(a), and five common indexable insert shapes with molded chipbreakers are shown in Figure 13.6(b).

13.4 LATHES AND LATHE COMPONENTS

Of the many standard and special types of turning machines that have been built, the most important, most versatile, and most widely recognized is the engine lathe. The standard engine lathe is not a high-production machine, but it can be readily tooled up for many one-piece or short-run jobs. It is also possible to modify the basic machine for many higher-production applications. The modern engine lathe provides a wide range of speeds and feeds, which allow optimum settings for almost any operation. There have been advances in headstock design to provide greater strength and rigidity. This allows the use of high-horsepower motors so that heavy cuts

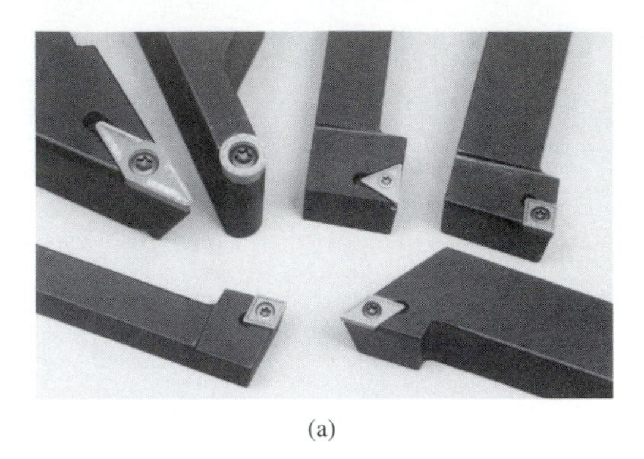

(a)

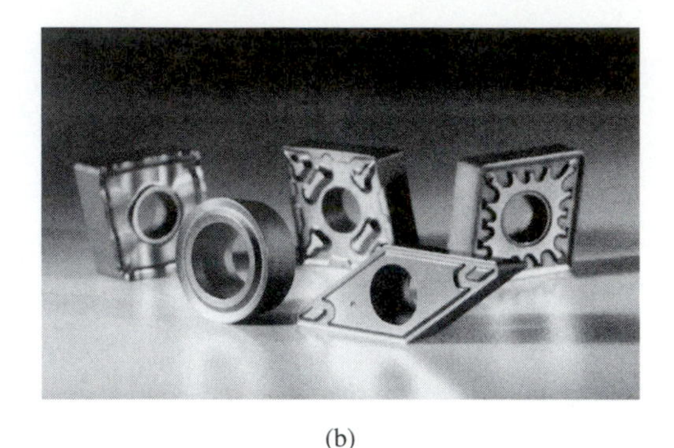

(b)

FIGURE 13.6 Common turning toolholders (a) and common indexable insert shapes (b) with molded chipbreakers are shown. (Courtesy Valenite Inc.)

with carbide tools are practical. To utilize this high power without losing accuracy, new lathes incorporate heavier **beds**, wider hardened **ways**, and deeper-sectioned **carriages**. A schematic illustration of the components of an engine lathe is shown in Figure 13.7 and described below:

Headstock: The headstock is the powered end and is always at the operator's left. This contains the speed-changing gears and the revolving, driving spindle, to which any one of several types of workholders is attached. The center of the spindle is hollow so that long bars may be put through it for machining.

Tailstock: The tailstock is nonrotating, but on hardened ways, it can be moved, to the left or right, to adjust to the length of the work. It can also be offset for cutting small-angle tapers.

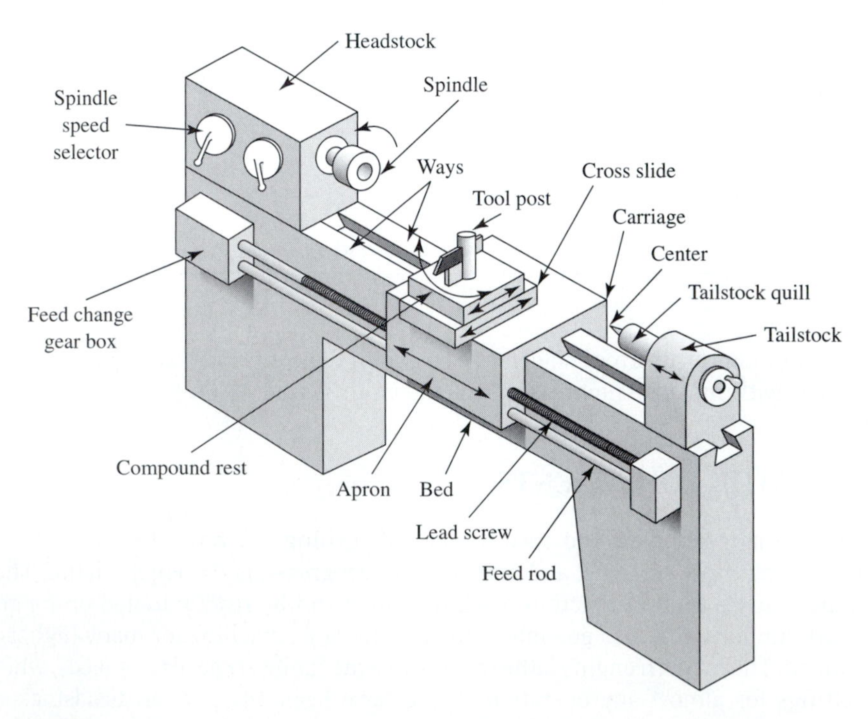

FIGURE 13.7 Schematic illustration of the components of a standard engine lathe.

Carriage: The carriage can be moved left or right either by handwheel or power feed. This provides the motion along the *Z*-axis. During this travel turning cuts are made.

Apron: The apron attached to the front of the carriage, holds most of the control levers. These include the levers that engage and reverse the feed lengthwise (*Z*-axis) or crosswise (*X*-axis), and the lever that engages the threading gears.

Cross Slide: The cross slide is mounted on the carriage and can be moved in and out (*X*-axis) perpendicular to the carriage motion. This is the part that moves when facing cuts are made with power feed, or at any time a cut must be made "square" with the *Z*-axis. This, or the compound, is also used to set the depth of cut when turning. The cross slide can be moved by its handwheel or by power feed.

Compound Rest: The compound rest, or compound for short, is mounted on the carriage. It can be moved in and out by its handwheel for facing or for setting the depth of cut. It can also be rotated 360° and fed by its handwheel at any angle. The compound does not have any power feed, but it always moves longitudinally with the cross slide and the carriage.

Tool Post: The tool post is mounted on the compound rest. This can be any of several varieties, but in its simplest form it is merely a slotted cylinder, which can be moved left or right in the T-slot in the compound and clamped in place. It can also be rotated so as to present the cutter to the work at whatever angle is best for the job.

Bed: The bed of the lathe is its "backbone." It must be rigid enough to resist deflection in any direction under load. The bed is made of cast iron or a steel weldment, in a box or I-beam shape, and is supported on legs, a cabinet, or a bench.

Ways: The ways of the lathe are the flat or V-shaped surfaces on which the carriage and the tailstock are moved left and right. Each has its separate pair of ways, often one flat surface, for stability, and one V-way for guidance in a perfectly straight line. These ways are hardened and scraped or ground to close tolerances. The basic accuracy of movement of the carriage depends on the ways. A typical toolroom engine lathe is shown in Figure 13.8.

Size: The size of a lathe is specified by two or three dimensions:

- The largest-diameter workpiece that will clear the bed of the lathe. The center is the headstock spindle center.

FIGURE 13.8 A typical toolroom engine lathe with face plate, square turret, follower, and steady rest. (Courtesy Summitt Machine Tool Manufacturing Corp.)

FIGURE 13.9 A more sophisticated 18-inch variable-speed engine lathe permits optimal cutting speed selection. (Courtesy Clausing Industries, Inc.)

- The largest-diameter workpiece that will clear the cross slide is sometimes also specified.
- The longest workpiece that can be held on centers between the headstock and the tailstock.

A larger, more sophisticated lathe is shown in Figure 13.9.

13.5 OPERATING CONDITIONS

Operating conditions control three important metal-cutting variables: metal removal rate, tool life, and surface finish. Correct operating conditions must be selected to balance these three variables and to achieve the minimum machining cost per piece, the maximum production rate, and/or the best surface finish, whichever is desirable for a particular operation.

The success of any machining operation is dependent on the **setup** of the workpiece and the cutting tool. Setup becomes especially important when the workpiece is not stiff or rigid and when the tooling or machine-tool components must be extended to reach the area to be machined.

Deflection of the workpiece, the cutting tool, and the machine is always present and can never be totally eliminated. This deflection is usually so minimal that it has no influence on an operation, and often goes unnoticed. The deflection only becomes a problem when it results in chatter, vibration, or distortion. It is therefore very important to take the necessary time and effort to ensure that the setup is as rigid as possible for the type of operation to be performed. This is especially important when making heavy or interrupted cuts.

Balancing should be considered when machining odd-shaped workpieces, especially those workpieces that have uneven weight distribution and those that are loaded off-center. An unbalanced situation can be a safety hazard and can cause work inaccuracies, chatter, and damage to the machine tool. While unbalance problems may not be apparent, they may exist at low-speed operations and will become increasingly severe as the speed is increased. Unbalance conditions most often occur when using turntables and lathe **face plates**.

As material is removed from the workpiece, the balance may change. If a series of roughing cuts causes the workpiece to become unbalanced, the problem will be compounded when the speed is increased to take finishing cuts. As a result, the reasons for problems in achieving the

FIGURE 13.10 Operating conditions become very important when machining very large parts. (Courtesy Sandvik Coromant Corp.)

required accuracy and surface finish may not be apparent until the machining operation has progressed to the finishing stage. Operating conditions become very important when machining very large parts as shown in Figure 13.10.

13.5.1 Work-Holding Methods

In lathe work the three most common work-holding methods are

- Held in a chuck
- Held between centers
- Held in a collet

Many of the various work-holding devices used on a lathe are shown in Figure 13.11.

FIGURE 13.11 Many of the various work-holding devices used on a lathe for turning operations. (Courtesy Kitagawa Div. Sumikin Bussan International Corp.)

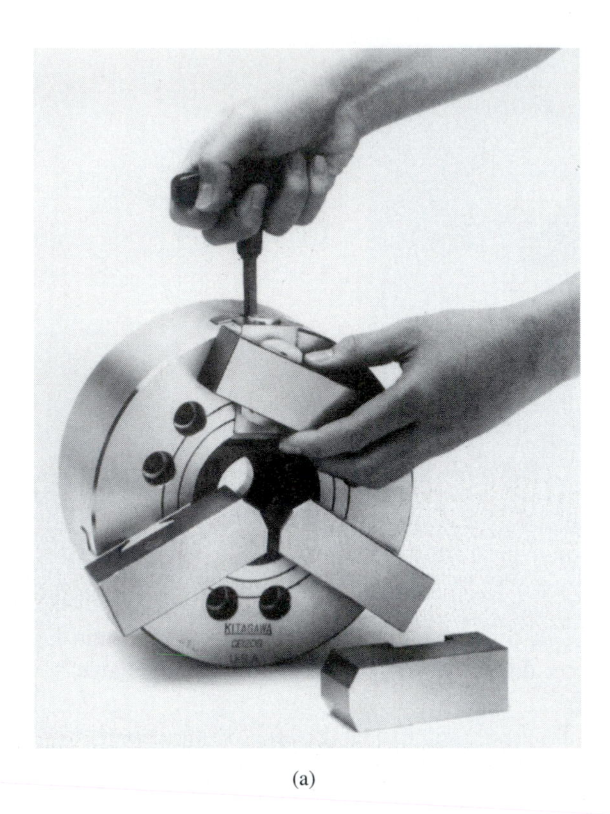

(a)

(b)

FIGURE 13.12 The most common method of work holding, the chuck, has either three jaws (a) or four jaws (b). (Courtesy Kitagawa Div. Sumikin Bussan International Corp.)

Chucks The most common method of work holding, the chuck, has either three or four jaws (Fig. 13.12) and is mounted on the end of the main spindle. A three-jaw chuck is used for gripping cylindrical workpieces when the operations to be performed are such that the machined surface is concentric with the work surfaces.

The jaws have a series of teeth that mesh with spiral grooves on a circular plate within the chuck. This plate can be rotated by the key inserted in the square socket, resulting in simultaneous radial motion of the jaws. Since the jaws maintain an equal distance from the chuck axis, cylindrical workpieces are automatically centered when gripped.

Three-jaw chucks, as shown in Figure 13.13, are often used to automatically clamp cylindrical parts using either electric or hydraulic power. With the four-jaw chuck, each jaw can be adjusted independently by rotation of the radially mounted threaded screws. Although accurate mounting of a workpiece can be time-consuming, a four-jaw chuck is often necessary for noncylindrical workpieces. Both three- and four-jaw chucks are shown in Figure 13.11.

Between Centers For accurate turning operations or in cases where the work surface is not truly cylindrical, the workpiece can be turned between centers. This form of work holding is illustrated in Figure 13.14. Initially the workpiece has a conical center hole drilled at each end to provide location for the lathe centers. Before supporting the workpiece between the centers (one in the headstock and one in the tailstock) a clamping device called a dog is secured to the workpiece. The dog is arranged so that the tip is inserted into a slot in the drive plate mounted on the main spindle, ensuring that the workpiece will rotate with the spindle.

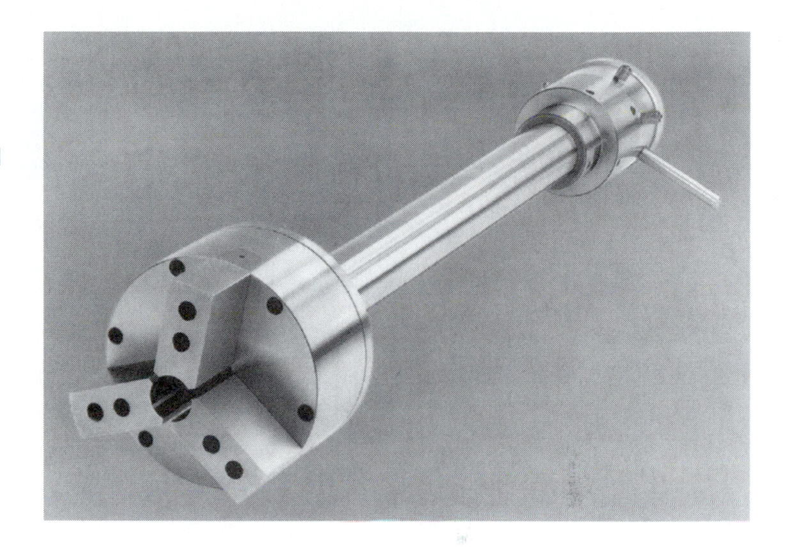

FIGURE 13.13 Three-jaw chucks are often used in automated machining systems to pneumatically or hydraulically clamp cylindrical parts. (Courtesy Royal Products)

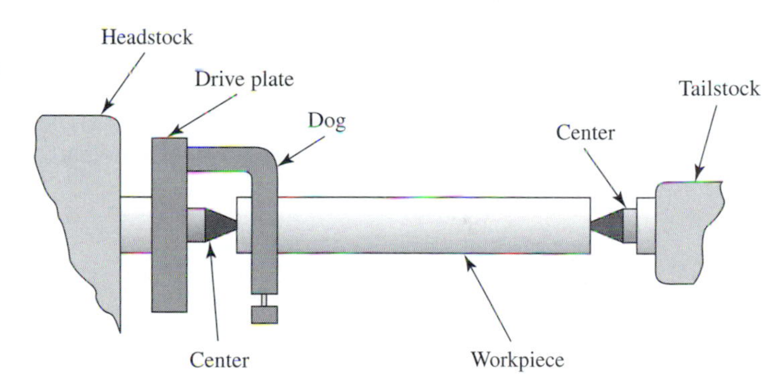

FIGURE 13.14 For accurate machining, cylindrical parts can be turned between centers.

Lathe centers support the workpiece between the headstock and the tailstock. The center used in the headstock spindle is called the "live" center. It rotates with the headstock spindle. The "dead" center is located in the tailstock spindle. This center usually does not rotate and must be hardened and lubricated to withstand the wear of the revolving work. Shown in Figure 13.15 are three kinds of dead centers. As shown in Figure 13.16, some manufacturers are making a roller-bearing or ball-bearing center in which the center revolves.

The hole in the spindle into which the center fits is usually of a Morse standard taper. It is important that the hole in the spindle be kept free of dirt and also that the taper of the center

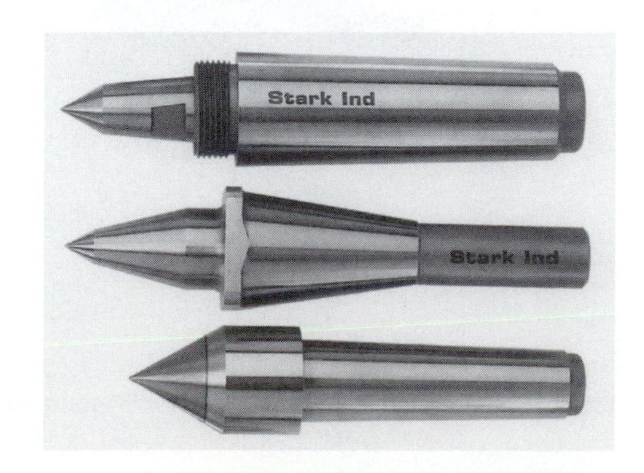

FIGURE 13.15 Hardened "dead" centers are mounted in the tailstock; they do not rotate with the workpiece and must be lubricated. (Courtesy Stark Industrial, Inc.)

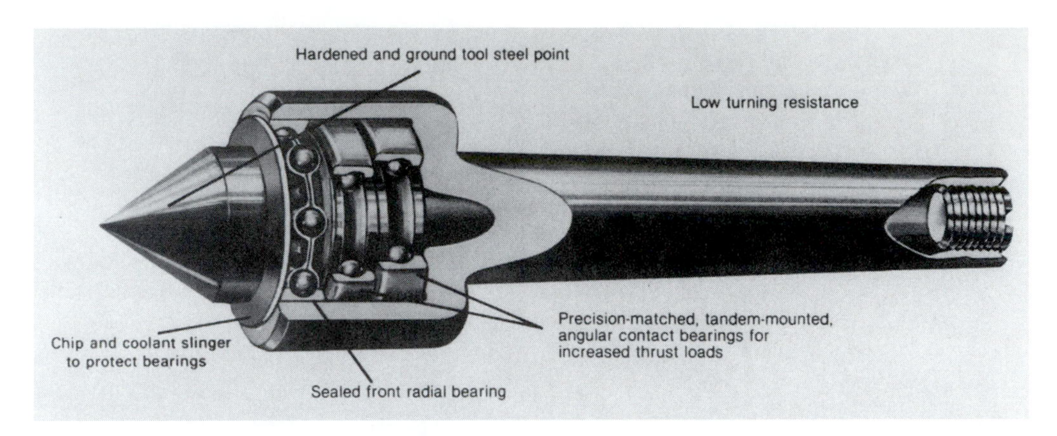

FIGURE 13.16 Hardened "live" centers are mounted in the tailstock; they rotate with the workpiece and do not need lubrication. (Courtesy Royal Products)

be clean and free of chips or **burrs.** If the taper of the live center has particles of dirt or a burr on it, it will not run true. The centers play a very important part in lathe operation. Since they give support to the workpiece, they must be properly ground and in perfect alignment with each other. The workpiece must have perfectly drilled and countersunk holes to receive the centers. The center must have a 60° point.

Collets **Collets** are used when smooth bar stock, or workpieces that have been machined to a given diameter, must be held more accurately than normally can be achieved in a regular three- or four-jaw chuck. Collets are relatively thin tubular steel **bushings** that are split into three longitudinal segments over about two thirds of their length (Fig. 13.17a). The smooth internal surface of the split end is shaped to fit the piece of stock that is to be held. The external surface at the split end is a taper that fits within an internal taper of a collet sleeve placed in the spindle hole. When the collet is pulled inward into the spindle, by means of the draw bar that engages threads on the inner end of the collet, the action of the two mating tapers squeezes the collet segments together, causing them to grip the workpiece (Fig. 13.17b).

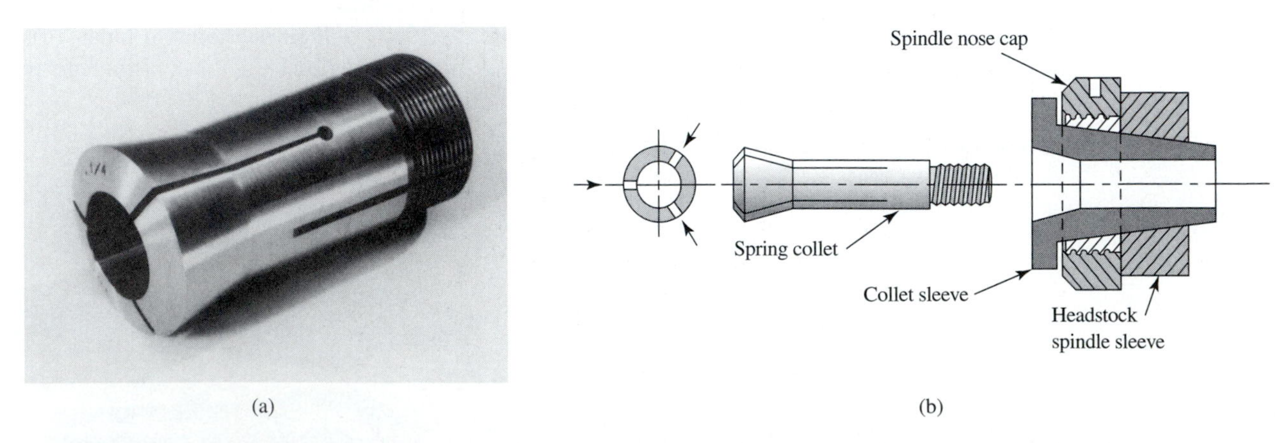

FIGURE 13.17 A collet (a) and a collet mounting assembly (b) are shown here. (Courtesy Lyndex Corp.)

Collets are made to fit a variety of symmetrical shapes. If the stock surface is smooth and accurate, collets will provide accurate centering; maximum **runout** should be less than 0.0005 inch. However, the work should be no more than 0.002 in. larger or 0.005 in. smaller than the nominal size of the collet. Consequently, collets are used only on drill rod, cold-drawn, extruded, or previously machined material.

Another type of collet has a size range of about $\frac{1}{8}$ in. Thin strips of hardened steel are bonded together on their sides by synthetic rubber to form a truncated cone with a central hole. The collet fits into a tapered spindle sleeve so that the outer edges of the metal strips are in contact with the inner taper of the sleeve. The inner edges bear against the workpiece. Pulling the collet into the adapter sleeve causes the strips to grip the work. Because of their greater size range, fewer of these collets are required than with the ordinary type.

13.5.2 Tool-Holding Devices

The simplest form of toolholder or post is illustrated in Figure 13.18(a) and is suitable for holding one single-point tool. Immediately below the tool is a curved block resting on a concave spherical surface. This method of support provides an easy way of inclining the tool so that its corner is at the correct height for the machining operation. In Figure 13.18(a) the toolpost is shown mounted on a compound rest. The rest is a small slideway that can be clamped in any angular position in the horizontal plane and is mounted on the cross slide of the lathe. The compound rest allows the tool to be hand fed at an oblique angle to the lathe bed and is required in operations like screw threading and the machining of short tapers or chamfers.

Another common form of toolpost, the square turret, is shown in Figure 13.18(b). It also is mounted on the compound rest. As its name suggests, this four-way toolpost can accommodate as many as four cutting tools. Any cutting tool can be quickly brought into position by unlocking the toolpost with the lever provided, rotating the toolpost, and then reclamping with the lever.

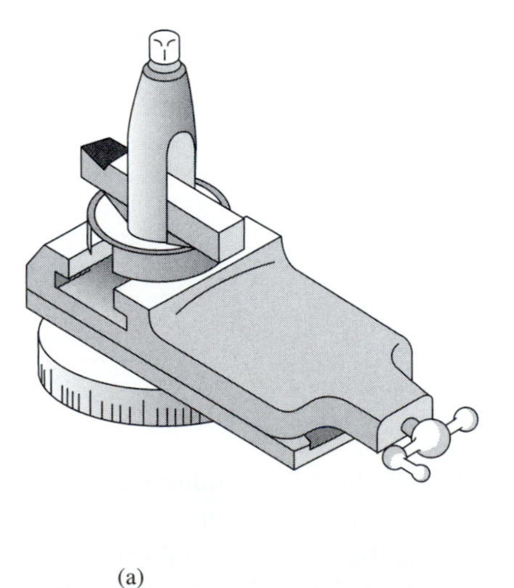

(a)

(b)

FIGURE 13.18 A toolpost for single-point tools (a) and a quick change indexing square turret, which can hold up to four tools (b). (Courtesy Dorian Tool)

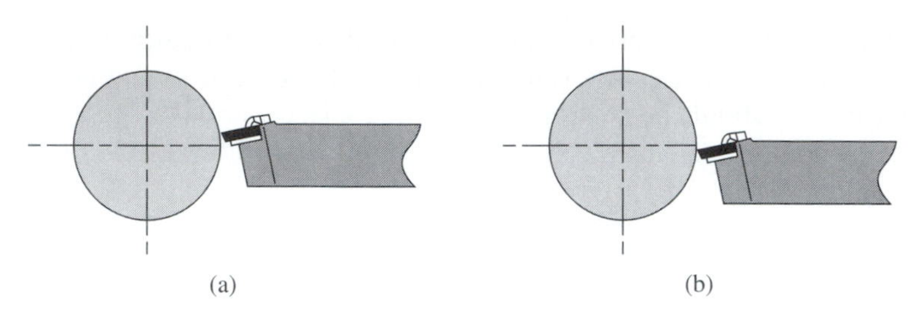

FIGURE 13.19 Cutting edge above workpiece centerline (a) and cutting edge below workpiece centerline (b). Both conditions result in poor performance.

All standard toolholders are designed to cut with the cutting point located on the **centerline** of the machine and workpiece. If the cutting point is not on the centerline, rake and clearance angles will not be correct as specified and problems will occur.

If the cutting edge is above the workpiece centerline, as shown in Figure 13.19(a), the clearance angle between the toolholders and the workpiece will be reduced. The lack of clearance will lead to poor tool life and poor surface finish. It will also force the workpiece away from the tool when working with small diameters.

On the other hand, if the cutting edge is positioned below the centerline, as shown in Figure 13.19(b), the rake angle becomes more negative. Very high cutting forces will be generated, and the chip will be directed into a tight curl. Insert fracture can very easily occur, and a small-diameter workpiece can even climb over the top of the tool and be torn from the machine.

Occasionally, however, moving the cutting point off centerline can solve a problem. An example is in situations when machining flimsy parts or when deep grooving chatter is a constant threat, even when a positive-rake tool is used. Moving the tool slightly above centerline (2 to 4% of the workpiece diameter) will change the rake angle slightly, and this in turn will reduce cutting forces and make chatter less of a danger.

Interrupted cuts present special problems, particularly when machining large-diameter workpieces. It is best to position the cutting point slightly below the centerline to present the insert in a stronger cutting position. A lead angle should also be used whenever possible. Moving the cutting point slightly below the centerline and using a lead angle allows the workpiece to contact the tool on a stronger part of the insert, behind the nose.

13.6 CUTTING CONDITIONS

After deciding on the machine tool and the cutting tool, the following main cutting conditions have to be considered:

- Cutting speed
- Depth of cut
- Feedrate

The choice of these cutting conditions will affect the productivity of the machining operation in general and the following factors in particular: the life of the cutting tool, the surface finish of the workpiece, the heat generated in the cutting operation (which in turn affects the life of the tool and the surface integrity of the machined parts), and the power consumption.

Cutting Speed Cutting speed refers to the relative surface speed between tool and work, expressed in surface feet per minute (**SFPM**). Either the work, the tool, or both, can move during cutting. Because the machine tool is built to operate in revolutions per minute, some means must then be available for converting surface speeds into revolutions per minute (**RPM**). The common formula for conversion is:

$$\text{SFPM} = \frac{D \times \pi \times (\text{RPM})}{12}$$

where D is the diameter (in inches) of the workpiece or the rotating tool, π equals 3.1416, and RPM is a function of the speed of the machine tool in revolutions per minute. For example, if a lathe is set to run at 250 RPM and the diameter of the workpiece is 5 in., then

$$\text{SFPM} = \frac{3.1416 \times 5 \times 250}{12} = \frac{3927}{12}$$

$$\text{Answer} = 327.25 \text{ or } 327 \text{ SFPM}$$

The RPM can be calculated when another cutting speed is desired. For example, to use 350 SFPM with a workpiece that is 4 in. in diameter, this formula is employed

$$\text{RPM} = \frac{12 \times \text{SFPM}}{\pi \times D} \quad \text{RPM} = \frac{12 \times 350}{3.1416 \times 4} = \frac{4200}{12.57}$$

$$\text{Answer} = 334.13 \text{ or } 334 \text{ RPM}$$

All tool materials are meant to run at a certain SFPM when machining various work materials. The SFPM range recommendations for tool and work materials are given in many reference publications.

Depth of Cut The depth of cut relates to the depth the tool cutting edge engages the work. The depth of cut determines one linear dimension of the area of cut. For example: To reduce the outside diameter (OD) of a workpiece by .500 in., the depth of cut would be .250 in.

Feedrate The feedrate for lathe turning is the axial advance of the tool along the work for each revolution of the work expressed as inches per revolution (**IPR**). The feed is also expressed as a distance traveled in a single minute or **IPM** (inches per minute). The following formula is used to calculate the feed in IPM:

$$\text{IPM} = \text{IPR} \times \text{RPM}$$

Feed, speed, and depth of cut have a direct effect on productivity, tool life, and machine requirements. Therefore, these elements must be carefully chosen for each operation. Whether the objective is rough cutting or finishing will have a great influence on the cutting conditions selected.

Roughing Cuts When roughing, the goal is usually maximum stock removal in minimum time with minor consideration given to tool life and surface finish. There are several important points to keep in mind when rough cutting.

The first is to use a heavy feed because this makes the most efficient use of power and, with less tool contact, tends to create less chatter. There are some exceptions where a deeper cut is more advantageous than a heavy feed, especially where longer tool life is needed. Increasing the depth of cut will increase tool life over an increase in feed rate, but, as long as it is practical and chip formation is satisfactory, it is better to choose a heavy feed rate.

A heavy feed or deeper cut is usually preferable to higher speed, because the machine is less efficient at high speed. When machining common materials, the unit horsepower (HP) factor is reduced in the cut itself, because the cutting speed increases up to a certain critical value. The machine inefficiencies will overcome any advantage when machining heavy workpieces.

Even more important, tool life is greatly reduced at high cutting speeds unless coated carbide or other modern tool materials are used, and these also have practical speed limits. Tool life is decreased most at high speeds, although some decrease in tool life occurs when feed or depth of cut is increased. This stands to reason, because more material will be removed in less time. It becomes a choice then, between longer tool life and increased stock removal. Since productivity generally outweighs tool costs, the most practical cutting conditions are usually those that first, are most productive, and second, will achieve reasonable tool life.

Finishing Cuts When taking finishing cuts, feedrate and depth of cut are of minor concern. The feedrate cannot exceed that which is necessary to achieve the required surface finish, and the depth of cut will be light. However, the rule about speed will still apply. The speeds will generally be higher for finish cuts, but they must still be within the operating speed of the tool material.

Tool life is of greater concern for finish cuts. It is often better to strive for greater tool life at the expense of material removed per minute. If tool wear can be minimized, especially on a long cut, greater accuracy can be achieved, and matching cuts, that result from tool changes can be avoided.

One way to minimize tool wear during finishing cuts is to use the maximum feedrate that will still produce the required surface finish. The less time the tool spends on the cut, the less tool wear can occur. Another way to minimize tool wear during a long finishing cut is to reduce the speed slightly. Coolant, spray mist, or air flow will also extend tool life because it reduces the heat of the tool.

13.7 PRODUCTION TURNING MACHINES

The standard engine lathe is versatile, but it is not a high-production machine. When production requirements are high, more-automated turning machines must be used. The turret lathe, some automatic lathes, and the computer-controlled lathe are discussed here. A thorough coverage of high-production machines, machining centers, and automation has been presented in Chapter 11.

13.7.1 Turret Lathes

The turret lathe represents the first step from the engine lathe toward the high-production turning machines. The turret lathe is similar to the engine lathe except that tool-holding turrets replace the tailstock and the toolpost compound assembly. These machines possess special features that adapt them to production. The "skill of the worker" is built into these machines, making it possible for inexperienced operators to reproduce identical parts. In contrast, the engine lathe requires a skilled operator and requires more time to produce parts that are dimensionally the same.

The principal characteristic of turret lathes is that the tools for consecutive operations are set up for use in the proper sequence. Although skill is required to set and adjust the tools properly, once they are correct, less skill is required to operate the turret lathe. Many parts can be produced before adjustments are necessary. These machines are normally used for small to medium-sized production runs when the engine lathe is too slow but the additional production rate desired does not warrant a special machine. A schematic illustration of the components of a turret lathe is shown in Figure 13.20.

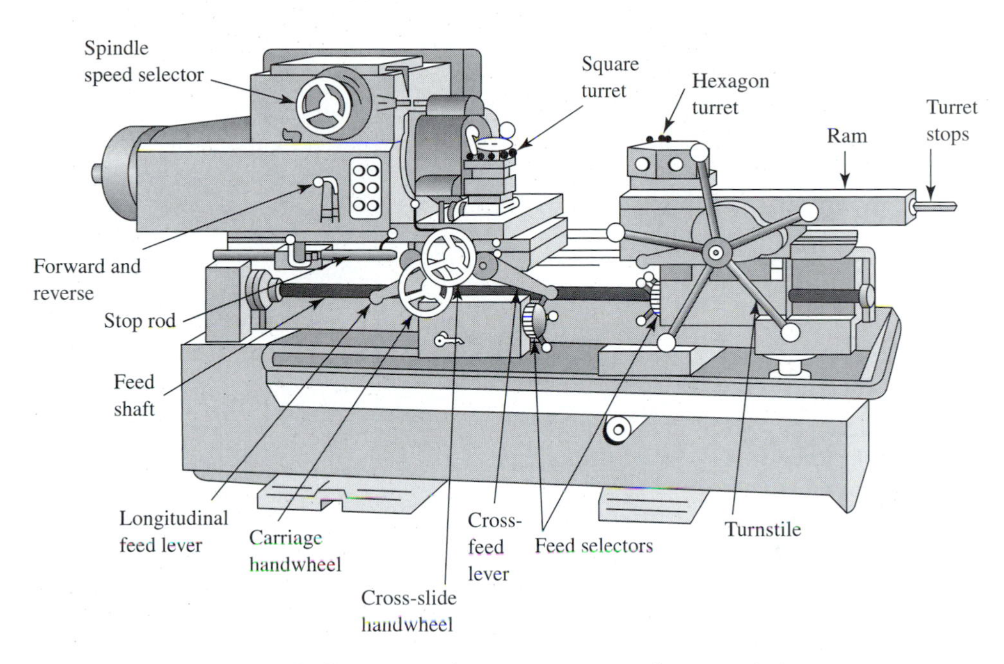

FIGURE 13.20 Schematic illustration of the components of a turret lathe.

Square and Hex Turrets A square turret is mounted on the top of the cross slide and is capable of holding four tools (Fig. 13.18b). If several different tools are required, they are set up in sequence and can be quickly indexed and locked in correct working position. So that cuts can be duplicated, the slide is provided with positive stops or feed trips. Likewise, the longitudinal position of the entire assembly may be controlled by positive stops on the left side of the apron. Cuts may be taken with square turret tools and with tools mounted on the hexagon turret simultaneously.

An outstanding feature is the turret in place of the tailstock. This turret, mounted on either the sliding ram or the **saddle**, or on the back of the structure, carries anywhere from 4 to 18 tool stations. The tools are preset for the various operations. The tools are mounted in proper sequence on the various faces of the turret so that as the turret indexes between machining operations, the proper tools are engaged into position. For each tool there is a stop screw or electric/electronic transducer, which controls the distance the tool will feed and cut. When this distance is reached, an automatic trip lever stops further movement of the tool by disengaging the drive clutch.

Like the engine lathe, the modern turret lathe provides fast spindle speeds, wide speed and feed ranges, high power, and great rigidity. The machine is operated in the high end of its speed range more than the engine lathe is, partly because the tools placed in the turret often work on small diameters on the workpiece, but also because the operator is more production conscious.

Horizontal Turret Lathes Horizontal turret lathes are made in two general designs and are known as the ram and saddle types. The ram-type turret lathe shown in Figure 13.21(a) has the turret mounted on a slide or ram, which moves back and forth on a saddle clamped to the lathe bed. The saddle-type turret lathe shown in Figure 13.21(b) has the turret mounted directly on a saddle that moves back and forth with the turret.

Vertical Turret Lathes A vertical turret lathe resembles a vertical **boring mill**, but it has the characteristic turret arrangement for holding the tools. It consists of a rotating chuck or table in the horizontal position with the turret mounted above on a cross rail. In addition, there is at least one side head provided with a square turret for holding tools. All tools mounted on the

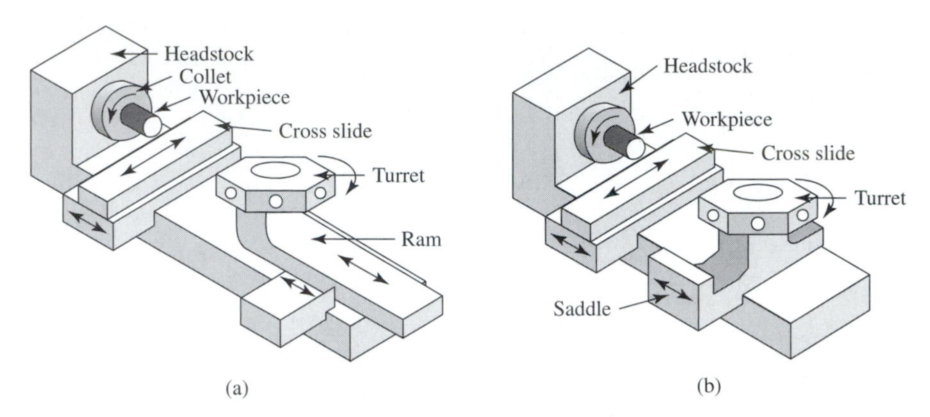

FIGURE 13.21 Ram-type horizontal turret lathe (a), and saddle-type horizontal turret lathe (b).

turret or side head have their respective stops set so that the length of cuts can be the same in successive machining cycles. It is, in effect, the same as a turret lathe standing on the headstock end, and it has all the features necessary for the production of duplicate parts. This machine was developed to facilitate mounting, holding, and machining of large-diameter heavy parts. Only chucking work is done on this kind of machine. In Figure 13.22 a vertical turning center is shown machining a heavy part.

A vertical turret lathe, shown in Figure 13.23, is provided with two cutter heads: the swiveling main turret head and the side head. The turret and side heads function in the same manner as the hexagonal and square turrets on a horizontal lathe. To provide for angle cuts both the ram and turret heads may be swiveled 30° right or left of center.

The machine can be provided with a control that permits automatic operation of each head including rate and direction of feed, change in spindle feed, indexing of turret, starting, and stopping. Once a cycle of operations is preset and tools are properly adjusted, the operator need only load, unload, and start the machine. Production rate is increased over those manually operated machines, because they operate almost continuously and make changes from one operation to another without hesitation or fatigue. Reducing the handling time and making the cycle automatic allows an operator to attend more than one machine.

The turret lathe normally has a jawed chuck to hold the workpiece; however, a collet may be more suitable when producing parts from bar stock. A turning machine equipped with a collet and a turret is called a screw machine, but it is actually a special turret lathe. The special features of screw machines are aimed primarily at reducing idle time on the parts being machined, thereby increasing productivity, as discussed in Chapter 11.

Advantages of Turret Lathes The difference between the engine and turret lathes is that the turret lathe is adapted to quantity production work, whereas the engine lathe is used primarily for miscellaneous jobbing, toolroom, or single-operation work. These are the features of a turret lathe that make it a quantity production machine:

- Tools may be set up in the turret in the proper sequence for the operation.
- Each station is provided with a feed stop or feed trip so that each cut of a tool is the same as its previous cut.
- Multiple cuts can be taken from the same station at the same time, such as two or more turning and/or boring cuts.

FIGURE 13.22 A vertical turning center machining a heavy part. (Courtesy Giddings & Lewis, LLC)

FIGURE 13.23 Vertical turning lathes are used for machining large-diameter and heavy parts.

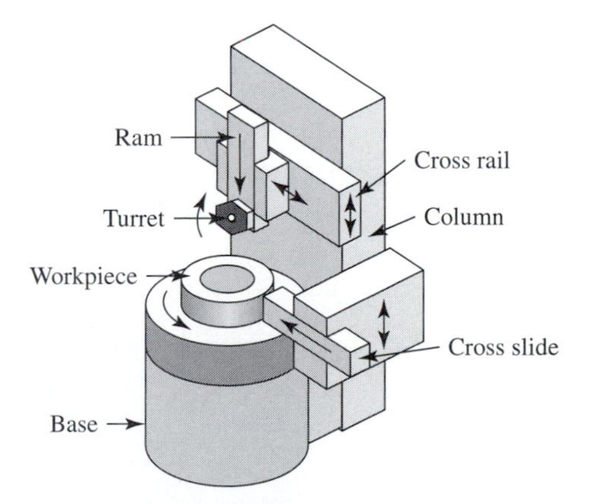

- Combined cuts can be made; tools on the cross slide can be used at the same time that tools on the turret are cutting.
- Rigidity in holding work and tools is built into the machine to permit multiple and combined cuts.
- Turret lathes can also have attachments for taper turning, thread chasing and duplicating, and can be tape controlled.

13.7.2 Automated Equipment

Some turning machines allow automatic chucking, indexing, feeding, spindle speed changes, and other work that has to be done by the operator on the engine lathe. These automatic lathes represent a refinement of the turret lathe, and they are particularly suitable for long-run, mass-production applications.

Automatic lathes may be made up as single-spindle or multiple-spindle machines. Generally, single-spindle machines provide for turning the workpiece, which is held in a collet or chucked on the headstock. Multiple-spindle automatic lathes usually provide means for indexing the workpiece to tools mounted on the various spindles. These tools might include drills, countersinks, boring bars, and other rotating cutters. Both single- and multiple-spindle automatics may be made up with vertical as well as horizontal spindle alignment. Single- and multiple-spindle automation was discussed in Chapter 11.

As far as the machining processes on an automatic lathe are concerned, the fundamental considerations are the high speeds desired for good productivity, the economics of the cutting process, and the balancing of speeds on various phases of the operation to obtain the desired rate of wear on each cutting tool.

Computer-Controlled Lathes In the most advanced lathes, movement and control of the machine and its components are actuated by computer numerical controls (CNCs). These lathes are usually equipped with one or more turrets. Each turret is equipped with a variety of tools and performs several operations on different surfaces of the workpiece. These machines are highly automated, the operations are repetitive and maintain the desired accuracy. They are suitable for low to medium volumes of production. A high-precision CNC lathe and some of the parts produced on such a computer-controlled lathe are shown in Figures 13.24 and 13.25 respectively.

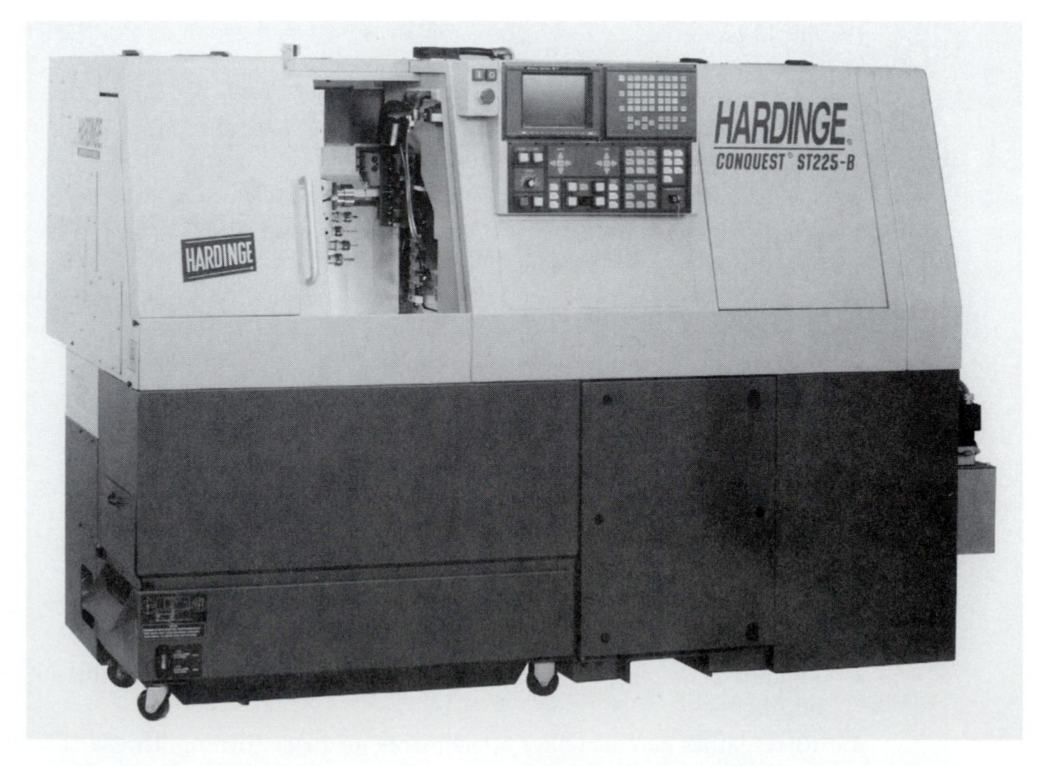

FIGURE 13.24 A high production computer-controlled Swiss Type lathe produces quality parts such as those shown in Figure 13.25. (Courtesy Hardinge Inc.)

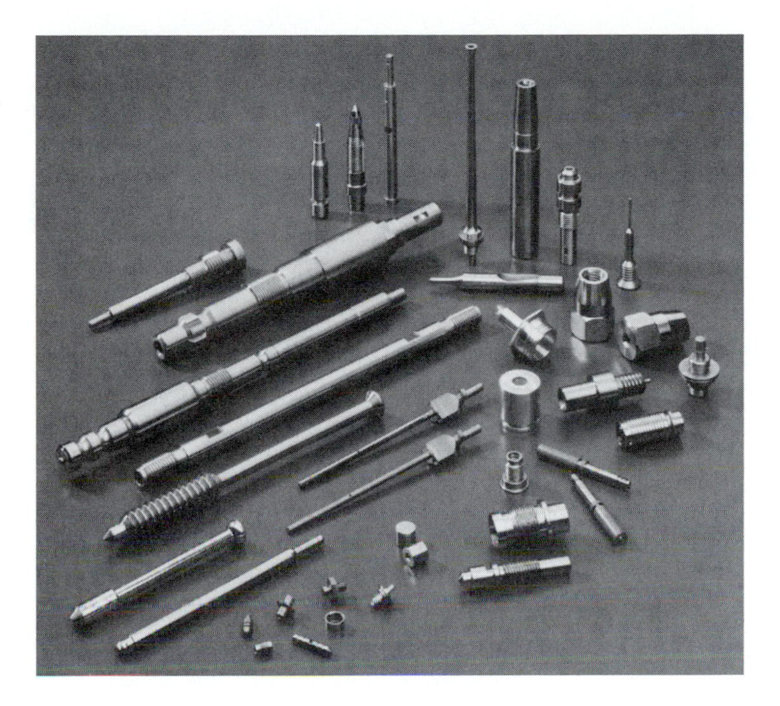

FIGURE 13.25 Various shaped parts are produced by high-production computer-controlled Swiss Type lathes such as the one shown in Figure 13.24. (Courtesy Hardinge Inc.)

13.8 REVIEW QUESTIONS AND PROBLEMS

1. Explain the differences between turning facing and profiling.
2. Discuss the effect of speed during turning and facing operations.
3. What other operations are related to turning?
4. Explain how to identify a left-hand from a right-hand tool.
5. Explain this statement: "the larger the included angle on the insert corner, the greater the insert strength."
6. Discuss the main components of a standard engine lathe.
7. Explain how the size of a lathe is determined.
8. What are the three important metal-cutting variables?
9. Discuss how deflection affects operating conditions.
10. Discuss how balancing affects operating conditions.
11. Explain the difference in function between three-jaw and four-jaw chucks.
12. Explain the difference in function between a dead center and a live center.
13. Describe the clamping mechanism of collets.
14. Explain the advantages of using a square turret toolholder.
15. Why is cutting "on center" so important and what are some exceptions?
16. Name the three main cutting conditions and explain their effect on productivity.
17. Define SFPM for turning.
18. Define IPR and IPM.
19. What are the main considerations in rough-cutting machining conditions?
20. What are the most important considerations in making finishing cuts?
21. Explain the principal characteristics of a turret lathe.
22. Why can a turret lathe be operated by a less-skilled worker?
23. What features make the turret lathe a quantity production machine?
24. What additional automated features are available for lathes?

CHAPTER FOURTEEN

SHAPING AND PLANING

CHAPTER OVERVIEW

- Introduction
- The Shaper
  - Drive Mechanisms
  - Vertical Shapers
- The Planer
  - Comparison of Shapers and Planers
- Review Questions and Problems

14.1 INTRODUCTION

Both the shaper and the planer are single-point tools and cut only in straight lines. They both make the same types of cuts. The **shaper** handles relatively small work. The **planer** handles work weighing up to several tons. The cutting stroke of the shaper is made by moving the tool bit attached to the ram. The cutting stroke of the planer is achieved by moving the work past a stationary tool bit.

The types of cuts that can be made with either machine are shown in Figure 14.1. Both the shaper and the planer usually cut only in one direction, so the return stroke is lost time. However, the return stroke is made at up to twice the speed of the cutting stroke.

14.2 THE SHAPER

The shaper is a relatively simple machine. It is used fairly often in the toolroom or for machining one or two pieces for prototype work. Tooling is simple, and shapers do not always require operator attention while cutting. The horizontal shaper is the most common type, and its principal components and motions are shown in Figure 14.2 and described as follows:

Ram: The ram slides back and forth in dovetail or square ways to transmit power to the cutter. The starting point and the length of the stroke can be adjusted.

Toolhead: The toolhead is fastened to the ram on a circular plate so that it can be rotated for making angular cuts. The toolhead can also be moved up or down by its hand crank for precise depth adjustments.

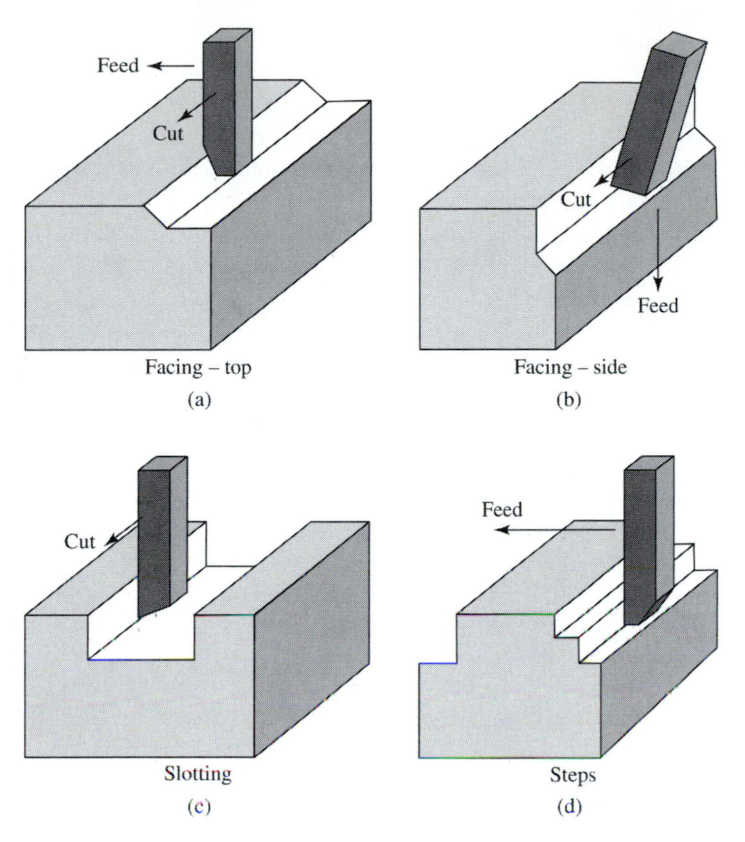

FIGURE 14.1 Typical cuts made by both shapers and planers.

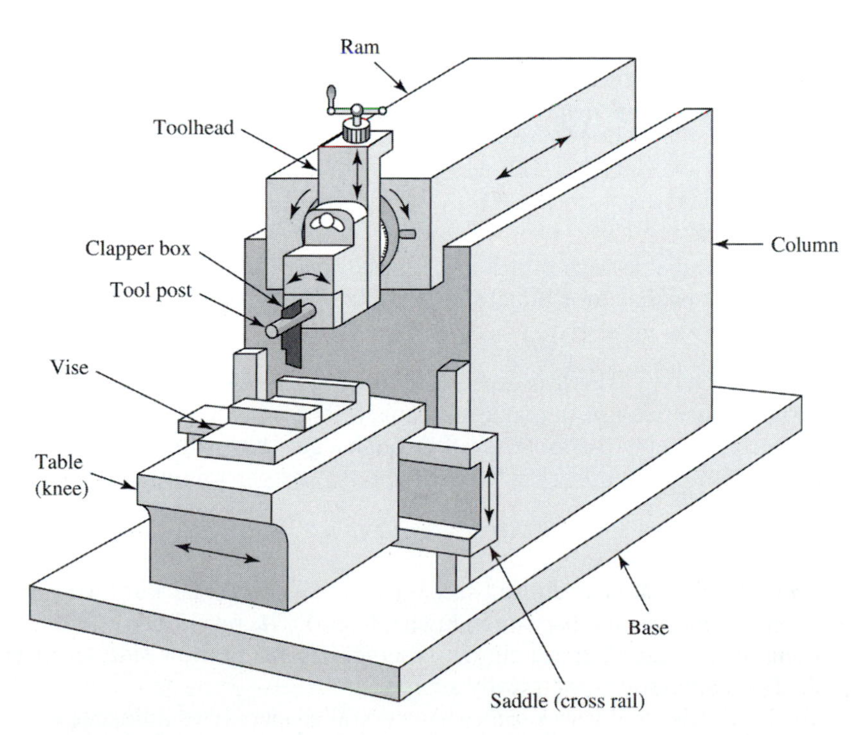

FIGURE 14.2 The horizontal shaper is the most common type. Its principal components and motions are shown here.

Attached to the toolhead is the tool-holding section. This has a toolpost very similar to that used on the engine lathe. The block holding the toolpost can be rotated a few degrees so that the cutter may be properly positioned in the cut.

Clapper Box: The clapper box is needed because the cutter drags over the work on the return stroke. The clapper box is hinged so that the cutting tool will not dig in. Often this clapper box is automatically raised by mechanical, air, or hydraulic action.

Table: The table is moved left and right, usually by hand, to position the work under the cutter when setting up. Then, either by hand or more often automatically, the table is moved sideways to feed the work under the cutter at the end or beginning of each stroke.

Saddle: The saddle moves up and down (*Y*-axis), usually manually, to set the rough position of the depth of cut. Final depth can be set by the hand crank on the toolhead.

Column: The column supports the ram and the rails for the saddle. The mechanism for moving the ram and table is housed inside the column.

Toolholders: Toolholders are the same as the ones used on an engine lathe, though often larger in size. The cutter is sharpened with rake and clearance angles similar to lathe tools, though the angles are smaller because the work surface is usually flat. These cutters are fastened into the toolholder, just as in the lathe, but in a vertical plane.

Work Holding: Work holding is frequently done in a vise. The **vise** is specially designed for use in shapers and has long ways, which allow the jaws to open up to 14 in. or more; therefore quite large workpieces can be held. The vise may also have a swivel base so that cuts may be made at an angle. Work that, due to size or shape, cannot be held in the vise, is clamped directly to the shaper table in much the same way as parts are secured on milling machine tables.

Shaper Size: The size of a shaper is the maximum length of stroke it can take. Horizontal shapers are most often made with strokes from 16 to 24 in. long, though some smaller and larger sizes are available. These shapers use motors from 2 to 5 HP to drive the head and the automatic feed.

Shaper Width: The maximum width that can be cut depends on the available movement of the table. Most shapers have a width capacity equal to or greater than the length of the stroke. The maximum vertical height available is about 12 to 15 in.

14.2.1 Drive Mechanisms

Shapers are available with either mechanical or hydraulic drive mechanisms. Figures 14.3(a) and 14.3(b) show diagrams of both shaper drive mechanisms.

Mechanical Drive

The less-expensive shaper, the one most often purchased, uses a mechanical drive. This drive uses a crank mechanism (Fig. 14.3a). The bull gear is driven by a **pinion**, which is connected to the motor shaft through a gear box with four, eight, or more speeds available. The RPM of the bull gear becomes the strokes per minute (sometimes abbreviated SPM) of the shaper.

Cutting Speed. The cutting speed of the tool across the work will vary during the stroke as shown by the velocity diagram in Figure 14.3(a). The maximum is at the center of the stroke. However, if the cutting speed chosen is somewhat on the slow side, the average speed may be used, and computations are greatly simplified.

Although the ratio varies somewhat, several shapers have a linkage using 220° of the cycle for the cutting stroke and 140° for the return stroke. This is close to a 3:2 ratio.

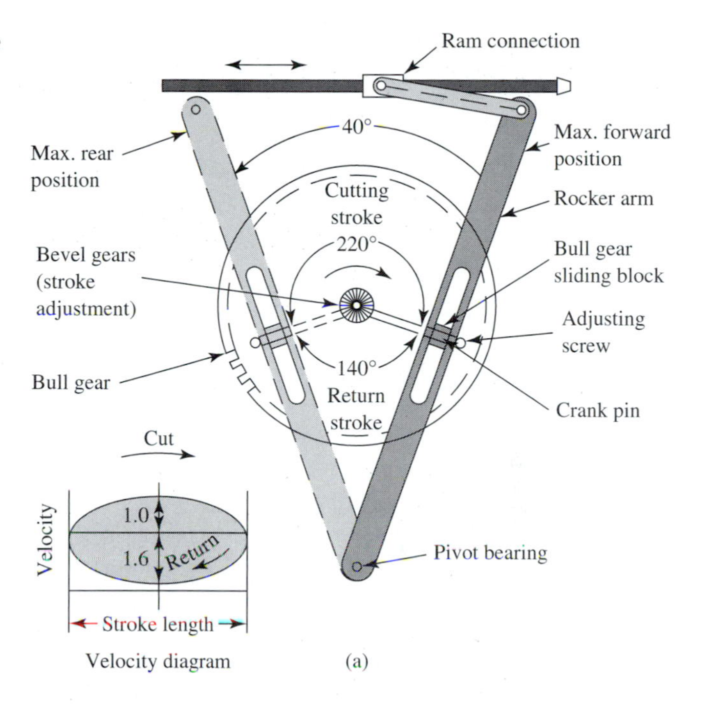

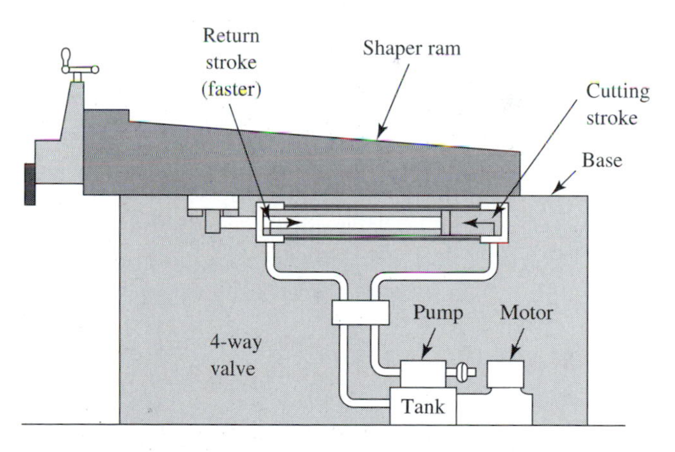

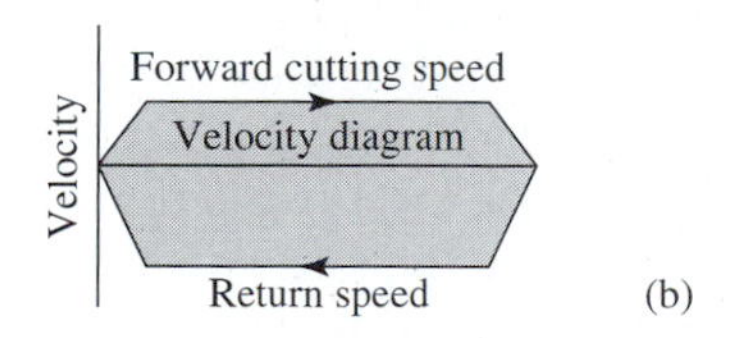

FIGURE 14.3 Shapers are available with either (a) mechanical drive mechanisms or (b) hydraulic drive mechanisms.

In setting up a mechanically operated shaper, the length of cut (in inches) is known and the cutting speed (in feet per minute) is selected according to the kind of metal being cut. It is then necessary to compute the number of strokes per minute because that is how the shaper speed is controlled. Such calculations are beyond the scope of this text.

The number of strokes per minute available on a shaper will vary according to the size of the shaper. The larger shapers will have lower speeds. A 16-in. shaper may have speeds of 27 to 150 strokes per minute, while a 24-in. shaper will have available speeds of 10 to 90 strokes per minute.

Cutting Feed. Feed per stroke on a shaper is comparable to the feed per revolution on a lathe. Coarse feeds for roughing range up to .100 in. per stroke (sometimes abbreviated as IPS) and finish cuts from .005 to .015 in. per stroke. Finish would also depend on the nose radius of the cutting tool.

Hydraulic Drive The hydraulic shaper (Fig. 14.3b) has the same major parts as the mechanical one; however, the ram is driven by a hydraulic cylinder as shown in the simplified sketch. These shapers use 5- to 10-HP motors.

Cutting Speed and Feed. The cutting speed of the hydraulic shaper is infinitely variable by means of hydraulic controls, as is the cross feed. The reverse stroke is made faster than the power stroke because of the smaller area in the return side of the cylinder, if a constant volume pump is used. Another method is to have the rate of fluid flow increased to speed up the return stroke.

Speed and feed on a hydraulic shaper are often controlled by simple dials. Speed is read directly in feet per minute, and feed is read directly in decimal inches. The cutting speed remains nearly constant through the full stroke.

14.2.2 Vertical Shapers

The vertical shaper, sometimes called a slotter, has a vertical ram, with table and saddle similar to the horizontal shaper. If a rotary table is mounted on the regular table, a number of slots can be made at quite accurately spaced intervals. This machine can work either outside or inside a part, provided that the interior opening is larger than the toolhead. A schematic illustration of a vertical shaper is shown in Figure 14.4.

14.3 THE PLANER

A planer makes the same types of cuts as a shaper. However, it is a production-type machine for certain types of work. It can machine any flat or angular surface, including grooves and slots, in medium-sized and large workpieces (see Fig. 14.1). Typical work would be machine beds and columns, marine diesel engine blocks, and bending plates for sheet metal work. These parts are usually large iron castings or steel weldments and may weigh a few hundred pounds or several tons. The most frequently used type of planer is the double-housing planer, shown in Figure 14.5, with the following components:

Frame: The frame is basically two heavy columns fastened together at the top with a large bracing section and fastened at the bottom to the machine bed. This creates a very strong, rigid structure that will handle heavy loads without deflection.

Crossrail: The crossrail is also a heavy box or similar construction. It slides up and down on V- or flat ways, controlled by hand or by power-operated screws. These crossrails are so heavy that they are counterweighted, with either cast-iron weights or hydraulic cylinders, in order that they may be moved easily and positioned accurately. After being positioned, they are clamped in place.

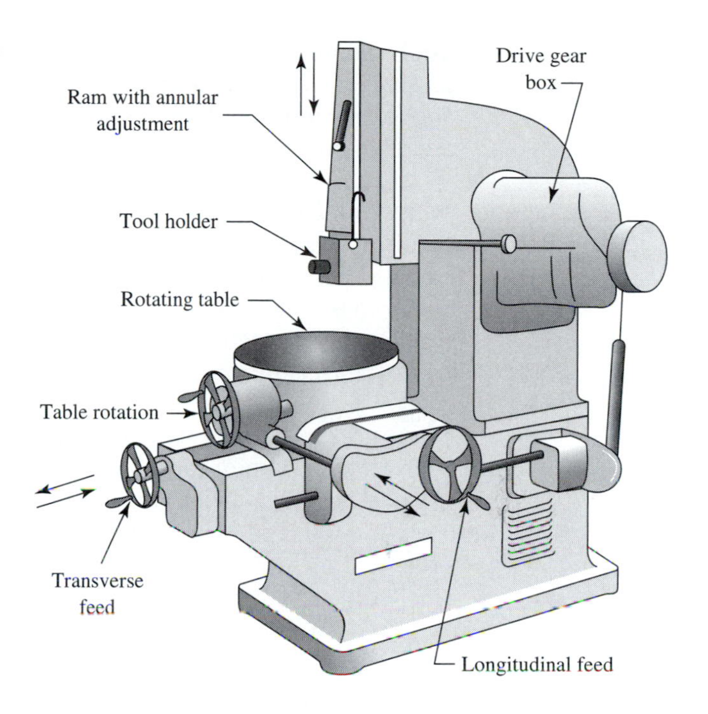

FIGURE 14.4 Schematic illustration of a vertical shaper, also called a slotter.

Railheads: The two railheads can be moved left or right across the crossrail, each controlled by a separate **lead screw**, which can be turned by hand but usually by power feed. The railheads can be rotated and vertically adjusted for depth of cut, the same as the shaper heads. They also have a clapper box (often with power lift) like the shaper.

Sideheads: The sideheads are independently moved up or down by hand or by power feed, and can also be rotated and moved in or out for depth of cut.

Table: The table is a heavy casting that carries the work past the cutting heads. It runs on V- or flat ways. The table is driven either by a very long hydraulic cylinder or by a pinion gear driving a rack that is fastened under the center of the table. The motor driving the pinion gear is the reversible type with variable speed.

Bed: The bed of the planer must be a weldment or casting twice as long as the table. Thus a 12-ft table requires a 24-ft bed. The gearing of hydraulic cylinders for driving the table is housed under the bed.

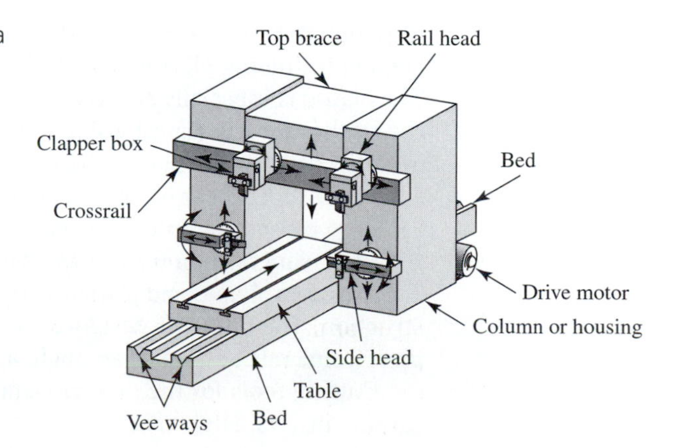

FIGURE 14.5 Schematic illustration of a double-housing planer.

Toolholders Planers use high-speed steel or carbide-tipped cutting tools similar to those used on shapers. However, because planers make heavy cuts, their tools are much larger. Rake relief angles are similar to those used on lathes for cutting cast iron or steel, although relief angles are often only 3 to 5°, because all cuts are on flat surfaces.

Work Holding Holding the work while machining such heavy cuts at 60 to 100 feet per minute requires considerable force; therefore, the workpieces must be solidly fastened to the table. Because the reversal of direction occurs quite rapidly, the work must be especially well braced at the ends. The table has **T-slots**, both lengthwise and across, in which heavy bolts and clamps may be used. Sometimes holes are drilled in the table so that large pins can be used to prevent the workpiece from going off the table when the machine reverses.

Planer Size Planers are often referred to as 30-in. planers or 60-in. planers. This specifies the approximate width of the table, which ranges from 30 to 72 in. A more complete specification is width of table × height under rail × length of table (e.g., 48 in. × 48 in. × 14 ft).

The width and height are usually, but not always, the same. Table length is often made to order and may be as short as 8 ft, or as long as 20 ft or more. The drive may be 15 HP on the smaller planers, and 100 HP or more on the larger models.

Mechanical and hydraulic power can be used for planers. Uniform cutting speed is attained throughout the cutting stroke. Acceleration and deceleration of the table takes place in a short distance of travel and does not influence the time to machine.

Double-Housing Planers Double-housing planers consist of a long heavy base on which the table reciprocates. The upright housing near the center on the side of the base supports the crossrail on which the tools are fed across the work. Figure 14.5 illustrates how the tools are supported both above and on the sides, and their adjustment for angle cuts. They are fed by power in either a vertical or a crosswise direction.

Open-Sided Planers Open-sided planers have the housing on one side only. The open side permits machining wider workpieces. Most planers have one flat and one double V-way, which allows for unequal bed and platen expansions. Adjustable dogs at the side of the bed control the stroke length of the platen. Planers are often converted to planer-mills, as explained in Chapter 25, under the heading "Special-Purpose Milling Machines."

14.3.1 Comparison of Shapers and Planers

Although both the planer and the shaper are able to machine flat surfaces, there is little overlapping in their application. They differ greatly in construction and in the method of operation. The planer is especially adapted to large work; the shaper can do only small work. On the planer the work is moved against a stationary tool; on the shaper the tool moves across the work, which is stationary. On the planer the tool is fed into the work; on the shaper the work is usually fed across the tool. The drive on the planer table is either by gears or by hydraulic means. The shaper ram also can be driven in this manner, but many times a quick-return link mechanism is used.

Most planers differ from shapers in that they approach more constant-velocity cutting speeds. Tools used in shaper and planer work are single point as used on a lathe, but are heavier in construction. The holder is designed to secure the tool bit near the centerline of the holder or the pivot point rather than at an angle as is customary with lathe toolholders.

Cutting tools for the planer operation are usually tipped with high-speed steel, cast alloy, or carbide inserts. High-speed steel or cast alloys are commonly used in heavy roughing cuts and carbides for secondary roughing and finishing.

Cutting angles for tools depend on the tool used and the workpiece material. They are similar to angles used on other single-point tools, but the end clearance does not exceed 4°. Cutting speeds are affected by the rigidity of the machine, how the work is held, tool, material, and the number of tools in operation. Worktables on planers and shapers are constructed with T-slots to hold and clamp parts that are to be machined.

14.4 REVIEW QUESTIONS AND PROBLEMS

1. Explain the difference between a shaper and a planer.
2. Describe the function and operation of the clapper box.
3. Describe the two types of shaper drives.
4. Explain speed and feed for mechanically driven shapers.
5. Explain speed and feed for hydraulically driven shapers.
6. How does the vertical shaper differ from the horizontal shaper in design and function?
7. What are the principal components of a planer?
8. Explain the two table drives for planers.
9. Describe and explain the special work-holding methods required on planer operations.
10. Compare the use of shapers and planers in manufacturing.

CHAPTER FIFTEEN

DRILLING OPERATIONS

CHAPTER OVERVIEW

15.1 INTRODUCTION

Drilling is the process most commonly associated with producing machined holes. Although many other processes contribute to the production of holes, including boring, **reaming**, **broaching**, and internal **grinding**, drilling accounts for the majority of holes produced in the machine shop. This is because drilling is a simple, quick, and economical method of hole production. The other methods are used principally for more accurate, smoother, larger holes. They are often used after a drill has already made the pilot hole.

Drilling is one of the most complex machining processes. The chief characteristic that distinguishes it from other machining operations is the combined cutting and extrusion of metal at the chisel edge in the center of the drill. Metal under the chisel edge is first extruded by the high thrust force caused by the feeding motion. Then it tends to shear under the action of a negative-rake-angle tool. Drilling of a single hole is shown in Figure 15.1, and high-production drilling of a plate component is shown in Figure 15.2.

FIGURE 15.1 Drilling accounts for the majority of holes produced in industry today. (Courtesy Valenite Inc.)

FIGURE 15.2 Holes can be drilled individually as shown in Figure 15.1, or many holes can be drilled at the same time as shown here. (Courtesy Sandvik Coromant Co.)

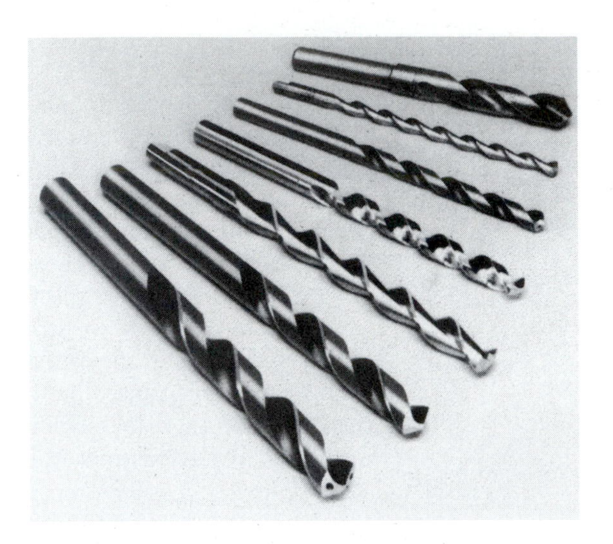

FIGURE 15.3 Many of the drills used in industry are shown here and described in this chapter. (Courtesy Cleveland Twist Drill Greenfield Industries)

The cutting action along the lips of the drill is not unlike that in other machining processes. Due to variable rake angle and inclination, however, there are differences in the cutting action at various radii on the cutting edges. This is complicated by the constraint of the whole chip on the chip flow at any single point along the lip. Still, the metal-removing action is true cutting, and the problems of variable geometry and constraint are present, but because it is such a small portion of the total drilling operation, it is not a distinguishing characteristic of the process. Many of the drills discussed in this chapter are shown in Figure 15.3.

The machine settings used in drilling reveal some important features of this hole-producing operation. Depth of cut, a fundamental dimension in other cutting processes, corresponds most closely to the drill radius. The undeformed chip width is equivalent to the length of the drill lip, which depends on the point angle as well as the drill size. For a given setup, the undeformed chip width is constant in drilling. The feed dimension specified for drilling is the feed per revolution of the spindle. A more fundamental quantity is the feed per lip. For the common two-flute drill, it is half the feed per revolution. The undeformed chip thickness differs from the feed per lip depending on the point angle.

The spindle speed is constant for any one operation, while the cutting speed varies all along the cutting edge. Cutting speed is normally computed for the outside diameter. At the center of the chisel edge the cutting speed is zero; at any point on the lip it is proportional to the radius of that point. This variation in cutting speed along the cutting edges is an important characteristic of drilling.

Once the drill engages the workpiece, the contact is continuous until the drill breaks through the bottom of the part or is withdrawn from the hole. In this respect, drilling resembles turning and is unlike milling. Continuous cutting means that steady forces and temperatures may be expected shortly after contact between the drill and the workpiece.

15.2 DRILL NOMENCLATURE

The most important type of drill is the twist drill. The important nomenclature listed here and illustrated in Figure 15.4 applies specifically to these tools.

Drill: A drill is an end-cutting tool for producing holes. It has one or more cutting edges, and flutes to allow fluids to enter and chips to be ejected. The drill is composed of a shank, body, and point.

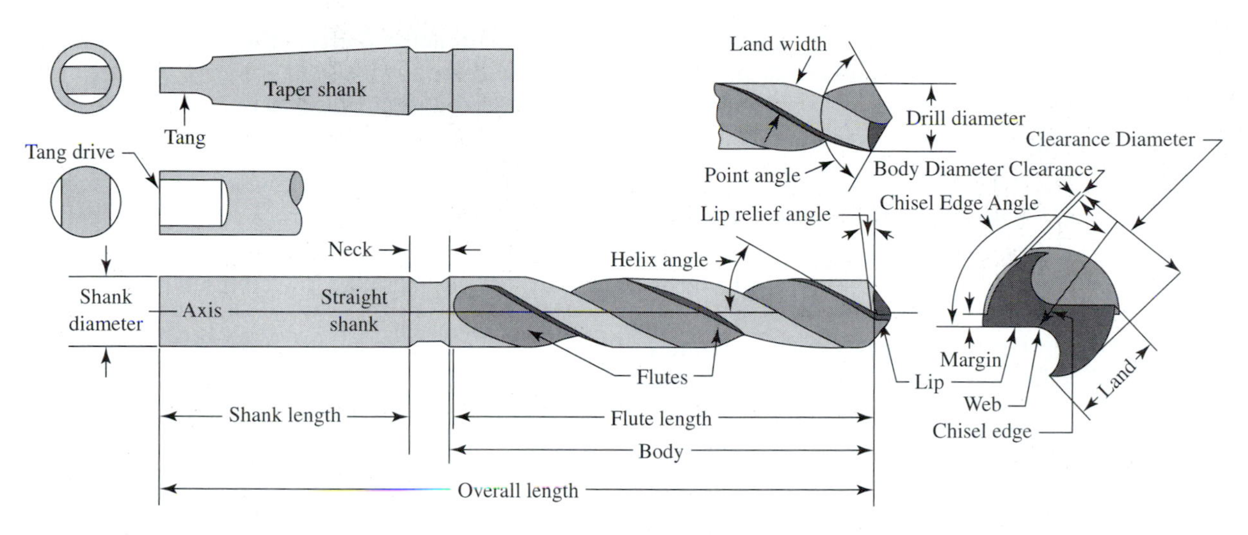

FIGURE 15.4 Nomenclature of a twist drill shown with taper and tang drives.

Shank: The shank is the part of the drill that is held and driven. It may be straight or tapered. Smaller-diameter drills normally have straight shanks. Larger drills have shanks ground with a taper and a tang to ensure accurate alignment and positive drive.

Tang: The tang is a flattened portion at the end of the shank that fits into a driving slot of the drill holder on the spindle of the machine.

Body: The body of the drill extends from the shank to the point and contains the flutes. During sharpening, it is the body of the drill that is partially ground away.

Point: The point is the cutting end of the drill.

Flutes: Flutes are grooves that are cut or formed in the body of the drill to allow fluids to reach the point and chips to reach the workpiece surface. Although straight flutes are used in some cases, they are normally **helical.**

Land: The land is the remainder of the outside of the drill body after the flutes are cut. The land is cut back somewhat from the outside drill diameter in order to provide clearance.

Margin: The margin is a short portion of the land not cut away for clearance. It preserves the full drill diameter.

Web: The web is the central portion of the drill body that connects the lands.

Chisel Edge: The edge ground on the tool point along the web is called the chisel edge. It connects the cutting lips.

Lips: The lips are the primary cutting edges of the drill. They extend from the chisel point to the periphery of the drill.

Axis: The axis of the drill is the centerline of the tool. It runs through the web and is perpendicular to the diameter.

Neck: Some drills are made with a relieved portion between the body and the shank. This is called the drill neck.

In addition to the preceding terms, which define the various parts of the drill, a number of terms apply to the dimensions of the drill, including the important drill angles. Among these terms are the following:

Length: Along with its outside diameter, the axial length of a drill is listed when the drill size is given. In addition, shank length, flute length, and neck length are often used (Fig. 15.4).

Body Diameter Clearance: The height of the step from the margin to the land is called the body diameter clearance.

Web Thickness: The web thickness is the smallest dimension across the web. It is measured at the point unless otherwise noted. Web thickness will often increase in going up the body away from the point, and it may have to be ground down during sharpening to reduce the size of the chisel edge. This process is called web thinning. Web thinning is shown later, in Figure 15.13.

Helix Angle: The angle that the leading edge of the land makes with the drill axis is called the helix angle. Drills with various **helix angles** are available for different operational requirements.

Point Angle: The included angle between the drill lips is called the point angle. It is varied for different workpiece materials.

Lip Relief Angle: Corresponding to the usual relief angles found on other tools is the lip relief angle. It is measured at the periphery.

Chisel Edge Angle: The chisel edge angle is the angle between the lip and the chisel edge, as seen from the end of the drill.

It is apparent from these partial lists of terms that many different drill geometries are possible.

15.3 CLASSES OF DRILLS

There are different classes of drills for different types of operations. Workpiece materials may also influence the class of drill used, but they usually determine the point geometry rather than the general type of drill best suited for the job. It has already been noted that the twist drill is the most important class. Within the general class of twist drills are a number of drill types made for different kinds of operations. Many of the special drills discussed here are shown in Figure 15.5.

High-Helix Drills: This drill has a high helix angle, which improves cutting efficiency but weakens the drill body. It is used for cutting softer metals and other low-strength materials.

Low-Helix Drills: A lower than normal helix angle is sometimes useful to prevent the tool from "running ahead" or "grabbing" when drilling brass and similar materials.

Heavy-Duty Drills: Drills subject to severe stresses can be made stronger by such methods as increasing the web thickness.

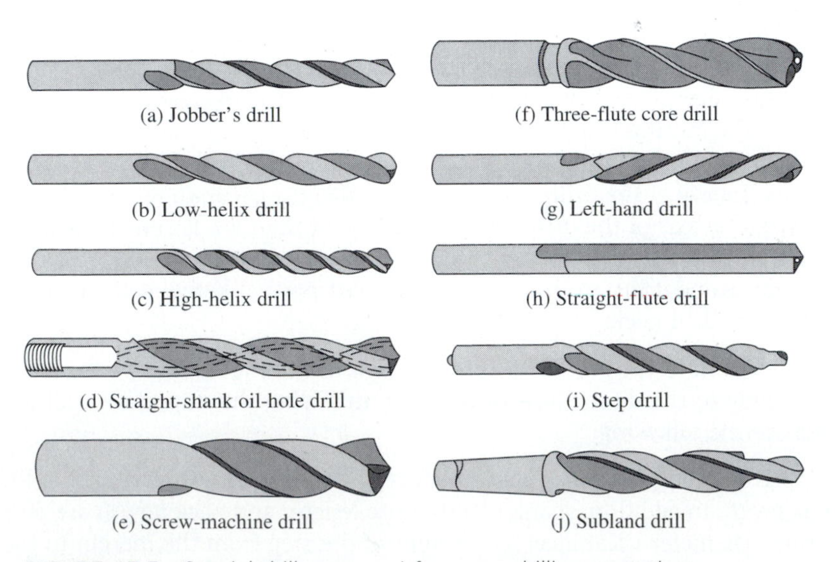

FIGURE 15.5 Special drills are used for some drilling operations.

Left-Hand Drills: Standard twist drills can be made as left-hand tools. These are used in multiple drill heads, where the head design is simplified by allowing the spindle to rotate in different directions.

Straight-Flute Drills: Straight-flute drills are an extreme case of low-helix drills. They are used for drilling brass and sheet metal.

Crankshaft Drills: Drills that are especially designed for crankshaft work have been found to be useful for machining deep holes in tough materials. They have a heavy web and helix angle that is somewhat higher than normal. The heavy web prompted the use of a specially notched chisel edge which has proven useful on other jobs as well. The crankshaft drill is an example of a special drill that has found wider application than originally anticipated and has become standard.

Extension Drills: The extension drill has a long, tempered shank to allow drilling in surfaces that are normally inaccessible.

Extra-Length Drills: For deep holes, the standard long drill may not suffice, and a longer-bodied drill is required.

Step Drill: Two or more diameters may be ground on a twist drill to produce a hole with stepped diameters.

Subland Drill: The subland or multicut drill does the same job as the step drill. It has separate lands running the full body length for each diameter, whereas the step drill uses one land. A subland drill looks like two drills twisted together.

Solid Carbide Drills: For drilling small holes in light alloys and nonmetallic materials, solid carbide rods may be ground to standard drill geometry. Light cuts without shock must be taken because carbide is quite brittle.

Carbide-Tipped Drills: Carbide tips may be used on twist drills to make the edges more wear resistant at higher speeds. Smaller helix angles and thicker webs are often used to improve the rigidity of these drills, which helps to preserve the carbide. Carbide-tipped drills are widely used for hard, abrasive nonmetallic materials such as masonry.

Oil-Hole Drills: Small holes through the lands, or small tubes in slots milled in the lands, can be used to force oil under pressure to the tool point. These drills are especially useful for drilling deep holes in tough materials.

Flat Drills: Flat bars may be ground with a conventional drill point at the end. This gives very large chip spaces, but no helix. Their major application is for drilling railroad track.

Three- and Four-Fluted Drills: There are drills with three or four flutes that resemble standard twist drills except that they have no chisel edge. They are used for enlarging holes that have been previously drilled or punched. These drills are used because they give better productivity, accuracy, and surface finish than a standard drill would provide on the same job.

Drill and Countersink: A combination drill and countersink is a useful tool for machining "center holes" on bars to be turned or ground between centers. The end of this tool resembles a standard drill. The countersink starts a short distance back on the body.

A double-ended combination drill and countersink, also called a **center drill**, is shown in Figure 15.6.

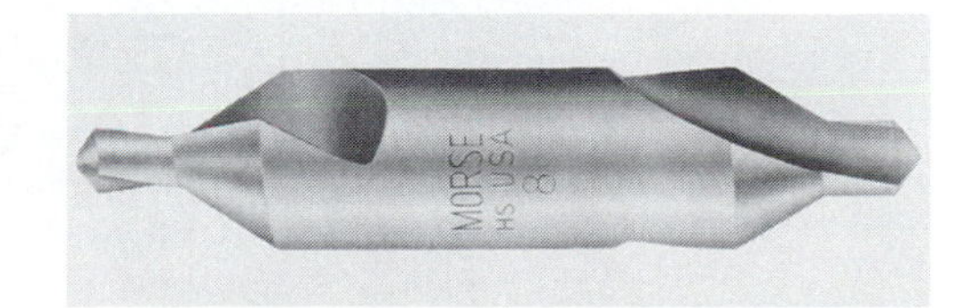

FIGURE 15.6 A double-ended combination drill and countersink, also called a center drill. (Courtesy Morse Cutting Tools)

15.4 RELATED DRILLING OPERATIONS

Several operations are related to drilling. In the following list, most of the operations follow drilling except for centering and spotfacing, which precede drilling. First a hole must be made by drilling, and then the hole is modified by one of the other operations. Some of these operations are described here and illustrated in Figure 15.7.

Reaming: A reamer is used to enlarge a previously drilled hole, to provide a higher tolerance and to improve the surface finish of the hole.

Tapping: A tap is used to provide **internal threads** on a previously drilled hole.

Reaming and tapping are more involved and complicated than counterboring, countersinking, centering, and spotfacing, and are therefore discussed in Chapter 16.

Counterboring: Counterboring produces a larger step in a hole to allow a bolt head to be seated below the part surface.

Countersinking: Countersinking is similar to counterboring except that the step is angular to allow flat-head screws to be seated below the surface.

Counterboring tools are shown in Figure 15.8(a), and a countersinking tool with two machined holes is shown in Figure 15.8(b).

Centering: Center drilling is used for accurately locating a hole to be drilled afterwards.

Spotfacing: Spotfacing is used to provide a flat machined surface on a part.

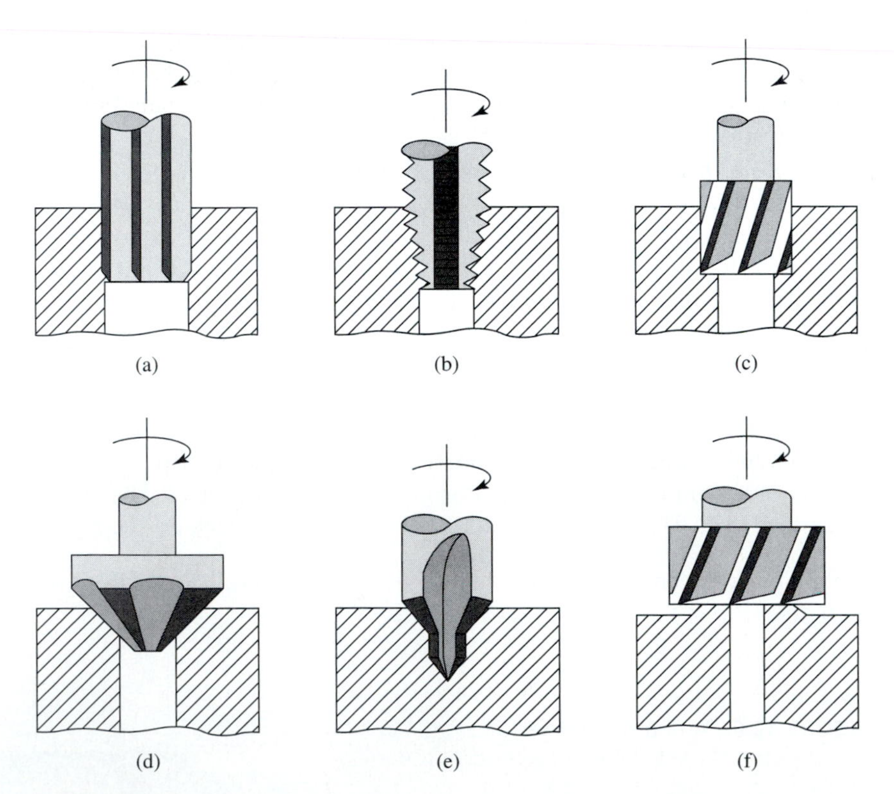

FIGURE 15.7 Related drilling operations: (a) reaming, (b) tapping, (c) counterboring, (d) countersinking, (e) centering, (f) spotfacing.

(a)

(b)

FIGURE 15.8 Counterboring tools (a) and a countersinking operation (b) are shown here. (Courtesy The Weldon Tool Co.)

15.5 OPERATING CONDITIONS

The varying conditions under which drills are used make it difficult to give set rules for speeds and feeds. Drill manufacturers and a variety of reference texts provide recommendations for proper speeds and feeds for drilling a variety of materials. General drilling speeds and feeds will be discussed here, and some examples will be given.

Drilling Speed Cutting speed may be referred to as the rate that a point on a circumference of a drill will travel in 1 minute. It is expressed in surface feet per minute (SFPM). Cutting speed is one of the most important factors that determines the life of a drill. If the cutting speed is too slow, the drill might chip or break. A cutting speed that is too fast rapidly dulls the cutting lips. Cutting speeds depend on the following seven variables:

- The type of material being drilled—the harder the material, the slower the cutting speed.
- The cutting tool material and diameter—the harder the cutting tool material, the faster it can machine the material, and the larger the drill, the slower the drill must revolve.
- The types and use of cutting fluids allow an increase in cutting speed.
- The rigidity of the drill press.
- The rigidity of the drill (the shorter the drill, the better).
- The rigidity of the work setup.
- The quality of the hole to be drilled.

Each variable should be considered prior to drilling a hole. Each variable is important, but the work material and its cutting speed are the most important factors. To calculate the revolutions per minute (RPM) rate of a drill, the diameter of the drill and the cutting speed of the material must be considered.

The formula normally used to calculate cutting speed is as follows:

$$\text{SFPM} = (\text{drill circumference}) \times (\text{RPM})$$

where SFPM = surface feet per minute, or the distance traveled by a point on the drill periphery in feet each minute
Drill circumference = the distance around the drill periphery in feet
RPM = revolutions per minute

In the case of a drill, the circumference is

$$\text{Drill circumference} = (\pi/12) \times d = .262 \times d$$

where Drill circumference = the distance around the drill periphery in feet
$\pi = 3.1416$
d = the drill diameter in inches

By substituting for the drill circumference, the cutting speed can now be written as

$$\text{SFPM} = .262 \times d \times \text{RPM}$$

This formula can be used to determine the cutting speed at the periphery of any rotating drill. For example, given a .25-in. drill, what is the cutting speed (SFPM) drilling cast iron at 5000 RPM?

$$\text{SFPM} = .262 \times d \times \text{RPM}$$
$$\text{SFPM} = .262 \times .25 \times 5000$$

$$\text{Answer} = 327.5 \text{ or } 327 \text{ SFPM}$$

RPM can be calculated as follows: Given a .75-in. drill, what is the RPM drilling low-carbon steel at 400 SFPM?

$$\text{RPM} = \frac{\text{SFPM}}{.262 \times d} = \frac{400}{.262 \times .75} = \frac{400}{.1965}$$

$$\text{Answer} = 2035.62 \text{ or } 2036 \text{ RPM}$$

Drilling Feed Once the cutting speed has been selected for a particular workpiece material and condition, the appropriate feedrate must be established. Drilling feedrates are selected to maximize productivity while maintaining chip control. Feed in drilling operations is expressed in inches per revolution, or IPR, which is the distance the drill moves in inches for each revolution of the drill. The feed may also be expressed as the distance traveled by the drill in a single minute, or IPM (inches per minute), which is the product of the RPM and IPR of the drill. It can be calculated as follows:

$$\text{IPM} = \text{IPR} \times \text{RPM}$$

where IPM = inches per minute
IPR = inches per revolution
RPM = revolutions per minute

For example, to maintain a .015 IPR feedrate on the .75-in. drill discussed earlier, what would the IPM feedrate be?

$$\text{IPM} = \text{IPR} \times \text{RPM}$$
$$\text{IPM} = .015 \times 2036$$

$$\text{Answer} = 30.54 \text{ or } 31 \text{ IPM}$$

The selection of drilling speed (SFPM) and drilling feed (IPR) for various materials to be machined often starts with recommendations in the form of application tables from manufacturers or by consulting reference books.

15.5.1 Twist Drill Wear

Drill wear starts as soon as cutting begins, and instead of progressing at a constant rate, the wear accelerates continuously. Wear starts at the sharp corners of the cutting edges and, at the same time, works its way along the cutting edges to the chisel edge and up the drill margins. As wear progresses, clearance is reduced. The resulting rubbing causes more heat, which in turn causes faster wear.

Wear lands behind the cutting edges are not the best indicators of wear, because they depend on the lip relief angle. The wear on the drill margins actually determines the degree of wear and is not nearly as obvious as wear lands. When the corners of the drill are rounded off, the drill has been damaged more than is readily apparent. Quite possibly the drill appeared to be working properly even while it was wearing. The margins could be worn in a taper as far back as an inch from the point. To restore the tool to new condition, the worn area must be removed. Because of the accelerating nature of wear, the number of holes per inch of drill can sometimes be doubled by reducing, by 25%, the number of holes drilled per grind.

15.5.2 Drill Point Grinding

It has been estimated that about 90% of drilling troubles are due to improper grinding of the drill point. Therefore, it is important that care be taken when resharpening drills. A good drill point will have both lips at the same angle to the axis of the drill, both lips the same length, the correct clearance angle, and the correct web thickness.

Lip Angle and Lip Length When the two cutting edges are ground, they should be equal in length and have the same angle with the axis of the drill as shown in Figure 15.9(a). Figure 15.9(b) shows two ground drill points.

For drilling hard or alloy steels, angle *C* (Fig. 15.9a) should be 135°. For soft materials and for general purposes, angle *C* should be 118°. For aluminum, angle *C* should be 90°.

If lips are not ground at the same angle with the axis, the drill will be subjected to an abnormal strain, because only one lip comes in contact with the work. This will result in unnecessary breakage and also cause the drill to dull quickly. A drill so sharpened will drill an oversized hole. When the point is ground with equal angles, but has lips of different lengths, a condition as shown in Figure 15.10(a) is produced.

A drill having cutting lips of different angles and of unequal lengths will be laboring under the severe conditions shown in Figure 15.10(b).

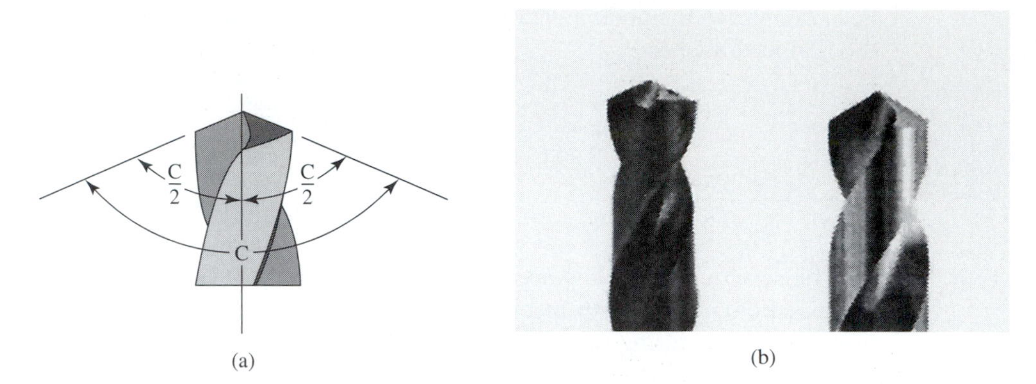

FIGURE 15.9 The included lip angle varies between 90 and 135 degrees (a); two drill points are shown in (b). (Courtesy Cleveland Twist Drill Greenfield Industries)

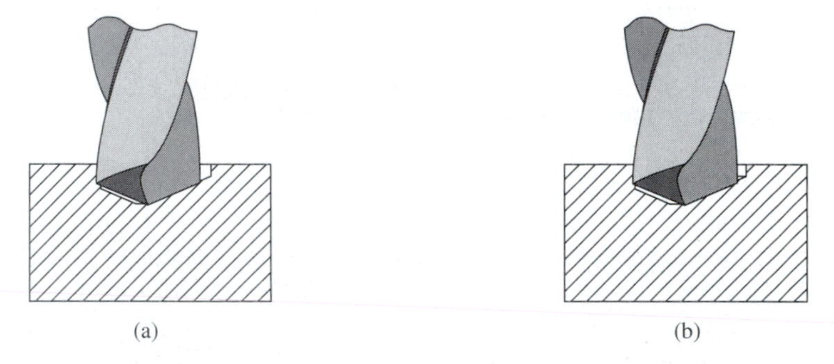

FIGURE 15.10 Drill with equal lip angle but unequal lip length (a), and drill with unequal lip angle and unequal lip length (b).

Lip Clearance Angle The clearance angle, or "backing-off" of the point, is the next important thing to consider. When drilling steel this angle A (Fig. 15.11a) should be from 6 to 9°. For soft cast iron and other soft materials, angle A may be increased to 12° (or even 15° in some cases).

This clearance angle should increase gradually as the center of the drill is approached. The amount of clearance at the center of the drill determines the chisel point angle B (Fig. 15.11b).

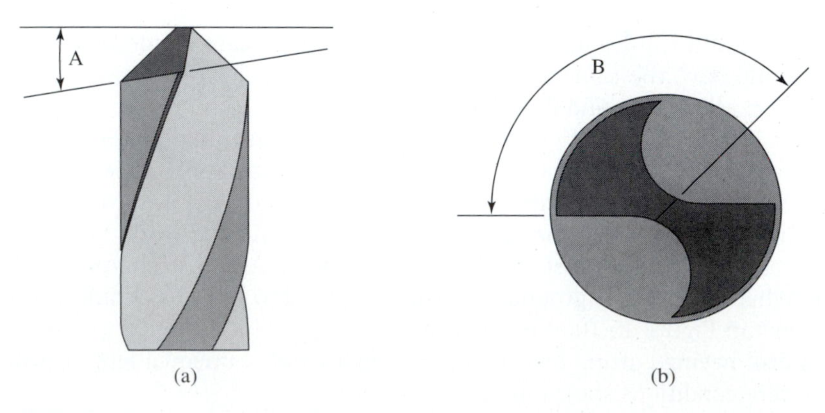

FIGURE 15.11 Drill lip clearance angle (a) and drill chisel point angle (b).

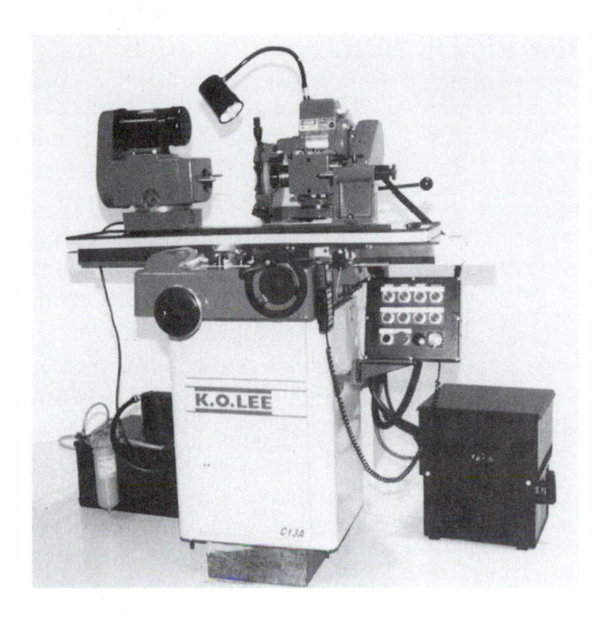

FIGURE 15.12 Tool and cutter grinders are used to properly sharpen drills and other cutting tools. (Courtesy K. O. Lee Co.)

The correct combination of clearance and chisel point angles should be as follows: When angle A is made to be 12° for soft materials, angle B should be made approximately 135°; when angle A is 6 to 9° for harder materials, angle B should be 115 to 125°.

While insufficient clearance at the center is the cause of drills splitting up the web, too much clearance at this point will cause the cutting edges to chip.

To maintain the necessary accuracy of point angles, lip lengths, lip clearance angle, and chisel edge angle, the use of machine point grinding is recommended. There are many commercial drill point grinders available today, which will make the accurate repointing of drills much easier. Tool and cutter grinders such as the one shown in Figure 15.12 are often used.

Twist Drill Web Thinning The tapered web drill is the most common type manufactured. The web thickness increases as this type of drill is resharpened. This requires an operation called web thinning to restore the tool's original web thickness. Without the web thinning process, more thrust would be required to drill, resulting in additional generated heat and reduced tool life. Figure 15.13 illustrates a standard drill before and after the web thinning process. Thinning is accomplished with a radiused wheel and should be done so the thinned section tapers gradually from the point. This prevents a blunt wedge from being formed that would be detrimental to chip flow. Thinning can be done by hand, but because point centrality is important, thinning by machine is recommended.

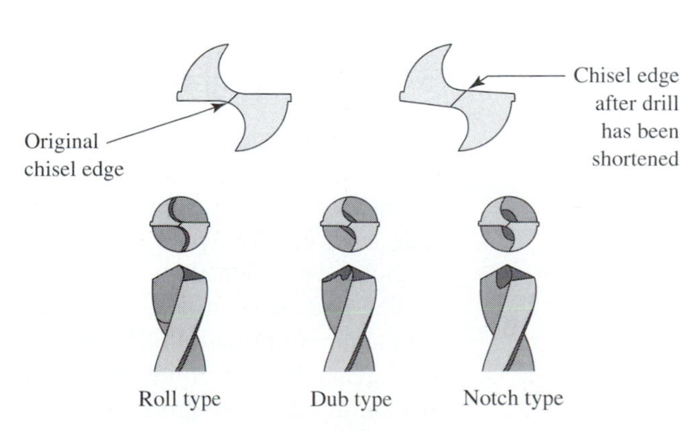

FIGURE 15.13 Web thinning restores proper web thickness after sharpening twist drills; three methods are shown.

15.6 SPADE DRILLS

The tool generally consists of a cutting blade secured in a fluted holder (Fig. 15.14). Spade drills can machine much larger holes (up to 15 in. in diameter) than twist drills. Spade drills usually are not available in diameters smaller than 0.75 in. The drilling depth capacity of spade drills, with length-to-diameter ratios over 100 to 1 possible, far exceeds that of twist drills. At the same time, because of their much greater feed capability, the penetration rates for spade drills exceed those of twist drills by 60 to 100%. However, hole finish generally suffers because of this. Compared to twist drills, spade drills are much more resistant to chatter under heavy feeds once they are fully engaged with the workpiece. Hole straightness is generally improved (with comparable size capability) by using a spade drill. However, these advantages can only be gained by using drilling machines of suitable capability and power.

The spade drill is also a very economical drill due to its diameter flexibility. A single holder will accommodate many blade diameters as shown in Figure 15.14. Therefore, when a diameter change is required, only the blade needs to be purchased, which is far less expensive than buying an entire drill.

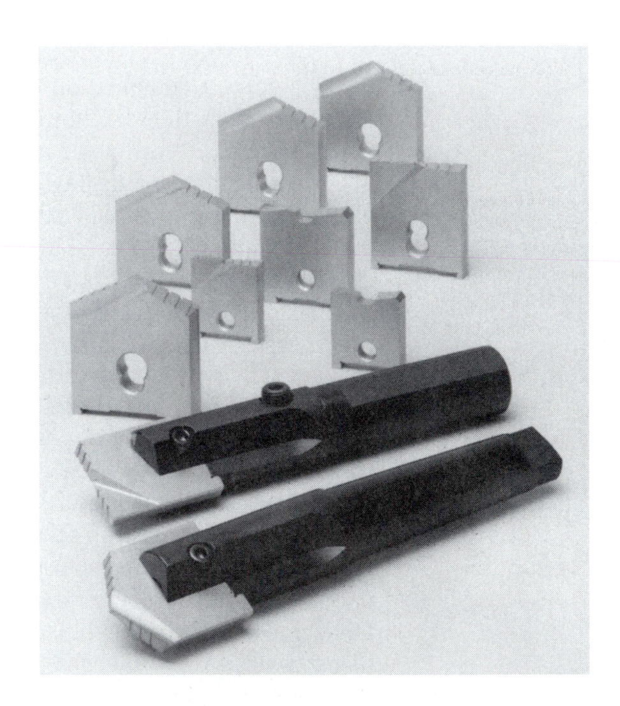

FIGURE 15.14 Spade drills with various cutting blades. (Courtesy Kennametal Inc.)

15.6.1 Spade Drill Blades

The design of spade drill blades varies with the manufacturer and the intended application. The most common design is shown in Figure 15.15. The locator length is ground to a precision dimension that, in conjunction with the ground thickness of the blade, precisely locates the blade in its holder. When the seating pads properly contact the holder, the holes in the blade and holder are aligned and the assembly can be secured with a screw.

The blade itself, as shown in Figure 15.15, possesses all the cutting geometry necessary. The point angle is normally 130° but may vary for special applications. In twist drill designs, the helix angle generally determines the cutting rake angle, but because spade drills have no helix, the rake surface must be ground into the blade at the cutting edge angle that produces the proper

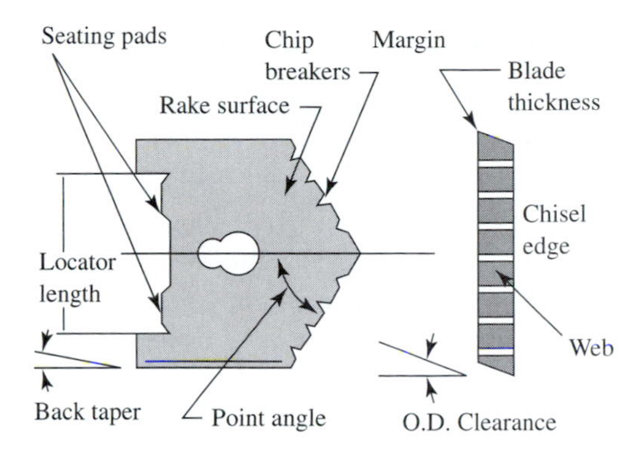

FIGURE 15.15 Spade drill cutting blade shows geometry specifications.

web thickness. The cutting edge clearance angle is a constant type of relief, generally 6 to 8°. After this clearance is ground, the chipbreakers are ground, about 0.025 in. deep, in the cutting edge.

These chipbreakers are necessary on spade drill blades and not optional as with twist drills. These notches make the chips narrow enough to flush around the holder. Depending on the feed rate, the grooves can also cause a rib to form in the chip. The rib stiffens the chip and causes it to fracture or break more easily, which results in shorter, more easily removed chips. Margins on the blade act as bearing surfaces once the tool is in a bushing or in the hole being drilled. The width of the margins will vary from $\frac{1}{16}$ to $\frac{3}{16}$ in., depending on the tool size. A slight back taper of 0.004 to 0.006 in. is normally provided, and outside diameter clearance angles are generally 10°.

15.6.2 Spade Drill Blade Holders

The blade holder makes up the major part of the spade drill. The blade holder is made of heat-treated alloy steel and is designed to hold a variety of blades in a certain size range as shown in Figure 15.14. Two straight chip channels or flutes are provided for chip ejection.

The holder shank designs are available in straight, Morse taper, and various other designs to fit the machine spindles. The holders are generally supplied with internal coolant passages to ensure that coolant reaches the cutting edges and to aid chip ejection.

When hole position is extremely critical and requires the use of a starting bushing, holders with guide strips are available. These strips are ground to fit closely with the starting bushing to support the tool until it is fully engaged in the workpiece. The strips may also be ground to just below the drill diameter to support the tool in the hole when the setup lacks rigidity.

15.6.3 Spade Drill Feeds and Speeds

The cutting speed for spade drills is generally 20% less than for twist drills. However, the spade drill feed capacity can be twice that of twist drills. The manufacturers of spade drills and other reference book publishers provide excellent recommendations for machining rates in a large variety of metals. These published rates should generally be observed. Spade drills work best under moderate speed and heavy feed. Feeding too lightly will result in either long, stringy chips or chips reduced almost to a powder. The drill cutting edges will chip and burn because of the absence of the thick, heat-absorbing, C-shaped chips. Chips can possibly jam and pack, which can break the tool or the workpiece. If the machine cannot supply the required thrust to maintain the proper feed without severe deflection, a change in tool or machine may be necessary.

15.7 INDEXABLE CARBIDE DRILLS

Indexable drilling has become so efficient and cost effective that in many cases it is less expensive to drill the hole than to cast or forge it. Basically, the indexable drill is a two-fluted, center-cutting tool with indexable carbide inserts. Indexable drills were introduced using square inserts (Fig. 15.16). Shown in Figure 15.17(a) are indexable drills using the more popular trigon insert (Fig. 15.17b). In most cases two inserts are used, but as size increases, more inserts are added, with as many as eight inserts in very large tools. Figure 15.18 shows six inserts being used.

Indexable drills have the problem of zero cutting speed at the center even though speeds can exceed 1000 SFPM at the outermost inserts. Because speed generally replaces feed to some degree, thrust forces are usually 25 to 30% of those required by conventional tools of the same size. Indexable drills have a shank, body, and multiedged point. The shank designs generally available are straight, tapered, and number 50 V-flange.

The bodies have two flutes, which are normally straight but may be helical. Because no margins are present to provide bearing support, the tools must rely on their inherent stiffness and on the balance in the cutting forces to maintain accurate hole size and straightness. Therefore, these tools are usually limited to length-to-diameter ratios of approximately 4 to 1.

The drill point is made of pocketed carbide inserts. These inserts are usually specially designed. The cutting rake can be negative, neutral, or positive, depending on holder and insert design. Coated and uncoated carbide grades are available for drilling a wide variety of work materials. Drills are sometimes combined with indexable or replaceable inserts to perform more than one operation, such as drilling, counterboring, and countersinking.

As shown in Figure 15.19(a) and (b), body-mounted insert tooling can perform multiple operations. More examples will be shown and discussed in Chapter 17, "Boring Operations."

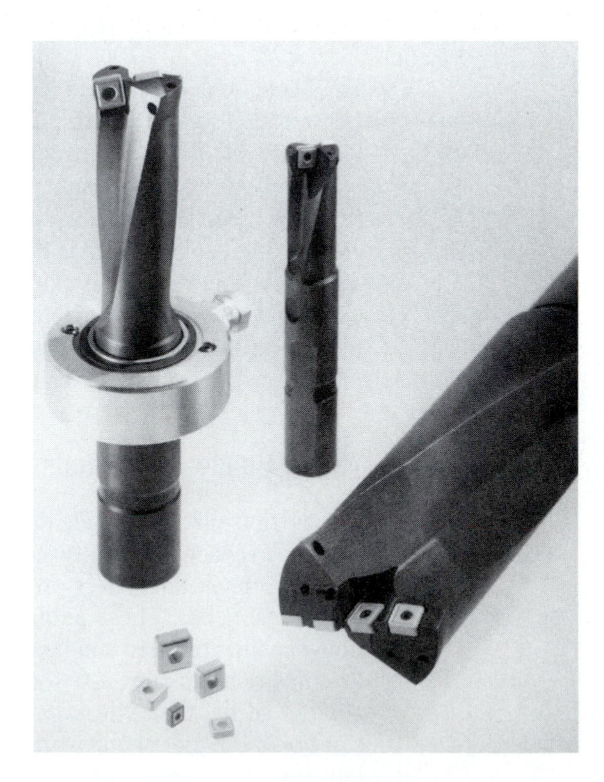

FIGURE 15.16 Indexable drills were introduced using square inserts; three sizes are shown here. (Courtesy Kennametal Inc.)

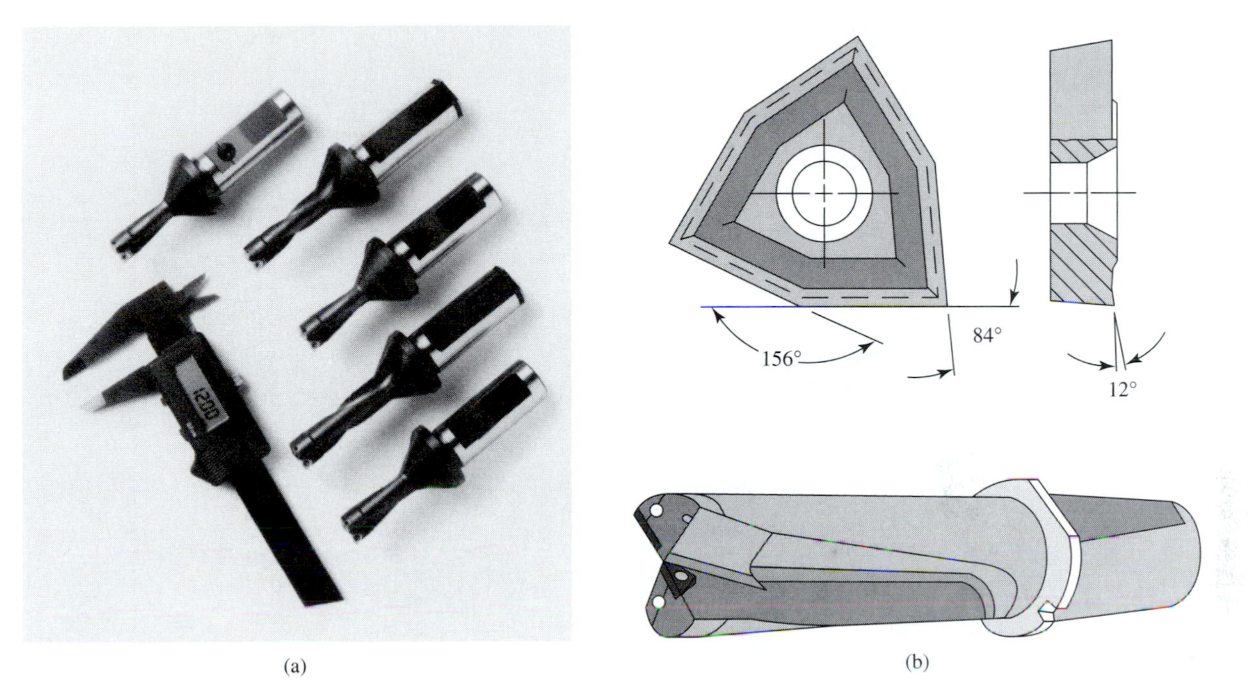

FIGURE 15.17 (a) Indexable drills using Trigon inserts. (b) A Trigon insert and a holder. (Courtesy Komet of America, Inc.)

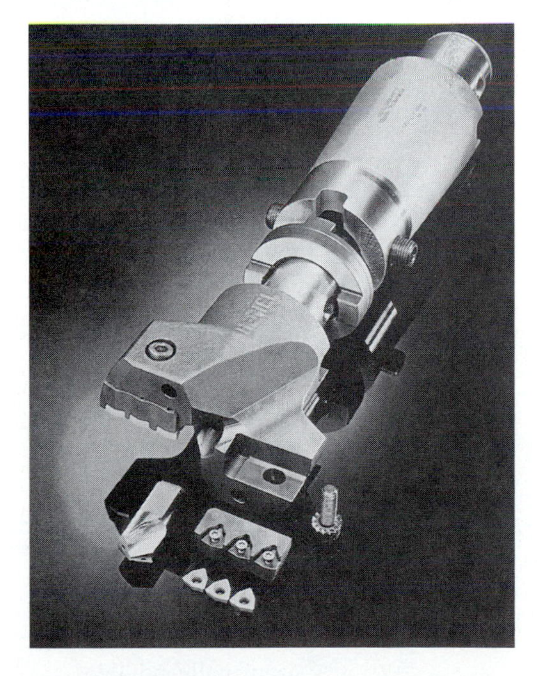

FIGURE 15.18 Indexable drill using six Trigon inserts for drilling large holes. (Courtesy Kennametal Inc.)

The overall geometry of the cutting edges is important to the performance of indexable drills. As mentioned earlier, there are no supporting margins to keep these tools on line, so the forces required to move the cutting edges through the work material must be balanced to minimize tool deflection, particularly on starting, and to maintain hole size.

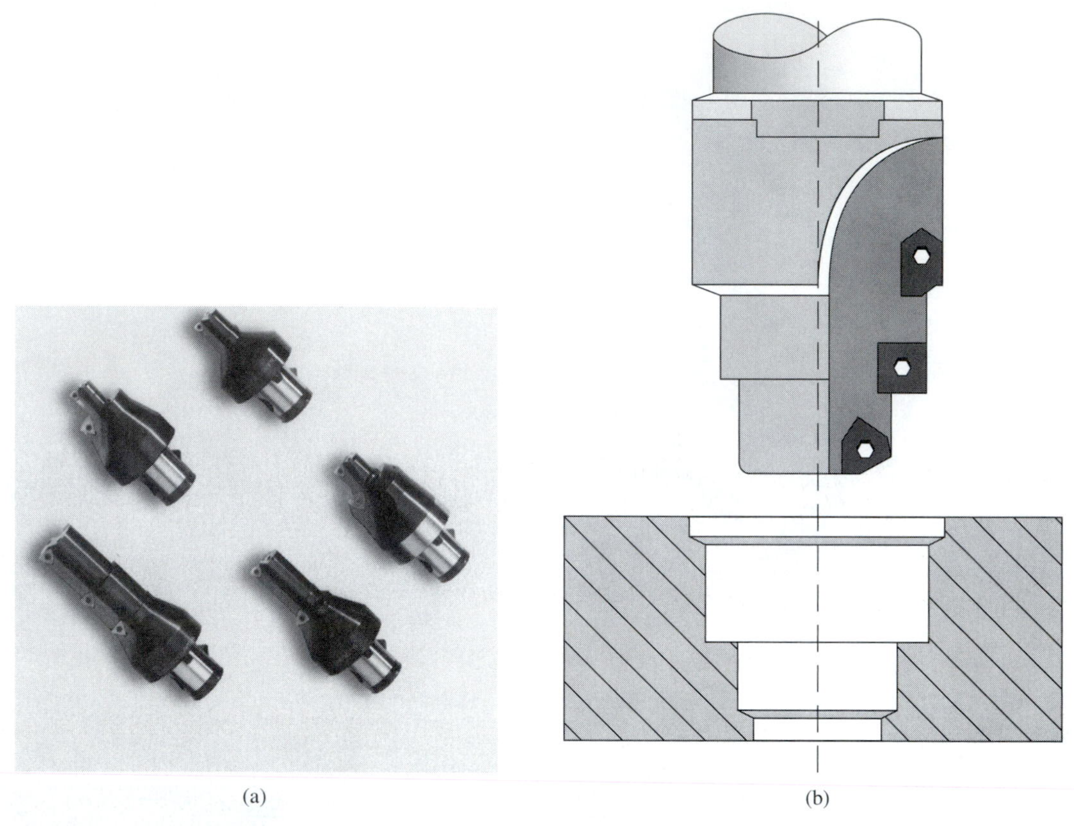

FIGURE 15.19 Body-mounted insert tooling can perform multiple operations. (Courtesy Komet of America, Inc.)

While they are principally designed for drilling, some indexable drills, as shown in Figure 15.20, can perform facing, and boring in lathe applications. How well these tools perform in these applications depends on their size, rigidity, and design.

15.7.1 Indexable Carbide Drill Operation

When used under the proper conditions, the performance of indexable drills is impressive. However, the manufacturer's recommendations must be carefully followed for successful applications.

Setup accuracy and rigidity is most important to tool life and performance. Chatter will destroy drilling inserts just as it destroys turning or milling inserts. If the inserts fail when the tool is rotating in the hole at high speed, the holder and workpiece will be damaged. Even if lack of rigidity has only a minor effect on tool life, hole size and finish will be poor. The machine must be powerful, rigid, and capable of high speed. **Radial drill presses** do not generally meet the rigidity requirements. Heavier lathes, horizontal boring mills, and NC machining centers are usually suitable.

When installing the tool in the machine, the same good practice followed for other drill types should be observed for indexable drills. The shanks must be clean and free from burrs to ensure good holding and to minimize run out. Run out in indexable drilling is dramatically amplified because of the high operating speeds and high penetration rates.

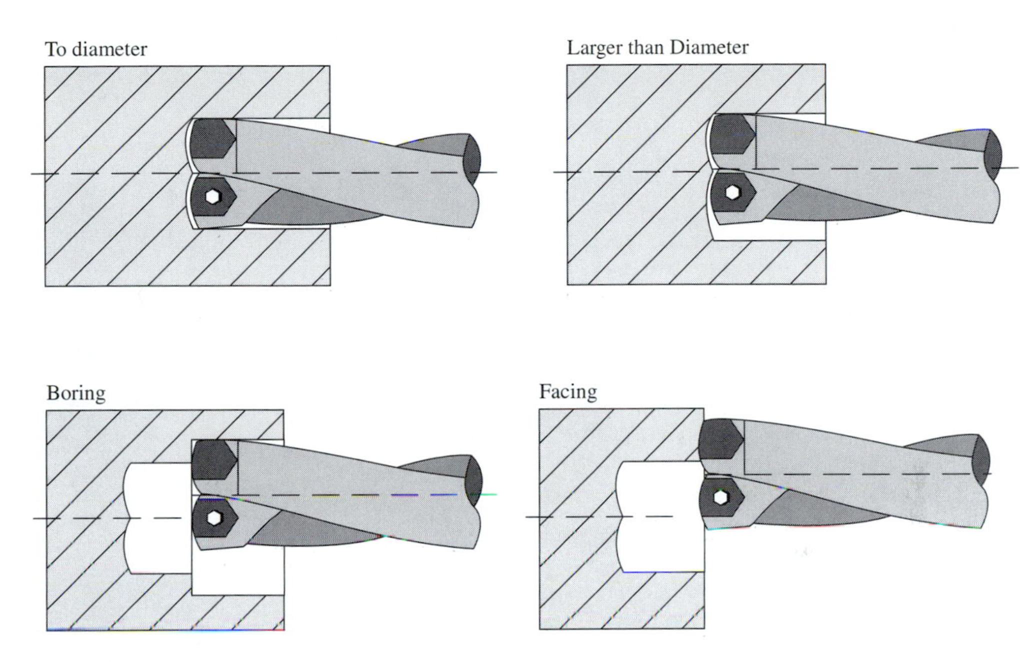

FIGURE 15.20 In addition to drilling, indexable drills can perform boring and facing operations.

When indexing the inserts is necessary, make sure that the pockets are clean and undamaged. A small speck of dirt, a chip, or a burr will cause stress in the carbide insert and result in a microscopic crack, which in turn will lead to early insert failure.

15.7.2 Indexable Drill Feeds and Speeds

Indexable drills are very sensitive to machining rates and work materials. The feed and speed ranges for various materials, as recommended by some manufacturers of these tools, can be very broad and vague, but can be used as starting points in determining exact feed and speed rates. Choosing the correct feed and speed rates, as well as selecting the proper insert style and grade, requires some experimentation. Chip formation is a critical factor and must be correct.

In general, soft low-carbon steel calls for high speed (650 SFPM or more) and low feed (.004/.006 IPR). Medium- and high-carbon steels, as well as cast iron, usually react best to lower speed and higher feed. The exact speed and feed settings must be consistent with machine and setup conditions, hole size and finish requirements, and chip formation for the particular job.

15.8 TYPES OF DRILL PRESSES

Figure 15.21 shows a schematic diagram of a standard vertical drill press as well as a schematic diagram of a three-axis computer numerical control (CNC) turret drilling machine. Described below are these and other types of drill presses and drilling operations such as sensitive and radial drills and gang and multispindle operations.

Base: The base is the main supporting member of the machine. It is a heavy gray cast iron or ductile iron casting that supports the column and the head.

Column: The column supports the table and the head of the drilling machine. The outer surface is machined to function as a precision way for aligning the spindle with the table.

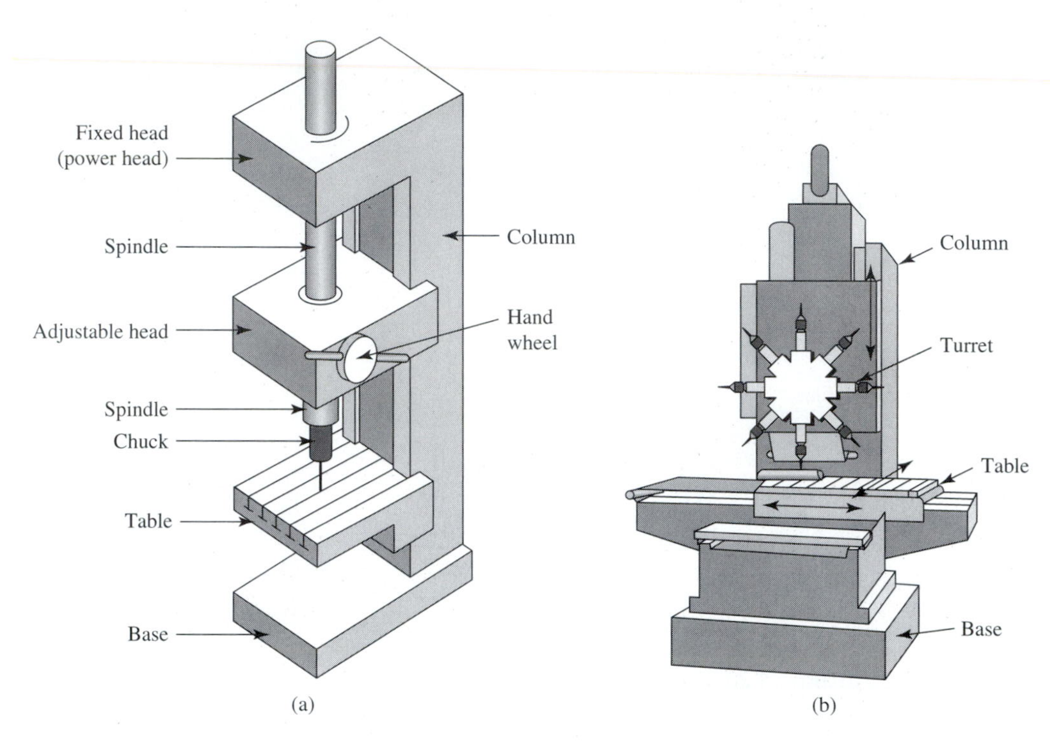

FIGURE 15.21 Schematic illustration of (a) vertical drill press, (b) CNC turret drilling machine.

Table: The table can be adjusted up or down the column to the proper heights. Most work-tables have slots and holes for mounting vises and other work-holding accessories.

Head: The head houses the spindle, motor, and feed mechanism. The motor drives the spindle and the spindle turns the drill.

Size: The size of a drill press is the distance from the center of the spindle to the column, that is, the radius of the largest circular part that can be drilled at its center. Speed ranges between 2000 and 4000 RPM, and feed is usually by hand.

15.8.1 Simple Drill Press

A simple drill press (Fig. 15.22) may be floor mounted as shown or have a shorter main post and be mounted on a bench. The motions of this machine are very simple. The table on a floor model can be raised or lowered and rotated around the machine column. The spindle rotates and can be raised and lowered, with a stroke of 4 to 8 in. Stops can be set to limit and regulate the depth.

15.8.2 Sensitive Drill Press

The term "sensitive" is used to indicate that the feed is hand operated and that the spindle and drilling head are counterbalanced so that the operator can "feel" the pressure needed for efficient cutting.

This drill press has the same motions as the previous one plus a telescoping screw for raising and lowering the table and a sliding "drill head." These two features allow easier handling of parts of varying heights.

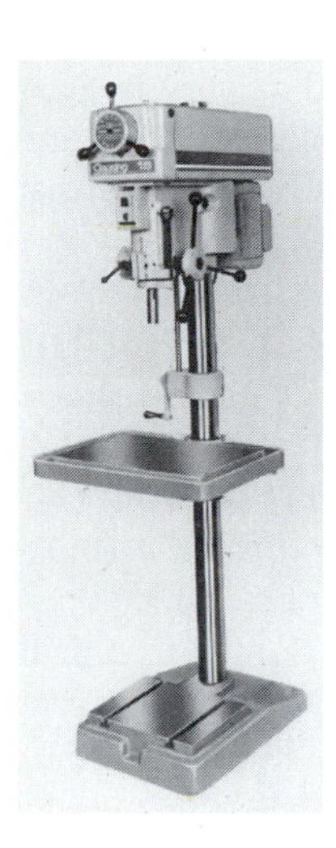

FIGURE 15.22 Simple 15-inch free-standing drill press. (Courtesy Clausing Industries, Inc.)

15.8.3 Radial Drill

For handling medium-sized to very large castings, weldments, or forgings, radial drills are ideal. Their size is specified by the length of the arm along which the spindle housing rides. This arm can be from 3 to 12 ft long. The column that holds the arm may be from 10 to 30 in. in diameter. A radial drill is shown in Figure 15.23.

FIGURE 15.23 Radial drills are used to machine large castings, weldments or forgings. (Courtesy Summit Machine Tool Manufacturing Corp.)

For very large work, the arm may be rotated 180° and the work placed on the shop floor. Speeds and feeds are dialed in by the machine operator and are the same as for other drill presses. Drilling is either hand or power feed.

15.8.4 Turret Drill

Turret drills (Fig. 15.21b) with either six or eight spindles enable the operator to use a wide variety of cutters and yet move the workpiece only a few inches, according to the hole spacing. The turret can be rotated (indexed) in either direction, then lowered, by hand or automatically, to make the cut.

Some turret drills have automatic, hydraulically controlled spindles. Speeds, feeds, and depths of cut can be preset for fast production. Figure 15.21(b) shows an automatic machine. These machines are also made with the entire operation computer controlled (CNC turret drill) so that the operator merely has to load and unload the parts.

15.8.5 Multispindle Drilling

This type of drilling can be done on drill presses by using special attachments. The spindle locations are adjustable, and the number of spindles may be from two to eight. Drills, reamers, countersinks, and so on can be used in the spindles. The RPM and feedrate of all spindles in one drill head are the same, and the horsepower needed is the sum of the power for all cutting tools used. In this type of machine, a large number of holes may be drilled at one time. Several different diameters of drills may be used at the same time.

15.8.6 Gang Drilling

An economical way to perform several different operations on one piece is by gang drilling as shown in Figure 15.24. This might include drilling two or more sizes of holes, reaming, tapping, and countersinking. The work is held in a vise or special **fixture** and is easily moved along the steel table from one spindle to the next. The drill presses usually run continuously, so the operator merely lowers each spindle to its preset stop to perform the required machining operation.

FIGURE 15.24 Gang drilling machines permit economical ways to perform several different operations. (Courtesy Clausing Industries, Inc.)

15.9 OPERATION SETUP

In drilling operations the three most common work-holding methods are

- Vises
- Angle plates
- Drill jigs

Vises Vises are widely used for holding work of regular size and shape, such as flat, square, and rectangular pieces. Parallels are generally used to support the work and protect the vise from being drilled. Figure 15.25 shows a typical vise. Vises should be clamped to the table of the drill press to prevent them from spinning during operation. Angular vises tilt the workpiece and provide a means of drilling a hole at an angle without tilting the table.

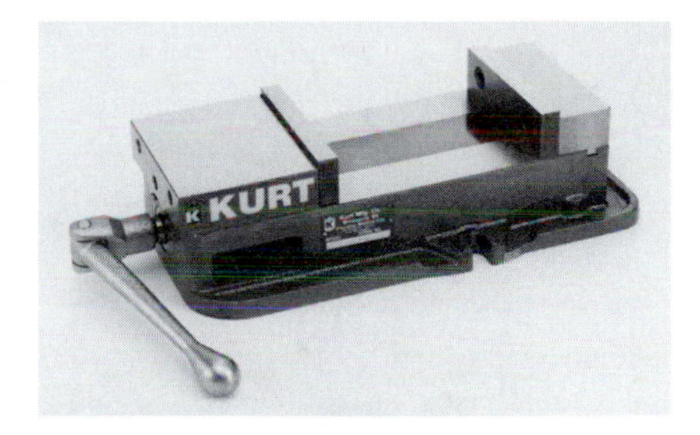

FIGURE 15.25 Vises should be clamped to the table of drill presses to prevent them from spinning. (Courtesy Kurt Manufacturing Co.)

Angle Plates An angle plate supports work on its edge. Angle plates accurately align the work perpendicular to the table surface, and they generally have holes and slots to permit clamping to the table and holding of the workpiece.

Drill Jigs A **drill jig** is a production tool used when a hole, or several holes, must be drilled in a large number of identical parts. Figure 15.26 shows a diagram of a typical drill jig. The drill jig has several functions. First, it is a work-holding device, clamping the work firmly. Second, it locates work in the correct position for drilling. The third function of the drill jig is to guide the drill straight into the work. This is accomplished by use of drill bushings.

15.9.1 Tool-Holding Devices

Some cutting tools used in drilling can be held directly in the spindle hole of the machine. Others must be held with a drill chuck, collet, sleeve, socket, or one of the many tool-holding devices shown in Figure 15.27.

Drill Chucks: Cutting tools with straight shanks are generally held in a drill chuck. The most common drill chuck uses a key to lock the cutting tool.

Sleeves: Cutting tools with tapered shanks are available in many different sizes. When a cutting tool that has a smaller taper than the spindle taper is used, a sleeve must be fitted to the shank of the cutting tool.

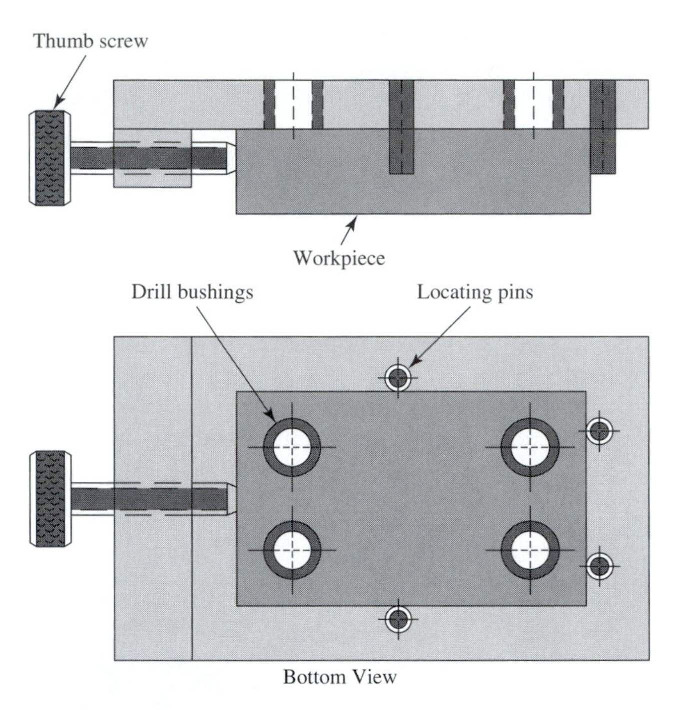

FIGURE 15.26 Drill jigs locate and clamp workpieces, and guide the drill through a drill bushing.

FIGURE 15.27 Various tool-holding devices such as chucks, collets, sleeves, and sockets are shown. (Courtesy Lyndex Corp.)

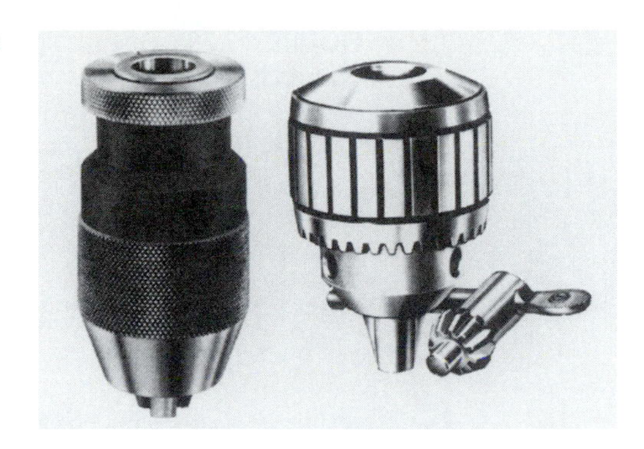

FIGURE 15.28 Key and keyless chucks are used to hold drills for hole making operations. (Courtesy Bridgeport Machine, Inc.)

Sockets: If the cutting tool has a tapered shank larger than the spindle taper, a socket is used to reduce it to the correct size. Drill chucks, both with key and keyless, are shown in Figure 15.28.

15.10 DEEP-HOLE DRILLING

The term "deep holes" originally referred to hole depths of over 5 times the diameter. Today, deep-hole drilling is a collective name for methods for the machining of both short and deep holes. Deep-hole drilling is the preferred method for drilling hole depths of more than 10 times the diameter, but because of the method's high metal removal capacity and precision, it is also competitive for small holes down to 2 times the diameter.

During drilling, it is important that the chips be broken and that they can be transported away without jamming and affecting the drilled surface. In deep-hole drilling, cutting fluid supply and chip transport have been provided for by the development of three different systems that permit trouble-free machining of hole depths of more than 100 times the diameter. The three systems are called the gun drilling system, the Ejector system (two-tube system), and the Single-Tube System (STS). Some of the tools used in deep-hole drilling are shown in Figure 15.29. The **gun drills** were manufactured by Hyper Tool, and the indexable tools were manufactured by Sandvik.

15.10.1 Gun Drilling Systems

The gun drill system uses the oldest principle for cutting fluid supply. The cutting fluid is supplied through a duct inside the drill and delivers coolant to the cutting edge, after which it removes the chips through a V-shaped chip flute along the outside of the drill. Due to the V-groove, the cross section of the tube occupies three fourths of its circumference. Figure 15.30 shows a gun drilling system and its component parts.

Gun Drills Gun drills belong to the pressurized coolant family of holemaking tools. They are outstanding for fast, precision machining regardless of hole depth. As a rule, a gun drill can hold hole straightness within .001 in. per inch (IPI) of penetration, even when the tool is reasonably dull. For most jobs a gun drill can be used to cut from 500 to 1000 in. in alloy steel before resharpening is necessary. In aluminum, it might be 15,000 in.; in cast iron it is usually around 2000 in. Figure 15.31(a) shows a gun drilling tool and Figure 15.31(b) shows the gun drilling process.

FIGURE 15.29 Deep-hole drilling tools; the gun drills were manufactured by Hyper Tool and the indexable tools were manufactured by Sandvik. (Courtesy TechniDrill Systems, Inc.)

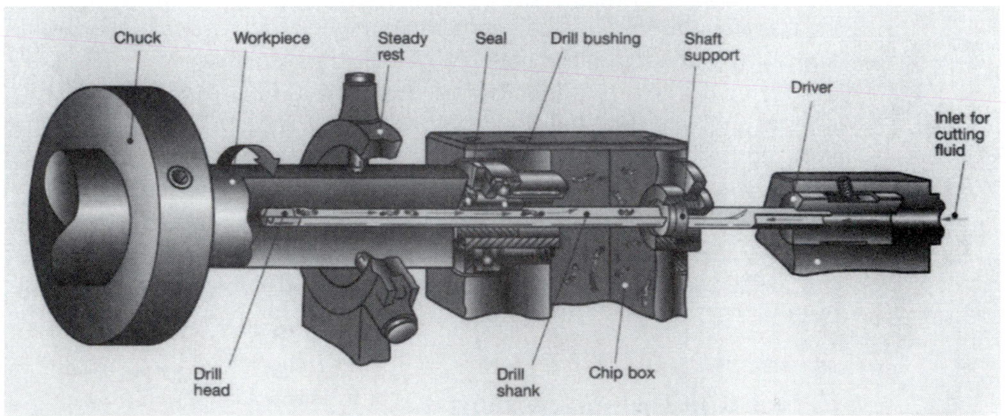

FIGURE 15.30 Schematic diagram of a gun drilling system with major components. (Courtesy Sandvik Coromant Co.)

Depending on the tool's diameter, a gun drill is seldom run at feedrates exceeding 0.003 in. per revolution (IPR). This is extremely light compared to twist drill feeds, which typically range from 0.005 IPR to 0.010 IPR. But gun drilling does use a relatively high speed compared to high-speed-steel (HSS) twist drilling. This accounts for the high metal removal rates associated with the process. In aluminum, speeds may be 600 surface feet per minute (SFPM), in steels from 400 SFPM to 450 SFPM.

Speeds and feeds for gun drilling are based on the workpiece material and shop floor conditions. Published charts only provide starting points. On-the-floor experimentation is critical to determine the right combination for maximum tool life.

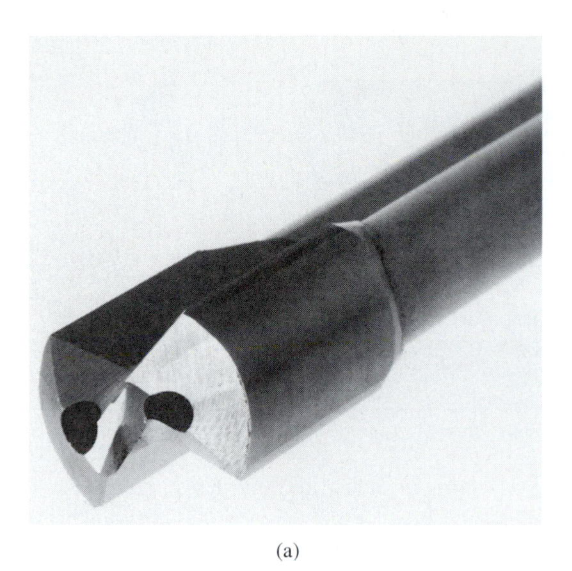

(a)

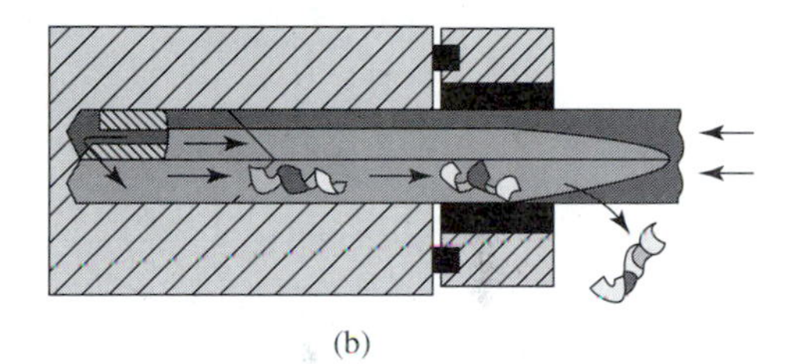

(b)

FIGURE 15.31 (a) Gun drilling head. (b) A drawing of a gun drilling process. (Courtesy Star Cutter Co.)

Gun Drill Body The body of a gun drill is typically constructed from 4120 aircraft quality steel tubing that is heat treated to between 35 to 40 Rc. A 4140 steel driver is brazed to one end of the tube, and a carbide tool tip is brazed to the other end.

There are two body styles for multiple-flute tools: milled and crimped. The former is a thick-walled tubular shaft with the flutes milled into the body. The latter is a thin-walled tubular shaft that has the flutes swaged into it. The number of flutes depends on the material being cut. When drilling in a material that breaks easily into small chips, such as cast iron, a two-flute tool is the choice. On the other hand, for a material such as D2 tool steel a single-flute design is preferred. In this case, chips tend to be stringy and a single-flute tool will minimize the chance of jamming as they are removed from the hole.

Figure 15.32 shows both a crimp-style gun drill body with two flutes produced by swaging and a conventional milled-style gun drill. The coolant holes in the crimped body have an irregular shape that permits carrying a much larger volume of coolant than comparable holes in a conventional equivalent-diameter tool body. Also, the flutes that are formed are much deeper than milled tools because allowance does not have to be made for wall thickness between flute and coolant hole. These deeper flutes improve the chip-removal efficiency of the tool.

Gun Drill Tip A conventional gun drill has a hole in its carbide tip underneath the cutting edge. Pressurized cutting fluid is pumped through the tool's body and out the hole (Fig. 15.31a). The fluid serves a threefold purpose: It lubricates and cools the cutting edge, it forces the chips back along the flute in the tool body, and it helps to stiffen the shank of the tool.

A new design has one hole in the top of the tool tip that effectively directs fluid at the cutting edge. The other hole, which is in the conventional location, helps to provide the chip ejection function. Total flow of cutting fluid is doubled with this two-hole arrangement. More important, the design produces chips about half the size of a conventional gun drill of the same diameter using the same speed and feedrate, so that packing of chips along the tool's shank is avoided in most materials.

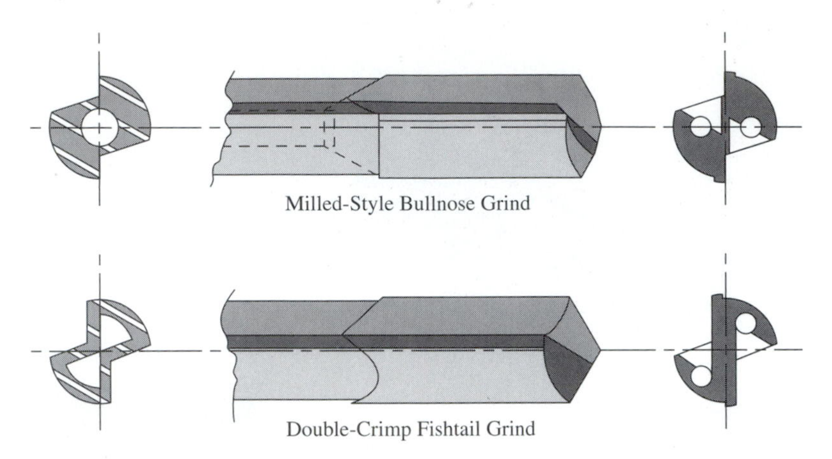

FIGURE 15.32 There are two body styles of multifluted gun drills: milled style and double-crimp style.

The most common tool tip material is C2 carbide, which is one of the harder grades and is generally associated with cast-iron applications. Because excessive tool wear is a major problem when cutting steel, a hard grade such as C2 is recommended, even though C5 carbide is labeled as the steel-machining grade in most text books. C5 carbide is a shock-resistant grade, not a wear-resistant grade, so that it is not as suitable for a gun drill tool tip. C3 carbide is harder than C2, and is used for certain applications; however, greater care must be taken when resharpening this material because it is easier to heat check the cutting edge.

Recently, coatings such as titanium nitride are being applied to gun drill tips to extend tool life. Physical Vapor Deposition (PVD) is the only practical process for depositing coatings on precision tools such as gun drills, but the results have not been encouraging. Unlike coating high-speed-steel tools, PVD coating of a carbide gun drill tip does not seem to form a good metallurgical bond. The coating wipes off during the metal-cutting process. Using Chemical Vapor Deposition (CVD) will form a metallurgical bond between the coating and carbide substrate, but the high heat required by the process distorts the tool. Hopefully these problems will be resolved in the near future.

15.10.2 The Ejector System

The ejector system consists of drill head, outer tube, inner tube, connector, collet, and sealing sleeve. The drill head is screwed to the drill tube by means of a four-start square thread. The inner tube is longer than the outer tube. The drill tube and the inner tube are attached to the connector by means of a collet and a sealing sleeve. The collet and sealing sleeve must be changed for different diameter ranges. Figure 15.33 shows the ejector system and its components.

15.10.3 The Single-Tube System (STS)

The single-tube system is based on external cutting fluid supply and internal chip transport. As a rule, the drill head is screwed onto the drill tube. The cutting fluid is supplied via the space between the drill tube and the drilled hole. The cutting fluid is then removed along with the chips through the drill tube. The velocity of the cutting fluid is so high that chip transport takes place through the tube without disturbances. Since chip evacuation is internal, no chip flute is

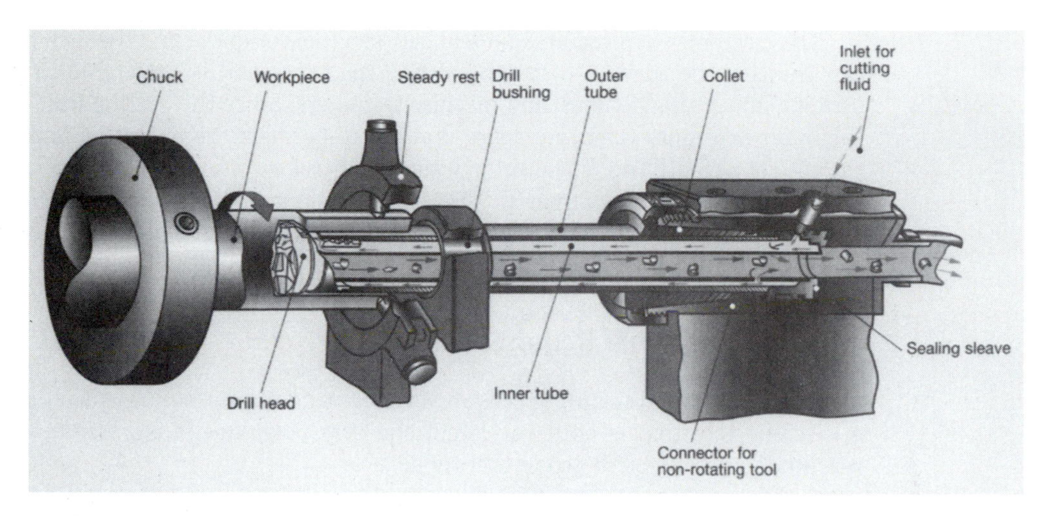

FIGURE 15.33 The ejector system and its major components. (Courtesy Sandvik Coromant Co.)

required in the shank, so tip cross section can be made completely round, which provides much higher rigidity than the gun drill system. Figure 15.34 shows the single-tube system and its components.

15.10.4 Comparison of STS and Ejector Systems

Both the single-tube system and the ejector system have wide ranges of application, but there are times when one system is preferable to the other. The STS is preferable in materials with poor chip formation properties such as stainless steel, low-carbon steel, and materials with an uneven structure, when chip-breaking problems exist. The STS is also more advantageous for long production runs, for uniform and extremely long workpieces, and for hole diameters greater than 7.875 in.

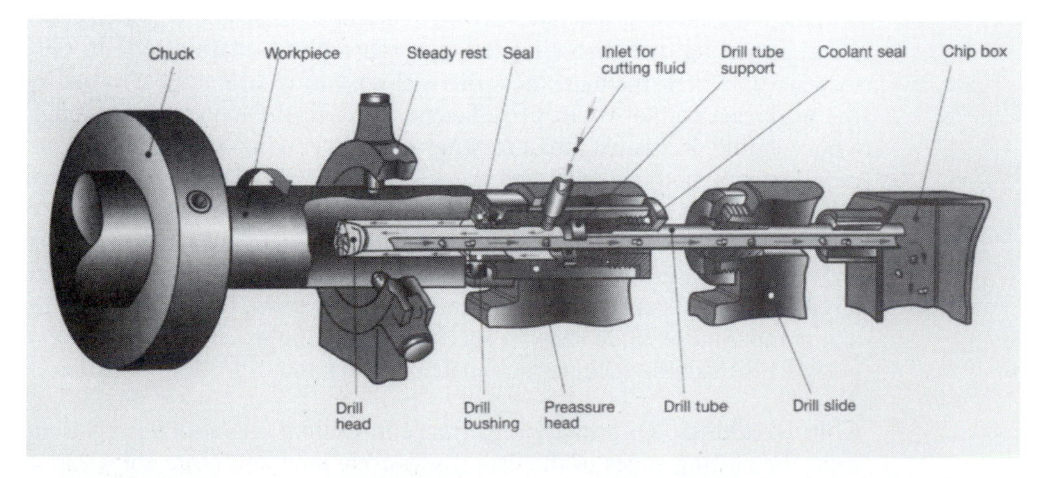

FIGURE 15.34 The single-tube system (STS) and its major components. (Courtesy Sandvik Coromant Co.)

The ejector system requires no seal between the workpiece and the drill bushing. The system can therefore be adapted easily to existing machines and is preferable in NC lathes, turning centers, universal machines, and machining centers. Since the cutting fluid is supplied between the outer and inner tubes, no space is required between the drill tube and the hole wall as in the case of STS drilling. The ejector system is therefore often used for machining in workpieces where sealing problems can arise. The ejector system can be used to advantage when it is possible to use a predrilled hole instead of a drill bushing for guidance, for example in machining centers.

15.10.5 Operational Requirements

Machining with high cutting speeds and high demands on surface finishes and tolerances requires a machine tool that is both very rigid and very powerful. It is possible to use conventional machines with sufficient power and rigidity.

Machine Requirements The high feed speeds that characterize deep-hole drilling impose high demands on available power. To achieve good precision, the machine must be rigid and the spindle bearings free of play. Good chip breaking often requires high feed, and the feed must be constant, otherwise the chip breaking may vary, leading to chip jamming. The best possible chip breaking can be obtained with infinitely adjustable feed.

It is important that the machine be equipped with safety devices to protect the machine, the tool, and the workpiece. The purpose of the safety device is to stop the machine automatically in the event of overloading. The machine spindle should not be able to start until the pressure of the cutting fluid has reached a preset minimum. The temperature and quantity of the cutting fluid should also reach a correct level before the machine starts.

Best are overload protections that are connected to the feed pressure. It is extremely important that the overload limits be set no more than 10 to 13% above the actual drill pressure for each drill diameter and feed. The feed will then be able to stop before the drill is damaged.

Machine Types The design of deep-hole drilling machines varies. The lengths of the machines are adapted to the special diameter ranges and lengths of the workpiece. A special very long machine is shown in Figure 15.35.

Deep-hole drilling machines are often designed to permit a choice between a rotating workpiece, a rotating tool, or both rotating workpiece and rotating tool. In the machining of asymmetric workpieces, the machine works with a rotating drill and a nonrotating workpiece because the workpiece cannot rotate at sufficient speed. In the machining of long, slender workpieces, a nonrotating drill is fed into a rotating workpiece. When the hole must meet high straightness requirements, both the drill and the workpiece rotate. The direction of rotation of the drill is then opposite to that of the workpiece.

The single-tube system is difficult to adapt to standard machines, whereas ejector drilling and, in some cases, gun drilling, can be done relatively simply in conventional machines. The largest extra costs are then for the cutting fluid system, chip-removal arrangement, filter tank, and pump. Figure 15.36 shows a special gun drilling machine to drill six camshafts simultaneously. This machine includes automated loading and unloading of parts.

Chip Breaking Of primary importance in drilling operations is transporting the chips away from the cutting edges of the drill. Excessively long and large chips can get stuck in the chip ducts. A suitable chip is as long as it is wide. However, the chips should not be broken harder than necessary, because chip breaking is power consuming, and the heat that is generated increases wear on the cutting edges. Chips with a length 3 to 4 times their width can be acceptable,

FIGURE 15.35 The length of a deep-hole drilling machine depends on the diameter and the length of the workpiece. (Courtesy Sandvik Coromant Co.)

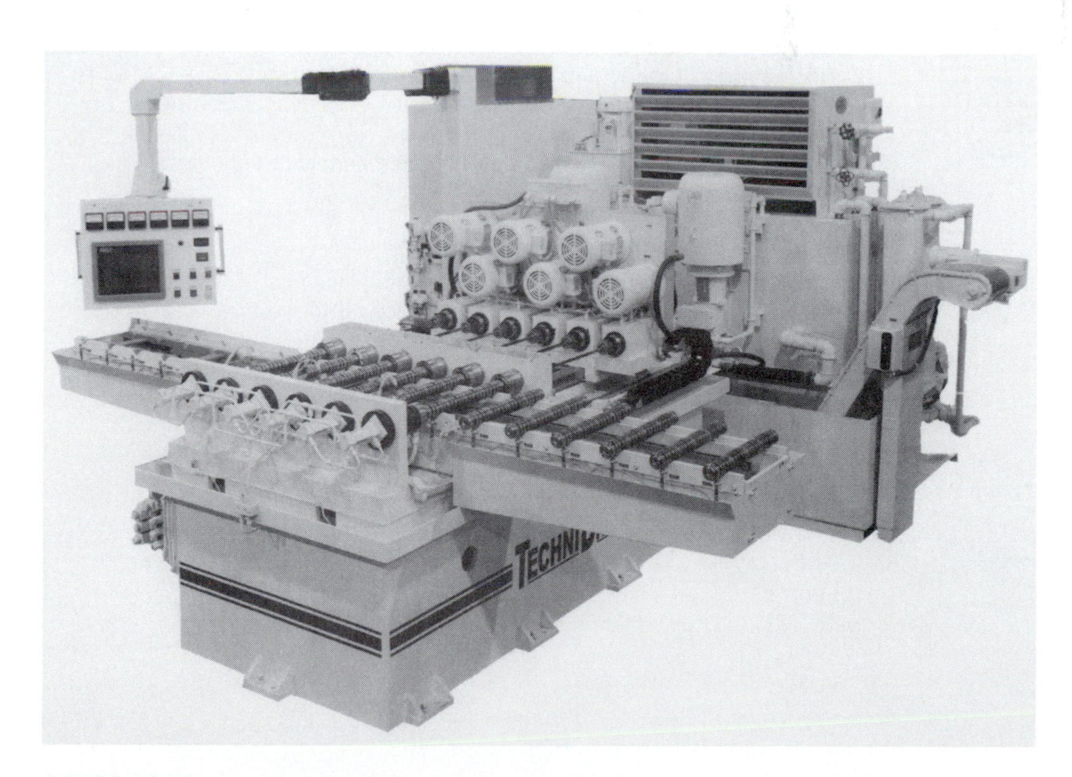

FIGURE 15.36 Special gun drilling machine is shown drilling six camshafts simultaneously. This machine includes automatic loading and unloading of parts. (Courtesy TechniDrill Systems, Inc.)

provided that they can pass through the chip duct and drill tube without difficulties. Chip formation is affected by the work material, chipbreaker geometry, cutting speed, feed, and choice of cutting fluid.

Coolant System The purposes of the coolant in a drilling system are

- Support and lubrication of the pads
- Improvement of the tool life
- Dissipation of heat
- Flushing of chips

The coolant system has to provide an adequate supply of clean coolant to the tool at the correct pressure and temperature.

15.11 REVIEW QUESTIONS AND PROBLEMS

1. How does drilling differ from other machining operations?
2. What makes drilling such a complicated machining process?
3. Explain web thickness and how it is affected by sharpening of the drill.
4. Why has the specially designed crankshaft drill become a standard drill?
5. Define SFPM for drilling.
6. Define IPR and IPM for drilling.
7. Explain the wear mechanisms of drills.
8. Discuss the importance of proper resharpening of drills.
9. Why is web thinning necessary as part of the drill resharpening process?
10. Why are spade drills often more economical than twist drills?
11. Discuss chip formation and breaking as it applies to spade drills.
12. Discuss some of the advantages of indexable carbide drills.
13. Give some reasons for the special care needed in using indexable drills.
14. Define and describe a "sensitive" drill press.
15. How does a radial drill operation differ from a normal drilling operation?
16. Explain the gang drilling process.
17. Explain the function of a drill jig.
18. How does a gun drill differ from a twist drill?
19. Discuss the advantages of a two-hole gun drill point.
20. Explain the differences between gun and ejector drill systems.

CHAPTER SIXTEEN

REAMING AND TAPPING

CHAPTER OVERVIEW

- Introduction
- Reaming
  - Reamer Nomenclature
  - Types of Reamers
  - Operating Conditions
  - Reaming Operations
- Tapping
  - Tap Nomenclature
  - Types of Taps
  - Operating Options
  - Tapping Operations
- Review Questions and Problems

16.1 INTRODUCTION

Twist drills do not make accurately sized or good finish holes; a reamer of some type is often used to cut the final size and finish. A reamer will not make the original hole; it will only enlarge a previously drilled or bored hole. It will cut to within +.0005 in. of tool size and give finishes to 32 μin.

Reamers are usually made of HSS, although solid carbide and carbide-tipped reamers are made in many sizes and styles. Regular chucking reamers are made in number and letter sizes, in fractional inch sizes, and in millimeter sizes. They can be purchased ground to any desired diameter.

Screw threads are used for a variety of purposes and applications in the machine tool industry. They are used to hold or fasten parts together (screws, bolts, and nuts), and to transmit motion (the lead screw moves the carriage on an engine lathe). Screw threads are also used to control or provide accurate movement (the spindle on a micrometer), and to provide a mechanical advantage (a screw jack raises heavy loads).

When defining a screw thread, one must consider separate definitions for an external thread (screw or bolt) and an internal thread (nut). An external thread is a cylindrical piece of material that has a uniform helical groove cut or formed around it. An internal thread is defined as a piece of material that has a helical groove around the interior of a cylindrical hole. This chapter will discuss internal threads and tapping, the operation that produces such threads.

16.2 REAMING

Reaming has been defined as a machining process that uses a multiedged fluted cutting tool to smooth, enlarge, or accurately size an existing hole. Reaming is performed using the same types of machines as drilling.

A reamer is a rotary cutting tool with one or more cutting elements, used for enlarging to size and contour a previously formed hole. Its principal support during the cutting action is obtained from the workpiece. A typical reaming operation is shown in Figure 16.1.

16.2.1 Reamer Nomenclature

The basic construction and nomenclature of reamers is shown in Figure 16.2. This shows the most frequently used style for holes up to 1 in., called a chucking reamer.

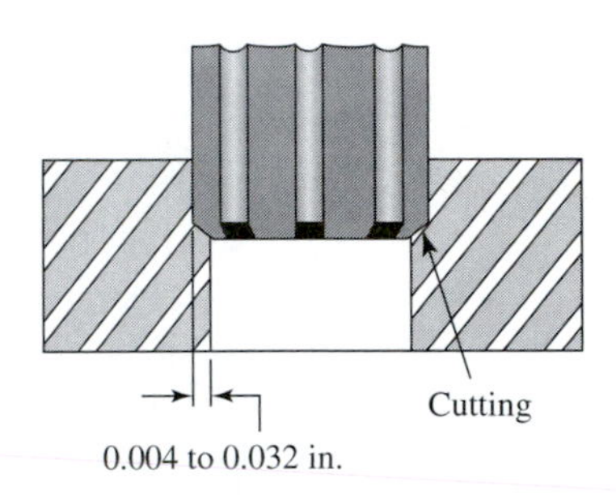

FIGURE 16.1 A typical reaming operation removes 0.004 to 0.032 in. of stock.

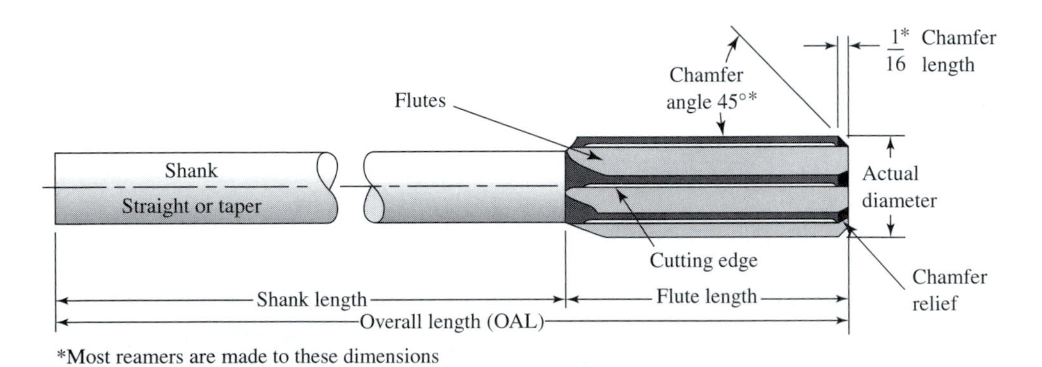

FIGURE 16.2 Construction and nomenclature of a straight-fluted machining reamer.

Solid reamers do almost all their cutting with the 45° chamfered front end. The flutes guide the reamer and slightly improve the finish. Therefore, reamers should not be used for heavy stock removal.

Axis: The axis is the imaginary straight line that forms the longitudinal centerline of a reamer, usually established by rotating the reamer between centers.

Back Taper: The back taper is a slight decrease in diameter, from front to back in the flute length of reamers.

Body: The body is (1) the fluted full diameter portion of a reamer, inclusive of the chamfer, starting taper, and bevel, or (2) the principal supporting member for a set of reamer blades, usually including the shank.

Chamfer: The chamfer is the angular cutting portion at the entering end of a reamer.

Chamfer Length: The chamfer length is the length of the chamfer measured parallel to the axis at the cutting edge.

Chamfer Relief Angle: The chamfer relief angle is the axial relief angle at the outer corner of the chamfer. It is measured by projection into a plane tangent to the periphery at the outer corner of the chamfer.

Clearance: Clearance is the space created by the relief behind the cutting edge or margin of a reamer.

Cutting Edge: The cutting edge is the leading edge of the land in the direction of rotation for cutting.

Flutes: The flutes are longitudinal channels formed in the body of the reamer to provide cutting edges, permit passage of chips, and allow cutting fluid to reach the cutting edges.

Flute Length: Flute length is the length of the flutes not including the cutter sweep.

Land: The land is the section of the reamer between adjacent flutes.

Margin: The margin is the unrelieved part of the periphery of the land adjacent to the cutting edge.

Neck: The neck is a section of reduced diameter connecting shank to body or connecting other portions of the reamer.

Overall Length: The overall length is the extreme length of the complete reamer from end to end, but not including external centers or expansion screws.

Shank: The shank is the portion of the reamer by which it is held and driven.

Straight Shank: A straight shank is a cylindrical shank.

Taper Shank: A taper shank is a shank made to fit a specified (conical) taper socket.

16.2.2 Types of Reamers

Reamers are made with three shapes of flutes, and all are standard.

Straight-Flute: Straight-flute reamers are satisfactory for most work and the least expensive, but should not be used if a keyway or other interruption is in the hole.

Right-Hand Spiral: Right-hand **spiral** fluted reamers give freer cutting action and tend to lift the chips out of the hole. They should not be used on copper or soft aluminum because these reamers tend to pull down into the hole.

Left-Hand Spiral: Left-hand spiral fluted reamers require slightly more pressure to feed but give a smooth cut and can be used on soft, gummy materials, because they tend to be pushed out of the hole as they advance. It is not wise to use these in **blind holes,** because they push the chips down into the hole.

All reamers are used to produce smooth and accurate holes. Some are turned by hand; others use machine power. The method used to identify left-hand and right-hand reamers is shown in Figure 16.3.

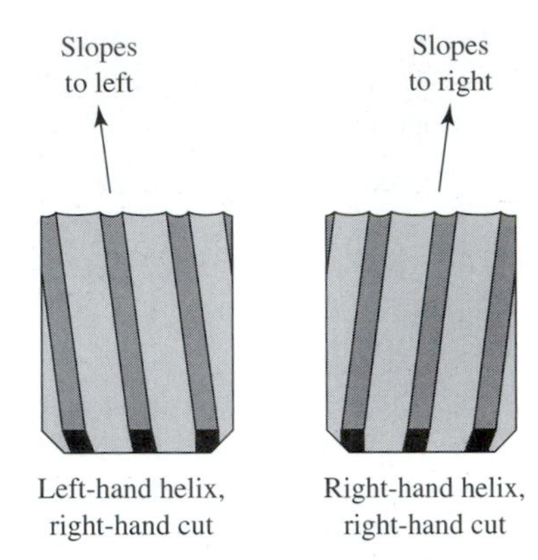

FIGURE 16.3 Method of identifying left-hand and right-hand reamers.

Machine Reamers Machine reamers are used on both drilling machines and lathes for roughing and finishing operations. Machine reamers are available with tapered or straight shanks, and with straight or helical flutes. Tapered-shank reamers (see Fig. 16.4) fit directly into the spindle, and the straight-shank reamer, generally called the chucking reamer, fits into a drill chuck.

Rose Reamers. Rose reamers are machine reamers that cut only on a 45° chamfer (bevel) located on the end. The body of the rose reamer tapers slightly (about 0.001 in. per inch of length) to prevent binding during operation. This reamer does not cut a smooth hole and is generally used to bring a hole to a few thousandths undersize. Because the rose reamer machines a hole 0.001 to 0.005 in. under a nominal size, a hand reamer is used to finish the hole to size. All hand reamers have a square shank and cannot be used and operated with machine power.

Fluted Reamers. Fluted reamers are machine reamers used to finish drilled holes. This type of reamer removes smaller portions of metal than the rose reamer. Fluted reamers have more cutting edges than rose reamers and therefore cut a smoother hole. Fluted reamers cut on the chamfered end as well as the sides. They are also available in solid carbide or have carbide inserts for cutting teeth.

Shell Reamers. Shell reamers (Fig. 16.5) are made in two parts: the reamer head and the arbor. In use, the reamer head is mounted on the arbor. The reamer head is available with either a rose or flute type, with straight or helical flutes. The arbor is available with either straight or tapered shank. The shell reamer is considered to be economical because only the reamer is replaced when it becomes worn or damaged.

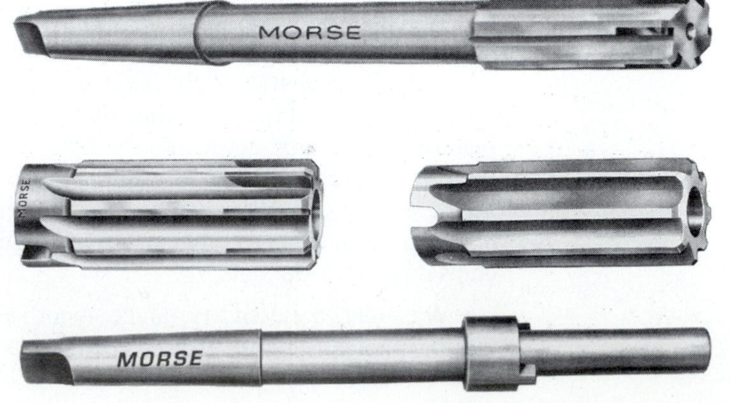

FIGURE 16.4 Carbide-tipped straight-fluted tapered-shank reamer. (Courtesy Morse Cutting Tools)

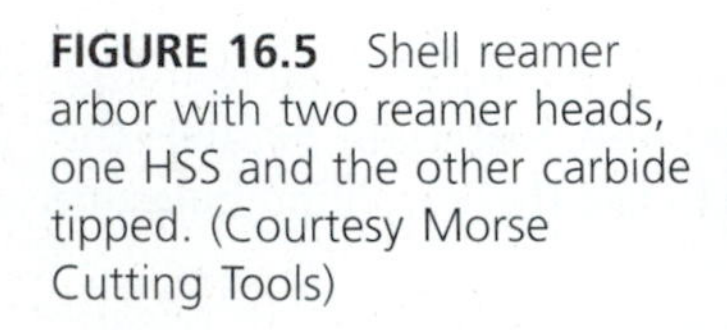

FIGURE 16.5 Shell reamer arbor with two reamer heads, one HSS and the other carbide tipped. (Courtesy Morse Cutting Tools)

FIGURE 16.6 Left-hand-helix hand reamer; square-shanked hand reamers cannot be power driven. (Courtesy Cleveland Twist Drill Greenfield Industries)

Hand Reamers Hand reamers are finishing reamers distinguished by the square on their shanks (Fig. 16.6). They are turned by hand with a tap wrench that fits over this square (Fig. 16.7). This type of reamer cuts only on the outer cutting edges. The end of the hand reamer is tapered slightly to permit easy alignment in the drilled hole. The length of taper is usually equal to the reamer's diameter.

Hand reamers must never be turned by machine power, and must be started true and straight. They should never remove more than 0.001 to 0.005 in. of material. Hand reamers are available from 1/8 to over 2 in. in diameter and are generally made of carbon steel or high-speed steel.

Taper Hand Reamers. Taper hand reamers are hand reamers made to ream all standard size tapers. They are made for both roughing and finishing tapered holes. Similar to the straight hand reamer, this taper should be used carefully, and never with machine power.

Adjustable Reamers (Fig. 16.8a). Adjustable reamers are used to produce any size hole within the range of the reamer. Their size is adjusted by sliding the cutting blades to and from the shank. These blades are moved by the two adjusting nuts located at each end of the blades. Adjustable hand reamers are available in sizes from $\frac{1}{4}$ to over 3-in. in diameter. Each reamer has approximately $\frac{1}{64}$-in. adjustment above and below its nominal diameter.

Expansion Hand Reamers (Fig. 16.8b). Expansion hand reamers are like the adjustable reamers, but have a limited range of approximately 0.010 in. adjustment. Expansion reamers have an adjusting screw at the end of the reamer. When turned, this adjusting screw forces a tapered plug inside the body of the reamer, expanding its diameter. Expansion reamers are also available as machine reamers.

Care of Reamers. Because reamers are precision finishing tools, they should be used with care:

- Reamers should be stored in separate containers or spaced in the tooling cabinet to prevent damage to the cutting edges.
- Cutting fluids must always be used during reaming operations, except with cast iron.

FIGURE 16.7 Tap wrenches are also used to hold hand reamers to finish drilled holes. (Courtesy Cleveland Twist Drill Greenfield Industries)

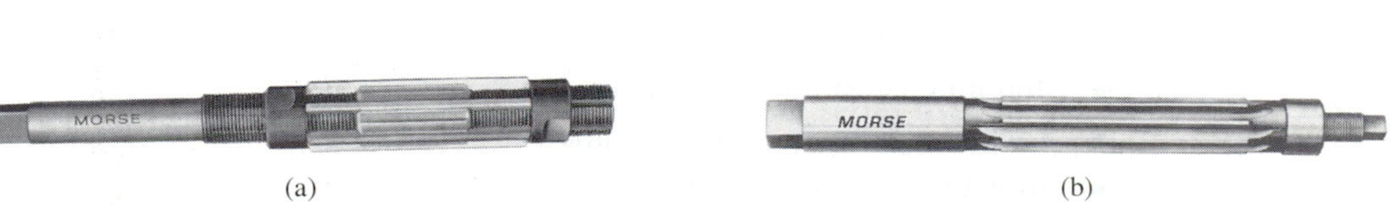

FIGURE 16.8 (a) Adjustable hand reamer. (b) A square-shanked expansion reamer. (Courtesy Morse Cutting Tools)

FIGURE 16.9 A typical automated tapping operation with self-reversing unit. (Courtesy Tapmatic Corp.)

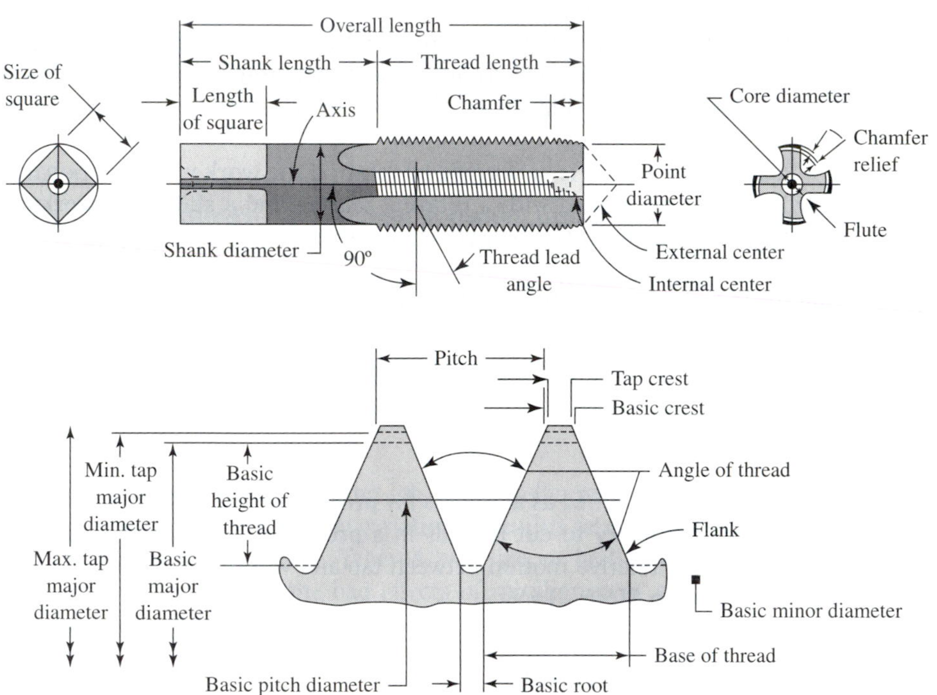

FIGURE 16.10 Tap and thread nomenclature.

Crest: Crest is the surface of the thread that joins the flanks of the thread and is farthest from the cylinder or cone from which the thread projects.

Flank: Flank is the part of a helical thread surface that connects the crest and the root, and is theoretically a straight line in an axial plane section.

Flute: Flute is the longitudinal channel formed in a tap to create cutting edges on the thread profile and to provide chip spaces and cutting fluid passage.

Hook Angle: The hook angle is the angle of inclination of a concave face, usually specified either as chordal hook or as tangential hook.

Land: The land is one of the threaded sections between the flutes of a tap.

FIGURE 16.6 Left-hand-helix hand reamer; square-shanked hand reamers cannot be power driven. (Courtesy Cleveland Twist Drill Greenfield Industries)

Hand Reamers

Hand reamers are finishing reamers distinguished by the square on their shanks (Fig. 16.6). They are turned by hand with a tap wrench that fits over this square (Fig. 16.7). This type of reamer cuts only on the outer cutting edges. The end of the hand reamer is tapered slightly to permit easy alignment in the drilled hole. The length of taper is usually equal to the reamer's diameter.

Hand reamers must never be turned by machine power, and must be started true and straight. They should never remove more than 0.001 to 0.005 in. of material. Hand reamers are available from 1/8 to over 2 in. in diameter and are generally made of carbon steel or high-speed steel.

Taper Hand Reamers. Taper hand reamers are hand reamers made to ream all standard size tapers. They are made for both roughing and finishing tapered holes. Similar to the straight hand reamer, this taper should be used carefully, and never with machine power.

Adjustable Reamers (Fig. 16.8a). Adjustable reamers are used to produce any size hole within the range of the reamer. Their size is adjusted by sliding the cutting blades to and from the shank. These blades are moved by the two adjusting nuts located at each end of the blades. Adjustable hand reamers are available in sizes from $\frac{1}{4}$ to over 3-in. in diameter. Each reamer has approximately $\frac{1}{64}$-in. adjustment above and below its nominal diameter.

Expansion Hand Reamers (Fig. 16.8b). Expansion hand reamers are like the adjustable reamers, but have a limited range of approximately 0.010 in. adjustment. Expansion reamers have an adjusting screw at the end of the reamer. When turned, this adjusting screw forces a tapered plug inside the body of the reamer, expanding its diameter. Expansion reamers are also available as machine reamers.

Care of Reamers. Because reamers are precision finishing tools, they should be used with care:

- Reamers should be stored in separate containers or spaced in the tooling cabinet to prevent damage to the cutting edges.
- Cutting fluids must always be used during reaming operations, except with cast iron.

FIGURE 16.7 Tap wrenches are also used to hold hand reamers to finish drilled holes. (Courtesy Cleveland Twist Drill Greenfield Industries)

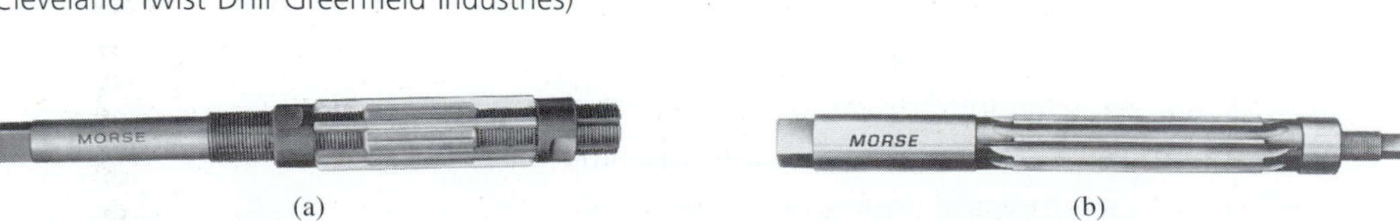

FIGURE 16.8 (a) Adjustable hand reamer. (b) A square-shanked expansion reamer. (Courtesy Morse Cutting Tools)

- A reamer must never be turned backward or the cutting edges will be dulled.
- Any burrs or nicks on the cutting edges must be removed with an oilstone to prevent cutting oversize holes.

16.2.3 Operating Conditions

In reaming speed and feed are important; stock removal and alignment must be considered in order to produce chatter-free holes.

Reaming Speeds Speeds for machine reaming may vary considerably depending in part on the material to be reamed, type of machine, and required finish and accuracy. In general most machine reaming is done at about two-thirds the speed used for drilling the same material.

Reaming Feeds Feeds for reaming are usually much higher than those used for drilling, often running 200 to 300% of drill feeds. Too low a feed may result in excessive reamer wear. At all times it is necessary that the feed be high enough to permit the reamer to cut rather than to rub or **burnish**. Too high a feed may tend to reduce the accuracy of the hole and may also lower the quality of the finish. The basic idea is to use as high a feed as possible and still produce the required finish and accuracy.

Stock to Be Removed For the same reason, insufficient stock for reaming may result in a burnishing rather than a cutting action. It is difficult to generalize about this phase because it is tied in closely with type of material, feed, finish required, depth of hole, and chip capacity of the reamer. For machine reaming, .010 in. on a $\frac{1}{4}$-in. hole, .015 in. on a $\frac{1}{2}$-in. hole, and up to .025 inch on a $1\frac{1}{2}$-in. hole seems a good starting point. For hand reaming, stock allowances are much smaller, partly because of the difficulty in forcing the reamer through greater stock. A common allowance is .001 in. to .003 in.

Alignment In the ideal reaming job, the spindle, reamer, bushing, and hole to be machined are all in perfect alignment. Any variation from this tends to increase reamer wear and detracts from the accuracy of the hole. Tapered, oversize, or **bell-mouthed** holes should call for a check of alignment. Sometimes the bad effects of misalignment can be reduced through the use of floating or adjustable holders. Quite often if the user will grind a slight back taper on the reamer, it will also be of help in overcoming the effects of misalignment.

Chatter The presence of chatter while reaming has a very bad effect on reamer life and on the finish in the hole. Chatter may be the result of one of several causes, some of which are listed here:

- Excessive speed
- Too much clearance on reamer
- Lack of rigidity in jig or machine
- Insecure holding of work
- Excessive overhand of reamer or spindle
- Too light a feed

Correcting the cause can materially increase both reamer life and the quality of the reamed holes. In reaming the emphasis is usually on finish, and a coolant is normally chosen for this purpose rather than for cooling.

16.2.4 Reaming Operations

Reaming operations can be performed on lathes, drills, and machining centers.

Lathe Reaming Reaming on a lathe can only be done by holding the reamer in the tailstock position either in a drill chuck for straight-shank reamers, or directly in the tailstock **quill** for tapered-shank reamers (Fig. 16.4). Work to be reamed can either be held in a chuck or mounted onto the face plate. In case of a turret lathe, the reamer can only be used in the hex turret.

Sometimes reamers are held in **floating** holders in the tailstock. These holders allow the reamer to center itself on the previously drilled hole. Deep holes (over three times the diameter of the drill) tend to "run out." The reamer will not correct this condition, and the hole must be bored if alignment is important.

Drill Press Reaming Reaming on a drill press also requires the reamer to be held in the spindle with a drill chuck for straight-shank machining reamers, or directly in the spindle for tapered-shank reamers (Fig. 16.4). The work to be reamed is usually held in a vise and centered on the drill table.

Reaming on a lathe is performed by rotating the work with a stationary reamer; reaming on a drill press is performed with a rotating reamer and a stationary workpiece. "Floating" heads can be used on drill presses as well as on lathes.

Machining-Center Reaming Reaming on a machining center is common. Reamers are usually held in the hex turret or in an automatic tool magazine. The setups are usually more complicated, and speeds and feeds are preprogrammed.

16.3 TAPPING

Tapping has been defined as a process for producing internal threads using a tool (**tap**) that has teeth on its periphery to cut threads in a predrilled hole. Threads are formed by a combined rotary and axial relative motion between tap and workpiece. A typical automated tapping operation is shown in Figure 16.9.

16.3.1 Tap Nomenclature

Screw threads have many dimensions. It is important in modern manufacturing to have a working knowledge of screw thread terminology. A right-hand thread is a screw thread that requires right-hand or clockwise rotation to tighten it. A left-hand thread is a screw thread that requires left-hand or counterclockwise rotation to tighten it. Thread fit is the range of tightness or looseness between external and internal mating threads. Thread series are groups of diameter and pitch combinations that are distinguished from each other by the number of threads per inch applied to a specific diameter. The two common thread series used in industry are the coarse and fine series, specified as UNC and UNF. Tap nomenclature is shown in Figure 16.10.

Chamfer: Chamfer is the tapering of the threads at the front end of each land of a chaser, tap, or die by cutting away and relieving the crest of the first few teeth to distribute the cutting action over several teeth.

FIGURE 16.9 A typical automated tapping operation with self-reversing unit. (Courtesy Tapmatic Corp.)

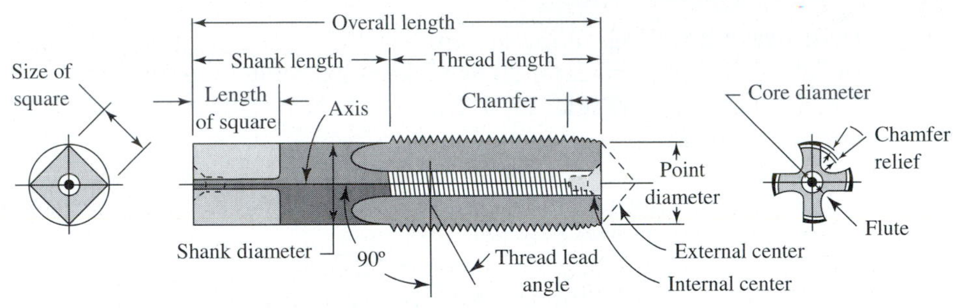

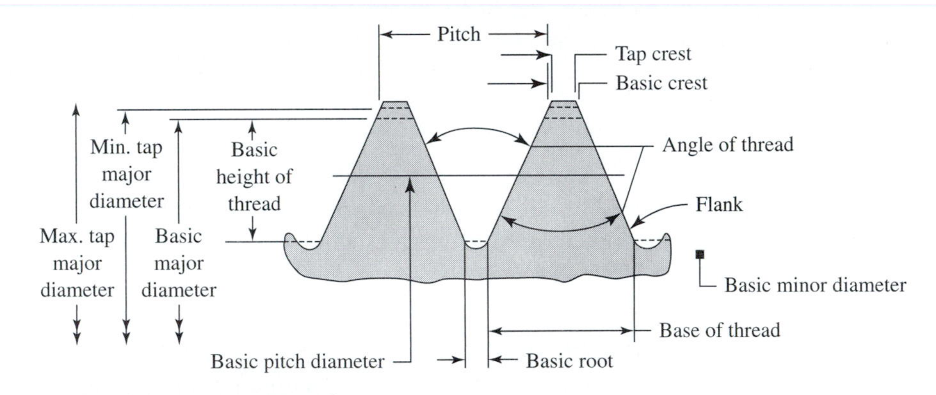

FIGURE 16.10 Tap and thread nomenclature.

Crest: Crest is the surface of the thread that joins the flanks of the thread and is farthest from the cylinder or cone from which the thread projects.

Flank: Flank is the part of a helical thread surface that connects the crest and the root, and is theoretically a straight line in an axial plane section.

Flute: Flute is the longitudinal channel formed in a tap to create cutting edges on the thread profile and to provide chip spaces and cutting fluid passage.

Hook Angle: The hook angle is the angle of inclination of a concave face, usually specified either as chordal hook or as tangential hook.

Land: The land is one of the threaded sections between the flutes of a tap.

Lead of Thread: The lead of thread is the distance a screw thread advances axially in one complete turn. On a single-start tap the lead and pitch are identical. On a multiple-start tap the lead is the multiple of the pitch.

Major Diameter: This is the diameter of the major cylinder or cone, at a given position on the axis, that bounds the crests of an external thread or the roots of an internal thread.

Minor Diameter: The minor diameter is the diameter of the minor cylinder or cone, at a given position on the axis, that bounds the roots of an external thread or the crests of an internal thread.

Pitch Diameter: Pitch diameter is the diameter of an imaginary cylinder or cone at a given point on the axis of such a diameter and location of its axis such that its surface would pass through the thread so as to make the thread ridge and the thread groove equal. As such, it is located equidistant between the sharp major and minor cylinders or cones of a given thread form. On a theoretically perfect thread, these widths are equal to half of the basic pitch (measured parallel to the axis).

Spiral Point: A spiral point is the angular fluting in the cutting face of the land at the chamfered end. It is formed at an angle with respect to the tap axis of opposite hand to that of rotation. Its length is usually greater than the chamfer length, and its angle with respect to the tap axis is usually made great enough to direct the chips ahead of the tap. The tap may or may not have longitudinal flutes.

Square: The four driving flats parallel to the axis on a tap shank form a square or square with round corners.

16.3.2 Types of Taps

Taps are manufactured in many sizes, styles, and types. Figure 16.11 shows some of the taps discussed here.

Hand Taps

Today the hand tap is used both by hand and in machines of all types. This is the basic tap design: four straight flutes, in taper, plug, or bottoming types. The small, numbered machine screw sizes are standard in two and three flutes depending on the size.

If soft and stringy metals are being tapped, or if horizontal holes are being made, either two- or three-flute taps can be used in the larger sizes. The flute spaces are larger, but the taps are weaker. The two-flute especially has a very small cross section.

The chips formed by these taps cannot get out; thus, they accumulate in the flute spaces. This causes added friction and is a major cause of broken taps.

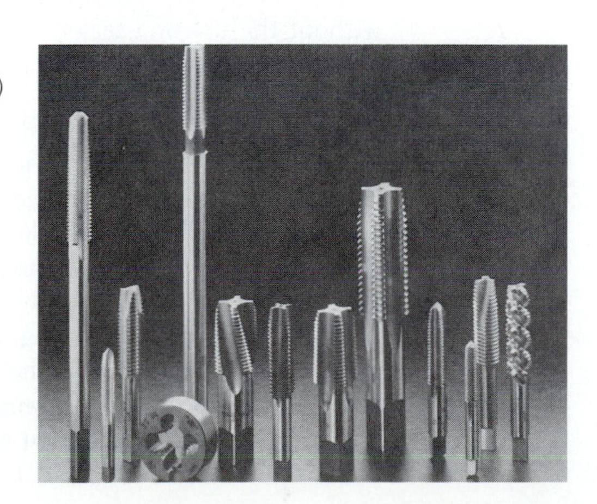

FIGURE 16.11 Some of the many styles and shapes of taps. (Courtesy Greenfield Industries)

FIGURE 16.12 (a) Spiral-point taps have replaced 'standard' taps in many cases. (b) A spiral-fluted bottoming tap. (Courtesy Morse Cutting Tools)

Spiral-Point Tap The spiral-point or gun tap (Fig. 16.12a) is made the same as the standard hand tap (Fig. 16.10) except at the point. A slash is ground in each flute at the point of the tap. This accomplishes several things:

- The gun tap has fewer flutes (usually three), and they are shallower. This means a stronger tap.
- The chips are forced out ahead of the tap instead of accumulating in the flutes as they will with a plug tap.
- Because of these two factors, the spiral-point tap can often be run faster than the hand tap, and tap breakage is greatly reduced.

The gun tap has, in many cases, replaced the standard style in industry, especially for open-ended trough holes in mild steel and aluminum. Both regular and spiral-point taps are made in all sizes including metric.

Spiral-Flute Tap The spiral-flute bottoming tap (Fig. 16.12b) is made in regular and fast spirals, that is, with small or large helix angles. These taps are sometimes called helical-fluted taps. The use of these taps has been increasing because they pull the chip up out of the hole and produce good threads in soft metals (such as aluminum, zinc, and copper), yet also work well in Monel metal, stainless steel, and cast steel. They are made in all sizes up to $1\frac{1}{2}$ in. and in metric sizes up to 12 mm.

The standard taps will efficiently do most work, but if a great deal of aluminum, brass, cast iron, or stainless steel is being tapped, the manufacturer can supply standard specials that will do a better job.

Pipe Taps General-purpose pipe taps are used for threading a wide range of materials, both ferrous and nonferrous. All pipe taps are supplied with a $2\frac{1}{2}$- to $3\frac{1}{2}$-thread chamfer. The nominal size of a pipe tap is that of the pipe fitting to be tapped, not the actual size of the tap.

Ground-thread pipe taps are standard in American Standard Pipe Form (NPT) and American Standard Dryseal Pipe Form (NPTF). NPT threads require the use of a sealer such as Teflon tape or pipe compound. Dryseal taps are used to tap fittings that will give a pressure-tight joint without the use of a sealer. Figure 16.13 shows straight and spiral-fluted pipe taps as well as a T-handle tap wrench.

Fluteless Taps Fluteless taps (Figure 16.14) do not look like taps, except for the spiral threads. These taps are not round. They are shaped so that they cold-form the metal out of the wall of the hole into the thread form with no chips. The fluteless tap was originally designed for use in aluminum, brass, and zinc alloys. However, it is being successfully used in mild steel and some stainless steels. Thus, it is worth checking for use where BHN is under 180. Fluteless taps are available in most sizes, including metric threads.

These taps are very strong and can often be run up to twice as fast as other styles. However, the size of the hole drilled before tapping must be no larger than the pitch diameter of the thread. The cold formed thread often has a better finish and is stronger than a cut thread. A cutting oil must be used, and the two ends of the hole should be countersunk because the tap raises the metal at all ends.

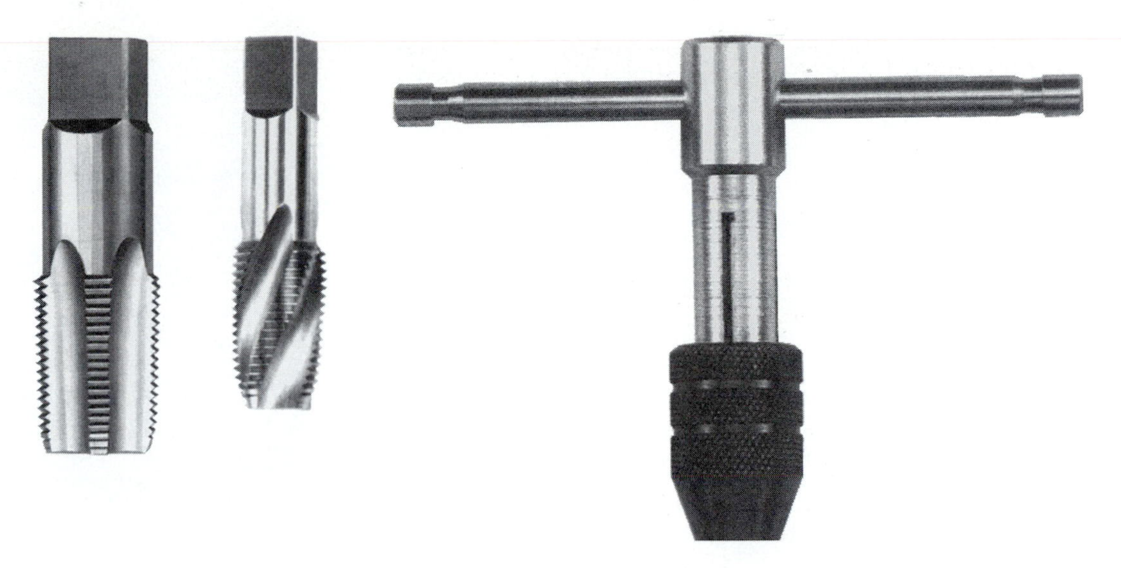

FIGURE 16.13 Straight and spiral-fluted pipe taps and a T-handle tap wrench. (Courtesy Morse Cutting Tools)

FIGURE 16.14 Fluteless taps are used to 'cold form' threads. (Courtesy The Weldon Tool Co.)

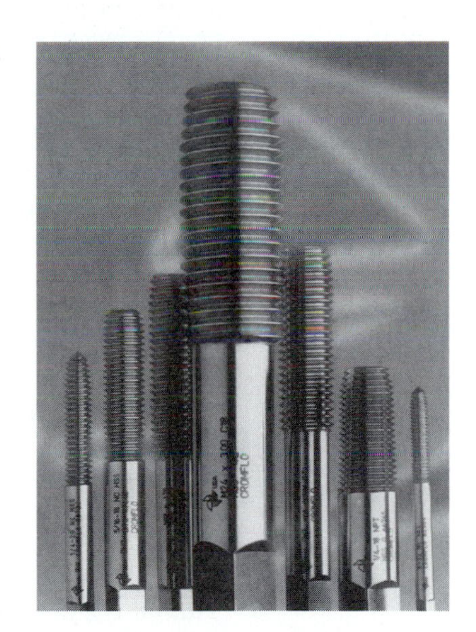

Collapsing Taps Collapsing taps (Fig. 16.15) collapse to a smaller diameter at the end of the cut. Thus, when used on lathes of any kind, they can be pulled back rapidly. They are made in sizes from about 1 in. up, in both machine and pipe threads. They use three to six separate chasers, which must be ground as a set. The tap holder and special dies make this assembly moderately expensive, but it is economical for medium- and high-production work.

16.3.3 Operating Options

Some threads, both external and internal, can be cut with a single-point tool as previously shown (Fig. 13.2). However, most frequently a die or tap of some type is used because it is faster and generally more accurate.

FIGURE 16.15 Collapsing tap assemblies are more expensive, but economical for medium- and high-production runs. (Courtesy Greenfield Industries)

Taps are made in many styles, but a few styles do 90% of the work. Figure 16.10 shows the general terms used to describe taps. The cutting end of the tap is made in three different tapers.

The taper tap is not often used today. Occasionally, it is used first as a starter if the metal is difficult to tap. The end is tapered about 5° per side, which makes eight partial threads.

The plug tap is the style used probably 90% of the time. With the proper geometry of the cutting edge and a good lubricant, a plug tap will do most of the work needed. The end is tapered 8° per side, which makes four or five incomplete threads.

The bottoming tap (Fig. 16.12b) is used only for blind holes where the thread must go close to the bottom of the hole. It has only $1\frac{1}{2}$ to 3 incomplete threads. If the hole can be drilled deeper, a bottoming tap may not be needed. The plug tap must be used first, followed by the bottoming tap. All three types of end tapers are made from identical taps. Size, length, and all measurements except the end taper are the same. The material used for taps is usually high-speed steel in the M1-, M2-, M7-, and sometimes the M40-series cobalt high-speed steels. A few taps are made of solid tungsten carbide.

Most taps today have ground threads. The grinding is done after hardening and makes much more accurate cutting tools. Cut thread taps are available at a somewhat lower cost in some styles and sizes.

16.3.4 Tapping Operations

Just like reaming operations, tapping can be performed on lathes, drills, and machining centers; a multihole tapping operation on a round part is shown in Figure 16.16.

Tap Drills It is quite obvious that the taps shown here cannot cut their own opening. Thus, a hole of the proper size must be made before the tap can be used. Usually this hole is drilled. A tap drill is not a special kind of drill. A tap drill is merely a convenient way to refer to the proper size drill to be used before using a tap. Tap drill sizes based on 75% of thread are given in reference tables. The trend today in many factories, in order to save taps, time, and rejects, is to use 60 to 65% of thread to determine tap drill sizes. Drills and drilling operations were discused in Chapter 15. A combination drill and tap is shown in Figure 16.17 and used to drill and tap in one pass.

FIGURE 16.16 An automated multihole tapping operation on a round part. (Courtesy Tapmatic Corp.)

FIGURE 16.17 Combination drill and tap tools are used for one-pass drilling and tapping. (Courtesy Morse Cutting Tools)

The deeper the hole is threaded, the longer it takes to drill and tap, and the more likely it is that the tap will break. Yet if there are too few threads holding the bolt, the threads will strip. Somewhere in between is a depth of thread engagement that is the minimum that will hold enough so that the bolt will break before the threads let go. This is called the optimum depth.

Tap drilling must be deep enough in blind holes to allow for the two to five tapered threads on the tap, plus chip clearance, plus the drill point.

Toolholders Toolholders for hand tapping are called tap wrenches. They are the same for taps and for reamers (Figs. 16.7 and 16.13), because most taps have a square shank. Tap wrenches are adjustable and can be used on several sizes of taps.

When taps are used in drill presses or machining centers, a special head with a reversing, slip-type clutch is used. These tapping heads (Fig. 16.18) can be set so that if a hard spot is met in the metal, the clutch slips and the tap will not break. They are constructed so that when the hand-feed lever or the automatic numerical control machine cycle starts upward, the rotation reverses (and often goes faster) to bring the tap safely out of the hole.

Workholding Workholding for tapping is the same as for any drill press or lathe work: clamps, vises, fixtures, and so on as needed. It is necessary to locate the tap centrally and straight in the hole. This is difficult in hand tapping but relatively easy in machine tapping.

Numerical control is especially efficient, as it will locate over a hole, regardless of when it was drilled, if it was drilled from the same tape and on the same setup.

Single-point threading was shown in Figure 13.2. **Thread chasing** or the manufacturing of outside threads is also performed with dies and self-opening die stocks. Figure 16.19 shows a number of die heads and die chasers used in the manufacturing of threads.

FIGURE 16.18 Various special tap heads with reversing, slip-type clutches are used in drill pressed and machining centers. (Courtesy Tapmatic Corp.)

FIGURE 16.19 Thread 'chasing,' or the manufacturing of outside threads, is performed with dies and self-opening die stocks. (Courtesy Greenfield Industries)

Lubrication The cutting edges on both taps and dies are buried in the material, so lubrication is quite necessary. For aluminum, light lard oil is used; other metals require a sulfur-based oil, sometimes chlorinated also. Figure 16.20 shows a tapping operation with an automated fluid dispensing system for machining centers. The Automiser unit shown here dispenses a lubricant/coolant through the tapping head automatically, while the head is in the machine spindle.

FIGURE 16.20 Multihole tapping operation with automatic coolant/lubrication system. (Courtesy Tapmatic Corp.)

Copper alloys are stained by sulfur, so mineral oils or soluble oil must be used. Cast iron is often threaded without any lubricant. There are several synthetic tapping fluids on the market today. They are somewhat more expensive but may save their cost in better threads and fewer broken taps.

16.4 REVIEW QUESTIONS AND PROBLEMS

1. Define the reaming process.
2. Discuss the use of straight-flute and spiral-flute reamers.
3. What distinguishes a hand reamer from a machine reamer?
4. Why are shell reamers considered economical?
5. Explain the difference between adjustable and expansion reamers.
6. Discuss reaming speeds and feeds as compared to drilling.
7. Discuss some of the causes of chatter in reaming operations.
8. Define the tapping process.
9. Discuss some of the advantages of the spiral-point or gun tap.
10. Explain the difference between fluteless and regular taps.
11. What are collapsing taps, and how do they differ from standard taps?
12. Explain the purpose and function of a tap drill.

CHAPTER SEVENTEEN

BORING OPERATIONS

CHAPTER OVERVIEW

- Introduction
- Boring Operations
- Boring Rigidity
- Boring Bars
- Boring Machines
    - Horizontal Boring Machines
    - Vertical Boring Machines
    - Jig Borers
- Review Questions and Problems

17.1 INTRODUCTION

Boring, also called internal turning, is used to increase the inside diameter of a hole. The original hole is made with a drill, or it may be a cored hole in a casting. Boring achieves three things:

Sizing: Boring brings the hole to the proper size and finish. A drill or reamer can only be used if the desired size is standard or if special tools are ground. The boring tool can work to any diameter, and it will give the required finish by adjusting speed, feed, and nose radius.

Straightness: Boring will straighten the original drilled or cast hole. Drills, especially the longer ones, may wander off center and cut at a slight angle because of eccentral forces on the drill, occasional hard spots in the material, or uneven sharpening of the drill (Fig. 15.10). Cored holes in castings are almost never completely straight. The boring tool being moved straight along the ways with the carriage feed will correct these errors.

Concentricity: Boring will make the hole concentric with the outside diameter within the limits of the accuracy of the chuck or holding device. For best concentricity, the turning of the outside diameter and the boring of the inside diameter are done in one setup, that is, without moving the work between operations. The basics discussed in Chapter 13, "Turning Operations," also apply to boring. However, with boring a number of limitations must be taken into account in order to reach a high stock removal rate combined with satisfactory accuracy, surface finish, and tool life. Therefore, in this chapter the limitations that distinguish internal turning from external turning will be discussed in greater detail. A typical boring operation is shown in Figure 17.1.

FIGURE 17.1 Typical horizontal boring operation. (Courtesy Sandvik Coromant Co.)

17.2 BORING OPERATIONS

Most of the turning operations that occur with external turning are also to be found in boring. With external turning, the tool overhang is not affected by the length of the workpiece and the size of the toolholder can be chosen so that it withstands the forces and stresses that arise during the operation. However, with internal turning, or boring, the choice of tool is very much restricted by the component's hole diameter and length.

A general rule, which applies to all machining, is to minimize the tool overhang in order to obtain the best possible stability and thereby accuracy. With boring the depth of the hole determines the overhang. The stability is increased when a larger tool diameter is used, but even then the possibilities are limited because the space allowed by the diameter of the hole in the component must be taken into consideration for chip evacuation and radial movements.

The limitations in regard to stability in boring mean that extra care must be taken with production planning and preparation. Understanding how cutting forces are affected by the tool geometry and the cutting data chosen, and also understanding how various types of boring bars and tool clampings will affect stability, can keep deflection and vibration to a minimum.

Insert Geometry

The geometry of the insert has a decisive influence on the cutting process. A positive insert has a positive rake angle. The insert's edge angle and clearance angle together will equal less than 90°. A positive rake angle means a lower tangential cutting force. However, a positive rake angle is obtained at the cost of the clearance angle or the edge angle. If the clearance angle is small, there is a risk of abrasion between the tool and workpiece, and the friction can give rise to vibration. In those cases where the rake angle is large and the edge angle is small, a sharper cutting edge is obtained. The sharp cutting edge penetrates the material more easily, but it is also more easily changed or damaged by edge or other uneven wear.

Edge wear means that the geometry of the insert is changed, resulting in a reduction in the clearance angle. Therefore, with finish machining it is the required surface finish of the workpiece that determines when the insert must be changed. Generally, the edge wear should be between .004 and .012 in. for finishing and between .012 and .040 in. for rough machining.

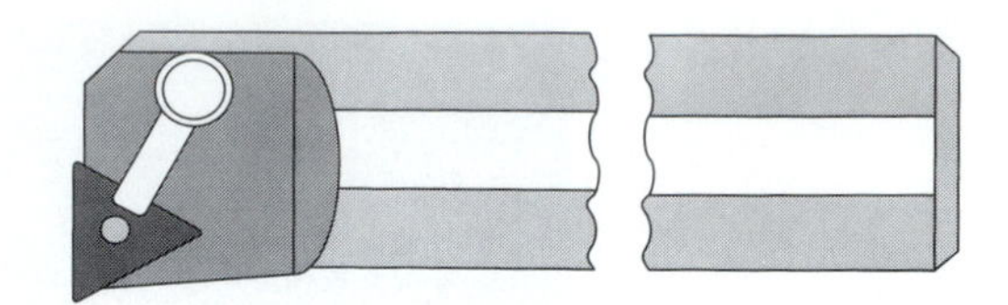

FIGURE 17.2 Typical indexable insert boring bar with 0° lead angle.

Lead Angle The lead angle affects the axial and radial directions of the cutting forces. A small lead angle produces a large axial cutting force component; a large lead angle results in a larger cutting force in the radial direction. The axial cutting force has a minimal negative effect on the operation because the force is directed along the boring bar. To avoid vibrations, it is consequently advantageous to choose a small lead angle but, because the lead angle also affects other factors such as the chip thickness and the direction of the chip flow, a compromise often has to be made.

The main disadvantage of a small lead angle is that the cutting forces are distributed over a shorter section of the cutting edge than with a large lead angle. Furthermore, the cutting edge is exposed to abrupt loading and unloading when the edge enters and leaves the workpiece. Since boring is done in most cases in a premachined hole and is designated as light machining, small lead angles generally do not cause a problem. Lead angles of 15° or less are normally recommended. However, at a lead angle of 15° the radial cutting force will be virtually double that of the cutting force with a 0° lead angle. A typical indexable insert boring bar with a 0° lead angle is shown in Figure 17.2.

Nose Radius The distribution of cutting forces is also affected by the nose radius of the insert: the greater the nose radius, the greater the radial and tangential cutting forces, and the greater the emergence of vibration. However, this is not the case with radial cutting forces. The deflection of the tool in a radial direction is instead affected by the relationship between the cutting depth and the size of the nose radius. If the cutting depth is smaller than the nose radius, the radial cutting forces will increase with increased cutting depth. If the cutting depth is equal to or greater than the size of the nose radius, the radial deflection will be determined by the lead angle. Therefore, it's a good idea to choose a nose radius that is somewhat smaller than the cutting depth. In this way the radial cutting forces can be kept to a minimum while utilizing the advantages of the largest possible nose radius, leading to a stronger cutting edge, better surface finish, and more even pressure on the cutting edge.

17.3 BORING RIGIDITY

Part geometries can have external turning operations as well as internal operations. Internal single-point turning is referred to as boring, and can be utilized for either a roughing or finishing operation. Single-point boring tools consist of a round shaft with one insert pocket designed to reach into a part hole or cavity to remove internal stock in one or several machine passes. Figure 17.3 shows various sizes and styles of boring bars.

The key to productivity in boring operations is the tool's rigidity. Boring bars are often required to reach long distances into parts to remove stock (Fig. 17.4). Hence, the rigidity of the machining operation is compromised because the diameter of the tool is restricted by the hole size and the need for added clearance to evacuate chips. The practical overhang limits for steel boring bars is four times their shank diameter. When the tool overhang exceeds this limit, the metal removal rate of the boring operation is compromised significantly due to lack of rigidity and the increased possibility of vibration.

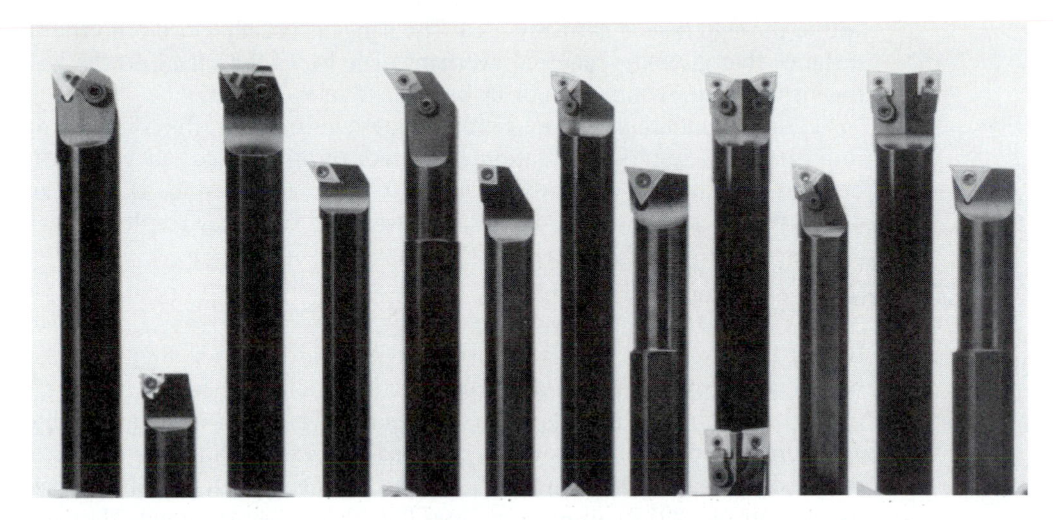

FIGURE 17.3 Various sizes and styles of boring bars. (Courtesy Dorian Tool)

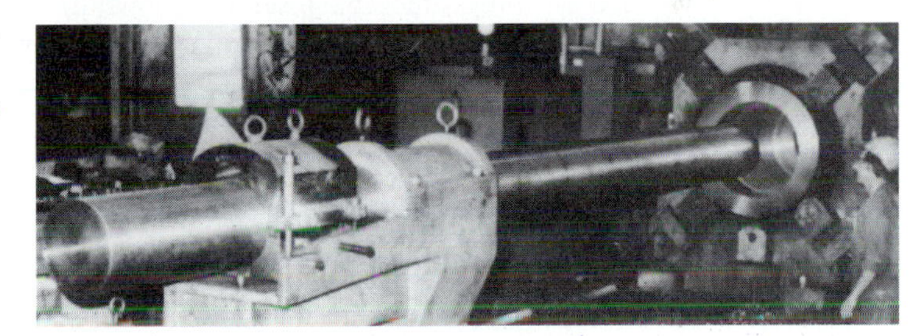

FIGURE 17.4 Boring bars are often required to reach long distances into parts to remove stock. (Courtesy Sandvik Coromant Co.)

Boring Bar Deflection The size of the boring bar's deflection is dependent on the bar material, the diameter, the overhang, and the size of the radial and tangential cutting forces. Boring bar deflection can be calculated, but such calculations are beyond the scope of this text.

This deflection can be counteracted by increasing the diameter of the tool to create an increased moment of inertia. Deflection can also be counteracted by choosing a boring bar made of a material that has a higher coefficient of elasticity. Since steel has a lower coefficient of elasticity than cemented carbide, cemented carbide boring bars are better for large overhangs.

Boring Bar Clamping The slightest amount of mobility in the fixed end of the boring bar will lead to deflection of the tool. The best stability is obtained with a holder that completely encases the bar. This type of holder is available in two styles: a rigid (Fig. 17.5a) or flange-mounted bar, or a divided block (Fig. 17.5b) that clamps when tightened. With a rigidly mounted bar, the bar is either preshrunk into the holder and/or welded in. With flange mounting, a flange

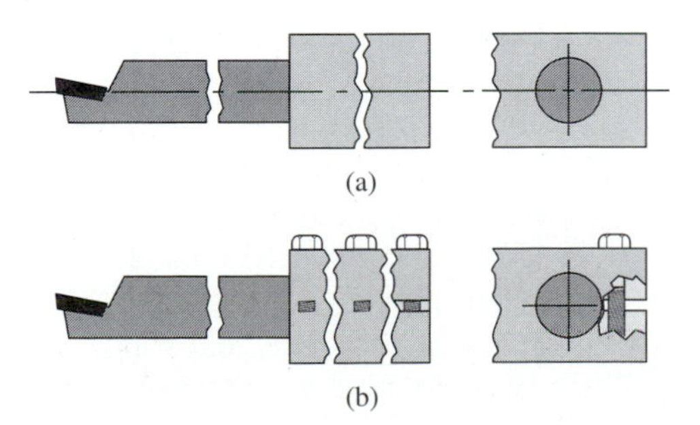

FIGURE 17.5 Two proper boring bar clamping methods.

with a through hole is normally used. The flange is usually glued onto the shank of the bar at a distance that gives the required overhang. The bar is then fed into the holder and clamped by means of a screw connection or by being held in the turret.

Less efficient are those tool-clamping methods in which the screw clamps onto the bar. This form generally results in vibration and is not recommended. Above all, this method must not be used for clamping cemented carbide bars. Cemented carbide is more brittle than steel, and cracks will occur as a result of vibration, which in turn may result in breakage.

17.4 BORING BARS

Boring bars are made in a wide variety of styles as shown in Figure 17.3. Single-point boring bars (Fig. 17.6) are easily ground but difficult to adjust when they are used in turret and automatic lathes and machining centers, unless they are held in an adjustable holder (Fig. 17.7).

More expensive boring bars are provided with easily adjustable inserts. These bars are made in standard sizes, with a range of $\frac{1}{4}$ to $\frac{1}{2}$ in. on the diameter. A fine adjustment is included in increments of 0.001 in. or in some cases 0.0001 in. They are standard up to about 6 in. in diameter. A boring bar with adjustments is shown in Figure 17.8. A different style of adjustable boring bar with two indexable inserts is shown in Figure 17.9. Standard boring bars with interchangeable heads to permit various internal operations such as turning, profiling, grooving, and threading are shown in Figure 17.10.

Many times it may be economical to order special bars with two or more diameters set at the proper distance apart. These special bars cost more and are generally only used when large quantities make their use economical. Sometimes this may be the only way to hold the required tolerances and concentricity. Such special boring bars, sometimes called boring heads, are designed with replaceable cartridges. A twin-cutter adjustable boring tool is shown in Figure 17.11. Various replaceable cartridges for special boring heads are shown in Figure 17.12.

FIGURE 17.6 Single-point boring bar. (Courtesy Morse Cutting Tools)

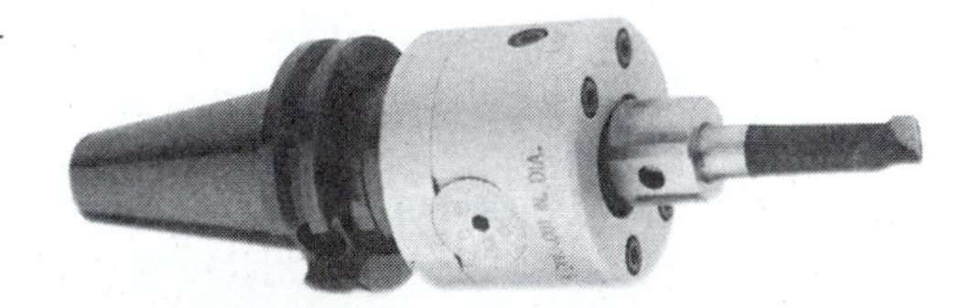

FIGURE 17.7 Adjustable boring head for single-point boring tools. (Courtesy Kennametal Inc.)

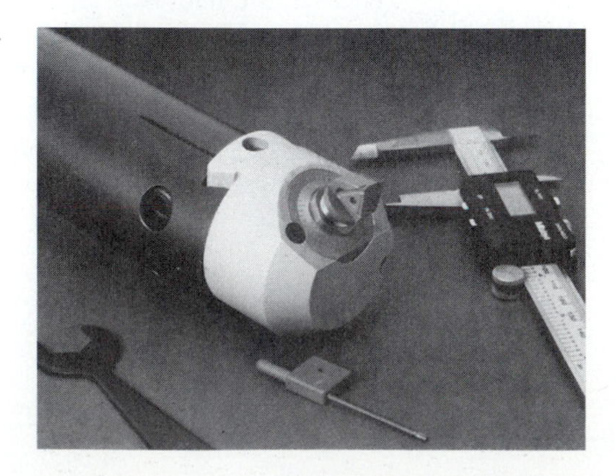

FIGURE 17.8 Adjustable boring bar with fine-tuning adjustment. (Courtesy Valenite Inc.)

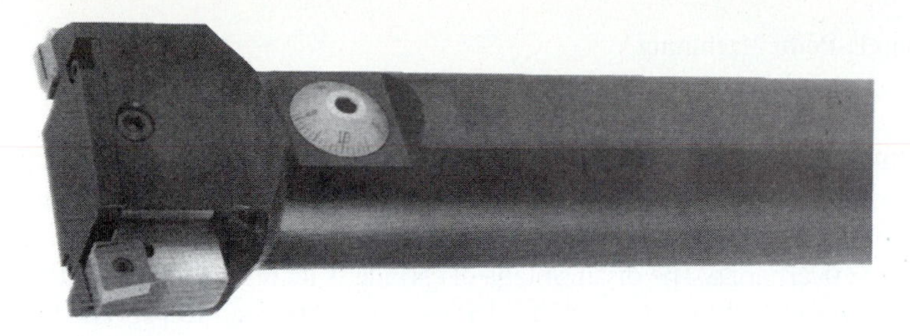

FIGURE 17.9 Adjustable boring bar with two indexable inserts. (Courtesy Kennametal Inc.)

FIGURE 17.10 Standard boring bar with interchangeable heads for various internal operations such as turning, profiling, grooving, and threading. (Courtesy Valenite Inc.)

FIGURE 17.11 A twin-cutter adjustable boring head with indexable Trigon inserts. (Courtesy Komet of America, Inc.)

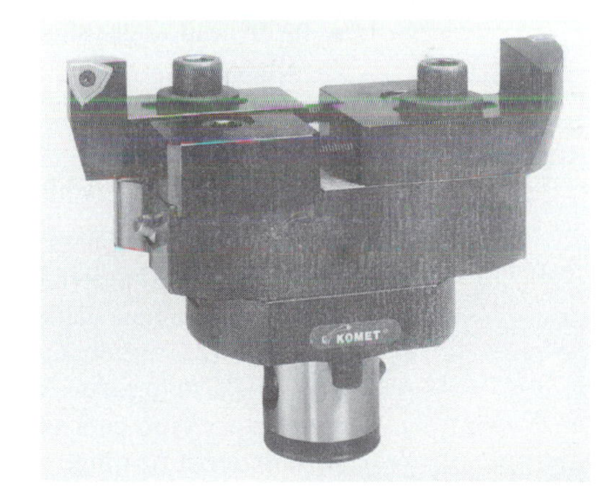

FIGURE 17.12 Various indexable and replaceable cartridges used in special boring heads. (Courtesy Valenite Inc.)

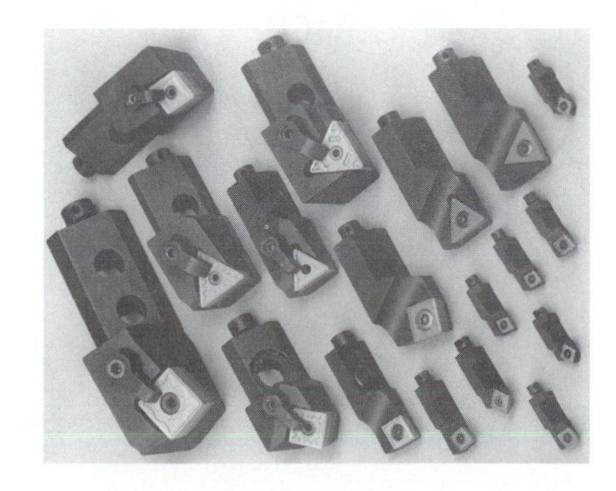

Boring Bar Types Boring bars are available in steel, solid carbide, and carbide-reinforced steel. The capacity to resist deflection increases as the coefficient of elasticity increases. Since the elasticity coefficient of carbide is three times larger than that of steel, carbide bars are preferred for large overhangs. The disadvantage of carbide is its poor ability to withstand tensile stresses. For carbide-reinforced bars, the carbide sleeves are prestressed to prevent tensile stresses.

Boring bars can be equipped with ducts for internal cooling, which is preferred for internal turning. An internal coolant supply provides efficient cooling of the cutting edge, plus better chip breaking and chip evacuation. In this way a longer tool life is obtained and quality problems, which often arise due to chip jamming, are avoided.

Boring Bar Choice When planning production, it is very important to minimize cutting forces and to create conditions where the greatest possible stability is achieved so that the tool can withstand the stresses that always arise. The length and diameter of the boring bar will be of great significance to the stability of the tool. Since the appearance of the workpiece is the decisive factor when selecting the minimum overhang and maximum tool diameter that can be used, it is important to choose the tool, tool clamping, and cutting data that minimize, as much as possible, the cutting forces that arise during the operation. The following recommendations should be followed to obtain the best possible stability:

- Choose the largest possible bar diameter, but at the same time ensure that there is enough room for chip evacuation.
- Choose the smallest possible overhang, but at the same time ensure that the length of the bar allows the recommended clamping lengths to be achieved.
- A 0° lead angle should be used. The lead angle should under no circumstances be more than 15°.
- The carbide grade should be tougher than for external turning to withstand the stresses to which the insert is exposed when chip jamming and vibration occur.
- Choose a nose radius that is smaller than the cutting depth.

Modern boring bars are designed to take into account the demands that must apply because the operation is undertaken internally and the dimensions of the tool are determined by the hole depth and the hole diameter. With a positive insert geometry, less material deformation and low cutting forces are obtained. The tool should offer good stability to resist the cutting forces that arise and also to reduce deflection and vibration as much as possible. Due to space requirements, satisfactory chip control and good accessibility are also properties of greater importance than with external turning.

17.5 BORING MACHINES

Boring operations can be performed on other than boring machines, such as lathes, milling machines, and machining centers. A typical boring operation performed on a lathe is shown in Figure 17.13. A steady rest is being used to provide support for the part being machined. Boring machines, like most other machine tools, can be classified as horizontal or vertical.

17.5.1 Horizontal Boring Machines

The horizontal boring machine (HBM) is made to handle medium-sized to very large parts, but these parts are usually somewhat rectangular in shape, though they may be asymmetrical or irregular. The size of cut is only limited by the available cutting tools, the rigidity of the spindle, and the available horsepower. There are two types of horizontal boring machines: table and floor.

FIGURE 17.13 A typical boring operation performed on a lathe; a steady rest is being used to provide support for the part being machined. (Courtesy Sandvik Coromant Co.)

Table-Type Horizontal Boring Machines

The table-type HBM shown in Figure 17.14 is built on the same principles as the horizontal-spindle milling machines. The base and column are fastened together, and the column does not move. The tables are heavy, ribbed castings, which may hold loads up to 20,000 pounds. Figure 17.15 shows a large part being machined on a table-type horizontal boring machine.

Size of HBM. The basic size of an HBM is the diameter of the spindle. Table-type machines usually have spindles from 3 to 6 in. in diameter. The larger sizes will transmit more power, and, equally important, the spindle will not sag or deflect as much when using a heavy cutting tool while extended. The size is further specified by the size of the table. Although each machine has a standard size table, special sizes may be ordered. The principal parts of the horizontal boring machine are shown in Figure 17.16.

Work Holding. Work holding is with clamps, bolts, or fixtures, the same as with other machines. Rotary tables allow machining of all four faces of a rectangular part or various angle cuts on any shape of part. Rotary tables up to 72 in. square or round are used for large work. If large, rather flat work is to be machined, an angle plate is used. The workpiece is bolted or clamped onto the angle plate so that the flat face is toward the spindle. Figure 17.17 shows a five-axis ram-style machining center. Parts can be clamped to the table and numerically (NC or CNC) positioned to perform a boring operation.

Cutting Tools. Cutting tools are held in the rotating spindle by a tapered hole and a drawbar. To speed up the process of tool changing, either or both of two things are done:

- The drawbar (which pulls the tapered toolholder tightly into the spindle hole) can be power operated. Thus, the holder is pulled tight or ejected very quickly.

FIGURE 17.14 Table-type horizontal boring machine (HBM) (Courtesy Summit Machine Tool Manufacturing Corp.)

FIGURE 17.15 Large part being machined on a table-type horizontal boring machine. (Courtesy WMW Machinery Co., Inc.)

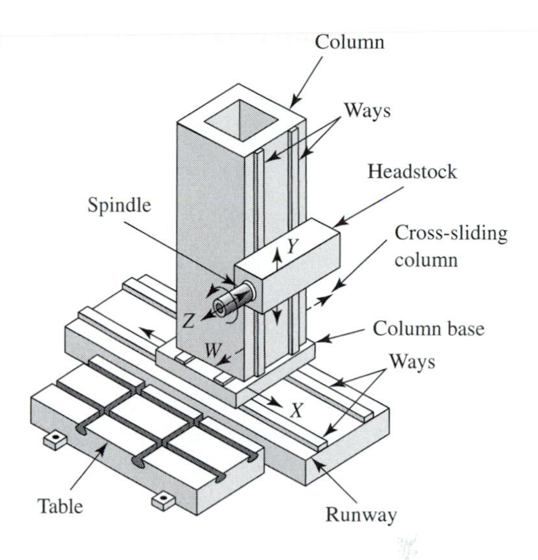

FIGURE 17.16 Principal parts of a floor-type horizontal boring machine (HBM).

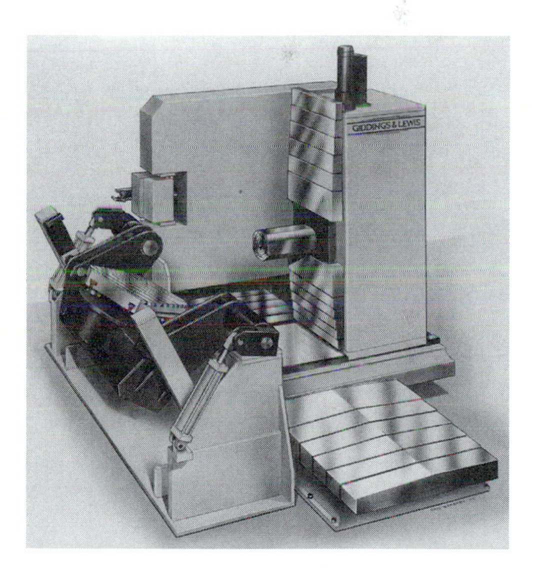

FIGURE 17.17 Five-axis ram-style machining center. (Courtesy Giddings and Lewis, LLC)

- Quick-change tooling is used. A basic holder is secured in the spindle. It has a taper into which tools may be secured by a quarter to half turn of the locking collar. Thus, the operator can change preset tools in 10 to 30 seconds.

Toolholders and quick-change toolholders in particular will be discussed in Chapter 25, "Milling Operations."

Speeds and Feeds. Speeds and feeds cover a wide range because of the wide variety of cutters which may be used on the HBM. Speeds from 15 to 1500 RPM and feedrates from 0.1 to 40 IPM are commonly used.

Floor-Type Horizontal Boring Machine The floor type HBM (Fig. 17.16) is used for especially tall or long workpieces. The standard 72-in. runway can be made almost any length required

FIGURE 17.18 Large floor-type horizontal boring machine. (Courtesy WMW Machinery Co., Inc.)

for special jobs. Lengths of 20 ft are in use today. The height of the column, which is usually 60 to 72 in., can be made to order up to twice this height if the work requires it. Figure 17.18 shows a large floor-type horizontal boring machine.

HBM Table. The table is separate from the boring machine, though it is, of course, fastened to the floor. It may be bolted to the runway. The entire column and column base move left and right (the *X*-axis) along special ways on the runway (Fig. 17.16). The runway must be carefully aligned and leveled when it is first installed, and then checked at intervals as the machine is used.

HBM Headstock. The headstock can be moved accurately up and down the column (the *Y*-axis). The 6- to 10-in.-diameter spindle rotates to do the machining. It is moved in and out (the *Z*-axis) up to 48 in. for boring cuts, drilling, setting the depth of milling cuts, and so on. As in the table-type HBM, the spindle diameter and table size specify the machine size.

Cutting Tools. Cutting tools are the same as those used on the table-type machine. Work holding is also the same, and angle plates are frequently used.

17.5.2 Vertical Boring Machines

A general description of a vertical boring machine (VBM) would be that it is a lathe turned on end with the headstock resting on the floor. This machine is needed because even the largest engine lathes cannot handle work much over 24 in. in diameter. A vertical boring machine is shown in Figure 17.19. Today's VBMs are often listed as turning and boring machines. If facing is added to that name, it pretty well describes the principal uses of this machine. Just like any lathe, these machines can make only round cuts plus facing and contouring cuts.

Figure 17.20 shows the general construction and the motions available on the VBM. The construction is the same as that of the double-housing planer, except that a round table has been substituted for the long reciprocating table, and the toolholders are different because the VBM does not need clapper boxes.

The size of a vertical boring machine is the diameter of the revolving worktable. The double-housing VBM is most often made with table diameters from 48 in. to 144 in. Larger machines have been made for special work.

FIGURE 17.19 Vertical boring machine (VBM). (Courtesy Summit Machine Tool Manufacturing Corp.)

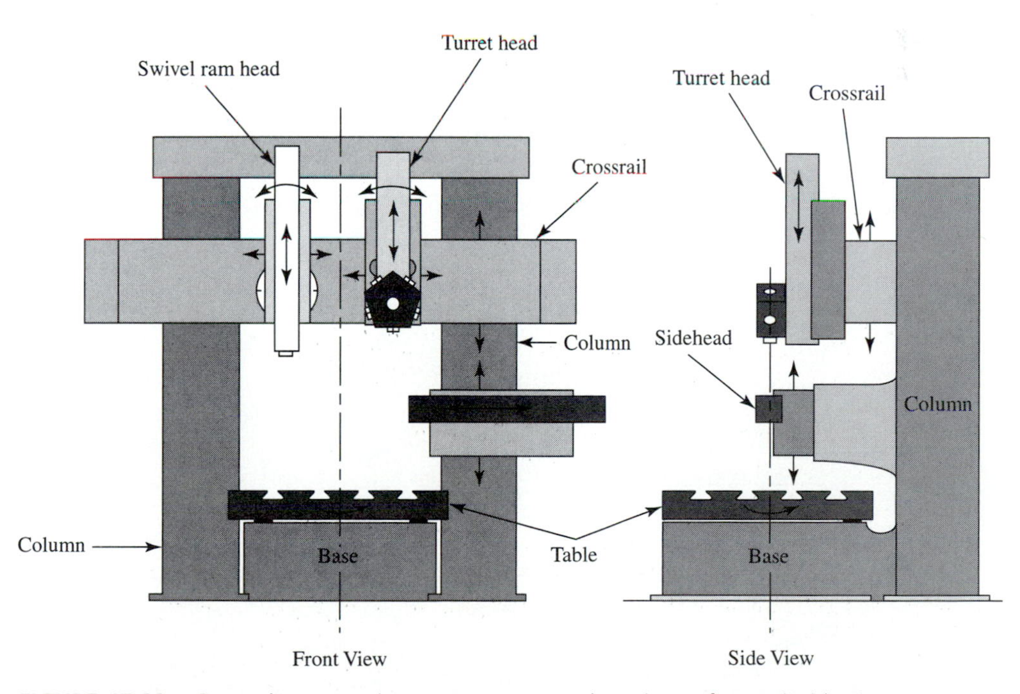

FIGURE 17.20 General construction, components and motions of a vertical boring machine (VBM).

17.5.3 Jig Borers

Jig borers are vertical boring machines with high-precision bearings. They are available in various sizes and used mainly in toolrooms for machining jigs and fixtures. Many jig borers are now being replaced by more versatile numerically controlled machines.

17.6 REVIEW QUESTIONS AND PROBLEMS

1. What three things does boring achieve?
2. What causes drills to "walk" or "wander" off center?
3. Why is tool overhang a very important consideration in boring?
4. Why are positive-rake inserts preferred for boring operations?
5. Discuss the importance of nose radius in boring operations.
6. Explain overhang limits and how they affect boring operations.
7. Why are carbide boring bars able to operate with a larger overhang?
8. Discuss the two types of horizontal boring machines.
9. What precautions are necessary when installing and using a floor-type HBM?
10. How is the vertical boring machine similar to a lathe?

PART V

METALWORKING PROCESSES

INTRODUCTION

Metalworking is a chipless process of producing parts from sheet metal and strip stock using tooling activated by the motion and pressure of a press. Tooling and materials used for this type of production are described and illustrated. Specific types of dies and die components are described and illustrated. Calculations for determining die and press forces are given. Empirical formulas for determining blank sizes for parts are provided, and examples are provided and illustrated.

Cross-section-changing processes utilize dies and presses similar to those used for conventional metalworking processes, but, as the name implies, dimensional thickness changes and cross-sectional variations are accomplished. Equipment, tooling required, and parts produced by this process are illustrated.

CHAPTER EIGHTEEN

METALWORKING MATERIALS

CHAPTER OVERVIEW

- Introduction
- Metalworking Materials
  - Material Form
  - Material Specifications
  - Material Receiving Inspection and Testing
- Testing
  - Olsen Tester
  - Erichsen Tester
  - Fuqui Cup Test
- Review Operations and Problems

18.1 INTRODUCTION

More parts are produced with dies and presses than with any other single metalworking process. The process, known as pressworking, produces parts ranging from extremely small watch and instrument components to very large parts required by the building, transportation, and machine-building industries. Production quantities range from prototype and sample parts to literally millions, each, of countless different parts annually.

Pressworking Defined Pressworking is the formation of a part from sheet metal with pressure exerted through a die by a press. Variations involving materials, die types, and presses exist and are covered in appropriate areas.

Comparison with Machining Pressworking manufacturing planning, facilities, and tooling parallel corresponding activities of machining and metal removal; however, the education, training, and especially the experience required for proficiency in each area is completely different. Pressworking's die engineer, die designer, die maker, and die setter function in similar roles to those of machining-tool engineer, tool designer, toolmaker, and job setter with the expertise required in each area comparable, specialized, and unique.

A machined part is produced by the orderly removal of chips from a casting, forging, or bar stock until the desired shape results; a stamped part is produced from sheet or strip stock by a chipless process of bending, forming, stretching, and piercing.

Tooling for machined parts is fully designed, detailed, and toleranced, and the tooling is manufactured to the design with full confidence it will work as planned. Also, any detail can be replaced by a like detail in the event of wear or damage.

Most tooling for die designs can be similarly designed and detailed but with an exception: Die working surfaces with severe draws or abrupt changes must be finally developed, barbered, and polished to suit the part shape and material at final die tryout. The final development of blank size and shape, die lubricant selection, and blank-holder pressures is also completed at that time. Alternatively, a great many dies have been made to the original design entirely and have produced satisfactory parts from the first start-up. Such dies have been designed by designers who had a great deal of experience and/or were based on a die that had produced similar parts satisfactorily.

18.2 METALWORKING MATERIALS

18.2.1 Material Form

Metalworking materials are supplied in coil (strip) and sheet forms directly from the mill in a wide range of widths, thicknesses, and alloys. Nonstandard widths and thicknesses are also supplied whenever the volume economically justifies the run. Specialty houses supply blanks to specific sizes and coil stock slit to desired widths.

Material can be purchased galvanized on one or two sides; phosphate, zinc, lead (terne), and aluminum coated; and also pre-coated with materials to provide color, wear, and weathering resistance.

Coil and sheet stock is available in aluminum brass, copper, magnesium, steel, zinc, titanium, and almost any other metal—and alloy. They can be ordered by alloy, formability, and hardness in hot- or cold-rolled states.

18.2.2 Material Specifications

Metalworking materials of aluminum, brass, bronze, copper, and steel are well covered in existing standards; most manufacturers use SAE standards for

- Composition
- Mechanical properties
- Thickness tolerances
- Heat treating specifications
- General applications

It is beyond the scope of this book to duplicate standards data, but it is a prime responsibility of the manufacturing engineer to obtain all data and specifications applying to the material to be processed. The material in this chapter is for the *applications* of the metalworking materials.

Product Engineering specifies the most economical material that satisfies the product's requirements. This practice is generally satisfactory for flat parts; however, both chemistry and formability specifications of material must often be modified to make possible the production of certain parts manufactured by bending, forming, and—in particular—drawing processes.

Manufacturing Engineering aids in selecting material for problem parts with recommendations based on similar part configuration/part process experience. Manufacturing Engineering must also make certain that all variables of the process such as punch and die clearances and radii, blank-holder forces, punch speeds, and die lubricants are optimized. Forming, drawing, and deep drawing are areas where, as yet, no substitute for experienced die designers and diemakers exists.

18.2.3 Material Receiving Inspection and Testing

After suitable material has been decided on, with sample parts having been satisfactorily produced and approved, it is imperative that all future shipments of material for that part duplicate the original tested material.

- **Material chemistry** must be confirmed. Incoming material is generally impounded while a coupon, or sample, is sent to the lab for analysis. The material is not released to production until chemistry and all other required tests are approved.
- **Material hardness** must be checked and verified. The Rockwell Hardness Test is usually used to check the hardness of most incoming materials. This test is not entirely accurate for sheet metal—especially in the thinner gages, where the indicator penetrates too far into the metal and the reading is affected by the hardness of the anvil. However, the reading, though not entirely accurate, does act as a comparison with acceptable steel shipments and their readings.
- **Formability** must be confirmed. Several methods of testing formability (previously known as ductility) are used. All are based on the adage, "The best way to determine whether a metal will draw is to draw it." All concerned users of material that require a high degree of formability draw a sample of the material in a test die employing a ball or hemispherically shaped punch. The depth of the draw is determined at the moment of rupture. Testing equipment used for this test is discussed next.

18.3 TESTING

18.3.1 Olsen Tester

The Olsen Tester shown in Figure 18.1 is used for formability testing by clamping a metal sample in the die, then reading both the pressures required and the depths at the point of rupture. Interchangeable cupping dies and punches are used to test samples of different thicknesses and widths.

18.3.2 Erichsen Tester

The Erichsen Tester, like the Olsen, is used to determine formability of material by clamping a sample in a die and drawing the metal to the point of rupture. The depth of rupture, but not the pressure required, is measured in the Erichsen. Illustrated in Figure 18.2, the Erichsen also has dies and punches of varying sizes to test different material thickness.

18.3.3 Fuqui Cup Test

The Fuqui Cup Test determines formability of metal with a die that does not utilize a blank holder. A disk (blank) of the strip metal to be tested is inked with a grid pattern and then drawn to the point of rupture. Flowline data can be deduced from examining the distorted grid pattern of the drawn and ruptured disk.

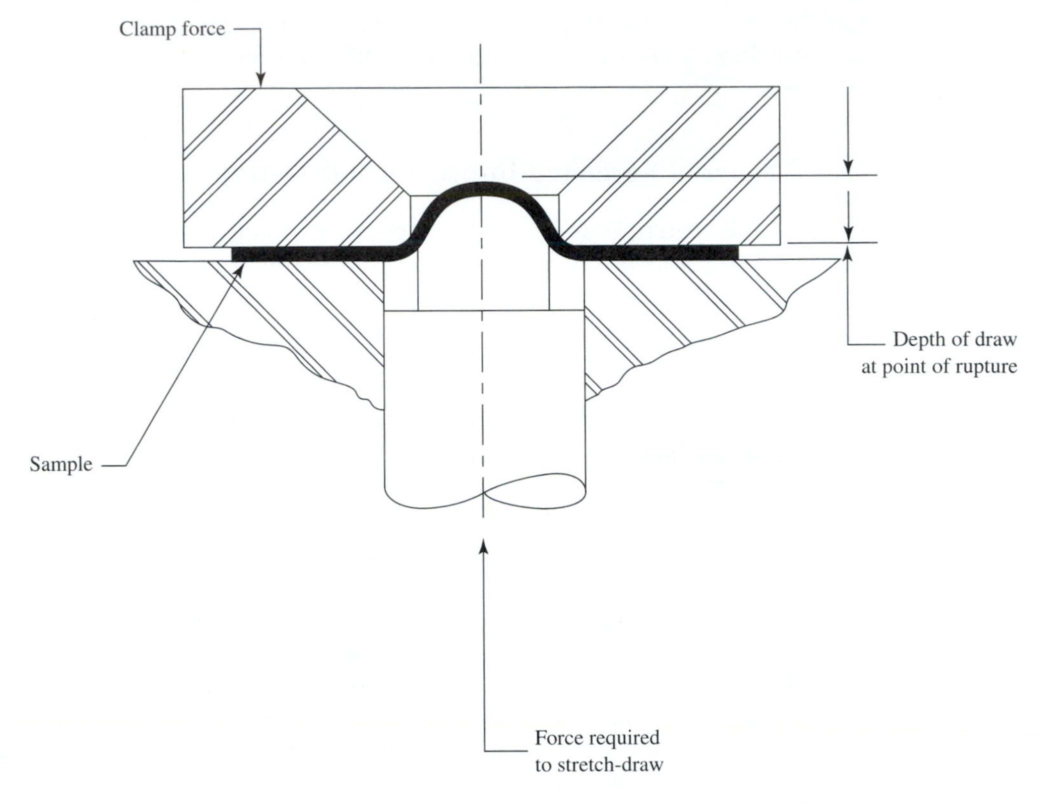

FIGURE 18.1 Olsen tester measures force required and depth to point of rupture.

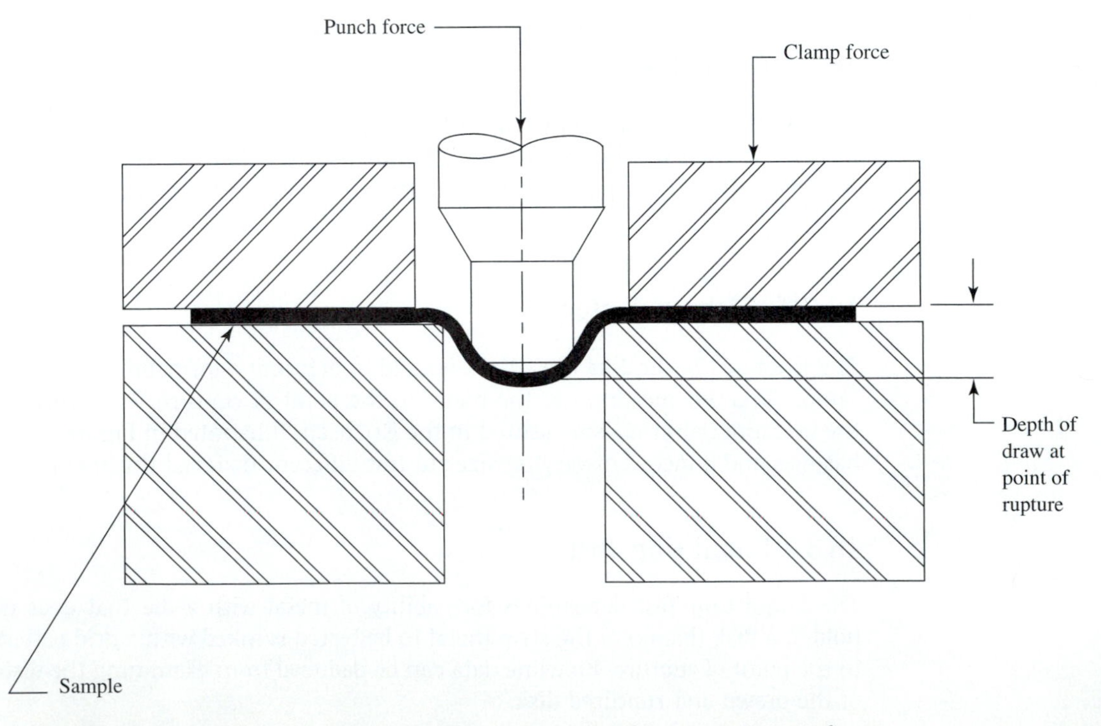

FIGURE 18.2 Erichsen cup tester measures only depth of cup at moment of rupture.

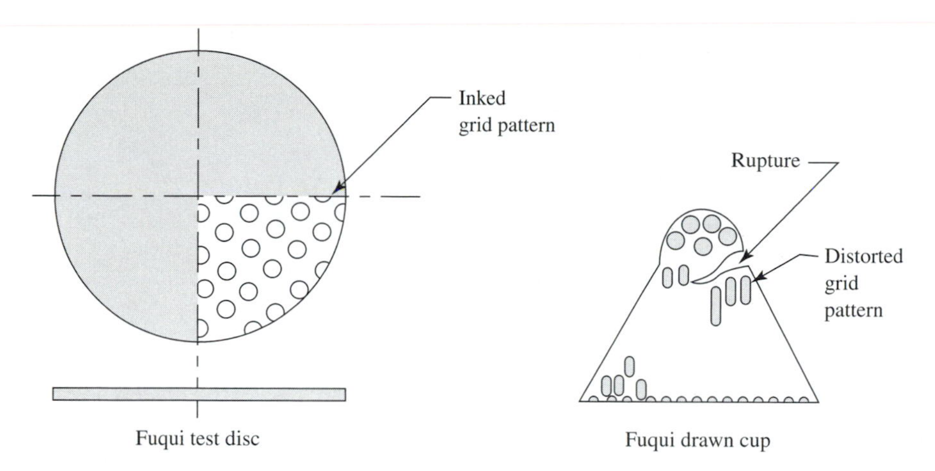

FIGURE 18.3 Fuqui cup test.

A Fuqui test disk and a drawn cup are shown in Figure 18.3. A polished hemispherical punch is used in a die with no blank holder to draw the cup. Wrinkles are avoided by utilizing proper ratios of blank diameter, blank thickness, punch diameter, and die diameter. Blank holders are used in production dies to prevent wrinkles by compressing the metal as it is drawn, creating the friction and resistance to metal flow that controls wrinkles.

A Caveat Pressworking can and does alter a metal's physical properties of hardness. This condition may require correcting with subsequent operations or processes such as annealing—or it may be feasible to factor the condition into a part by starting with softer material and allowing work hardening to bring the metal hardness up to specification.

18.4 REVIEW QUESTIONS AND PROBLEMS

1. Define "pressworking."
2. In what ways is pressworking similar to machining?
3. In what ways do they differ?
4. Discuss the principal forms in which metalworking material is supplied as compared to machining material.
5. What does an Olsen Tester determine?
6. What does an Erichsen Tester determine?
7. Discuss the use of a Fuqui Cup Test. How is it different from the Olsen and Erichsen testers?

CHAPTER NINETEEN

DIE COMPONENTS

CHAPTER OVERVIEW

- Introduction
- Die Components
  - Die Sets
  - Stock Guides
  - Stock Stops
  - Pilots
  - Punches
  - Die Blocks
  - Pressure Pads and Blank Holders
  - Stock Strippers
  - Knockouts and Shedders
- Review Questions and Problems

19.1 INTRODUCTION

Most dies are provided through the efforts of the Manufacturing Engineering department, whose responsibility is to have dies designed, made, and tried out ready for production. The term "die" is the generic name of tooling used, in conjunction with a press, to work metal into a desired form by performing a specific process or processes. Such processes include bending, piercing, forming, drawing, shaving, embossing, and numerous others.

Dies are generally mounted in a press with the tang of the punch holder clamped to the press ram and the die holder bolted to the press bed at the bolster plate. The press, selected by the manufacturing engineer for the specific application, stroke, and tonnage required, provides motion and pressure to the punch holder.

Die details, which are designed and made to suit the specific part design, are assembled in the die set that has been machined and altered to accept the details. Figure 19.1 illustrates the arrangement.

Very few "standard" dies exist that can be purchased off the shelf. Many die components, however, have been standardized, and suppliers exist for die sets, springs, punches, bushings, pilots and other parts, some of which are illustrated in Figure 19.2. This chapter is devoted to

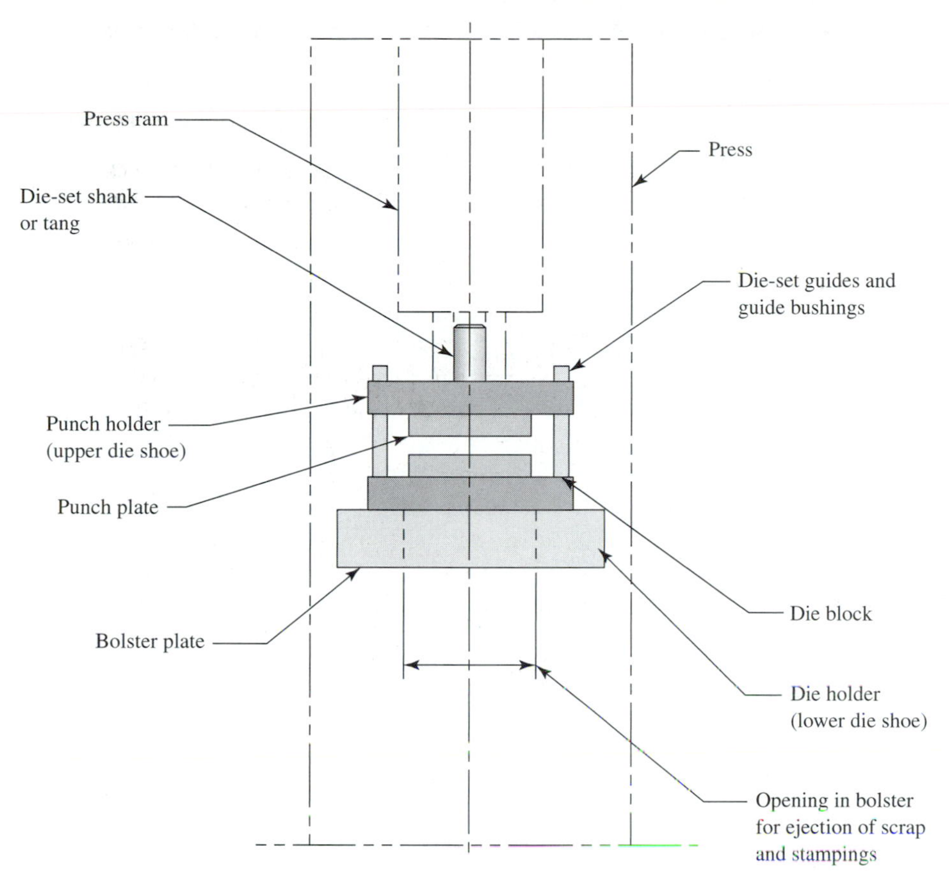

FIGURE 19.1 Arrangement of die in press.

FIGURE 19.2 Standard die makers supplies. (Courtesy Danly Die Set)

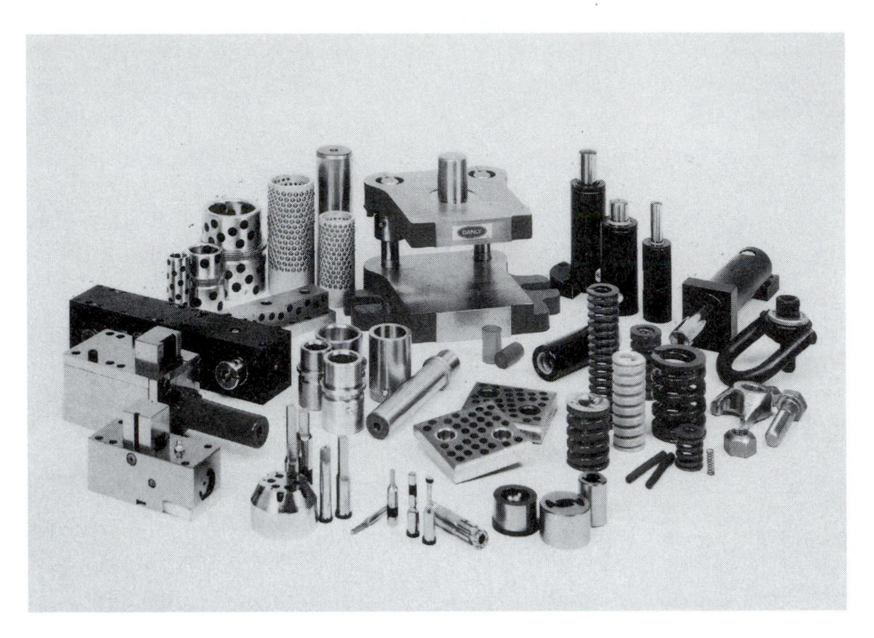

FIGURE 19.2 Continued. (Courtesy Danly Die Set)

the description and function of die components. Other chapters cover the function and description of the different complete dies and processes.

19.2 DIE COMPONENTS

19.2.1 Die Sets

Die sets composed of punch holder and shank (upper shoe), die holder (lower shoe), guideposts, and bushings are available in two-post back, center and diagonal series as well as four-post series. Illustrated in Figure 19.3, they are available in many sizes and levels of quality.

Die sets are very accurately and ruggedly constructed to provide continuing dimensional accuracy and form integrity to the part. Indeed, the heavy guideposts of the die sets not only

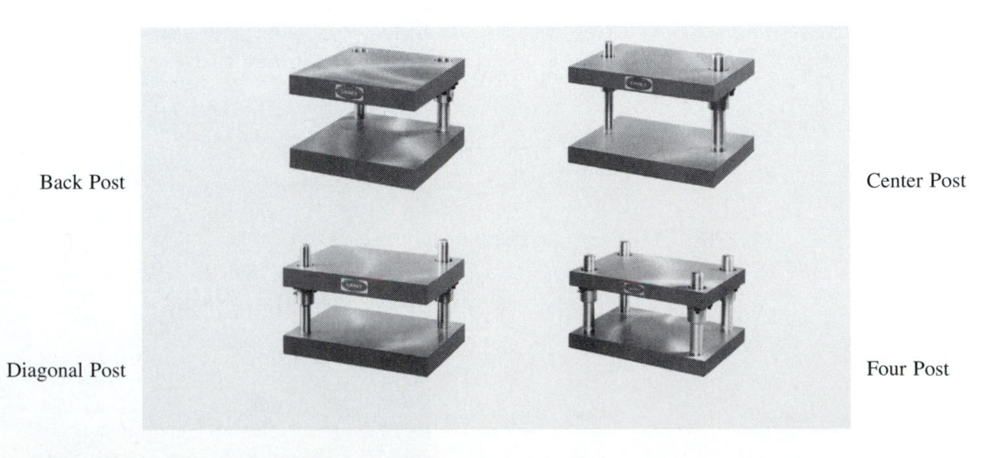

FIGURE 19.3 Standard die sets. (Courtesy Danly Die Set)

maintain alignment between die and punch but, in too many instances, also provide alignment for loosely fitted or worn press ram gibs.

19.2.2 Stock Guides

Stock guides, sometimes called "gages," guide strip stock into and through a die. Many dies, especially those used for short runs and low production, are fed by the operator manually pushing strip stock along and through a guide to a stop in the die as shown in Figures 19.4 and 19.5. The guide must be sufficiently long to mechanically aid aligning the metal strip with the back edge. Some guides are provided with rollers or other devices to force the metal strip into contact with the rear edge of the guide. Die construction dimensions relate to this surface.

Guides can be used to strip stock off punches for certain parts and conditions. In use, the operator pushes the strip stock through the guide slot or tunnel to a stop and activates the press. The punch pierces the metal on the downstroke, then (by friction) lifts it to contact the under surface of the tunnel cover. The metal is retained and stripped from the punch as the press completes the upstroke, as illustrated in Figure 19.5.

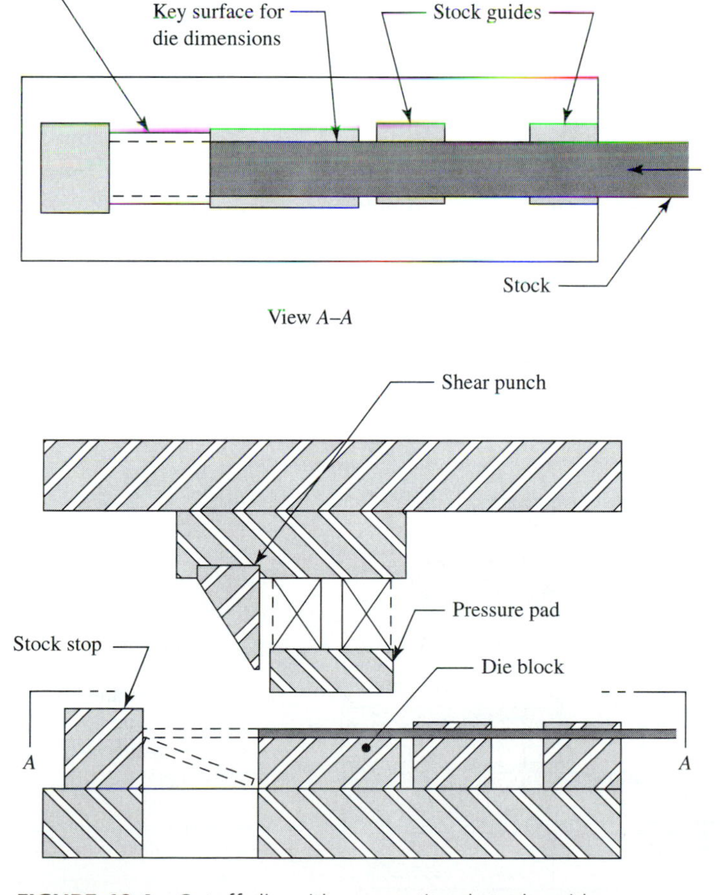

FIGURE 19.4 Cutoff die with conventional stock guides.

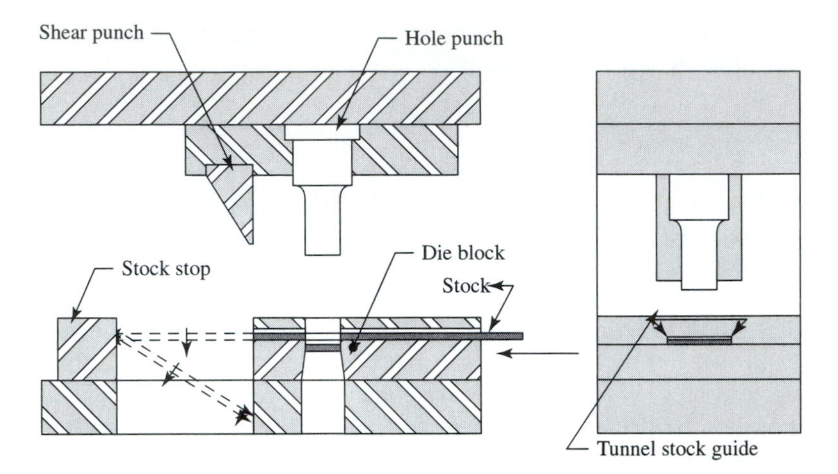

FIGURE 19.5 Pierce and cutoff die with tunnel stock guide.

19.2.3 Stock Stops

Stock stops are devices used to stop the motion of and locate strip stock moving through a die. Several types of stops are employed, depending on the part design and rate of production. Stops function by registering in a previously pierced hole or notch or against an end face so as to interfere with and stop lateral motion of the strip during the die stroke. There are several types of stops:

- Pin. This is the simplest form of stop and can be used with the tunnel stripper. It functions by registering in a previously pierced hole or notch and is perhaps the fastest and most convenient form of stop for an operator to use manually. It is shown in Figure 19.6.

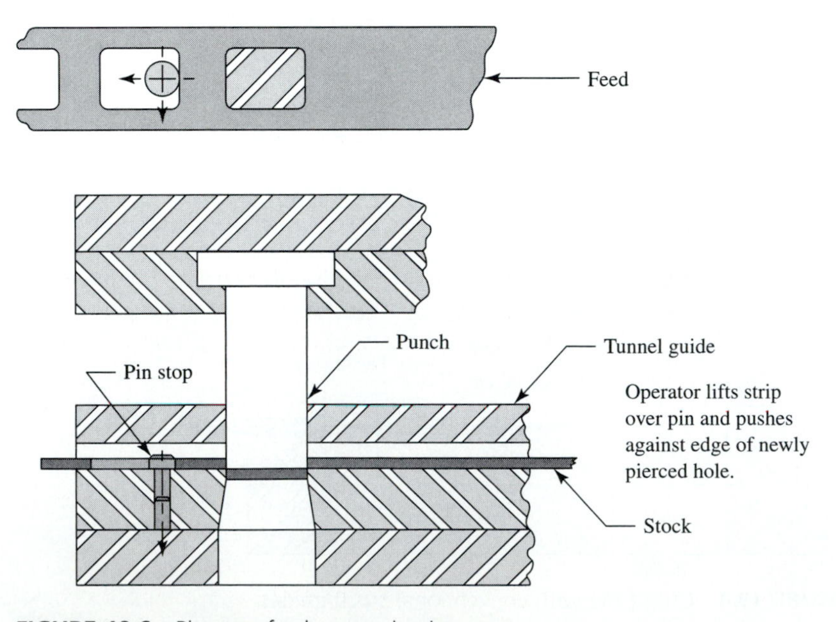

FIGURE 19.6 Pin stop for low production runs.

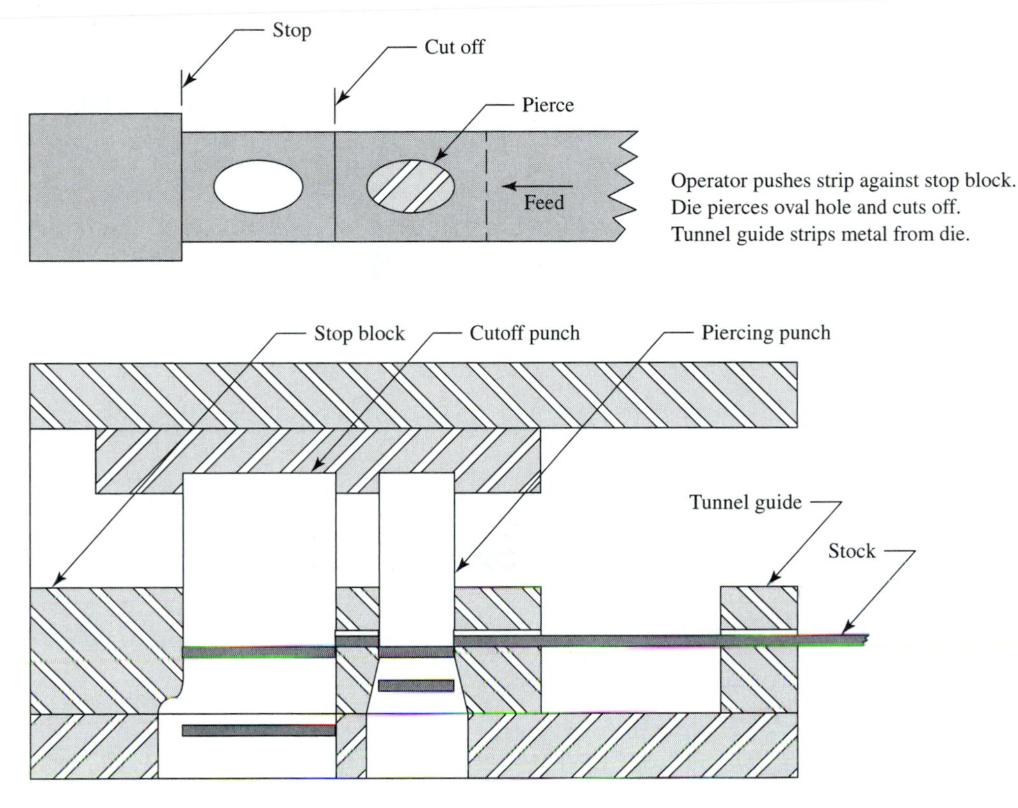

FIGURE 19.7 Block stop for low and medium production.

- **Block.** Another stop design provides a block that registers against the end face of a metal strip. After a subsequent shearing of the part from the strip, the stock strip is free to again advance to the stop. This stop is illustrated in Figure 19.7.
- **Automatic Stop.** One stop design features a spring loaded member that is forced automatically into a register hole or notch. It is released by a push rod during a portion of the die stroke, allowing the strip to advance one increment. This design is illustrated in Figure 19.8.

19.2.4 Pilots

Strip stock that is power fed through a progressive die must be rapidly and accurately positioned in relation to each stroke of the press. For this purpose, a pilot hole is pierced in the first stage of the die. This hole may or may not survive as a feature of the part design due to further part development at a succeeding stage of the die.

If the use of an integral hole in the part is not feasible, a hole, known as an indirect pilot, may be pierced along the edge of the strip. A pilot, located in the second stage of the die, registers in the pierced hole to locate the stock strip as shown in Figure 19.9.

Pilots may be solid or spring loaded and operating in a bushing. Some components are commercially available, but the pilot must be designed to be an integral part of the die. It is good practice to treat a pilot as a punch. If the pilot hole is missing due to a broken punch or any other reason, the pilot will punch a hole in the strip, and provisions must be made to dispose of the resulting slug. Not to do so risks wrecking the die caused by a buildup of slugs.

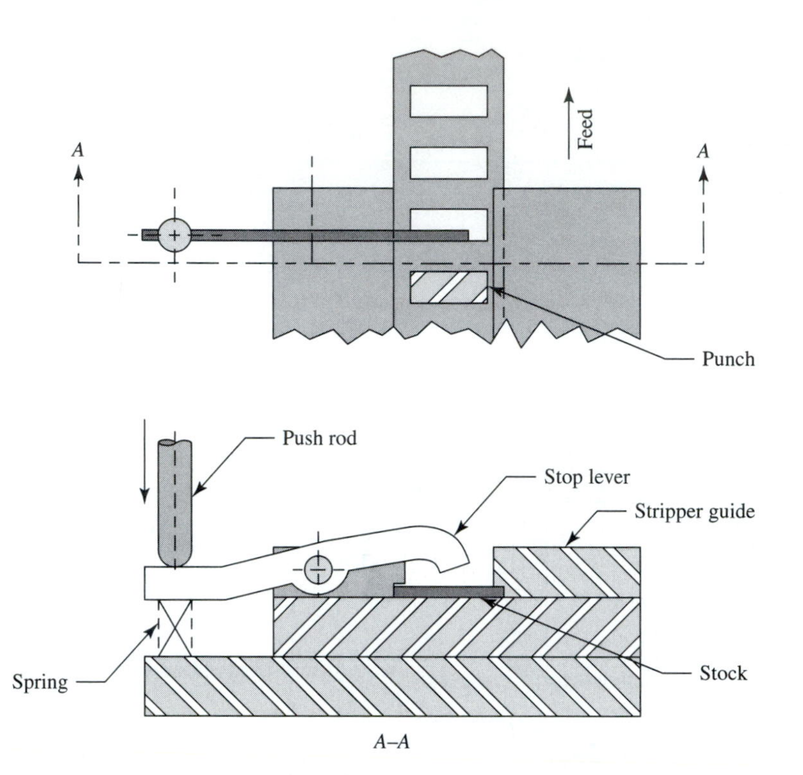

FIGURE 19.8 Automatic stop for medium and high production.

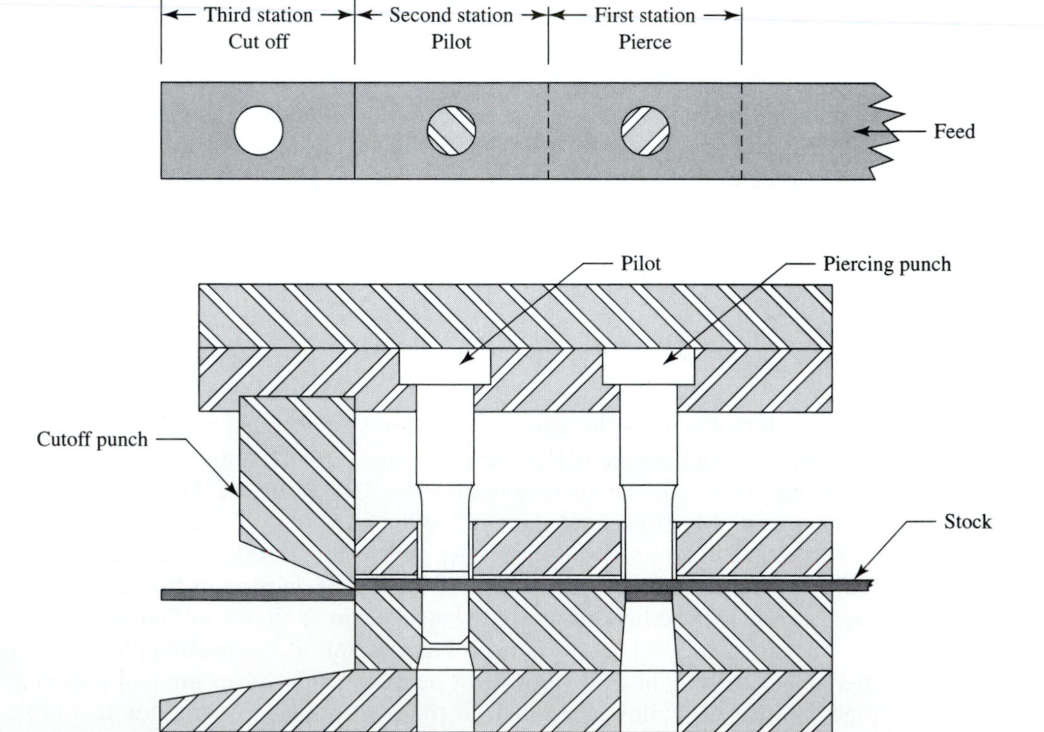

FIGURE 19.9 Use of pilot in simple progressive die.

19.2.5 Punches

The working (movable) upper portion of a die set is known as the punch holder. Details known as punches may form the strip material against a correspondingly contoured die block, or they may be so designed to pierce holes (round, geometrically shaped, or irregular) in the metal.

Punches used to pierce round holes are commercially available in many diameters, lengths, and methods of retaining. Special sizes, lengths, and unique shapes must be designed and made to suit the part design. Punches are sharpened by grinding their end faces. One is shown in Figure 19.10.

19.2.6 Die Blocks

Die blocks from which dies are cut are generally mounted on the lower shoe of a die set. Some inverted die designs place the dies on the upper shoe for special reasons, which will be covered later. Dies are designed to match and accommodate piercing punches, pilots, and forming punches located and held in the upper punch holder.

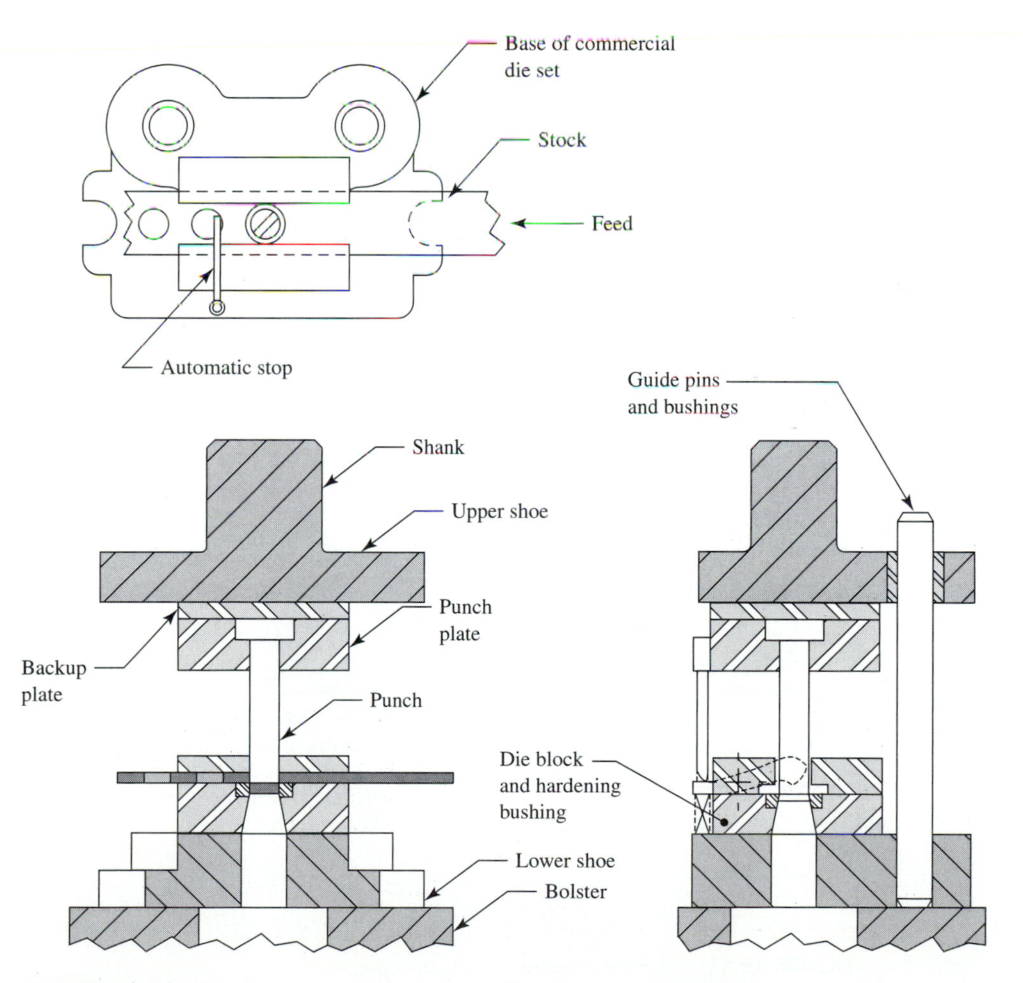

FIGURE 19.10 Punch and die relationships with other components.

Die blocks used in piercing operations may have hardened bushings or hardened top wear plates included for purposes of wear and maintenance. They are sharpened by grinding their top surfaces. Figure 19.10 illustrates the punch and die relationships with other components.

19.2.7 Pressure Pads and Blank Holders

The terms **pressure pad** and **blank holder** are often used interchangeably—and, in certain instances of practice, they are one unit. A pressure pad is illustrated in Figure 19.11.

A pressure pad is a die component that utilizes air pressure, hydraulic pressure or the forces of compressed die springs or rubber to function. A pressure pad can be used as

- A component of a pad-type form die (Fig. 19.11a)
- A hold down in a bending die (Fig. 19.11b), and cutoff die (Fig. 19.11c)
- A part stripper (Fig. 19.11d)
- A blank holder in a draw die (Fig. 19.11e)

A blank holder, used to locate and apply pressure and friction to a blank, controls the flow of metal as it is drawn into a cup by a punch.

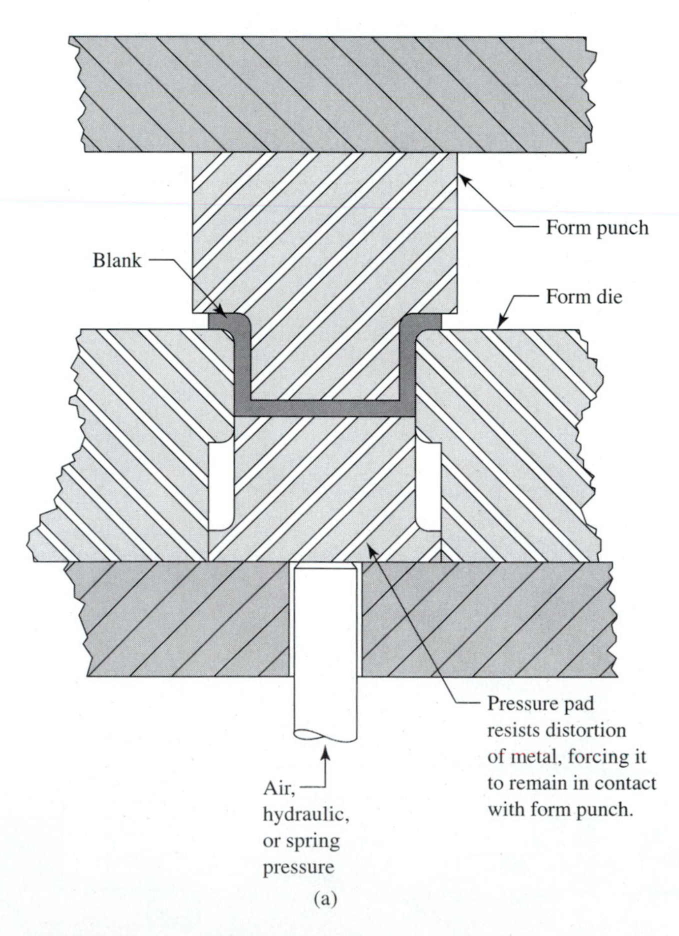

FIGURE 19.11 (a) Pressure-pad use in form die.

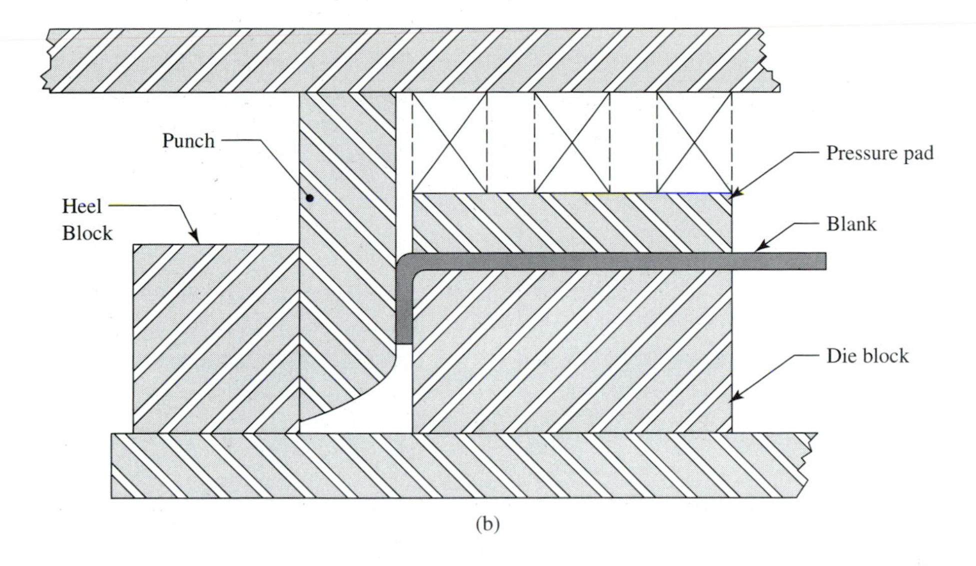

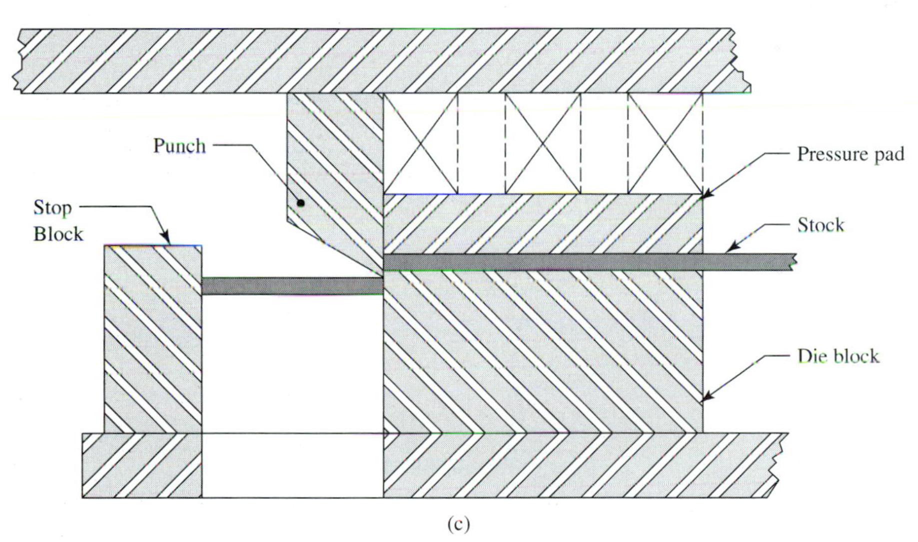

FIGURE 19.11 Continued (b) Pressure-pad use in bending die. (c) Pressure-pad use in cutoff die.

19.2.8 Stock Strippers

Two general types of strippers are used to strip parts from a punch or die:

- Fixed or passive
- Pressure or spring operated

Passive strippers in the form of tunnels and hooks function by retaining a part as the punch withdraws. Figure 19.12 shows a stock guide designed with a top cover, or plate, to form a tunnel.

After completion of the piercing operation, the punch begins its upstroke carrying the part up to contact the under surface of the cover; it is stripped off as the punch completes its travel.

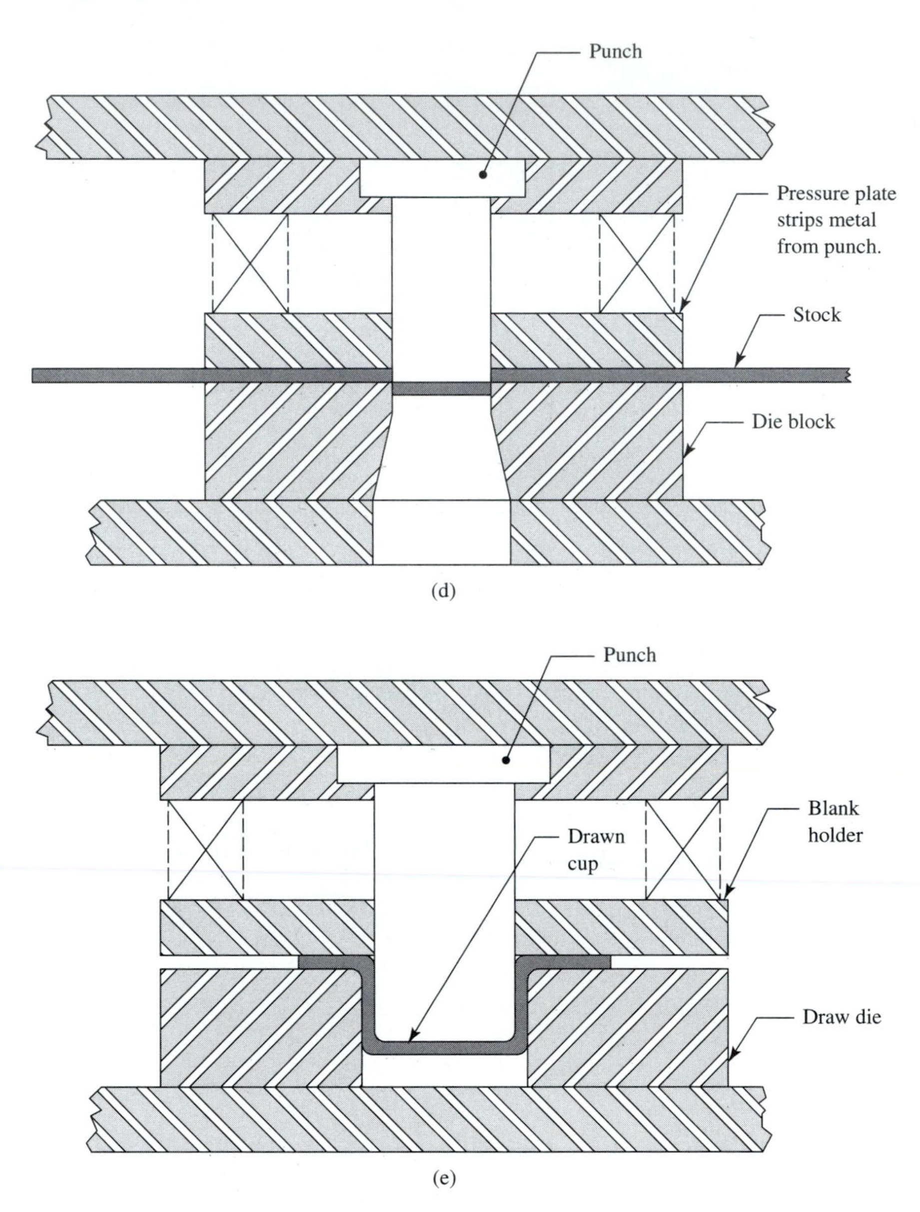

FIGURE 19.11 Continued (d) Pressure-plate use as stripper in piercing die. (e) Pressure plate as blank holder in draw die.

The hooks shown in Figure 19.13 function in a similar manner but are used on heavier gage metal (#13 gage and heavier) where distortion will not be a problem.

Pressure strippers are designed to have the required motion and pressure to positively remove, or strip, a part from a punch or die. Strippers used to eject parts out of a die are called knockouts.

Most pressure strippers are designed with commercial die springs having calculated quantities, grades, and lengths that, when compressed, provide the required pressure to force the part from the punch. Calculations of the very significant stripping forces involved are covered in the section 20.8, "Pressures and Forces." Figure 19.11(d) illustrates the function and arrangement of a typical pressure die stripper.

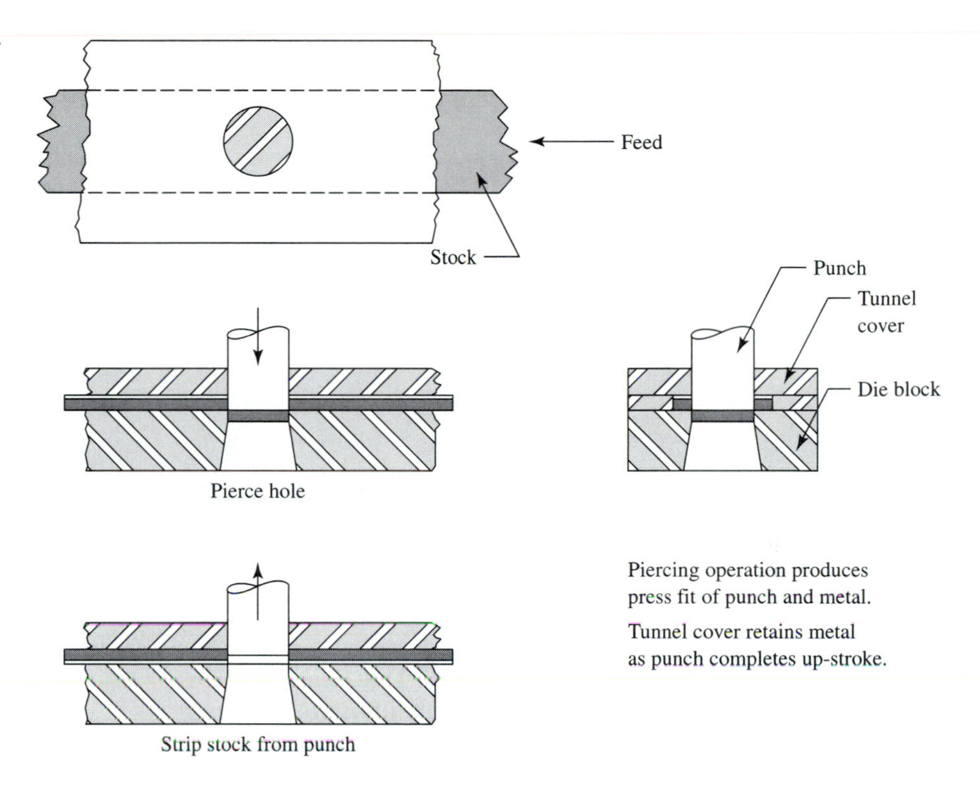

FIGURE 19.12 Tunnel stripper operation.

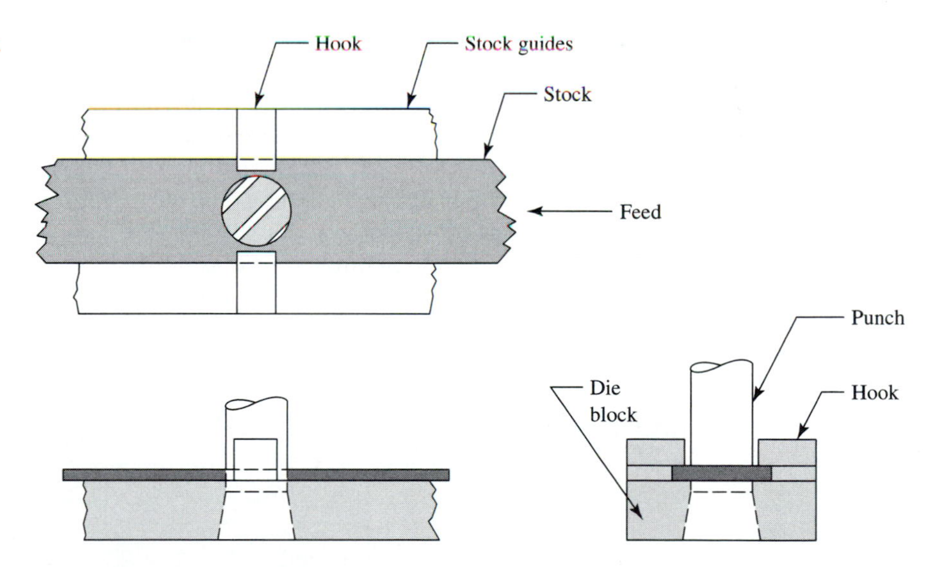

FIGURE 19.13 Hook-type stripper.

19.2.9 Knockouts and Shedders

Knockouts Knockouts, devices used to eject blanks retained in a die by metal pressure, are operated by spring pressure or mechanically.

Spring-operated units have springs that are compressed on the downstroke of the press and expand on the upstroke, forcing a blank out of a die. The action is similar to that of a die spring-operated stripper.

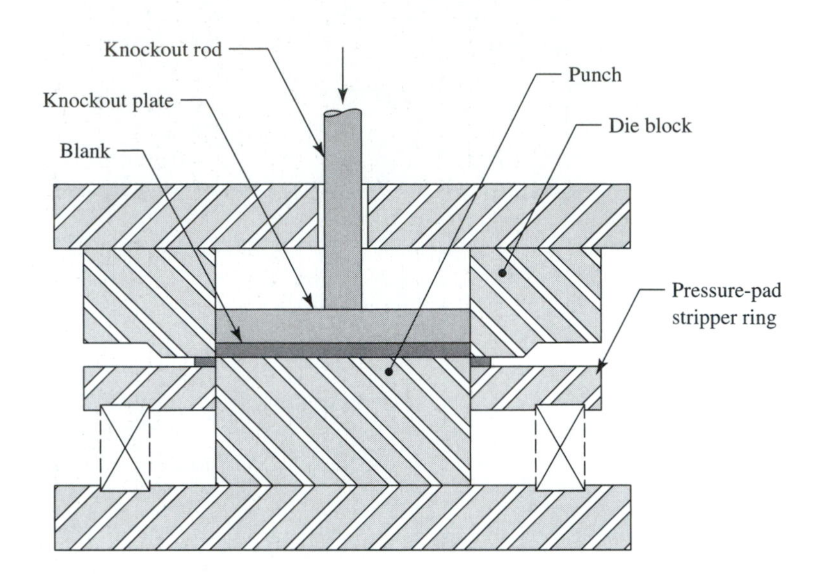

FIGURE 19.14 Mechanically operated knockout.

A **mechanically operated unit**, sometimes called a positive knockout, is operated by a mechanical link as motion to dependably eject a blank from a die.

The inverted piercing die illustrated in Figure 19.14 utilizes a positive knockout to eject a blank from the die, which, in this design, is mounted to the upper shoe. Where a mechanical knockout can be used in a die design, the press tonnage or pressure requirements are lowered, because there are no die springs to compress. Mechanical knockouts are not feasible on all die designs, however, due to space or accessibility considerations.

Shedders Similar in design to knockouts, but normally of much lighter construction, a device known as a shedder strips blanks that are held to the face of a punch by an oily film. Such parts, blanked from light-gage metal, are carried up with the punch. They must be removed or they will stack and accumulate, with eventual damage to the die and possibly the press. One arrangement for this device is shown in Figure 19.15.

These devices, also sometimes known as push-off pins or oil seal breakers, are also used to elevate strip stock prior to indexing in a progressive die. This is done to ease partially formed parts in the strip development over forming die edges as the metal advances to the next station.

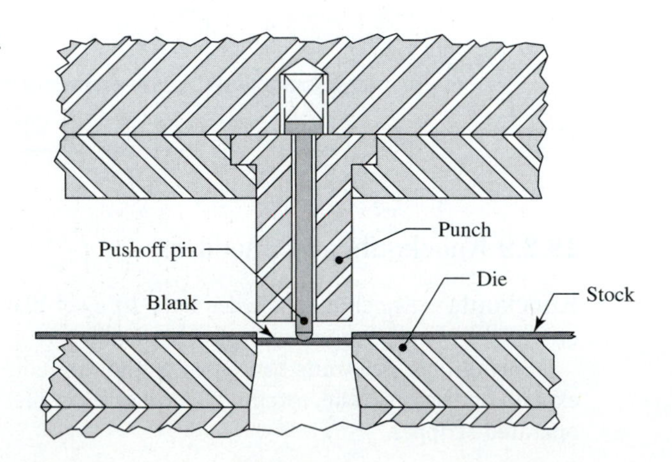

FIGURE 19.15 Shedder.

19.3 REVIEW QUESTIONS AND PROBLEMS

1. Which component of a die set is movable; that is, which is clamped to and moved by the press?
2. What are the major components of a die set?
3. Which surface of a stock guide is considered a "key" surface? Why?
4. How many different general types of stock stops are there? What are their functions?
5. How does a pilot function?
6. What operations do punches perform?
7. What is a pressure pad? Describe it. What functions does it perform?
8. How are die blocks used?
9. What does a stock stripper do? How many general types of stock strippers are there, and how does each work?
10. What is a knockout? What is a shedder, and how does it differ from a knockout in its function?

CHAPTER TWENTY

DIE PROCESSES

CHAPTER OVERVIEW

- Introduction
- Metal-Cutting Dies—Blanking and Piercing
  - Shearing Action
  - Fineblanking
  - Blanking Die—Conventional Design
  - Blanking Die—Inverted Design
  - Piercing Die
- Bending Dies
- Forming Dies
  - Solid Form Die
  - Pressure-Pad Forming Die
- Progressive Dies
- Drawing Dies
- Calculating Blank Sizes, Reduction of Diameters (Severity of Draw), and Cup Heights for Cylindrical Shells
- Pressures and Forces
  - Center of Pressure
  - Cutting, Blanking, and Shearing Forces
  - Bending Forces
  - Drawing Forces
  - Blank-Holder Force
  - Stripping Forces
  - Press Tonnage
- Roll Forming
- Review Questions and Problems

20.1 INTRODUCTION

Manufacturing Engineering develops processing, specifies tooling, and calculates die forces required to produce sheet metal parts.

More than 20 different dies are recognized as unique to perform specific metalworking operations. Some dies, termed "progressive," are capable of performing more than two operations in sequence as the metal is advanced. All are special. Each is designed to satisfy part designs, metal specifications, and rates of production. The more common and widely used dies are metal cutting, bending, forming, drawing, and progressive. Each is illustrated and functionally described in this section.

Die forces must be calculated to determine the required tonnage of presses as well as to provide guidance to the die designer for determining die sections and materials. Formulas of die force calculations are provided as well as examples of drawing, blank development and forces.

20.2 METAL-CUTTING DIES—BLANKING AND PIERCING

In pressworking, the term "blank" refers to a shape cut, or sheared, from strip or sheet stock ready for subsequent operations; the term **blanking** refers to the action of producing the blank. A blanking die cuts, or shears, a shape from strip or sheet stock. The die may be single purpose, or it may be a station in a progressive die.

20.2.1 Shearing Action

The mechanics of shearing are illustrated in Figure 20.1(a) and (b). Figure 20.1(a) shows a blank deformed under pressure between punch and die elements. As the operation continues, the punch penetrates to a depth, depending on metal hardness and punch clearances, where the metal fractures, as shown in Figure 20.1(b), producing a slug.

The resulting pierced hole has a rounded or deformed entry edge followed by an accurate burnished diameter (cut band) and then a fractured breakout. The width of the cut band depends on the softness of the material, provided that the punch clearance is optimal. Softer materials provide wider cut bands. The cut band section of the hole is the only portion that is to the size of the punch and as specified by product design. Many stamping designs specify "die edge of blank," which must be observed when designing the die.

A burr is formed on both the blank and slug. Burr heights increase with greater ductility of metal, increased clearances, and developing dullness of punches. Burrs must be controlled for general appearance and quality reasons, but especially where they will detrimentally affect following operations such as beading and flanging.

20.2.2 Fineblanking

Fineblanking is an operation that, where applicable, can be used to blank parts and provide smooth, full-length edges with virtually no burrs. Illustrated in Figure 20.1(c), the operation utilizes a circular V-shaped stinger, which locks the metal to prevent distortion and forces parallel penetration of the punch. The punch has extremely small clearances. A pressure pad, compressed by punch stroke, helps "lock in" material.

Fineblanking requires a triple-action hydraulic press where the movements of the stinger ring, punch, and pressure pad are independently controlled. Punch speed is very low. The part design must be suitable for the operation.

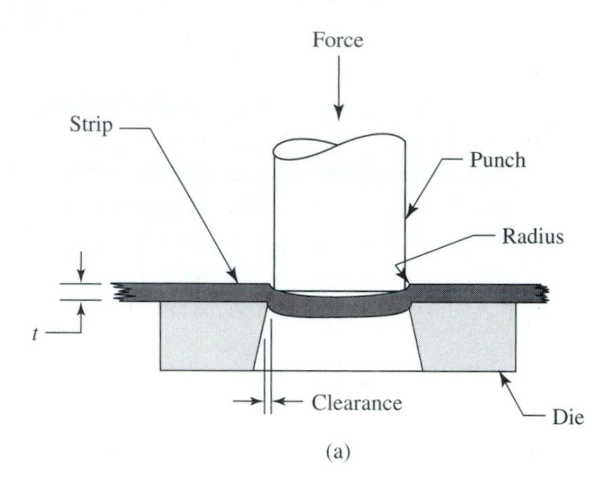

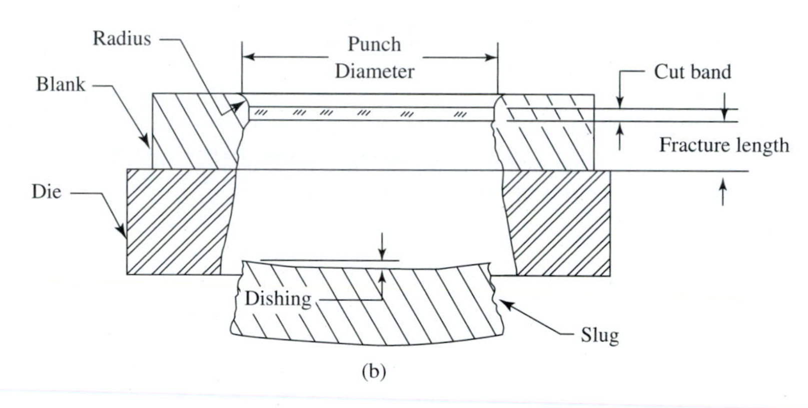

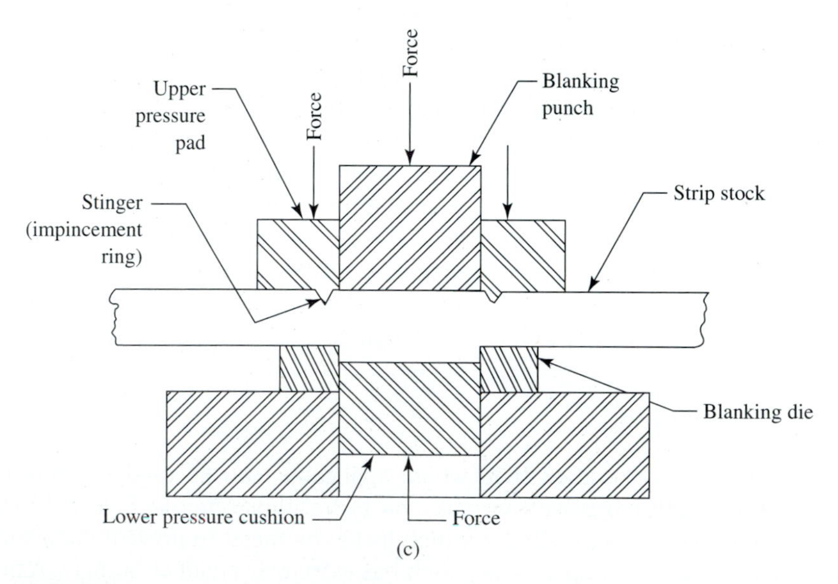

FIGURE 20.1 (a) Shearing process schematic. (b) Pierced hole and slug characteristics. (c) Fineblanking process.

20.2.3 Blanking Die—Conventional Design

A **blanking die** conventionally arranged, with the punch fastened to the upper shoe and the die located on the lower, is limited to smaller-size blanks. Blanks produced by this design of die are ejected through the opening in the bolster plate and are therefore limited to the size of that opening. A conventionally designed die for small blanks is shown in Figure 20.2(a) and (b).

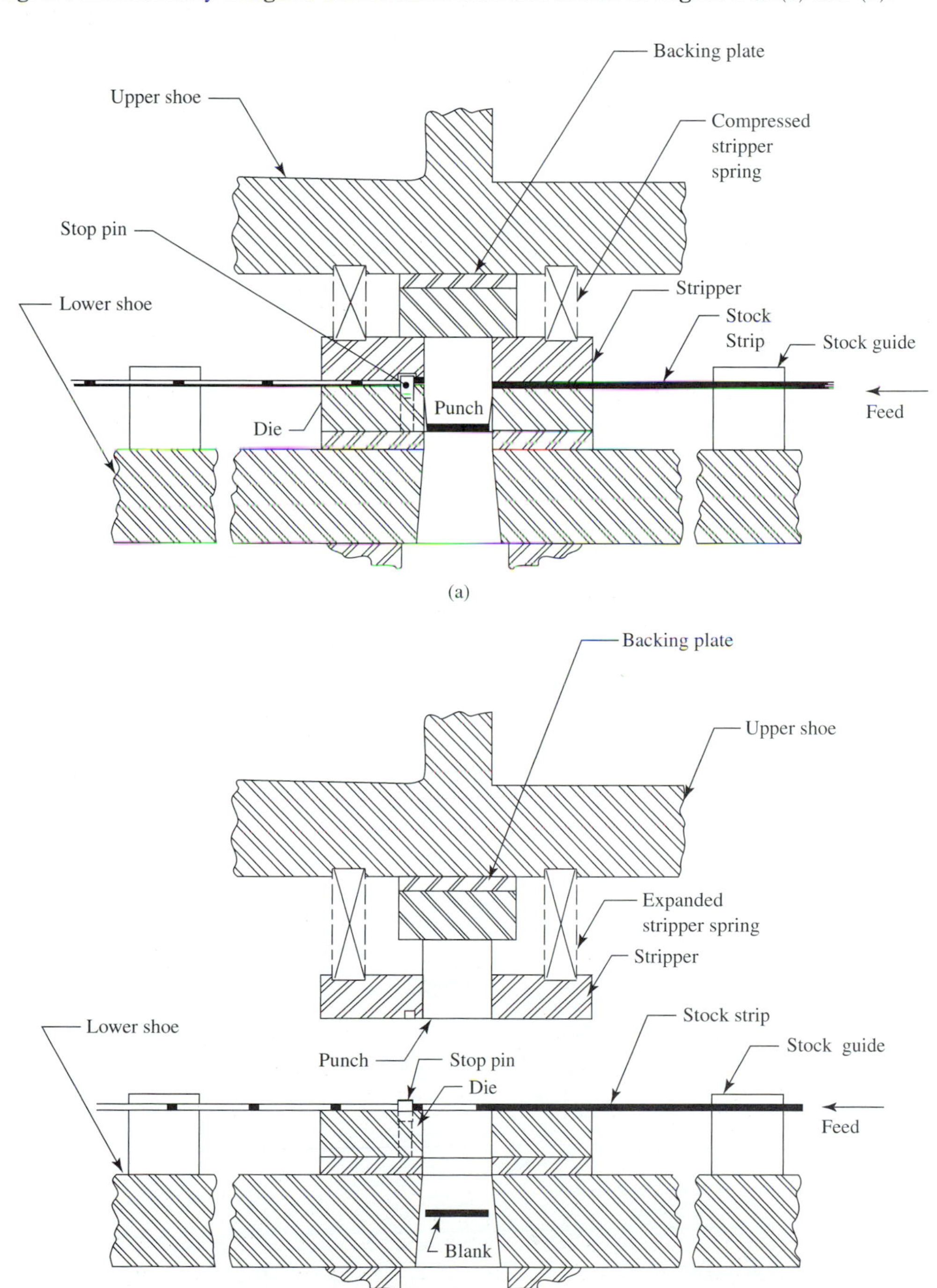

FIGURE 20.2 Conventional blanking dies for small parts: (a) closed, and (b) open.

Functions During One Cycle of Die (Fig. 20.2)

- Strip stock is advanced to stop pin.
- Press is cycled.

Downstroke (Fig. 20.2a)

- Punch shears blank into die.
- Stripper springs are compressed.

Upstroke (Fig. 20.2b)

- Stripper retains stock as springs expand.
- Punch withdraws.
- Blank ejects through bolster plate opening.

20.2.4 Blanking Die—Inverted Design

A blanking die designed for larger parts reverses the arrangement of punch and die locations as shown in Figure 20.3.

Functions During One Cycle of Die (Fig. 20.3)

- Strip stock is advanced to stop.
- Press is cycled.

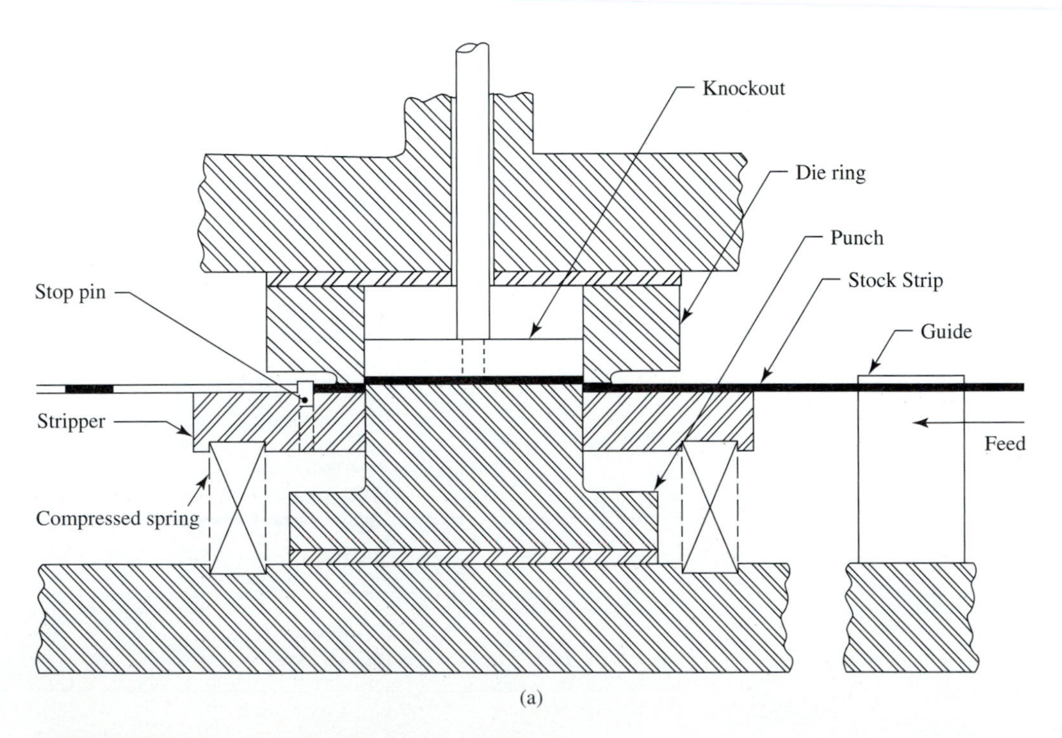

FIGURE 20.3 Inverted blanking die for larger parts.

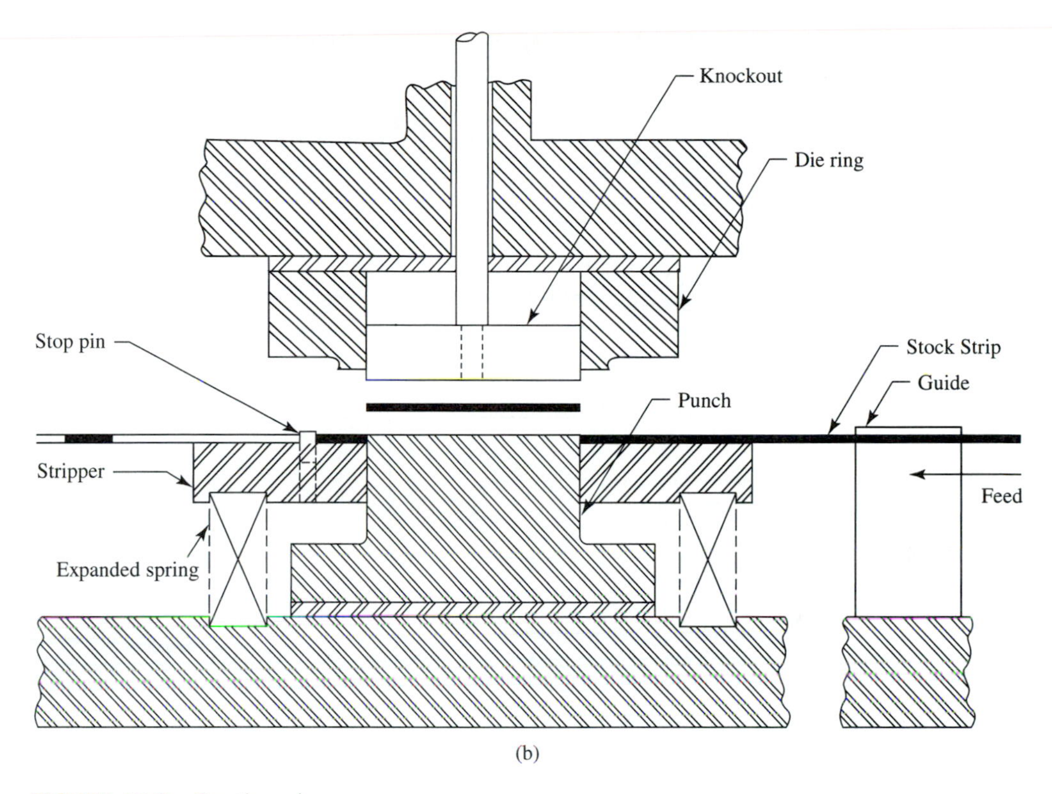

FIGURE 20.3 Continued

Downstroke (Fig. 20.3a)

- Die ring shears stock over punch.
- Stripper springs are compressed.

Upstroke (Fig. 20.3b)

- Stripper pushes stock off punch as springs expand.
- Knockout ejects blank from die ring.

20.2.5 Piercing Die

Piercing is the process of cutting holes, openings, and slots in sheet metal strip or blanks. Similar in design to a blanking die, a piercing die can be used to provide secondary operations such as those illustrated in Figure 20.4. The single-station die illustrated in Figure 20.5 is designed to pierce two holes in an oval blank.

Functions During One Cycle of Die (Fig. 20.5)

- Blank is loaded in nest.
- Press is cycled.

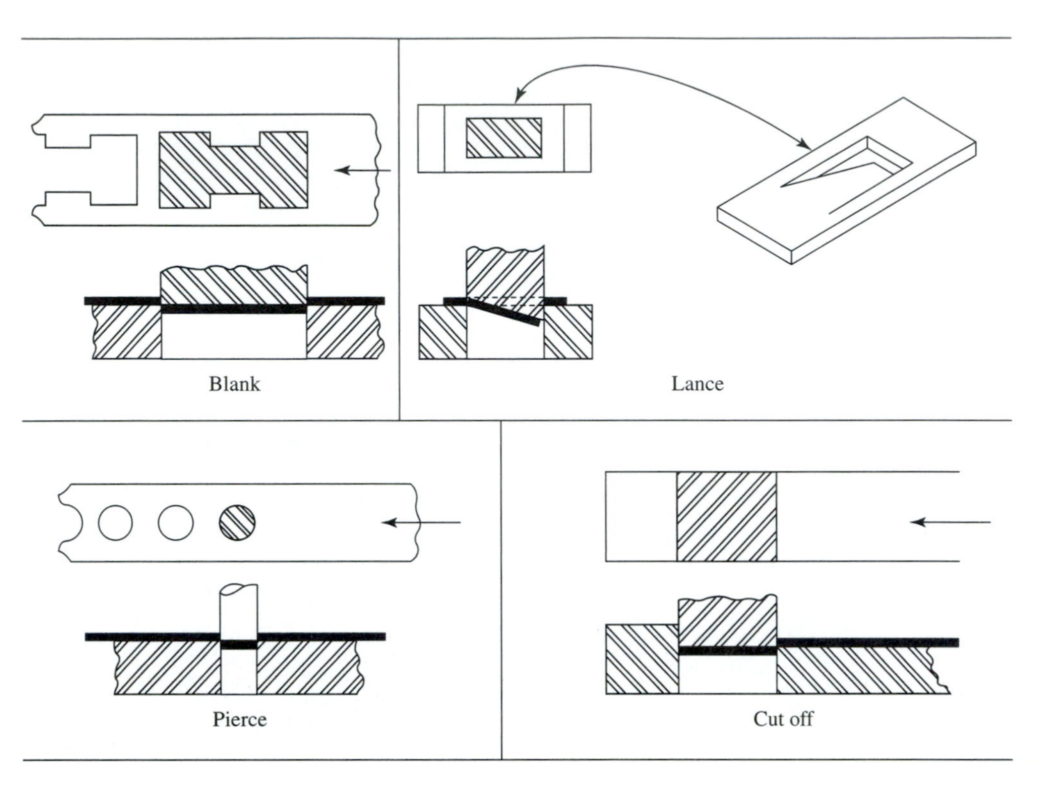

FIGURE 20.4 Piercing die-cutting operations.

Downstroke (Fig. 20.5a)

- Punches pierce two holes.
- Stripper springs are compressed.

Upstroke (Fig. 20.5b)

- Stripper ejects blank from punches as springs expand.
- Slugs are disposed of through bolster opening.

20.3 BENDING DIES

Bending dies deform metal along a straight axis to an angle, or a radius, but do not reproduce the exact shape of the punch and die. Where exact reproduction of a die shape is required, a forming operation should be used.

Three variables affect the bending process and must be provided, or allowed for:

1. **Bending Radii.** The *minimum* bending radius of each metal is different but, in general, annealed metals can be bent to a minimum radius equal to their thickness.

2. **Bending Allowances for Length.** The overall length of a metal piece increases with bending. If this development affects a part's function, it must be allowed for. The length of bent metal can be calculated from the formula

$$B = \frac{A}{360} \times 2\pi(R_i + Kt)$$

FIGURE 20.5 Piercing die: Pierce two holes. (a) Closed. (b) Open.

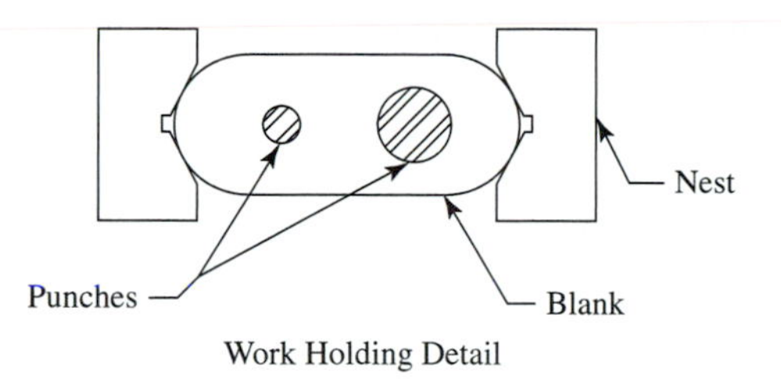

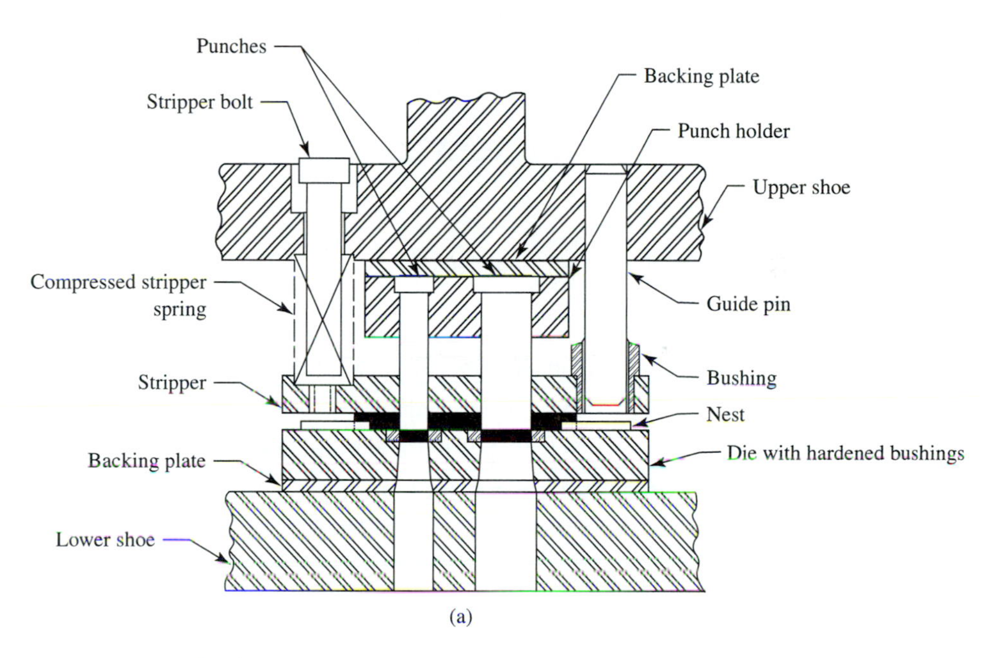

(a)

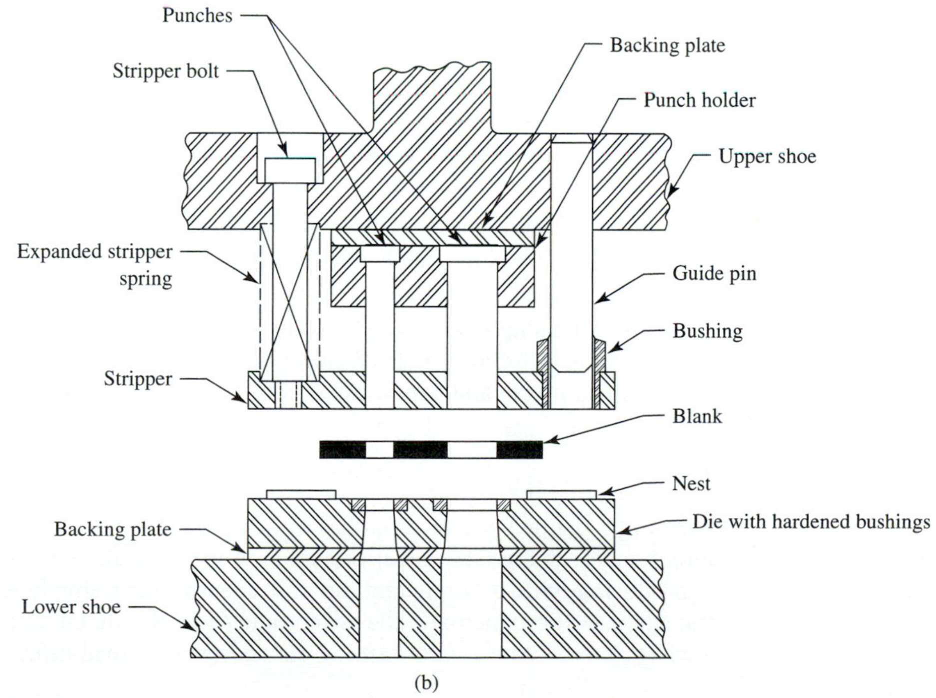

(b)

20.3 BENDING DIES (Continued)

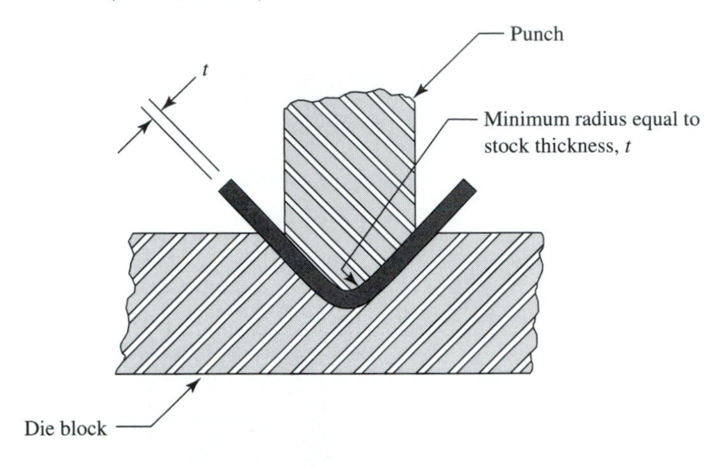

FIGURE 20.6 V-bending method.

where B = Bend allowance in inches (along neutral axis)
A = Bend angle in degrees
R_i = Inside Radius of Bend in Inches
K = Neutral axis constant
0.33 where R_i is less than $2t$
0.50 where R_i is greater than $2t$
t = Metal thickness in inches

3. **Springback.** When die clamping and bending pressures on metal are released, the material's bend angles decrease and corner radii increase as built-up stresses are released. The amount of this condition, known as springback, varies with the different kinds of materials and their hardnesses. Springback can be compensated for in a V-bending die with punches and dies that have reduced included angles and radii.

Little or no springback is experienced with edge bending because the metal is "ironed" and formed by the punch, normalizing the work strains.

- **The V-bending method,** illustrated in Figure 20.6, is accomplished by forcing a metal strip or sheet into a die block which has the required V-angle, by a wedge-shaped die with the same angle.
- **Edge bending** is performed by clamping the workpiece to the die block so that the portion to be bent extends in a cantilever fashion. The punch then moves down, contacts the extended metal, and wipes it down the side of the die block as shown in Figure 20.7.

20.4 FORMING DIES

Forming is defined as a process that makes a change in the shape of a metal piece but does not intentionally change the metal thickness. A forming die is really a bending die that is capable of producing more accurate and intricate shapes than a simple V. A forming die is one in which the shape of the punch and die is reproduced in the metal with little or no metal flow. There are two general types: solid-forming die and pressure-pad forming die.

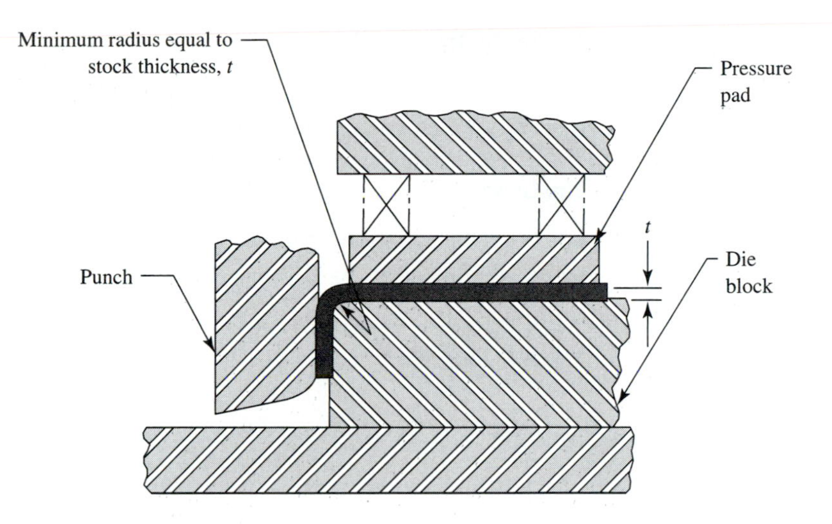

FIGURE 20.7 Edge-bending method.

20.4.1 Solid Form Die

This die adheres to the definition of a forming die by having both the punch and the die follow the part design precisely. It is the simplest form die to design and construct and is acceptably productive if kept within its limitations. Solid form dies are mainly used for parts with generous radii and open angles that are easily removed from the die after forming. A typical application is shown in Figure 20.8.

Function During One Cycle of Die (Fig. 20.8)

- Blank is loaded in nest against stop.
- Press is cycled.

Downstroke

- Punch forces blank into contour of die.

Upstroke

- Shedder pushes part out of die.
- Knockout ejects part from punch.

20.4.2 Pressure-Pad Forming Die

A pressure-pad forming die is used for parts with more severe forming requirements than can be accommodated in a solid form die: small corner radii, close tolerances, 90° bends, and the elimination of springback. A pressure-pad forming die is illustrated in Figure 20.9.

Function During One Cycle of Die (Fig. 20.9)

- Blank is loaded in the nest.
- Press is cycled.

Downstroke

- Punch forces blank into die, compressing springs under pressure pad.
- Pressure-pad resistance forces metal to form to bottom of punch.

Upstroke

- Pressure pad forces part out of die as springs expand.
- Knockout ejects part from punch.

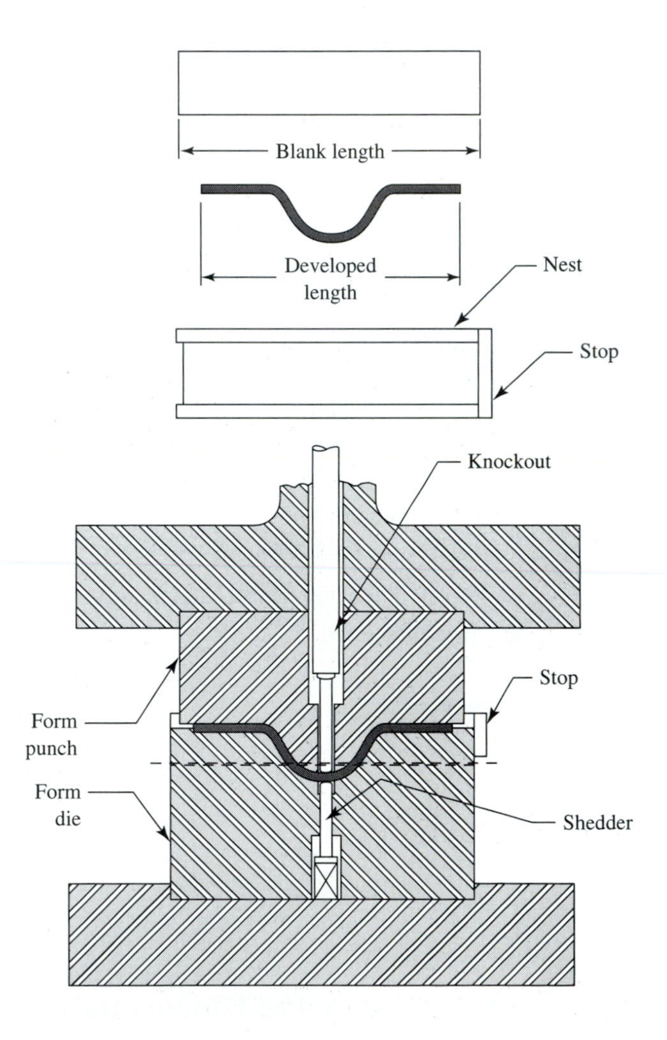

FIGURE 20.8 Solid form die.

20.5 PROGRESSIVE DIES

A **progressive die,** sometimes called a "follower," "gang," or "cut and carry" die is one that performs several individual operations of a part sequentially as strip stock is advanced station by station through the die.

The strip, moved by means of power roller or shuttle-type stock feeders, is accurately positioned at each station by means of pilots. One or more operations are performed at each station, and the sum of all operations yields a completed part; the strip must progress through all stations to complete the part.

To illustrate the process, the simple progressive die shown in Figure 20.10(a) produces a spacer (Fig. 20.10c) by a progression illustrated with a scrap strip (Fig. 20.10b).

The sequence of operations is

Station 1: Pierce holes.

Station 2: Pilot strip (use large pierced hole).

Station 3: Blank part from strip.

The completed blanked part is disposed of downward through the bolster plate opening, as are the slugs from the pierced holes. Provision is also made to dispose of any slugs pierced by the pilot in the event of a broken piercing punch.

Many considerations and decisions must be resolved when processing a stamped part for production—the first being which type of die to use (in this instance, whether to use a progressive die or several individual dies).

When to Use a Progressive Die Progressive dies are efficient, reducing direct labor costs and providing significant savings in metal. They produce a better-quality product because the part is not loaded and removed from several individual dies—each with the potential for mislocating and damage.

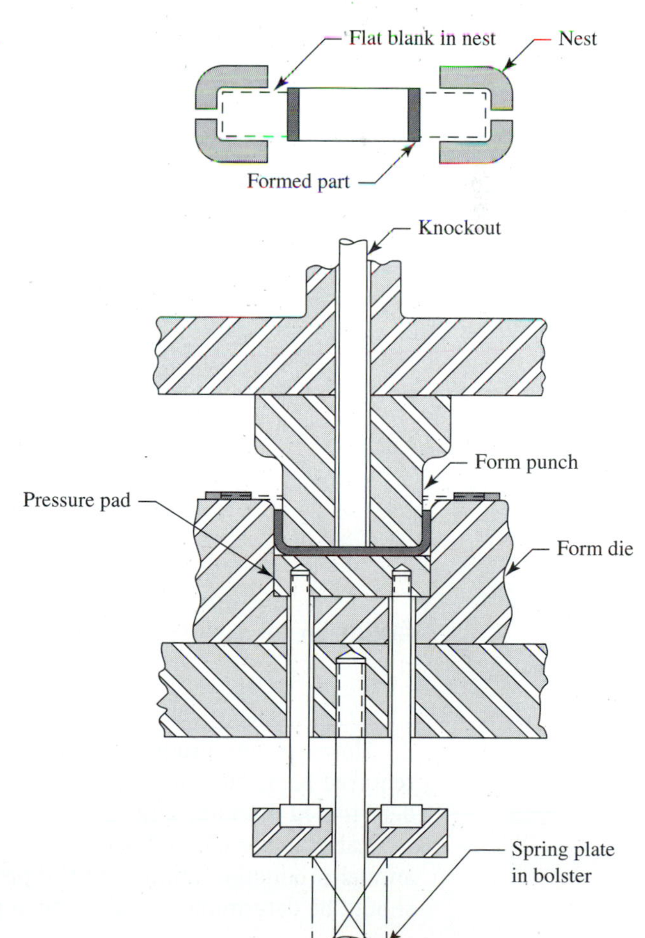

FIGURE 20.9 Pressure-pad forming die.

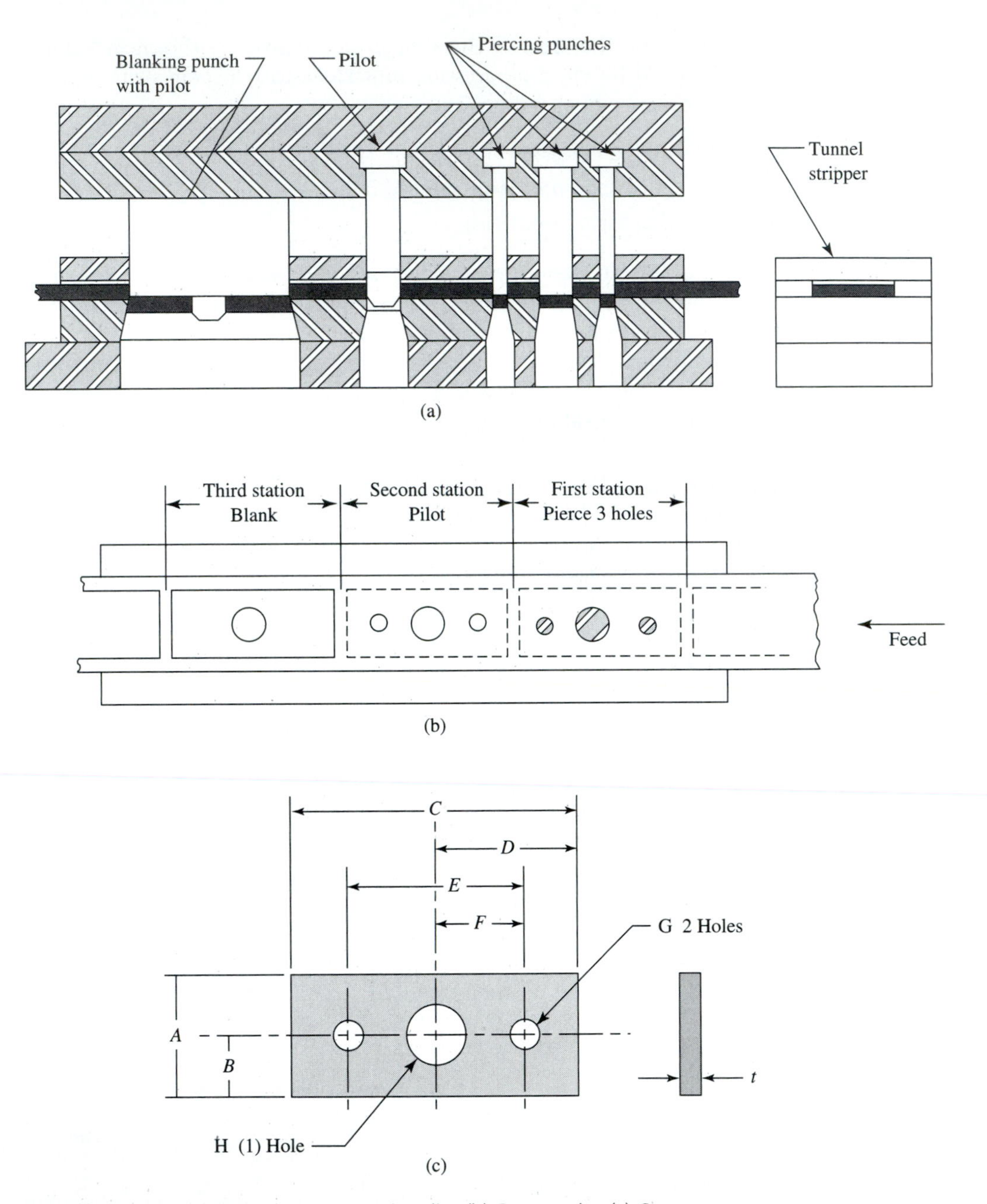

FIGURE 20.10 (a) 3-Station progressive die. (b) Scrap strip. (c) Spacer.

However, one progressive die can prove more costly than several individual dies, and the required press facilities can be much more costly. Justification of a progressive die should be determined economically.

Labor, material, and scrap costs should be compared for both processes for the projected annual production and life of the product. Length of individual runs with related setup costs should be determined. Availability of presses may be a factor.

Progressive Die Development—General When the decision is to produce a part with a progressive die, two steps must precede the actual design of the die:

1. **Process Planning.** All operations required to produce a part (piercing, forming, blanking, etc.) and the order in which they must occur must be determined. A formal plan must be made listing operations, sequence, and special conditions such as close tolerances, flatness requirements, or any other.
2. **Scrap Strip Development.** A **scrap strip** must be laid out to scale to include
   a. Processing as per the formal plan.
   b. Efficient use of material.
   c. Manual stops to aid starting new strips.
   d. Automatic stop for production runs.
   e. Pilot requirements.
   f. Stock strippers where required.
   g. Special conditions, flatness, tolerances, etc.
   h. Part disposal.
   i. Scrap disposal.

Progressive Die Design Example The part shown in Figure 20.11 is to be produced from cold-rolled strip stock on a progressive die. A process plan and a scrap strip development follow.

Process Plan

- Pierce one $\frac{1}{2}$-in.-diameter hole (use for piloting).
- Pierce 2 slots.
- Pierce four $\frac{1}{4}$-in.-diameter holes.
- Blank part.

Pay special attention to close tolerances of slot locations and widths.

Scrap Strip Development (Fig. 20.12)

1. Provide 4 working stations.
2. Since the part is rectangular, the most efficient use of material will be to arrange the parts end to end with minimum separation.
3. Provide manual stops to aid starting new strips through first three stations.
4. Provide an automatic stop for station 4 and thereafter.
5. Use the $\frac{1}{2}$-in.-diameter hole for piloting, providing pilots on each punch to maintain the relationship and tolerances of slots to holes.
6. Provide a tunnel stripper for piercing punches.
7. See #5.
8. Dispose of blanked part through the lower shoe and bolster plate openings.
9. Dispose of slugs through the lower shoe and bolster plate openings. Provide for possible slugs from pilot in the event of piercing punch break. Dispose of scrap strip with scrap cutter downstream from die.

Die Designs. In addition to the foregoing, the die designer must also resolve any potential thin sections of die resulting from specified operation sequence.

FIGURE 20.11
Energizer plate.

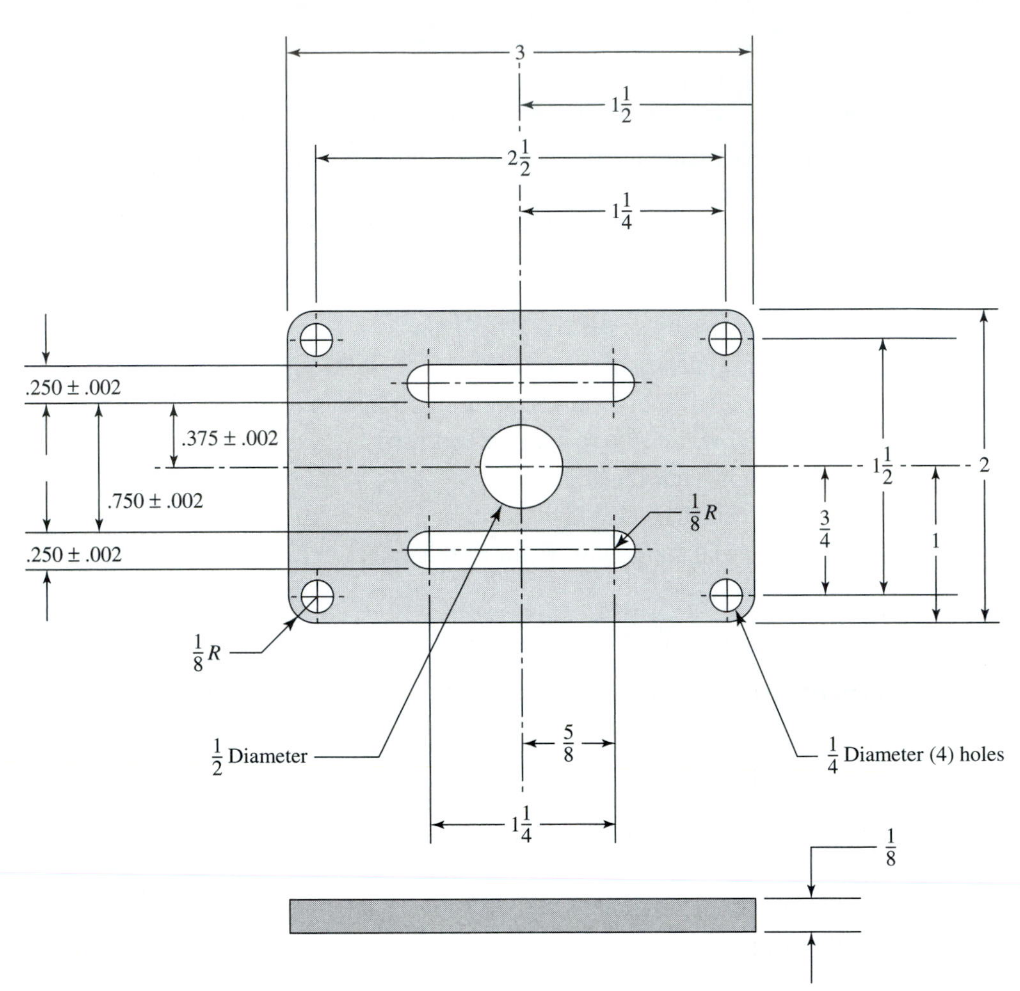

Material: CRS
Tolerances
Fractional ±.016
Decimals as shown
Production
400/Month—one run
12 Setups per year

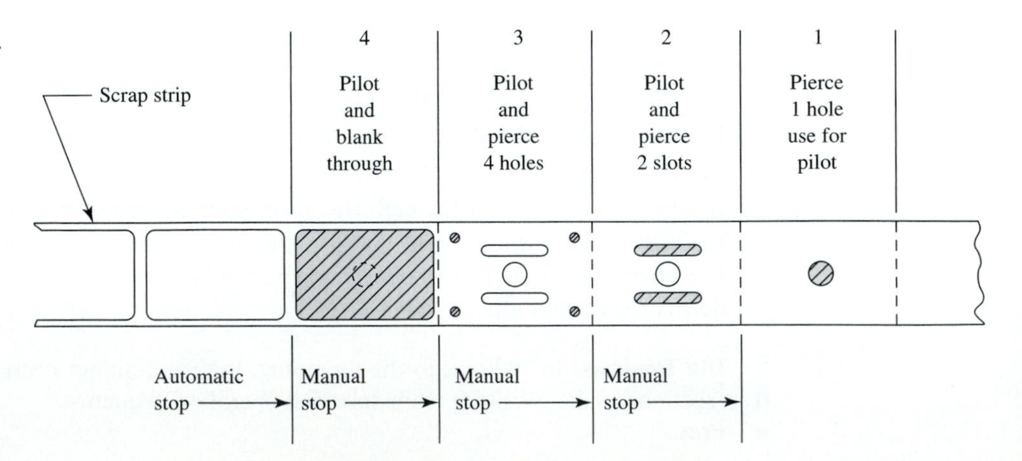

FIGURE 20.12 Scrap strip.

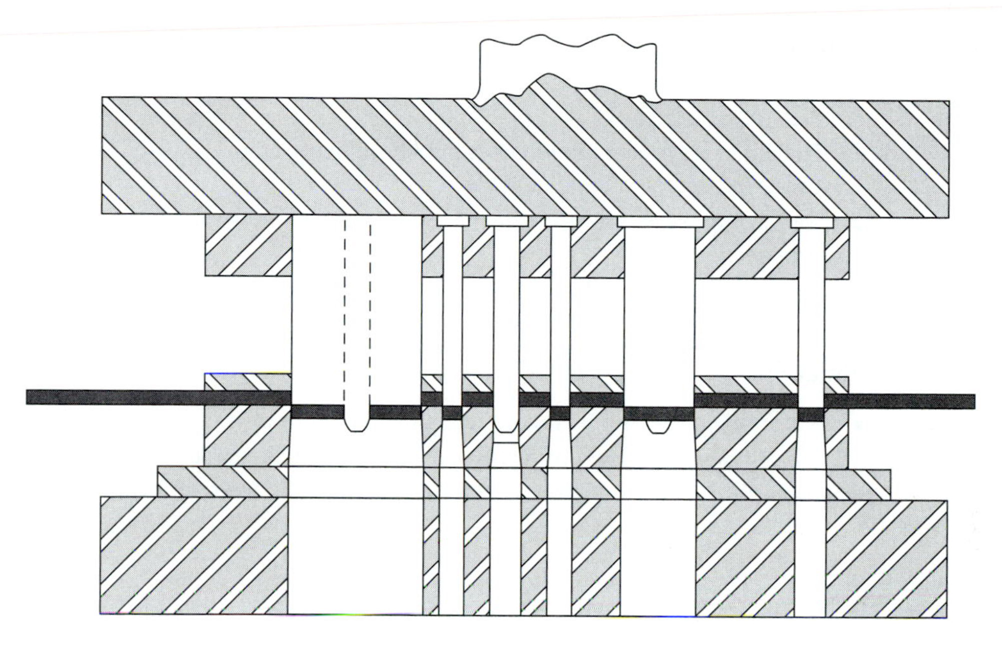

FIGURE 20.13 Front sectional view of progressive die for energizer plate.

Die Design Cross Section. Progressive die design (Fig. 20.13) highlights the components and features required and their arrangement relative to the scrap strip (Fig. 20.12). The die functions as follows:

Station #1

- New strip is fed to first manual stop.
- Press is cycled one cycle.
- Half-inch-diameter hole is pierced.

Station #2

- New strip is fed to second manual stop.
- Press is cycled one cycle.
- Two slots are pierced with punch piloting in $\frac{1}{2}$-in.-diameter hole.

Station #3

- New strip is fed to third manual stop.
- Press is cycled one cycle.
- Four $\frac{1}{4}$-in.-diameter holes are pierced with punch piloting in $\frac{1}{2}$-in.-diameter hole.

Station #4

- Strip is fed to automatic stop.
- Press is set to cycle continuously.
- Part is blanked with punch piloting in $\frac{1}{2}$-in.-diameter hole.
- Press continues to cycle automatically.

20.6 DRAWING DIES

Definitions

- **Drawing** is a process in which a punch, under pressure, causes flat metal to flow into a die cavity to assume a cuplike shape.
- **Redrawing** refers to the second and additional operation, in which a cuplike shell is deepened.
- **Deep drawing** is used to define a drawn cup that has a depth greater than its radius.

Metal Flow Metal flow causes, or is caused by, internal forces set up as shown in Figure 20.14.

- Little or no metal flow takes place in the bottom of the cup.
- Metal flows uniformly into the cup walls from the blank under suasion of the punch, with the wall thickness retaining the original blank thickness but increasing slightly at the open end of the cup.

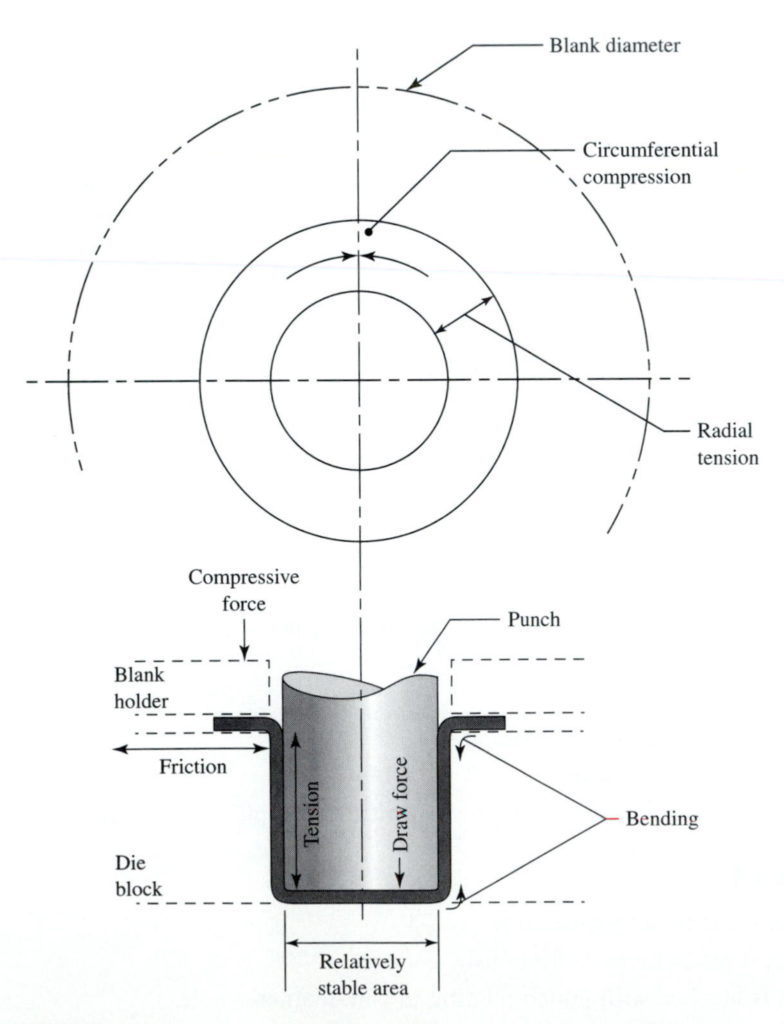

FIGURE 20.14 Forces affecting plastic flow of metal during draw.

- The remaining blank, at the flange area, tends to wrinkle and must be retained and compressed by pressure from a blank holder. The amount of pressure or compressive force required is determined for each part design by trial. Some heavier-gage metals can be successfully drawn without the use of a blank holder or compressive pressure.

Metal flow for symmetrical shells and cups is uniform and predictable whereas that of non-cylindrical shapes requires experimentation in developing the shape of a blank.

Different metals, thicknesses, and hardnesses have different drawing qualities, requiring differing punch and die clearances, radii, and drawing speeds.

Sheet steel of #14 gage (.078 in.) and heavier can usually be drawn without a punch holder and without wrinkling. SAE 1010 steel is considered deep-drawing quality. Of the nonferrous metals, copper draws best, followed by brasses and aluminum. All metals may require annealing where processing calls for successive operations with subsequent work hardening.

Die Schematic Figure 20.15 illustrates drawing action in a schematic view of a drawing die.

View *A*

- Blank is loaded in die.
- Blank holder forces blank against die, supplying compressive force to resist wrinkling.

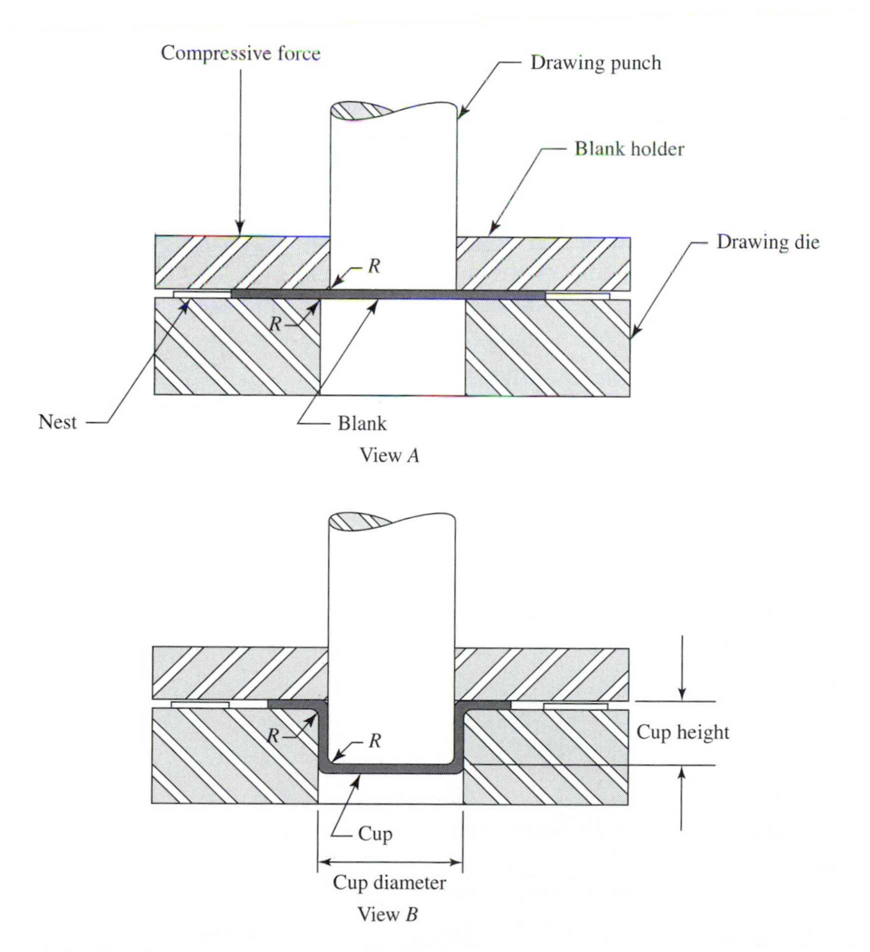

FIGURE 20.15 Schematic of drawing die.

View *B*

- Punch, at bottom of stroke, has drawn metal from blank into walls of cup. Compressive force of blank holder has forced smooth flow of metal into die.
- A stripper or shedder and a knockout, not shown, would be required to eject the cup from the die.

Determining the Number of Drawing Operations Required The number of drawing and redrawing operations required to produce a specific part must be determined during process planning to provide for the required tooling.

A generally used method is to calculate the **reduction of diameter** and compare with recommended empirical allowances for the amount of draw that can be done in one operation for the specified metal.

If the amount of reduction exceeds the maximum recommended draw for the metal, a redrawing operation must be added—perhaps more than one. Many parts do require multiple redrawing operations—depending on the severity of the draw.

Recommended Maximum Reductions of Drawing The maximum recommended reductions of the most commonly used metals are shown in Figure 20.16 and are based on optimum conditions of punch radii, diameter ratios, and metal hardnesses. The practical maximum reduction per draw is approximately 50%, provided that optimum conditions of tools and metals prevail.

Work hardening, detrimental but unavoidable die radii, and clearance conditions can reduce the maximum amount of draw possible. Corrective measures of annealing and/or additional drawing operations must be used to avoid metal rupture during draw.

The term "reduction of diameter" relates to the percentage relationship between the diameter of a blank and the diameter of the resulting draw cup or shell. For example, where the blank diameter is 12 in. and that of the draw cup is 9 in., the reduction of shell diameter is 25%. Redrawing a shell from 10 in. diameter to 6 in. is a 40% reduction.

Blank Development for Cylindrical Shells A blank of sufficient diameter, yet economically sized to control scrap, must be provided for use in the drawing operation. The size may be determined with the use of graphics or mathematics. The following formulas are in common

Metal	Reduction in Diameter, %	Reduction in Area, %
Aluminum alloys	45	28
Aluminum heat-treatable	40	23
Brasses	50	33
Bronze, tin	50	33
Copper	45	28
Steel, low carbon	45	28
Steel, stainless	50	33
Zinc	40	23

FIGURE 20.16 Maximum recommended reductions for drawing.

use for this purpose. The ratio of shell diameter to the punch corner radius (d/r) affects the blank diameter and must be taken into consideration.

The blank should be developed early-on for use in any new drawing die to permit timely designing and making of the die and, if required, the designs of redrawing dies. Due to variables of material thickness, hardness, and some irregularities in the top edge of the cup, it is common practice to add sufficient metal to the basic blank diameter to permit a trimming operation.

Blank Development for Noncylindrical Shapes Metal does not draw uniformly in noncylindrical shapes (cubes, prisms, rectangular solids, etc.). A combination of bending, side flow, drawing, and stretch forming creates problems of folds, tears, excessive thickness in some sections, and considerable difficulty in forming corners.

Some calculations, some experience based on drawing similar parts in the past, and some trial and error—all directed by a die engineer with years of experience in this area—are the methods used to develop an operational, economical blank. Modifications of the product designs, especially in the corners, are often required to make production feasible by the drawing process.

20.7 CALCULATING BLANK SIZES, REDUCTION OF DIAMETERS (SEVERITY OF DRAW), AND CUP HEIGHTS FOR CYLINDRICAL SHELLS

Determining blank sizes for draw parts requires that calculations be made in a progressive order employing empirical formulas.

1. Determine the shell diameter to corner radius ratio, d/r.
2. Calculate blank size utilizing the relevant formula:

$$\text{For } \frac{d}{r} \text{ 20 or greater,} \qquad D = \sqrt{d^2 + 4dh}$$

$$\text{For } \frac{d}{r} \text{ 10 to 20,} \qquad D = \sqrt{d^2 + 4dh} - 0.5r$$

$$\text{For } \frac{d}{r} \text{ less than 10,} \qquad D = \sqrt{(d - 2r)^2 + 4d(h - r) + 2\pi r(d - 0.74)}$$

where D = Blank diameter
d = Shell diameter
h = Shell height
r = Corner radius of punch

3. Calculate Severity of Draw (reduction of diameter)
   - Determine the reduction of diameter with the following formula:

$$R_c = 100\left(1 - \frac{d}{D}\right) = \text{Percentage reduction}$$

   - Compare R_c, percentage reduction calculated, with the maximum recommended reduction shown in Figure 20.16.
   - Provide additional redrawing operations for part processing, where R_c exceeds maximum recommended amount.

4. Determine the height of a cup wall (or depth of cup) with the following empirical formula:

$$h = \frac{(D^2 - d^2)}{4d}$$

where h = Cup height
D = Blank diameter
d = Cup diameter

This formula assumes the cup to have sharp corners and no differences of thickness in cup walls and bottom.

EXAMPLE 1

Determine the blank size required and reduction of diameter for the part shown in Figure 20.17.

Solution

1. Calculate blank size:

   Determine d/r:

$$\frac{6}{.250} = 24 \text{ (ratio over 20)}$$

   Use the appropriate formula

$$\begin{aligned} D &= \sqrt{d^2 + 4dh} \\ &= \sqrt{6^2 + 4(6)(1.25)} \\ &= \sqrt{66} \\ &= 8.12 \text{ in.} \end{aligned}$$

2. Calculate reduction of diameter:

   Use the formula.

$$R_c = 100\left(1 - \frac{d}{D}\right) = 100\left(1 - \frac{6}{8.12}\right) = 100(.2611) = 26\%$$

Comparing 26% reduction with the maximum recommended reduction from Figure 20.16 shows the part, as designed, can be successfully drawn in a single operation from low-carbon steel.

FIGURE 20.17 Problem: Determine blank diameter and percent reduction.

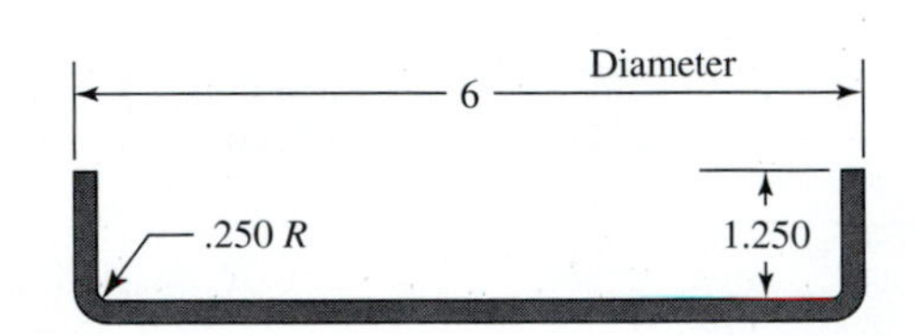

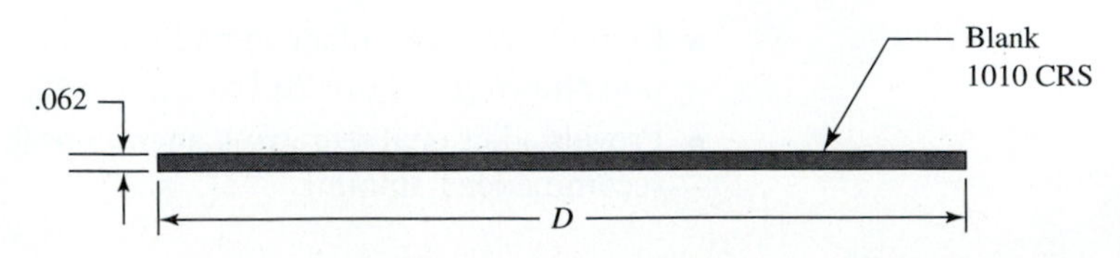

EXAMPLE 2

Determine the blank size required for the cup shown in Figure 20.18(a) and processed with three drawing operations as shown in Figure 20.18(b). Also calculate cup heights and reduction of diameter for each draw.

Solution

1. Calculate blank size. The third and last draw produces the cup to size for which a blank must be developed.

Determine $\frac{d}{r}$.

$$\frac{2}{.187} = 10.69 \text{ (ratio from 10 to 20)}$$

Use the appropriate formula.

$$D = \sqrt{d^2 + 4dh} - 0.5r = \sqrt{(4)^2 + 4(2)(6)} - .5(.187) = \sqrt{64} - .09 = 8 - .09$$

Use 8.00 in.

With the blank size established, cup heights and reductions of diameters can be calculated for all draws.

2. First draw

Cup height:

$$h_1 = \frac{(D^2 - d^2)}{4d} = \frac{(8)^2 - (4.5)^2}{4(4.5)} = 2.43 \text{ in.}$$

Reduction of diameter:

$$R_c = 100\left(1 - \frac{d}{D}\right) = 100\left(1 - \frac{4.5}{8}\right) = 44\% \text{ (acceptable)}$$

3. Second draw

Cup height:

$$h_2 = \frac{(D^2 - d^2)}{4d} = \frac{(8)^2 - (2.625)^2}{4(2.625)} = 5.43 \text{ in.}$$

Reduction of diameter

$$R_{c_2} = 100\left(1 - \frac{d}{D^*}\right) = 100\left(1 - \frac{2.625}{4.5^*}\right) = 42\% \text{ (acceptable)}$$

4. Third Draw

Cup height: The original part design in Figure 20.18 established the cup height to be 6 in.

Reduction of diameter:

$$R_{c_3} = 100\left(1 - \frac{d}{D^{**}}\right) = 100\left(1 - \frac{2}{2.625^{**}}\right) = 24\% \text{ (acceptable)}$$

*Used from the first drawing operation.

**Used from the second drawing operation.

FIGURE 20.18 (a) Problem: Determine blank diameter, reductions, and cup heights. (b) Solutions: Blank diameter, reductions, and cup heights.

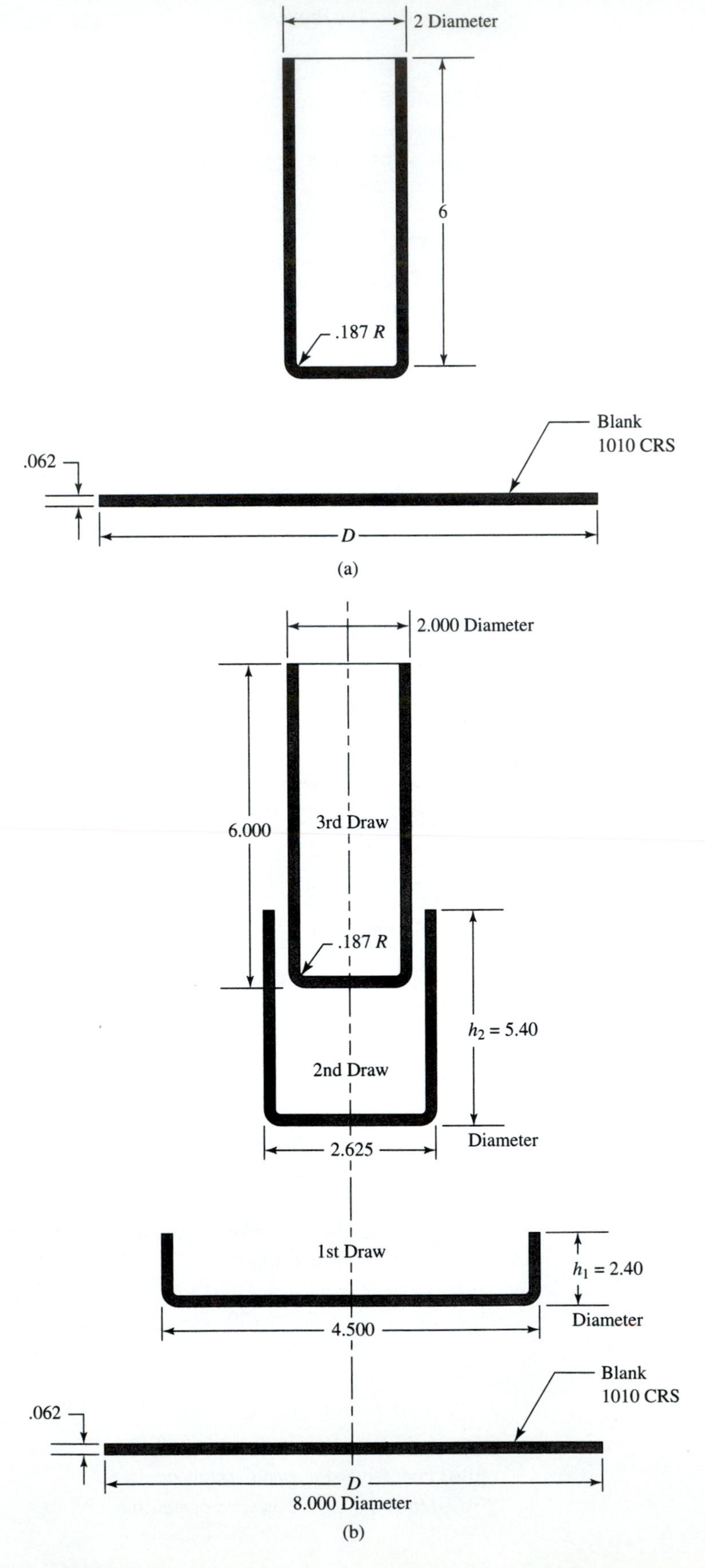

Drawing Operation Illustration An inverted drawing die is shown in cross section in Figure 20.19. The die is fastened to the ram with the punch and pressure-plate mounted on the bolster plate. Pressure-pad springs are located below the bolster plate.

Die Function

Figure 20.19(a)

- Blank is loaded in nest on top surface of pressure plate (also stripper pad).

Figure 20.19(b)

- Press is at bottom of stroke.
- Die draw ring has drawn blank over punch, compressing springs in pressure pad with same movement.

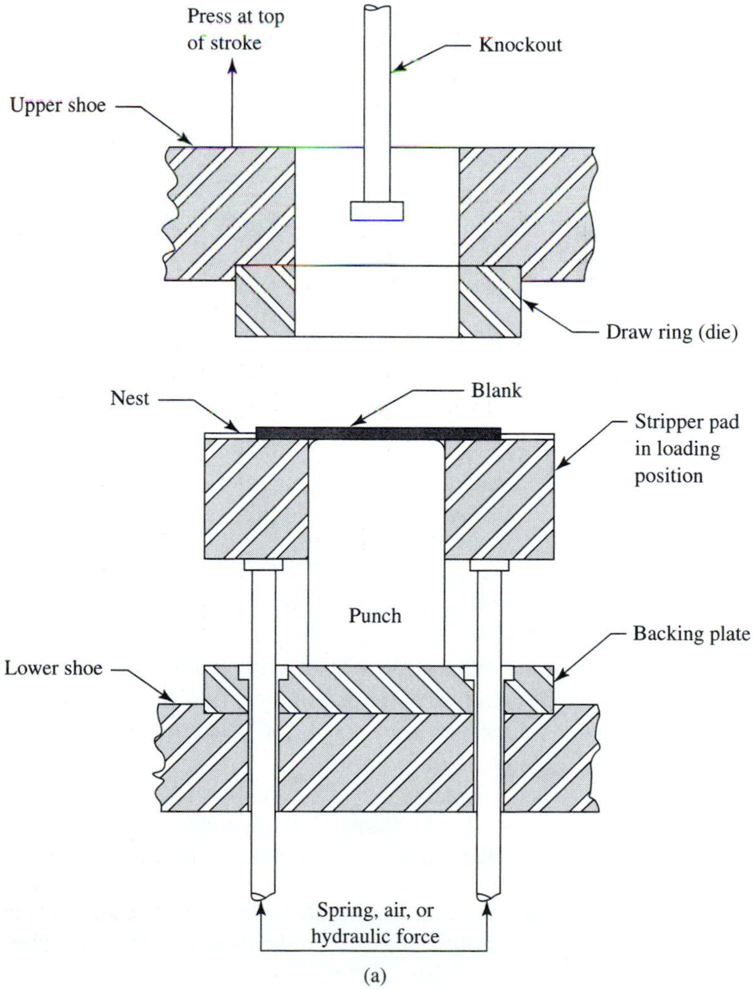

FIGURE 20.19
(a) Inverted drawing die—blank is loaded.

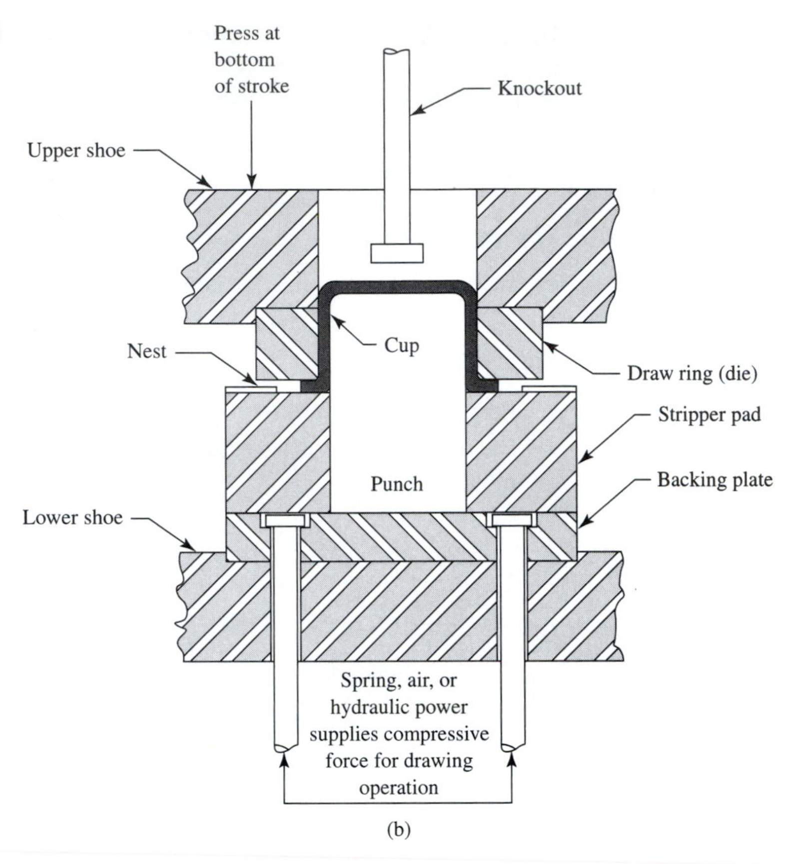

FIGURE 20.19 Continued (b) Inverted drawing die—cup is drawn.

Figure 20.19(c)

- Press is at top of stroke.
- Pressure pad has stripped cup from punch.
- Knockout has ejected cup from die.

20.8 PRESSURES AND FORCES

The force or pressure required for each press operation must be calculated for the type of operation, part design, and metal characteristics. Forces so calculated determine the required press tonnage and also provide guidance to the die designer for determining die and press materials and sections.

Where more than one operation is performed in a die—as in a progressive die—the pressures for each operation are additive in determining the required press size. When strippers and spring pads are used, the force to compress the springs must be calculated and included in the total required pressure.

Shearing operations generate the greatest required forces of all sheet metal operations; measures can be taken to reduce peak pressures and spread the load over more length of die stroke so that the size of the press can be reduced:

- A shear angle can be added to the cutting edges of a die or a punch, changing a total impact condition to a progressive operation as shown in Figure 20.20.

FIGURE 20.19 Continued (c) Inverted draw die—cup is ejected.

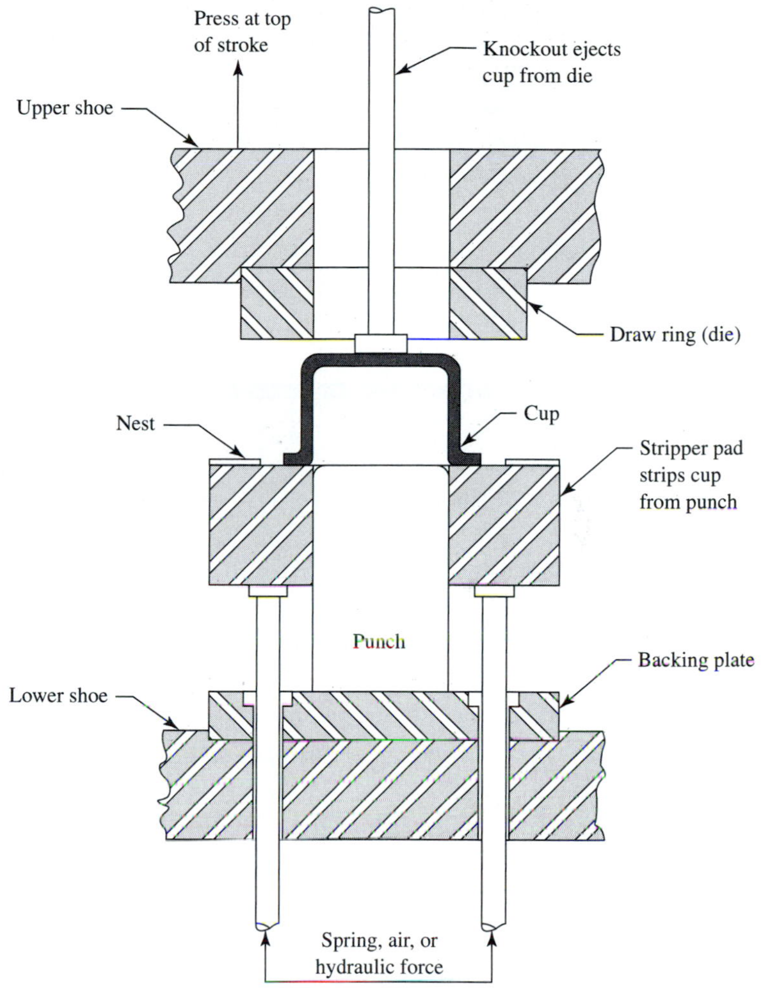

Press at top of stroke
Knockout ejects cup from die
Upper shoe
Draw ring (die)
Nest
Cup
Stripper pad strips cup from punch
Punch
Backing plate
Lower shoe
Spring, air, or hydraulic force
(c)

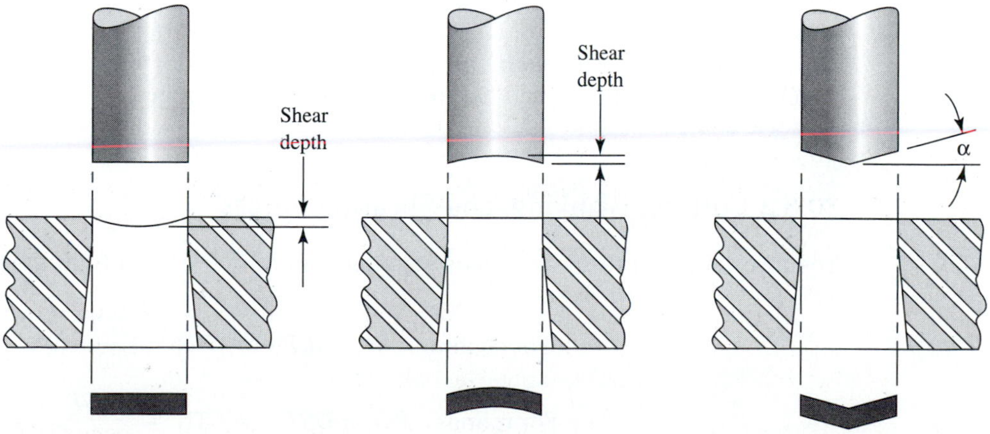

Shear depth
Shear depth
α

FIGURE 20.20 Shear as applied to die components.

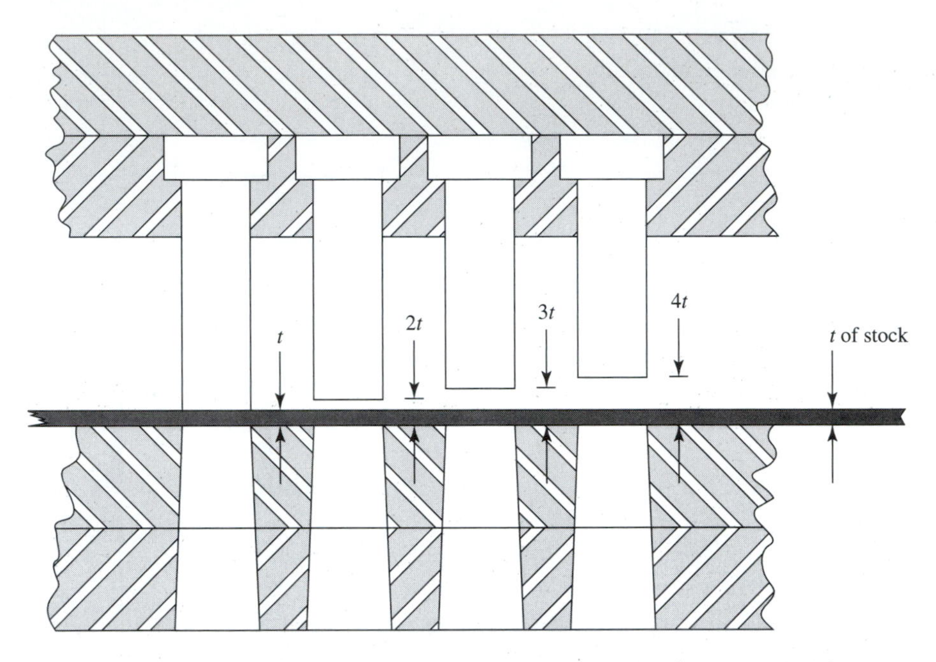

FIGURE 20.21 Stepped punches–4 holes pierced with same force required for 1.

- Punches can be stepped so that all do not impact together but rather in a progressive order as illustrated in Figure 20.21. The arrangement of the four punches will reduce the peak load to 25% over a longer portion of the die stroke.

20.8.1 Center of Pressure

The center of pressure of a die is the point where all bending loads of a blank cancel out and where the pressure at the central axis of a press ram can act vertically with no deflection. It is the point where all shearing forces balance.

It has been said that it is the center of gravity of a *line* that is the perimeter of a blank. It is *not* the center of gravity of an *area*. A center of pressure can be determined mathematically but is usually found with a wire model of the part's outline. Trial and error locate the point at which the model balances and has proven sufficiently accurate.

20.8.2 Cutting, Blanking, and Shearing Forces

The force required to shear various materials can be determined from the following formulas:

$$\text{For contours} \quad P = SLT \quad \text{or} \quad \text{TN} = \frac{SLT}{2000}$$

$$\text{For rounds} \quad P = \pi DST \quad \text{or} \quad \text{TN} = \frac{\pi DST}{2000}$$

where P = Cutting force in pounds
TN = Cuting force in tons
D = Diameter in inches
L = Length of cut in inches
S = Shear strength in pounds per square inch
T = Material thickness in inches

20.8.3 Bending Forces

Bending forces can be determined from the empirical formula for V-bends:

$$P = \frac{KLST^2}{W}$$

where P = Bending force in tons
K = Die opening factor
1.20 for die opening 16 times metal thickness
1.80 for die opening 8 times metal thickness
L = Length of part in inches
S = Tensile strength in tons per square inch
W = Width of die in inches
T = Metal thickness in inches

Note: The bending force for a U-bend die will be approximately two times that calculated for a V-bend die. Edge bending requires approximately half of the force calculated for a V-bend die.

20.8.4 Drawing Forces

The force required for a draw die can be determined from the following formula:

$$F + \pi\, d\, T\, \mathrm{Sy}\left(\frac{D}{d} - C\right)$$

where F = Drawing force in pounds
D = Diameter of blank in inches
d = Diameter of punch in inches
T = Metal thickness in inches
Sy = Yield strength of material in PSI
C = Empirical constant allowance for friction and bending ranging from .6 to .7

Note: Blank-holder force is not included in the formula and must be added where used.

20.8.5 Blank-Holder Force

The amount of blank holder force required is developed by trial and error. It can range from zero for certain thick sections that require no blank holder to as much as half the die force for thinner metal.

20.8.6 Stripping Forces

The substantial forces required to strip metal from punches can be determined from the following formulas:

$$\text{For contours} \quad P_s = 3500LT \quad \text{or} \quad \text{TN}_s = \frac{3500LT}{2000}$$

$$\text{For rounds} \quad P_s = 3500\pi DT \quad \text{or} \quad \text{TN}_s = \frac{3500\pi DT}{2000}$$

where P_s = Stripping force in pounds
TN_s = Stripping force in tons
D = Punch diameter in inches
L = Length of cut in inches
T = Material thickness in inches

Shearing forces for any appreciable length of cut are high and are exerted over a short portion of the die stroke. Total forces resulting from a large punch diameter, a long punch perimeter, and/or a group of several punches can result in undesirable, excessive tonnage requirements; they can be detrimental to the life of the die and, indeed, may exceed the capacity of an available press. Several methods are used to reduce and spread peak load over more length of die stroke.

20.8.7 Press Tonnage

The tonnage required for a press to perform an operation is determined by the total forces involved during one stroke of the press. For instance, a progressive die may pierce, bend, and blank during one stroke of the press, plus compress springs in a die stripper. Pressures required for each must be determined and added. It is good practice to add a safety factor when specifying the press size in order to cover such problems as heavier or harder strip stock than specified, as well as the dulling of dies.

The capacity of a press in terms of pieces per hour or strokes per minute does not affect the tonnage rating. Tonnage is the total force required to complete one cycle of the press.

20.9 ROLL FORMING

Roll forming changes the shape of flat strip stock as it advances between a series of rolls to the desired configuration. Cross-sectional area does not change.

One station, or set of rolls, is illustrated in Figure 20.22 for forming stainless steel molding. The number of stations, or rolls, required to produce a particular shape depends on the material, material thickness, and size and intricacy of the shape.

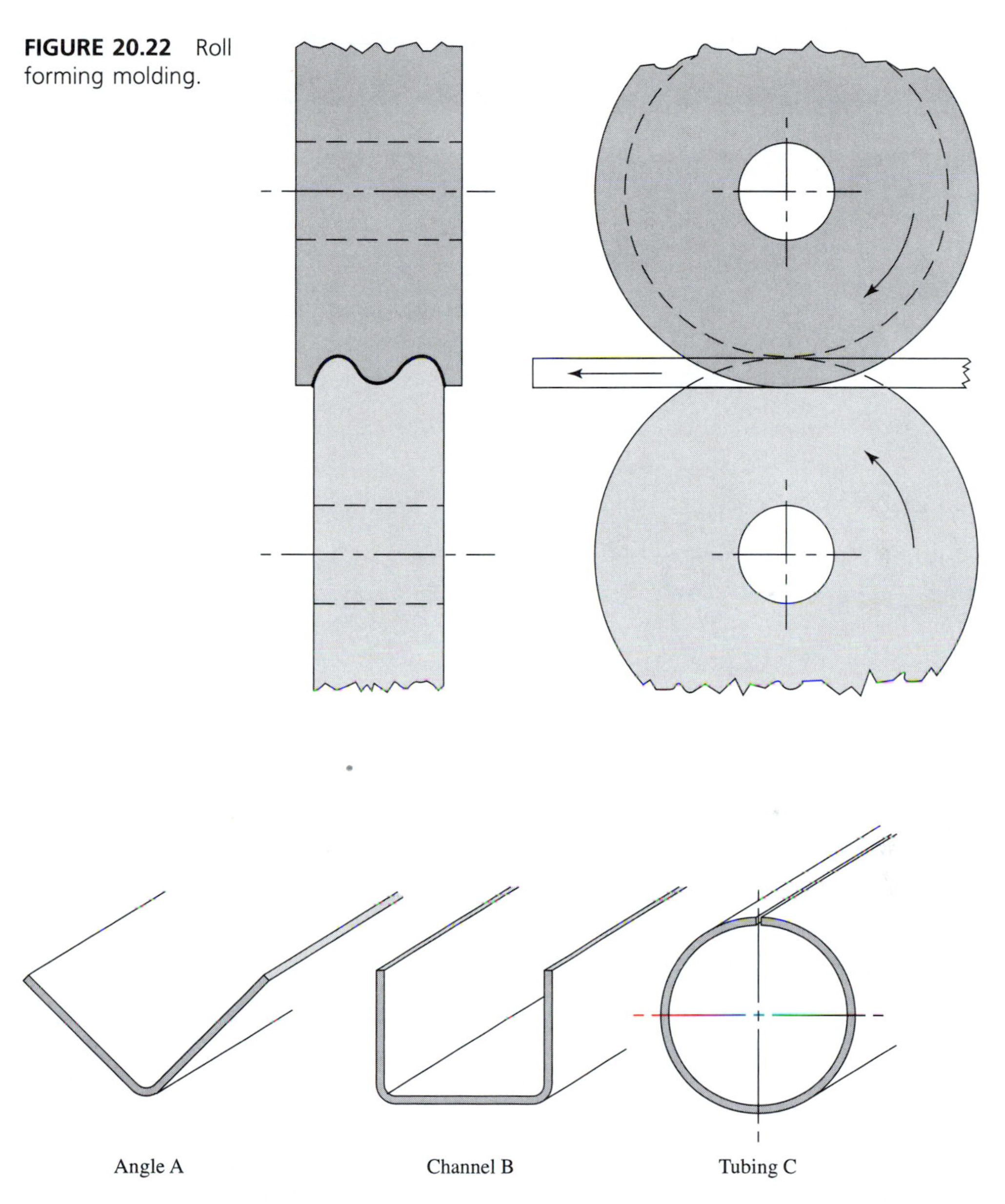

FIGURE 20.22 Roll forming molding.

FIGURE 20.23 Roll-formed shapes.

The angle *A* shown in Figure 20.23 required three sets of rolls; the channel *B* illustrated in the figure was produced on four sets of rolls. Both were formed from .032-in.-thick aluminum strip stock.

The steel tubing *C* shown in Figure 20.23 is in production to provide automobile shock absorber main sections. The .062-in.-thick steel strip requires 11 roll passes to form the strip into 2.50-in.-diameter tubing. The roll forming equipment has four additional stations where the rolls do no forming but serve only to guide and straighten the tubing as it is formed.

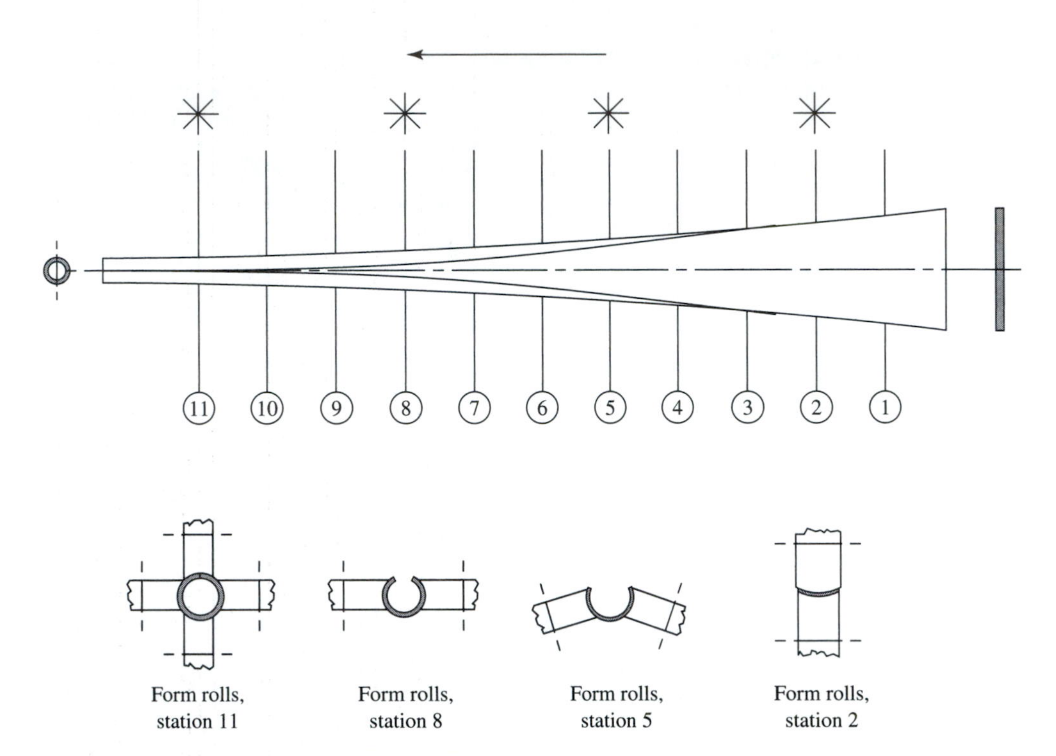

FIGURE 20.24 Roll form flat steel strip into tubing for shock absorbers.

Figure 20.24 diagrams the shock absorber tubing forming operation and several of the rolls used in the process. The tubing seam is welded as it leaves the last set of rolls, and the tubing is then cut to length ready for machining on secondary operations.

20.10 REVIEW QUESTIONS AND PROBLEMS

1. What operations do metal cutting dies perform? List them.
2. What is breakout?
3. What part of a sheared edge, or of a pierced hole, is accurate dimensionally?
4. Define a blank.
5. What is the principle of fineblanking?
6. What are two methods of bending metal? Describe the processes.
7. How sharp can the radius be in metal bent by either process?
8. Describe a forming die.
9. Describe a progressive die.
10. What would limit the number of stations in a progressive die?
11. Generally speaking, which metalworking process or die requires the most pressure to operate?
12. Discuss how proper die design can lower total or peak pressures of the shearing operation.

13. Calculate blank requirements, reduction of diameter(s), and cup height(s) for the part shown in Figure 20.25.

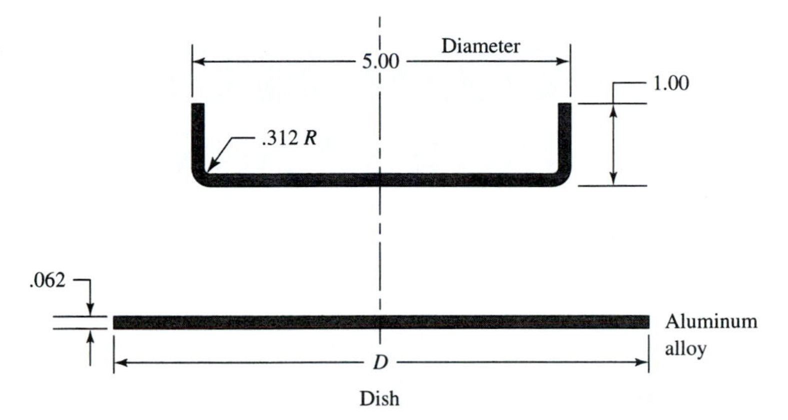

FIGURE 20.25 Problem: Determine blank diameter and percent reduction.

14. Calculate blank requirements, reduction of diameter(s), and cup height(s) for the part shown in Figure 20.26.

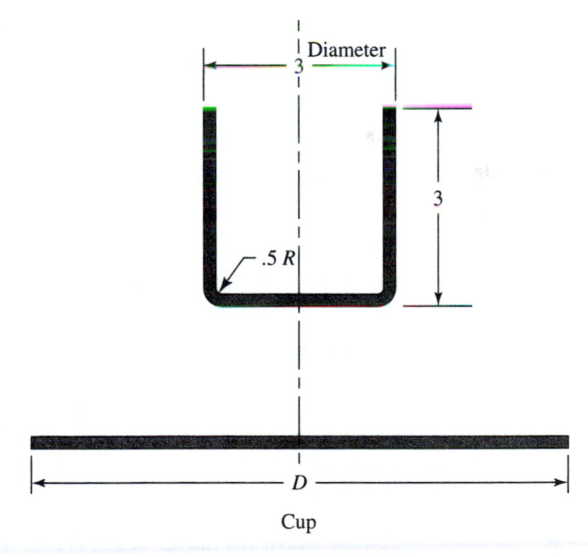

FIGURE 20.26 Problem: Determine blank diameter and percent reduction.

15. What is the center of pressure?

CHAPTER TWENTY-ONE

CROSS-SECTION-CHANGING PROCESSES

CHAPTER OVERVIEW

21.1 INTRODUCTION

Cross-section-changing processes are used to increase or decrease cross-sectional areas of material or workpieces. Direct material for parts and workpieces produced by cross-section-changing processes—and, indeed, for all metalworking processes—is provided by the **rolling process.** Rolling, which is an industry by itself, provides bar, strip, sheet, and structural shapes from cast metal ingots.

Heated ingots of coarse-grained, brittle, and porous metal are passed between rollers, as shown in Figure 21.1, compressing and elongating them into wrought material of finer grain and ductility. Repeated passes reduce and elongate the cast ingot to final thickness and shape with further increased ductility and finer graining.

Rolling equipment is massive, specialized, and costly, requiring large continuous production runs to justify the required investment in space, inventory, and equipment costs. Shapes and sizes are standardized to provide fewer changes and longer runs. Special shapes and sizes can

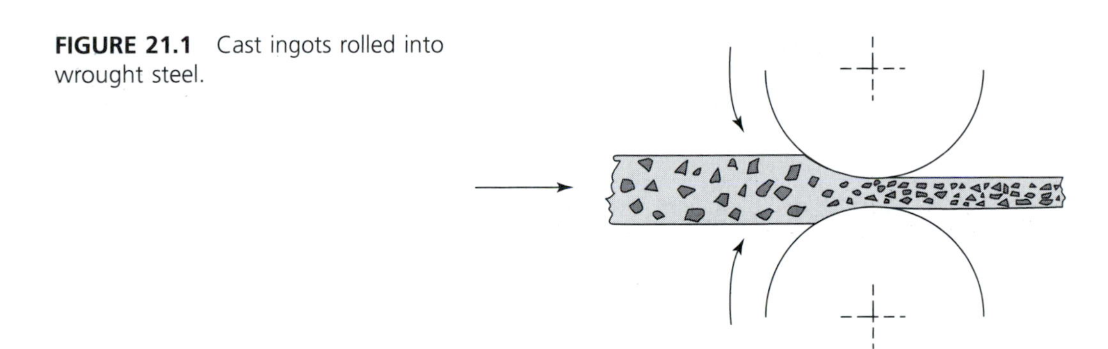

FIGURE 21.1 Cast ingots rolled into wrought steel.

be and are produced where the volume warrants the run. Forging utilizes many of the shapes produced by rolling, including round, square, hexagonal, and rectangular cross sections.

Hot-rolled material is generally free from internal residual stresses and has little distortional properties. The surface finish is not entirely smooth and is covered with hard mill scale (oxide). **Cold-rolled material** has smoother surfaces than hot-rolled material, more accurately controlled dimensions, and no mill scale. It is available in four levels of hardness, with each having physical properties peculiar to that grade. Cold-rolled metal also has internal stresses and more directional properties than hot-rolled metal. When cold-rolled material is heated, the internal stresses are relieved, and a part made from this material will distort.

Forging, extruding, heading, and swaging can be considered as secondary operations to rolled material. These processes provide shape to workpieces, can eliminate many machining operations, and further develop ductility and finer-grained patterns as illustrated in Figure 21.2.

Parts designed for cross-section-changing processes are usually assigned to machining manufacturing engineers for processing rather than metalworking engineers, because the secondary operations are almost always machining operations.

21.2 FORGING

A forging is generally considered to be a workpiece shaped by plastic deformations caused by compressive forces. Compressive forces may be provided by a press or by impacts from hammer blows. The forging operation is usually performed hot but in some instances may be accomplished at room temperature.

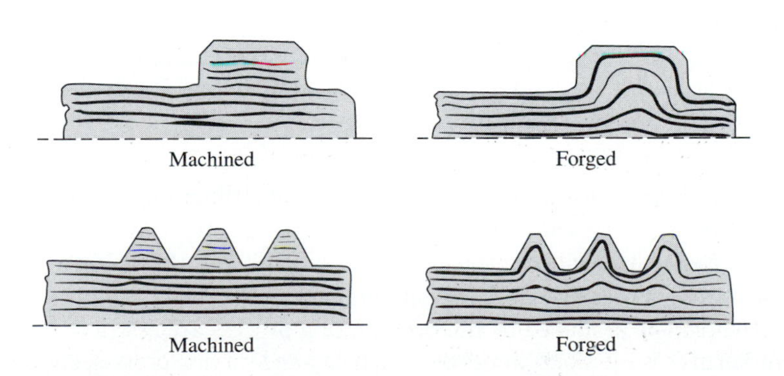

FIGURE 21.2 Grain patterns: forged versus machined.

21.2.1 Open Die Forging

A blacksmith operation comprised of forge, hammer, and anvil is the simplest form of open die forging:

- A forge brings the metal up to working temperature.
- The heated metal is placed on, and is supported by, the anvil.
- A hammer blow compresses the metal between anvil and hammer, displacing metal as required.

Double-acting steam and air hammers are used in production, with gas, oil, and elecric furnaces providing heat. Simple, open top dies may be used to confine and aid in shaping the metal.

Large open die forgings such as heavy-duty truck axle beams, railroad parts, and so on may require several men working together to lift, move, heat, operate a power hammer, reheat, restrike, and dispose of a forging. Aids in the form of lifts, counterbalances, and hoists may be used, but the quality of the forging will depend on the experience, judgment, and skill of the workers. This method can be used selectively to produce certain forgings complete, but it is more often used to preshape a billet, readying it for final machining operations.

21.2.2 Drop Forging

Drop forgings are so named because they are formed by the weighted upper die of a closed die set falling—dropping—onto a heated billet of metal located in the lower die, forcing metal into the die cavity.

Most drop-forging dies require several consecutive cavities to completely form a part. The heated metal billet is transferred from cavity to cavity, being struck a blow (drop) at each cavity, which forces the metal progressively into its final configuration. At the end of the progression, the completed forging is ready for trimming the flash. A drop-forging progression is illustrated in Figure 21.3.

A typical drop-forging press and die arrangement is illustrated in Figure 21.4. In this view the lower die is secured to the base or anvil of the press while the upper die is attached to the press ram. The anvil is usually supported by a large casting known as a sow block, which, in turn, rests on a bed of timbers in a pit underneath the press.

Drop-Forging Process Sequence

- The press ram and upper die, which have a total specific weight, are elevated to a height sufficient to produce the required impact. (*Note*: Drop height is calculated or is determined by trial and error and is adjusted during setup.)
- A heated metal billet is positioned in lower die.
- The ram and upper die are released, dropping and striking the heated metal, forcing it into a die cavity.

Methods used to elevate the ram in gravity presses are shown in Figure 21.5.

Parts that can be produced by the drop-forging process are restricted to those with moderate and uniform cross sections. Forgings with heavy and varying cross-section thicknesses usually must be produced on a press, where squeezing can penetrate the entire mass of metal, forcing a uniform flow.

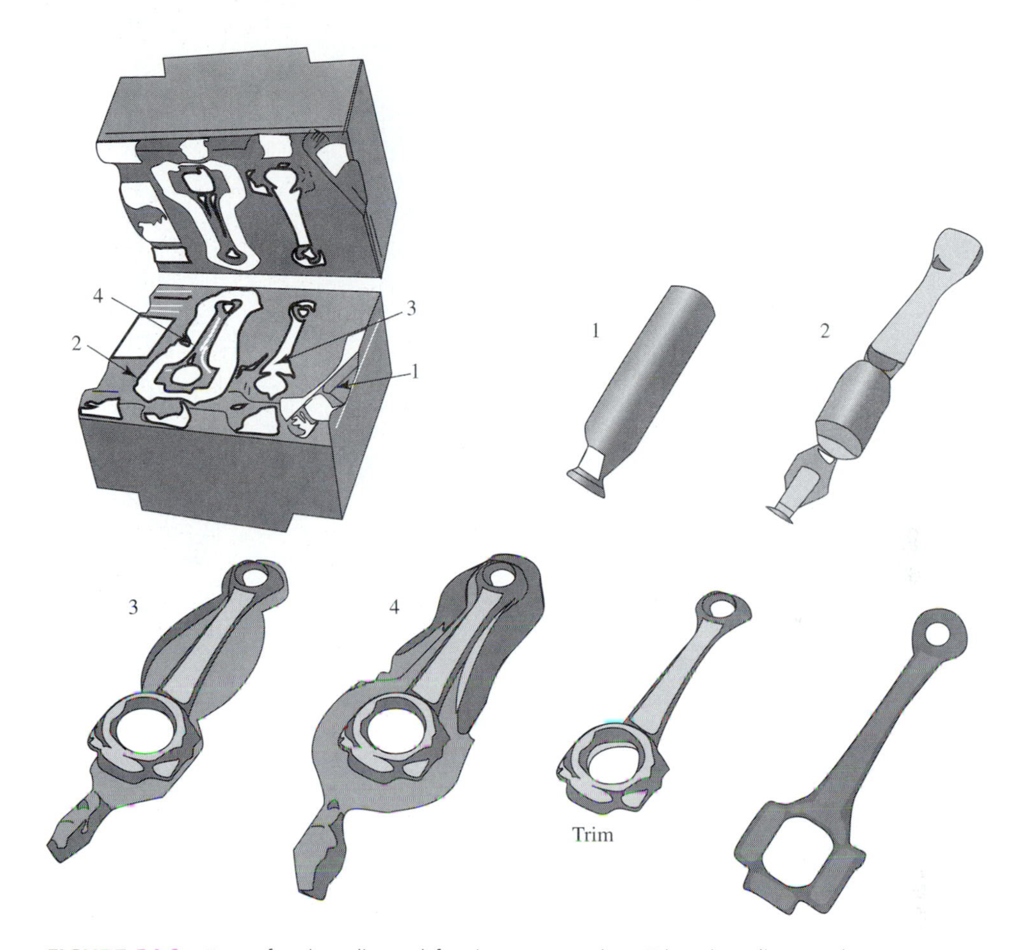

FIGURE 21.3 Drop-forging die and forging progression. Trimming die not shown.

All commercially produced wrought metals can be forged by the drop-forging method with the exception noted regarding cross sections. Dies must be designed with optimum draft angles, parting lines, and radii to suit each type of metal and design of part.

21.2.3 Press Forging

Press forgings are produced by a squeezing action, forcing heated metal into closed die cavities. Mechanical and hydraulic presses are used to provide the high pressures required for the process. Figure 21.6(a) illustrates the rugged proportions of a well-engineered hydraulic forging press with robotic transfer. A 4000-ton mechanical forging press is shown in production in Figure 21.6(b). A precision hydraulic forging press is illustrated in Figure 21.6(c).

Originally, press forgings could be accomplished only on forgings with uniform cross sections; present practice includes forgings of any shape and section, and results from years of experience in designing dies, billets, and preforms.

The accuracy, or dimensional tolerance, of a press forging can be held closer than that of a drop forging due to the more positive control of the process—especially in the amount of flash material required. Some excess material (which will become flash) is required to assure complete die fills with no voids left in the part.

FIGURE 21.4 Drop-forging press die and arrangement.

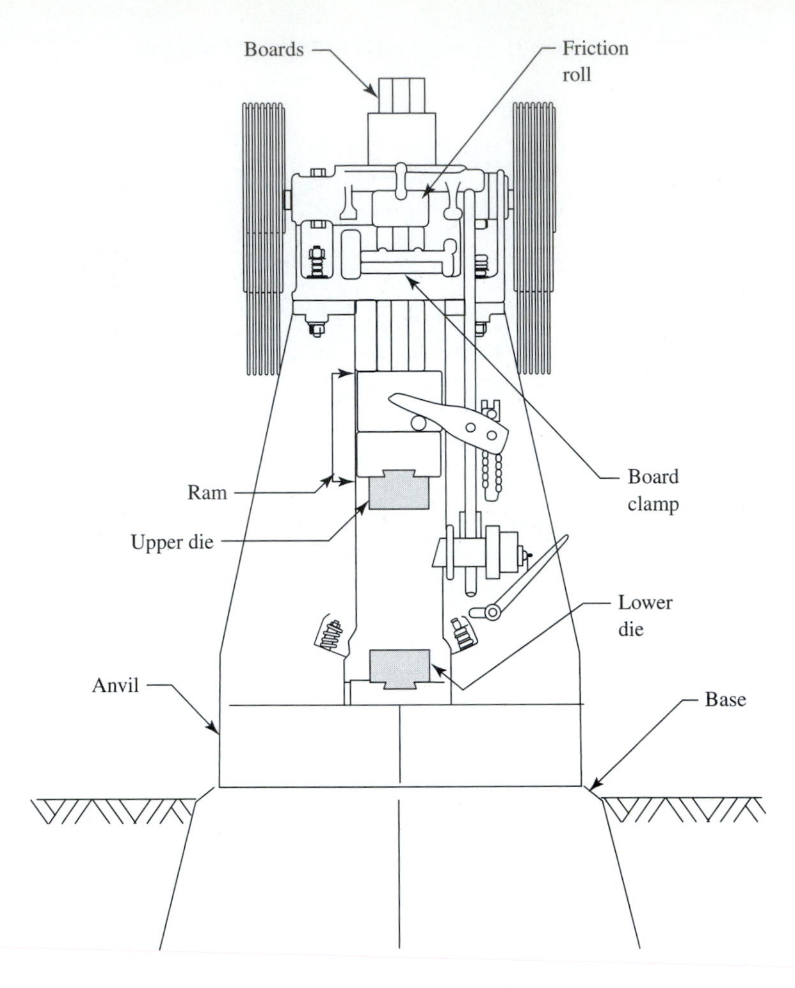

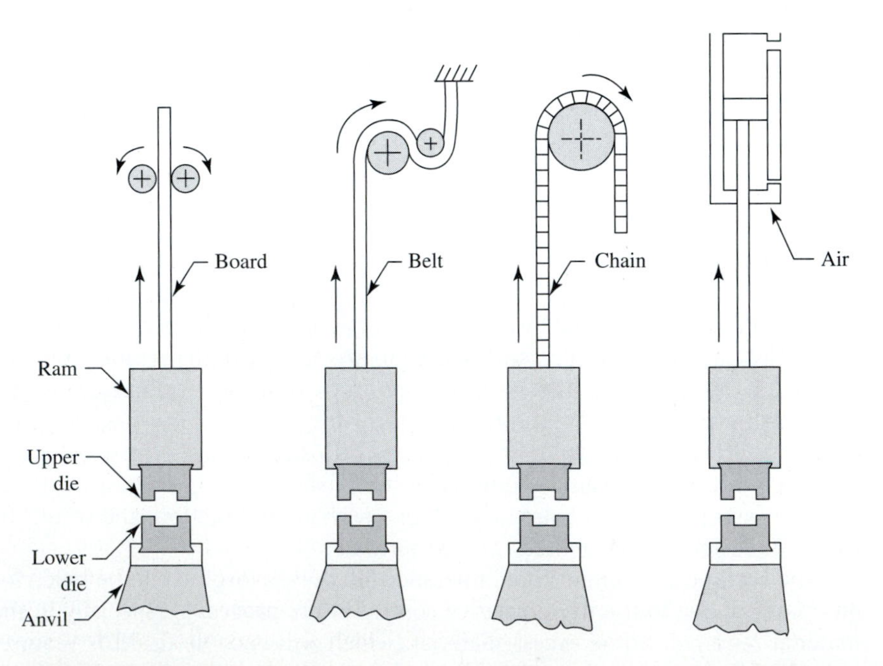

FIGURE 21.5 Methods used to elevate ram and upper die in gravity press.

(a)

(b)

(c)

FIGURE 21.6 (a) Shell forging system (60-155 mm projectile 800/800/400-ton hydraulic press with robotic transfer). (b) 4000-Ton mechanical press. (c) 1500-Ton hydraulic precision forging press. (Courtesy Erie Press Systems)

One major problem inherent in this process is that—due to the longer time required for the operations—heat is lost from the material, which then becomes harder and more resistant to metal flow. Some heated dies are used, but die life is shortened. The majority of press forgings are completed in one press cycle, other than flash trimming.

21.2.4 Forging Dies

Material Forging dies must withstand wear, heat, and—in the case of drop forgings—shock. Medium- to high-carbon tool steels—alloys with chromium, tungsten, and molybdenum—are used; AISI-H-26, AISI-D-2, and AISI-M-10 are representative. Great care must be taken in heat treating to obtain uniform hardness and toughness with a minimum of distortion.

Draft Angles Forging-die draft angles range from 3° to 10°, with most designs specifying 7°. A die for a thin, uniform section can utilize a 3° draft angle; a deep, thick-sectioned forging may require as much as 10° to assure self-releasing.

Die sections that are encircled by the part may require 10° of draft angle because, as the metal cools, interference fits can create problems of die release. This condition is illustrated in Figure 21.7.

Forgings must be planned to have flat locating surfaces, without draft angles or flash where required for use in subsequent machining operations. Figure 21.8(a) illustrates a forged steel control knob that requires a drilling and tapping operation to complete. Machining will be accomplished in a self-centering pump jig while located on the large, flat, as-forged end face. The forging will be trimmed of flash before machining. The die design shown in Figure 21.8(b) provides the required locating surface. The die design in Figure 21.8(c) does not provide an acceptable locating surface but does present an awkward, possibly unacceptable, flash trimming operation.

Fillets and Corner Radii The amount of die fillet radii and corner radii provided in a forging die significantly affect the flow of metal filling a die cavity. It must be noted that "fillet radii" refers to the forging; a die fillet radius is one that provides the fillet of the *part*.

The optimum amount of die fillet radius permits a smooth, progressive flow of metal into a forging die cavity whereas a radius that is too sharp causes hot metal to flow across rather than

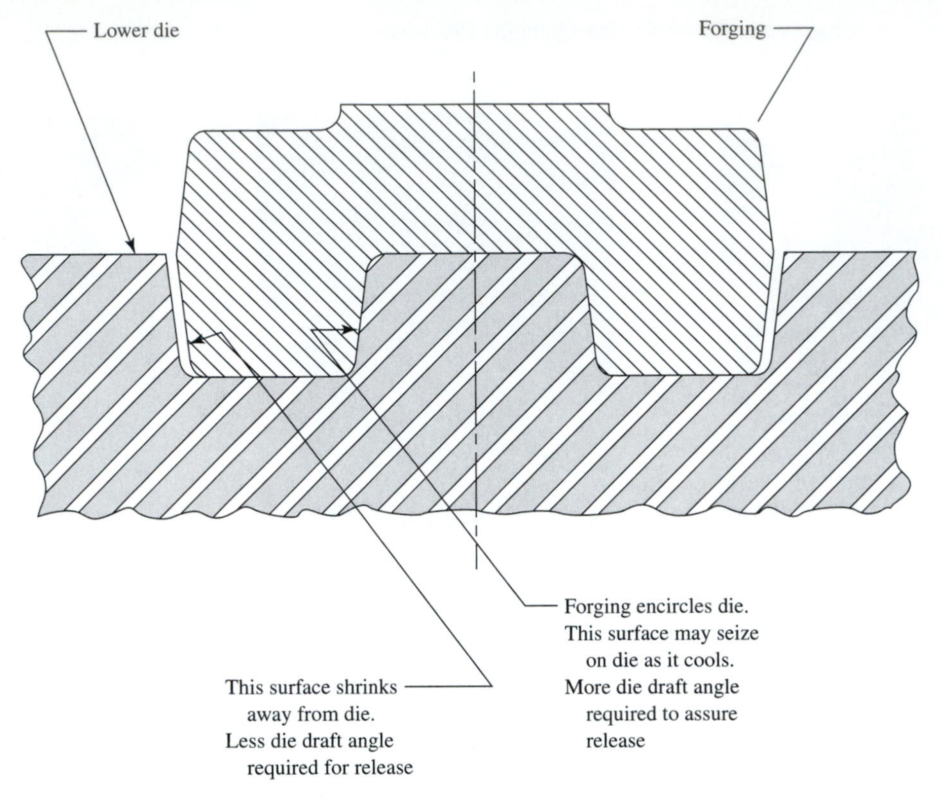

FIGURE 21.7 Forging-die draft-angle considerations.

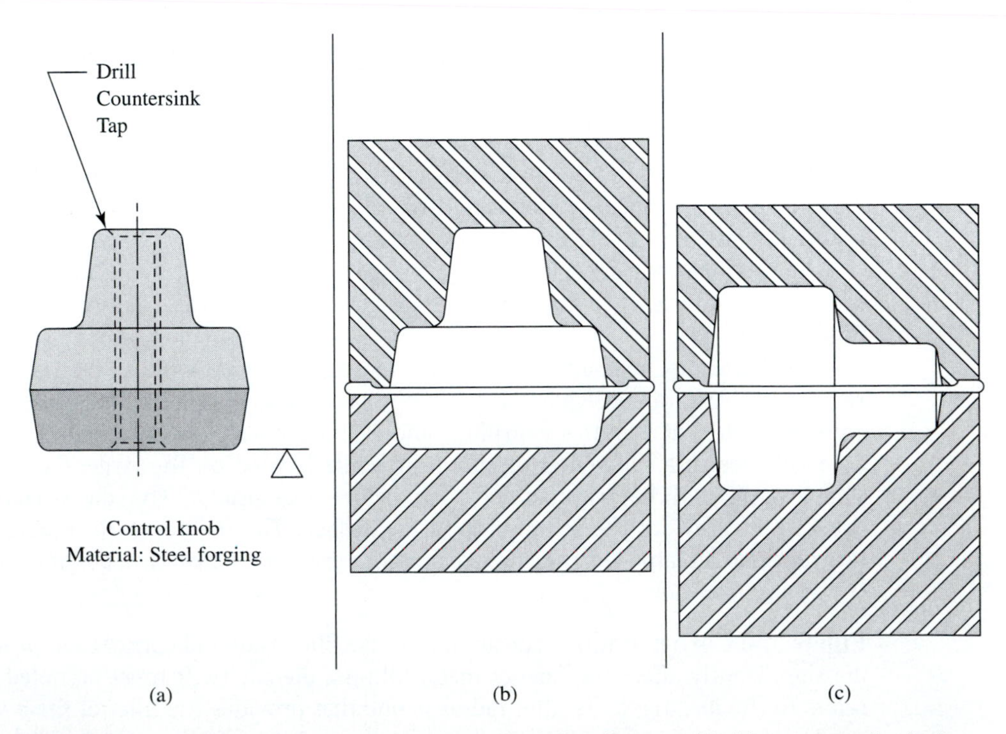

FIGURE 21.8 (a) Finished part. (b) Correct die design for part (a). (c) Incorrect die design for part.

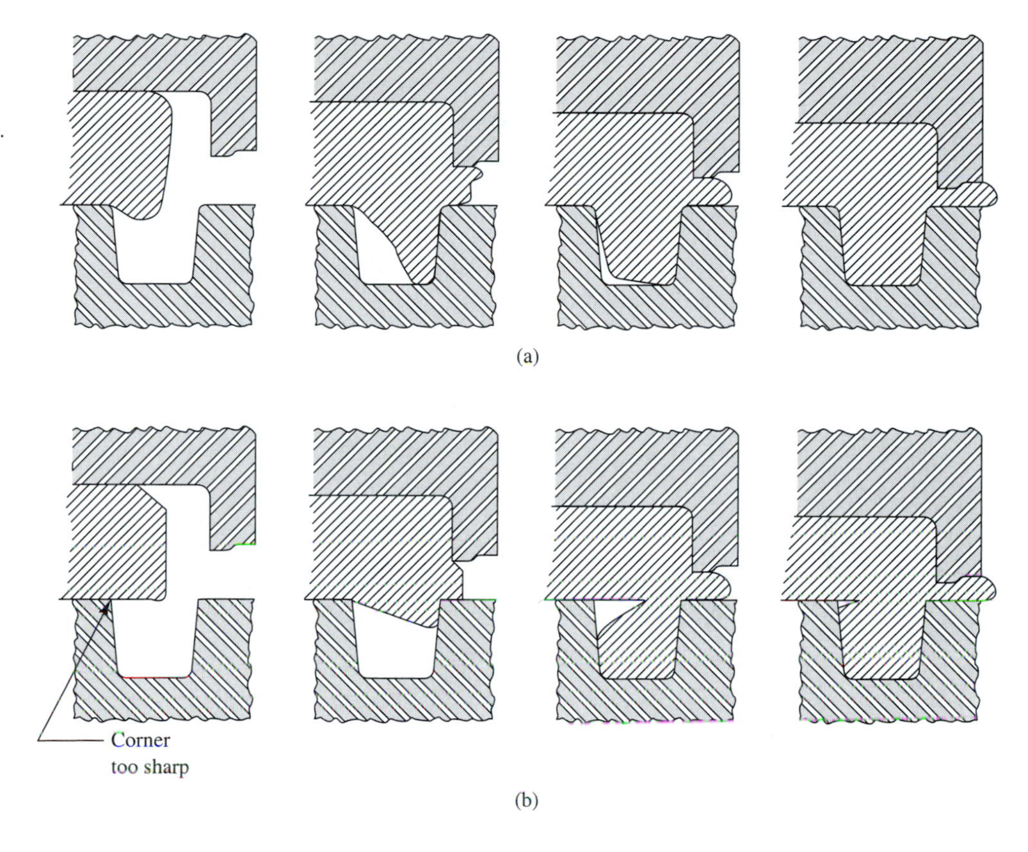

FIGURE 21.9 (a) Die cavity fill with optimum fillet radius. (b) Die cavity fill with sharp fillet radius.

down into it. The metal then backfills or folds back as shown in Figure 21.9(b). The folding action causes an incomplete section, or flow, called a fold or lap.

The Drop Forging Association's recommended standards for fillets and corner radii are shown in Figure 21.10. The die cavity should also form a radius on the corner of a forging; a sharp internal corner in the die cavity traps air, oil, and foreign material, resulting in incompletely filled and faulty forgings. Forging-die cavities must be provided with draft angles that will assure

FIGURE 21.10 Forging die fillet and corner radii practice.

Maximum Radii of Fillets and Corners, inches		
Net Weights in Pounds up to—	Commercial	Close
.3	$\frac{3}{32}$	$\frac{3}{64}$
1	$\frac{1}{8}$	$\frac{1}{16}$
3	$\frac{5}{32}$	$\frac{5}{64}$
10	$\frac{3}{16}$	$\frac{3}{32}$
30	$\frac{7}{32}$	$\frac{7}{64}$
100	$\frac{1}{4}$	$\frac{1}{8}$

Draft-Angle Tolerances—Permissable Variations from Standard or Nominal Draft Angles							
Drop Hammer Forgings				Upset Forgings			
Location of Surface	Nominal Angle, Degrees	Commercial Limits	Close Limits	Location of Surface	Nominal Angle, Degrees	Commercial Limits	Close Limits
Outside	7°	0°–10°	0°–8°	Outside	3°	0°–5°	0°–4°
Holes and depressions	10°	10°–13°	—	Holes and depressions	5°	0°–8°	0°–7°
	7°	—	0°–8°				

FIGURE 21.11 Forging-die draft-angle standards.

release of the forging. The Drop Forging Association's recommended draft angles are shown in Figure 21.11.

Forging Dies for Copper and Bronze Forgings for copper and bronze are produced with dies similar in design to those used for steel forgings. One major difference is the higher quality required for die surface finishes. Surfaces must be extremely smooth and free of any tool marks and scratches because the soft metal will be driven into any imperfection. Draft angles, fillets, and radii are virtually the same as those used for steel forging dies.

Tolerances The Drop Forging Association's recommended tolerances apply to forgings produced to either commercial or close standards. Commercial standard tolerances apply to general forging practice. Close standard tolerances apply when forgings must be held to closer tolerances than standard. Additional care and costs apply to this grade.

- The thickness tolerances shown in Figure 21.12(a) apply to forging dimensions that cross parting lines of dies.
- The width and length tolerances shown in Figure 21.12(b) are identical and apply to overall widths and overall lengths of forgings, but do not apply across parting lines.

Trimming Forging flash is removed—trimmed—on a press with a trimming die. The process is usually performed hot, although some smaller steel forgings and many aluminum forgings are trimmed cold.

Hot-trimming dies are made of a special grade of steel known as hot-trimming die stock. This is a medium-carbon steel and may be faced with Stellite—a wear material that maintains hardness at elevated temperatures. Conventional tool steels are not used for hot trimming because the edges of such dies check, crack, and break away at higher temperatures.

Cold-trimming dies and punches are made of tough alloy tool steels hardened to a medium hardness for shock resistance. Punches are designed to have some float and are given a lower hardness than the die so that, in the event of contacting the die, the edge of the punch will shear, protecting the more costly die. A trimming die cross section is illustrated in Figure 21.13.

Thickness Tolerances, Inch*									
Net Weights, Pounds, up to —	Commercial		Close		Net Weights, Pounds, up to —	Commercial		Close	
	–	+	–	+		–	+	–	+
.2	.008	.024	.004	.012	20	.026	.078	.013	.039
.4	.009	.027	.005	.015	30	.030	.090	.015	.045
.6	.010	.030	.005	.015	40	.034	.102	.017	.051
.8	.011	.033	.006	.018	50	.038	.114	.019	.057
1	.012	.036	.006	.018	60	.042	.126	.021	.063
2	.015	.045	.008	.024	70	.046	.138	.023	.069
3	.017	.051	.009	.027	80	.050	.150	.025	.075
4	.018	.054	.009	.027	90	.054	.162	.027	.081
5	.019	.057	.010	.030	100	.058	.174	.029	.087
10	.022	.066	.011	.033					

*Thickness tolerances apply to the overall thickness. For drop-hammer forgings, they apply to the thickness in a direction perpendicular to the main or fundamental parting plane of the die. For upset forgings, they apply to the thickness in the direction parallel to the travel of the ram, but only to such dimensions as are enclosed by and actually formed by the die.

Adopted, 1937, by Drop Forging Association for forgings under 100 pounds each

(a)

Shrinkage		Plus	Die Wear			Mismatching		
Lengths or Widths up to — in.	Commercial + or –	Close + or –	Net wt. up to — lbs.	Commercial + or –	Close + or –	Net Weights, Pounds, up to —	Commercial	Close
1	.003	.002	1	.032	.016	1	.015	.010
2	.006	.003	3	.035	.018	7	.018	.012
3	.009	.005	5	.038	.019	13	.021	.014
4	.012	.006	7	.041	.021	19	.024	.016
5	.015	.008	9	.044	.022	25	.027	.018
6	.018	.009	11	.047	.024	31	.030	.020

For each additional inch under shrinkage, add 0.003 to the commercial tolerance and 0.0015 to the close tolerance. For example, if length or width is 12 inches, the commercial tolerance is plus or minus 0.036 and the close tolerance plus or minus 0.018.

For each additional 2 pounds under die wear, add 0.003 to the commercial tolerance and 0.0015 to the close tolerance. Thus, if the net weight is 21 pounds, the die wear commercial tolerance is 0.062 plus or minus, and the close tolerance 0.031 plus or minus.

For each additional 6 pounds under mismatching, add 0.003 to the commercial tolerance and 0.002 to the close tolerance. Thus, if the net weight is 37 pounds, the mismatching commercial tolerance is 0.033 and the close tolerance 0.022.

Adopted, 1937, by Drop Forging Association for forgings under 100 pounds each

(b)

FIGURE 21.12 (a) Standard forging thickness tolerances. (b) Standard forging length and width tolerances. (Courtesy Drop Forging Association)

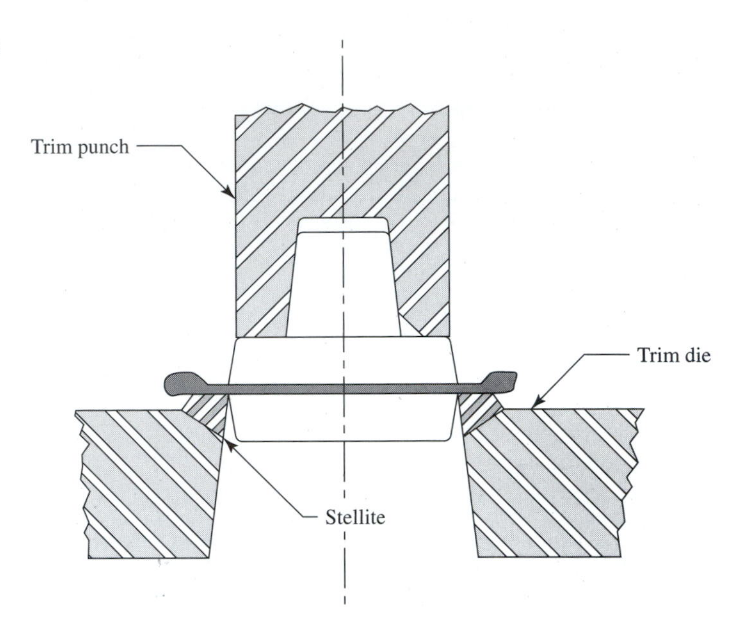

FIGURE 21.13 Trimming die cross section.

21.3 EXTRUSION

First developed in the late eighteenth century, the **extrusion process** was used to produce lead pipe. Then, as now, the process consisted of placing a billet of metal in a closed chamber and forcing it with pressure from a press through an opening in a die.

Process Term Definitions

- Metal, extruded in continuous solid shape and also hollow, is said to be produced by the extrusion process.
- Metal parts of finite shapes and lengths are generally termed cold extrusions.
- Impact extrusions are cold, reverse extrusions produced at a high rate of production with the ram and punch contacting the metal rapidly.

21.3.1 Extrusion Process

Paste forced through the opening of a toothpaste tube demonstrates the extrusion process, and the tube itself is an extrusion. In metal parts production, the process consists of positioning a billet in a die chamber, where it is confined between a ram and a die as shown in Figure 21.14. Pressure applied to the billet from the press forces the metal to flow plastically through a shaped opening in the die.

Forward Extrusion Metal can be extruded in this process utilizing the forward extrusion arrangement whereby the metal flows in the same direction as the motion of the ram. Sometimes known as direct extrusion, forward extrusion is also illustrated in Figure 21.14.

Backward Extrusion Metal can also be extruded with this process utilizing the backward extrusion arrangement, in which the metal flows in the direction opposite to the motion of the

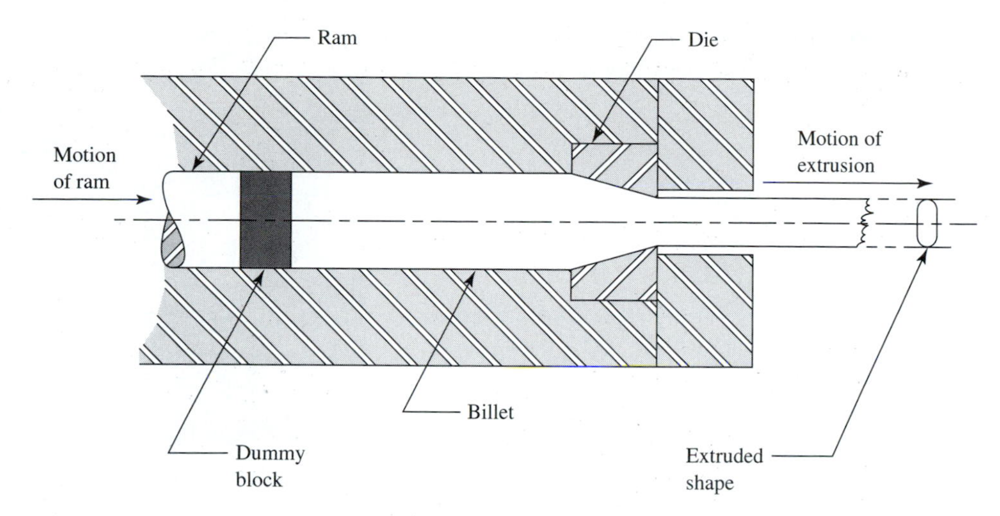

FIGURE 21.14 Forward (direct) extrusion.

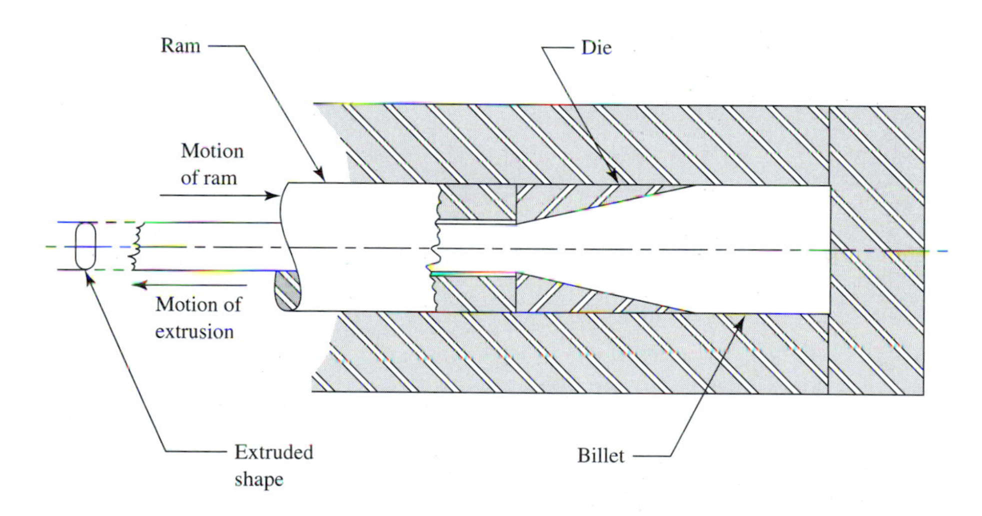

FIGURE 21.15 Backward (indirect) extrusion.

ram as shown in Figure 21.15. This category, also known as indirect extrusion, has two limiting conditions, which cause the more predominant use of forward extrusion:

- Length of product is restricted
- Tooling is more costly

A 1000-ton hydraulic vertical extrusion press is illustrated in Figure 21.16.

Hydrostatic Extrusion In this variation of the extrusion process, the ram applies pressure to a fluid, which forces the billet through a die by hydraulic pressure as shown in Figure 21.17. Since the billet is surrounded by a fluid, no direct contact and friction exists between the billet and chamber walls. Also, the fluid, which is a lubricant, lubricates the metal as it flows into the die opening.

FIGURE 21.16 1000-Ton hydraulic vertical extrusion press. (Courtesy Erie Press Systems)

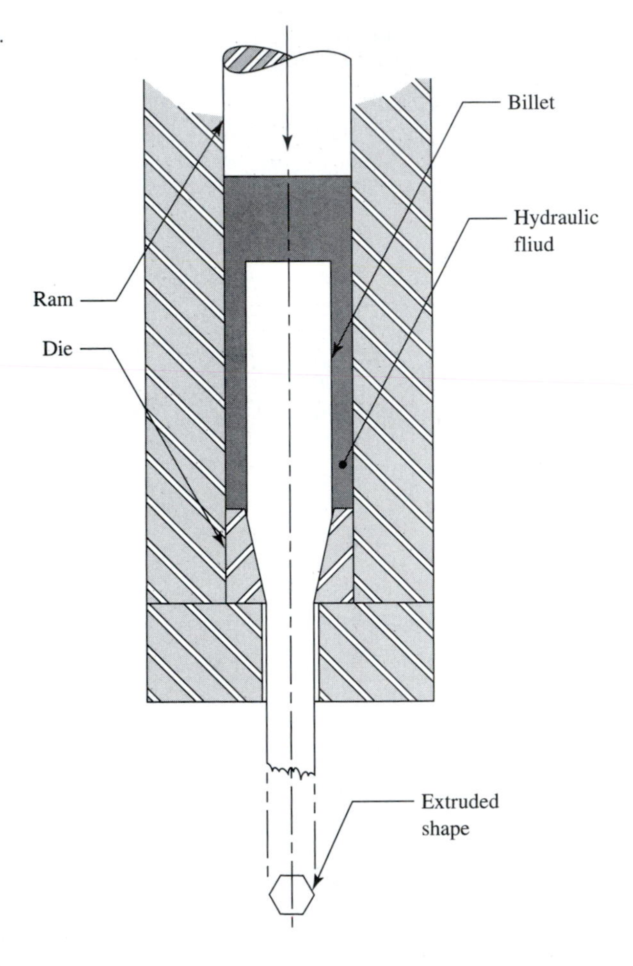

FIGURE 21.17 Hydrostatic extrusion.

The process is accomplished cold using oil and can be done at elevated temperatures with molten glass or a heat-resistant fluid. Brittle, as well as ductile, metals can be extruded with this process. However, the process is not used extensively due to the slow rate of extrusion and high cost and complexity of tooling.

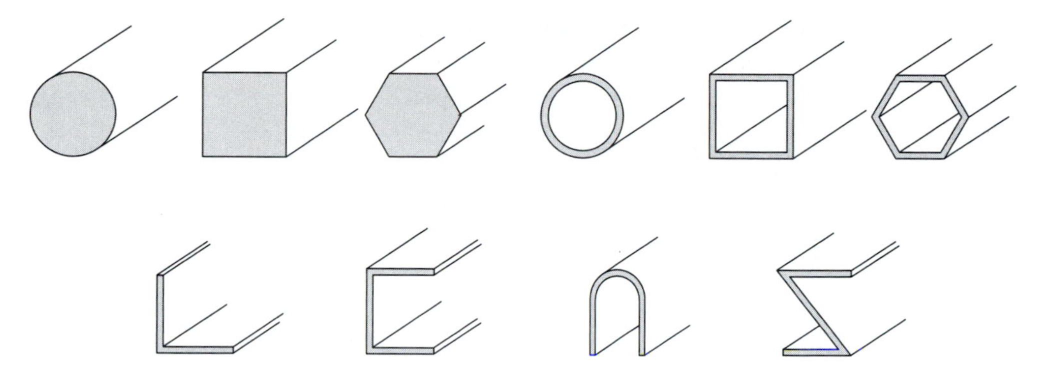

FIGURE 21.18 Extruded shapes used as structural components.

Metals Extruded Aluminum, copper, lead, and magnesium are easily extruded due to their relatively low yield strengths. Very little steel is extruded cold in the continuous process; the extremely high pressure required and the heat from friction cause problems with the metal welding to the chamber and die. Considerable quantities of hot steel extrusions are produced. Heated billets must be used in conjunction with lubricants of heated glass, graphite, and molybdenum disulfide.

Examples Some extruded shapes, as illustrated in Figure 21.18, are used as extruded for structural members and components of other assemblies. Many standard geometrical cross-sectional forms as well as special shapes are extruded and then cut to length to form individual parts, some of which require additional secondary operations to complete. Examples of this processing are shown in Figure 21.19.

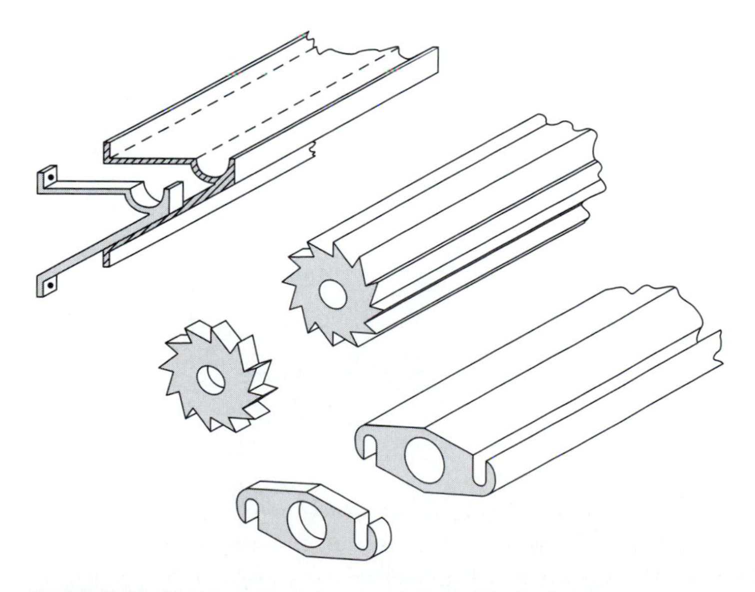

FIGURE 21.19 Extruded shapes sectioned to form parts.

Advantages of the Extrusion Process

- Many shapes including hollow ones can be produced.
- No draft angles are required on surfaces.
- Very good-quality surface finishes are produced.
- Inherent dimensional tolerance control is good.
- Net shapes save material and labor.
- Tooling costs are moderate.

21.3.2 Cold Extrusion—Cross-Sectional Shapes

A cold extrusion is produced by placing a slug or billet in a cylindrically shaped die, then applying enough pressure through a punch to cause the metal to conform to the die by plastic flow. Illustrated in Figure 21.20 is a forward extrusion of a billet—so named because the metal flow is forward, or in the same direction as the punch movement.

A backward extrusion, shown in Figure 21.21, is produced with the metal flowing in the direction opposite to the punch movement. Parts are also produced in combined extrusion operations as illustrated in Figure 21.22. Forward, backward, and combined extrusions are produced at a fair rate of production; quantities to 100 pieces per hour are typical, depending on the alloy and percentage of reduction required.

21.3.3 Impact Extrusion

A version of cold extrusion termed impact extrusion is used to get a higher rate of production. The process is used to produce cylindrical parts of thin wall thicknesses where wall thickness is

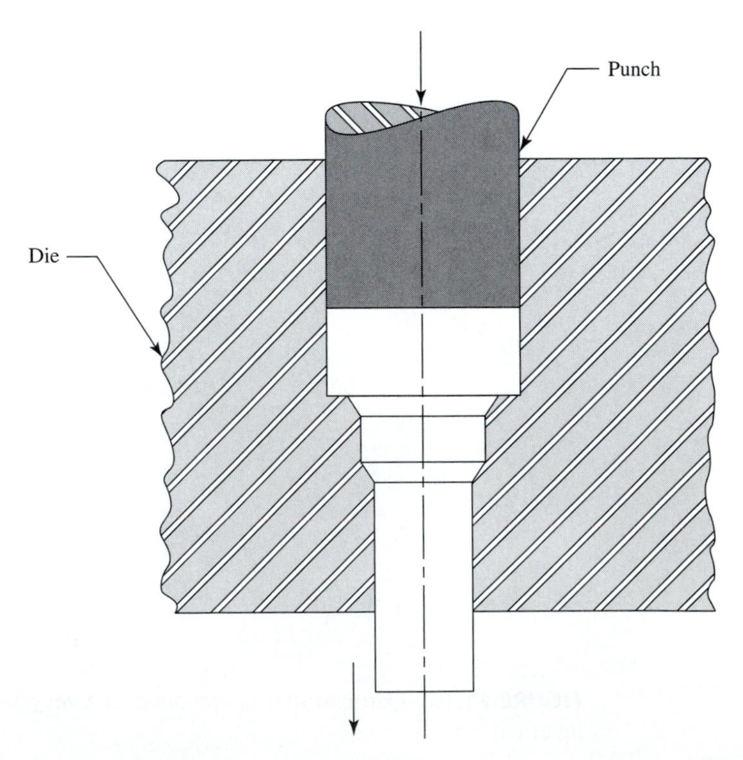

FIGURE 21.20 Forward extrusion.

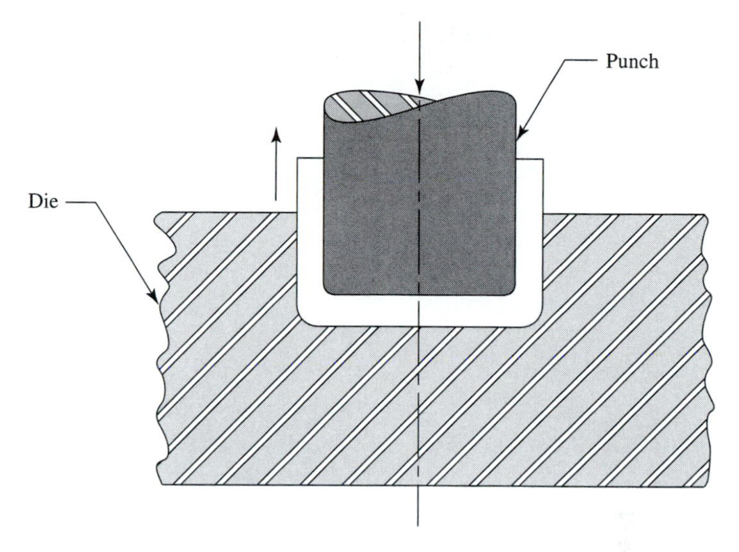

FIGURE 21.21 Backward extrusion.

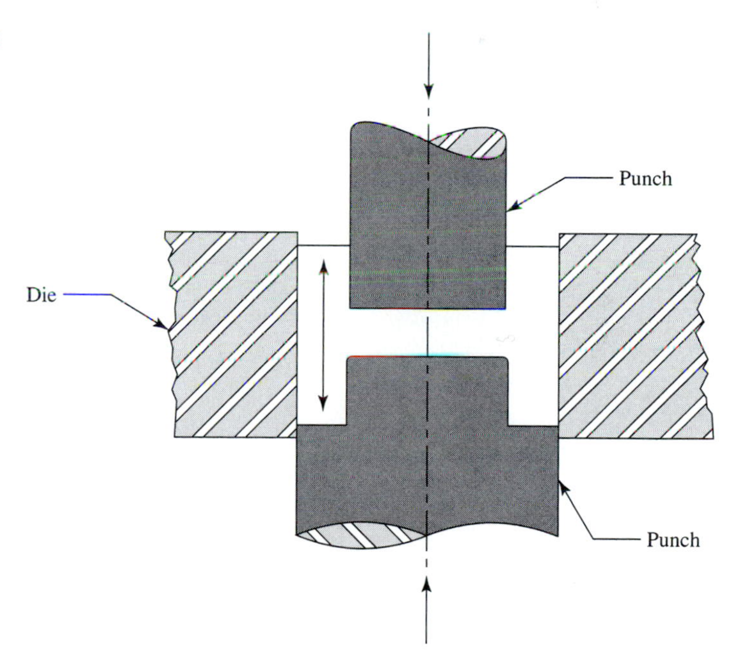

FIGURE 21.22 Combination extrusion.

small compared to the part's diameter. Examples are beverage cans, collapsible tubes (toothpaste), electronic shields and cans, and cigar packaging tubes.

In this process, illustrated in Figure 21.23, a slug is positioned in the die cavity and the press is cycled, bringing the ram and punch into contact with the metal very rapidly and, forcing the metal upward (reverse extrusion). This operation is used on softer, nonferrous metals and is normally highly automated.

One of the suppliers of basic aluminum bars, plate, and sheet also produces aluminum containers. Manufacturing this product proves especially efficient for them because all offal, scrap, and trimmings are remelted at the site with very low handling costs. A schematic view of the layout is shown in Figure 21.24.

FIGURE 21.23 Impact extrusion.

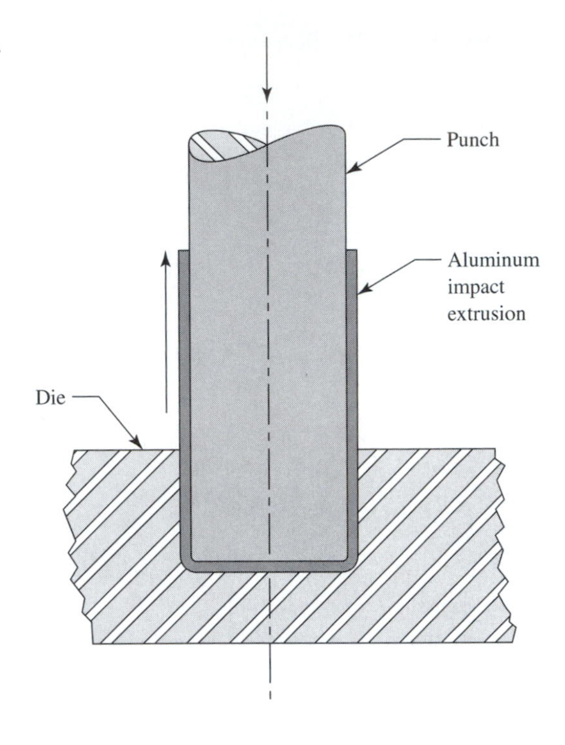

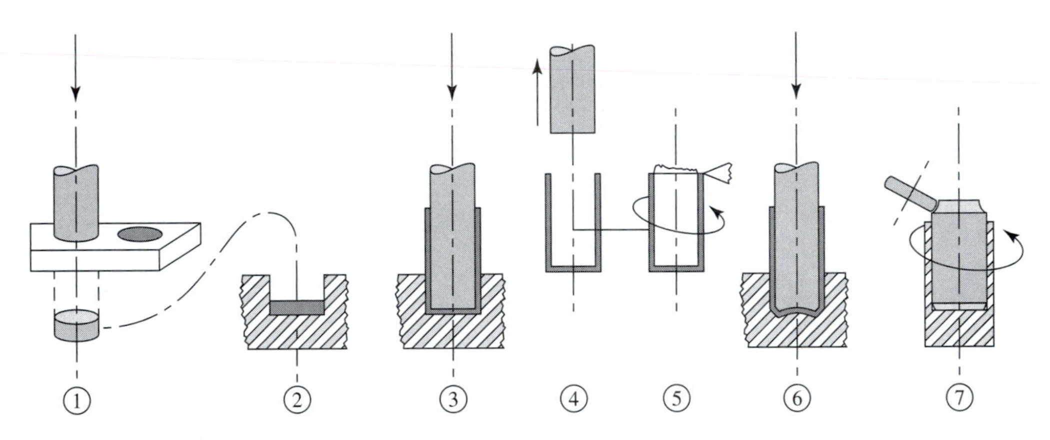

1. Punch slug from aluminum strip.
2. Feed slug into die.
3. Reverse extrude.
4. Strip from punch.
5. Trim end.
6. Form bottom.
7. Swedge (roll) top.
8. Wash.
9. Apply logo.

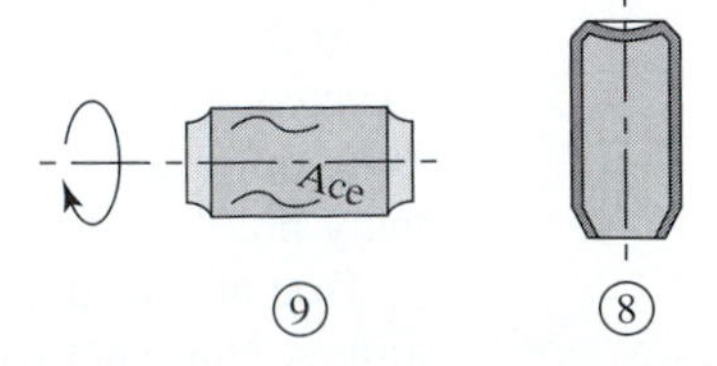

FIGURE 21.24 Impact extrusion—beverage can process layout.

	Pressure (tons/in^2)						
Alloy	450°F	500°F	550°F	600°F	650°F	700°F	750°F
AZ31B	33	33	30	27	26	25	23
AZ61A	35	34	33	32	31	30	29
AZ80A	36	35	34	33	32	31	30
ZK60A	34	33	32	31	29	27	26

FIGURE 21.25 Impact extrusion temperature/pressure requirements for magnesium alloys.

Basic Steps in Impact Extrusion Process for Cans

1. Punch slug from aluminum strip.
2. Feed slug automatically into extrusion die.
3. Reverse extrude part.
4. Strip part from punch and transfer to trimmer.
5. Trim end.
6. Form bottom.
7. Swage neck.
8. Paint logo.

Some magnesium alloys can be impact extruded, but the slug must be heated and precoated with a very uniform coating of lubricant such as colloidal graphite. The extruding die is also usually heated. Figure 21.25 shows the temperatures and pressures required to extrude magnesium alloys.

Extruding Pressures Pressures required for extruding depend on the yield point of the specified metal plus the amount of force required to overcome friction between the metal and the die during plastic flow. A general range of extruding pressures for the more commonly extruded metals is shown in Figure 21.26.

21.4 SWAGING

Swaging reduces the diameter of wire, bar, and tubing. Two principal means of accomplishing the operation are with the use of dies and by rotary or radial hammering. The process is used primarily for steel parts, but it can be used for any ductile metal. However, the softer metals such as aluminum and bronze are generally more economically produced by extruding, spinning, or rolling processes.

FIGURE 21.26 Extrusion pressures for common metals.

Material	Pressure Range (tons/in^2)
Pure aluminum—extrusion grade	40–70
Brass—soft	30–50
Copper—soft	25–70
Steel C1010—Extrusion grade	50–165
Steel C-1020—Spheroidized	60–200

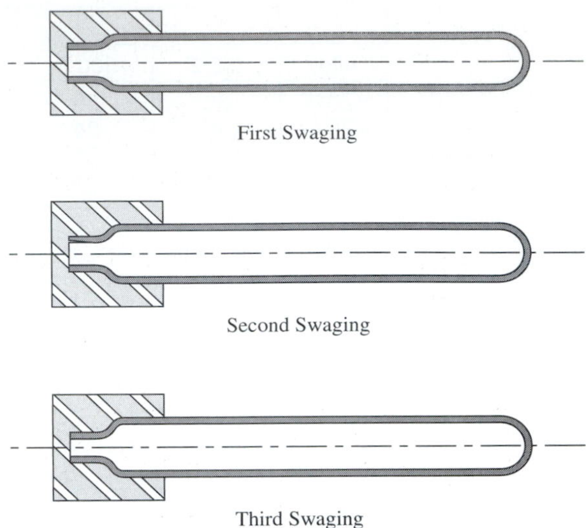

FIGURE 21.27 Swaging process to neck end of cylinder.

Die Swaging Die swaging is accomplished by forcing or hammering a tube or rod into a shaped die to reduce and re-form the diameter. The die often is the hammer. The operation is generally done hot, especially with thicker sections. Illustrated in Figure 21.27 is a series of dies required to form the end, or neck, of a heavy-walled cylinder.

21.5 UPSET FORGING (HEADING)

Upset forging, also known as **cold heading**, is accomplished by transferring a cutoff length of steel wire progressively through multiple stages of a die—each of which does incremental work—until the part is completed and ejected into a container or fed into secondary equipment such as drilling or thread rolling.

Advantages of the process are extremely high attainable rates of production, good surface finishes, and very low scrap loss. More total parts are purchased by heading than by all other upset processes combined. Cold-headed parts are produced from steel wire of up to 1.000 in. in diameter supplied in coils.

Heading can be done hot or cold; most parts are produced cold. Parts suitable for production by this process include fasteners of all kinds, bolts, pins, shafts, small hubs and gear blanks, and the like. Figure 21.28 illustrates a cold-heading operation used to produce bolt blanks with hexagonal heads ready for the threading operation.

21.6 WIRE DRAWING

Drawn wire, which itself is produced by a cross-section-changing process, is the raw or direct material for other such processes—heading in particular. Steel wire may be purchased in the desired alloy and diameter in coils ready for use in heading machines. Many producers elect to purchase wire in this condition.

Other producers of headed parts purchase steel wire drawn to nominal sizes and commercial tolerances, then draw (and lubricate) the metal to final precise desired sizes and tolerances. Savings are effected in purchase costs and reduced inventory requirements. Figure 21.29 illustrates a typical wire drawing line used by several producers of cold-headed parts. A brief description of the system's operation follows:

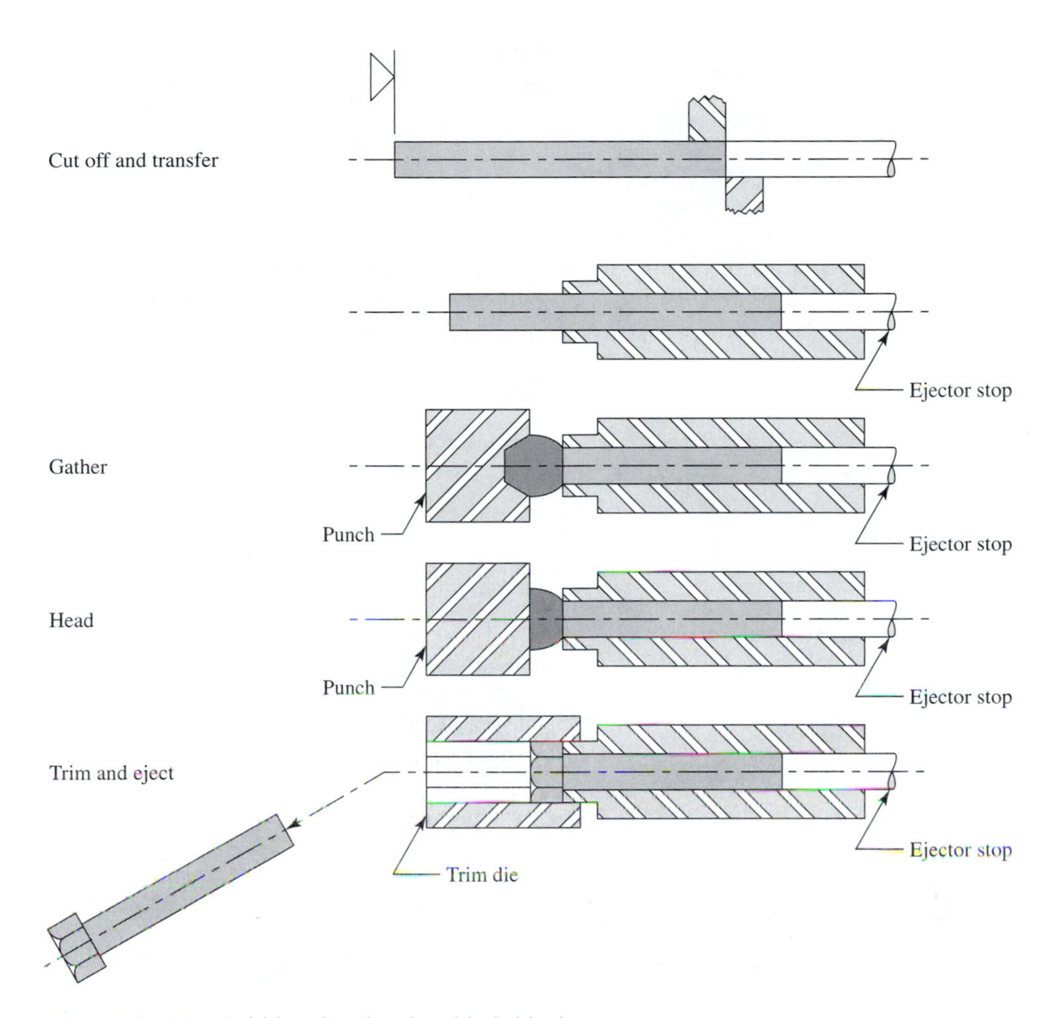

FIGURE 21.28 Cold heading hex head bolt blank.

- A coil of wire of nominal size is loaded onto one station ① of the capstan ②. One end of the steel wire is welded to the face end of the previously loaded coil ③. Welding flash is removed and the joint is ground smooth.
- When the previously loaded coil of wire is consumed by drawing, the capstan is unlocked and rotates 180°, bringing the new coil into production with no delay for reloading.
- Most of the mill scale coating the wire is removed by drawing the wire around relatively small-diameter wire guide wheels ④, which causes the inelastic scale to fracture and split off.
- The wire is drawn through a cleaning tank ⑤, where a caustic solution, in conjunction with additional small-diameter wire guide wheels that flex the wire severely, removes all remaining scale, readying the metal surface for coating with lubricant.
- As the wire passes through the lubricating tank ⑥, powdered metalic soap clings to the mildly etched surface, lubricating it for drawing to size in carbide die ⑦.
- The wire is drawn to size. The very considerable effort required to draw the wire is provided by an electric motor-driven bull-wheel wire puller ⑧. Several turns of wire around grooves in the rim of the bull-wheel provide a friction drive.
- Wire flows through a guide as it leaves the bull-wheel rim, forming it into a specific diameter and feeding onto a revolving wire reel ⑨ known as a "top hat" due to its unique shape.

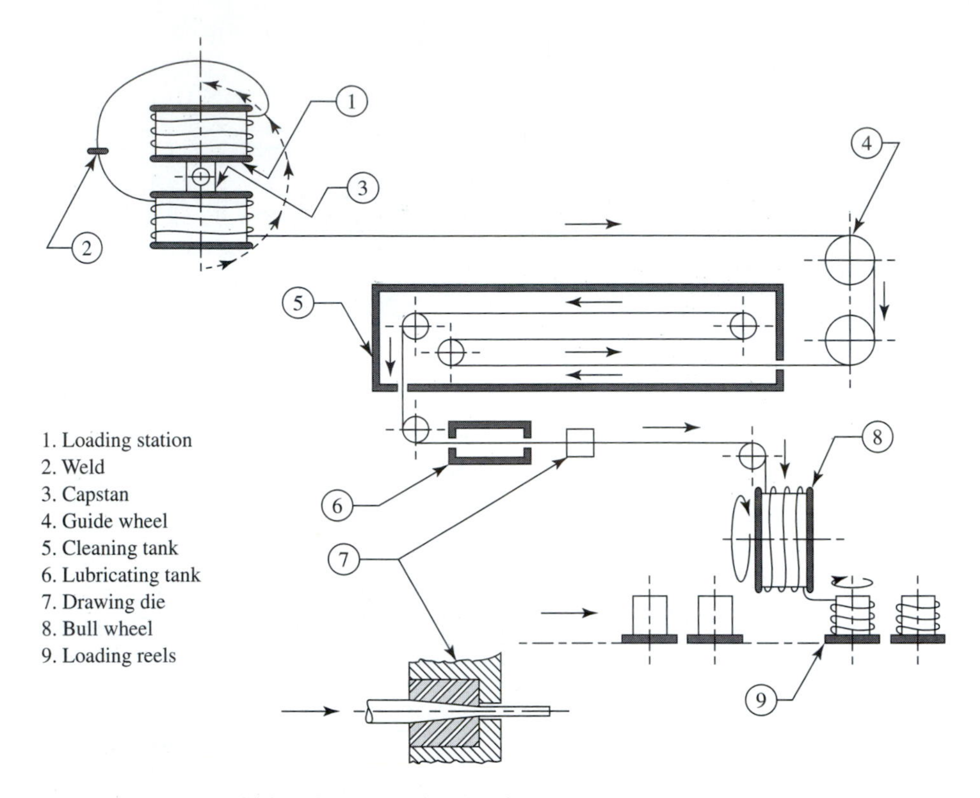

FIGURE 21.29 Cold heading wire drawing line.

- As wire reels fill to capacity, the wire is cut, the loaded reel (top hat) is sent to the headers, and an empty reel begins accumulating wire.
- Top hats are designed not only to accumulate and store drawn wire from the wire drawing line, but also to load directly into the heading machines, feeding the wire directly into the dies.

Wires drawn range up to 1.00 in. in diameter.

21.7 REVIEW QUESTIONS AND PROBLEMS

1. Describe the rolling process.
2. What is an open die forging?
3. What force or form of energy is used in the drop-forging process?
4. What is one part design characteristic that is unsuitable for the drop-forging process?
5. In what ways are parts produced by the press-forging process superior?
6. Why must forging dies for copper and bronze materials have a better surface finish than dies used for steel?
7. Describe, in general terms, the extrusion process.
8. Assume a part can be processed by either machining or extruding. What must be considered, and what are the advantages of either process?
9. Explain the difference between cold extrusion and impact extrusion.
10. What is swaging?
11. Describe the cold-heading process.
12. What kind of parts are produced by the cold-heading process?
13. What material is used in the cold-heading process?
14. How is the direct material produced for the cold-heading process? Describe the process.

PART VI

METALWORKING PROCESS PLANNING

INTRODUCTION

Process planning for metalworking follows the same basic procedure used to develop a process plan for machined parts. Both require that complete information and data be obtained before an attempt is made to process.

One element not used in metalworking processing is a tolerance chart; final part dimensions and tolerances are established by the die or die set and are not subject to tolerance accumulations from succeeding operations. Stock removal is not calculated as such; the part is usually blanked from sheet or strip stock. Trimming and shaving operations require only a simple calculation of material to be allowed for these operations.

Die lubricants are a vital part of a die process. They are required to prevent metal tearing, die pickup and welding, and die wear. Specific lubricants must be selected for optimum lubricity for each die/material combination and specified for use on the operation sheets. Data, charts, and graphs are provided for die lubricant selection in Chapter 23.

Press types and sizes of presses along with the required accompanying equipment such as stock feeders, reels, stock straighteners, scrap cutters, and so on are illustrated and explained in Chapter 24.

Also, special machines for metalworking such as multislides, turret presses, plasma cutters, and laser cutters are explained and illustrated.

CHAPTER TWENTY-TWO

METALWORKING PROCESS PLANNING

CHAPTER OVERVIEW

22.1 INTRODUCTION

A manufacturing engineer developing a metalworking process plan for a stamping will follow the same basic procedure as that used to develop a machined part process plan (see Chapter 6). Both must initially obtain complete pertinent information and data; both require analysis of specific processes required to satisfy the part design; and both require analysis of the equipment capacity required to satisfy production requirements.

All required tooling, machines, and gages must be determined and provided. Equipment and tooling for ancillary, auxiliary, and support operations must be determined and provided for. Operation sheets must be written for the plan and distributed to all concerned.

Although the same basic procedure is used to develop process plans for metal worked and also machined parts, the knowledge, training, and experience required by engineers producing either plan is very different.

This chapter highlights metalworking processing as it differs from machining processing. The time required to become proficient in either discipline is so great that few engineers become expert in both areas. Indeed, most major manufacturers make a clear distinction between "die engineer" and "tool engineer." Separate departments are maintained: Die Engineering and Tool Engineering are usually supervised by a chief die engineer and a chief tool engineer, with both normally reporting to the head of Manufacturing Engineering.

Many manufacturing engineers within each area were typically promoted from the skilled trades. A beginning engineer with a journeyman diemaker certificate and an education in mechanical or manufacturing engineering could, after some years experience of increasingly complex metalworking process planning, enjoy a satisfying and financially rewarding career. The same can be said for an individual following a similar career path in tool engineering.

22.2 PART RELEASE FOR PRODUCTION

A part release is the formal document issued to notify all concerned of the effective date of production for which a part of specified design and part number will be required. Releases for certain parts are issued to Purchasing to obtain competitive bids; other parts are considered proprietary. Releases for proprietary automotive components such as transmissions, engines, and bodies are issued directly to corporate-owned transmission, engine, and stamping plants.

Releases must be issued early-on to provide sufficient lead time for process planning and acquisition and setup of tooling and equipment by the start-up date. Also, time should be allowed for possible engineering changes (see Section 22.3).

22.3 DETERMINE DESIGN FEASIBILITY

Product Engineering's responsibility is to assure that designs of released parts are functional and producible. Part designs are checked and rechecked; sample, prototype, and preproduction parts are tested to destruction. Still, Manufacturing Engineering must perform a final check of *feasibility*; that is, a determination must be made as to whether a part, as designed, can be produced with available technology and equipment. If it cannot, the part design must be altered, more capable equipment obtained, or the part resourced to a manufacturer with facilities capable of producing the part as designed. Areas of review in a feasibility study must include the following:

- Determine if the part's designed form, shape, or contour can be duplicated in practice to satisfy the specifications.
- Determine if the part's dimensional and surface finish tolerances are attainable.
- Determine if the size of the stamping is within available press size capacity and that available presses have sufficient tonnage.
- Determine if a quality part can be produced without tears, cracks, or distortion, or whether a higher grade of metal may be required.
- Determine if additional operations such as annealing will be required.

Any changes required to make production of released parts feasible must be communicated with Product Engineering promptly to permit the changes to be made with no holdup of planned production dates. Also, any known change of design that would enhance quality or reduce cost should be made at this point.

22.4 PART ANALYSIS AND TENTATIVE PROCESSING

To develop a process plan for a specific part, a manufacturing engineer must study and become thoroughly familiar with the designs, specifications, and function of the part. All design features such as form, holes, notches, and so on must be identified and suitable processes provided to accomplish production.

22.4.1 Analyze Part Design

Each significant element of a part that requires a manufacturing operation must be determined and entered in a list for consideration. These considerations should include

- Material
- Material hardness and heat treat specifications
- Gross size, metal thickness, and finished part weight
- Close tolerance dimensions, critical dimensions, and surface finish tolerances
- Description of each part feature for which an operation is required
- Mating parts: Determine needed clearances, fits, and mating surfaces of parts that will become details of an assembly.

Examples An analysis of the Slide Bracket design in Figure 22.1 determines that eight operations as well as inspections and material handling are required for production. These operations are shown in Figure 22.2.

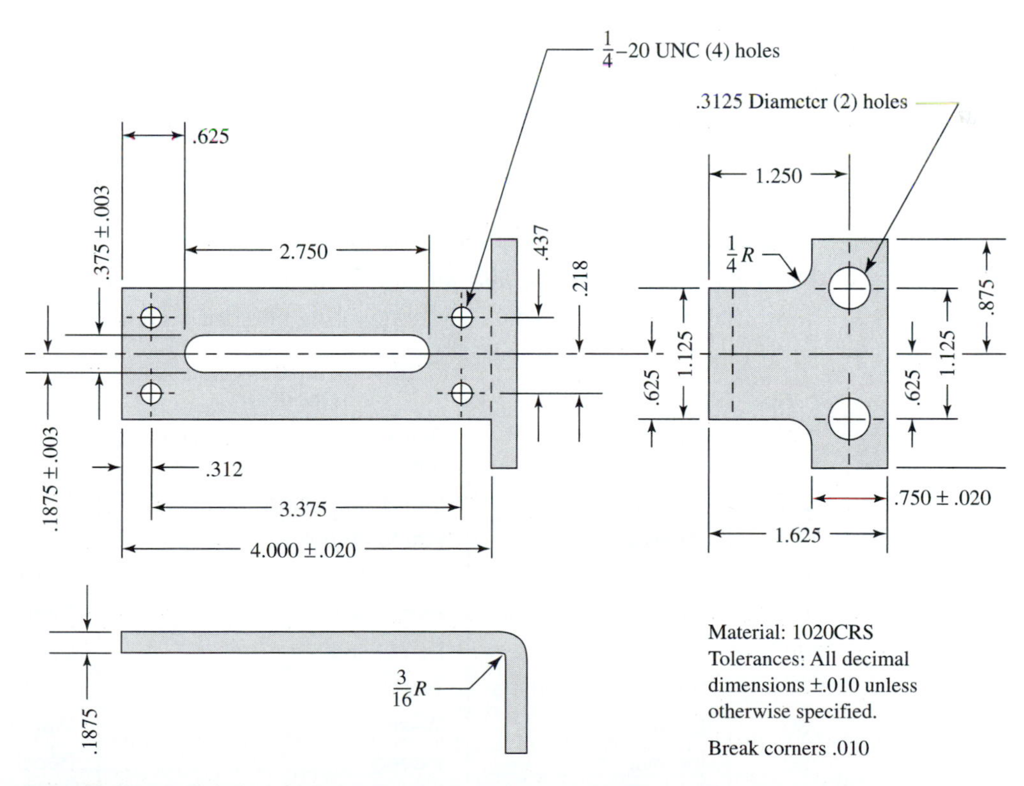

FIGURE 22.1 Slide bracket.

FIGURE 22.2 Part design features.

No.	Feature	Operation Required
1	Material: 1020CRS	
2	Blank size and shape	✓
3	Bend	✓
4	Slot	✓
5	(2) Holes	✓
6	(4) Holes	✓
7	Thread (4) holes	✓
8	Break corners	✓
9	Wash	✓
10	Inspect	
11	Material handling	

22.4.2 Develop Tentative Processing

Each processing requirement as determined from analyzing the part design must be considered and a means to produce them planned. Following these determinations, a tentative process sequence must be developed that will satisfy the part design and production quantity requirements.

Figure 22.2 illustrates the individual operations required to produce the Slide Bracket in Figure 22.1. Figure 22.3 illustrates a tentative process sequence and the tooling required to produce it.

22.5 ANALYZE FACTORS AFFECTING MANUFACTURING COST

22.5.1 Material

- Material specified for slide bracket is 1010 CRS.
- No viable substitute is available for cost reduction
- Cost of blank might be reduced by eliminating notched sections

Op. No.	Part Feature	Operation	Tooling
10	Blank	Blank part to developed length and shape	Blanking die
20	(2) Holes	Pierce (2) holes	Piercing die
30	(4) Holes	Pierce (4) holes	Piercing die
40	Thread (4) holes	Tap (4) holes	Tapping machine
50	Slot	Pierce slot	Piercing die
60	Bend	Form bend	Bending die
70	Break sharp corners	Tumble	Tumbling barrel
80	Clean	Wash	Washing machine
90	Inspect	Inspect	Bench

FIGURE 22.3 Tentative process sequence.

22.5.2 Equipment Availability

- The plant has a wide range of press types, sizes, and tonnages available. From that standpoint no new equipment will be required.

22.5.3 Production Quantities Required

- Production is scheduled for 250,000 annually for a model run of 5 years. Parts may be scheduled for monthly runs or for a single yearly run and parts banked.
- Adequate capacity exists in the presses for the sizes likely to be used.

22.5.4 Alternative Part Design

An alternative part design that eliminates the notches would save material and enhance processing.

- The developed blank for the part in Figure 22.1, as designed, is shown in Figure 22.4(a).
- An alternative blank design that eliminates the notches is shown in Figure 22.4(b). This design will reduce the amount of material required by 26% and will also be compatible with processing on a progressive die for a subsequent labor savings.

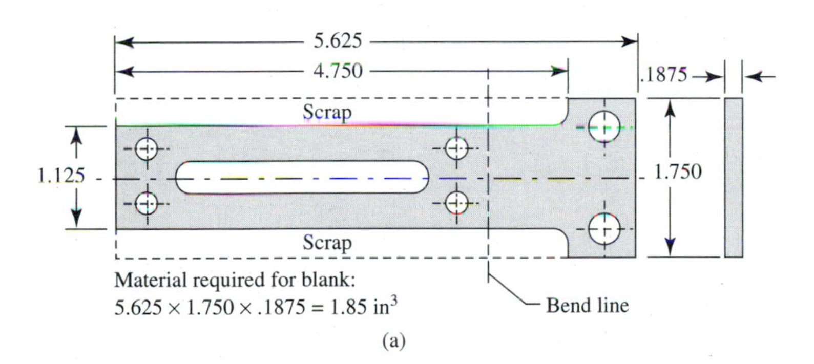

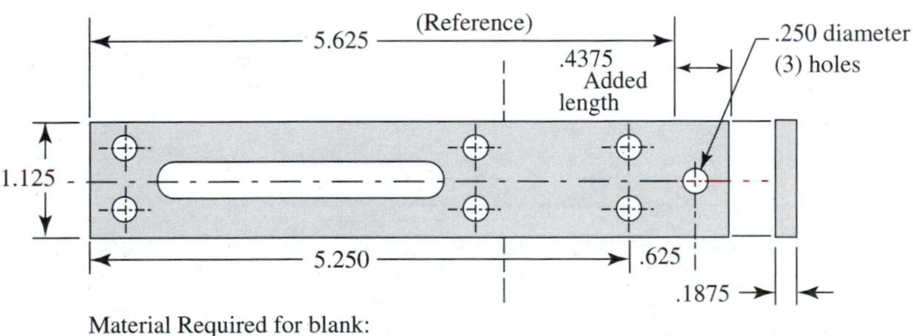

FIGURE 22.4 (a) Developed length and shape of blank as designed. (b) Developed length and shape of alternative design.

22.6 DEVELOP FINAL PROCESS SEQUENCE

After approval of the modified part design (Figure 22.4b) is received, the final processing must be developed based on the tentative processing, considered factors, and the now approved alternate design features. The developed final process sequence is illustrated in Figure 22.5, including process pictures of each operation.

22.6.1 Check Final Process Plan

A check of all design features, specifications, and tolerances must be made to ensure that the final process plan satisfies the part design and specifications.

Example The process plan *check sheet* in Figure 22.6 illustrates one method of checking the operations of a process plan against the part design details for conformance.

22.6.2 Develop Operation Sheets

Operation sheets must be written and distributed to include

- Operation numbers for each process
- Operation sequence
- Operation description

Oper No		Operation	Tooling
10	Station 1	Pierce (4) holes	4-Station Progressive die
	Station 2	Pierce slot	
	Station 3	Pierce (3) holes	
	Station 4	Cut off	

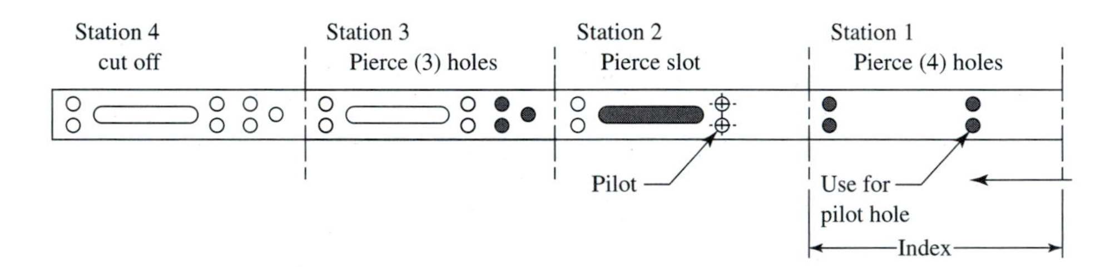

20		Tap (4) holes	Tapping machine	
30		Bend	Bending die	
40		Break corners	Tumbling barrel	
50		Wash	Washer	
60		Inspect	Bench—gages	

FIGURE 22.5 Develop final process sequence.

Item #	Part Design Feature	Dimensions and Tolerances of Design Features	How Produced and How Controlled
1	(4) Holes	Hole size for subsequent tapping, .196–.207 diameter. Location dimensions of .312, 3.375, .218, and .437 have ±.010 tolerances.	Holes are to be pierced simultaneously in one station of progressive die. Hole and location tolerances are well within die capability.
2	Slot	Slot location and length dimensions of .625 and 2.750 have ±.010 tolerances. Slot width dimension of .375 and centrality dimensions of .1875 each have a tolerance of ±.003.	Slot will be pierced in one station of progressive die. All tolerances are well within die capability.
3	(3) Holes	.250 diameter holes have a tolerance of ±.010. Location dimensions of .625 and 5.250 have ±.10 tolerances.	Holes are to be pierced simultaneously in one station of progressive die. Hole and location tolerances are well within die capability.
4	Length	5.625 length has a tolerance of ±.040 (4.000 ± .020 + 1.625 ± .020).	Blank will be cut to length in last station of progressive die. (Actual working tolerance will probably be ±.002)
5	Bend	1.625 and 4.000 dimensions each have ±.010 tolerance. Bend is assumed as 90°.	Bend of 90° will be made in a bending die. All tolerances are attainable with correct die setting.

FIGURE 22.6 Process plan check sheet.

- Equipment and machining required
- Tooling required
- Gages required

Also to be included are

- Heat treating, plating, and finishing operations
- Quality Control functions
- Material-handling facilities required

Example Figure 22.7 illustrates the final operation sheets developed to produce the Slide Bracket. In practice, copies of operation sheets, the process plan, are distributed to Industrial Engineering, Quality Control, and Production. These become the bible and are the official documents used to produce parts.

All companies have their own unique process sheet forms. Figure 22.7 is typical. Process sheet forms are augmented by corresponding standards sheets issued by the Industrial Engineering department. Standards sheets show time elements of direct labor, setup, and relief time allowances for each part.

Oper. No.	Dept. No.	Operation Description	Equipment	Tooling	Gages
10	200	Pierce (4) holes Pierce slot Pierce (3) holes Cut off	OBI press, 30-ton capacity, 90 spm	4-Station progressive die (D-1)	.192–.207 D.E. Plug gage (G-1) .240–.260 D.E. Plug gage (G-2) .372–.378 Spec width gage (G-3) Hole location gage (G-4) Template gage (G-5)
20	250	Tap (4) holes	Tapping machine	Holding fixture (F-1)	Thread plug gage (G-6)
30	200	Bend	Gap press, 10-ton capacity, 60 spm	Bending die (D-2)	Template gage (G-7)
40	260	Break sharp corners	Tumbling barrel		
50	260	Wash	Washing machine		
60	300	Inspect	Bench		.240–.260 D.E. Plug gage (G-2) .372–.378 Spec. width gage (G-3) Thread plug gage (G-6) Template gage (G-7) Functional relationship gage (G-8)

FIGURE 22.7 Process plan Tool and Operation Sheets.

22.7 COMMENTS

Metalworked parts, like machined parts, can usually be processed several different ways. The final process selection will boil down to the most economical processing of the part for the company planning to produce it. However, in metalworking, the part design dictates the process required to produce it more so than in machining.

In machining, a slot may be produced by one of many different processes—torch cut, sawed, shaped, broached, milled—the choice depending at least partially on specified tolerances and finishes. In metalworking, a hole is usually pierced, a bent part is bent, and a drawn part must be drawn.

Financial analysis of all elements of feasible processes will reveal the difference in costs of one process over another—everything else being equal—but everything else is usually not equal: The final choice will probably be dictated by some overriding factor such as availability of equipment, labor costs, required quantities, and so on. The final selection must be made by knowledgeable engineers guided by experienced supervisors and approved by management.

Metalworking process planning of regular geometrical shapes is more or less straight forward, as illustrated by the work contained in this chapter. Designs with more complex geometrical shapes require engineers possessing more actual experience in processing. Metalworking process planning of nonuniform, nonregular geometrical shapes such as automotive doors, hoods, and body panels requires yet more expertise and is beyond the scope of this textbook. Indeed, many independent specialty die engineering shops exist whose function is to supply major manufacturing companies with their services in this area. A die engineer interested in pursuing a career in this area would do well to apprentice or become employed in such a shop at an entry level.

22.8 REVIEW QUESTIONS AND PROBLEMS

1. Why is Manufacturing Engineering involved in assuring a metalworking part's design feasibility for manufacturing?
2. What areas of a part's design must be studied for feasibility of manufacture?
3. What must be considered when developing a tentative process plan?
4. How can a final process plan be checked and verified?
5. What is the function of operation sheets?
6. Assume the proposal to eliminate the notches is not acceptable and the part must be produced as originally designed.
   a. What effect would this have on the processing as developed for this example?
   b. Develop a tentative process sequence for the original configuration.
7. Assume the yearly production requirements for the part are too low to justify a progressive die. Develop a tentative process sequence for the part that is shown on the progressive die, but with individual operations run on simple, individual dies.
8. Assume that 10 pieces of the part as originally designed are required as soon as possible with no time available to produce dies of any kind. Develop a tentative process to produce 10 parts by other means.

CHAPTER TWENTY-THREE

DIE LUBRICANTS

CHAPTER OVERVIEW

- Introduction
- Basic Factors Affecting Selection of Die Lubricants
- Forms of Lubricants
- Components and Types of Die Lubricants
    - Die Lubricant Components and Treatments
    - Extreme-Pressure Lubricants
    - Pigmented Die Lubricants
    - Nonpigment Lubricants
    - Soap-Type Lubricants
- Die Lubricants for Pressworking Various Metals
    - Steel
    - Stainless Steel
    - Aluminum
    - Brass
    - Magnesium
- Lubricants for Cold Extrusion
- Die Lubricant Applications Summary
- Review Questions and Problems

23.1 INTRODUCTION

Lubricants are indispensable in pressworking operations, where pressures of up to several hundred thousand pounds per square inch develop in a sliding condition of one metal contacting another—the die and the workpiece.

A die lubricant is a substance that is applied to a workpiece to reduce the friction between the die and the workpiece. It is used in many forms and is applied by several different methods.

Manufacturing engineers must select and specify die lubricants that will reduce metal-to-metal friction, which:

- Makes an operation possible; that is, prevents tears, galling and/or welding to the die.
- Reduces die wear.
- Lowers tonnage requirements of the press.

Die lubricants function in two different ways:

1. As a tenacious lubricant that resists removal from a surface and that has properties capable of reducing the coefficiency of friction between two metals. This is known as a **polar lubricant.**

2. As a lubricant that reacts chemically between two rubbing surfaces to prevent welding and has properties capable of lowering the coefficiency of friction; this is known as an **extreme pressure lubricant.** Typical chemically active ingredients are chlorine, phosphorus, and sulfur.

The coefficient of friction between any two specified metals is, of course, the point from which die lubricants must reduce friction to a workable level.

Figure 23.1 illustrates the more common combinations of steel (die) and the workpiece metals. No definite or exact amount of reduction can be predicted, but most applications are perhaps in the 20% to 30% range of the original amount of friction.

Some metalworking companies custom mix and compound their own lubricants to formulas developed for specific applications. Other companies utilize commercial premixed products they have found, by testing, are suitable for their applications. The use of premixed commercial products does away with a messy mixing operation but may incur a cost penalty. The use of precoated metal in sheet and strip form has gained wide use. It provides a superior, more uniform coating.

Many lubricants are applied by dipping, others by spraying, oil-coated rollers, electroplating, and so on. Parts must be washed after metalworking to remove die lubricant. This is especially critical ahead of painting operations. Different lubricants present different levels of difficulty in removal. Graphite is difficult to remove, and all traces must be removed before annealing as well as painting operations.

Data for the use and application of die lubricants has been grouped in this chapter in the following order:

First. Factors effecting lubricant selection as related to the metal and as related to lubricant forms are shown in Sections 23.2 and 23.3.

Second. Components of lubricants are identified for general information and especially for engineers involved in the mixing of lubricants for in-house use. These are shown in Section 23.4.

Third. The severity of pressworking operations is defined and die lubricant applications are shown for various metals and severity of operations in Section 23.5.

Fourth. Die lubricant recommendations for steel and aluminum cold extrusions are given in Section 23.6.

Fifth. A summary sheet of die lubricant recommendations for various metals and for different severity of operations is given in Section 23.7.

FIGURE 23.1 Coefficient of friction for various metal combinations.

Metal Combination	Coefficient of Friction with No Lubrication
1. Steel–aluminum	.8
2. Steel–bronze	.4
3. Steel–copper	.36
4. Steel–lead	.35
5. Steel–steel	.7

Recommendations. Sources of valuable help are the die lubricant manufacturers and suppliers. They are specialists in the field and often have developed—or will develop—the exact product required.

23.2 BASIC FACTORS AFFECTING SELECTION OF DIE LUBRICANTS

The selection of a die lubricant must include considerations of the metal's physical specifications as well as the general characteristics of the lubricant to be used.

Factors to Consider Relating to *Metal Specifications*

1. Yield strength
2. Coefficient of friction
3. Work-hardening rate
4. Chemical reaction with lubricant
5. Surface oxides

Factors to Consider Relating to *Selected Lubricant*

1. Purchase prepared commercial product or custom mix on site
2. Form of lubricant
3. Application and removal methods
4. Wetting properties
5. Odor and toxicity
6. Corrosion or staining possibilities
7. Difficulty of removal
8. Cost

23.3 FORMS OF LUBRICANTS

Die lubricants are used in several different forms or modes. Some have infinite variations of mix due to the differing requirements of any given application. The more common forms of lubricant are

1. Mineral oils
2. Synthetic oils
3. Soap and water
4. Soap dispersions, fatty acids and fatty oils
5. Low-melting-temperature solid lubricants
6. High-melting-temperature solid lubricants
7. Oxide lubricants
8. Metalic lubricants
9. Phosphate coatings
10. Resins including Teflon

23.4 COMPONENTS AND TYPES OF DIE LUBRICANTS

23.4.1 Die Lubricant Components and Treatments

1. Mineral oils
   - Nondetergent motor oils
   - Paraffin
   - Kerosene
2. Fats, fatty oils, and waxes
   - Tallow
   - Lard
   - Fish oils
   - Cottonseed
   - Castor
   - Waxes
3. Fatty acids
   - Stearic
   - Oleic
4. Chlorinated oils
   - Fatty oils
   - Paraffin
   - Waxes

   Chemically treated with chlorine
5. Sulfurized oils
   - Mineral oils
   - Fatty oils

   Chemically treated with sulfur at elevated temperatures
6. Sulfonated oils
   - Caster oils
   - Cottonseed oil

   Chemically treated with sulfuric acid and neutralized with an alkali
7. Soluble oils
   - Mineral oils — Chemically treated with an additive making oils emulsifiable in water
8. Soaps
   - Fatty acids — Compounded with a base of oleates or stearates
9. Metallic soaps
   - Fatty acids — Compounded with a base such as aluminum, lead or zinc stearates (insoluble) or potassium and sodium (soluble)
10. Pigments
    - China clay — Additives of finely ground particles
    - Graphite
    - Lithopone
    - Resins
    - Talc
    - White lead

23.4.2 Extreme-Pressure Lubricants

- Sulfurized fatty oils and paraffin waxes, concentrated
- Sulfurized fatty oils and paraffin waxes, mixed with mineral oil
- Chlorinated fatty oils and paraffin waxes, concentrated
- Chlorinated fatty oils and paraffin waxes, mixed with mineral oil

23.4.3 Pigmented Die Lubricants

1. Emulsion compounds
   - Fatty oils
   - Pigments
   - Emulsifier

   } Used concentrated or mixed with water and/or mineral oils
2. Oil compounds
   - Fatty oils
   - Pigments
   - Sulfonated oil

   } Used concentrated or mixed with water and/or mineral oils
3. Precoated
   - Selected lubricants applied and dried on metal surface before metalworking operations

23.4.4 Nonpigment Lubricants

1. Emulsion compounds
   - Fatty oils
   - Emulsifier
   - Water
2. Fats, fatty acids, fatty oils
   - Mineral oils
   - Emulsifier
   - Water
3. Mineral oils and greases
4. Soluble oils
   - Water

23.4.5 Soap-Type Lubricants

1. Dry powders
   - Sodium
   - Phosphate
   - Graphite
2. Precoat or dried film
   - Soluble soaps
   - Soluble fillers
   - Waxes
   - Wetting agents
   - Metallic soaps

   Metal dipped or sprayed, then dried previous to metalworking

23.5 DIE LUBRICANTS FOR PRESSWORKING VARIOUS METALS

One basis of die lubricant selection is the severity of the operation on specific metals.

- A mild operation is defined as a shallow draw on low-carbon steel.
- A medium operation is defined as a deep draw on low-carbon steel.
- A severe operation is a cartridge-case-type draw.

Recommendations in this section are grouped for metals in terms of the levels of severity of operation.

23.5.1 Steel

1. Mild operations
   - Mineral oil
   - Soap solutions
   - Fat, fatty oil, and mineral oil emulsions
   - Fat, fatty oil, and mineral oil in soap-base emulsions
   - Lard oil
2. Medium operations
   - Fat or oil in soap-base emulsions with pigments such as lithium compounds
   - Fat or oil in soap-base emulsions containing sulfurized oils
   - Fat or oil in soap-base emulsions containing pigments and sulfurized oils
   - Phosphate coating plus emulsion lubricants in soap solution
3. Severe operations
   - Precoat of soap or wax film, phosphate, or other metallic coatings
   - Precoat as above but also including pigments and sulfurized oils
   - Emulsions or lubricants containing sulfur and sulfides
   - Oil-based sulfurized blends containing pigments

23.5.2 Stainless Steel

1. Mild operations
   - Corn oil or castor oil
   - Castor oil and emulsified soap
   - Waxed or oiled sheets of paper
2. Medium operations
   - Powdered graphite coating
   - Journal bearing lubricant emulsion
   - Wax precoat
3. Severe operations
   - Lithium compounds and boiled linseed oil
   - White lead and boiled linseed oil

23.5.3 Aluminum

1. Mild operations
   - Mineral oil
   - Fatty oil blends in mineral oil

2. Medium operations
   - Tallow and paraffin
   - Sulfurized fatty oil blends
3. Severe operations
   - Soap film
   - Wax film
   - Mineral oil or fatty oil blends plus pigments
   - Fat emulsions in soap water plus pigments

23.5.4 Brass

1. Mild operations
   - Soap solution
   - Fat or oil emulsions and emulsifier
   - Lard oil blends in mineral oil
2. Medium operations
   - Soap solution, melted tallow, and stearic acid
   - Fatty oil emulsions with soap emulsions
   - Lard oil blends
3. Severe operations
   - Soap solution, tallow, and stearic acid
   - Lard oil blends
   - Dried soap films

23.5.5 Magnesium

1. Mild, 300°F or less
   1. Neutral soap solution
   2. Beeswax
   3. Tallow and colloidal graphite
2. Medium, 300°F to 500°F
   1. Soap solutions
   2. Tallow and colloidal graphite
3. Severe, 500°F to 750°F
   1. Colloidal graphite in solvent sprayed or rolled on metal
   2. Colloidal molybdenum disulfide in solvent sprayed or rolled on metal

23.6 LUBRICANTS FOR COLD EXTRUSION

Steel is lubricated in two steps prior to the cold extrusion operation.

1. The workpiece or slug is phosphate coated.
2. A soap-type emulsion is applied to the phosphate coating.

Step 1 provides a uniform base coating of phosphate by spraying or dipping followed by drying. This coating acts as a base to absorb and hold the subsequent lubricant emulsion in a uniform film. The phosphate coating also acts as a lubricant for any areas of thin or noncoverage of emulsion, preventing metal-to-metal contact.

Step 2 provides a fatty acid soap emulsion, which is applied at approximately 150°F and allowed to dry. Aluminum workpieces or slugs to be cold extruded are lubricated with one or more (a combination) of

- Tallow
- Waxes
- Soaps
- Fatty acids
- Lanolin

Copper and brass workpieces or slugs to be cold extruded are lubricated with

- Tallow
- Waxes
- Beeswax

Lubricant is applied to workpieces or slugs with the use of tumbling barrels in some instances and by dipping with subsequent drying in other instances, depending on the consistency of the lubricant.

23.7 DIE LUBRICANT APPLICATIONS SUMMARY

A summary of recommended die lubricants for pressworking various metals is shown in Fig. 23.2.

Die Lubricant	Metal				
	Carbon Steels	Alloy Steels	Stainless Steels	Aluminum	Copper Brass Bronze
Soluble oil	A				A
Soap solution				A	A B
Soap–fat compound (non-pigmented)	A B	A			A B C
Soap–fat compound (pigmented)	A C	B C	A		C
Dry-film soap	C	C			
Oil-soluble types:					
Straight mineral oil	A			A	A
Straight fatty oil				B C	C
Straight chlorinated oil		B	B C		
Straight sulfurized oil (non corrosive)					C
Mineral oils					
Oiliness blend	A B			A B	A B
Sulfurized oil blend (non corrosive)					A B
Sulfurized oil blend (corrosive)	A B	A			
Chlorinated oil blend	A B C	A C	A B C	C	C

A Flat Stamping
B Shallow Draw
C Deep Draw

FIGURE 23.2 Recommended lubricants for metalworking operations of various metals.

23.8 REVIEW QUESTIONS AND PROBLEMS

1. What are the functions of a die lubricant?
2. How do die lubricants accomplish their functions?
3. Name some of the ways die lubricants are applied.
4. What factors must be considered, *other than those relating to the die lubricant,* when selecting a lubricant?
5. What factors *relating to the lubricant itself* must be considered when selecting a lubricant?
6. Name some of the forms in which die lubricants are used.
7. Select two die lubricants that do not require mixing of components to be used for a mild metalworking operation of a steel part.
8. Select a die lubricant to be used for a severe pressworking operation of an aluminum part where wetting cannot be tolerated.
9. Summarize the lubricating process of steel workpieces that are to be extruded.
10. Define a die lubricant.

CHAPTER TWENTY-FOUR

PRESSES, EQUIPMENT, AND SPECIAL MACHINES

CHAPTER OVERVIEW

- Introduction
- Presses
  - OBI Presses
  - Gap Presses
  - Horning and Adjustable-Bed Presses
  - Power Press Brake
  - Straight-Side Press
  - Dieing Machine
  - Knuckle Press
  - Hydraulic Press
  - Double- and Triple-Action Presses
- Equipment
  - Die Cushions
  - Material Handling—Coil and Strip Stock
  - Material Handling—Sheet Stock and Blanks
  - Safety Requirements
- Special Machines and Machine Concepts
  - Turret Presses
  - Laser Beam Machining
  - Water-Jet Cutting Machines
  - Automation
  - Plasma Arc Machining
- Review Questions and Problems

24.1 INTRODUCTION

Manufacturing Engineering must select presses from the many available types and sizes to suit parts designed for production by the stamping and metalworking process.

This chapter describes presses, their types, and the types of parts processed on each; also, equipment is described that is required for loading parts, feeding stock, transferring, unloading, cutting scrap, and operator safety.

Presses are devices that provide motion and power to tooling for shaping, shearing, and forming metal. A press may be described as a machine designed with a stationary table and having a slide, or ram, positioned at right angles to the table or bed.

An upper die section attached to the ram mates with a lower die section mounted on the press bed. Activating the press causes the ram to move toward the bed—the working stroke—and back to its original position to complete the cycle.

One planned operation is performed on a metal workpiece or section of strip stock during each complete cycle of the press, after which the workpiece is removed or ejected and a new blank is loaded in the die—or the strip stock is advanced.

Major components of a typical press are illustrated in Figure 24.1 and enumerated in this section:

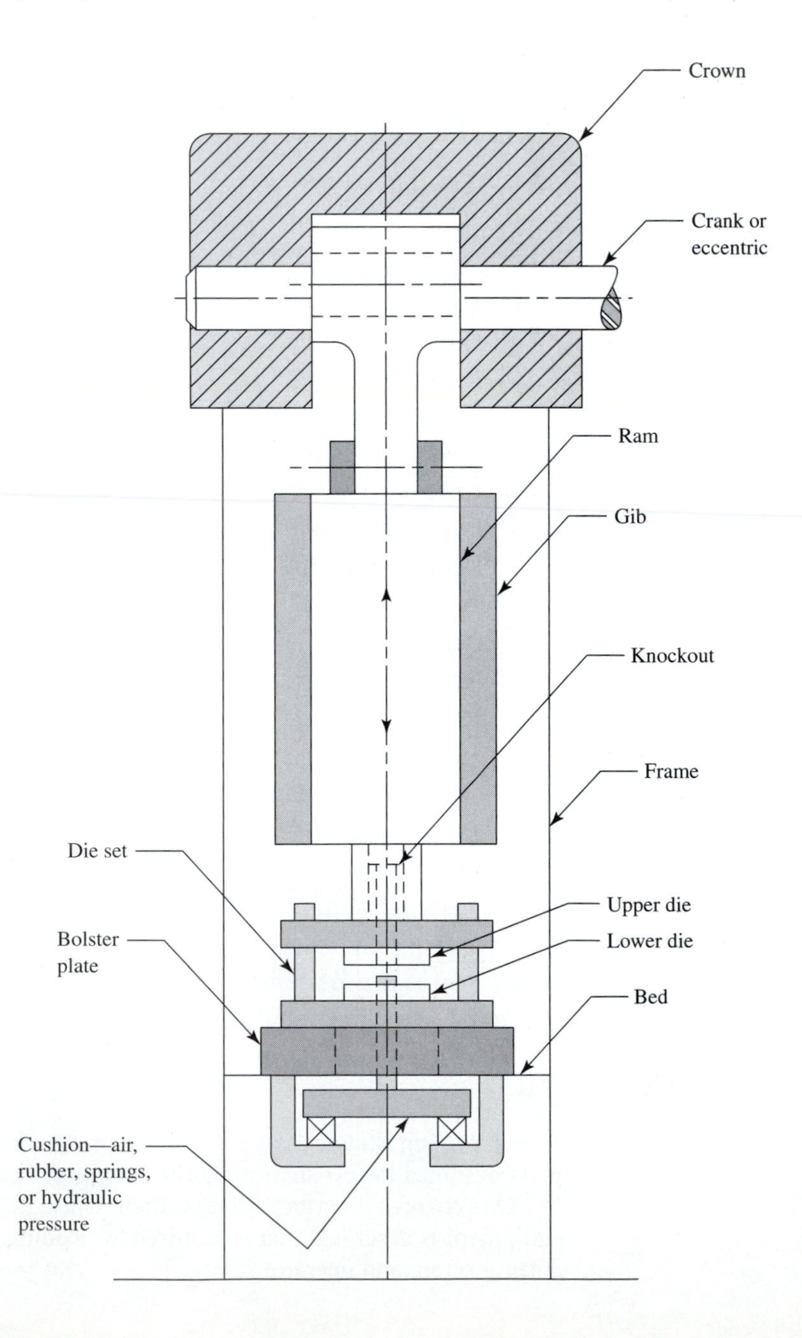

FIGURE 24.1 Press component arrangement.

1. **Press Frame**—the main structure of a press. Frames may be castings or weldments.
2. **Press Bed**—the base of the frame, usually open in the center to permit parts or slug ejection, or to enable the mounting of cushion devices below. Tooling is sometimes mounted directly on the frame.
3. **Bolster Plate**—a flat steel plate bolted to the top face of the bed upon which tooling and accessories are mounted.
4. **Ram**—the movable member of a press to which tooling is attached and whose movement produces the work or operation.
5. **Gibs**—retain position and guide the ram throughout its stroke.
6. **Knockout**—a mechanism that operates on a press up-stroke to eject a workpiece from a die—not used on all dies.
7. **Cushion**—an accessory located under the press bed or bolster plate to provide an upward force and motion. It is powered variously with air, rubber under compression, or hydraulic pressures—not used on all dies.

24.2 PRESSES

24.2.1 OBI Presses

The Open-Backed Inclinable (OBI) press shown in Figure 24.2 is a C-frame press, so called because its frame is designed with a C configuration. The frame is a casting or a weldment, pivot mounted on a base to permit inclining the press from the vertical.

FIGURE 24.2 OBI press. (Courtesy The Minster Machine Co.)

Advantages of the OBI press are that it may be operated vertically or run at an inclined angle to aid ejection of parts by gravity through the open back. Also, the C-frame makes the die accessible from three sides for purposes of loading and unloading workpieces, blanks or strip stock. The open back also permits strip stock to be fed through the die from front to back.

General design parameters of a C-frame press include superior accessibility to dies, low to moderate tonnages, relatively short strokes, and low cost or purchasing prices. Capacities range up to 200 tons.

C-frame presses are used for shearing, bending, forming, and drawing within their stroke limitations. Very high rates of production are attainable for small parts run on progressive dies.

24.2.2 Gap Presses

The press illustrated in Figure 24.3, known as a gap press, is a C-frame design but is not inclinable. Gap presses are built in larger sizes and tonnages than OBIs, are used for similar work, but can produce larger, heavier parts on larger dies. Capacities range up to 300 tons.

24.2.3 Horning and Adjustable-Bed Presses

Figure 24.4 illustrates a horning and adjustable-bed press. This press is a fixed-base C-frame design with special provisions or features:

1. **Horn provision**—a bore in the frame is provided to accommodate a post, or horn, to be used with tubular parts and parts that require internal support during a press working operation.

FIGURE 24.3 Gap press. (Courtesy The Minster Machine Co.)

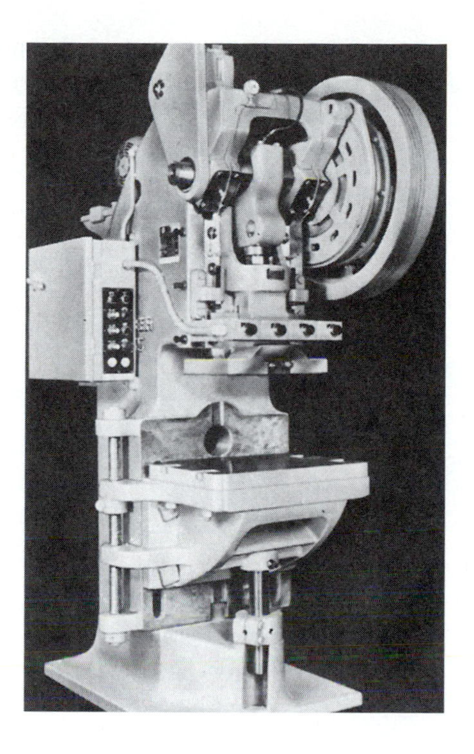

FIGURE 24.4 Horning and adjustable-bed press. (Courtesy The Minster Machine Co.)

2. **Adjustable bed height**—a bed that can be raised or lowered to suit different die heights is provided. The press is built in a horn-only design, with adjustable bed only, or with both features. This design of press is versatile, is suitable for prototype work, and is used for low-volume production work.

24.2.4 Power Press Brake

A power press brake, illustrated in Figure 24.5 is a C-frame design with a long, narrow bed and ram. It is used to bend, shear, and form long workpieces. It can also be used to produce several shorter workpieces simultaneously or progressively on a series of dies set up in-line on the press bed.

Press brakes are used for low-production and prototype part applications. Simple, low-cost dies can be used where long tool life is not a requirement. Labor costs, however, are proportionally higher because the workpieces are transferred manually from one die to the next. Provisions for the safety of the operator must be provided for all presses, with special attention to this one.

24.2.5 Straight-Side Press

A straight-side press is designed with columns or uprights across each end to form a boxlike structure. Windows, or openings, are provided in the ends to permit feeding stock to the die as well as for unloading, ejecting, and handling workpieces and scrap strips. Figure 24.6 illustrates a straight-side press.

Because the straight-side press is designed to withstand high pressures without deflection or loss of alignment, it is used for longer, heavier parts. Tonnages range up to 2000, and bed sizes of 250 in. of length are feasible.

FIGURE 24.5 Power press brake. (Courtesy The Minster Machine Co.)

FIGURE 24.6 Straight-side press. (Courtesy The Minster Machine Co.)

Reciprocating rates are lower than those of C-frame presses due to the inertia of the larger, heavier components. Strokes of 5 to 60 per minute are typical of the larger sizes, but rates of 300 strokes per minute are attainable on the smaller sizes.

Straight-side presses are used with single dies where the workpiece is loaded directly into the die either manually or automatically. They are also used with transfer dies, Figure 24.8b, progressive dies, Figure 24.7, and in transfer press lines, Figure 24.8a.

Transfer Die Use **Transfer dies** have a device shown in Figure 24.8b that feeds workpieces from a magazine or hopper into the first station and first operation. From there the workpiece is picked up, moved, and loaded into and out of each succeeding station until all operations are completed and the part is unloaded. The press is equipped with attachments that mechanically move transfer fingers of the loading device through the die.

Progressive Die Use A typical **progressive die** arrangement in a straight-side press is shown in Figure 24.7. Also, Figure 24.16 shows the application of material-handling equipment to a progressive die operation. In this illustration, a coil of stock on a reel ① is uncoiled and straightened in rollers ② then fed into a progressive die ③.

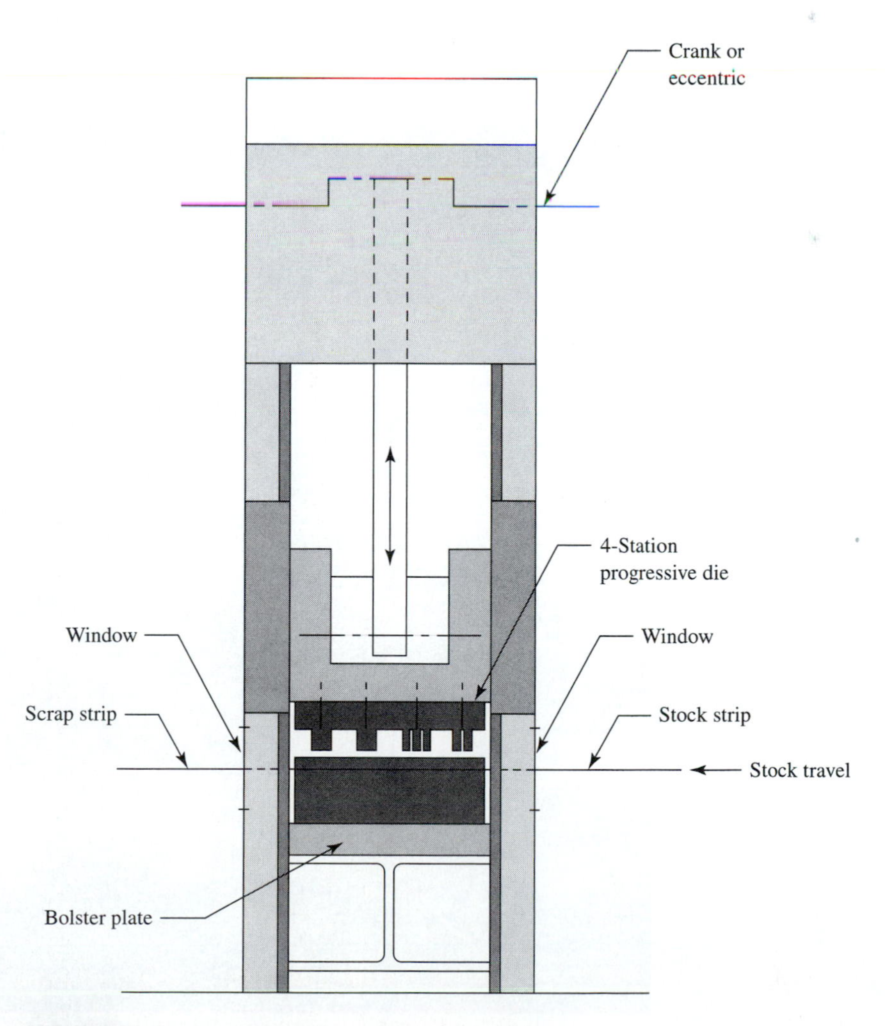

FIGURE 24.7 Progressive die use in straight-side press.

FIGURE 24.8 (a) Plan view, straight-side press transfer press line. (b) Transfer die on straight-side press. (Courtesy The Minster Machine Co.)

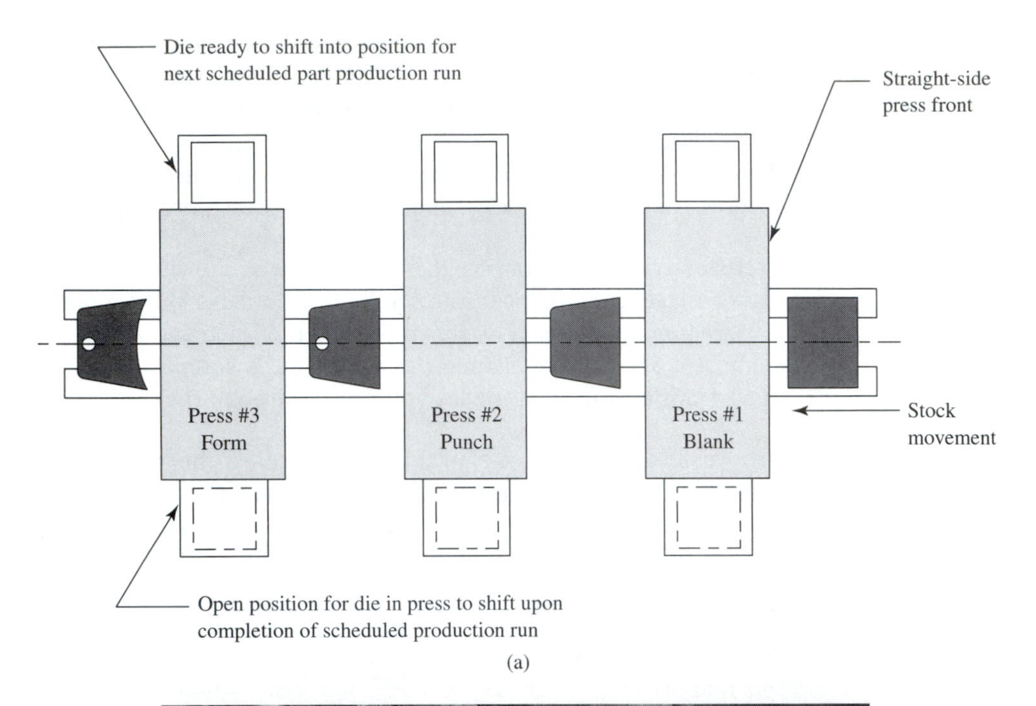

(a)

(b)

The strip advances through the die with the finished part ejected at ④ and the scrap strip emerging from the die at ⑤. The scrap strip is cut off into segments and disposed of at ⑥. A sensing arm ⑦ monitors and provides a loop of stock by activating the powered reel ① and control ② intermittently.

Transfer Press Line Use Straight-side presses are used for high production of large workpieces such as automobile body panels, doors, tops, and floorpans. The presses are arranged in tandem (in-line), with workpieces transferred out of the back of the first press into the front of the second press and so on until the finished part emerges from the final press.

Most press lines are completely automated. Many lines are provided with quick-change die stations where succeeding dies for the next run can be positioned. A corresponding vacant die station is also provided for the die that will be deactivated. Significant reductions of downtime due to die setting are effected with the quick-change technique. Figure 24.8a illustrates the quick-change transfer die line concept.

24.2.6 Dieing Machine

A **dieing machine**, illustrated in Figure 24.9 has the drive mechanism located under the bed. The upper die of this type of press is fastened to the ram, which is pulled, as opposed to the ram pushing in a conventional press. The lower die is mounted on the press bed.

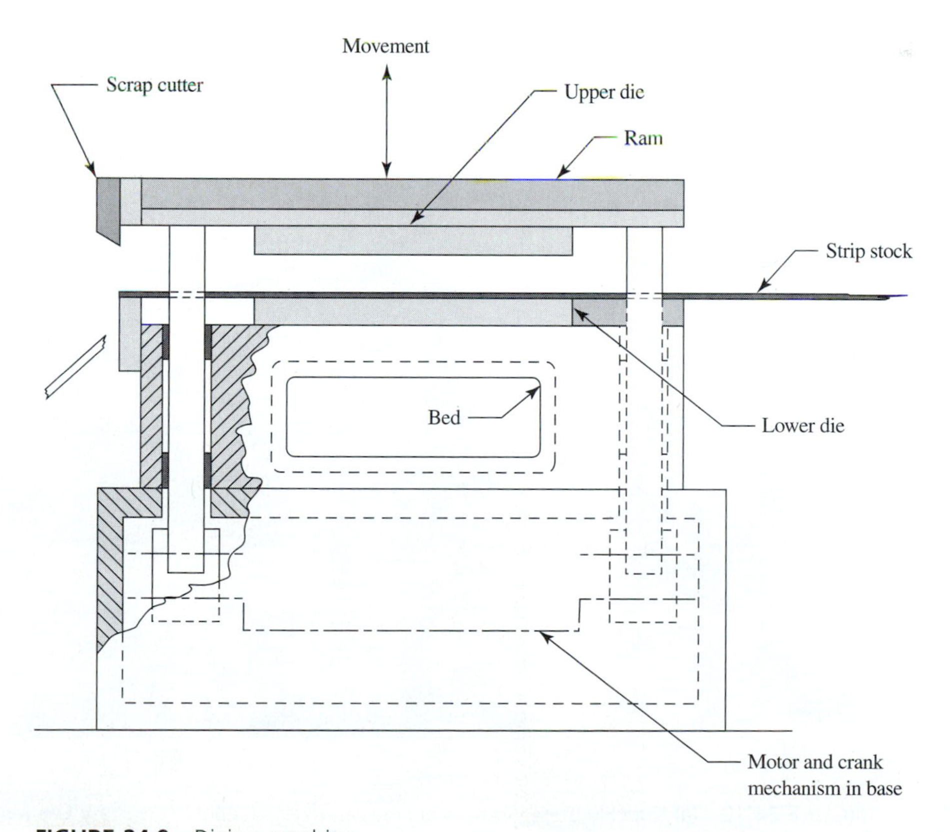

FIGURE 24.9 Dieing machine.

A high rate of strokes per minute can be realized due to the low weight, mass, or inertia of the ram. Some machines can run in excess of 600 strokes per minute with capacities ranging up to 400 tons.

These machines are ideal for the production of large quantities of small parts such as electric motor laminations, torque converter impeller blades, transformer core laminations, and so forth on progressive dies with mechanized coil stock feed.

Some dieing machines are installed in pits to bring the operating height of the bed down to an operator's level without the use of a platform. However, such arrangements create problems when the drive mechanism requires maintenance.

24.2.7 Knuckle Press

A knuckle press, sometimes known as a **knuckle joint press** or a **toggle press** is illustrated in Figure 24.10. Very high tonnages can be obtained with the use of a knuckle, as illustrated in Figure 24.11.

In action, a crankshaft alternately folds and realigns the links into a straight line through a connecting rod. As the links move into alignment in a straight line, the high leverage ratio makes very high tonnages of capacity attainable—some in excess of 1000 tons.

The press slide and ram are connected to the lower link of the lower link of the knuckle; the upper end of the link is secured to the press crown. The resulting movement of the slide is a short, increasingly powerful stroke. As it nears the bottom, the motion slows and dwells before the up-stroke occurs. These characteristics, coupled with small required bed areas (as compared to an equivalent hydraulic press), make it ideal for use with coining, embossing, flattening, forging, and certain extruding operations.

One limitation is the short stroke of up to only 8 in. available.

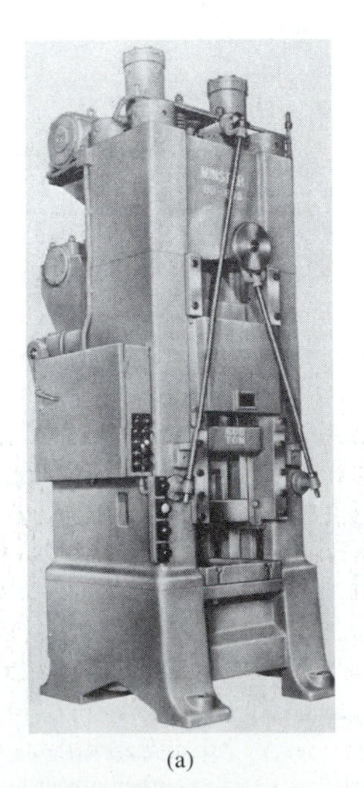

(a)

(b)

FIGURE 24.10 (a) Knuckle press. (b) Knuckle presses in production. (Courtesy The Minster Machine Co.)

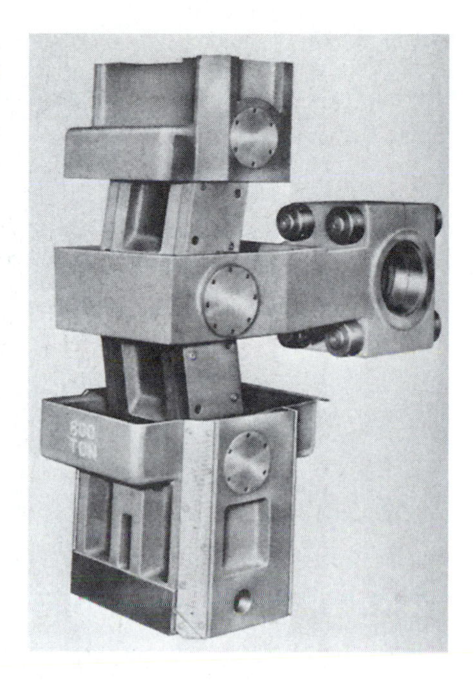

FIGURE 24.11 Detail of knuckle. (Courtesy The Minster Machine Co.)

24.2.8 Hydraulic Press

Hydraulic presses are used primarily for forming operations. Due to their slower motion, and high attainable pressures, they are ideal for workpieces requiring a squeezing action. Hydraulic presses are extremely flexible with adjustable working pressures, lengths of stroke, and ram speeds. These presses are made in all desired sizes and tonnages. A hydraulic press is shown in Figure 24.12.

24.2.9 Double- and Triple-Action Presses

Double-action and **triple-action presses** are used for deep drawing of large parts such as certain automobile and truck panels, tubs, and sinks.

A double-action press has two independent rams, or slides, moving within one another. In operation, the outer rams operate the ram holder while the inner ram operates the punch holder. During the operating cycle, the blank holder contacts the metal first and applies the necessary pressure for the punch holder to draw the part.

A triple-action press has the same inner and outer ram construction and action as a double-acting press, but also has a third ram in the press bed that moves up, enabling a reverse draw to be made during the press cycle. Triple-action presses are not common.

24.3 EQUIPMENT

Presses must be provided with certain accessories and auxiliary units—devices termed generally as "equipment," to function and operate at planned levels of production.

Production of parts on a press tooled with a simple manually loaded die requires little more additionally than containers for blanks and finished parts. Other presses are provided with

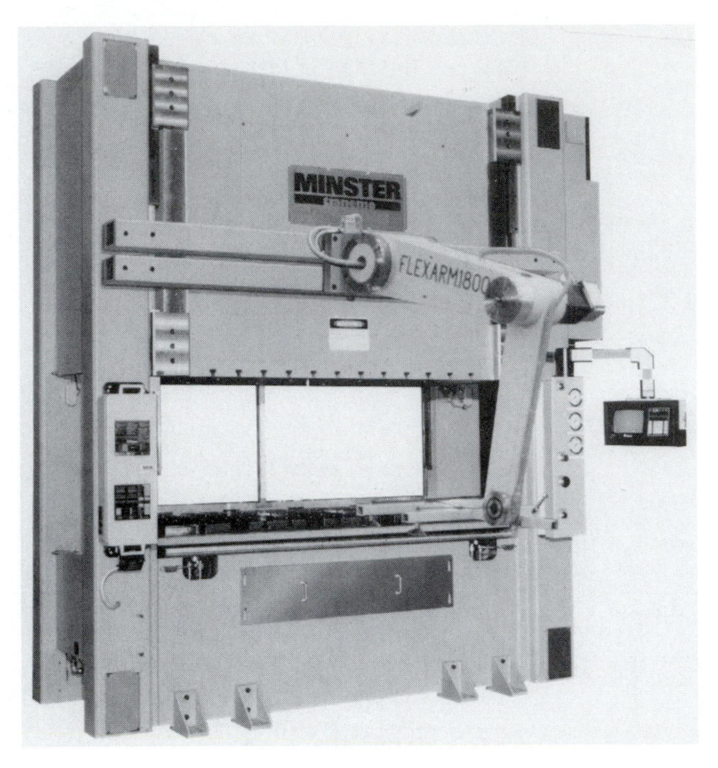

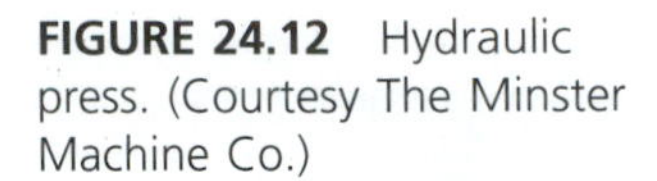
FIGURE 24.12 Hydraulic press. (Courtesy The Minster Machine Co.)

sophisticated material-handling systems to operate at high planned production rates. Some operations require that a press have a cushion for certain elements of die function. All presses must be safe to operate. Provision of safely designed tooling, guarding, controls, and processes is mandatory.

24.3.1 Die Cushions

A **die cushion** is an accessory for a press that provides the resistive force and motion required for some operations such as blank holding, drawing and redrawing, maintaining uniform pressure on a workpiece, and knocking out or stripping. Also called pads or jacks, die cushions are usually mounted in or under the press bed; they are also used in or on a slide.

Blanking Die Use Figure 24.13 illustrates the use of a die cushion with a **blanking die**—a fineblanking die operation in this example. The cushion provides resistance to the metal slug as it is forced into the blanking die, keeping it flat and generally minimizing distortion.

Forming Die Use The pressure pad in a **forming die**, or form die (see Figure 24.14) provides upward pressure to keep the bottom of the part flat as it is formed. A spring-operated cushion fixed to the bottom of the bolster supplies pressure to the pad.

Drawing Die Use The **drawing die** shown in Figure 24.15 requires a cushion to apply pressure to the stripper pad which acts as a blank holder during the draw operation, Figure 24.15(a) and

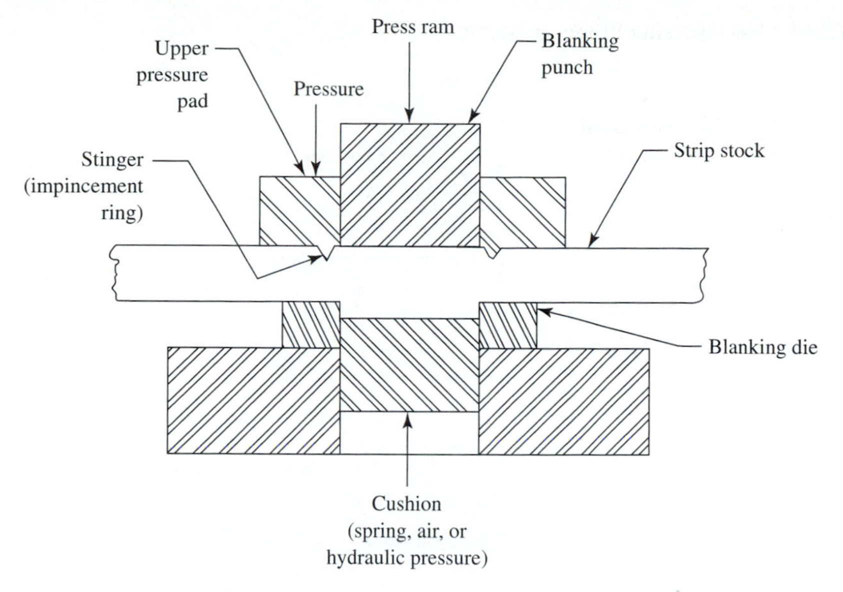

FIGURE 24.13 Fineblank die with cushion.

FIGURE 24.14 Form die with cushion.

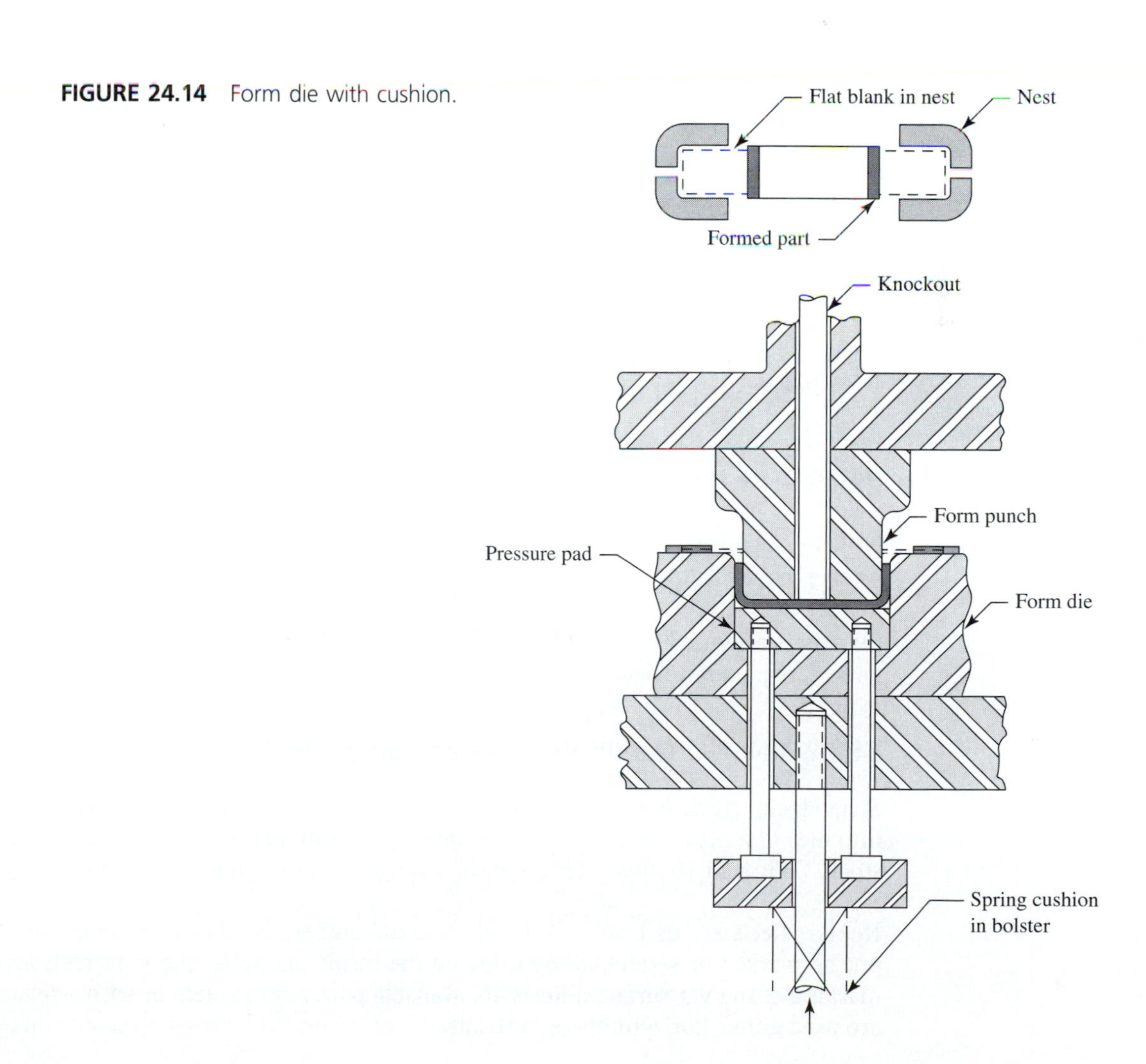

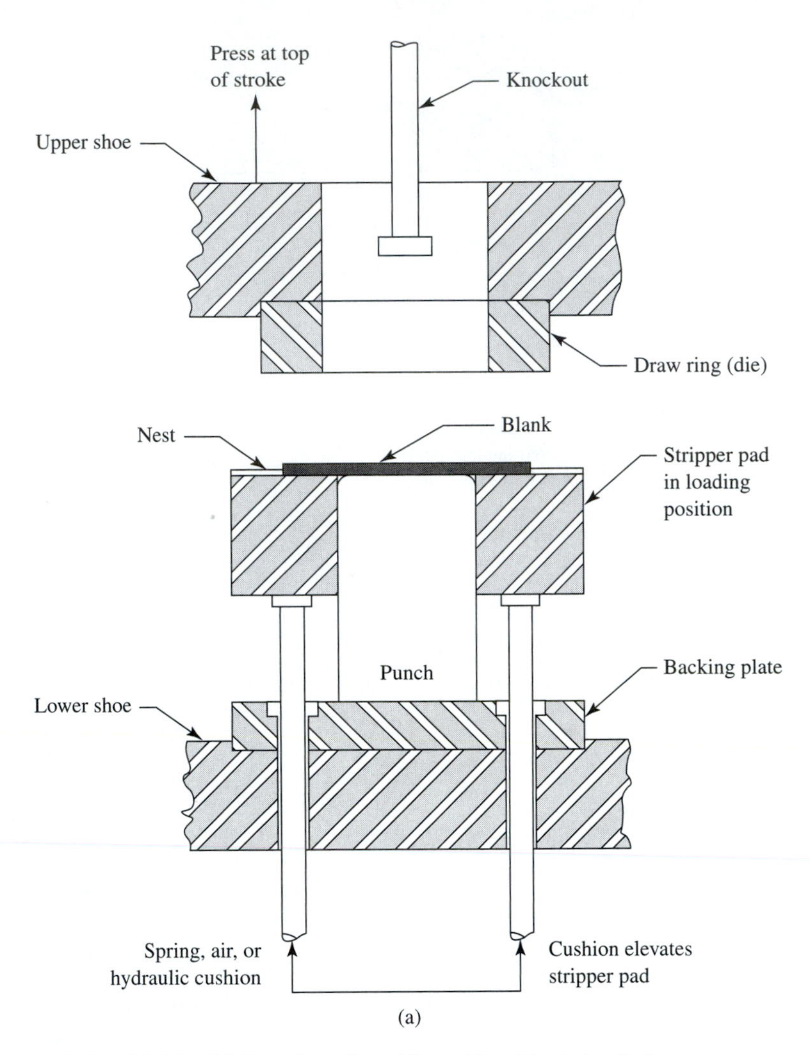

FIGURE 24.15 (a) Drawing die with cushion blank loaded.

24.3.1 Die Cushions (Continued)

(b), and is also used to strip the part from the punch, Figure 24.15(c). A spring, air, or hydraulic force is required as shown.

24.3.2 Material Handling—Coil and Strip Stock

Strip stock, which is received in coils, must be uncoiled and straightened just prior to feeding into dies of a press. Most facilities required to accomplish this are standard and available from stock. Figure 24.16 illustrates a typical arrangement of material-handling units.

Reels Reels are used to uncoil coils of metal and are used in particular where surfaces must not be marked or scratched. By gripping the inside diameter of a coil, reels avoid contact and marking of the flat surfaces. Reels are available powered to rotate or with a retarding brake, and are used either horizontally or vertically.

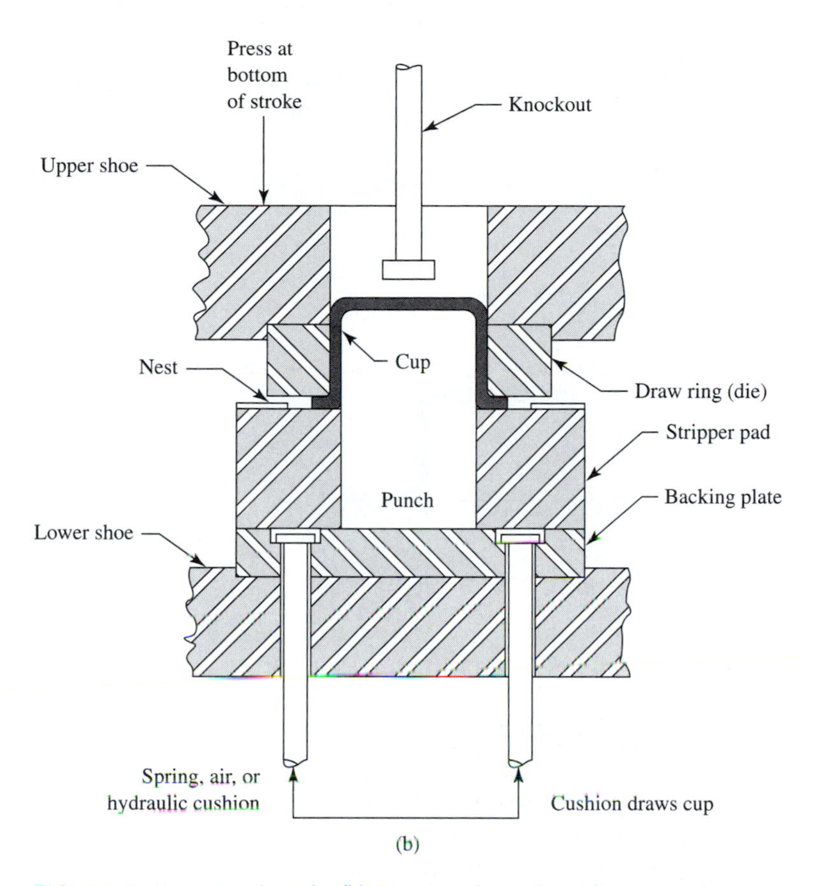

FIGURE 24.15 Continued (b) Drawing die with cushion cup drawn.

24.3.2 Material Handling—Coil and Strip Stock (Continued)

Care must be taken in selecting reels to ensure adequate capacity and ability to handle the largest sizes of coil weight and unwinding tendencies that may be encountered in the future use of any line. Conventional reels are illustrated in Figure 24.17.

Cradles Cradles are used to uncoil heavy coils of steel, in particular where slight marking or scratching of the surfaces can be tolerated. As shown in Figure 24.18, a coil of metal's outside diameter rests on rollers. The rollers are available powered or free.

Pallet Decoilers Pallet decoilers permit uncoiling and use of a wide range of metal widths, thicknesses, and forms, including wire, directly from the shipping pallets. Not only is there a time and labor saving, but this method of coil handling is safer, eliminating the need for workers to lift and load heavy coils onto a reel or into a cradle. Figure 24.19 illustrates a pallet decoiler in line with a powered stock straightener, press, and scrap cutter.

Stock Feeders The force required to uncoil strip stock and move it through straightening rolls and a die can be supplied by powered straightening rolls, powered reels, cradles and pallet decoilers, or "grip and shuttle" units, which are usually air operated. Also, some press lines use a ratchet-type feed powered by the motion of the press ram. Feed methods are illustrated in Figure 24.20.

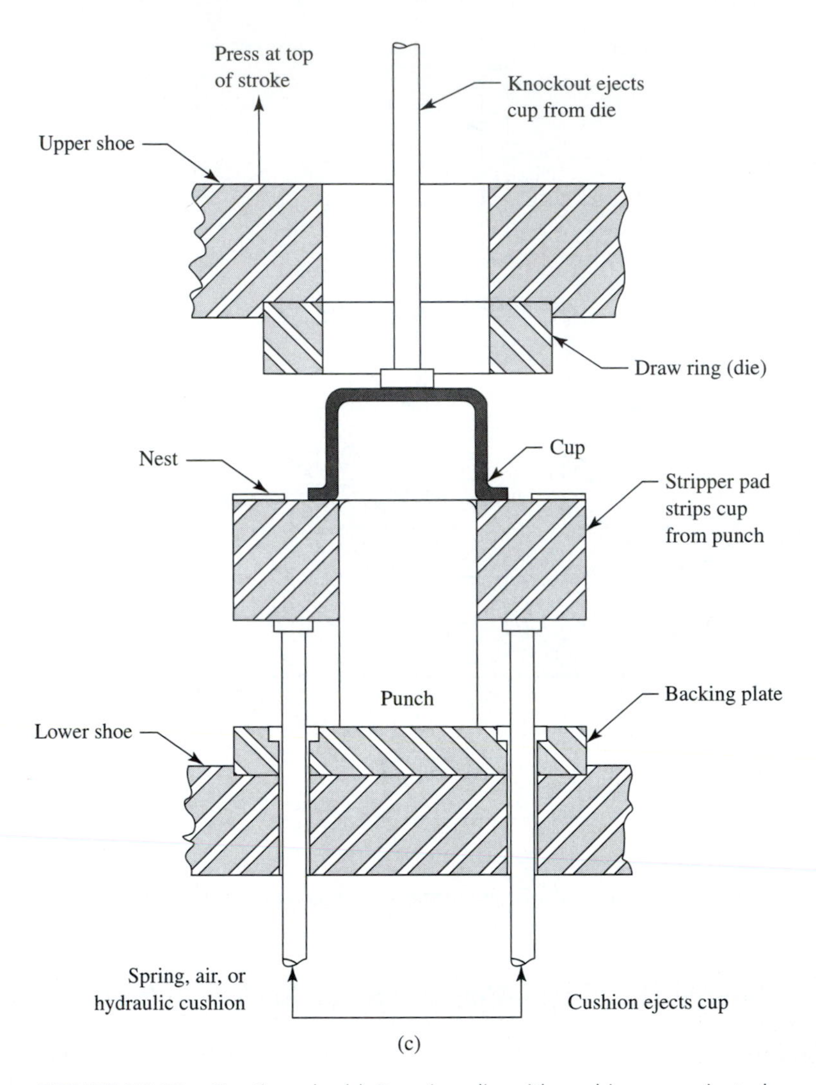

FIGURE 24.15 Continued (c) Drawing die with cushion cup ejected.

24.3.2 Material Handling—Coil and Strip Stock (Continued)

Controls Sensors, switches, and related circuitry are required to control the amount of metal uncoiled and fed into a die; metal is fed into a die at an interrupted rate to suit the press rate of strokes per minute.

In practice a loop of material is maintained to provide a "bank" of material, and some flexibility, from which metal can be drawn. The amount of metal in the loop is maintained by sensing the loop's height and activating powered rolls accordingly.

Controls are available in mechanically operated switch versions and also in sensing noncontact proximity switch types as shown in Figure 24.21 for loop control.

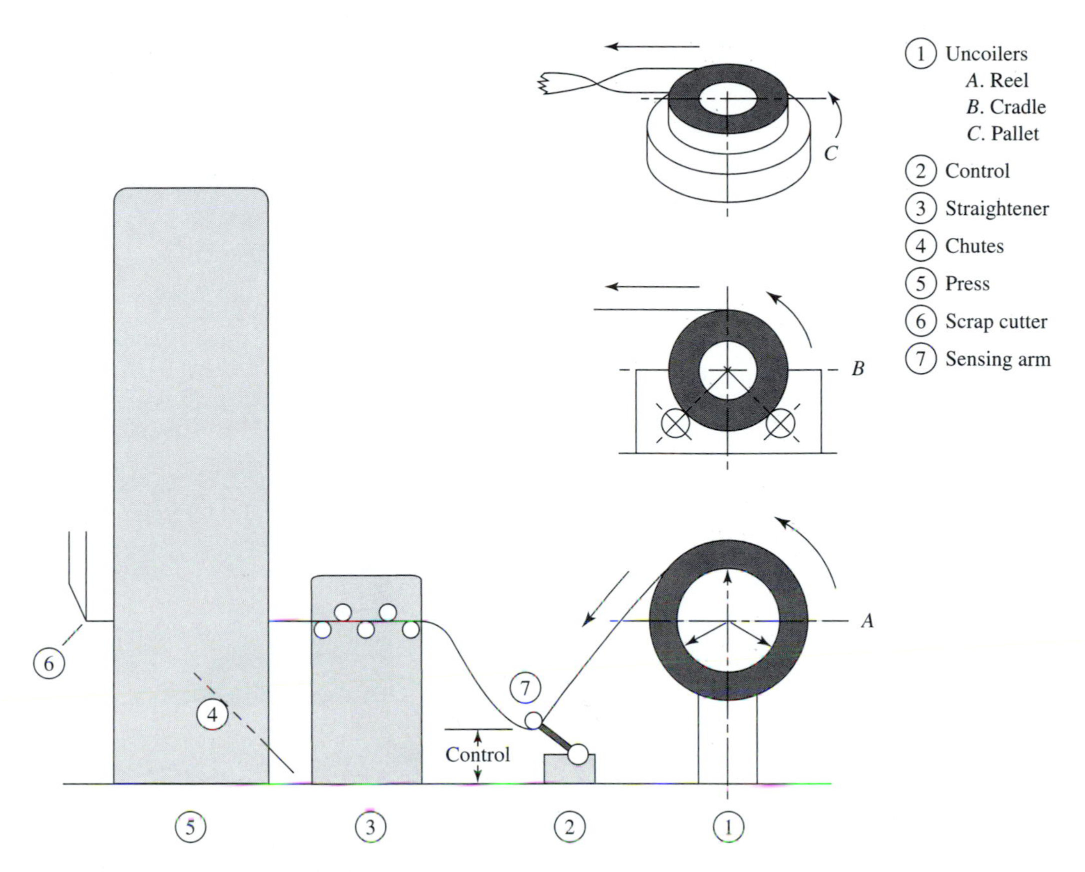

FIGURE 24.16 Coil stock material-handling equipment. As used with a progressive die in a straight-side press.

FIGURE 24.17 Stock reel. (Courtesy The Minster Machine Co.)

FIGURE 24.18 Coil car (cradle). (Courtesy The Minster Machine Co.)

FIGURE 24.19 Pallet decoilers. (Courtesy The Minster Machine Co.)

(a)

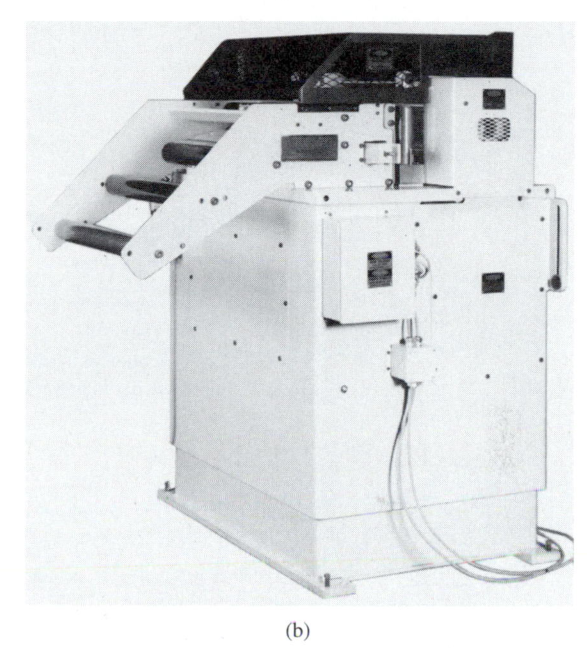

(b)

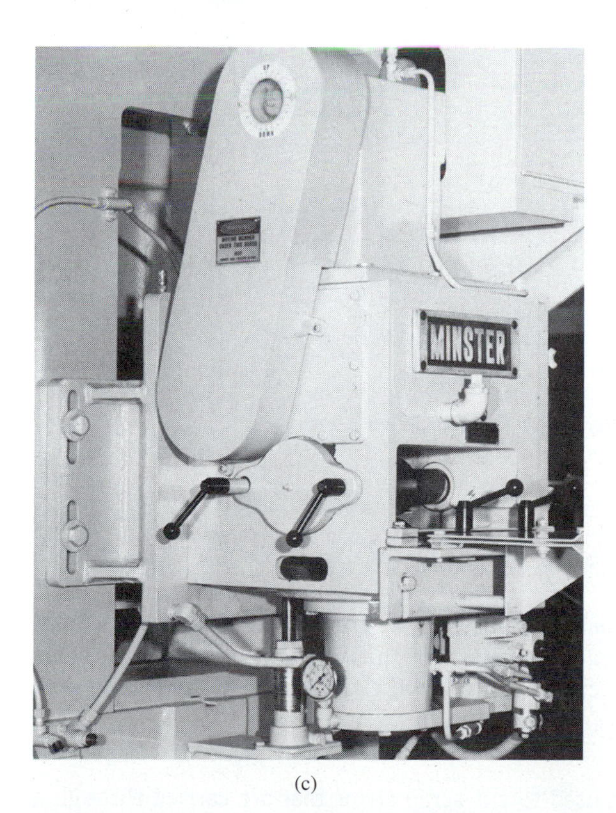

(c)

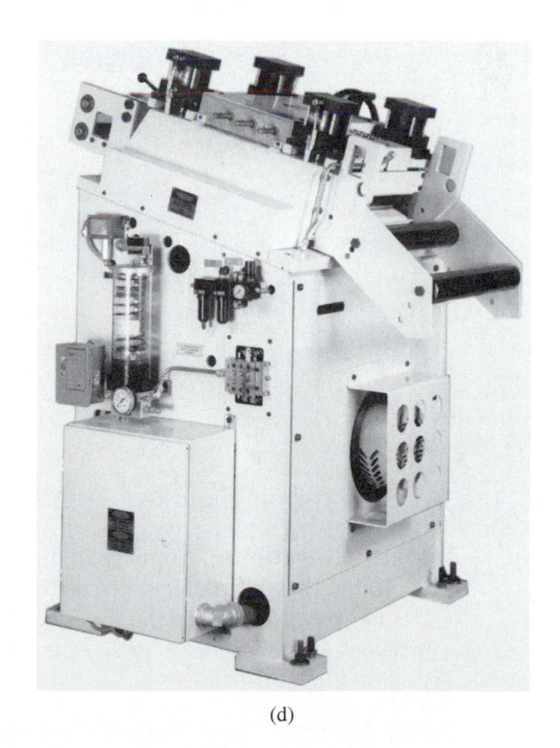

(d)

FIGURE 24.20 (a) Press control operated feed (rack and pinion). (b) Electric servo feed (power feed rolls). (c) Cam feed. (d) Stock straightener and feed rolls. (Courtesy The Minster Machine Co.)

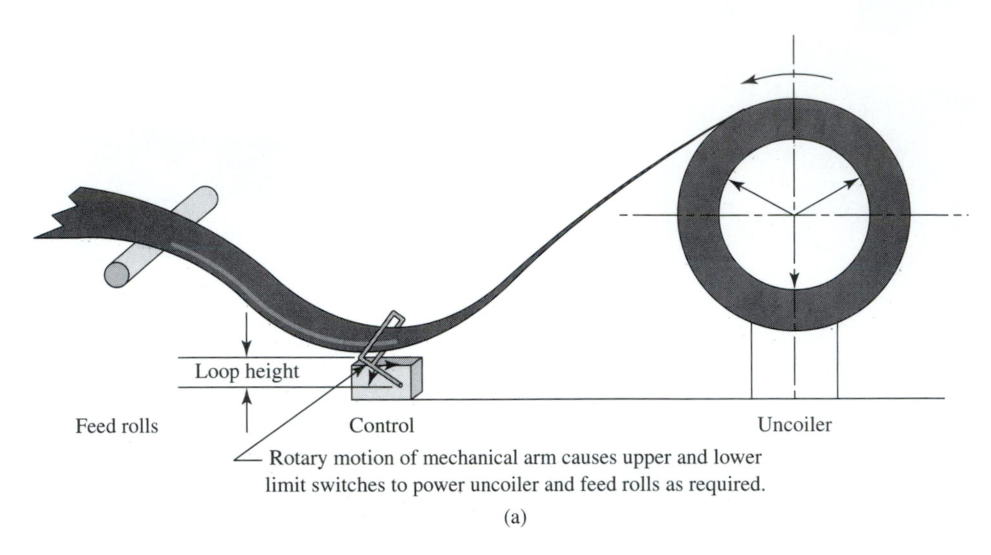

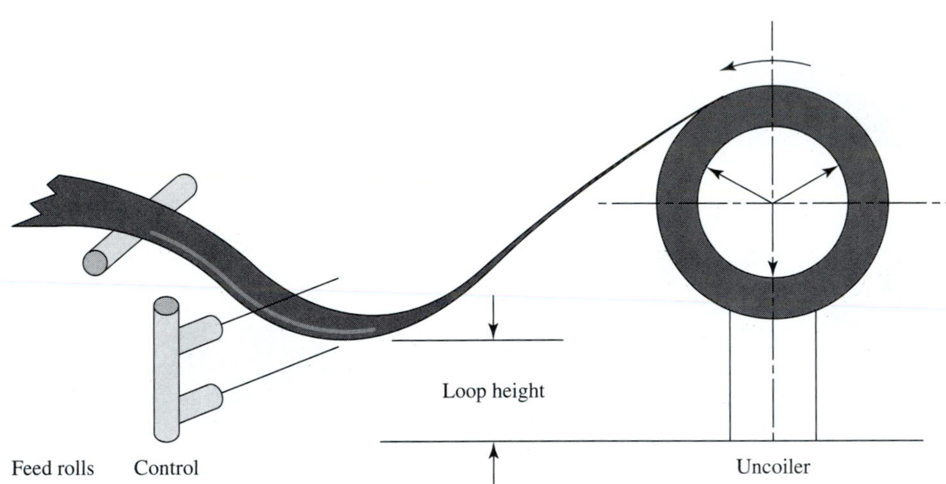

FIGURE 24.21 (a) Loop control—mechanical/electrical limit switches. (b) Loop control—antenna contact.

24.3.2 Material Handling—Coil and Strip Stock (Continued)

Die Considerations Some form of material handling must be provided for slugs punched out and ejected through a die as well as trimmings and material resulting from notching. Provisions must be made to handle the final workpiece; this can be in the form of containers, chutes to conveyors, or conveyors.

Scrap Cutters Scrap cutters are devices used to cut scrap strips that are carried through a progressive die into convenient, disposable lengths. Two types of scrap cutters, self-powered and powered by the press ram, are shown in Figure 24.22.

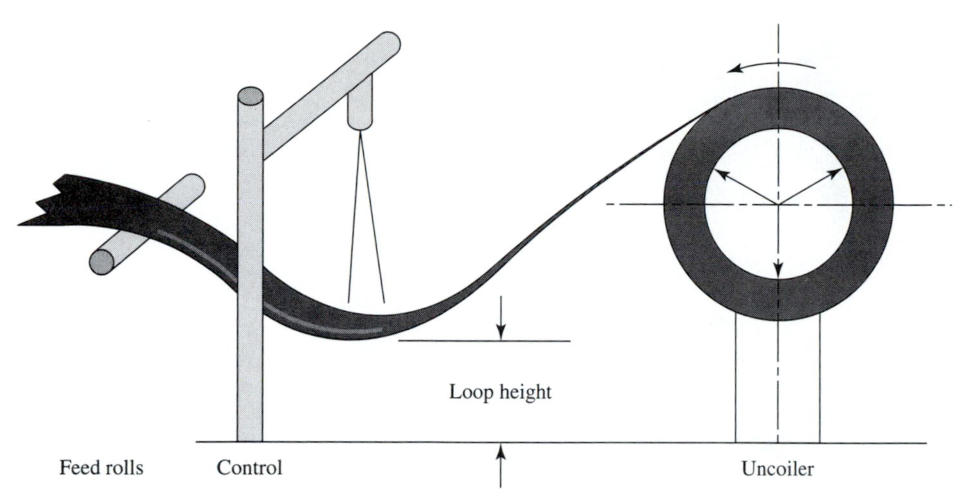

FIGURE 24.21 Continued (c) Loop control—sonic no contact.

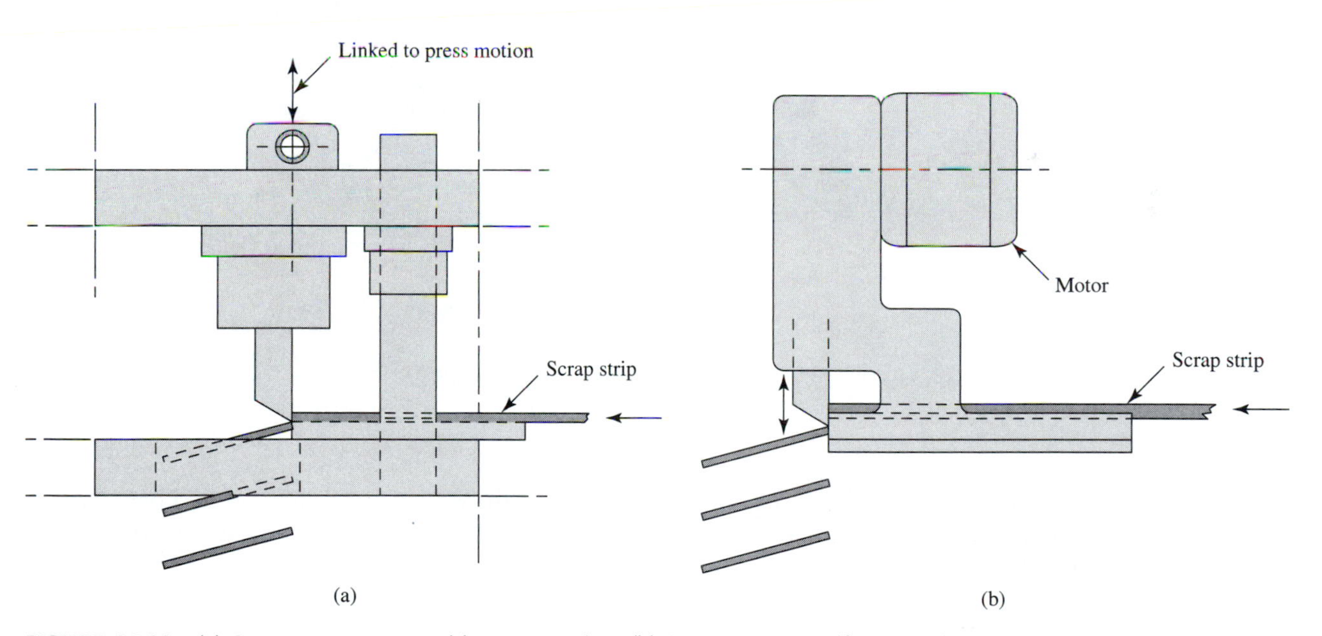

FIGURE 24.22 (a) Scrap cutter powered by press action. (b) Scrap cutter—self powered.

24.3.3 Material Handling—Sheet Stock and Blanks

Large stamped parts and workpieces processed from sheet stock and blanks require means of loading into and out of presses. Where more than one press is used in-line, a means of transfer between presses is also needed.

Loaders Loading a large blank into a press can be accomplished manually with the use of up to four operators, or with mechanical aids: slides, chutes, robots, and specially designed loading

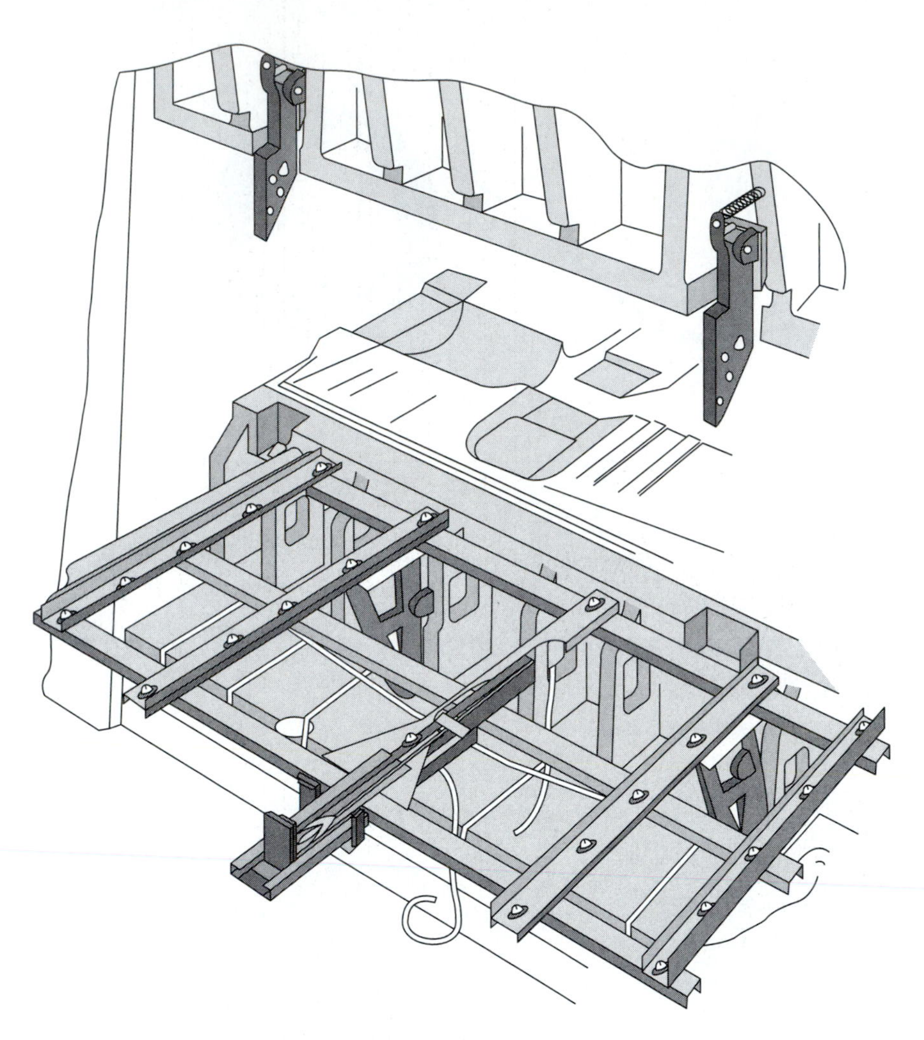

FIGURE 24.23 Blank loader. (Courtesy The Minster Machine Co.)

fixtures. Selection of the type of loader depends on the workpiece design, size and type of press and die, and required rate of production. One specially designed loading fixture is shown in Figure 24.23.

Unloaders The type of press loader selected usually governs the type of unloader used. The workpiece can be unloaded manually or with the help of a mechanical aid. Special unloaders are illustrated in Figure 24.24.

Transfer Devices Material-handling equipment must be provided to move workpieces from one press to a succeeding press in tandem or in-line layouts. Slides are provided and parts are moved manually in low-volume lines, but where production rates justify, efficiency and operator safety are both better served with a special powered mechanism. Figure 24.25 illustrates a specially designed powered transfer mechanism.

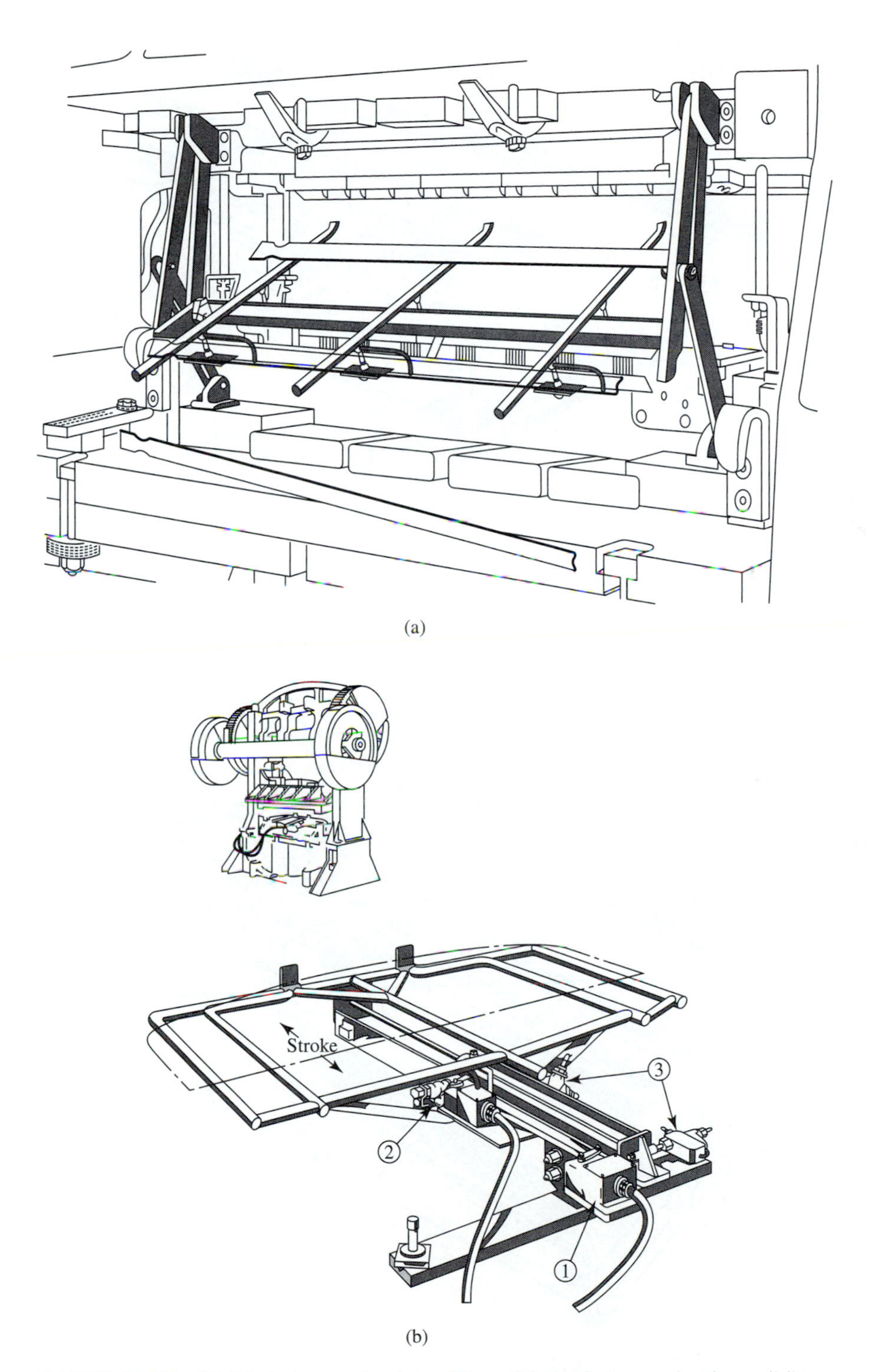

FIGURE 24.24 (a) Workpiece unloader—tilting. (b) Workpiece unloader—sliding. (Courtesy The Minster Machine Co.)

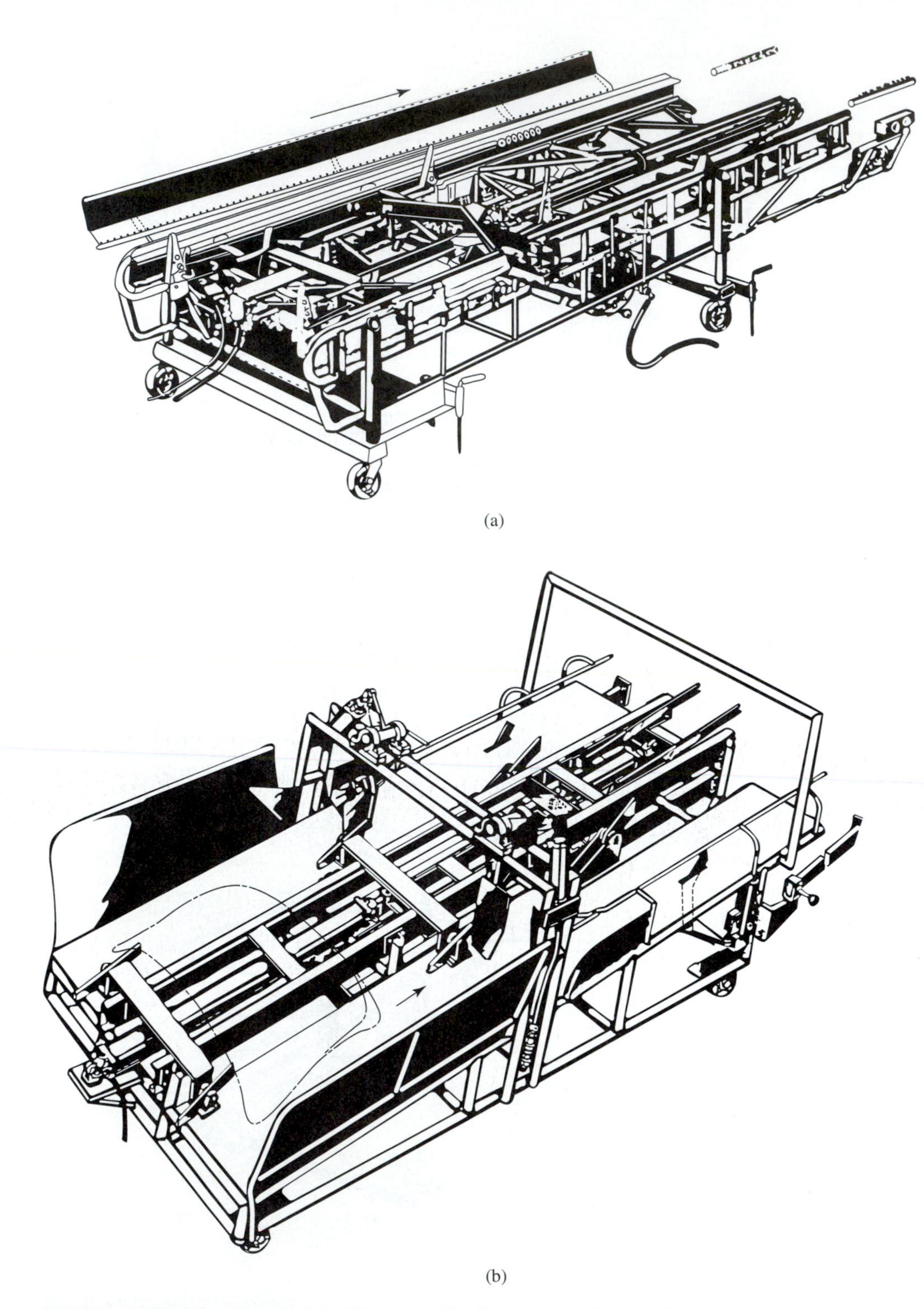

FIGURE 24.25 Transfer devices. (Courtesy The Minster Machine Co.)

24.3.4 Safety Requirements

Press work can and must be made safe for operators and maintenance personnel. Manufacturing Engineering must provide safe tooling and equipment and, working with plant safety personnel, must assure that safe practices of operating equipment are followed by monitoring the same. Included in considerations for press safety must be the following:

- **Design of Tooling.** The design of all tooling must be checked for elimination of sharp edges, pinch points, or other potentially injurious condition. Any element of a process that would require dangerous motions or actions of the operator must be eliminated.
- **Controls.** All presses must be equipped with two-hand anti–tie down push buttons to control the activation of press movements. All in-line and tandem press lines must have emergency stop facilities to control the entire line.
- **Guards.** Fixed guarding must be provided for all possible hazardous areas. Light screens (motion detectors) or other effective guarding must be provided to shield an operator's hands from hazardous areas. The light screen shown in Figure 24.26 stops the press action instantly if penetrated by any object including an operator's hand.

(a)

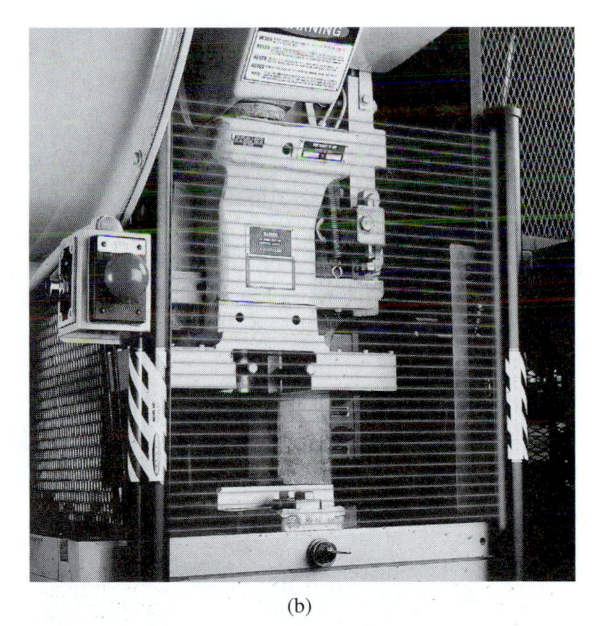

(b)

FIGURE 24.26 Light screen motion detectors. (Courtesy Banner Engineering Corp.)

- **Lockout Systems.** A system requiring all maintenance personnel to deactivate, by turning off power and physically locking out electrical controls of any press before working in any associated hazardous area must be in effect. Lockout systems are formal, enforced plant operating procedures. Details of actions required to comply are posted in conspicuous places. Individual locks, usually padlocks, are issued to affected persons and supervision.

24.4 SPECIAL MACHINES AND MACHINE CONCEPTS

24.4.1 Turret Presses

Punching machines, also known as turret presses (Fig. 24.27), used in flexible sheet metal processing have a punching power of approximately 25 tons and, for the most part, employ stan-

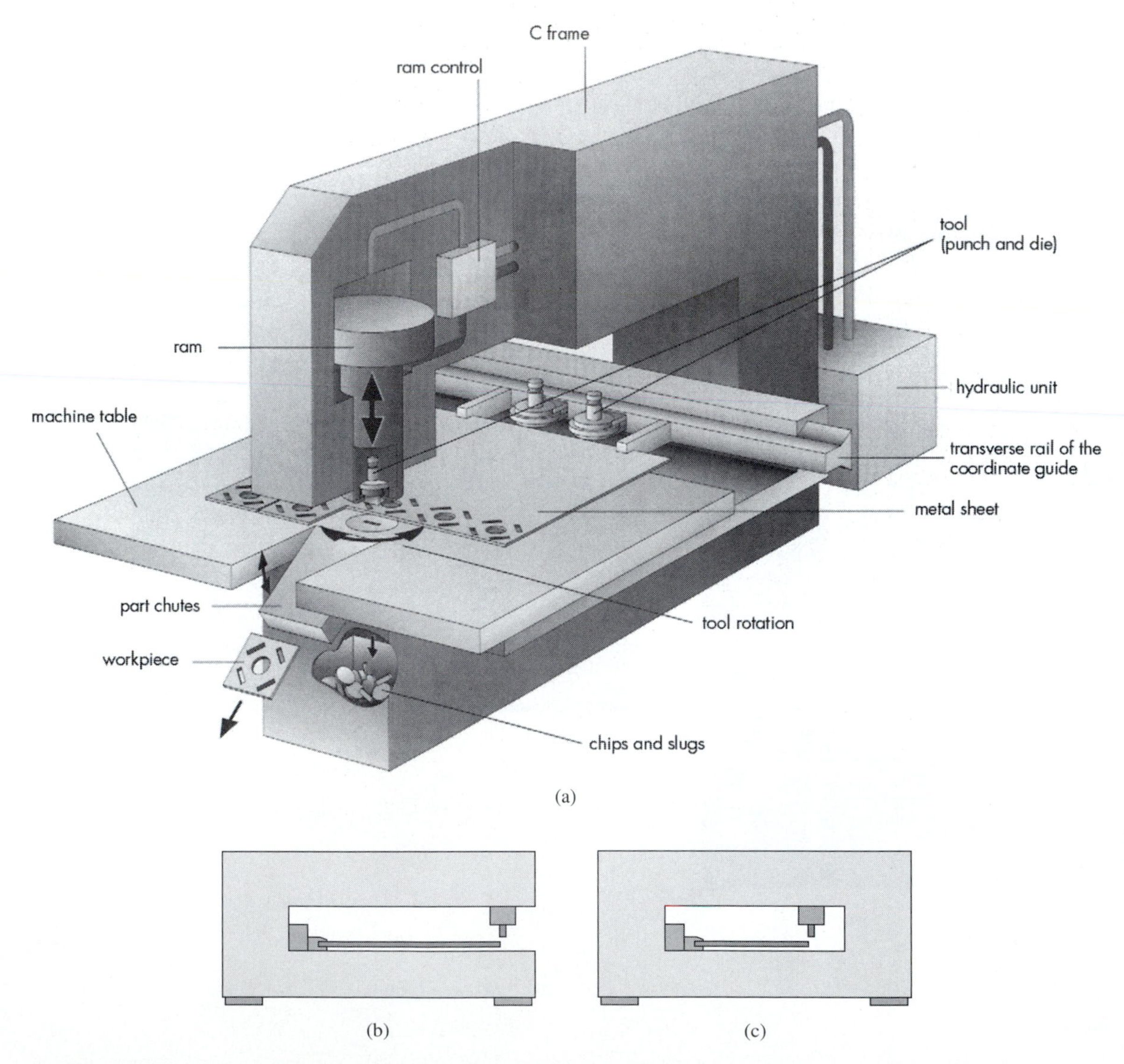

FIGURE 24.27 (a) The schematic diagram displays the most important features of a punching machine. (b and c) A machine frame has the shape of a "C" or "O." (Courtesy of TRUMPF USA)

dardized tools. Punching machines of this type are particularly suited for small and medium-sized lots due to their short setup times, quick tool-changing times, easy-to-use tool and program administration, and automation components for removing and sorting pieces.

Machine Frame The machine frame of a punching machine absorbs the forces required for punching. It is either shaped as a C open at the front or as a closed O. The C-frame allows you to access the work area easily and to process oversize sheets by simply turning them.

Ram Drive The up-and-down motion of the ram is produced either electrohydraulically with a hydraulic cylinder or electromechanically with an eccentric drive.

In an eccentric drive the eccentric connection of the ram (connecting rod) to the rotating shaft causes the ram stroke. A flywheel is located on the shaft, whose rotational energy is transferred to the punch by means of a quick-engaging clutch.

Oil under pressure moves the hydraulic cylinder or punching head. A hydraulic pump creates the required pressure (up to c 3380 pounds per square inch), which is then directed into the cylinder. The result of this transfer of force is the generation of a punching stroke.

Hydraulic Punching Head A hydraulic punching head (Fig. 24.28) works in a way that has been optimized for power and energy. The required punching power level is selected automati-

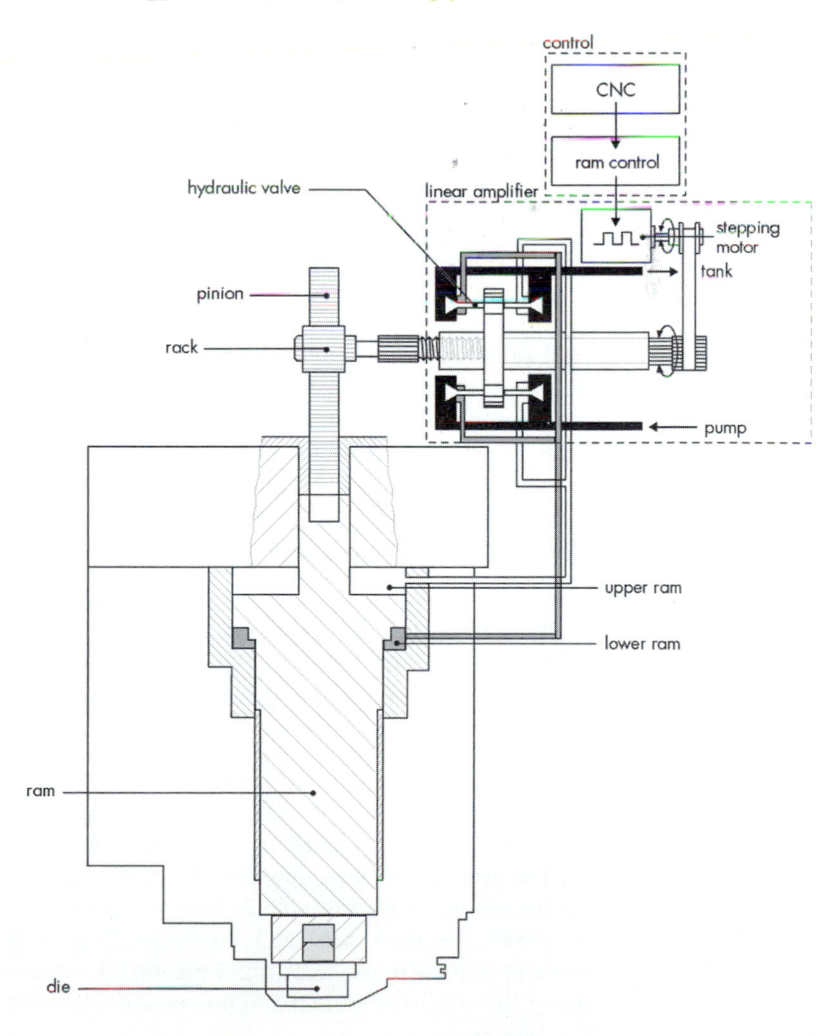

FIGURE 24.28 Schematic illustration of hydraulic punching head. (Courtesy of TRUMPF USA)

cally, depending on the required punching power. Thin sheet metal only requires low oil pressure, which results in less punching power. This in turn reduces the punching machine's energy requirements.

The hydraulic ram has its own NC-controlled axis, which regulates the start and reversal points of the stroke movement of the ram according to the metal thickness. Unnecessary stroke lengths are thus eliminated, increasing the number of strokes and ensuring a perfect penetrating depth for the punch into the die. This leads to a significant increase in tool life.

By regulating the ram speed precisely, a hydraulic punching head can reduce noise up to 80%. The pressure increase in the punching head is measured and evaluated when the punch strikes the workpiece. A temporary reduction of the ram speed can lead to a more quiet punching of the workpiece.

Principle of the Hydraulic Punching Head

1. **Setting the Ram Position.** The control regulates the distance between the ram and the surface of the sheet metal. A stepping motor regulates this distance. In this initial position prior to the punching stroke the lower ram surface is under hydraulic pressure.

2. **Executing the Punching Stroke.** The punching stroke is initiated by an additional signal from the control. The hydraulic valve of the linear amplifier is opened and conducts a flow of oil onto the upper ram surface upon which the ram is pressed downward. When the ram makes contact with the sheet, the hydraulic pressure on the upper ram surface automatically increases until the punch is executed. If a high degree of punching power is required, the hydraulic pressure on the lower ram surface is also cut off.

3. **Return Ram Stroke.** After the punch, the ram reaches its lowest position. The hydraulic valve is then closed by the mechanical rack and pinion. The ram is returned to its initial position by increasing the hydraulic pressure on the lower ram surface while decreasing the pressure on the upper ram surface.

Tool Adapter

In a punching machine we distinguish between a tool adapter, in which a complete tool set is hydraulically tightened so that it cannot move, and a turret, which serves as tool storage and a guide.

With hydraulically tightened tools, the ram as the upper tool adapter holds the punch with the stripper. The lower tool adapter holds the die. The upper and lower tool adapters have the same centerpoint. The long, hydraulic ram guide provides highly precise tool coordination and allows extreme eccentric loads on the tools (for example during nibbling and notching).

Tool Rotation. The tool adapter can also be implemented as a rotational axis allowing all the tools to be rotated. The upper and lower tool adapters rotate synchronously and along the shortest route around the programmed rotational angle.

Workpiece Clamping

Workpiece clamps hold the sheet metal and guide it as it is processed. They are hydraulically or pneumatically activated. The clamps can either be mounted to set positions or variably placed on the transverse rail.

The area of the sheet upon which the clamps are located remains unprocessed, because the clamps would otherwise collide with the punching head. If this collision area is also to be processed, the sheet must be repositioned. The sheet is then clamped with a stripper, and the coordinate guide moves in a program-controlled manner with open clamps. Afterward, the clamps clamp the sheet once again and processing continues.

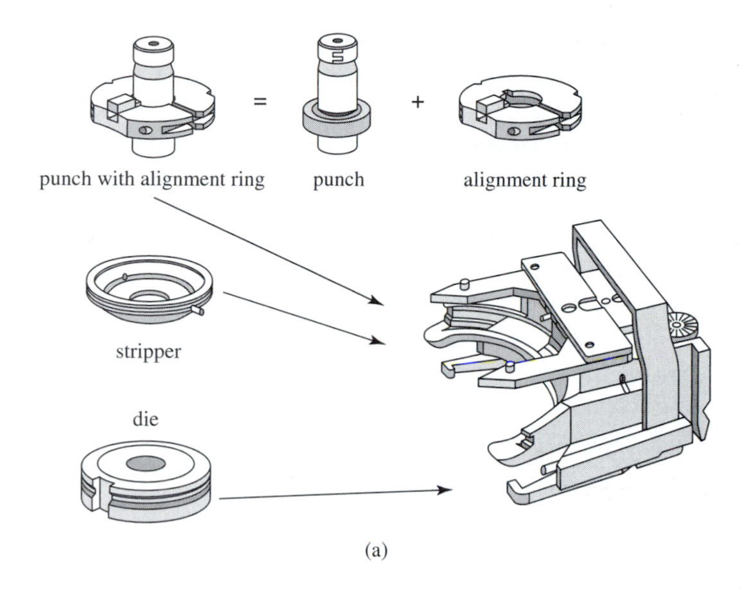

(a)

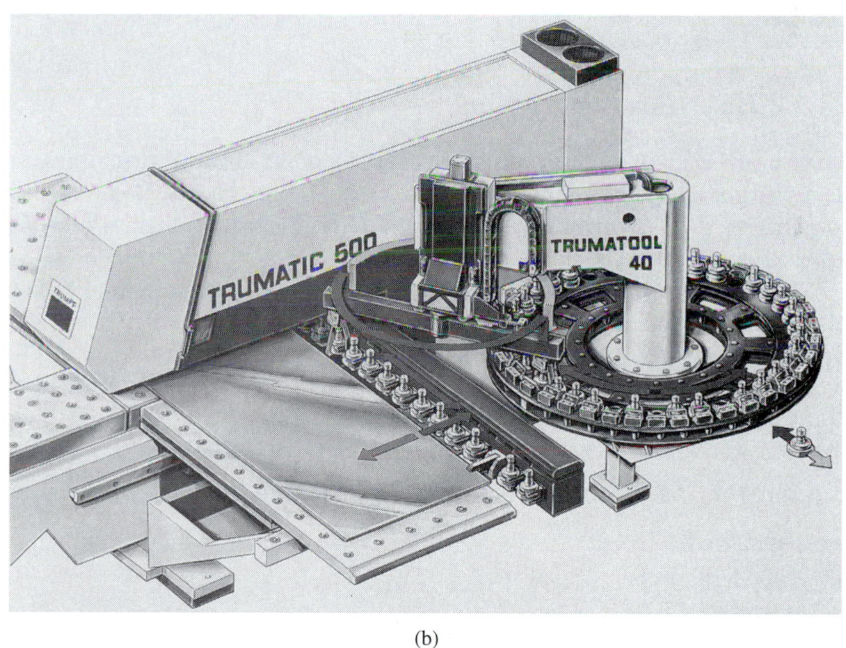

(b)

FIGURE 24.29 (a) Punching tools. (b) External tool storage can supply the machine automatically with additional tools. (Courtesy of TRUMPF USA)

A punching tool (Fig. 24.29a) consists of a punch and die. An alignment ring and stripper complete the tool set. A tool cassette holds the tool set. This means that the complete tool set can be inserted into either the tool adapter or the tool storage in a single step. Tool storage is shown in Figure 24.29(b).

Punch with Alignment Ring. The punch is manufactured from high-speed steel (HSS). For specific applications, punches with coated or oxidized surfaces can be used. The alignment ring is a clamping ring, which fastens the forming punch (e.g., a rectangular tool) in the zero position. It transfers punching power from the ram to the punch.

Stripper/Holding-Down Clamp. The stripper or holding-down clamp can perform three functions:

Stripping Function. The stripper separates the sheet metal from the punch during the return stroke. This prevents the metal from being pulled up with the punch. During processing, the stripper is situated approximately 0.5 to 1 mm above the workpiece surface.

Holding-Down Function. The hydraulic control lowers the stripper onto the sheet before the punching stroke and holds it securely to prevent warping. This function is particularly important when processing thin sheet metal. Self-separating tools are used as an alternative to strippers and holding-down clamps.

Clamping the Workpiece. To rework oversized sheets, the stripper lowers itself onto the sheet and clamps it down while the coordinate guide moves with opened clamps for reworking.

Die. Punch and die function as shearing tools that pass alongside each other in a scissorlike motion and cut the sheet. The resulting punching slug falls through the die and is usually extracted.

24.4.2 Laser Beam Machining

Laser beam machining, previously considered a nontraditional machining process, has emerged as a well-developed and increasingly used machining process. Some applicable data developed for Chapter 29 is repeated here along with descriptions and illustrations of current commercial applications.

The **laser beam machining** process is applicable to any material, hardness, and condition of electrical conductance. It is especially capable of producing very small holes (.001 in.) in steel up to .100-in. thick but is not practical for heavy metal cuts of over .200-in. thickness. Laser beam machining has potentially hazardous areas that must be controlled by proper safeguards and techniques for the safety of the operator.

Equipment Of the many types of lasers developed, the most practical for metal cutting are

- Saphire lens
- Ruby lens
- Yttrium aluminum garnet
- CO_2

The ruby lens laser produces the most powerful output but is the most costly. CO_2 lasers are the most efficient.

Lasers can be programmed to cut, slot, drill, and scribe. Numerical control is used on many applications to position and move material in relation to the beam.

Applications

- Any material
- Any hardness
- Thicknesses of cut to .400 in.

- Drilling
- Cutting
- Scribing

Laser Safety The laser beam machining process has several inherent conditions that can be dangerous and affect the safety of the operator or bystanders. They include the following:

- **High Electrical Power.** High voltages and high currents (1200 V and 1000 A) combine to create a potentially hazardous condition for operators. Proper insulation, covers, interlocks and correct operating techniques must be provided and monitored for compliance with safe operating conditions.
- **X-Rays.** Lead shielding must be provided and, in some instances, operators must be issued film badges to monitor exposure to X-rays.
- **Vision.** Light-absorbing shields must be provided, and operators must be issued eye protection goggles to prevent any accidental exposure during setup, part loading, and so on. Closed circuit TV systems are available and are in use for the operator to safely view and conduct such tasks as focusing the laser beam.

Advantages

- Versatile—processes any material, any hardness
- Nondistorting
- Can produce sharp corners
- No cutting tools required
- Simple fixturing required
- Adaptable to any type of control system

Disadvantages

- Limited ability to cut thick metal
- Potentially dangerous
- Equipment relatively costly

Laser Cutting-Machine Concepts The various applications for laser technology in sheet metal processing have given rise to the development of various system and machine concepts. Two-dimensional laser systems are used primarily for flat workpieces and 3D lasers for spatial components. One-dimensional systems and laser robots play a rather subordinate role in sheet metal processing. Figure 24.30 provides an overview of the most common types of machines. The possibility of moving the workpiece or the focusing optics available are important differentiating features. A typical commercial machine is shown in Figure 24.31.

2D Systems. Two-dimensional laser systems have proven themselves in applications requiring a high degree of flexibility for geometrical changes in the workpiece. Two-dimensional machine processing is characterized by high precision and high processing speed. Typical cutting applications are slitting steel, high-grade steel, and nonferrous metals in automotive, machine, tool, and equipment construction as well as air conditioning and aerospace technology. The machine concept best suited for 2D processing depends on the requirements of the tasks.

Variant 1	Variant 2	Variant 3	Variant 4	Variant 5
2D laser only machine with flying optics; flatbed machine	2D laser only machine with stationary optics	3D laser only machine with cantilever	3D laser only machine with rotation axis on the machine body	3D laser only machine with portal design
flying optics	—	flying optics	flying optics	flying optics
—	movable work piece	—	movable work piece	movable work piece
optics: 3 axes	optics: 3 axes	optics: 5 axes	workpiece: 1 axis optics: 5 axes	workpiece: 1 axis optics: 4 axes

FIGURE 24.30 Table of various machine types. (Courtesy of TRUMPF USA)

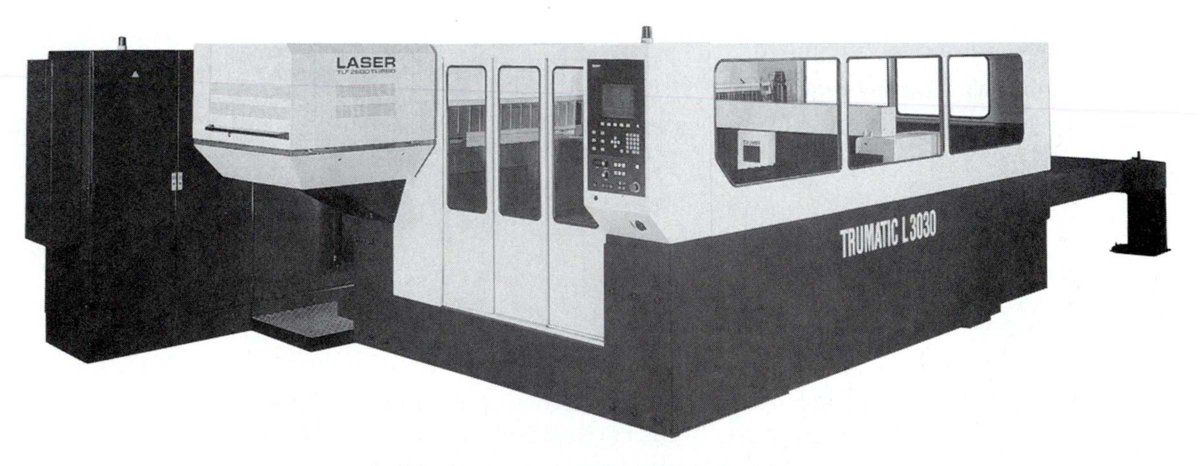

FIGURE 24.31 Typical laser cutting machine. (Courtesy of TRUMPF USA)

Flatbed machines with a fixed laser but movable laser cutting head can cut large sheets up to 6000 × 2000 mm and thicknesses of up to 20 mm without the need for any machine modifications. The dimensions of the raw materials can vary greatly.

Overshot and scratching are of no consequence when processing with movable, noncontact optics. Furthermore, a machine that only moves its cutting head requires less space than one that has to position the workpiece.

The path that the laser beam must follow from the laser device to the cutting optics varies greatly in laser cutting systems with movable optics, the result being that the focusing position and spot size are not constant throughout the entire work area. By employing adaptive optics,

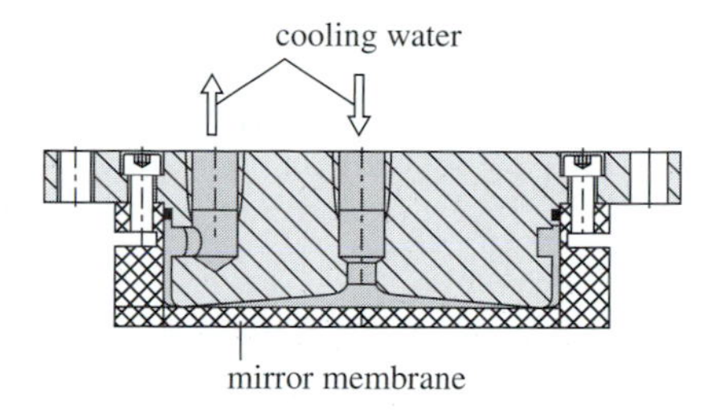

FIGURE 24.32 Adaptive mirror. (Courtesy of TRUMPF USA)

these effects can be rectified: The mirror membrane of these optics is changed by the pressure of the cooling water (Fig. 24.32).

The principle of a movable workpiece is only practical up to a particular workpiece weight, because the machine cannot position an arbitrary mass quickly and precisely.

Laser cutting systems with a fixed laser cutting head are advantageous because they have a constant beam path, ensuring uniform beam characteristics for processing a complete sheet. It is easier and more efficient to extract particles produced as waste in systems with a fixed laser head than in systems in which particles have to be extracted along the entire work area.

3D Systems. Three-dimensional laser systems are used for processing 3D pieces. As is the case with 2D processing, the laser head is positioned vertically to the workpiece surface (Fig. 24.33). In order to ensure this with a stationary workpiece, a 5-axis beam delivery is needed, as shown in Figure 24.34.

The appropriate 3D system concept is determined primarily by the spectrum of workpieces and their respective dimensions and weights. As a consequence, flexible machine concepts have proven themselves in a modular construction form in reference to the work area, the focusing

FIGURE 24.33 Cutting head processing a 3D part. (Courtesy of TRUMPF USA)

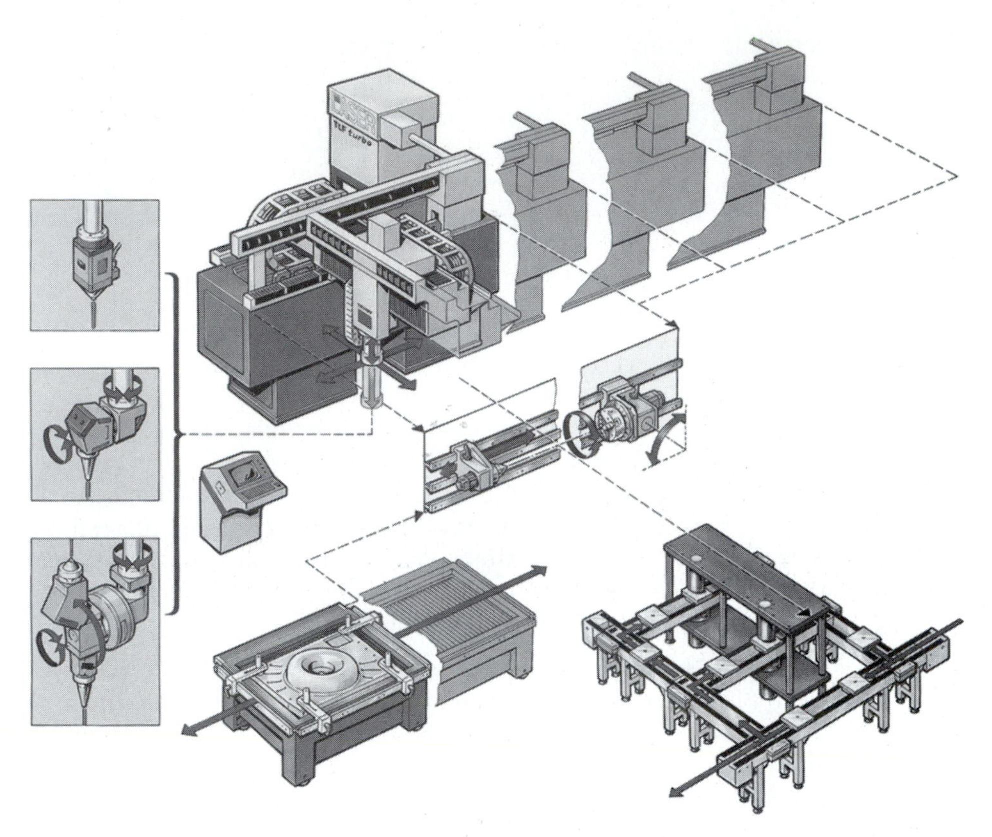

FIGURE 24.34 5-axis beam delivery shown as part of a 3D modular laser processing cell. (Courtesy of TRUMPF USA)

optics, the CNC control, and the laser device. This concept allows a customer-specific, job-oriented system to be realized.

A typical 3D application is processing preformed steel sheets in which cutouts and openings have to be made or edges need to be cut.

Processing Pipes. Laser technology is gaining increasing importance as a means for processing pipes and profile sections with 2D and 3D machines. Normally, an additional rotary axis is used on the machine, which in combination with the machine axes allows a rotating piece to be processed as shown in Figure 24.35. A schematic plan view of a flatbed laser machine used to process flat sheet and plate is shown in Figure 24.36.

To take full advantage of the machine, 2D laser machines frequently employ two pallets, which are fed into the machine alternately like chest drawers. While a sheet is being processed in the machine, the second rest outside of the machine can be cleared and loaded with a new sheet.

Extraction and Filtering. An extraction system located under the pallet extracts particles and gases caused by processing and delivers them to a filter system. Extraction and filtering play an important role in keeping the workplace clean and free of contamination.

Protective Enclosure. A protective enclosure is required as a means of protection against stray radiation from the machine. The enclosure protects the operator from light radiation caused by straying or reflection during processing. The protective enclosure is usually made of steel sheets and transparent polycarbonate (macrolon) for the windows.

FIGURE 24.35 Module for pipe processing. (Courtesy of TRUMPF USA)

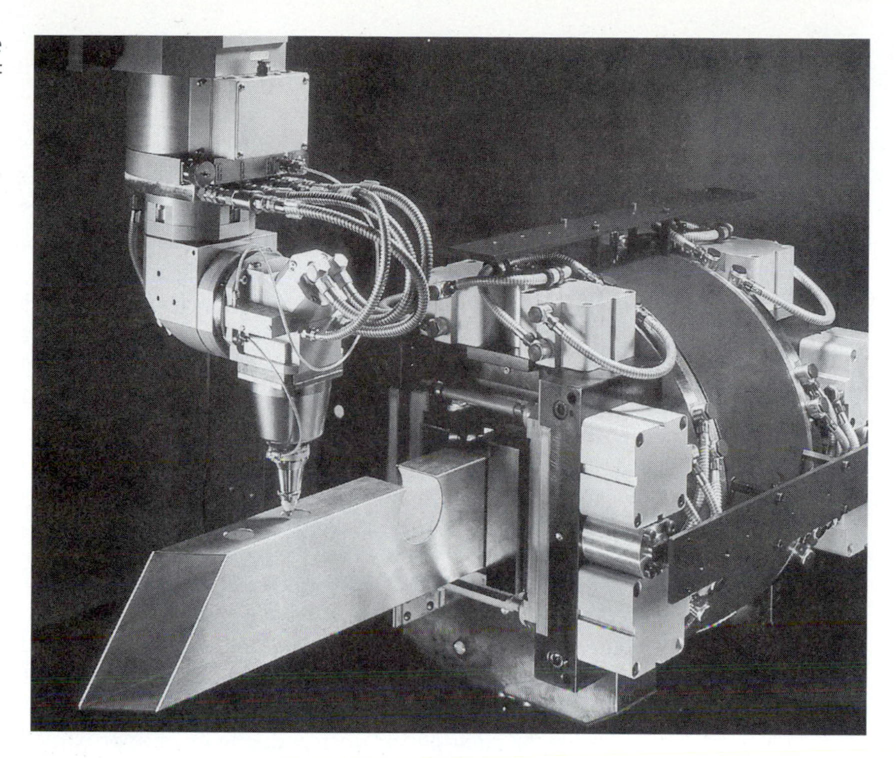

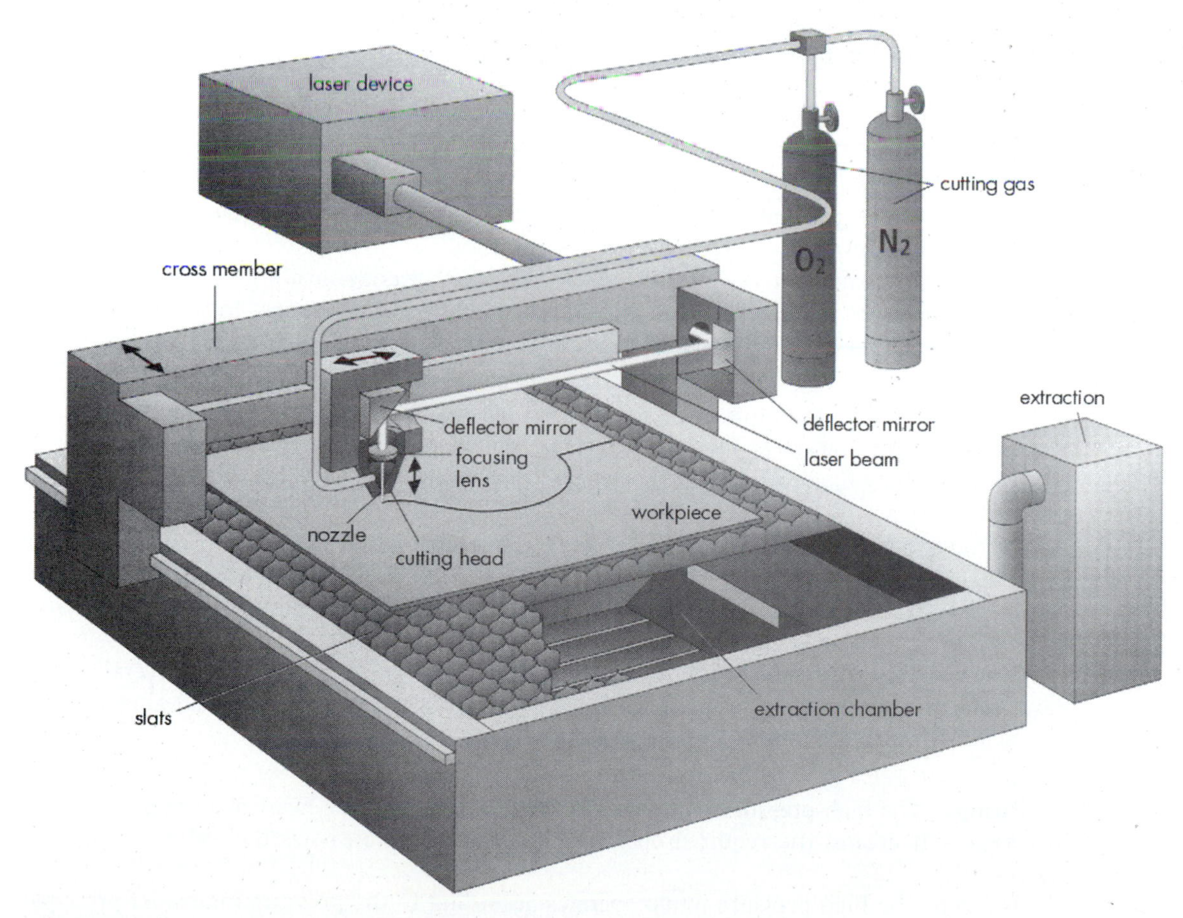

FIGURE 24.36 Structure of flatbed laser machine. (Courtesy of TRUMPF USA)

Safety. Safety has been an important factor for laser processing from the very beginning. For that reason, numerous measures exist that deal with the dangers of lasers. Generally there are two types of danger groups in laser technology: First are the dangers that emanate from the laser itself such as light radiation and electricity. Second, particles and gas emissions caused by processing are a source of danger. As a result of the establishment of universally applicable laser classes in 1977 and their related safety concepts, radiation accidents have been practically eliminated in laser cutting.

Modern extraction and filter systems have brought dangerous materials produced during processing under control. Laser cutting is thus not only highly suitable for industry, but also a very safe production process.

24.4.3 Water-Jet Cutting Machines

How a Water-Jet Cutting Installation Works When starting a cutting procedure, the high-pressure pump is activated and creates a very high pressure of up to 4000 bar. The delivery system conducts the water under pressure from the pump to the cutting head. The cutting head is positioned over the piercing point in a very quick motion. The water jet is regulated by a valve located over the cutting head.

A sapphire or ruby nozzle with a diameter of approximately .08 to .5 mm is at the end of the high-pressure line. The water pressure is relieved in the nozzle and a water jet is created. The abrasive material is added to the water jet in the mixing chamber of the cutting head. The jet is then focused in the second nozzle—the abrasive nozzle.

The cutting head makes a circular or pendular motion over the piercing point. The abrasive water jet then cuts a path through the material. As soon as the jet has penetrated the material, the cutting head moves along its programmed contour. The abrasive water jet, guided by the cutting head, forms the contour in the material.

When it reaches the end of the contour, the cutting head lifts itself and moves to the next piercing point. A new processing step then begins.

The energy remaining in the jet is absorbed by the catcher underneath the workpiece. Cutting water and sludge are transported away from the machine's processing area to the disposal unit. A schematic plan view of a water-jet cutting machine is shown in Figure 24.37.

Machine Concepts Machines designed for a specific application are just as varied as the fields of application in water-jet technology. These concepts range from systems with fixed cutting heads to machines equipped with an *XY* guiding unit and to robot-guided systems that process with pure water.

Machines designed in the form of a portal, shown in Figure 24.38, are used primarily in sheet metal processing. This construction combines high positioning precision with a very dynamic freedom of movement. Furthermore, machines with a portal design are accessible from three sides.

A water-jet cutting machine can be equipped with one or more cutting heads. Several cutting heads offer the advantage of processing several workpieces in parallel—a big advantage in the race against time and money.

Pumps. The high-pressure pump shown in Figure 24.39 is the heart of a water-jet cutting installation. It creates the required operating pressure and transports the water through the system.

As a rule, the high-pressure pump operates according to the pressure intensifier principle: A hydraulic pump creates the oil flow required for powering the single or double pressure inten-

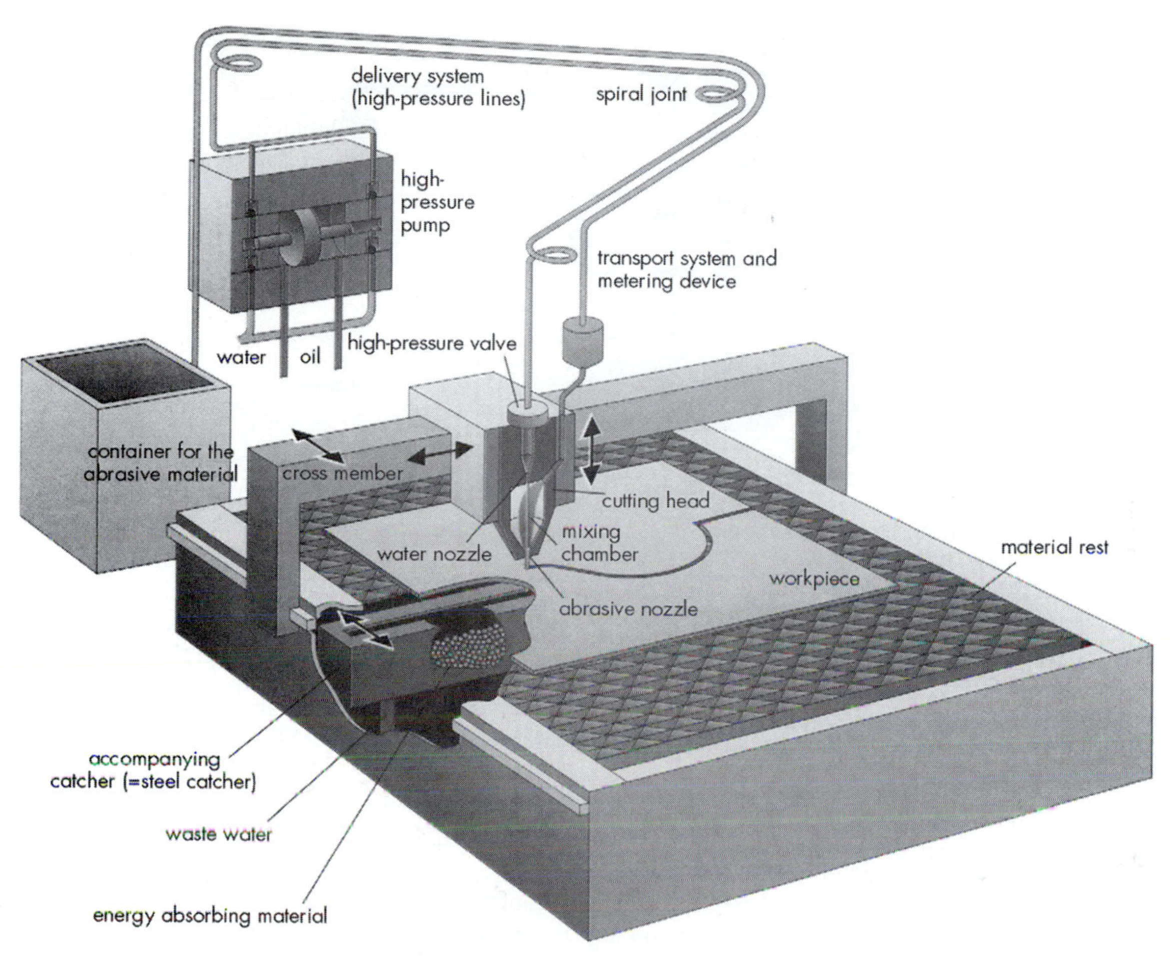

FIGURE 24.37 Schematic illustration shows the most important features of a water jet cutting installation. (Courtesy of TRUMPF USA)

FIGURE 24.38 An abrasive water jet cutting installation with a portal design and two cutting heads. (Courtesy of TRUMPF USA)

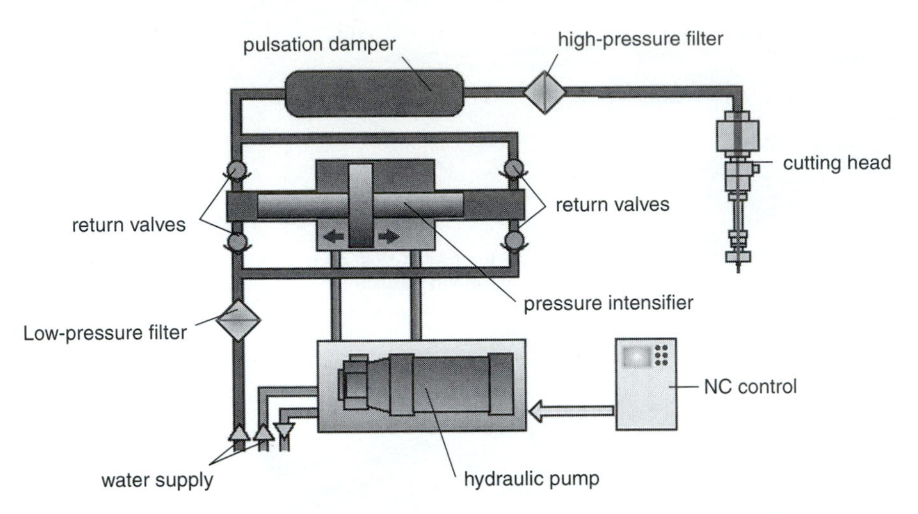

FIGURE 24.39 Structure of a high-pressure pump displayed schematically. (Courtesy of TRUMPF USA)

sifier. The maximum oil pressure is 200 bar. The transmission ratio between the low-pressure hydraulics (oil) and the high-pressure hydraulics (water) is 1:20.

The pressure pumps for water-jet cutting machines are high-performance pumps. They create a pressure of up to 4000 bar. In comparison, the pressure in household water pipes amounts to approximately 6 bar.

The back and forth motion of the pressure intensifier causes pressure fluctuations in the piston's end of travel. For that reason, pumps for this pressure are usually equipped with a pulsation damper to smooth out pressure fluctuations.

During processing the desired pressure is set automatically through input made at the control panel or by the program system. It can also be set manually at the pump.

High-Pressure Delivery System. The high-pressure delivery system safely conducts the water under pressure from the pump to the cutting head and hence to the processing area. The high-pressure delivery system consists of thick-walled high-pressure pipes made of strain-hardened, austenite steel.

On the one hand, the high-pressure pipes have to be stable enough that they can withstand the immense high pressure of 4000 bar. On the other hand, the delivery system as a whole must be flexible enough that it can follow the motions of the cutting head.

In the machine depicted in the Figure 24.39, cutting water is supplied to the cutting head over defined high-pressure pipes without turnable knuckle joints. This type of delivery system has proven itself to be highly wear resistant and reliable.

Cutting Head and Nozzles. Aside from the high-pressure pump, the cutting head is the most important component of a water-jet cutting machine. It is responsible for creating the water jet.

If a workpiece is processed without abrasive materials, the cutting head (Fig. 24.40) then consists of a high-pressure pipe—the collimation pipe—which is terminated by a water nozzle. If a workpiece is to be processed with abrasive materials, as in Figure 24.41, an abrasive cutting head is installed as well.

The water nozzle—sapphire, ruby, or diamond—forms a very thin and fast water jet with a diameter of approximately .08 to .5 mm. The water is pressed out of the nozzle at a speed of

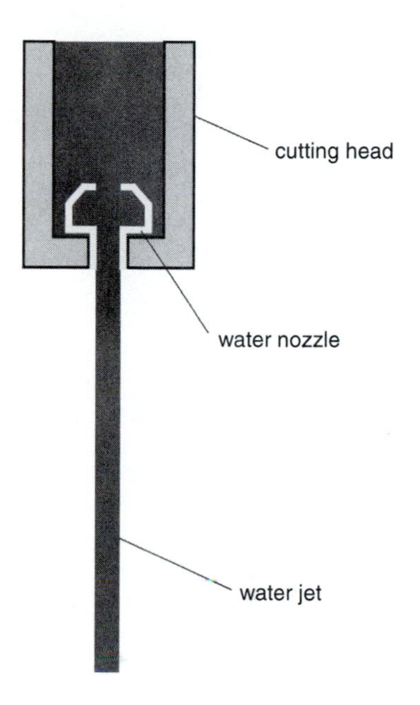

FIGURE 24.40 Cutting head arranged to cut with water jet. (Courtesy of TRUMPF USA)

approximately 900 m/s—nearly three times the speed of sound. The high speed of the water jet creates a partial vacuum in the mixing chamber so that abrasive material and air are sucked in and flushed away by the water jet. This is known as the injector principle.

The water–abrasive material–air mixture is then led through the abrasive nozzle—a 50- to 100-mm-long nozzle made of hard metal. It has a diameter of approximately .8 to 1.2 mm and two functions: First, it accelerates the abrasive particles, and second, it focuses the jet. The result is the highly piercing, energy-laden abrasive water jet.

The water and abrasive nozzles are subject to high loads. The water nozzle is particularly affected due to the high water speed passing through it. The abrasive nozzle is subject to high

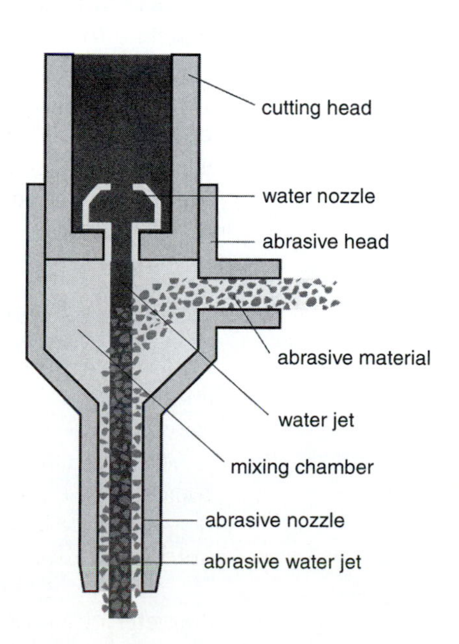

FIGURE 24.41 Cutting head arranged to cut with abrasive water jet. (Courtesy of TRUMPF USA)

FIGURE 24.42 Cutting head with metering device for abrasive material and height regulation. (Courtesy of TRUMPF USA)

wear because the abrasive material not only grinds the workpiece but also the nozzle itself. Both nozzles are thus wearing parts that have to be replaced after every 50 hours of operation. When it comes time to replace a nozzle, accessibility and adjustibility demonstrate the advantageous features of a cutting head. The distance between the workpiece and cutting head is approximately 1 to 2 mm. A uniform cutting quality along the entire contour can only be ensured if the distance between the nozzle and the surface of the material remains constant. This task is performed by the height regulator located on the cutting head. In cases where the surface is not completely level, the height regulator also prevents the cutting head from colliding with the workpiece. A device to control cutting height and abrasive material concentration is shown in Figures 24.42 and 24.43.

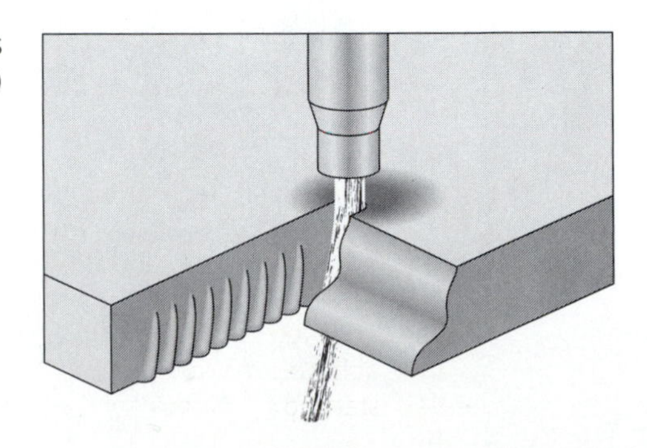

FIGURE 24.43 Abrasive water jet cutting process illustrated schematically. (Courtesy of TRUMPF USA)

Materials and Material Thicknesses. Nearly all materials can be cut with an abrasive water jet. Materials that are difficult to work with using other known processes are particularly suited for sheet metal processing:

- Highly reflective and heat-conducting metals such as aluminum, aluminum alloys or copper
- Titanium/titanium compounds
- Very thick metals
- Composite materials
- Laminates
- Coated metals

An abrasive water jet also cuts all kinds of synthetic materials, for example: fiber-reinforced plastics, stone (e.g., granite or marble), glass, acrylic glass, and foam rubber.

The maximum material thickness is essentially determined by the machine. In sheet metal processing, abrasive water jet cutting machines can normally cut material thicknesses up to 100 mm. Only in exceptional cases is it economically feasible to process thicker materials. The progressive development of a metal edge as it is cut by a water jet is shown in Figure 24.44.

Cutting Edges

Roughness. Cutting edges that have been processed with an abrasive water jet are classified into two groups: the smooth cut zone and the remaining surface.

The smooth cut edge is a surface with a totally stochastic structure, whereas the remaining surface has a rippling, fluted structure. This fluted structure is superimposed on the stochastic structure. The flutes, which run parallel to the cutting jet direction, are curved away from the cutting direction.

There is a gradual transformation of the smooth cut zone to the remaining surface. The size of both surfaces is primarily determined by the cutting speed. The following guidelines apply:

- The smooth cut portion increases in size as the cutting speed is reduced. This means that the average roughness of the entire cutting edge is reduced as the cutting speed is reduced.
- The following applies for metals: At a cutting speed that can just barely cut a metal (the slitting speed), the smooth cut zone makes up approximately one-third of the total surface.

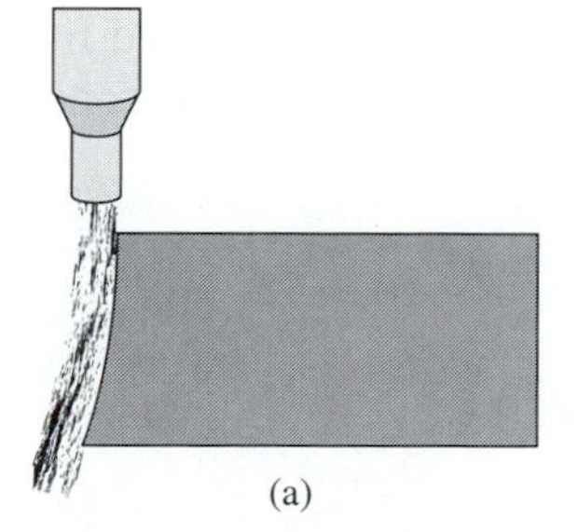

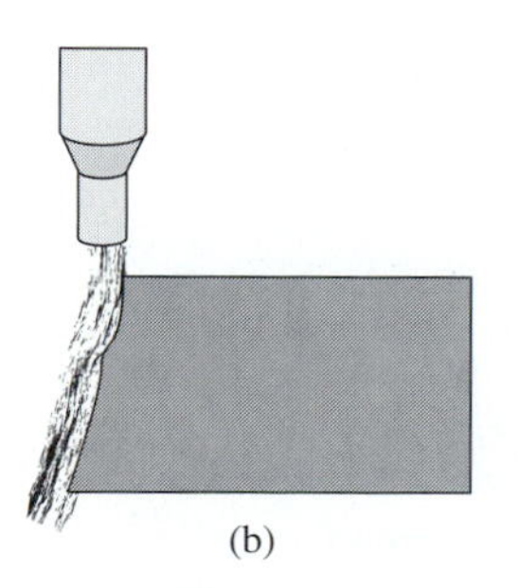

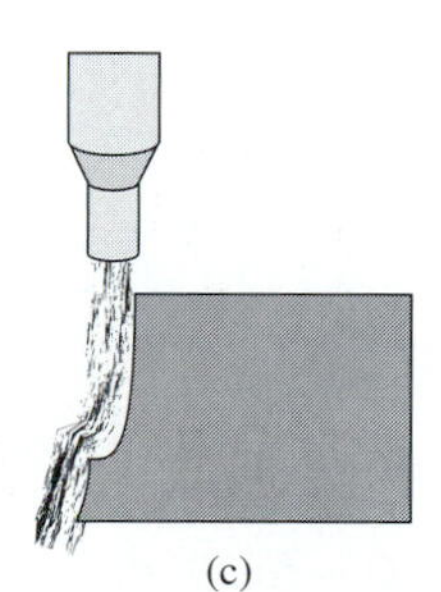

FIGURE 24.44 Edge development of water jet cutting process. (a) The angle between the undisturbed abrasive water jet and the front edge constantly increases. As a result the water jet is deflected away from the cutting direction. (b) During the cutting process a step is created. The angle of deflection on the step increases constantly. There is increasingly less abrasion below the step. (c) The step is ground out of the workpiece relatively quickly, until the colliding particles are no longer able to abrade the material. As the step moves downwards, a smooth front edge is created. (d) The starting condition has been reached again. (Courtesy of TRUMPF USA)

In practice a distinction is made between slitting cut, quality cut, and fine cut—according to the quality of the cutting edges, as shown in Figure 24.45.

The Width of the Slit and Perpendicularity. The form of the kerf also depends on the cutting speed. The following guidelines apply:

- The kerf on the top side of the workpiece has approximately the same width as the circumference of the abrasive nozzle (i.e., .8 mm).
- The diameter of the abrasive nozzle and hence the kerf on the top side of the workpiece increases as the nozzle wears.
- The actual kerf width must be taken into account when executing a program.

FIGURE 24.45 Cutting edge quality
Edge quality and perpendicularity depend on the cutting speed. In practice, a distinction is made between slitting cuts, quality cuts, and fine cuts. (Courtesy of TRUMPF USA)

Slitting cut

Slitting speed = maximum cutting speed
The slitting cut is a cut optimized for speed by which the processing parameters have been selected to allow a part to be processed as fast as possible.

Quality cut

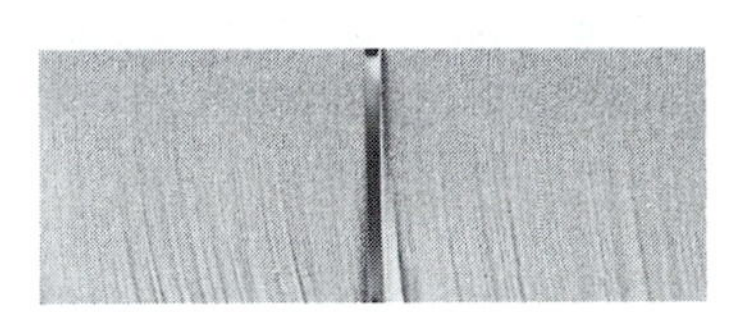

Half of the slitting speed
The quality cut is characterized by good edge and corner quality at an economically sensible cutting speed.

Fine cut

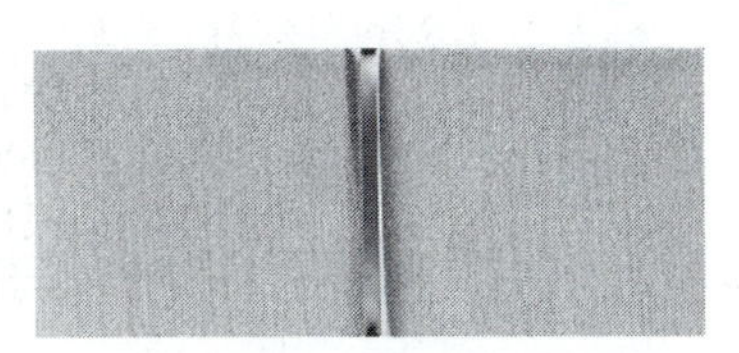

One fourth of the slitting speed
For a fine cut the processing parameters are set to produce a very good edge quality. The smooth cut part is very large and the edges are nearly perpendicular.

Form and width of the slit
The kerf width is determined by the diameter of the abrasive nozzle. On the top side of the workpiece the kerf has approx. the same width as the diameter of the abrasive nozzle (approx. 0.8 mm). The width on the bottom side of the workpiece depends on the cutting speed.

- On the underside of the workpiece the kerf width is dependent on the cutting speed. The kerf profile has a slight V shape.
- It can be said for all homogenous materials that the slower the cutting speed, the more perpendicular the cutting edge will be.

Burr Formation. Cutting edges produced with an abrasive water jet are for the most part free of burrs.

Corners. It must be possible to direct the abrasive water jet parallel to the edge before the cutting direction is changed at corners along the contour path. This would otherwise lead to contour deformation on the underside of the workpiece. For that reason, the cutting speed must be reduced before reaching a corner. After the jet has changed directions (i.e., after the corner) it can slowly be accelerated to the original cutting speed.

Starting at an Existing Contour. At the end of a cutting procedure, contours frequently have to be closed—for example, when processing a circular section. When starting at an existing contour, deformations can occur on the way around the circumference so that upon completion of cutting the circle, there can be a definite mismatch. To avoid or minimize this condition, maintenance of equipment, cutting heads, pressures, and cutting speeds must be kept to the optimum. Less than satisfactory maintenance will result in poor quality parts, rework, and possibly scrap.

24.4.4 Automation

What Is the Goal of Automated Manufacturing? Every industry is interested in continually increasing the

- productivity,
- flexibility, and
- profitability

of its manufacturing process. In practice, these three factors cannot be clearly distinguished from each other. Indeed, they are mutually dependent, as indicated in Figure 24.46.

Productivity. One goal of every production control process is to minimize the nonproductive times (i.e., the times when the machines are not producing). Loading and unloading the machine, clamping the workpiece, or setting up the tools can only be carried out manually directly at the machine. Reducing nonproductive times is therefore one of the most essential demands on a modern automation concept for sheet metal processing.

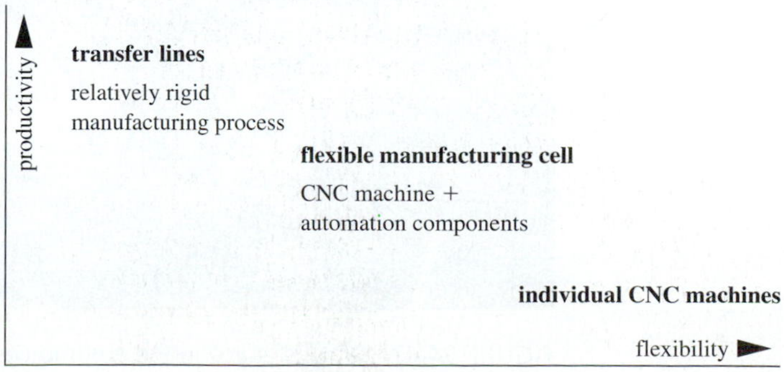

FIGURE 24.46 Production versus flexibility for key processes. (Courtesy of TRUMPF USA)

productivity
transfer lines
relatively rigid manufacturing process
flexible manufacturing cell
CNC machine + automation components
individual CNC machines
flexibility

Flexibility. Increasingly shorter product life cycles must be taken into account by increasing flexibility. Flexibility is required, however, not only with respect to the range of parts, but also in reference to the number of parts to be produced. The machine is supposed to produce in a profitable way, independent of the number of pieces. Whereas the constructional material sheet metal provides far-reaching possibilities for prototypes and trial pieces—which at times leads to the production of a single piece, the same machine must be able to produce a vendor part in the thousands for large-scale, industrial production.

Profitability. As the demand for variable lot sizes makes clear, any automation plans must not be looked at from the standpoint of the costs incurred. Fully automatic systems are expensive and must pay for themselves by producing a sufficient number of pieces or by altering the production organization.

Transfer Lines for Large-Scale, Industrial Production

Automation projects increase productivity. This applies to the large-scale production area as well as to smaller companies with individual operating machines.

In large-scale industrial production, automation concepts are still used that can be subsumed to the label transfer lines. This refers to a more or less fixed arrangement of machines, such as in Figure 24.47. These are connected to each other by a transfer system, which transports the pieces from one station to the next. Robots, which are more responsible for optimizing time than increasing flexibility, are installed at most of the stations. Nonetheless, products cannot be accelerated by minimizing the cycle time or reducing the time it takes to transfer a workpiece between two stations. If all cycle times remain the same, the best that can be produced is a material pileup, because every concept intended to reduce the processing time must take the combination of all components of the transfer line into account. Time optimization means, however, not only reducing the cycle times, but staying within the time frame. The less the total time for a production process varies, the more precisely the required material can be disposed and delivered "just in time." The achieved reduction of storage and transport times decides at least as

FIGURE 24.47 A transfer line in the automotive industry. (Courtesy of TRUMPF USA)

much about the efficiency of the system as the reduction of main machine times and nonproductive times.

Wherever product variations are to be manufactured, branching is planned in the production flow or more flexible individual components are interconnected. This, however, has the effect of making production control immensely more complex. Today, in the automotive industry you can find transfer lines that meet this challenge with individual programs: Every workpiece is equipped with a chip containing all production-relevant information. This information is read in at the processing stations. The workpiece is processed according to the input instructions and then transferred to the next processing station.

Making a Transfer Line More Flexible Transfer lines are fully automatic, high-production systems for relatively rigid manufacturing processes. They are usually designed for the production of a single product line including all variations thereof. Whether, in the final analysis, the construction of such a transfer line is profitable depends not only on the lot sizes and the number of variations but also on the rate of use. These are, however, parameters whose dynamics can only be calculated under certain conditions. As a result, the trend today is again moving away from large complete systems to linking standardized work stations. These systems are known as flexible production systems. Their initial cost is not only considerably lower than traditional production lines, but due to their increased flexibility, they are also instrumental in reducing costs. Systems that have to be modified several times during their service lives are not uncommon today. Nevertheless, during the planning phase for a new transfer line, each new production step will still be calculated very deliberately before it is integrated into the automated part of the production flow. In contrast to three or four years ago, today's user will, in case of doubt, opt for manual handling and flexible manufacturing cells which are time independent instead of attempting to overengineer a problem.

Automating a Punching Machine By imbedding production line machines in a system of automation components, they are made into flexible manufacturing cells. Automation within the manufacturing cell concerns tool handling and the material flow. In so far as they are not rigidly bound to a single, linear production process, they are considered autonomous. Time independency does not mean, however, that they are detached from the flow of materials. Flexible manufacturing cells are always intermediate stations for a product on the way from raw material storage to the shipping department.

The automation components of a flexible manufacturing cell can be standardized components or individually designed units. Figure 24.48 shows standardized components for automating a punching machine. These components support the entire range of processing possibilities regardless of how the raw sheets have been cut, the lot size and the dimensions of the finished parts. Thanks to a wheel-like storage system, approximately 60 different tools can be utilized without manual intervention. The finished parts are sorted onto pallets or into crates. The scrap skeleton is placed onto a separate stack.

Flexible Small-Scale Production To be competitive as a contractor of small lots, a sheet metal production company must rely on being able to handle orders in a flexible manner. Prolonged nonproductive times for resetting or programming the machine must not be allowed to occur between two different jobs. Workpieces that have already completed one phase of operation must be ready for the next phase. This level of flexibility in the material flow can only be achieved in large firms by means of a fully automatic buffer system—for example, with a high storage bay—which acts as the interface between the machines.

More flexibility in processing an order is often a consequence of measures designed primarily to increase productivity. Automatic loading and unloading devices and the integration of a

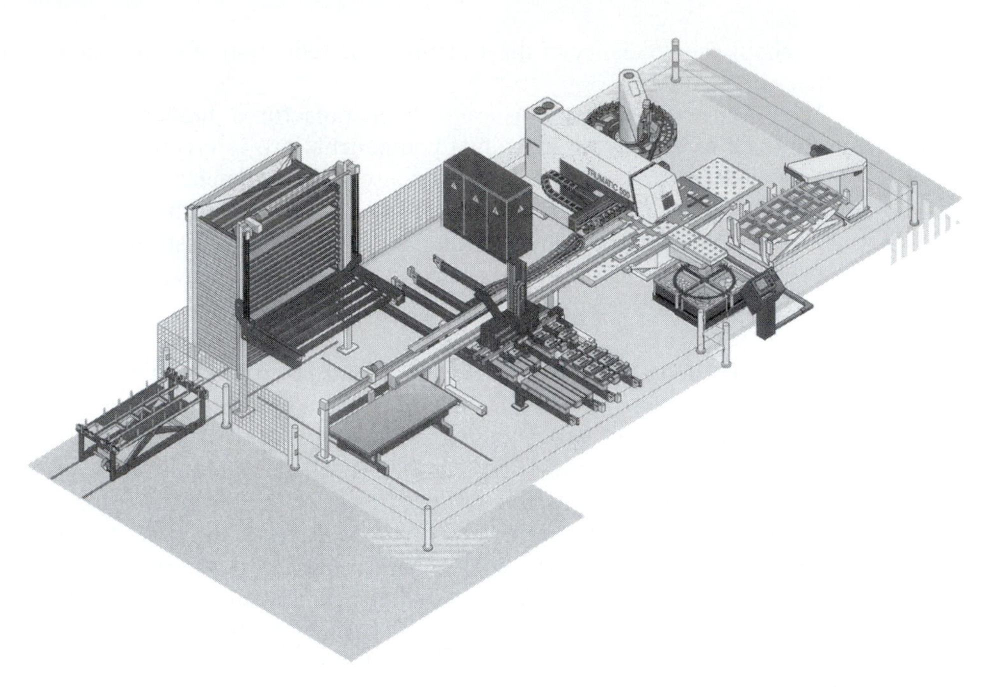

FIGURE 24.48 Punching machine as the core of a flexible manufacturing cell. (Courtesy of TRUMPF USA)

storage system make it possible to operate machines around the clock without manually intervening in the processing procedure. The required integration of the machines into the material and information flow places a big demand on the control system.

An NC program is created for every phase of a job operation, enabling various parts to be processed on a machine in any sequence. Job distribution to the individual machines is determined according to urgency and availability of the resources.

Providing the required resources at the scheduled time (materials and tools, providing the machines with the required NC programs) as well as the job plan and organization is monitored and controlled by a computer-aided control system. This system is frequently a part of a comprehensive CIM concept that coordinates all production-related actions.

Processing Tubes. Flexible manufacturing cells are not only used to process flat sheets, but also, for example, tubes and profile sections. The laser system for tubes depicted in Figure 24.49 is integrated into a handling system, which automatically feeds the machine with raw materials. Tubes up to 14 m in length are initially sectioned off automatically and then fed to the machine. The finished parts, which are ejected from the machine, are sorted according to their length and deposited into containers. Sensors monitor the individual processes to ensure trouble-free operation. The operator's tasks are for the most part limited to eliminating disturbances and maintaining the installation.

From Ordering to the Finished Product Despite very sophisticated automation and highly qualified personnel, the productivity of a manufacturing cell does not meet expectations if the machine cannot be integrated into the information flow of the entire company. The concept "information flow" should be understood as a comprehensive term; simply processing a sheet on an NC-controlled machine is not what is meant. The flexible manufacturing cell is rather

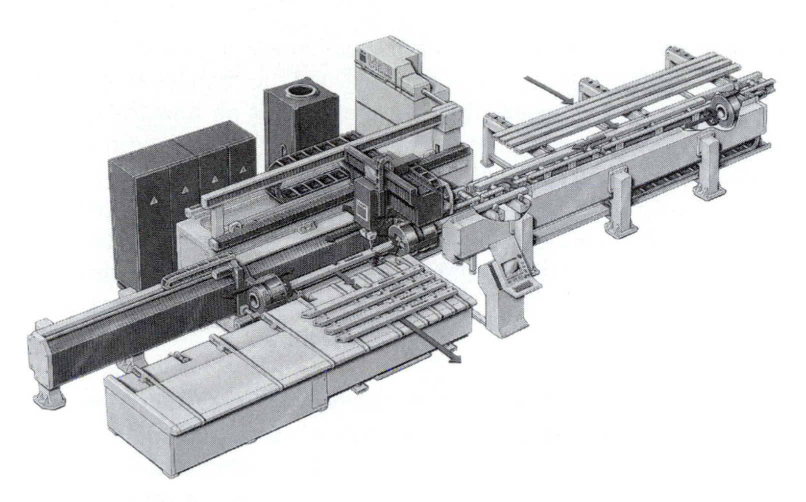

FIGURE 24.49 A laser cutting installation for processing pipes and profile sections. (Courtesy of TRUMPF USA)

embedded in a complex network of business functions on the one side and technical work preparation on the other.

If the material and information flow are organized in this network using computers, we then refer to it as Computer Integrated Manufacturing (CIM): CIM does not refer to a defined structure; it is a concept which, depending on the type and size of the company, can have a totally different form. Consequently, the workflow in a flexible manufacturing cell for sheet metal processing can only be described as an example.

Production Planning and Control. When an NC program is started on a machine for processing a workpiece, the production order will have already completed an in-depth planning stage that begins when the customer places an order. The designer then creates a drawing with a CAD system, which is input into the program system in order to create a NC program.

All the data from this technical work preparation that is relevant to production planning and control are adopted for a production and planning control system, which generates a work plan. It contains a sequence of all the individual work steps in which a product is produced. The work plan assigns the resources required to each individual work process, for example, tools, materials, work place, and NC program. The work plan is thus not the complete picture. It must be implemented. The machine or equipment must be operated to produce the part and the part or product must be inspected and finally shipped to the customer.

The use of resources is optimized automatically. For example, the sequence of the production jobs for a punching machine is arranged in such a way that it requires as few tool changes as necessary. The system calls on calculations performed for the number of required tool changes for each possible sequence by comparing the current setup of the machine with the tools required for the following NC programs.

Each manufacturing cell has a screen on which the current production plan can be viewed. This allows the machine operator to alter the sequence of jobs on short notice. Once a job has been selected, the NC program is then transferred on-line from the program archive to the machine control. The times when NC programs had to be input with punched tape are, as far as sheet metal processing is concerned, a remnant of the past.

Collecting Production Data. To correctly diagnose weak points and sources of errors, production data is collected continuously as soon as an operator has started work at a manufacturing cell. All interruptions in production are registered, processing times are measured, the number of produced pieces calculated, and the changes recorded in the storage system. The status information provided automatically depends on how the machines are equipped. If, for example, the storage and retrieval unit responsible for moving the pallets in a high storage bay back and forth is equipped with a weighing unit, material consumption and an inventory change in inventory are recorded without the operator having to enter these amounts. The collected production data can be evaluated statistically from various standpoints and displayed in diagrams, so that defective NC programs, material defects, tool wear or unusually long downtimes can be specifically analyzed. In this way, the production data flow back into production planning.

Interface High Storage Bays High-storage bays in automation concepts are not only intended for storing materials. They are also a central buffer in a combination of independent sheet metal processing centers, which take on the function of a logistic center.

In the foreground of Figure 24.50, punching machines form the core of the two manufacturing cells. The material flow is regulated in the following way: As soon as a job is started on the machine, the storage and retrieval unit withdraws the pallet with the raw materials and transfers it to a relocating station. The station cart delivers the pallet to an area where the automatic loading and unloading systems, the handling device, delivers individual sheets to the machine. The machine has a tool storage containing a sufficient number of tools to process several jobs in succession. In most cases, it is then no longer necessary to retool. The punched finished parts are then removed from the scrap skeleton with the handling device and sorted onto a pallet that is driven back to the high storage bay by the station cart. Waste parts and scrap skeletons are placed onto separate pallets or into waste crates.

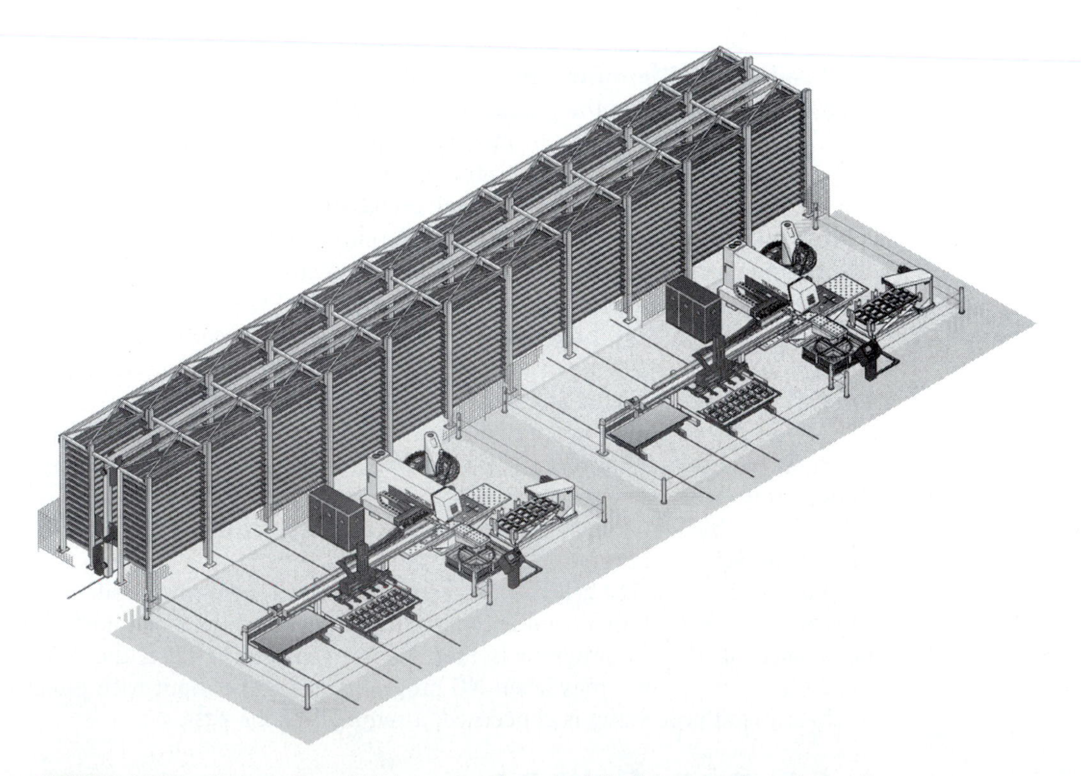

FIGURE 24.50 Integration of high storage bays and flexible manufacturing cells. (Courtesy of TRUMPF USA)

All processing steps are carried out without intervention on part of the operating personnel. The entire system is monitored by sensors and controlled by computer.

Are Skilled Workers No Longer Needed? When F.W. Taylor set about to revolutionize production organization in the automotive industry by introducing the principle of labor division at the assembly line, he was able to call on a seemingly inexhaustible reservoir of unskilled, inexpensive workers. At that time, the assembly line was considered the epitome of automation in the industrial production process. This view changed completely after the Second World War, due to two parallel developments: at first, wages increased and became a central factor for calculating costs. Second, the degree of automation for production procedures rose so quickly that the increase in wages was more than compensated for.

The more automation progressed, the more ambitious the goals became, even in sheet metal processing. This led to the notion harbored by many industrialists that, one day, fully automated equipment would even replace skilled workers. Experience showed, however, that the possibilities of automating manufacturing cells had been completely overestimated. Even when considering the high standards for process safety that have been achieved in the meantime, technical and organizational errors and failures cannot be completely eliminated. The scenario of fully automated production without human intervention is still only a vision, particularly for flexible processing centers. In practice, the personnel saved through the automating process are mainly used for process optimization.

The qualifications profile of a machine operator reveals the picture of a comprehensively qualified skilled worker who is not just competent in the world of NC programs and numerical control. It is his experience in the field of processing technology that contributes to the steady improvement of the manufacturing process and, in particular, increases the utilization ratio. The qualified operator is the one who guarantees high machine usage and long running times, without which expensive automated systems would not be profitable.

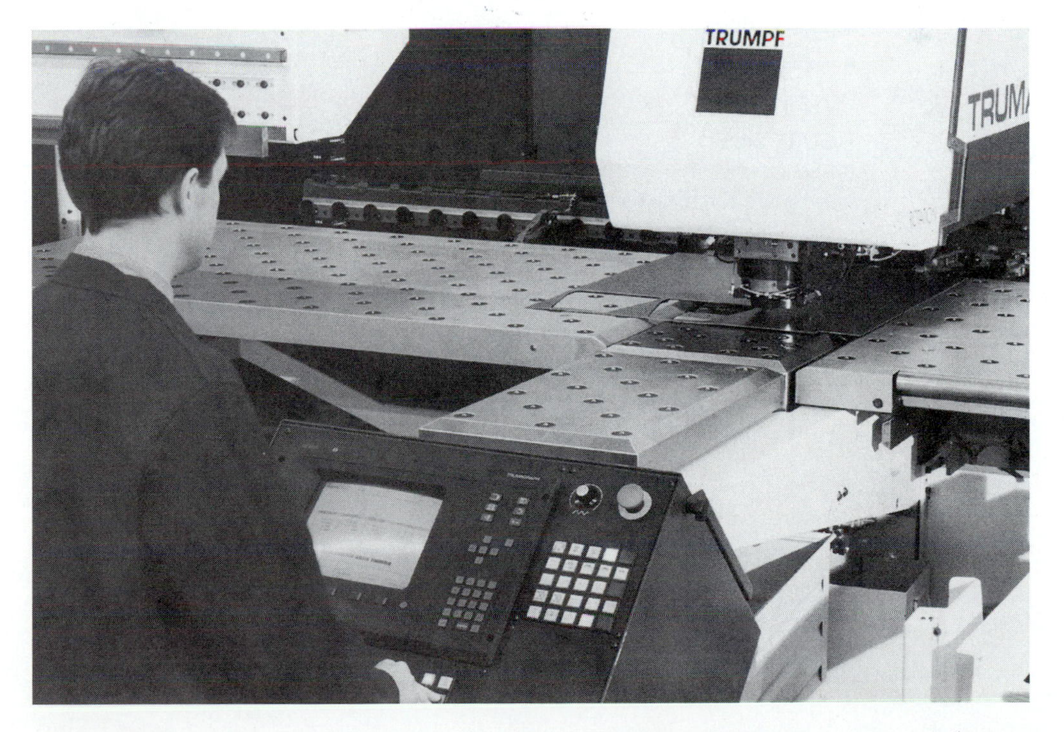

FIGURE 24.51 Highly skilled workers are required to operate the equipment. (Courtesy of TRUMPF USA)

24.4.5 Plasma Arc Machining

Plasma Arc Machining (PAM) is accomplished by constricting an electrical arc with a nozzle to increase and intensify the ensuing jet's voltage and temperature. A surrounding primary gas of nitrogen and argon is ionized, which develops a concentrated, high-velocity, intensely hot plasma jet. A secondary gas surrounds the plasma jet and acts as to shield the jet from impingement of air molecules.

The plasma jet raises metal temperatures to the melting point while the high-velocity gas stream blows the molten material away. Oxygen is injected through an orifice of the nozzle in some applications to add another source of energy and further speed the process.

Water injection is used to constrict the arc, in some applications, to produce a square-edged cut, narrower kerf, and increased nozzle life at the expense of cutting speed. Schematic views of a conventional plasma arc torch and also of oxygen and water-injection plasma torches are shown in Figure 24.52.

Equipment

Machines. Plasma arc torches are used with specially designed machines that resemble turret presses, having a rotating table for the workpiece over which is mounted the torch. Many conventional punch presses are also fitted with plasma torches.

Tools. Plasma arc torches, which are the cutting tools of the process, are available in several sizes. One development provides one size of torch head that is used with a variety of nozzles and electrodes to suit the application. Portable, handheld plasma arc torches are also available.

Applications

- Any metal
- Any hardness

FIGURE 24.52
(a) Schematic of plasma arc torch.

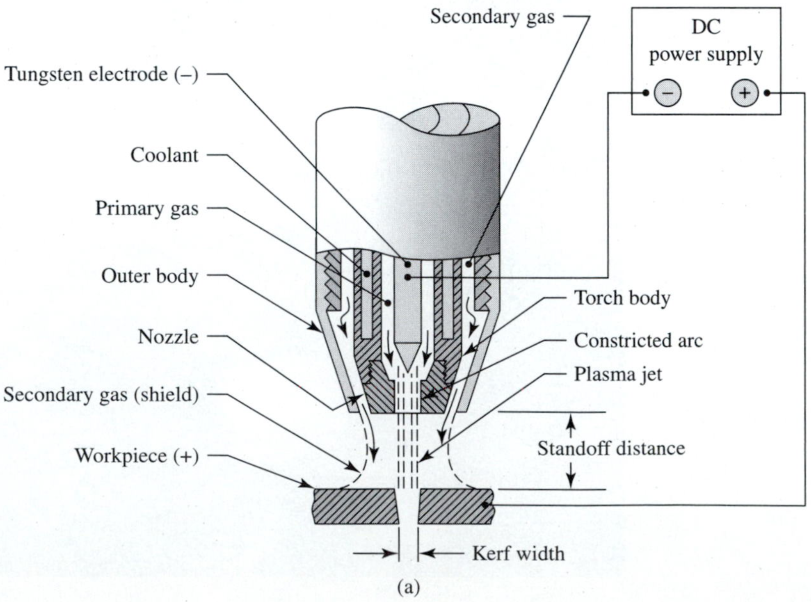

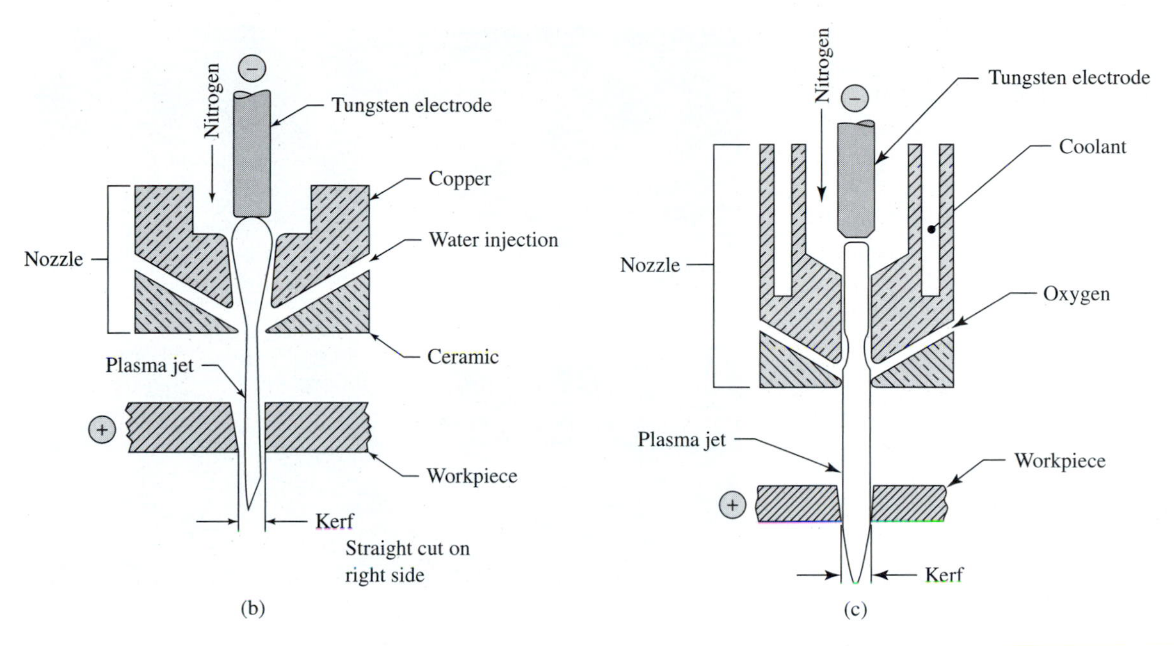

FIGURE 24.52 Continued (b) Water-injection plasma torch. (c) Oxygen injection increases total heat of plasma arc torch.

- Cutting process capabilities
  - Straight lines
  - Circles
  - Irregular shapes
  - Grooves
- Hole piercing
  - Repetitive
  - Good quality
- Stack cutting
  - Aluminum sheet
  - Stainless steel sheet
  - Carbon steels tend to weld.
- Edge preparation
  - Beveled edges can be provided.

Advantages

- Fast—highly productive
- Good quality
- Low operating costs
- Easily automated

FIGURE 24.53 Plasma arc torch in operation. (Courtesy ESAB Cutting Systems)

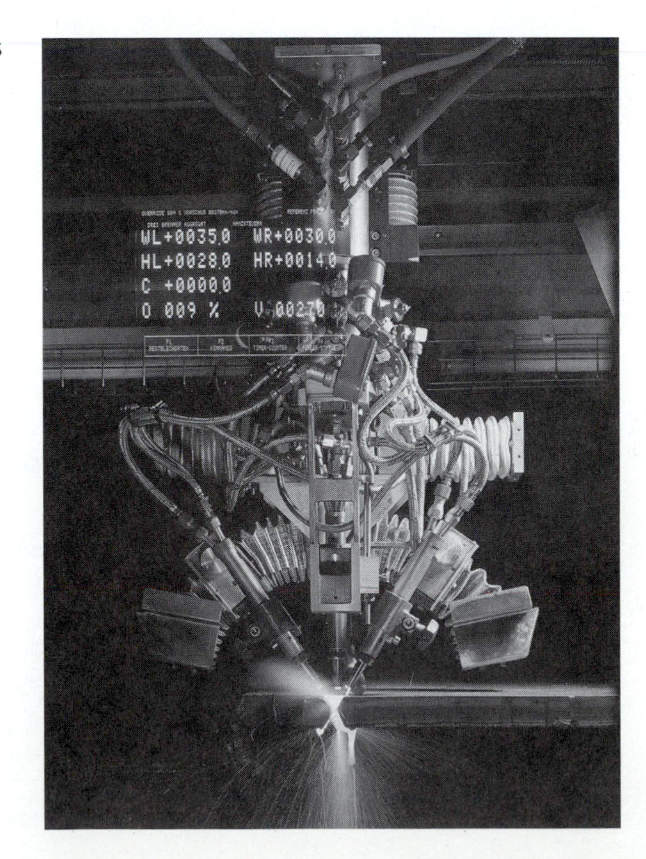

FIGURE 24.54 Triple torch station bevels edges of plate. (Courtesy ESAB Cutting Systems)

Disadvantages

- Metallurgical changes of material from heat (hardening of material and uneven top surface)

A plasma arc torch is shown in operation in Figure 24.53.

Torches and Applications The triple torch contour station shown in Figure 24.54 is ideal for preparing edge surfaces for welding. The cut surface can be contoured to the exact shape required for a good weld, eliminating the need for secondary operations. Lateral and angle setting of the outside torches can be controlled automatically. An example of the welded edge produced by a programmable oxyfuel triple torch contour station is shown in Figure 24.55.

Offering cutting widths up to 70 feet, specialized plasma cutting machines such as that illustrated in Figure 24.56 are well suited for industries such as shipbuilding that require extra-large cutting thicknesses and widths.

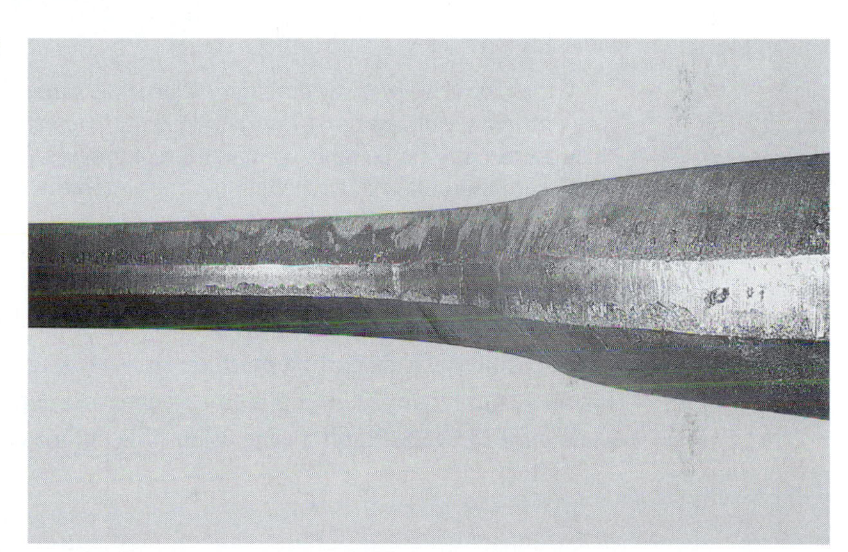

FIGURE 24.55 Example of plasma arc cut through welded part. (Courtesy ESAB Cutting Systems)

FIGURE 24.56 Plasma arc cutting machine capable of processing large sizes of heavy steel plate. (Courtesy ESAB Cutting Systems)

FIGURE 24.57 Plasma arc torch station combines cutting and marking processes. (Courtesy ESAB Cutting Systems)

A massive drive system combines AC brushless motors and planetary gearboxes to allow accurate cutting speeds up to 600 in. per minute. The extremely rugged structure enables them to carry almost any conceivable combination of process stations. The torch station shown in Figure 24.57 combines plasma, oxyfuel, and punch marking capabilities.

Gantry-Type Plasma Cutting Machines Most plasma arc cutting machines use a gantry-type system to program and move plasma arc torches along the paths required to produce the required shapes. Typical of gantrys in present use are the following:

- The gantry plasma cutting machine shown in Figure 24.58 is a heavy-duty 2-axis machine.
- Shown in Figure 24.59 is a high-speed plasma cutting machine for smaller parts.
- Figure 24.60 illustrates a general-purpose plasma cutting machine with shape cutting capabilities.

FIGURE 24.58 Heavy-duty 2-axis gantry plasma cutting machine. (Courtesy ESAB Cutting Systems)

FIGURE 24.59 High-speed gantry plasma cutting machine for smaller parts. (Courtesy ESAB Cutting Systems)

FIGURE 24.60 General-purpose plasma cutting machine with shape cutting capabilities. (Courtesy ESAB Cutting Systems)

24.5 REVIEW QUESTIONS AND PROBLEMS

1. Define a press.
2. Describe a press cycle.
3. Enumerate the advantages of an O.B.I. press.
4. What is a horning press, and how is it used?
5. List the uses and applications of power press brakes.
6. Describe the construction of a straight-side press.
7. Discuss the applications of a straight-side press.
8. How does a dieing machine differ from a conventional press?
9. What is the main feature of a knuckle press? What applications (processes) is it most suitable for?
10. State the principal uses of double-action and triple-action presses.
11. What are some of the applications of a die cushion?
12. Discuss the uses of
    a. Reels.
    b. Stock feeders.
    c. Scrap cutters.

13. Discuss areas where operator safety must be considered and appropriate plans and actions taken.
14. Describe the action of a turret press.
15. What is the main use of a turret press?
16. What are the advantages of a plasma cutting machine?
17. State the advantages of laser cutting.
18. What materials are most suitable for the water jet cutting process?

PART VII

MULTIPOINT MACHINING

INTRODUCTION

Machine tools are built to produce cylindrical surfaces, holes, plane surfaces, irregular contours, gear teeth, and so on. This part will discuss multipoint machining operations such as milling, broaching, sawing, and grinding, as well as finishing operations such as lapping and honing.

The more traditional multipoint machining operations are shown in Figure VII.1 and are described throughout this part. It has been shown throughout this text that machine tools can be designed for general-purpose, high-production, or special-purpose applications.

General-Purpose Machine Tools. These are designed to be adapted quickly and easily to a large variety of operations on many different kinds of parts. They are used extensively in jobbing shops, repair shops, and tool and die shops. Sometimes these machine tools are called job-shop machine tools.

Production Machine Tools. These tools are designed to perform an operation or a sequence of operations in a repetitive manner in order to achieve a rapid output of machined parts at minimum cost. They can be set up to machine a variety of different parts. However, their operation is economical only when the quantity of parts to be machined is relatively large.

Special-Purpose Machine Tools. These tools are designed to perform one operation or a sequence of operations repetitively on a specific part. These machines are often automatic, and one operator can sometimes be responsible for more than one machine. They are used in mass-production shops such as in the automotive industry.

Computers have enhanced automated manufacturing; machining is no exception. Numerical control (NC) of machine tools has been available for many years. Today typical machine tools of all types are equipped with their own computer numerical controls (CNC), and almost every machining process can now be efficiently automated with an exceptional degree of accuracy, reliability, and repeatability.

Not only machine tools have changed; new methods of material removal are quite different from the traditional chip-producing processes. They include the use of lasers, electrical energy, electrochemical processes, ultrasound, high-pressure water jets, and high-temperature plasma arcs as material removal tools. These new processes are discussed in detail in Chapter 29, "Nontraditional Machining."

Operation	Diagram	Type of Machines	Characteristics
Milling (horizontal)	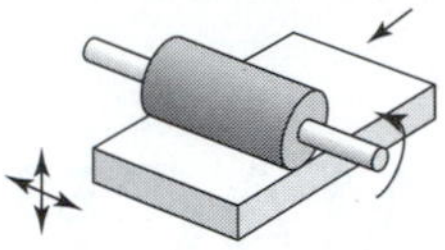	Horizontal milling machine	Cutter rotates and cuts on periphery. Work feeds into cutter and can be moved in these axes.
Face milling	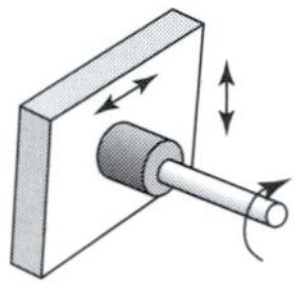	Horizontal mill, profile mill, and machining center	Cutter rotates to cut on its end and periphery of vertical workpiece.
Vertical (end) milling	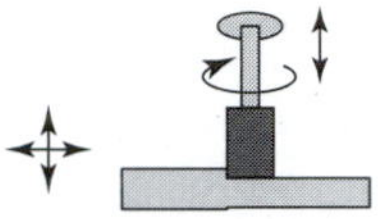	Vertical milling machine, die sinker, machining center	Cutter rotates to cut on its end and periphery, work moves on three axes for feed or position. Spindle also moves up or down.
Broaching	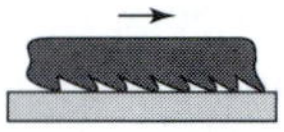	Vertical broaching machine, horizontal broaching machine	Workpiece is held stationary while a multitooth cutter is moved across the surface. Each tooth in the cutter cuts progressively deeper than the previous one.
Horizontal sawing (cutoff)	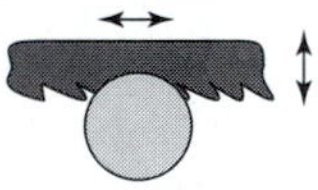	Horizontal bandsaw, reciprocating cutoff saw	Work is held stationary while the saw cuts either in one direction as in bandsawing or it reciprocates while being fed downward into the work.
Horizontal spindle surface grinding	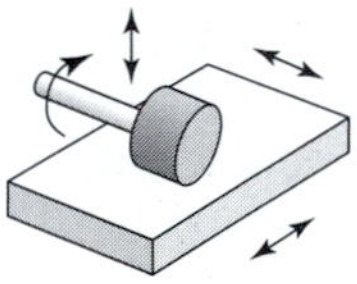	Surface grinders, specialized industrial grinding machines	The rotating grinding wheel can be moved up or down to feed into the workpiece. The table, which is made to reciprocate, holds the work and can also be moved crosswise.
Cylindrical grinding	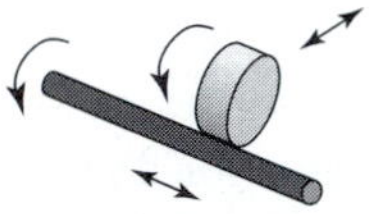	Cylindrical grinders, specialized industrial grinding machines	The rotating grinding wheel contacts a turning workpiece that can reciprocate from end to end. The wheelhead can be moved into the work or away from it.
Centerless grinding	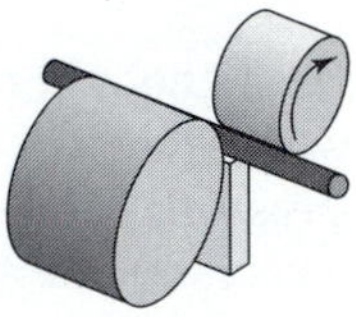	Centerless grinder	Work is supported by a workrest between a large grinding wheel and a smaller feed wheel.

FIGURE VII.1 The more traditional multipoint machining operations are shown.

CHAPTER TWENTY-FIVE

MILLING OPERATIONS

CHAPTER OVEVRVIEW

- Introduction
- Types of Milling Cutters
  - High-Speed Steel Milling Cutters
  - Milling Cutter Nomenclature
  - Indexable Milling Cutters
  - Milling Cutter Geometry
- Types of Milling Machines
  - Column and Knee Machines
  - Bed-Type Milling Machines
  - Special-Purpose Milling Machines
  - Machining Centers
- Milling Machine Attachments and Accessories
  - Special Milling Heads
  - Vises and Fixtures
  - Arbors, Collets, and Toolholders
- Basic Milling Operations
  - Operating Conditions
  - Direction of Milling Feed
- Types of Milling Operations
- Review Questions and Problems

25.1 INTRODUCTION

The two basic cutting-tool types used in the metalworking industry are of the single-point and multipoint design, although they may differ in appearance and in their methods of application. Fundamentally, they are similar in that the action of metal cutting is the same regardless of the type of operation. By grouping a number of single-point tools in a circular holder, the familiar **milling cutter** is created.

FIGURE 25.1 A typical milling operation; the on-edge insert design is being used. (Courtesy Ingersoll Cutting Tool Co.)

Milling is a process of generating machined surfaces by progressively removing a predetermined amount of material or stock from the workpiece, which is advanced at a relatively slow rate of movement or feed to a milling cutter rotating at a comparatively high speed. The characteristic feature of the milling process is that each milling cutter tooth removes its share of the stock in the form of small individual chips. A typical face milling operation is shown in Figure 25.1. The on-edge insert design is used (Fig. 2.30).

25.2 TYPES OF MILLING CUTTERS

The variety of milling cutters available for all types of milling machines helps make milling a very versatile machining process. Cutters are made in a large range of sizes and of several different cutting-tool materials. Milling cutters are made from High-Speed Steel (HSS), others are carbide tipped, and many are replaceable or indexable inserts. The three basic milling operations are shown in Figure 25.2. Peripheral and end milling cutters will be discussed below. **Face milling cutters** are usually indexable and will be discussed later in this chapter.

25.2.1 High-Speed Steel Milling Cutters

A High-Speed Steel (HSS) shell end-milling cutter is shown in Figure 25.3, and other common HSS cutters are shown in Figure 25.4 and briefly described here.

Periphery Milling Cutters **Periphery milling cutters** are usually arbor mounted to perform various operations.

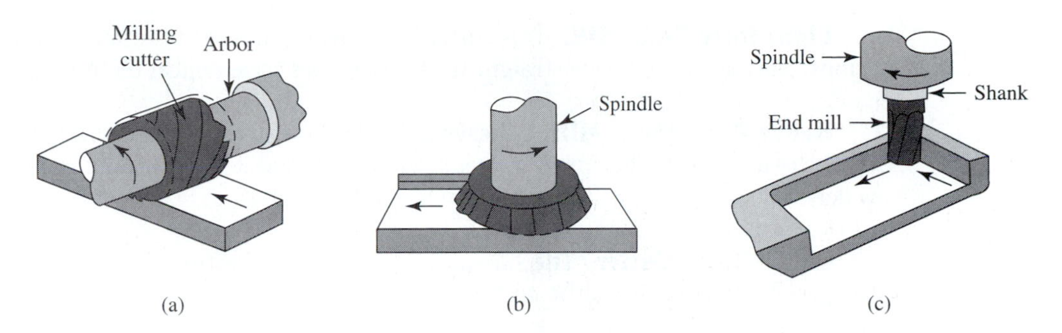

FIGURE 25.2 The three basic milling operations: (a) peripheral milling, (b) face milling, (c) end milling.

FIGURE 25.3 High-speed steel (HSS) shell end milling cutter. (Courtesy Morse Cutting Tools)

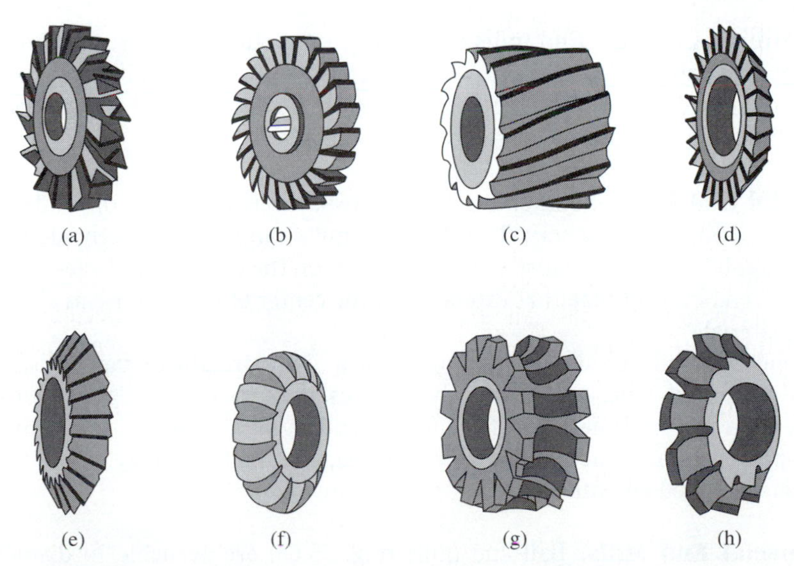

FIGURE 25.4 Common HSS milling cutters: (a) staggered-tooth cutter, (b) side milling cutter, (c) plain milling cutter, (d) single-angle milling cutter, (e) double-angle milling cutter, (f) convex milling cutter, (g) concave milling cutter, (h) corner rounded milling cutter.

Light-Duty Plain Mill. This cutter is a general-purpose cutter for peripheral milling operations. Narrow cutters have straight teeth; wide ones have helical teeth (Fig. 25.4c).

Heavy-Duty Plain Mill. A heavy-duty plain mill is similar to the light-duty mill except that it is used for higher rates of metal removal. To aid it in this function, the teeth are more widely spaced and the helix angle is increased to about 45°.

Side-Milling Cutter. The side-milling cutter has a cutting edge on the sides as well as on the periphery. This allows the cutter to mill slots (Fig. 25.4b).

Half-Side-Milling Cutter. This tool is the same as the one previously described except that cutting edges are provided on a single side. It is used for milling shoulders. Two cutters of this type are often mounted on a single arbor for straddle milling.

Stagger-Tooth Side Mill. This cutter is the same as the side-milling cutter except that the teeth are staggered so that every other tooth cuts on a given side of the slot. This allows deep, heavy-duty cuts to be taken (25.4a).

Angle Cutters. On angle cutters, the peripheral cutting edges lie on a cone rather than on a cylinder. A single or double angle may be provided (Fig. 25.4d and e).

Shell End Mill. The shell end mill has peripheral cutting edges plus face-cutting edges on one end. It has a hole through it for a bolt to secure it to the spindle (Fig. 25.3).

Form Mill. A form mill is a peripheral cutter whose edge is shaped to produce a special configuration on the surface. One example of this class of tool is the gear tooth cutter. The exact contour of the cutting edge of a form mill is reproduced on the surface of the workpiece (Fig. 25.4f, g, and h).

End-Milling Cutters

End mills can be used on vertical and horizontal milling machines for a variety of facing, slotting, and profiling operations. Solid end mills are made from high-speed steel or sintered carbide. Other types, such as shell end mills and **fly cutters**, consist of cutting tools that are bolted or otherwise fastened to adapters.

Solid End Mills. Solid end mills have two, three, four, or more flutes and cutting edges on the end and the periphery. Two-flute end mills can be fed directly along their longitudinal axis into solid material because the cutting faces on the end meet. Three- and four-fluted cutters with one end-cutting edge that extends past the center of the cutter can also be fed directly into solid material.

Solid end mills are double or single ended, with straight or tapered shanks. The end mill can be of the stub type, with short cutting flutes, or of the extra long type for reaching into deep cavities. On end mills designed for effective cutting of aluminum, the helix angle is increased for improved shearing action and chip removal, and the flutes may be polished. Various single- and double-ended end mills are shown in Figure 25.5.

Special End Mills. Ball end mills (Fig. 25.6a) are available in diameters ranging from 1/32 to $2\frac{1}{2}$ in. in single- and double-ended types. Single-purpose end mills such as Woodruff keyseat cutters, corner rounding cutters, and **dovetail cutters** (Fig. 25.6b) are used on both vertical and horizontal milling machines. They are usually made of high-speed steel and may have straight or tapered shanks.

FIGURE 25.5 Various single- and double-ended HSS end mills. (Courtesy The Weldon Tool Co.)

(a)

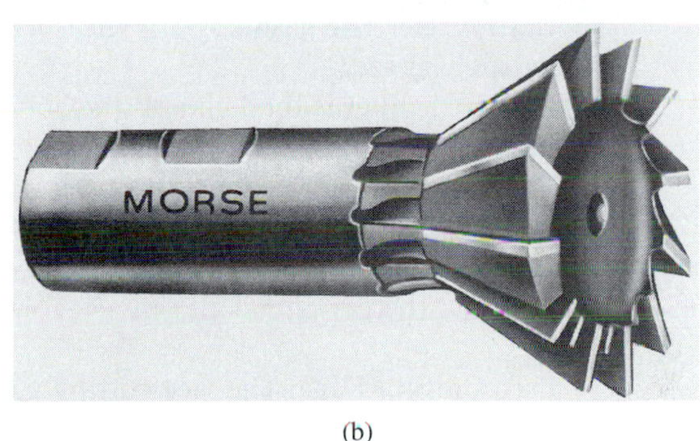

(b)

FIGURE 25.6 (a) Ball-nose end-milling cutters are available in diameters ranging from 1/32 to $2\frac{1}{2}$ inches. (Courtesy The Weldon Tool Co.) (b) HSS dovetail cutters can be used on both vertical and horizontal milling machines. (Courtesy Morse Cutting Tools)

25.2.2 Milling Cutter Nomenclature

As far as metal-cutting action is concerned, the pertinent angles on the tooth are those that define the configuration of the cutting edge, the orientation of the tooth face, and the relief to prevent rubbing on the land. The terms defined here and illustrated in Figure 25.7(a) and (b) are important and fundamental to milling cutter configuration.

Outside Diameter: The outside diameter of a milling cutter is the diameter of a circle passing through the peripheral cutting edges. It is the dimension used in conjunction with the spindle speed to find the cutting speed (SFPM).

Root Diameter: This diameter is measured on a circle passing through the bottom of the fillets of the teeth.

Tooth: The tooth is the part of the cutter starting at the body and ending with the peripheral cutting edge. Replaceable teeth are also called inserts.

Tooth Face: The tooth face is the surface of the tooth between the fillet and the cutting edge, where the chip slides during its formation.

Land: The area behind the cutting edge on the tooth, which is relieved to avoid interference, is called the land.

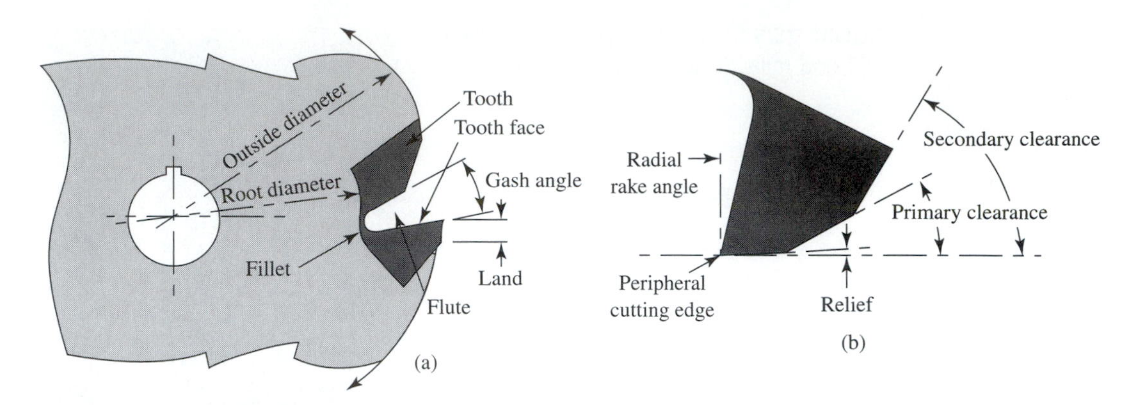

FIGURE 25.7 Milling cutter configuration: (a) plain milling cutter nomenclature; (b) plain milling cutter tooth geometry.

Flute: The flute is the space provided for chip flow between the teeth.

Gash Angle: The gash angle is measured between the tooth face and the back of the tooth immediately ahead.

Fillet: The fillet is the radius at the bottom of the flute, provided to allow chip flow and chip curling.

The terms defined here apply primarily to milling cutters, particularly to plain milling cutters. In defining the configuration of the teeth on the cutter, the following terms are important.

Peripheral Cutting Edge: The cutting edge aligned principally in the direction of the cutter axis is called the peripheral cutting edge. In peripheral milling, it is this edge that removes the metal.

Face-Cutting Edge: The face-cutting edge is the metal-removing edge aligned primarily in a radial direction. In side milling and face milling, this edge actually forms the new surface, although the peripheral cutting edge may still be removing most of the metal. It corresponds to the end-cutting edge on single-point tools.

Relief Angle: This angle is measured between the land and a tangent to the cutting edge at the periphery.

Clearance Angle: The clearance angle is provided to make room for chips, thus forming the flute. Normally two clearance angles are provided to maintain the strength of the tooth and still provide sufficient chip space.

Radial Rake Angle: The radial rake angle is the angle between the tooth face and a cutter radius, measured in a plane normal to the cutter axis.

Axial Rake Angle: The axial rake angle is measured between the peripheral cutting edge and the axis of the cutter, when looking radially at the point of intersection.

Blade Setting Angle: When a slot is provided in the cutter body for a blade, the angle between the base of the slot and the cutter axis is called the blade setting angle.

25.2.3 Indexable Milling Cutters

The three basic types of milling operations were introduced earlier. Figure 25.8 shows a variety of indexable milling cutters used in all three of the basic types of milling operations (Fig. 25.2).

Insert Clamping Methods There are a variety of clamping systems for indexable inserts in milling cutter bodies. The examples shown here cover the most popular methods now in use.

FIGURE 25.8 A variety of indexable milling cutters. (Courtesy Ingersoll Cutting Tool Co.)

Wedge Clamping. Milling inserts have been clamped using wedges for many years in the cutting-tool industry. This principle is generally applied in one of the following ways: Either the wedge is designed and oriented to support the insert as it is clamped, or the wedge clamps on the cutting face of the insert, forcing the insert against the milling body. When the wedge is used to support the insert, all of the force generated during the cut must be absorbed by the wedge. This is why wedge clamping on the cutting face of the insert is preferred, because this method transfers the loads generated by the cut through the insert and into the cutter body. Both of the wedge clamping methods are shown in Figure 25.9.

The wedge clamp system however, has two distinct disadvantages. First, the wedge covers almost half of the insert cutting face, thus obstructing normal chip flow while producing premature cutter body wear; second, high clamping forces, which cause clamping element and

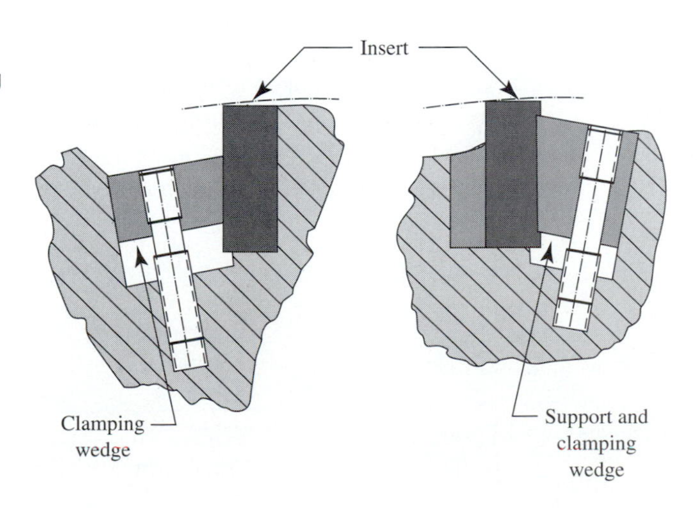

FIGURE 25.9 Two methods of wedge clamping indexable milling cutter inserts.

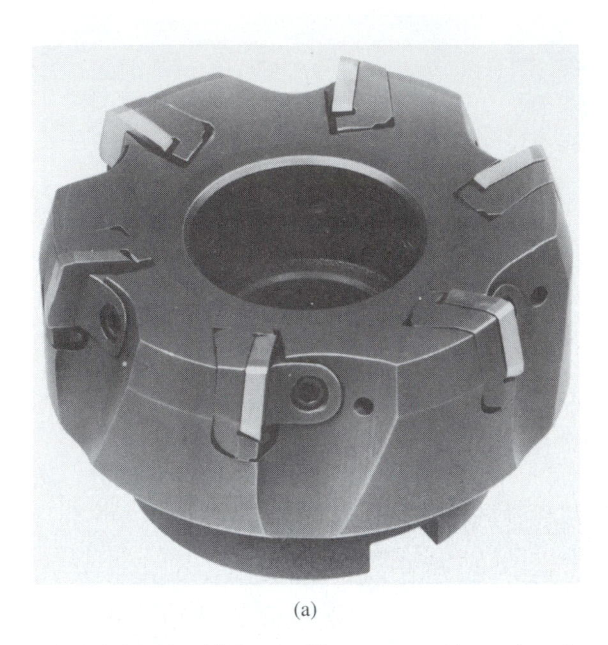

(a)

(b)

FIGURE 25.10 (a) Face milling cutter with wedge clamped indexable inserts. (Courtesy Iscar Metals, Inc.) (b) Face milling cutters with indexable inserts and wedge clamped milling cutter cartridges. (Courtesy Greenleaf Corp.)

cutter body deformation, can and often will result. The excessive clamping forces can cause enough cutter body distortion that in some cases when loading inserts into a milling body, the last insert slot will have narrowed to a point where the last insert will not fit into the body. When this occurs, several of the other inserts already loaded in the milling cutter are removed and reset. Wedge clamping can be used to clamp individual inserts (Fig. 25.10a) or indexable and replaceable milling cutter cartridges as shown in Figure 25.10b.

Screw Clamping. This method of clamping is used in conjunction with an insert that has a pressed countersink or counterbore. A torque screw is often used to eccentrically mount and force the insert against the insert pocket walls. This clamping action results either from offsetting the centerline of the screw toward the back walls of the insert pocket or from drilling and tapping the mounting hole at a slight angle, thereby bending the screw to attain the same type of clamping action. The screw clamping method for indexable inserts is shown in Figure 25.11.

Screw clamping is excellent for small-diameter end mills where space is at a premium. It also provides an open unhampered path for chips to flow free of wedges or any other obstructive hardware. Screw clamping produces lower clamping forces than those attained with the

FIGURE 25.11 Screw clamping method for indexable inserts.

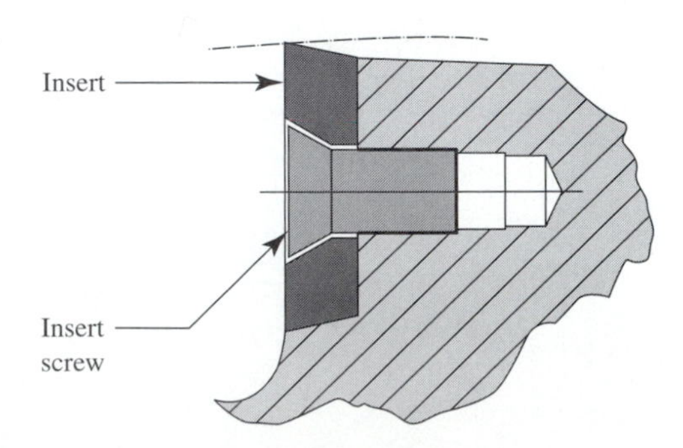

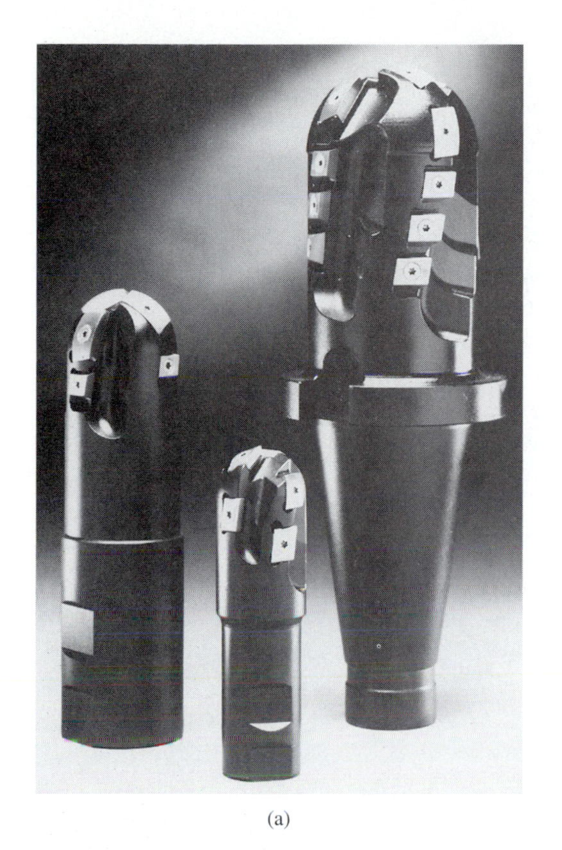

(a)

(b)

FIGURE 25.12 (a) Indexable-insert ball-nosed milling cutters using the screw clamping method. (Courtesy Ingersoll Cutting Tool Co.) (b) Slotting cutters and face milling cutters with screw-on-type indexable inserts. (Courtesy Duramet Corp.)

wedge clamping system. However, when the cutting-edge temperature rises significantly, the insert frequently expands and causes an undesirable retightening effect, increasing the **torque** required to unlock the insert screw. The screw clamping method can be used on indexable ball milling cutters (Fig. 25.12a) or on indexable-insert slotting and face milling cutters as shown in Figure 25.12(b).

25.2.4 Milling Cutter Geometry

There are three industry standard milling cutter geometries: double negative, double positive, and positive/negative. Each cutter geometry type has certain advantages and disadvantages, which must be considered when selecting the right milling cutter for the job. Positive-rake and negative-rake milling cutter geometries are shown in Figure 25.13.

Double-Negative Geometry A double-negative milling cutter uses only negative inserts held in a negative pocket. This provides cutting-edge strength for roughing and severe interrupted cuts. When choosing a cutter geometry it is important to remember that a negative insert tends to push the cutter away, exerting considerable force against the workpiece. This could be a problem when machining flimsy or lightly held workpieces, or when using light machines. However, this tendency to push the work down, or push the cutter away from the workpiece, may be beneficial in some cases because the force tends to "load" the system, which often reduces chatter.

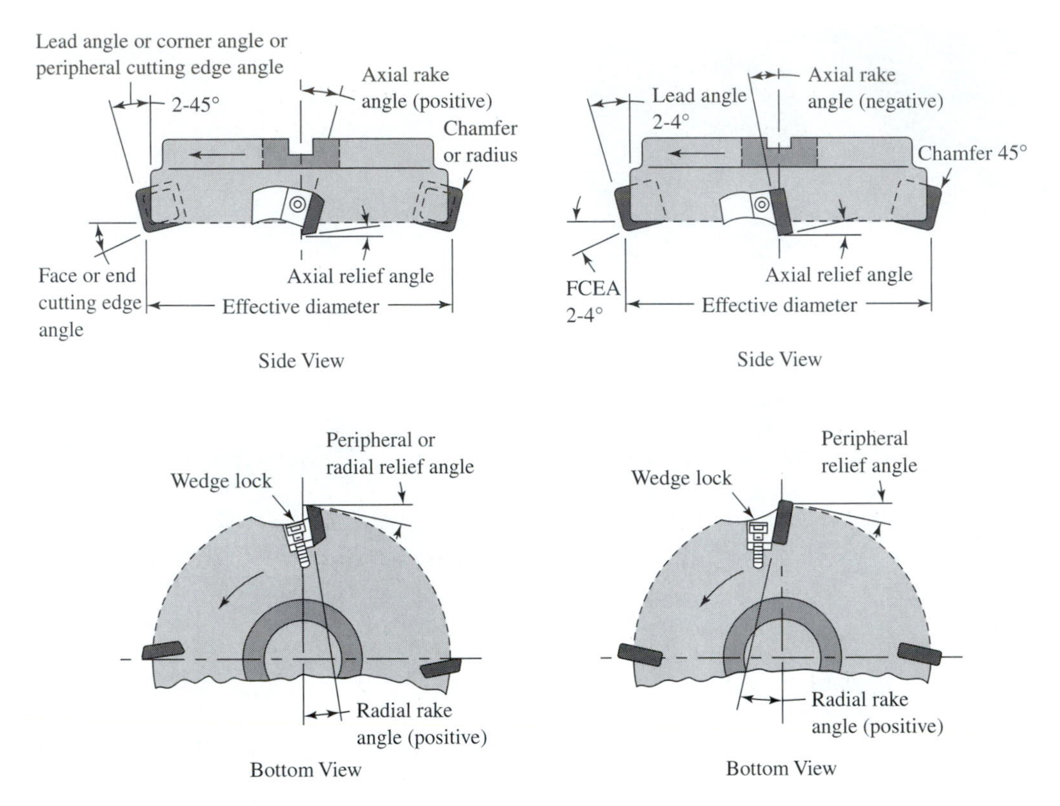

FIGURE 25.13 Positive-rake and negative-rake face milling cutter nomenclature.

Double-Positive Geometry Double-positive cutters use positive inserts held in positive pockets to provide the proper clearance for cutting. Double-positive cutter geometry provides for low-force cutting, but the inserts contact the workpiece at their weakest point, the cutting edge. In positive-rake milling, the cutting forces tend to lift the workpiece or pull the cutter into the work. The greatest advantage of double-positive milling is free cutting. Less force is exerted against the workpiece, so less power is required. This can be especially helpful with machining materials that tend to work harden.

Positive/Negative Geometry Positive/negative cutter geometry combines positive inserts held in negative pockets. This provides a positive axial rake and a negative radial rake, and as with double-positive inserts, this provides the proper clearance for cutting. In the case of positive/negative cutters, the workpiece is contacted away from the cutting edge in the radial direction and on the cutting edge in the axial direction. The positive/negative cutter can be considered a low-force cutter because it uses a free-cutting positive insert. On the other hand, the positive/negative cutter provides contact away from the cutting edge in the radial direction, the feed direction of a face mill.

In positive/negative milling, some of the advantages of both positive and negative milling are available. Positive/negative milling combines the free cutting or shearing away of the chip of a positive cutter with some of the edge strength of a negative cutter.

Lead Angle The lead angle (Fig. 25.14) is the angle between the insert and the axis of the cutter. Several factors must be considered to determine which lead angle is best for a specific operation. First, the lead angle must be small enough to cover the depth of cut. The greater the

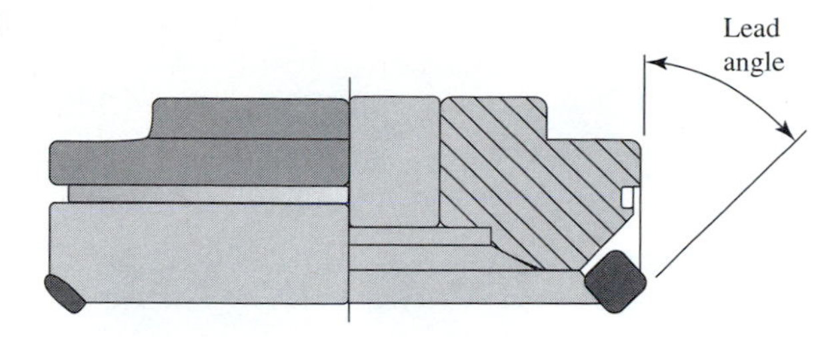

FIGURE 25.14 Drawing of a positive lead angle on an indexable-insert face milling cutter.

lead angle, the less the depth of cut that can be taken for a given size insert. In addition, the part being machined may require a small lead angle in order to clear a portion or form a certain shape on the part. As the lead angle increases, the forces change toward the direction of the workpiece. This could cause deflections when machining thin sections of the part.

The lead angle also determines the thickness of the chip. The greater the lead angle for the same feed rate or chip load per tooth, the thinner the chip becomes. As in single-point tooling, the depth of cut is distributed over a longer surface of contact. Therefore, lead-angle cutters are recommended when maximum material removal is the objective. Thinning the chip allows the feedrate to be increased or maximized.

Lead angles can range from zero to 85°. The most common lead angles available on standard cutters are 0, 15, 30 and 45°. Lead angles larger than 45° are usually considered special, and are used for very shallow cuts for fine finishing or for cutting very hard work materials.

Milling cutters with large lead angles also have greater heat dissipating capacity. Extremely high temperatures are generated at the insert cutting edge while the insert is in the cut. Carbide, as well as other tool materials, often softens when heated, and when a cutting edge is softened it will wear away more easily. However, if more of the tool can be employed in the cut, as in the case of larger lead angles, the tool's heat dissipating capacity will be improved, which, in turn, improves tool life. In addition, as lead angle is increased, axial force is increased and radial force is reduced, an important factor in controlling chatter.

The use of large lead-angle cutters is especially beneficial when machining materials with scaly or work-hardened surfaces. With a large lead angle, the surface is spread over a large area of the cutting edge. This reduces the detrimental effect on the inserts, extending tool life. Large lead angles will also reduce burring and breakout at the workpiece edge.

The most obvious limitation on lead-angle cutters is part configuration. If a square shoulder must be machined on a part, a 0° lead angle is required. It is impossible to produce a 0° lead-angle milling cutter with square inserts because of the need to provide face clearance. Often a near square shoulder is permissible. In this case a 3° lead-angle cutter may be used.

Milling Insert Corner Geometry

Indexable-insert shape and size were discussed in Chapter 2. Selecting the proper corner geometry is probably the most complex element of insert selection. A wide variety of corner styles is available. The corner style chosen will have a major effect on surface finish and insert cost. Figure 25.15(a) shows various sizes and shapes of indexable milling cutter inserts.

Nose Radius. An insert with a nose radius is generally less expensive than a similar insert with any other corner geometry. A nose radius is also the strongest possible corner geometry because it has no sharp corners where two flats come together, as in the case of a chamfered corner. For these two reasons alone, a nose radius insert should be the first choice for any application where it can be used.

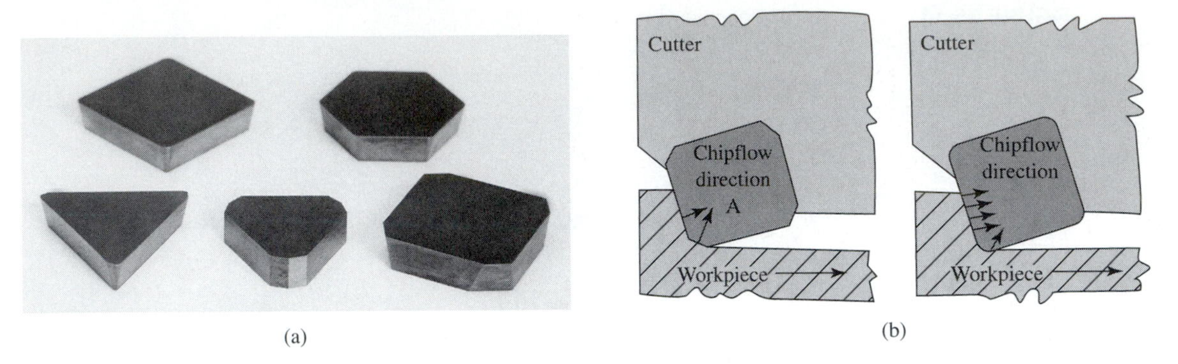

(a) (b)

FIGURE 25.15 (a) Various sizes and shapes of indexable milling cutter inserts. (Courtesy American National Carbide Co.) (b) Indexable milling cutter insert chip flow directions are shown.

Inserts with nose radii can offer tool life improvement when they are used in 0 to 15° lead-angle cutters, as shown in Figure 25.15(b). When a chamfer is used, as in the left drawing, the section of the chip formed above and below point *A*, will converge at point *A*, generating a large amount of heat at that point, which will promote faster than normal tool wear. When a radius insert is used, as shown in the right drawing, the chip is still compressed, but the heat is spread more evenly along the cutting edge, resulting in longer tool life.

The major disadvantage of an insert with a nose radius is that the surface finish it produces is generally not as good as other common corner geometries. For this reason, inserts with nose radii are generally limited to roughing applications and applications where a sweep wiper insert is used for the surface. A sweep wiper is an insert with a very wide flat edge or a very large radiused edge that appears to be flat. There is usually only one wiper blade used in a cutter, and this blade gets its name from its sweeping action, which blends the workpiece surface to a very smooth finish.

Inserts with nose radii are not available on many double-positive and positive/negative cutters because the clearance required under the nose radius is different from that needed under the edge. This clearance difference would require expensive grinding procedures, which would more than offset the other advantages of nose radius inserts.

Chamfer. There are two basic ways in which inserts with a corner chamfer can be applied. Depending both on the chamfer angle and the lead angle of the cutter body in which the insert is used, the land of the chamfer will be either parallel or angular (tilted) to the direction of feed, as shown in Figure 25.16(a).

Inserts that are applied with the chamfer angular to the direction of feed normally have only a single chamfer. These inserts are generally not as strong and the cost is usually higher than inserts which have a large nose radius. Angular-land chamfer inserts are frequently used for general-purpose machining with double-negative cutters.

Inserts designed to be used with the chamfer parallel to the direction of feed may have a single chamfer, a single chamfer and corner break, a double chamfer, or a double chamfer and corner break. The larger lands are referred to as primary facets and the smaller lands as secondary facets. The cost of chamfers, in relation to other types of corner geometries, depends upon the number of facets. A single-facet insert is the least expensive, and multiple-facet inserts cost more because of the additional grinding expense. Figure 25.16(b) shows two precision ground indexable milling cutter inserts. A face milling cutter with six square precision ground indexable milling cutter inserts was shown in Figure 25.10(a).

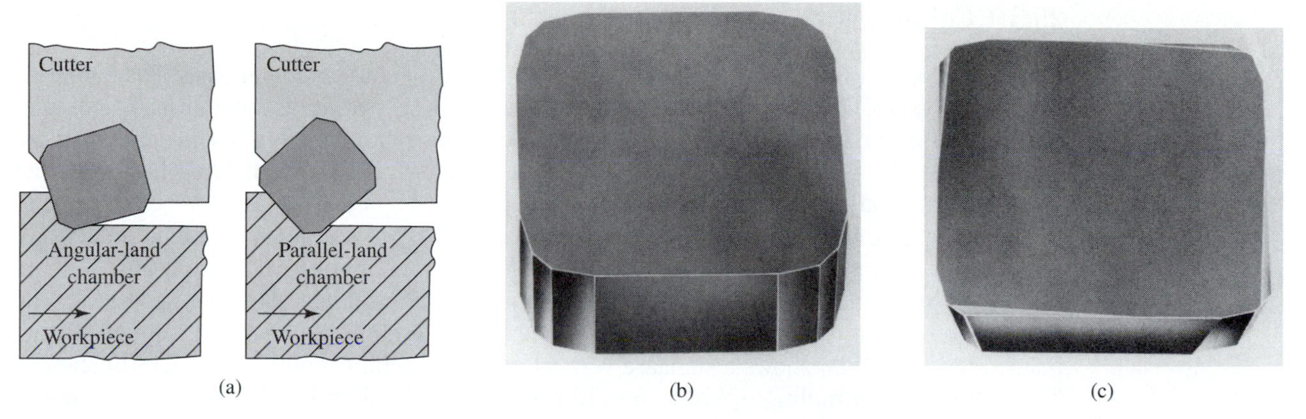

FIGURE 25.16 (a) Indexable milling cutter inserts with angular-land chamfer and parallel-land chamfer. (b and c) Two precision ground indexable milling cutter inserts. (Courtesy Iscar Metals, Inc.)

The greatest advantage of using inserts with the land parallel to the direction of feed is that, when used correctly, they generate an excellent surface finish. When the land width is greater than the advance per revolution, one insert forms the surface. This means that an excellent surface finish normally will be produced regardless of the insert face runout. Parallel-land inserts also make excellent roughing and general-purpose inserts for positive/negative and double-positive cutters. When a parallel-land chamfer insert is used for roughing, the land width should be as small as possible to reduce friction.

Sweep Wipers. Sweep wipers are unique in both appearance and application. These inserts have only one or two very long wiping lands. A single-sweep wiper is used in a cutter body filled with other inserts (usually roughing inserts) and is set approximately .003 to .005 in. higher than the other inserts, so that the sweep wiper alone forms the finished surface.

The finish obtained with a sweep wiper is even better than the excellent finish attained with a parallel-land chamfer insert. In addition, because the edge of the sweep wiper insert is exceptionally long, a greater advance per revolution may be used. The sweep wiper also offers the same easy setup as the parallel-land insert.

Sweep wiper inserts are available with both flat and crowned wiping surfaces. The crowned cutting edge is ground to a very large radius, usually from 3 to 10 in. The crowned cutting edges eliminate the possibility of sawtooth profiles being produced on the machined surface because the land is not exactly parallel to the direction of feed, a condition normally caused by spindle tilt. On the other hand, sweep wipers with flat cutting edges produce a somewhat better finish if the land is perfectly aligned with the direction of feed.

25.3 TYPES OF MILLING MACHINES

The many types of milling machines used in machine shops have been grouped into three general classes.

- Column and knee machines
- Bed-type milling machines
- Special-purpose milling machines

The common subtypes are also identified and discussed.

25.3.1 Column and Knee Machines

Column and knee milling machines are made in both vertical and horizontal types. The schematic diagrams (Fig. 25.17a and b) show both types of machines. Versatility is a major feature of knee and column milling machines. On a basic machine of this type, the table, saddle, and knee can be moved. Many accessories such as universal vises, rotary tables, and **dividing heads**, further increase the versatility of this type of machine.

Whether the machine is of the vertical or horizontal type, several components on all knee and column milling machines are similar, except for size and minor variations because of manufacturer's preference. These similarities are described in terms of general shape, geometric relationship to the rest of the machine, function, and the material from which the components are made. Figure 25.18(a) shows a standard vertical milling machine, and Figure 25.18(b) shows a 3-axis CNC vertical milling machine. A universal horizontal column and knee milling machine is shown in Figure 25.19.

Column The column, which is usually combined with the base as a single casting, is cast gray iron or ductile iron. The column houses the spindle, bearings, and the necessary gears, **clutches**, shafts, pumps, and shifting mechanisms for transmitting power from the electric motor to the spindle at the selected speed. The gears usually run in oil and are made of **carburized** alloy steel for long life. Some of the necessary controls are usually mounted on the side of the column.

The base is usually hollow and in many cases serves as a sump for the cutting fluid. A pump and filtration system can be installed in the base. The hole in the center of the base houses the support for the screw the raises and lowers the knee.

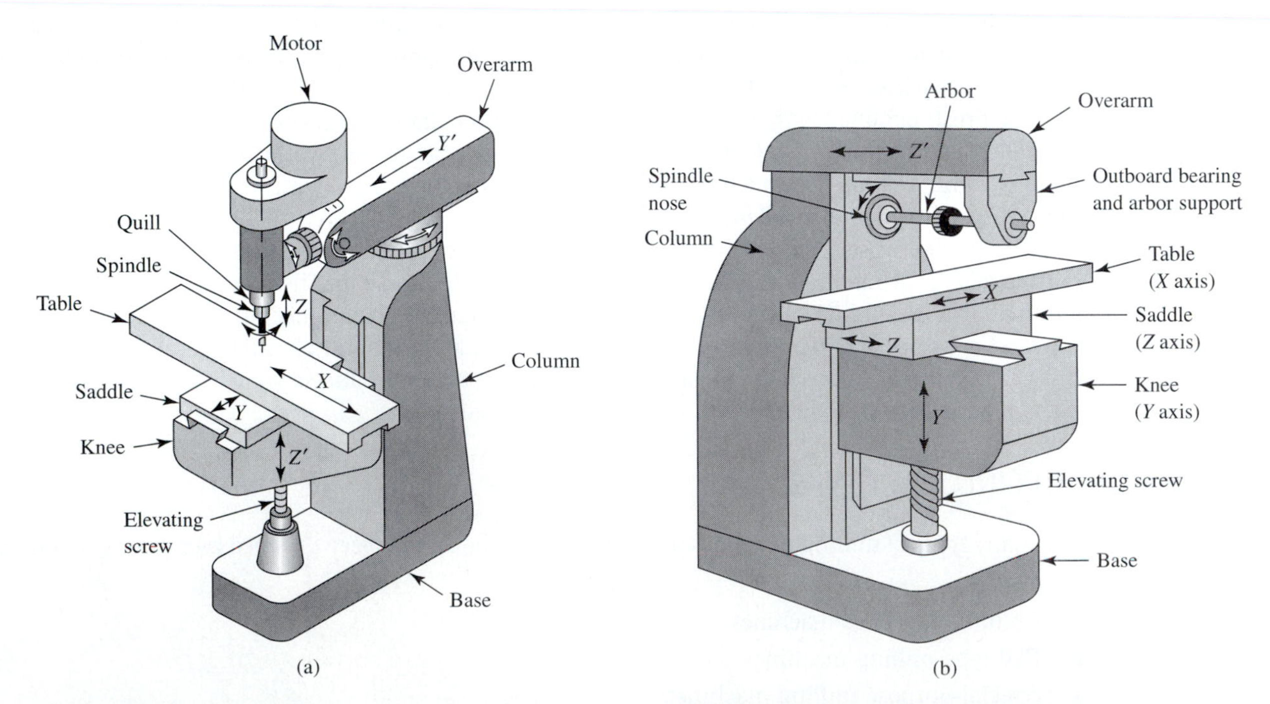

FIGURE 25.17 Schematic illustration of the motions and components (a) of a vertical-spindle column and knee type of milling machine and (b) of a horizontal-spindle column and knee type of milling machine.

(a)

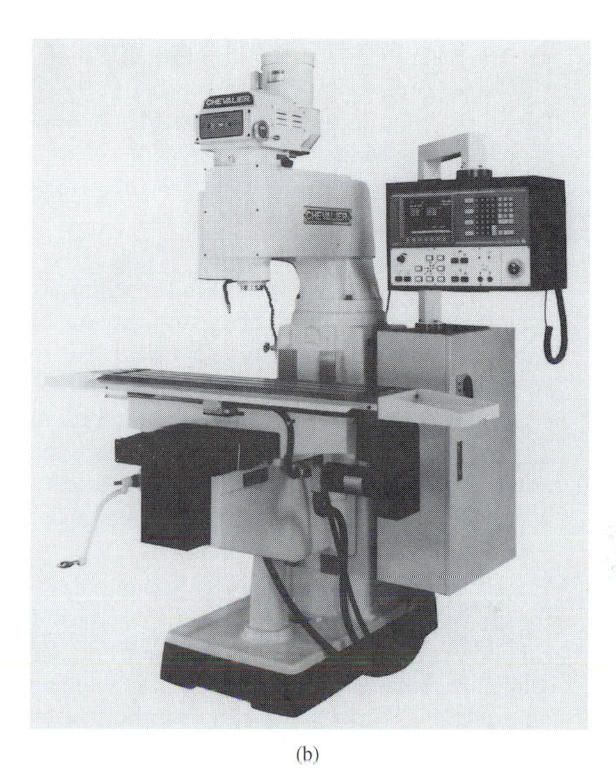

(b)

FIGURE 25.18 (a) Standard vertical-spindle column and knee type milling machine. (Courtesy Bridgeport Machinery, Inc.) (b) Three-axis CNC vertical-spindle column and knee type milling machine. (Courtesy Chevalier Machinery, Inc.)

FIGURE 25.19 A universal horizontal-spindle column and knee type milling machine. (Courtesy Summit Machine Tool Manufacturing Corp.)

The machined vertical slide on the front of the column may be of the square or dovetail type. The knee moves up and down on this slide. The slide must be machined at a 90° angle to the face of the column in both the lateral and vertical planes. The tolerances are very close and are usually expressed in minutes or seconds of arc. The large hole in the face of the column casting is for the spindle. The hole is very accurately bored perpendicular to the front slide in two planes and parallel to the upper slide.

Spindle On a horizontal milling machine, the spindle is one of the most critical parts. It is usually machined from an alloy steel forging and is heat treated to resist wear, vibration, thrust, and bending loads. The spindle is usually supported by a combination of ball and straight roller bearings, or by tapered roller bearings that absorb both radial loads and end thrust loads. Spindles are hollow so that a drawbar can be used to hold arbors securely in place.

The front of the spindle is machined to accept standard arbors. The actual driving of the arbor is done by the two keys that fit into corresponding slots in the arbor. The internal taper, which is accurately ground so that it is concentric with the spindle, locates the arbor.

Knee The knee is a casting that is moved up or down the slide on the front of the column by the elevating screw. Two dovetail or square slides are machined at 90° to each other. The vertical slide mates with the slide on the front of the column, and the horizontal slide carries the saddle. It contains the necessary gears, screws, and other mechanisms to provide power feeds in all directions. Various feed rates can be selected by the operator with the controls mounted on the knee.

Saddle The saddle for a plain milling machine is a casting with two slides machined at an exact 90° angle to each other. The lower slide fits the slide on the top of the knee, and the upper slide accepts the slide on the bottom of the table. The surfaces of the slides that make contact with the knee and the table are parallel to each other. Locks for both the cross slide and table are fitted to the saddle, along with the nuts that engage with the cross feed and table feed screws.

On a universal milling machine the saddle is made in two pieces and is more complex because it must allow the table to swivel through a limited arc. The lower part has a dovetail slide that fits the top of the knee, and a circular slide above it is graduated in degrees for a small portion of its periphery. The upper portion of the saddle consists of a circular face that fits against the lower circular slide, a central pivot point, and a dovetail slide that accepts the table. Locking bolts moving in a circular T–slot are provided so that the two parts of the saddle can be locked in any position.

Table Milling machine tables vary greatly in size, but generally they have the same physical characteristics. The bottom of the table has a dovetail slide that fits in the slide on top of the saddle. It also has bearings at each end to carry the table feed screw. The top of the table is machined parallel with the slide on the bottom and has several full-length T–slots for mounting vises or other work-holding fixtures.

A dial graduated in thousandths of an inch is provided to allow for accurate table movement and placement. The table feed screw usually has an **Acme thread.**

Milling machines with vertical spindles (see Figs. 25.17a and 25.18a) are available in a large variety of types and sizes. The head, which houses the spindle, motor, and feed controls, is fully universal and can be placed at a compound angle to the surface of the table. The ram, to which the head is attached, can be moved forward and back, and locked in any position. A turret on top of the column allows the head and ram assembly to swing laterally, increasing the reach of the head of the machine.

Some ram-type milling machines can be used for both vertical and horizontal milling. On ram-type vertical mills that have the motor in the column, power is transmitted to the spindle by gears and **splined shafts.** Some heavy-duty vertical mills have a spindle and head assembly that can be moved only vertically by either a power feed mechanism or manually. These are generally known as overarm-type vertical milling machines.

25.3.2 Bed-Type Milling Machines

High production calls for heavy cuts, and the rigidity of a knee and column type of milling machine may not be sufficient to take the high forces. A bed-type milling machine is often ideal for this kind of work. In this machine the table is supported directly on a heavy bed, and the column is placed behind the bed. A CNC bed-type milling machine is shown in Figure 25.20.

There are several advantages of the bed-type machine, particularly for production runs. Hydraulic table feeds are possible, the hydraulic components being housed in the bed casting. This allows very high feed forces, variable feedrates during any given cut, and automatic table cycling. The spindle may be raised or lowered by a **cam** and **template** arrangement to produce special contours. The basically heavier construction allows more power to be supplied to the spindle, which gives higher productivity through faster metal removal. Duplex bed-type milling machines have two columns and spindles for milling two surfaces on a part simultaneously. A large 5-axis CNC bed-type milling machine is shown in Figure 25.21.

The chief disadvantage of a bed-type milling machine compared to one of the knee and column type is that it is less versatile for machining small parts. Its advantages lie in its higher productivity, its adaptability to large machines, and its ease of modification to special applications.

25.3.3 Special-Purpose Milling Machines

As industrial products have become more complex, new and unusual variations of the more common milling machines have been developed. The objectives are to accommodate larger work, make many duplicate parts, locate holes and surfaces precisely, or do other unusual machining jobs.

FIGURE 25.20 CNC bed type milling machine. (Courtesy WMW Machinery Co., Inc.)

FIGURE 25.21 Large 5-axis CNC bed type milling machine. (Courtesy Ingersoll Milling Machine Co.)

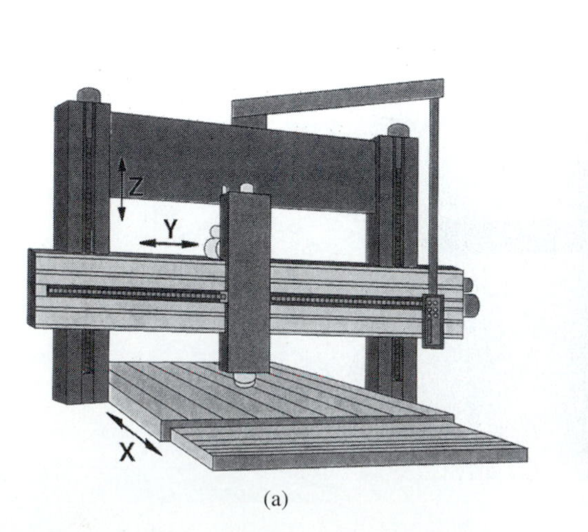

(a)

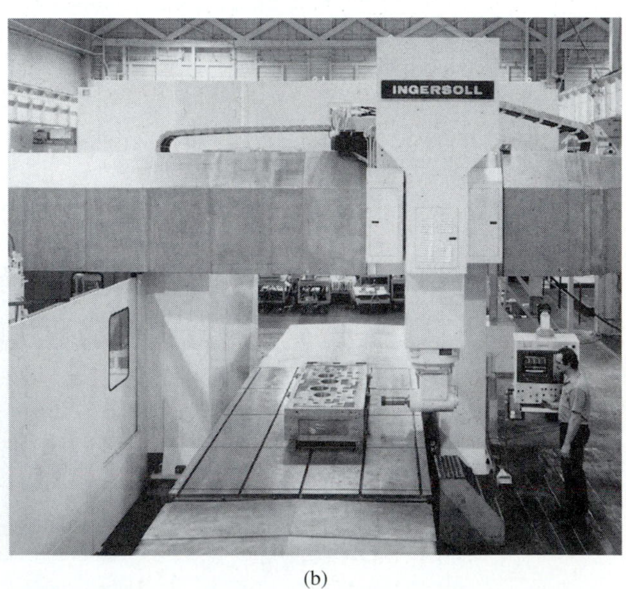

(b)

FIGURE 25.22 (a) Illustration of a planer-type milling machine. (Courtesy Kennametal Inc.) (b) Part being machined on a planer-type milling machine. (Courtesy Ingersoll Milling Machine Co.)

Planer-Type Milling Machines The general arrangement of these types of machines is similar to that for planers (Chapter 14), except that in place of individual tool bits, milling heads are installed. The table of the machine carries the work past the rotating cutter heads, which are individually powered and can be run at different speeds if necessary. As many as four cutter heads can be used, with two mounted on the cross rail and two on the vertical pillars. A planer-type milling machine is shown in Figure 25.22(a). A part being machined on a planer-type milling machine is shown in Figure 25.22(b). Planer-type machines are used mostly for machining parts like the bedways for large machine tools and other long workpieces that require accurate flat and angular surfaces or grooves.

Profile Milling Machines Two-dimensional profiling can be done by using a template or with a numerically controlled vertical milling machine. Some profilers have several spindles, and a number of duplicate parts can be produced in each cycle. Hydraulic-type profilers have a stylus that is brought into contact with the template to start the operation. The operator then moves the stylus along the template, causing hydraulic fluid under pressure to flow to the proper actuating cylinders. The table moves the work past the cutter, duplicating the shape of the template. A profile milling operation is shown, using a ball-nosed milling cutter, in Figure 25.23(a). A 5-axis, multispindle profile milling machine is shown in Figure 25.23(b).

Diesinking and other processes involving the machining of cavities can be done on three-dimensional profilers. An accurate pattern of the cavity is made of wood, plaster, or soft metal, and the stylus follows the contour of the pattern, guiding the cutter as it machines out the cavity. Numerically controlled milling machines can also be used for this type of work.

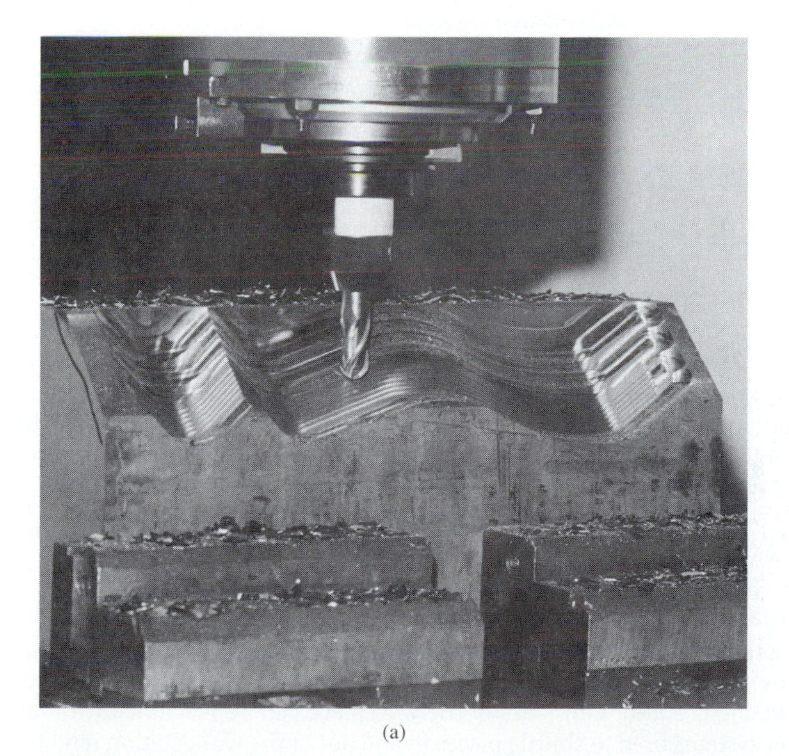

(a)

FIGURE 25.23 (a) Profile milling operation is shown, using a ball-nosed end milling cutter. (Courtesy Ingersoll Milling Machine Co.)

(b)

FIGURE 25.23 Continued (b) A large 5-axis multispindle profile milling machine. (Courtesy Cincinnati Machine)

25.3.4 Machining Centers

As shown in Chapter 11, the most versatile milling machine is now the machining center. Machining centers are designed and built to provide for flexible manufacturing. They can be used to machine just a few parts or large production runs. Programming can be relatively simple, and the use of "canned" cycles provides a great deal of versatility. An NC machining center by definition is able to perform milling, drilling, and boring cuts, and has either indexing turret toolholders or provides for automatic tool change. A toolholder magazine is shown in Figure 25.24.

Machining centers are built in either horizontal or vertical configuration. The relative merits of each will be discussed briefly.

Horizontal Machines Horizontal machines tend to be advantageous for heavy box-shaped parts, such as gear housings, which have many features that need to be machined on the side faces. The horizontal machine easily supports heavy workpieces of this type. If a rotary indexing worktable is added, four sides of the workpiece can be machined without refixturing.

Pallet systems used to shuttle pieces in and out of the work station tend to be easier to design for horizontal machines, where everything in front of the main column is open and accessible. A horizontal machining center with a pallet shuttling system is shown in Figure 25.25.

FIGURE 25.24 Preset machining center tools are stored in a toolholder magazine. (Courtesy Kennametal Inc.)

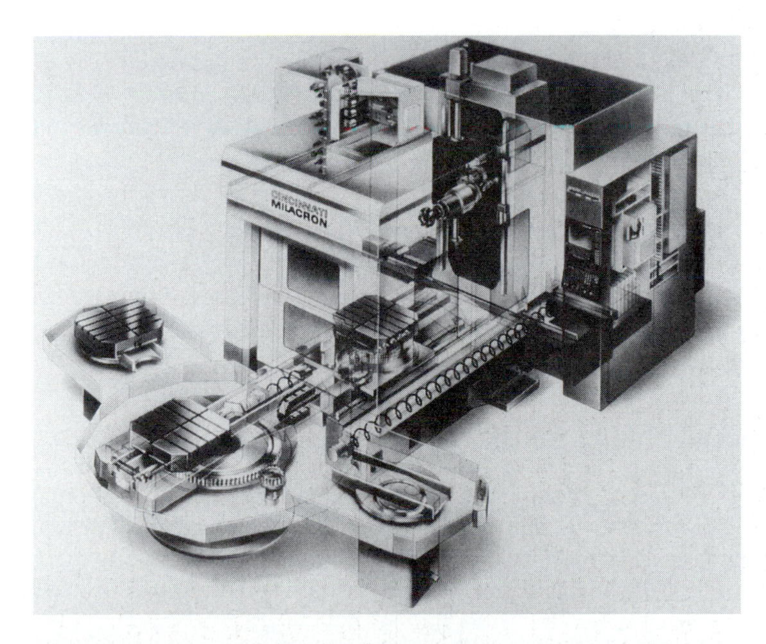

FIGURE 25.25 Horizontal machining center with pallet shuttling system. (Courtesy Cincinnati Machine)

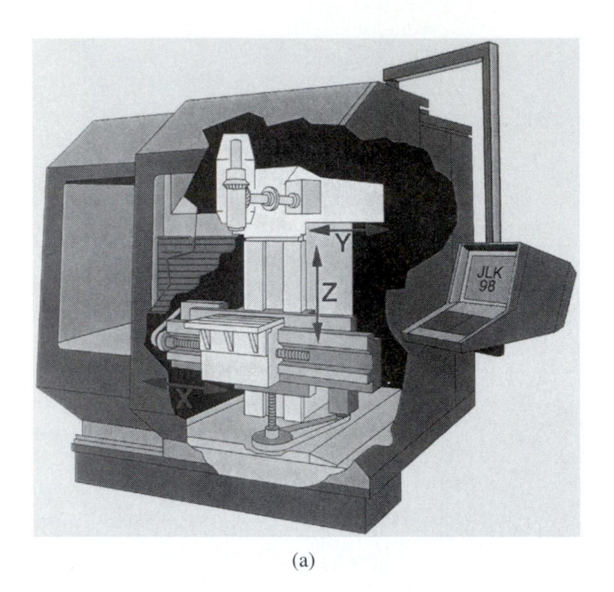

(a)

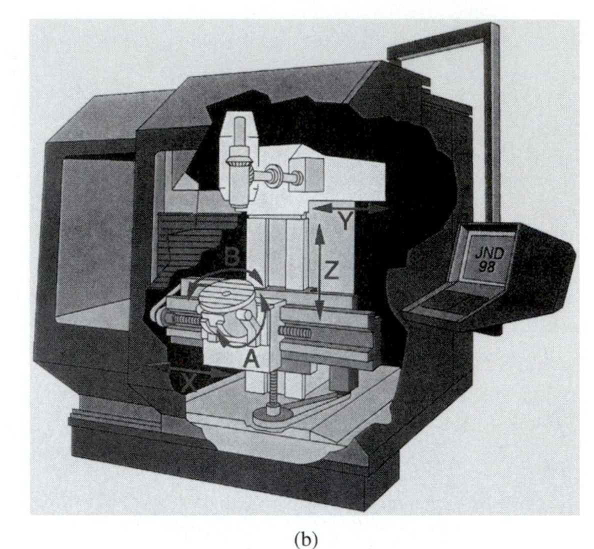

(b)

FIGURE 25.26 The motions of (a) a 3 axis CNC vertical machining center and (b) a 5-axis CNC vertical machining center. (Courtesy Kennametal Inc.)

Vertical Machines Vertical machining centers are often preferred for flat parts that must have through holes. Fixtures for these parts are more easily designed and built for a vertical spindle. Also, the thrust of the cut developed in drilling or in milling pockets can be absorbed directly by the bed of the machine. The motions of a 3- and 5-axis CNC vertical machining center are shown in Figure 25.26.

The vertical machine is preferred where 3-axis work is done on a single face as in mold and die work. The weight of the head of a vertical machine as it extends away from the column, particularly on large machines, can be a factor in maintaining accuracy, because there may be some tendency for it to drop, lose accuracy, and cause chatter. A vertical machining center is shown in Figure 25.27.

25.4 MILLING MACHINE ATTACHMENTS AND ACCESSORIES

Many accessories have been developed for milling machines. Some are specialized and can be used for only a few operations. Others, such as vises, arbors, and collets, are used in almost all milling operations.

25.4.1 Special Milling Heads

Several types of special heads have been developed for use on horizontal or vertical milling machines. These accessories increase the versatility of the machine. For example, a vertical head can be attached to a conventional horizontal column and knee milling machine, greatly increasing its usefulness, especially in small shops with a limited number of machines.

Vertical Heads Vertical heads are generally attached to the face of the column or to the overarm of a horizontal milling machine. The head is a semiuniversal type, which pivots only on the axis parallel to the center line of the spindle, or fully universal. Fully universal heads can

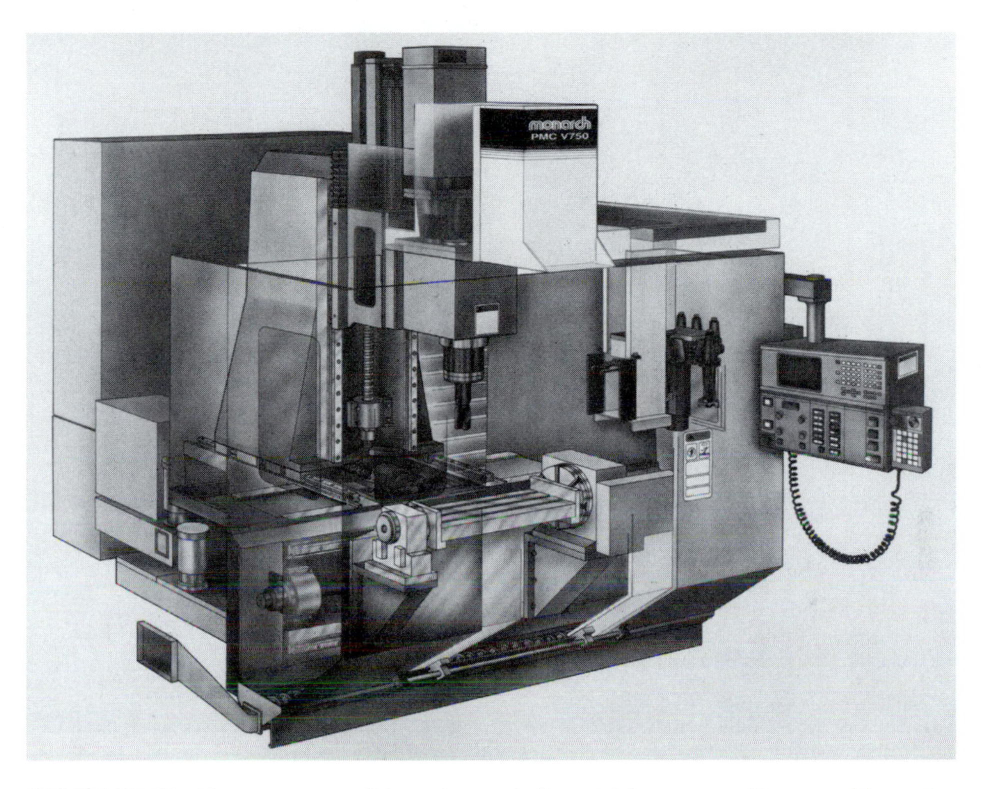

FIGURE 25.27 Transparent rendition of a vertical machining center. (Courtesy Monarch Machine Tool.)

be set to cut compound angles. Both types of heads are powered by the spindle of the milling machine and accept standard arbors and collets. A universal milling head attachment is shown in Figure 25.28(a). Figure 25.28(b) shows a universal milling head attachment being used on a horizontal milling machine.

Rack-Milling Attachment The rack-milling attachment bolts to the spindle housing of the milling machine. Its spindle is at a right angle to the main spindle of the machine. Both **spur** and helical **racks** can be milled with this attachment, and it can also be used to mill **worms.** Some rack-milling attachments have an outboard support for the spindle, which makes it possible to take heavier cuts.

Slotting Attachment This attachment, which is bolted to the column of a horizontal milling machine, can be swiveled 90° in either direction from the vertical position. It is used primarily in toolmaking and prototype work for cutting keyways, internal splines, and square or rectangular cavities. The crank that actuates the reciprocating slide is driven directly by the spindle, and the stroke is adjustable.

High-Speed Attachment When spindle speeds beyond the operating range of the machine are necessary, high-speed attachments can be placed on both horizontal and vertical milling machines. A **gear train** is generally used to step up the speed as much as 6 to 1, which allows more efficient use of small cutters.

(a)

(b)

FIGURE 25.28 (a) Universal milling head attachment. (b) A universal milling head attachment being used on a horizontal milling machine. (Courtesy WMW Machinery Co., Inc.)

25.4.2 Vises and Fixtures

In all milling operations, the work is held by fixtures, vises, or clamping arrangements. In most cases the work is held stationary in relation to the table while it is being machined, but work held in indexing heads and rotary tables can be moved in two planes while machining operations are in progress.

Plain Vise Plain milling vises (Fig. 25.29a) are actuated by an Acme threaded screw, and the movable jaw moves on either a dovetail or rectangular slide. The vises are usually cast of high-grade gray cast iron or ductile iron and can be heat treated. Steel keys are attached in slots machined into the bottom of the vise, parallel with and perpendicular to the fixed jaw to allow accurate placement on the milling table. The jaw inserts are usually heat treated alloy steel and are attached by cap screws. Vises of this type are classified by the jaw width and maximum opening. Cam-operated plain milling vises are widely used in production work because of the savings in time and effort and the uniform clamping pressure that can be achieved.

Swivel-Base Vise A swivel-base vise is more convenient to use than the plain vise, although it is somewhat less rigid in construction. The base, which is graduated in degrees, is slotted for keys that align it with the T-slots in the table. The upper part of the vise is held to the base by T-bolts that engage a circular T-slot. The swivel-base vise, when used on a milling machine with a semiuniversal head, makes it possible to mill compound angles on a workpiece.

Universal Vise A universal vise (Fig. 25.29b) is used mostly in toolroom diemaking and prototype work. The base of the vise is graduated in degrees and held to the table by T-bolts. The

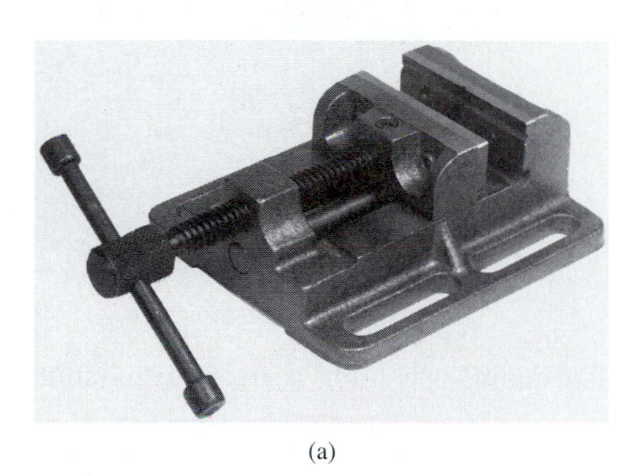

(a)

(b)

FIGURE 25.29 (a) A plain vise. (b) A universal vise, which is generally used in toolroom diemaking and prototype work. (Courtesy Palmgren Steel Products, Inc.)

intermediate part of the vise has a horizontal pivot upon which the vise itself can rotate 90°. Because there are several joints and pivots in the vise assembly, the universal vise is usually the least rigid of the various types of milling machine vises.

Angle Plates Several types of angle plates can be used to hold work or work-holding fixtures for milling (Fig. 25.30). Plain angle plates are available in T-slotted or blank form and are usually strong iron castings. Adjustable angle plates may tilt in one direction only or have a swivel base. They are very useful for milling workpieces that are irregular in shape and cannot be held easily in a vise.

Holding fixtures that are a combination of a simple angle plate and a collet are sometimes used to hold round or hexagonal work for milling. The collet holding fixture may be manually or air operated. Both fixtures can be bolted to the milling table in the vertical or horizontal position or attached to an adjustable angle plate for holding workpieces at simple or compound angles to the table or other reference surface.

Indexing Heads The indexing head, also known as the dividing head (Fig. 25.31), can be used on vertical and horizontal milling machines to space the cuts for such operations as making splines, gears, worm wheels, and many other parts requiring accurate division. It can also be

FIGURE 25.30 Adjustable-angle tables can be tilted in one or more directions to machine irregularly shaped parts. (Courtesy Palmgren Steel Products, Inc.)

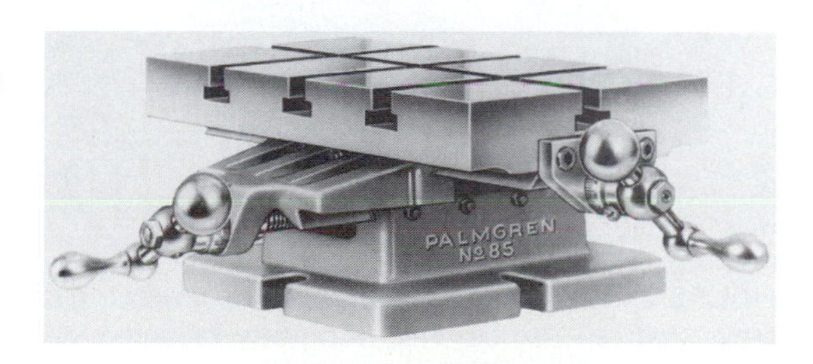

FIGURE 25.31 Indexing heads, also called dividing heads, can be used on vertical and horizontal milling machines. (Courtesy Cincinnati Machine)

geared to the table screw for helical milling operations such as cutting flutes in twist drills and making helical gears.

Indexing heads are of the plain or universal type. Plain heads cannot be tilted; universal heads can be tilted to the vertical or any intermediate position. The spindle of the indexing head can be fitted with a chuck, or with other work-holding devices, including collets or a center.

Most indexing heads have a worm and wheel reduction ratio of 40:1, requiring 40 turns of the hand crank to make the spindle revolve once. When the necessary index plates are available, all divisions up to and including 50 can be achieved by plain indexing. For some numbers above 50, differential indexing is necessary.

In recent years, programmable precision indexers have become fairly common in shops doing work that requires accurate spacing of complex hole patterns on surfaces. The indexer may be mounted with the axis of the chuck vertical or horizontal, and in some cases the chuck may be replaced with a specially made holding fixture or a faceplate. If necessary, a tailstock may be used to support the end of the workpiece. The controller is capable of storing a series of programs, each of which may incorporate as many as 100 operational steps or positions.

Rotary Table Rotary tables (Fig. 25.32) are available in a wide range of sizes and can be used on both vertical and horizontal milling machines. Most can also be clamped with the face at a

(a)

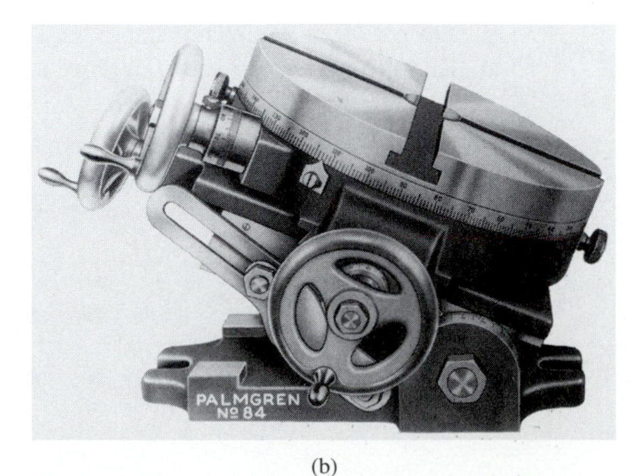

(b)

FIGURE 25.32 (a) Rotary tables can be used on both vertical and horizontal milling machines. (b) A rotary table, that can also be tilted. (Courtesy Palmgren Steel Products, Inc.)

90° angle to the surface of the milling machine table. The face of the rotary table has four or more T-slots and an accurately bored hole in the center, which is concentric with the axis about which the table rotates.

The base of the rotary table, which houses the worm drive mechanism, is graduated in degrees, and the handwheel can be graduated in increments as small as 5 minutes or $\frac{1}{12}^{\circ}$. On some rotary tables an index plate may be attached to the base. Rotary tables can also be geared to the table feed screw. When set up in this manner, the rotary table can be used to make plate cams and to generate a number of other irregular shapes.

25.4.3 Arbors, Collets, and Toolholders

Several basic types of arbors and collets are used to hold milling cutters and to transmit power from the spindle to the cutter. Regardless of type, they are usually precisely made of alloy steel and heat treated for wear resistance and strength.

Arbors Arbors for horizontal milling machines (Fig. 25.33) are available in three basic types: style A, style B, and style C. A drawbolt that goes through the spindle of the machine screws into the small end of the taper and draws the arbor tightly into the tapered hole in the milling machine spindle. Power is transmitted from the spindle to the arbor by two short keys that engage with the slots on the flange of the arbor.

Style A arbors consist of the tapered portion that fits the spindle, the shaft on which the cutter or cutters fit, the spacers, and the nut. The shaft has a keyway along its entire length. The outboard end of the arbor has a pilot that fits into a bronze bushing in the outboard support of the milling machine overarm (Fig. 25.17b). One or more cutters can be mounted on the arbor, either adjacent to each other or separated by spacers and shims. Style A arbors are used primarily for light- and medium-duty milling jobs.

Style B arbors are used for heavy milling operations, especially where it is necessary to provide support close to a milling cutter, such as in a straddle milling operation (Fig. 25.39a). One or more bearing sleeves may be placed on the arbor as near to the cutters as possible. An outboard bearing support is used for each bearing sleeve on the arbor.

Style C arbors are used to hold and drive shell end mills (Fig. 25.35) and some types of face milling cutters, and require no outboard support. In some cases, they can also be fitted with adapters for mounting other types of cutters.

Collets On some vertical milling machines the spindle is bored to accept a collet that has a partly straight and partly tapered shank. The collet is secured by a drawbar that is screwed into

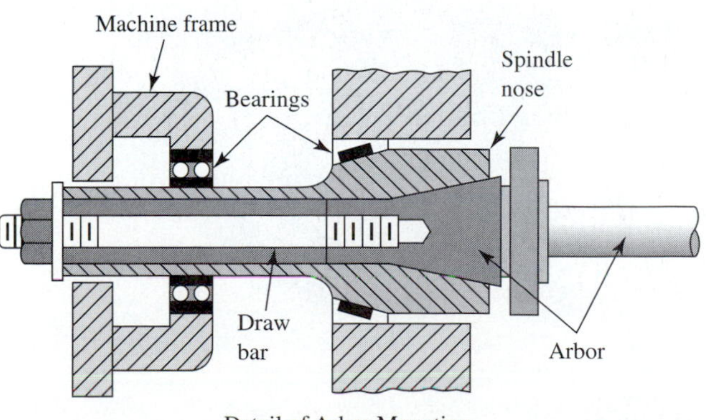

FIGURE 25.33 Schematic drawing of an arbor mounting with drawbolt used in horizontal milling machines.

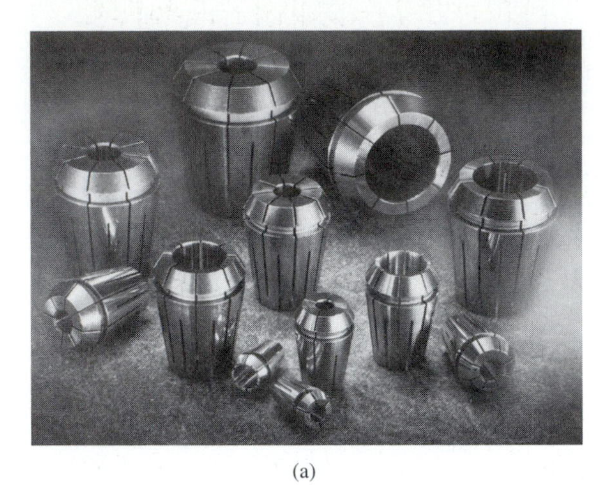

(a)

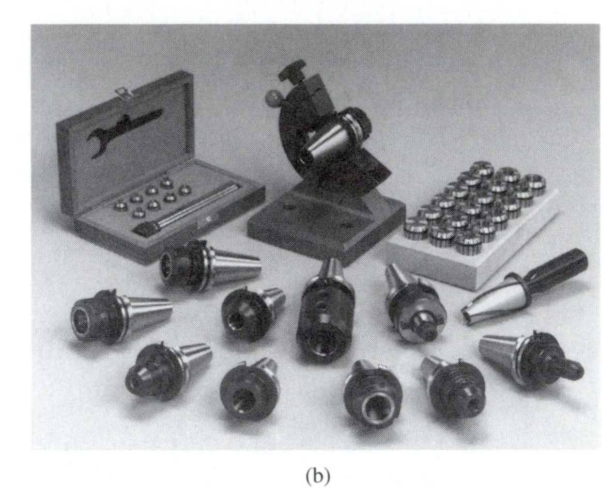

(b)

FIGURE 25.34 (a) Collets of various sizes. (Courtesy Lyndex Corp.) (b) Various toolholders with collets and a setup fixture. (Courtesy Valenite Inc.)

a tapped hole in the back of the collet and tightened from the top of the spindle. Some milling machine manufacturers offer collet arrangements that do not need a drawbar. Collets of this type can be closed with a lever-operated cam or with a large locking nut. Figure 25.34(a) shows various size collets and Figure 25.34(b) shows a number of toolholders including collets and a setup fixture.

Toolholders Standard toolholders are available for end mills and shell mills as shown in Figure 25.35(a) and (b) respectively. For some operations that require the use of tools with nonstandard shank sizes, chucks can be used to hold the tool. These chucks are available with Morse

(a)

(b)

FIGURE 25.35 (a) End milling cutter toolholders. (b) Shell end milling cutter toolholders. (Courtesy Lyndex Corp.)

FIGURE 25.36 Offset boring heads are often used in milling machines for boring, facing, and chamfering operations. (Courtesy Bridgeport Machine, Inc.)

taper or straight shanks. Either type can be used in milling machines when the proper adapters or collets are available.

Offset boring heads (Fig. 25.36) are often used in vertical milling machines for boring, facing, chamfering, and outside-diameter turning operations. They are available with straight, Morse taper or standard milling machine taper shanks, and usually have three mounting holes for boring bars. Two of the holes are usually parallel with the centerline of the tool, and one is perpendicular to the centerline. Some boring heads have two adjusting mechanisms, and the movable slide can be adjusted accurately in increments of .0001 in.

Flycutters can be used for facing operations. The tools in cutters of this type are adjusted so that both a roughing and finishing cut are taken in one pass.

25.5 BASIC MILLING OPERATIONS

Before any milling job is attempted, several decisions must be made. In addition to selecting the best means of holding the work and the most appropriate cutters to be used, the cutting speed and feedrate must be established to provide good balance between rapid metal removal and long tool life.

25.5.1 Operating Conditions

Proper determination of cutting speed and feedrate can be made only when the following six factors are known:

- Type of material to be machined
- Rigidity of the setup
- Physical strength of the cutter
- Cutting-tool material
- Power available at the spindle
- Type of finish desired

Several of these factors affect cutting speed only, and some affect both cutting speed and feedrate. The tables in reference handbooks provide approximate figures that can be used as starting points. After the cutting speed is chosen, the spindle speed must be computed and the machine adjusted.

Cutting Speed Cutting speed is defined as the distance in feet that is traveled by a point on the cutter periphery in one minute. Since a cutter's periphery is its circumference,

$$\text{Circumference} = \pi \times d$$

In the case of a cutter, the circumference is

$$\begin{aligned}\text{Cutter circumference} &= (\pi/12) \times d \\ &= .262 \times d\end{aligned}$$

Since cutting speed is expressed in surface feet per minute (SFPM),

$$\text{SFPM} = \text{Cutter circumference} \times \text{RPM}$$

by substituting for the cutter circumference, the cutting speed can be expressed as

$$\text{SFPM} = .262 \times d \times \text{RPM}$$

The concept of cutting speed (SFPM) was introduced in Chapter 13, "Turning Operations" and explained again in Chapter 15, "Drilling Operations." It has again been reviewed here without giving additional examples. However, because milling is a multipoint operation, feed needs to be explained in more detail than in previous chapters.

Feedrate Once the cutting speed is established for a particular workpiece material, the appropriate feedrate must be selected. Feedrate is defined in metal cutting as the linear distance the tool moves at a constant rate relative to the workpiece in a specified amount of time. Feedrate is normally measured in units of inches per minute or IPM. In turning and drilling operations the feedrate is expressed in IPR or inches per revolution.

When establishing the feedrates for milling cutters, the goal is to attain the fastest feed per insert possible, to achieve an optimum level of productivity and tool life, consistent with efficient manufacturing practices. The ultimate feedrate is a function of the cutting-edge strength and the rigidity of the workpiece, machine, and fixturing. To calculate the appropriate feedrate for a specific milling application, the RPM, number of effective inserts (N) and feed per insert in inches (IPT or **a.p.t.**) should be supplied.

The milling cutter shown in Figure 25.37 on the left (one insert cutter) will advance .006 in. at the cutter centerline every time it rotates one full revolution. In this case, the cutter is

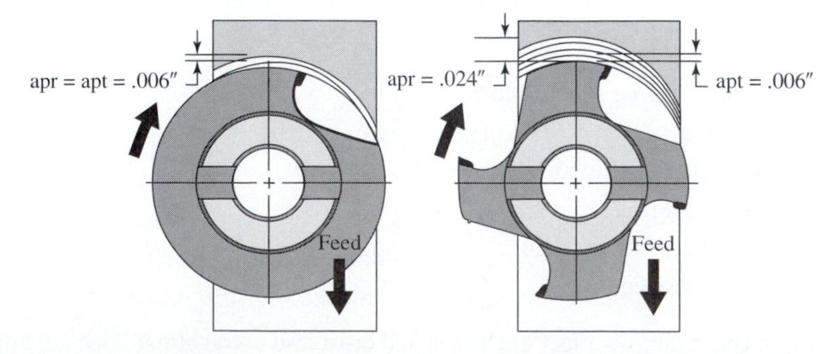

FIGURE 25.37 Drawing of a milling cutter showing the difference between advance per revolution (**apr**) and advance per tooth (**apt**).

said to have a feed per insert or an IPT (inches per tooth), **a.p.t.** (advance per tooth), and **a.p.r.** (advance per revolution) of .006 in. The same style of cutter with four inserts is shown in the right-hand drawing. However, to maintain an equal load on each insert, the milling cutter will now advance .024 in. at the centerline every time it rotates one full revolution. The milling cutter on the right is said to have an IPT and **a.p.t.** of .006 in., but an **a.p.r.** (advance per revolution) of .024 in. (.006 in. for each insert).

These concepts are used to determine the actual feedrate of a milling cutter in IPM (inches per minute) using one of the following formulas:

$$\text{IPM} = (\text{IPT}) \times (N) \times (\text{RPM})$$

or

$$\text{IPM} = (\text{a.p.t.}) \times (N) \times (\text{RPM})$$

where: IPM = Inches per minute
N = Number of effective inserts
IPT = Inches per tooth
a.p.t. = Advance per tooth
RPM = Revolutions per minute

For example, if you were milling automotive gray cast iron using a 4-in.-diameter face mill with 8 inserts at 400 SFPM and 30.5 IPM, what **a.p.r.** and **a.p.t.** would this be?

$$\text{RPM} = \frac{\text{SFPM}}{.262 \times d} = \frac{400}{.262 \times 4} = 382$$

$$\text{a.p.r.} = \frac{\text{IPM}}{\text{RPM}} = \frac{30.5}{382} = .080 \text{ in.}$$

$$\text{a.p.t.} = \frac{\text{a.p.r.}}{N} = \frac{.080}{8} = .010 \text{ in.}$$

$$= .080 \text{ in. a.p.r.}$$

$$= .010 \text{ in. a.p.t.}$$

When milling a 300M steel landing gear with a 6-in.-diameter 45° lead face mill (containing 10 inserts) at 380 SFPM and a .006 in. advance per tooth, what feedrate should be run in IPM?

$$\text{RPM} = \frac{\text{SFPM}}{.262 \times d} = \frac{380}{.262 \times 6} = 242$$

$$\text{IPM} = \text{apt} \times N \times \text{RPM} = .006 \times 10 \times 242 = 14.5$$

$$= 14.5 \text{ IPM}$$

The following basic list of formulas can be used to determine IPM, RPM, **a.p.t.**, **a.p.r.**, or N, depending on what information is supplied for a specific milling application:

IPM = Inches per minute
N = Number of effective inserts
a.p.r. = Inches of cutter advance every revolution
a.p.t. = Inches of cutter advance for each effective insert every revolution
RPM = Revolutions per minute

Find	Given	Using
IPM	**a.p.r.** RPM	IPM = **a.p.r.** × RPM
IPM	RPM, *N*, **a.p.t.**	IPM = **a.p.t.** × *N* × RPM
a.p.r.	IPM, RPM	**a.p.r.** = IPM/RPM
RPM	IPM, **a.p.r.**	RPM = IPM/**a.p.r.**
RPM	IPM, *N*, **a.p.t.**	$\text{RPM} = \dfrac{\text{IPM}}{N \times \textbf{a.p.t.}}$
N	IPM, RPM, **a.p.t.**	$N = \dfrac{\text{IPM}}{\text{RPM} \times \textbf{a.p.t.}}$
a.p.t.	IPM, *N*, RPM	$\textbf{a.p.t.} = \dfrac{\text{IPM}}{\text{RPM} \times N}$

Note: In these formulas IPT can be substituted for **a.p.t.** and IPR can be substituted for **a.p.r.**

Horsepower Requirements In metal cutting, the horsepower consumed is directly proportional to the volume (Q) of material machined per unit of time (cubic inches/minute). Metals have distinct unit power factors, which indicate the average amount of horsepower required to remove one cubic inch of material in a minute. The power factor (k*) can be used either to determine the machine size in terms of horsepower required to make a specific machining pass or the feedrate that can be attained once a depth and width of cut are established on a particular part feature. To determine the metal removal rate (Q) use the following:

$$Q = \text{D.O.C.} \times \text{W.O.C} \times \text{IPM}$$

where: D.O.C. = Depth of cut in inches
W.O.C. = Width of cut in inches
IPM = Feedrate, in inches/minute

The average spindle horsepower required for machining metal workpieces is as follows:

$$\text{HP} = Q \times k$$

where: HP = Horsepower required at the machine spindle
Q = The metal removal rate in cubic inches/minute
k = The unit power factor in HP/cubic inch/minute

For example, what feed should be selected to mill a 2-in.-wide by .25-in.-deep cut on aircraft aluminum, utilizing all the available horsepower on a 20-HP machine using a 3-in.-diameter face mill?

$$\text{HP} = Q \times k$$
$$k = .25\ \text{HP/in}^3\text{/min} \quad \text{for aluminum}$$

The maximum possible metal removal rate (Q), for a 20-HP machine running an aluminum part is

$$Q = \frac{\text{HP}}{k} = \frac{20}{.25} = 80\ \text{in}^3\text{/min}$$
$$Q = 80\text{in}^3\text{/min}$$

*k factors are available from reference books.

To remove 80 in^3/min, what feedrate will be needed?

$$Q = (\text{D.O.C.}) \times (\text{W.O.C.}) \times \text{IPM}$$

$$\text{IPM} = \frac{Q}{(\text{D.O.C.}) \times (\text{W.O.C.})} = \frac{80}{.25 \times 2} = 160$$

$$= 160 \text{ IPM}$$

25.5.2 Direction of Milling Feed

The application of the milling tool in terms of its machining direction is critical to the performance and tool life of the entire operation. The two options in milling direction are described as either **conventional** or **climb milling.** The concept of conventional and climb milling was introduced in Chapter 8 and described as it relates to cutting forces and deflection. Conventional and climb milling also affects chip formation and tool life as explained here. Figure 25.38 shows drawings of both conventional and climb milling.

Conventional Milling The term often associated with this milling technique is "up-cut" milling. The cutter rotates against the direction of feed as the workpiece advances toward it from the side where the teeth are moving upward. The separating forces produced between cutter and workpiece oppose the motion of the work. The thickness of the chip at the beginning of the cut is at a minimum, gradually increasing in thickness to a maximum at the end of the cut.

Climb Milling The term often associated with this milling technique is "down-cut" milling. The cutter rotates in the direction of the feed and the workpiece, and therefore advances toward the cutter from the side where the teeth are moving downward. As the cutter teeth begin to cut, forces of considerable intensity are produced, which favor the motion of the workpiece and tend to pull the work under the cutter. The chip is at a maximum thickness at the beginning of the cut, reducing to a minimum at the exit. Generally climb milling is recommended wherever possible. With climb milling a better finish is produced and longer cutter life is obtained. As each tooth enters the work, it immediately takes a cut and is not dulled while building up pressure to dig into the work.

Advantages and Disadvantages If the workpiece has a highly abrasive surface, conventional milling will usually produce better cutter life because the cutting edge engages the work below

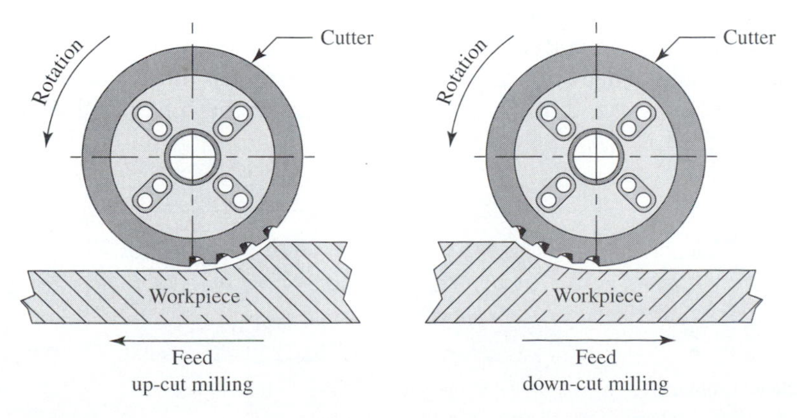

FIGURE 25.38 Conventional or up-milling as compared to climb or down-milling.

the abrasive surface. Conventional milling also protects the edge by chipping off the surface ahead of the cutting edge.

Limitations on the use of climb milling are mainly affected by the condition of the machine and the rigidity with which the work is clamped and supported. Since there is a tendency for the cutter to climb up on the work, the milling machine arbor and arbor support must be rigid enough to overcome this tendency. The feed must be uniform, and if the machine does not have a **backlash** eliminator drive, the table **gibs** should be tightened to prevent the workpiece from being pulled into the cutter. Most present-day machines are built rigidly enough. Older machines can usually be tightened to permit use of climb milling.

The downward pressure caused by climb milling has an inherent advantage in that it tends to hold the work and fixture against the table, and the table against the ways. In conventional milling, the reverse is true and the workpiece tends to be lifted from the table.

25.6 TYPES OF MILLING OPERATIONS

Milling cutters are used either individually or in combinations to machine various surfaces as described here and shown in Figure 25.39.

Plain Milling Plain milling is the process of milling a surface that is parallel to the axis of the cutter and basically flat. It is done on plain or universal horizontal milling machines with cutters of varying widths that have teeth only on the periphery. A plain milling cutter is shown in Figure 25.4(c).

Side Milling For side milling, a cutter is used that has teeth on the periphery and on one or both sides. When a single cutter is being used, the teeth on both the periphery and sides may be cutting. The machined surfaces are usually either perpendicular or parallel to the spindle. Angle cutters can be used to produce surfaces that are at an angle to the spindle for such operations as making external dovetails or flutes in reamers. A side milling cutter is shown in Figure 25.4(b).

Straddle Milling In a typical straddle milling setup (Fig. 25.39) two side milling cutters are used. A straddle milling operation is shown in Figure 25.40. The cutters are half-side or plain side milling cutters, and have straight or helical teeth. Stagger-tooth side milling cutters can also be used.

The cutters cut on the inner sides only, or on the inner sides and the periphery. If the straddle milling operation involves side and peripheral cuts, the diameter of the two cutters must be exactly the same. When cutters with helical teeth are used, the helix angles must be opposite.

Since straddle-milled surfaces must be parallel to each other and are usually held to close tolerances in terms of width, the condition and size of the collars and shims that separate the

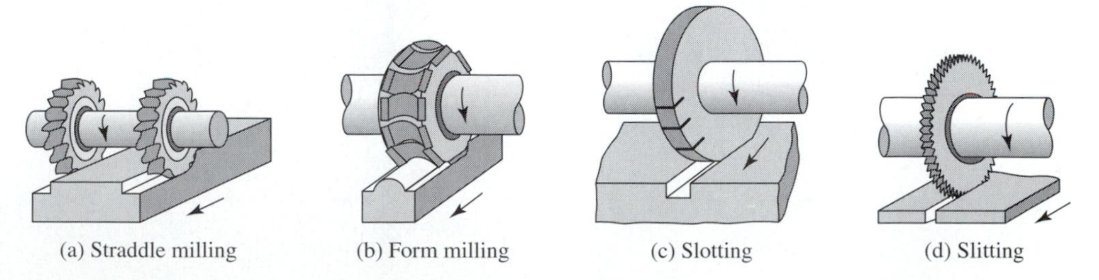

FIGURE 25.39 Types of milling operations: (a) straddle milling, (b) form milling, (c) slotting operation, and (d) slitting operation.

FIGURE 25.40 A straddle milling operation. (Courtesy Sandvik Coromant Co.)

cutters are important. The arbor must also turn as true as possible to avoid cutting the workpiece undersize.

Usually, a combination of collars and steel shims can be assembled to provide the correct spacing between cutters. For some production operations, a special collar can be made from alloy or medium-carbon steel, heat treated, and surface ground to length. The faces must be perpendicular to the bore and parallel to each other. The cutters should be keyed to the arbor, and the outboard bearing supports must be placed as close to the cutters as possible.

Gang Milling In gang milling, three or more cutters are mounted on the arbor, and several horizontal, vertical, or angular surfaces are machined in one pass (Fig. 25.41). When making a gang milling setup, several different types of cutters can be used, depending on the job to be done. Cutters used for producing vertical or angular surfaces must be of the side-cutting type; plain milling cutters of the proper width can be used for horizontal surfaces. In some cases face mills with the teeth facing inward can be used at one or both ends of the gang milling setup.

When only one wide plain helical milling cutter is used as part of a gang milling setup, the side thrust caused by that cutter should be directed toward the spindle of the machine. If possible, interlocking cutters with opposite helix angles should be used to eliminate side thrust and reduce the possibility of chatter.

Because of the time and effort involved in setting up the milling machine for gang milling, the process is used mainly for production work. Since all or almost all of the workpiece is being machined at one time, power and rigidity are very desirable features in the machine being used.

FIGURE 25.41 In gang milling operations, horizontal, vertical, or angular surfaces are machined in one path. (Courtesy Sandvik Coromant Co.)

(a)

(b)

FIGURE 25.42 (a) Slotting operation used in the manufacturing of rotors. (Courtesy Iscar Metals, Inc.) (b) Slitting operation to separate cast automotive parts. (Courtesy Sandvik Coromant Co.)

Every effort should be made to control vibration, including the use of support bars that are bolted to both the knee and the outboard bearing support.

Form Milling The number of parallel surfaces and angular relationships that can be machined by peripheral milling is limited almost only by cutter design. Form cutters are expensive, but often there is no other satisfactory means of producing complex contours as shown in Figure 25.39.

Slotting and Slitting Operations Milling cutters of either the plain or side-cutting type are used for slotting and slitting operations (Fig. 25.42a and b). Slotting and slitting are usually done on horizontal milling machines, but can also be done on vertical mills by using the proper adaptors and accessories. Metal slitting cutters of various diameters and widths are also used to cut slots. A number of identical cutters can also be mounted on the same arbor for cutting fins. When the thickness of the fins must be held to close tolerances, spacers are usually machined and surface ground to provide the necessary accuracy.

Face Milling Face milling (Fig. 25.43) can be done on vertical and horizontal milling machines. It produces a flat surface that is perpendicular to the spindle on which the cutter is mounted. The cutter ranges in size and complexity from a simple, single-tool flycutter to an inserted-tooth cutter with many cutting edges. Large face mills are usually mounted rigidly to the nose of the spindle. They are very effective for removing large amounts of metal, and the workpiece must be securely held on the milling table.

End Milling End milling (Fig. 25.44) is probably the most versatile milling operation. Many types of end mills can be used on both vertical and horizontal milling machines. End mills are available in sizes ranging from 1/32 to 6 in. (for shell end mills) and in almost any shape needed.

Milling cutters can be used individually or in pairs to machine various slots (Fig. 25.45a). Many milling cutters can also be mounted onto an arbor to perform in a gang milling operation of automotive engine blocks as shown in Figure 25.45(b).

FIGURE 25.43 Face milling operation. (Courtesy Valenite Inc.)

FIGURE 25.44 End milling operation. (Courtesy Valenite Inc.)

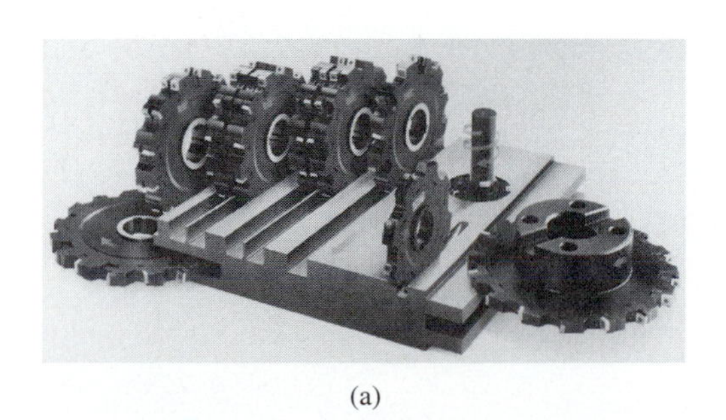

(a)

(b)

FIGURE 25.45 Milling operations: (a) side and face milling cutters used to machine various slots, (b) various indexable milling cutters used to machine an automotive engine block. (Courtesy Sandvik Coromant Co.)

25.7 REVIEW QUESTIONS AND PROBLEMS

1. Explain the milling process and how it differs from turning.
2. Explain the difference between light-duty and heavy-duty plain milling cutters.
3. Discuss the straddle milling cutter setup.
4. Describe the shell end mill.
5. Discuss the many applications of end mills.
6. Explain the advantages and disadvantages of the wedge clamping method.
7. Discuss the screw clamping method.
8. Compare double-negative and double-positive geometry.
9. Discuss the advantages of positive/negative geometry.
10. Discuss the importance of lead angles in milling operations.
11. Why should a nose radius corner geometry always be considered first?
12. Explain the function of a sweep wiper insert.
13. What are the advantages of a column and knee milling machine?
14. What are the advantages of a bed-type milling machine?
15. Explain the differences between a planer and a planer-type milling machine.
16. Describe the profiling process.
17. What is the function of a high-speed milling attachment?
18. Explain the function and list the uses of indexing heads.
19. Describe the A-style milling arbor and discuss its use.
20. How does the feedrate differ in milling as compared to turning?
21. Explain the difference between climb and conventional milling.
22. Compare straddle milling and gang milling.
23. Explain slotting and slitting setups.
24. Compare face milling and end milling.

CHAPTER TWENTY-SIX

BROACHING AND SAWING

CHAPTER OVERVIEW

26.1 INTRODUCTION

The broaching operation is similar to shaping with multiple teeth and is used to machine internal and external surfaces such as holes of circular, square, or irregular shapes, keyways, and teeth of internal gears. A **broach** is a long, multitooth cutting tool with successively deeper cuts. Each tooth removes a predetermined amount of material in a predetermined location. The total depth of material removed in one path is the sum of the depth of cut of each tooth. Broaching is an important production process and can produce parts with very good surface finish and dimensional accuracy. Broaching competes favorably with other processes such as boring, milling, shaping, and reaming. Although broaches tend to be expensive, the cost is justified because of their use for high-production runs. A typical broaching operation is shown in Figure 26.1.

Several types of power saws are also extensively used in machine shops. Once sawing was considered a secondary machining process, and saws were used mostly for cutting bar stock in preparation for other machining operations. In recent years, the development of new types of saws and better blade materials has made metal sawing a much more effective, versatile, and economical process. In many cases bandsaws are now being used as the primary means of shaping certain types of metal parts.

When the proper sawing machines and blades are used, sawing is one of the most economical means of cutting metal. The saw cut (**kerf**) is narrow, and relatively few chips are produced in making a cut. When a bandsaw is used for cutting the contours of complex shapes, only a small portion of the metal is removed in the form of chips. Therefore, the power used in removing large amounts of waste metal is at a minimum.

FIGURE 26.1 Typical broaching operation of an internal spline. (Courtesy Detroit Broach Co.)

26.2 BROACHING

Tooling is the heart of any broaching process. The broaching tool is based on a concept unique to the process—rough, semifinish, and finish cutting teeth combined in one tool or string of tools. A broach tool frequently can finish machine a rough surface in a single stroke.

For exterior surface broaching, the broach tool may be pulled or pushed across a workpiece surface, or the surface may move across the tool. Internal broaching requires a starting hole or opening in the workpiece so the broaching tool can be inserted. The tool or the workpiece is then pushed or pulled to force the tool through the starter hole. Almost any irregular cross section can be broached as long as all surfaces of the section remain parallel to the direction of broach travel.

26.2.1 Broaching Tools

A broach is like a single-point tool with many "points," each of which cuts like a flat-ended shaper tool, although some broaches have teeth set diagonally, called shear cutting. The principal parts of an internal broach are shown in Figure 26.2.

Broach Nomenclature

Front Pilot: When an internal-pull broach is used, the pull end and front pilot are passed through the starting hole. Then the pull end is locked to the pull head of the broaching machine. The front pilot assures correct axial alignment of the tool with the starting hole and serves as a check on the starting hole size.

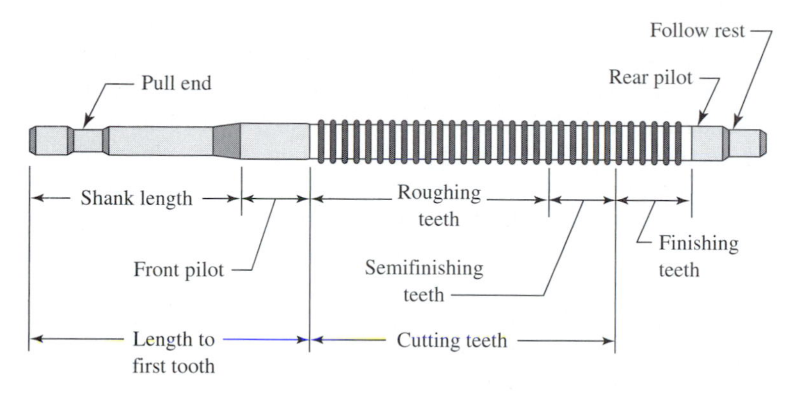

FIGURE 26.2 Principal parts of a round internal-pull broach.

Length: The length of a broach tool or string of tools is determined by the amount of stock to be removed and limited by the machine stroke.

Rear Pilot: The rear pilot maintains tool alignment as the final finish teeth pass through the workpiece hole. On round tools the diameter of the rear pilot is slightly less than the diameter of the finish teeth.

Broach tooth nomenclature and terminology are shown in Figure 26.3(a).

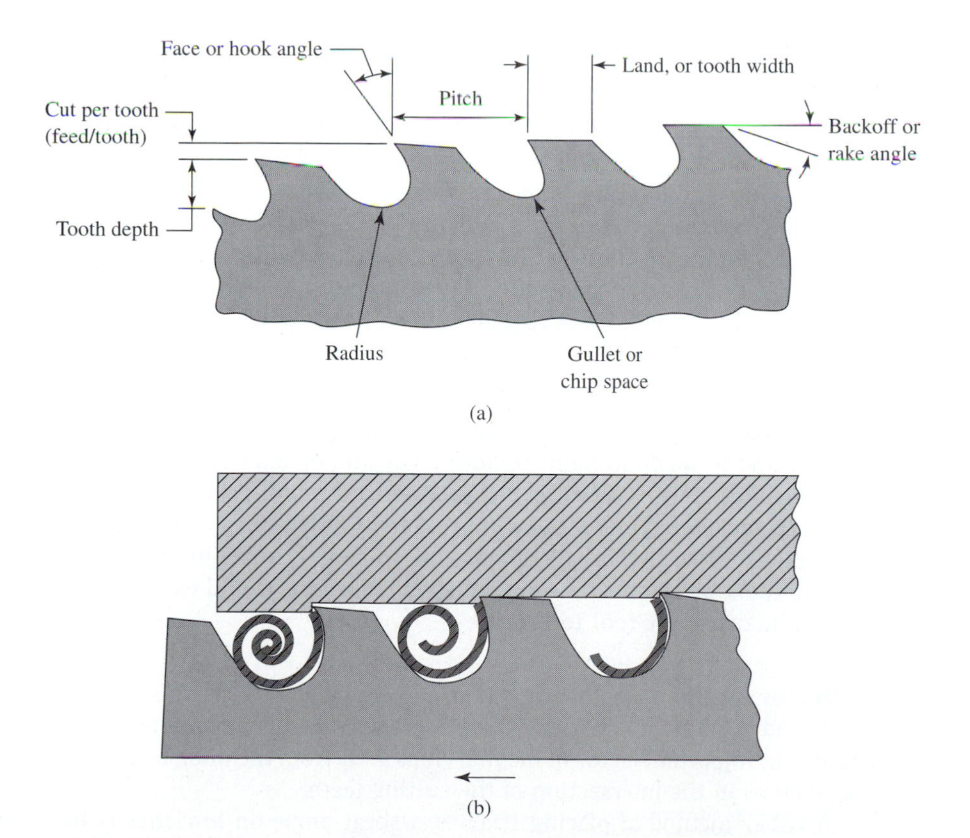

FIGURE 26.3 (a) Broach tooth nomenclature and terminology. (b) Illustration of how a chip fills the gullet during a broaching operation.

Cutting Teeth: Broach teeth are usually divided into three separate sections along the length of the tool: roughing teeth, semifinishing teeth, and finishing teeth (Fig. 26.2). The first roughing tooth is proportionately the smallest tooth on the tool. The subsequent teeth progressively increase in size up to and including the first finishing tooth. The difference in height between each tooth, or the tooth rise, is usually greater along the roughing section and less along the semifinishing section. All finishing teeth are the same size. The face is ground with a hook or face angle that is determined by the workpiece material. For instance, soft steel workpieces usually require greater hook angles, and hard or brittle steel pieces require smaller hook angles.

Tooth Land: The land supports the cutting edge against stresses. A slight clearance or backoff angle is ground onto the lands to reduce friction. On roughing and semifinishing teeth, the entire land is relieved with a backoff angle. On finishing teeth, part of the land immediately behind the cutting edge is often left straight, so that repeated sharpening (by grinding the face of the tooth) will not alter the tooth size.

Tooth Pitch: The distance between teeth, or **pitch**, is determined by the length of cut and influenced by type of workpiece material. A relatively large pitch may be required for roughing teeth to accommodate a greater chip load. Tooth pitch may be smaller on semifinishing teeth to reduce the overall length of the broach tool. Pitch is calculated so that preferably, two or more teeth cut simultaneously. This prevents the tool from drifting or chattering.

Tooth Gullet: The depth of the tooth **gullet** is related to the tooth rise, pitch, and workpiece material. The tooth root radius is usually designed so that chips curl tightly within themselves, occupying as little space as possible (Fig. 26.3b).

When designing broaches, attention must also be given to chip load, chipbreakers, shear angles and side relief.

Chip Load. As each tooth enters the workpiece, it cuts a fixed thickness of material. The fixed chip length and thickness produced by broaching create a chip load that is determined by the design of the broach tool and the predetermined feedrate.

This chip load feedrate cannot be altered by the machine operator as it can in most other machining operations. The entire chip produced by a complete pass of each broach tooth must be freely contained within the preceding tooth gullet (Fig. 26.3b). The size of the tooth gullet is a function of the chip load and the type of chips produced. However the form that each chip takes depends on the workpiece material and hook. Brittle materials produce flakes. Ductile or malleable materials produce spiral chips.

Chipbreakers. Notches, called chipbreakers, are used on broach tools to eliminate chip packing and to facilitate chip removal. The chipbreakers are ground into the roughing and semifinishing teeth of the broach, parallel to the tool axis. Chipbreakers on alternate teeth are staggered so that one set of chipbreakers is followed by a cutting edge. The finishing teeth complete the job. Chipbreakers are vital on round broaching tools. Without the chipbreakers, the tools would machine ring-shaped chips that would wedge into the tooth gullets and eventually cause the tool to break.

Shear Angle. Broach designers may place broach teeth at a shear angle to improve surface finish and reduce tool chatter. When two adjacent surfaces are cut simultaneously, the shear angle is an important factor in moving chips away from the intersecting corner to prevent crowding of chips in the intersection of the cutting teeth.

Another method of placing teeth at a shear angle on broaches is by using a herringbone pattern. An advantage of this design is that it eliminates the tendency for parts to move sideways in the work-holding fixtures during broaching.

Side Relief. When broaching slots, the tool becomes enclosed by the slot during cutting and must carry the chips produced through the entire length of the workpiece. Sides of the broach teeth will rub the sides of the slot and cause rapid tool wear unless clearance is provided. This is done by grinding a single relief angle on both sides of each tooth. Thus only a small portion of the tooth near the cutting edge, called the side land, is allowed to rub against the slot. The same approach is used for one-sided corner cuts and spline broaches.

26.2.2 Types of Broaches

Two major types of broaches are the push broach and the pull broach. A second division is internal and external broaches.

Push and Pull Broaches A push broach must be relatively short because it is a column in compression and will buckle and break under too heavy a load. Push broaches are often used with a simple arbor press if quantities of work are low. For medium- to high-volume production they are used in broaching machines.

Pull broaches (Fig. 26.2) are pulled either up, down, or horizontally through or across the workpiece, always by a machine. Flat or nearly flat broaches may be of the pull type, or the broach may be rigidly mounted, with the workpiece then pulled across the broaching teeth. Automobile cylinder blocks and heads are often faced flat by this method. Figure 26.4 shows various broach configurations both round and flat types. Figure 26.1 shows a vertical spline broaching operation; Figure 26.5 shows a large spline broaching operation using a horizontal broaching machine.

Internal Broaches Internal broaches are either pulled or pushed through a starter hole. The machines can range from fully automated multistationed verticals, to horizontal pull types, to simple presses. Figure 26.6 shows a variety of forms that can be produced by internal broaches.

Keyway Broach. Almost all keyways in machine tools and parts are cut by a keyway broach—a narrow, flat bar with cutting teeth spaced along one surface. Both external and internal keyways can be cut with these broaches. Internal keyways usually require a slotted bushing or horn to fit the hole, with the keyway broach pulled through the horn, guided by the slot.

If a number of parts, all of the same diameter and keyway size, are to be machined, an internal keyway broach can be designed to fit into the hole to support the cutting teeth. Only the cutting teeth extend beyond the hole diameter to cut the keyway. Bushings or horns are not required.

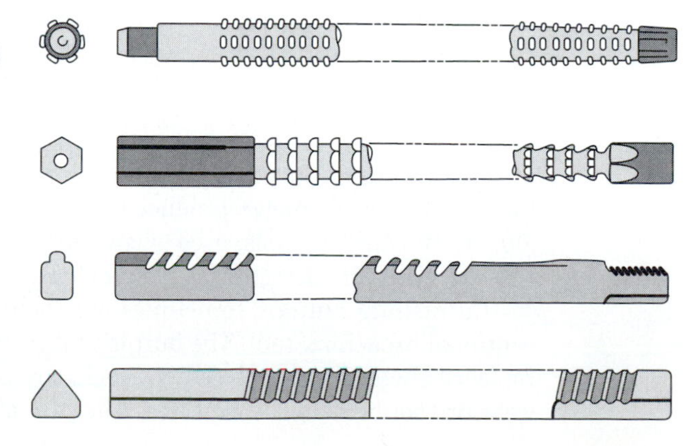

FIGURE 26.4 Various broach configurations, both round and flat types.

FIGURE 26.5 A large spline broaching operation using a horizontal broaching machine. (Courtesy US Broach & Machine Co.)

FIGURE 26.6 A variety of forms that can be produced with an internal broach.

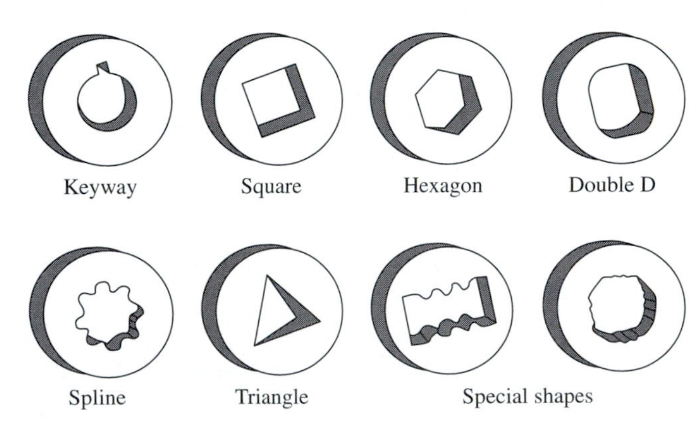

Burnishers. Burnishers are broaching tools designed to polish rather than cut a hole. The total change in diameter produced by a burnishing operation may be no more than .0005 to .001 in. Burnishing tools, used when surface finish and accuracy are critical, are relatively short and are generally designed as push broaches.

Burnishing buttons sometimes are included behind the finishing tooth section of a conventional broaching tool. The burnishing section may be added as a special attachment or easily replaced shell. These replacement shells are commonly used to reduce tooling costs when high wear or tool breakage is expected. They are also used to improve surface finish.

Shell Broaches. Shell broaches can be used on the roughing, semifinishing, and finishing sections of a broach tool. The principal advantage of a shell broach is that worn sections can be removed and resharpened or replaced, at far less cost than a conventional single-piece tool. When shells are used for the finishing teeth of long broaches, the teeth of the shell can be ground to far greater accuracy than those of a long conventional broach tool and the tool can continue to be used by replacing the shell. Shell broaches are similar to shell milling cutters, which were discussed in Chapter 25.

Surface Broaches The broaches used to remove material from an external surface are commonly known as surface broaches. Such broaches are passed over the workpiece surface to be cut, or the workpiece passes over the tool on horizontal, vertical, or chain machines to produce flat or contoured surfaces.

Some surface broaches are of solid construction, but most are of built-up design, with sections, inserts, or indexable tool bits that are assembled end to end in a broach holder or subholder. The holder fits on the machine slide and provides rigid alignment and support. A surface broach assembly is shown in Figure 26.7(a).

Sectional Broaches. Sectional broaches are used to broach unusual or difficult shapes, often in a single pass. The sectional broach may be round or flat, internal or external. The principle behind this tool is similar to that of the shell broach, but straight sections of teeth are bolted along the long axis of the broach rather than being mounted on an arbor. A complex broaching tool can be built up from a group of fairly simple tooth sections to produce a cut of considerable complexity.

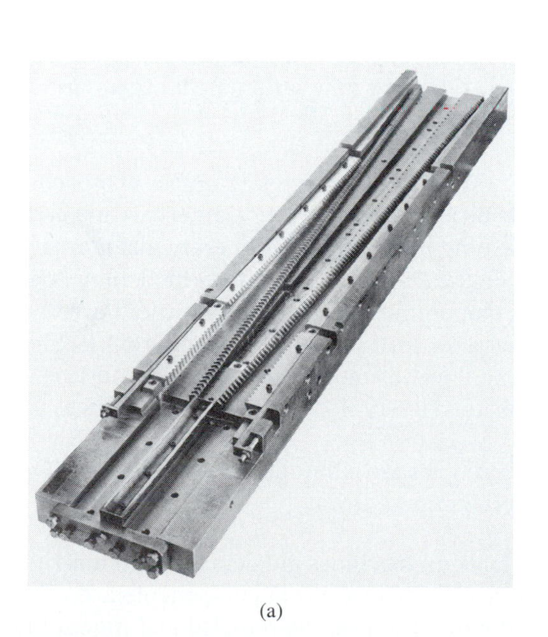

(a)

(b)

FIGURE 26.7 (a) A surface broach assembly. (Courtesy Detroit Broach Co.) (b) A surface broach assembly with indexable carbide inserts. (Courtesy Ingersoll Cutting Tools).

Carbide Broach Inserts. Broaching tools with brazed carbide broach inserts are frequently used to machine cast-iron parts. Present practice, such as machining automotive engine blocks, has moved heavily to the use of indexable inserts (Fig. 26.7(b)), and this has drastically cut tooling costs in many applications.

Slab Broaches. Slab broaches, simple tools for producing flat surfaces, come closest to being truly general-purpose broaches. A single slab broach can be used to produce flat surfaces having different widths and depths on any workpiece by making minor adjustments to the broach, fixture, and/or machine.

Slot Broaches. Slot broaches are for cutting slots, but are not as general purpose in function as slab broaches. Adjustments can easily be made to produce different slot depths, but slot widths are a function of the broach width. When sufficient production volume is required, however, slot broaches are often faster and more economical than milling cutters. In broaching, two or more slots can often be cut simultaneously.

26.2.3 Types of Broaching Machines

The type of broach cutting tool required for a given job is the single most important factor in determining the type of broaching machine to be used. Second in importance is the production requirement. Taken together, these factors usually determine the specific type of machine for the job.

The type of broach tool (internal or surface) immediately narrows down the kinds of machines that could be used. The number of pieces required per hour, or over the entire production run, will further narrow the field.

For internal broaching, the length of a broach in relation to its diameter may determine whether it must be pulled rather than pushed through the workpiece, for a broach tool is stronger in tension than in compression. This in turn helps determine the type of machine for the job.

The type of drive, hydraulic or electromechanical, is another important factor in machine selection. So are convertibility and automation. Some machine designs allow for conversion from internal to surface work. Some designs are fully automated; others are limited in scope and operate only with close operator supervision.

Vertical Broaching Machines About 60% of the total number of broaching machines in existence are vertical, almost equally divided between vertical internals and vertical surface or combination machines. Vertical broaching machines, used in every major area of metalworking, are almost all hydraulically driven. Figure 26.1 shows a vertical broaching operation.

One of the essential features that promoted their development, however, is beginning to turn into a limitation. Cutting strokes now in use often exceed existing factory ceiling clearances. When machines reach heights of 20 feet or more, expensive pits must be dug for the machine so that the operator can work at the factory floor level. A large vertical broaching machine is shown in Figure 26.8(a).

Vertical internal broaching machines are either table-up, pull-up, pull-down, or push-down, depending on their mode of operation.

Vertical Table-Up. Today table-up machines are demanded to meet the cell concept (flexible) manufacturing, where short runs of specialized components are required. Upon completion of short runs (1–2 years) the machines can be retooled and moved to another area of the plant without the problem of what to do with pits in shop floors. With this type of machine the

(a)

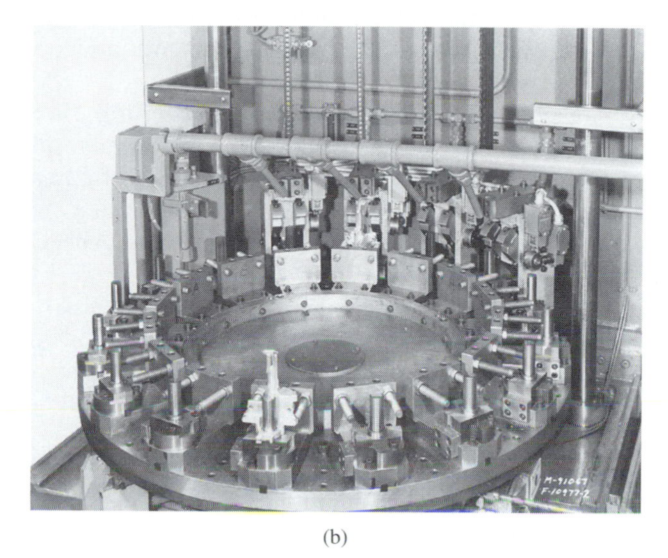

(b)

FIGURE 26.8 (a) A large vertical broaching machine. (b) Special multistation vertical broaching machine fixture. (Courtesy US Broach & Machine Co.)

part sits on a table, which moves up while the broach is stationary. Stroke lengths from 30 to 90 in. and capacities from 5 to 30 tons are the limits for this machine.

Vertical Internal Pull-Up. The pull-up type, in which the workpiece is placed below the worktable, was the first to be introduced. Its principal use is in broaching round and irregular shaped holes. Pull-up machines are now furnished with pulling capacities of 6 to 50 tons, strokes up to 72 in., and broaching speeds of 30 FPM. Larger machines are available; some have electromechanical drives for greater broaching speed and higher productivity.

Vertical Internal Pull-Down. The more sophisticated pull-down machines, in which the work is placed on top of the table, were developed later than the pull-up type. These pull-down machines are capable of holding internal shapes to closer tolerances by means of locating fixtures on top of the work table. Machines come with pulling capacities of 2 to 75 tons, 30- to 110-in. strokes, and speeds of up to 80 FPM.

Vertical Internal Push-Down. Vertical push-down machines are often nothing more than general-purpose hydraulic presses with special fixtures. They are available with capacities of 2 to 25 tons, strokes up to 36 in., and speeds as high as 40 FPM. In some cases, universal machines have been designed that combine as many as three different broaching operations, such as push, pull, and surface, simply through the addition of special fixtures. A special multistation vertical broaching machine fixture is shown in Figure 26.8(b).

Horizontal Broaching Machines The favorite configuration for broaching machines seems now to have come full circle. The original gear- or screw-driven machines were designed as horizontal units. Gradually, the vertical machines evolved as it became apparent that floor space could be much more efficiently used with vertical units. Now the horizontal machine, both hydraulically and mechanically driven, is again finding increasing favor among users because of its very long strokes and the limitation that ceiling height places on vertical machines. About 40% of all broaching machines are now horizontals. For some types of work such as roughing and finishing automotive engine blocks, they are used exclusively. A two-station internal horizontal broaching machine is shown in Figure 26.9(a).

Horizontal Internal Broaching Machines. By far the greatest amount of horizontal internal broaching is done on hydraulic pull-type machines for which configurations have become somewhat standardized over the years. Fully one third of the broaching machines in existence are this type, and of these nearly one fourth are over 20 years old. They find their heaviest application in the production of general industrial equipment but can be found in nearly every type of industry.

Hydraulically driven horizontal internal machines are built with pulling capacities ranging from $2\frac{1}{2}$ to 75 tons, the former representing machines only about 8 feet long, the latter machines over 35 feet long. Strokes up to 120 in. are available, with cutting speeds generally limited to less than 40 FPM.

Horizontal Surface Broaching Machines. This type accounts for only about 10% of existing broaching machines, but this is not indicative of the percentage of the total investment they represent or of the volume of work they produce. Horizontal surface broaching machines belong in a class by themselves in terms of size and productivity. Only the large continuous

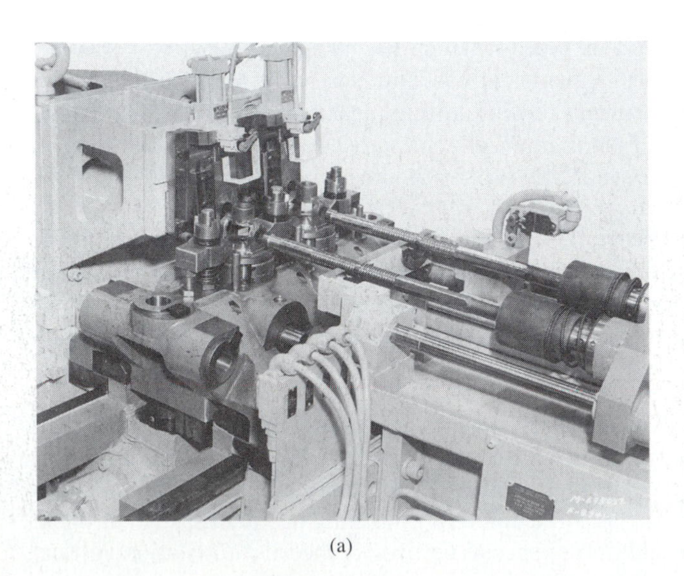

(a)

(b)

FIGURE 26.9 (a) Two-station internal horizontal broaching machine. (b) Gear broaching operation. (Courtesy Detroit Broach Co.)

horizontal units can match or exceed them in productivity. Horizontal surface units are manufactured in both hydraulically and electromechanically driven models, with the latter now becoming dominant. A gear broaching operation is shown in Figure 26.9(b).

The older hydraulically driven horizontal surface machines now are produced with capacities up to 40 tons, strokes up to 180 in., and normal cutting speeds of 100 FPM. These machines, a major factor in the automotive industry for many years, turn out a great variety of cast-iron parts. They use standard carbide cutting tools and have some of the highest cutting speeds used in broaching.

Electromechanically driven horizontal surface machines are taking over at an ever-increasing rate for certain applications, despite their generally higher cost. Because of their smooth ram motion and the resultant improvements in surface finish and part tolerances, these machines have become the largest class of horizontal surface broaching units built. They are available with pulling capacities in excess of 100 tons, strokes up to 30 feet, and cutting speeds, in some instances, of over 300 FPM.

Chain Broaching Machines. These have been the most popular type of machine produced for high-production surface broaching. The key to the productivity of a continuous horizontal broaching machine is elimination of the return stroke by mounting the workpieces, or the tools, on a continuous chain (Fig. 26.10a). Most frequently, the tools remain stationary, mounted in a tunnel in the top half of the machine, and the chain-mounted workpieces pass underneath them. A schematic of a chain broaching machine is shown in Figure 26.10(b).

26.3 SAWING

Sawing is a process where a narrow slit is cut into the workpiece by a tool consisting of a series of narrowly spaced teeth called a saw blade. Sawing is normally used to separate work parts into two or more pieces or to cut off an unwanted section of a part. These processes are often called cutoff operations and because many manufacturing projects require cutoff operations at some point of the production sequence, sawing is an important manufacturing process.

Sawing is a basically simple process. As the blade moves past the work, each tooth takes a cut. Depending on the thickness or diameter of the work, the number of teeth cutting at one time varies from 2 to 10 or more. Saws may be of the continuous cutting (band or rotary) or reciprocating type. A typical sawing operation is shown in Figure 26.11.

The cutting speeds and characteristics of the materials must be understood before the proper blades and operating conditions can be selected. Saws are an effective and efficient category of machine tools found in almost every type of machine shop.

26.3.1 Saw Blades

All saw blades have certain common characteristics and terminology. Some of these terms are shown in Figure 26.12, and others are explained here.

Rake Angles: Rake angles are 0° or neutral rake on most saw blades. Some have a positive rake angle as shown in Figure 26.12(a).

Width: The width of a saw blade is its total width including the teeth.

Set: The set of a saw blade means the offsetting of some teeth so that the back of the blade clears the cut. The "raker" set is most frequently used and is furnished with all hacksaws and band saws unless otherwise specified (see Fig. 26.12b).

Kerf: The kerf is the width of the cut made by the saw blade or the material cut away. The thickness of the blade is called the gage.

FIGURE 26.10 (a) Continuous chain broaching operation. (Courtesy US Broach & Machine Co.) (b) Schematic illustration of a continuous chain broaching machine.

(a)

(b)

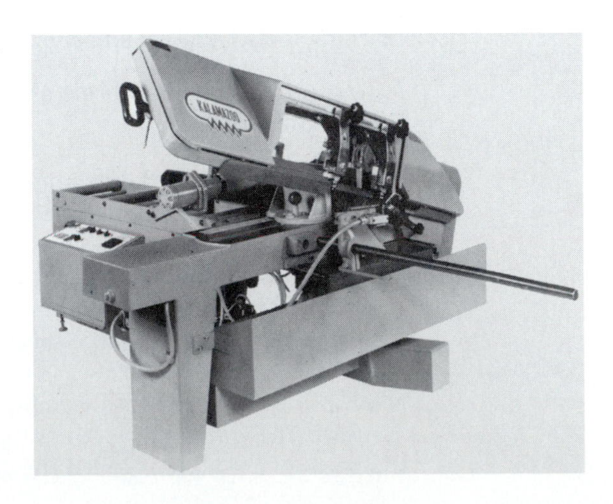

FIGURE 26.11 Typical sawing operation. (Courtesy Clausing Industries, Inc.)

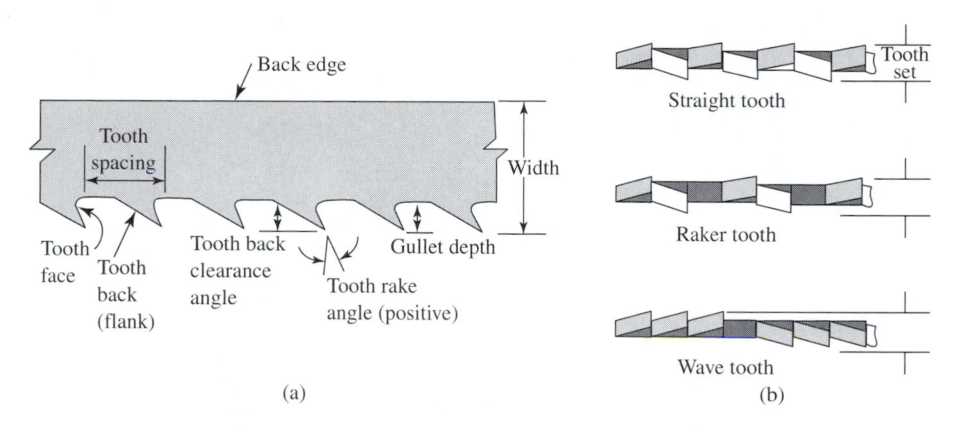

FIGURE 26.12 Saw blade characteristics and terminology.

Pitch: The pitch of a saw blade is the distance between the tops of two adjacent teeth. This is specified in teeth per inch.

Saw Blade Material Saw blades are made from various materials:

Carbon Steel: General utility for small-lot, low-speed work. The least expensive blade, these may have a hard "back" for greater wear.

High-Speed Steel: This costs two to three times as much as carbon steel, but it is much longer wearing and is a necessity for the "difficult-to-machine" metals.

High-Speed Edge: This is a carbon steel blade that has a narrow strip with HSS teeth welded on. This is a tough blade, intermediately priced, and widely used for most materials.

Tungsten Carbide Tipped Blades: These are available in a few sizes and are used only on large, very rigid sawing machines for high-production sawing of difficult materials.

26.3.2 Sawing Equipment

In most sawing operations, the work is held stationary and the saw blade is moved relative to it. As shown in Figure 26.13, there are three basic types of sawing operations, according to the saw blade motion involved.

Hacksawing Hacksawing involves a linear reciprocating motion of the saw against the workpiece. This method of sawing is often used in cutoff operations. Cutting only takes place on the forward stroke of the saw blade. Due to this intermittent cutting action, hacksawing is less efficient than other sawing methods. Hacksawing can be done manually or with a power hacksaw. A power hacksaw provides a drive mechanism to operate the saw blade at a desired speed and feed rate (Fig. 26.13a).

Power Hacksaw. The power hacksaw is the original and least expensive saw for the work. As shown in Figure 26.14(a), these saws work the same as a hand hacksaw: They cut on the forward stroke and then lift slightly so that the blade does not drag on the return stroke.

The size of a power hacksaw is the cross section of the largest piece of stock which it can cut. Typical sizes are 6 × 6 inches to 24 × 24 ins. The motors used will vary from 1 to 10 HP.

The speed of these saws is in strokes per minute. This may be from 30 strokes per minute, for large cuts with heavy saws on difficult materials, up to 165 strokes per minute on carbon steels and nonferrous materials. The hacksaw usually has four to six different speeds available.

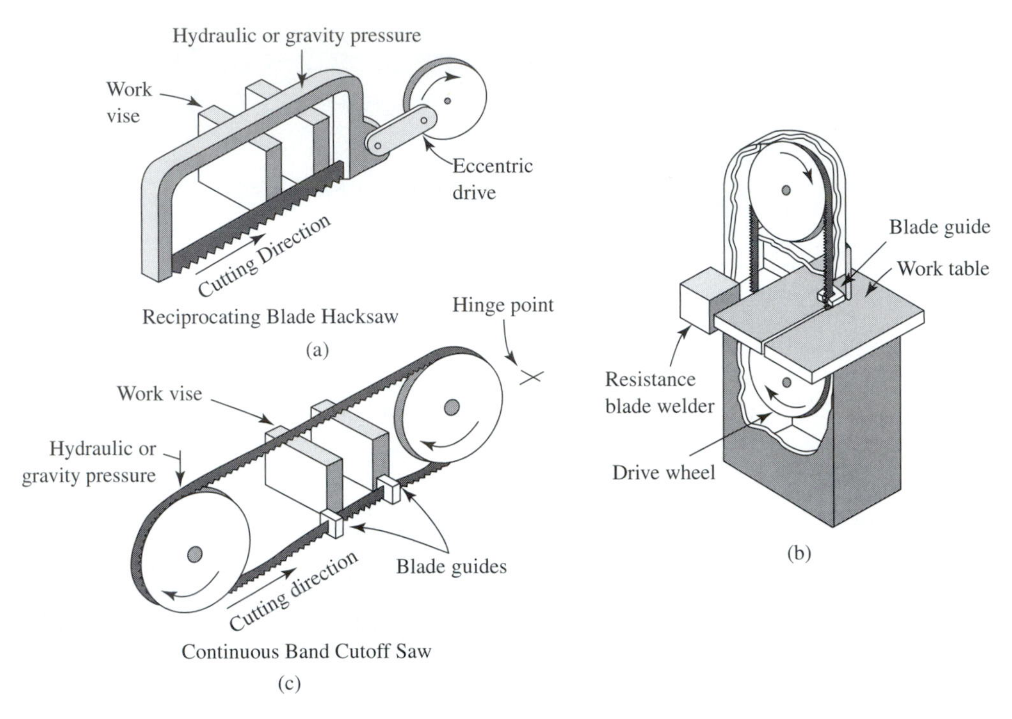

FIGURE 26.13 Three basic types of sawing operations: (a) hacksawing, (b) vertical bandsawing, (c) horizontal bandsawing.

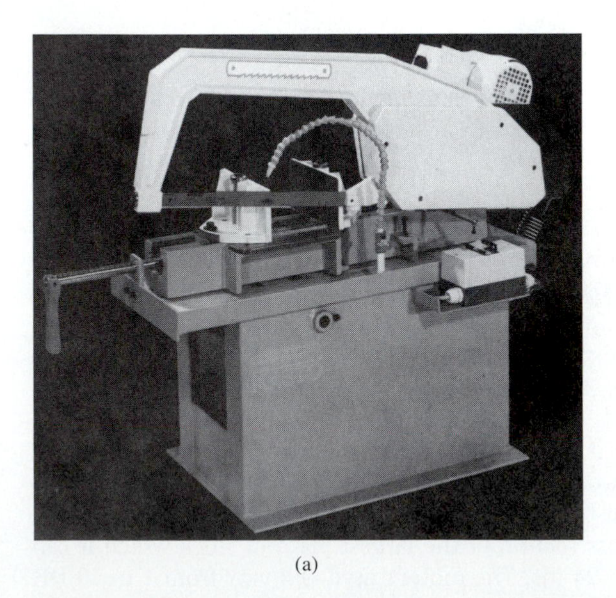

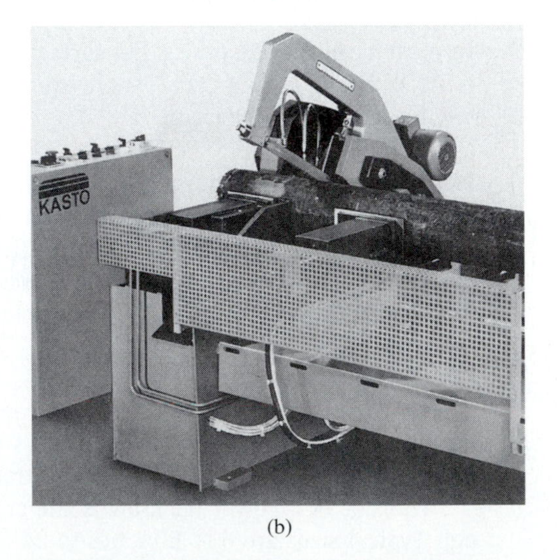

(a) (b)

FIGURE 26.14 (a) Semiautomatic power hacksaw. (b) Automatic power hacksaw used in high-production sawing. (Courtesy Kasto-Racine, Inc.)

Feed may be a positive advance per stroke or may be gaged by a friction or pressure drive. The smaller power hacksaws feed about .006 in. per stroke and the larger ones .012 to .030 in. per stroke. Feed pressures will be 450 to 750 pounds on the blades. Work is held in a built-in vise, which may be hand or power operated.

Automatic power hacksaws (Fig. 26.14b) will feed the stock a preset length, clamp the vise, cut off, and raise the saw for the next cut, all with preset gages and limit switches. These will cut accurate lengths to within .010 in. or less. They are, of course, expensive, so they would be used only if a large amount of work is to be done.

Bandsawing

Bandsawing involves a linear continuous motion, using a bandsaw blade made in the form of an endless loop. The band saw provides a **pulley**-like drive mechanism to continuously move and guide the bandsaw blade past the work. Bandsaws are classified as vertical or horizontal. This designation refers to the direction of saw blade motion during cutting. Vertical bandsaws are used for cutoff and other operations such as contouring and slotting. Horizontal bandsaws are normally used for cutoff operations as alternatives to power hacksaws (Fig. 26.13b and c).

Vertical Bandsaws. All vertical bandsaws, regardless of whether they are light-, medium-, or heavy-duty machines, are made up of certain basic components. Although these major parts of the machine may be made by different methods, depending on the manufacturer, their function is essentially the same. A typical vertical bandsaw is shown in Figure 26.15(a).

Vertical bandsaws are available in sizes and configurations ranging from light-duty hand-fed machines to heavy-duty machines with power-fed tables. The light-duty machines usually have two wheels and are driven through a variable-speed belt drive, V-belts and step pulleys, or some other type of speed-change mechanism. Blades ranging from $\frac{3}{16}$ in. to $\frac{5}{8}$ in. in width can be used on light-duty machines.

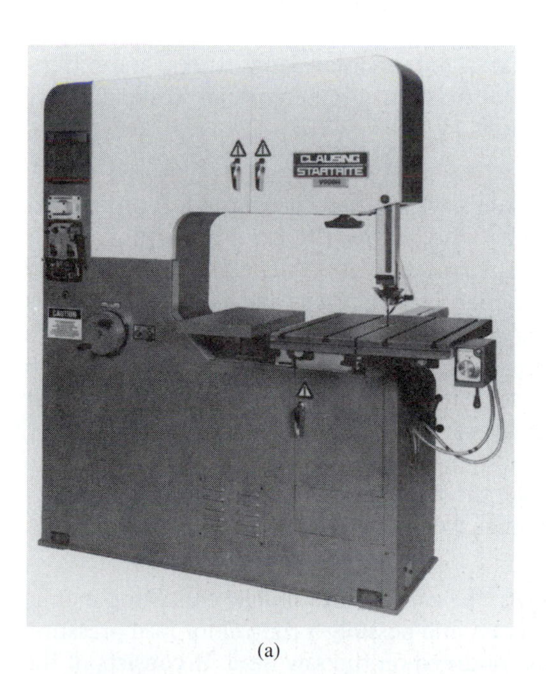

(a)

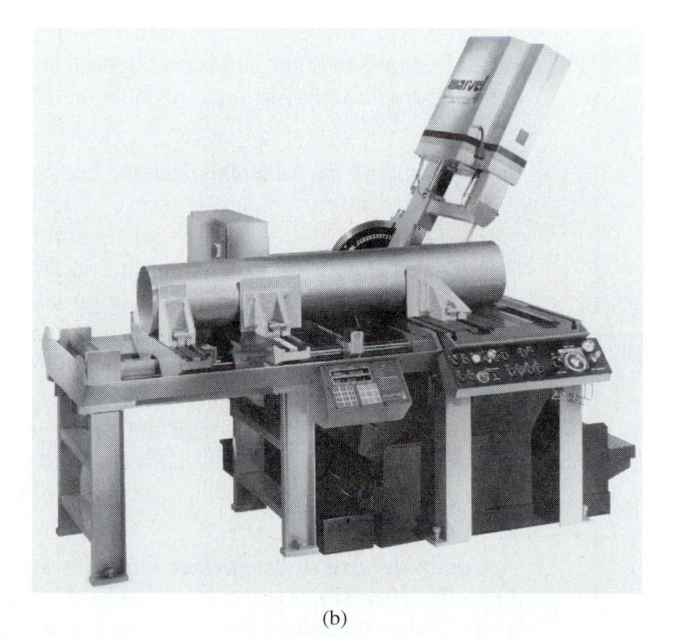

(b)

FIGURE 26.15 (a) Typical vertical bandsaw. (Courtesy Clausing Industries, Inc.) (b) High-production automated vertical bandsaw table machine. (Courtesy Armstrong-Blum Mfg. Co.)

***Table Types*:** The table of the vertical metal cutting bandsaw is usually made of cast iron and fitted with a tilting mechanism so that simple or compound angle cuts can be made. On fixed-table machines, the table does not move with the work, but can be tilted 45° to the right and 10° to the left on most machines. The work can be fed and guided manually, or a weight-operated feed mechanism can be used to supply the feed pressure.

Vertical bandsaws with power tables are generally heavy-duty machines. The feed pressure is provided by the mechanism that moves the table; the feedrate can be varied by the operator.

There is usually enough power available to make effective use of high-speed steel or tungsten carbide saw blades rather than the high-carbon steel blades used on light-duty machines. Coolant systems are also widely used on power table machines, thus allowing higher cutting speeds and higher feedrates along with longer blade life. Many types of fixtures can be used on power table machines, particularly when they are used for repetitive operations. A high-production automated table machine is shown in Figure 26.15(b).

***Accessories*:** Most bandsaws that do not have a coolant system have an air pump that directs a stream of air at the point where the blade is cutting the workpiece. This removes the chips, letting the operator see the layout lines clearly, and provides some cooling.

If the machine has a fluid coolant system, the tank and pump are usually located in the base. The pump is controlled by a separate switch. Coolant systems are usually found on medium- and heavy-duty vertical bandsaws.

Blade welding attachments, which are a specialized form of electric butt-welding machines, are a standard accessory on almost all bandsaws. The blade welder usually consists of cast copper or bronze blade clamps, a grinder, a saw thickness gage, and the necessary switches and operating levers.

Weight-operated feed devices can be used on bandsaws not fitted with power-feed attachments. This reduces operator fatigue and generally results in more uniform feedrates and longer blade life.

Other attachments such as fixtures for cutting arcs and circles, ripping fences, and miters, are used extensively on bandsaws. Special fixtures for holding specific types of workpieces are often designed for use in mass production applications.

Horizontal Bandsaws. Because horizontal bandsaws are used primarily for cutting bar stock and structural shapes, they are also known as cutoff saws. The band-type cutoff saw is widely used because it is easy to set up and takes a narrow saw cut, thus requiring less power to operate and wasting less material. The cutting action is continuous and rapid. The blade is supported close to either side of the material being cut, so the cut is accurate if the machine is properly adjusted and the blade is in good condition. A typical horizontal bandsaw is shown in Figure 26.16(a). Horizontal bandsaws range in capacity from small, fractional horsepower machines (Fig. 26.16a), to large heavy-duty industrial saws, as shown in Figure 26.16(b).

The saw guides are an important factor in accurate cutoff operations. The saw blade has to twist as it leaves the idler pulley, and the guides make the blade travel perpendicular to the material being cut. Tungsten carbide inserts help minimize wear.

***Controls and Accessories*:** On light-duty saws, the controls are simple, consisting mainly of an off–on switch, a means for changing blade speed, and possibly a control for feed pressure. On the larger machines a control panel is usually mounted on the saw head. It consists of the necessary switches, valves, and instruments that indicate blade speed in feet per minute, feedrate in inches per minute, and other factors, such as blade tension. Some machines used for

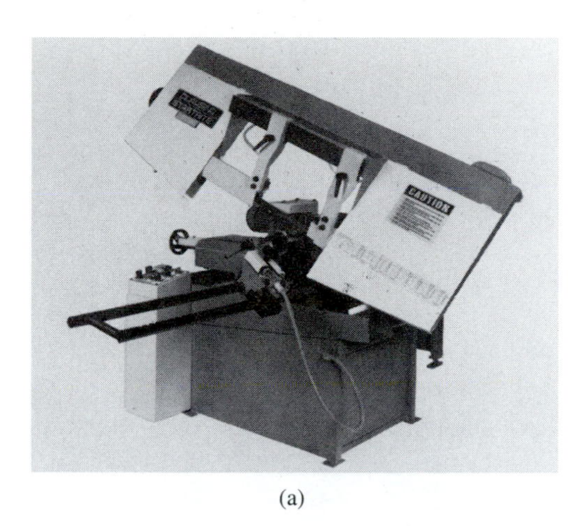

(a)

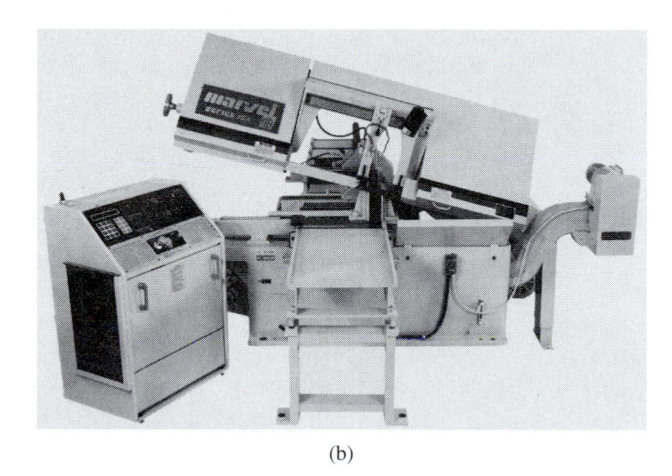

(b)

FIGURE 26.16 (a) Typical horizontal bandsaw. (Courtesy Clausing Industries, Inc.) (b) Large, heavy-duty industrial horizontal bandsaw. (Courtesy Armstrong-Blum Mfg. Co.)

production work are capable of fully automatic operation and can be preset to cut a given number of pieces of work. A counter is usually part of the instrumentation on semiautomatic and automatic machines.

There are coolant systems on almost all medium- and heavy-duty horizontal bandsaws. The coolant extends blade life and allows higher cutting speeds and metal removal rates. The rate of coolant flow is controlled by the operator. Solid lubricants such as wax or grease can also be used. Wax in stick form is usually applied manually to the blade on light-duty machines.

Circular Sawing

Circular sawing uses a rotating saw blade to provide a continuous motion of the tool past the work. Circular sawing is often used to cut long bars and tubes to specific lengths. The cutting action is similar to slot milling (Figs. 25.39c and 25.42a), except that the saw blade is thinner and contains more cutting teeth. Circular sawing machines have power spindles to rotate the saw blade and a feeding mechanism to drive the rotating blade into the work.

Cold Saws. Most cold saws, regardless of size, consist of a base, drive mechanism, blade arbor, vise, feed mechanism, and necessary guards and switches. On some small saws the blade is fed into the work by hand (Fig. 26.17). On larger machines the feed mechanism is pneumatically or hydraulically operated. The rate of feed is controlled by the operator (Fig. 26.18). The base of the machine or the vise can be swiveled to make angular cuts. In some cases two machines can be set up on a single work stand for production operations.

Cold Saw Blades: Blades smaller than 18 in. in diameter are cut directly in the rim of the saw disk. For cutting soft materials, the teeth are spaced farther apart, as in the case of bandsaw and power hacksaw blades, so that the gullet (the space between the teeth) will be large enough to accommodate large chips. When cutting thin tubing or other thin materials, use saw blades with closely spaced teeth to avoid chattering and tooth breakage. Cold saw blades with teeth cut directly on the periphery of the disk may be made of high-carbon or high-speed steel.

FIGURE 26.17 Typical cold saw. (Courtesy Clausing Industries, Inc.)

FIGURE 26.18 Large, heavy-duty industrial cold saw. (Courtesy Clausing Industries, Inc.)

Larger blades usually have segmented teeth. The body of the blade is made of a rough, resilient alloy steel, and the inserted teeth are made of high-speed steel or tungsten carbide. The individual teeth or segments of three or four teeth are wedged or riveted to the blade and can be easily replaced if a tooth is damaged or broken. Larger cold saw blades can cut a kerf as wide as $\frac{1}{4}$ in. and remove metal rapidly.

An operation similar to circular sawing is abrasive cutoff machining. In abrasive cutoff, an abrasive disk is used to cut hard materials. Abrasive cutoff will be covered in Chapter 27, "Grinding Operations."

26.4 REVIEW QUESTIONS AND PROBLEMS

1. Define the broaching process.
2. Describe the different sections of teeth on a broaching tool.
3. What is the purpose of the tooth gullet?
4. Explain the importance of side relief angles.
5. Discuss the differences between push and pull broaches.
6. Explain the burnishing process.
7. What are the advantages of shell broaches?
8. Compare vertical and horizontal broaching machines.
9. Explain how chain broaches differ from other broaching machines.
10. Define the sawing process.
11. Discuss the choices of saw blade materials.
12. Explain why hacksawing is not considered an economical process.
13. Compare vertical and horizontal bandsaws.
14. Discuss some of the accessories used with bandsaws.
15. Describe the circular sawing process.
16. Why is coolant necessary in sawing operations?

CHAPTER TWENTY-SEVEN

GRINDING OPERATIONS

CHAPTER OVERVIEW

- Introduction
- Grinding Wheels
  - Types of Abrasives
  - Types of Bonds
  - Abrasive Grain Size
  - Grinding Wheel Grade
  - Grinding Wheel Structure
- Grinding Wheel Specifications
  - Grinding Wheel Markings
  - Grinding Wheel Shapes and Faces
- Grinding Machines and Operations
  - Surface Grinding
  - Cylindrical Grinding
  - Centerless Grinding
  - Special Grinding Processes
- Grinding Wheel Selection
- Wheel Balancing, Dressing, and Truing
- Grinding Wheel Wear
- Grindability
- Review Questions and Problems

27.1 INTRODUCTION

Grinding, or **abrasive** machining, is one of the most rapidly growing metal removal processes in manufacturing. Many machining operations previously done on conventional milling machines, lathes, and shapers are now being performed on various types of grinding machines. Greater productivity, improved accuracy, reliability, and rigid construction characterize today's industrial grinding machines. A typical internal grinding operation is shown in Figure 27.1.

FIGURE 27.1 Typical internal grinding operation. (Courtesy Kellenberger, A Hardinge Co.)

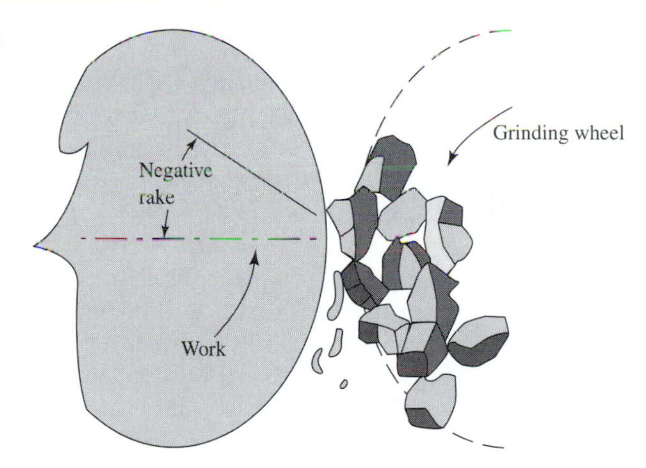

FIGURE 27.2 Abrasive grains cutting material during a grinding operation.

Grinding, or abrasive machining, is the process of removing metal in the form of minute chips by the action of irregularly shaped abrasive particles. These particles may be in bonded wheels, in coated belts, or simply loose. The way an abrasive grain cuts material is shown in Figure 27.2.

27.2 GRINDING WHEELS

Grinding wheels are composed of thousands of small abrasive grains held together by a bonding material. Some typical grinding products are shown in Figure 27.3. Each abrasive grain is a cutting edge. As the grain passes over the workpiece it cuts a small chip, leaving a smooth, accurate surface. As each abrasive grain becomes dull, it breaks away from the bonding material because of machining forces and exposes new, sharp grains.

FIGURE 27.3 Typical grinding products. (Courtesy Norton Company)

27.2.1 Types of Abrasives

Two types of abrasives are used in grinding wheels: natural and manufactured. Except for diamonds, manufactured abrasives have almost totally replaced natural abrasive materials. Even natural diamonds have been replaced in some cases by synthetic diamonds. The manufactured abrasives most commonly used in grinding wheels are aluminum oxide, **silicon carbide**, cubic boron nitride, and diamond.

Aluminum Oxide Aluminum oxide is made by refining bauxite ore in an electric furnace. The bauxite ore is first heated to eliminate any moisture, then mixed with coke and iron to form a furnace charge. The mixture is then fused and cooled. The fused mixture resembles a rocklike mass. It is washed, crushed, and screened to separate the various grain sizes.

Aluminum oxide wheels are manufactured with abrasives of different degrees of purity to give them certain characteristics for different grinding operations and applications. The color and toughness of the wheel are influenced by the degree of purity.

General-purpose aluminum oxide wheels, usually gray and 95% pure, are the most popular abrasives used. They are used for grinding most steels and other ferrous alloys. White aluminum oxide wheels are nearly pure and are very friable (able to break away from the bonding material easily). They are used for grinding high-strength, heat-sensitive steels.

Silicon Carbide Silicon carbide grinding wheels are made by mixing pure white quartz, petroleum coke, and small amounts of sawdust and salt, and firing the mixture in an electric furnace. This process is called synthesizing the coke and sand. As in the making of aluminum oxide abrasive, the resulting **crystalline** mass is crushed and graded by particle size.

Silicon carbide wheels are harder and more brittle than aluminum oxide wheels. There are two principal types of silicon carbide wheels: black and green. Black wheels are used for grinding cast irons; nonferrous metals such as copper, brass, aluminum, and magnesium; and nonmetallics such as ceramics and gem stones. Green silicon carbide wheels are more friable than the black wheels and are used for tool and cutter grinding of cemented carbide.

Cubic Boron Nitride Cubic Boron Nitride (CBN) is an extremely hard, sharp, and cool cutting abrasive. It is one of the newest manufactured abrasives and $2\frac{1}{2}$ times harder than aluminum oxide. It can withstand temperatures up to 2500°F. Cubic boron nitride is produced by high-temperature, high-pressure processes similar to those used to produce manufactured diamond and is nearly as hard as diamond.

Cubic boron nitride is used for grinding superhard high-speed steels, tool and die steels, hardened cast irons, and stainless steels. Two types of CBN wheels are used in industry today. One type is metal coated to promote good bond adhesion and is used in general-purpose grinding. The second type is an uncoated abrasive for use in electroplated metal and vitrified bond systems.

Diamond Two types of diamond are used in the production of grinding wheels: natural and manufactured. Natural diamond is a crystalline form of carbon and very expensive. In the form of bonded wheels, natural diamonds are used for grinding very hard materials such as cemented carbides, marble, granite, and stone.

Recent developments in the production of manufactured diamonds have brought their cost down and led to expanded use in grinding applications. Manufactured diamonds are now used for grinding tough and very hard steels, cemented carbide, and aluminum oxide cutting tools.

27.2.2 Types of Bonds

Abrasive grains are held together in a grinding wheel by a bonding material. The bonding material does not cut during a grinding operation. Its main function is to hold the grains together with varying degrees of strength. Standard grinding wheel bonds are vitrified, resinoid, silicate, shellac, rubber, and metal.

Vitrified Bond Vitrified bonds are used on more than 75% of all grinding wheels. Vitrified bond material is comprised of finely ground clay and fluxes with which the abrasive is thoroughly mixed. The mixture of bonding agent and abrasive in the form of a wheel is then heated to 2400°F to fuse the materials.

Vitrified wheels are strong and rigid. They retain high strength at elevated temperatures and are practically unaffected by water, oils, or acids. One disadvantage of vitrified bond wheels is that they exhibit poor shock resistance. Therefore, their application is limited where impact and large temperature differentials occur.

Resinoid Bond Resinoid bonded grinding wheels are second in popularity to vitrified wheels. Phenolic resin in powdered or liquid form is mixed with the abrasive grains in a form and cured at about 360°F. Resinoid wheels are used for grinding speeds up to 16,500 SFPM. Their main use is in rough grinding and cutoff operations. Care must be taken with resinoid bonded wheels because they will soften if they are exposed to water for extended periods of time.

Silicate Bond This bonding material is used when heat generated by grinding must be kept to a minimum. Silicate bonding material releases the abrasive grains more readily than other types of bonding agents. Speed is limited to below 4500 SFPM.

Shellac Bond Shellac is an organic bond used for grinding wheels that produce very smooth finishes on parts such as rolls, cutlery, camshafts, and crankpins. These wheels are not generally used on heavy-duty grinding operations.

Rubber Bond Rubber bonded wheels are extremely tough and strong. Their principal uses are as thin cutoff wheels and driving wheels in centerless grinding machines. They are also used when extremely fine finishes are required on bearing surfaces.

Metal Bond Metal bonds are used primarily as bonding agents for diamond abrasives. They are also used in **electrolytic grinding**, where the bond must be electrically conductive.

27.2.3 Abrasive Grain Size

The size of an abrasive grain is important because it influences stock removal rate, chip clearance in the wheel, and surface finish obtained.

Abrasive grain size is determined by the size of the screen opening through which the abrasive grits pass. The number of the nominal size indicates the number of the openings per inch in the screen. For example, a 60-grit-sized grain will pass through a screen with 55 openings per inch, but it will not pass through a screen size of 65. A low grain size number indicates a large **grit**, and a high number indicates a small grain.

Grain sizes vary from 6 (very coarse) to 1000 (very fine). Grain sizes are broadly defined as coarse (6 to 24), medium (30 to 60), fine (70 to 180), and very fine (220 to 1000). Figure 27.4 shows a comparison of three different grain sizes and the screens used for sizing. Very fine grits are used for polishing and **lapping** operations, fine grains for fine finish and small-diameter grinding operations. Medium grain sizes are used in high-stock-removal operations where some control of surface finish is required. Coarse grain sizes are used for billet conditioning and **snagging** operations in steel mills and foundries, where stock removal rates are important and there is little concern about surface finish.

27.2.4 Grinding Wheel Grade

The grade of a grinding wheel is a measure of the strength of the bonding material holding the individual grains in the wheel. It is used to indicate the relative hardness of a grinding wheel. Grade or hardness refers to the amount of bonding material used in the wheel, not to the hardness of the abrasive. A soft wheel has less bonding material than a hard wheel.

The range used to indicate grade is A to Z, with A representing maximum softness and Z maximum hardness. The selection of the proper grade of wheel is very improtant. Wheels that are too soft tend to release grains too rapidly, and wheel wear is great. Wheels that are too hard do not release the abrasive grains fast enough, and the dull grains remain bonded to the wheel, causing a condition know as **glazing**.

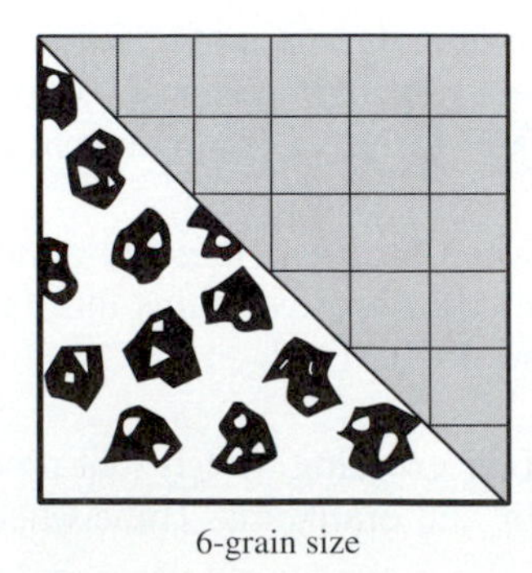

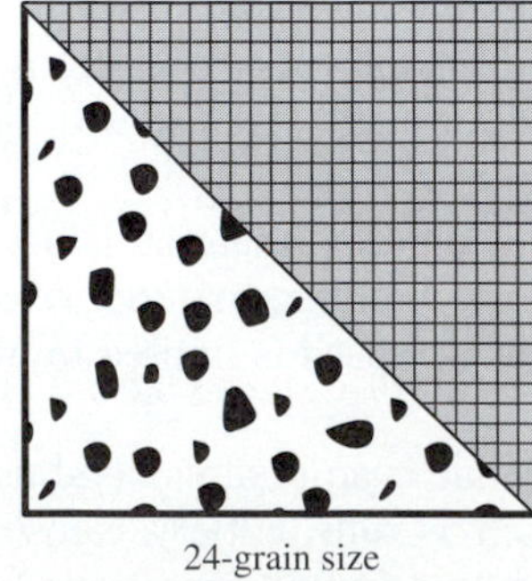

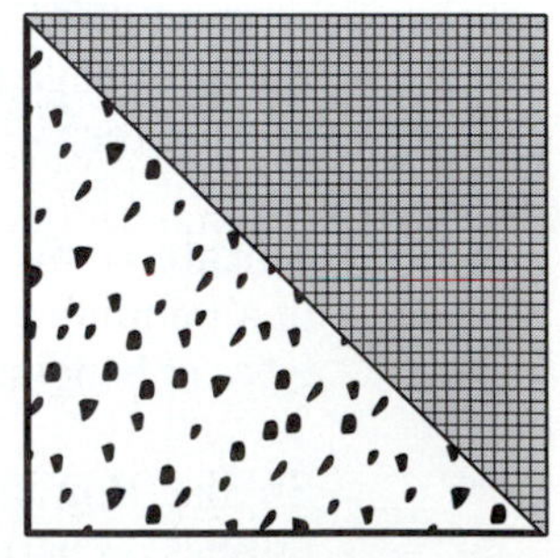

FIGURE 27.4 Comparison of three different grain sizes.

An open-structure grinding wheel

A medium-structure grinding wheel

A dense- or closed-structure grinding wheel

FIGURE 27.5 Comparison of three different grain structures.

27.2.5 Grinding Wheel Structure

The structure of a grinding wheel refers to the relative spacing of the abrasive grains; it is the wheel's density. There are fewer abrasive grains in an open-structure wheel than in a closed-structure wheel. Figure 27.5 shows a comparison of different structures used in a grinding wheel. The structure of a wheel is designated by a number from 1 to 15. The higher the number, the more open is the structure; the lower the number, the more dense the structure.

27.3 GRINDING WHEEL SPECIFICATIONS

Grinding wheel manufacturers have agreed to a standardization system to describe wheel composition as well as wheel shapes and faces.

27.3.1 Grinding Wheel Markings

Abrasive Grinding Wheels This marking system is used to describe the wheel composition as to type of abrasive, grain size, grade, structure, and bond type. Figure 27.6 illustrates this standard marking system.

CBN and Diamond Wheels The same standardization is applicable to CBN and diamond wheels. Some typical CBN and diamond wheels are shown in Figure 27.7. Wheel markings are a combination of letters and numbers as shown in Figure 27.8.

27.3.2 Grinding Wheel Shapes and Faces

Eight standard wheel shapes and 12 standard wheel faces have been adopted for general use by most grinding wheel manufacturers. Figure 27.9 shows the most common standard wheel shapes used on all types of grinders. Figure 27.10 illustrates the standard wheel faces used on most grinding wheel shapes.

STANDARD MARKING SYSTEM CHART

ANSI STANDARD B74.13 – 1970

Sequence: Prefix	1 Abrasive Type	2 Grain Size	3 Grade	4 Structure	5 Bond Type	6 Manufacturer's Record
51	A	36	L	5	V	23

Prefix	Abrasive Type	Abrasive (grain) Size				Grade			Structure		Bond Type	Manufacturer's Record
		Coarse	Medium	Fine	Very fine	Soft	Medium	Hard	Dense to Open			
Manufacturer's symbol indicating exact kind of abrasive (*Use optional*)	A • Aluminum Oxide C • Silicon Carbide	8	30	70	220	A E	I M	Q V	1	9	B • Resinoid	Manufacturer's private marking to identify when (*Use optional*)
		10	36	80	240	B F	J N	R W	2	10	BF • Resinoid Reinforced	
		12	46	90	280	C G	K O	S X	3	11	E • Shellac	
		14	54	100	320	D H	L P	T Y	4	12	O • Oxychloride	
		16	60	120	400			U Z	5	13	R • Rubber	
		20		150	500				6	14	RF • Rubber Reinforced	
		24		180	600				7	15	S • Silicate	
									8	16 etc.	V • Vitrified	
									(*Use optional*)			

FIGURE 27.6 ANSI standard marking system for abrasive grinding wheels. (Unified abrasives manufacturers' association.)

FIGURE 27.7 Typical Cubic Boron Nitride (CBN) and diamond grinding wheels. (Courtesy Norton Company)

STANDARD MARKING SYSTEM CHART
FOR DIAMOND AND CBN WHEELS

M	D	100 -	P	100 -	B		1/8
Prefix	Abrasive Type	Grit Size	Grade	Diamond Concentration	Bond	Bond Modification	Diamond Depth (in.)
Manufacturer's symbol to indicate type of diamond	B Cubic boron nitride D Diamond	20 | 1000	A (soft) to Z (hard)	25 (low) 50 75 100 (high)	B Resinoid M Metal V Vitrified	A letter or numeral or combination used here will indicate a variation from standard bond	1/16 1/8 1/4 Absense of depth symbol indicates solid diamond

FIGURE 27.8 Standard marking system for Cubic Boron Nitride (CBN) and diamond grinding wheels.

27.4 GRINDING MACHINES AND OPERATIONS

Grinding machines have advanced in design, construction, rigidity, and application far more in the last decade than any other standard machine tool in the manufacturing industry. Grinding machines fall into four categories:

- Surface grinders
- Cylindrical grinders
- Centerless grinders
- Special types of grinders

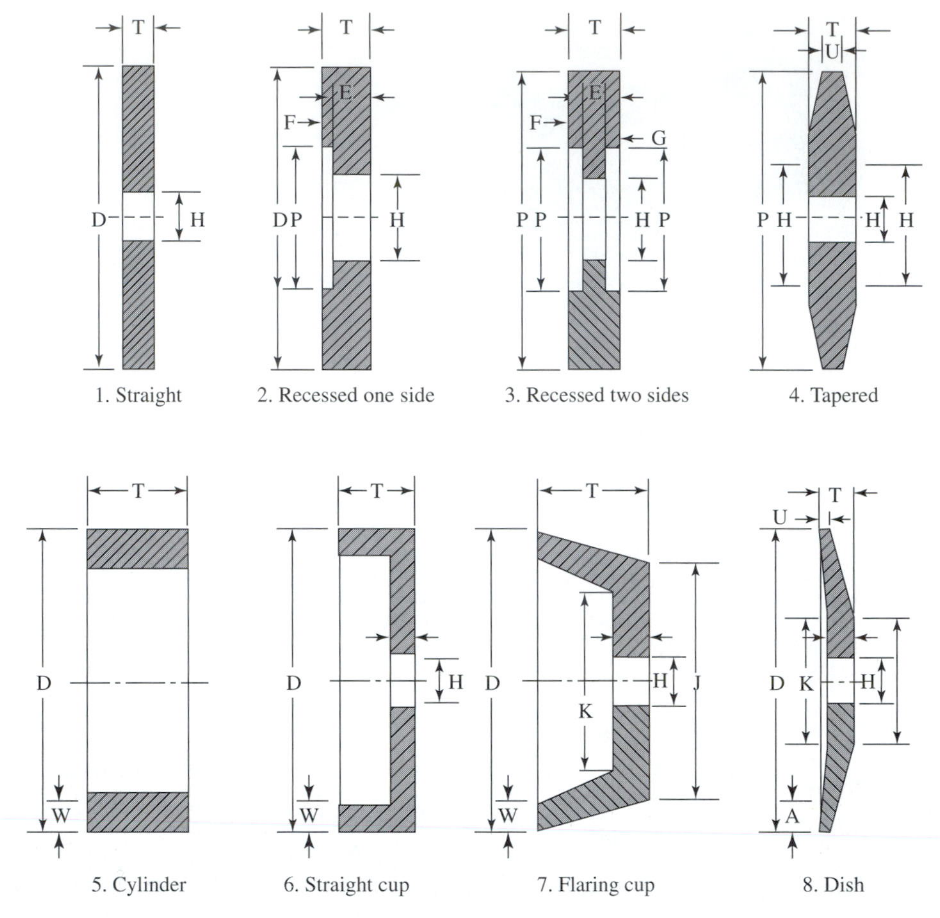

FIGURE 27.9 Eight standard grinding wheel shapes.

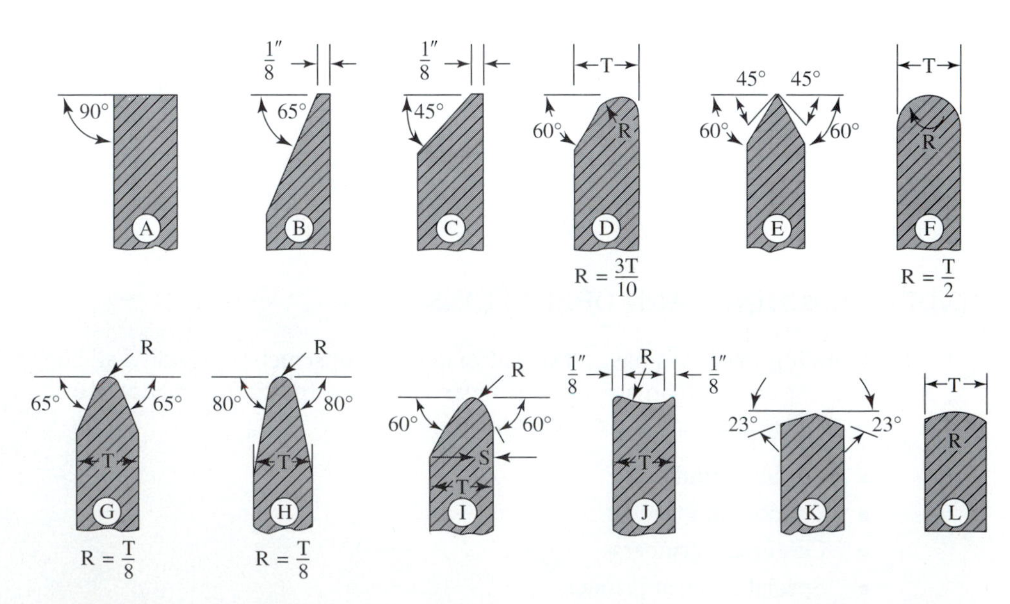

FIGURE 27.10 Twelve standard grinding wheel face contours.

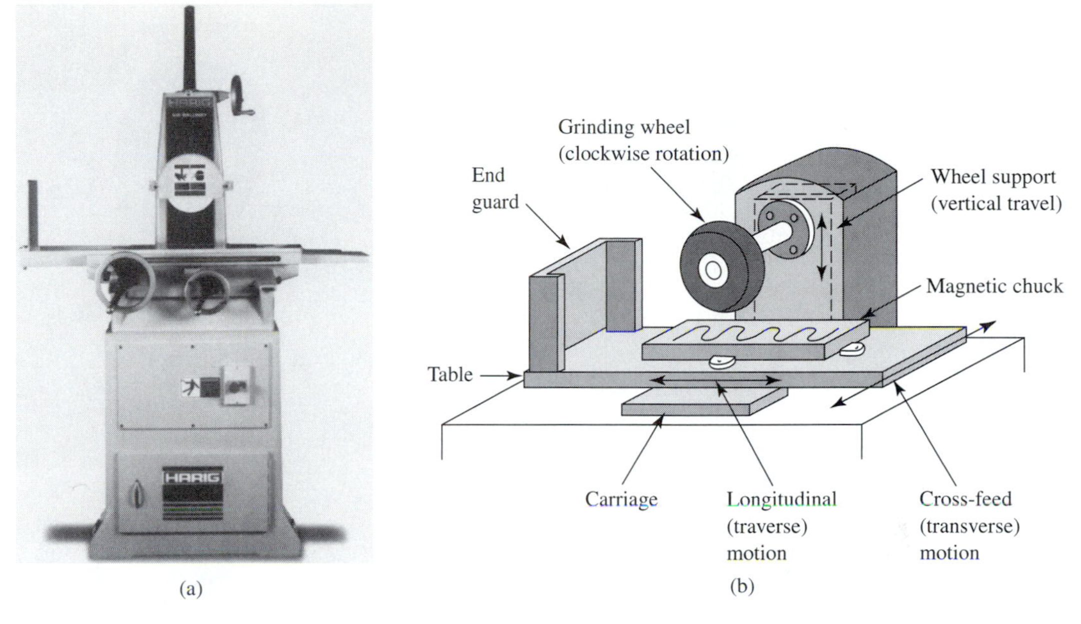

FIGURE 27.11 (a) Typical standard surface grinder. (Courtesy Bridgeport Machine, Inc.) (b) Schematic illustration of the basic components and motions of a surface grinder.

27.4.1 Surface Grinding

Surface grinders are used to produce flat, angular, and irregular surfaces. A typical surface grinder is shown in Figure 27.11(a). In the surface grinding process, the grinding wheel revolves on a spindle and the workpiece, mounted on either a reciprocating or rotary table, is brought into contact with the grinding wheel (Fig. 27.11b). A typical surface grinding operation is shown in Figure 27.12. Four types of surface grinders are commonly used in industry (Fig. 27.13).

Horizontal Spindle/Reciprocating Table This surface grinder is the most commonly used type in industry. A manual surface grinder was shown in Figure 27.11(a). A more sophisticated and automated surface grinder is shown in Figure 27.14. It is available in various sizes to accommodate large or small workpieces. With this type of surface grinder, the work moves back and

FIGURE 27.12 Typical surface grinding operation. (Courtesy Norton Company).

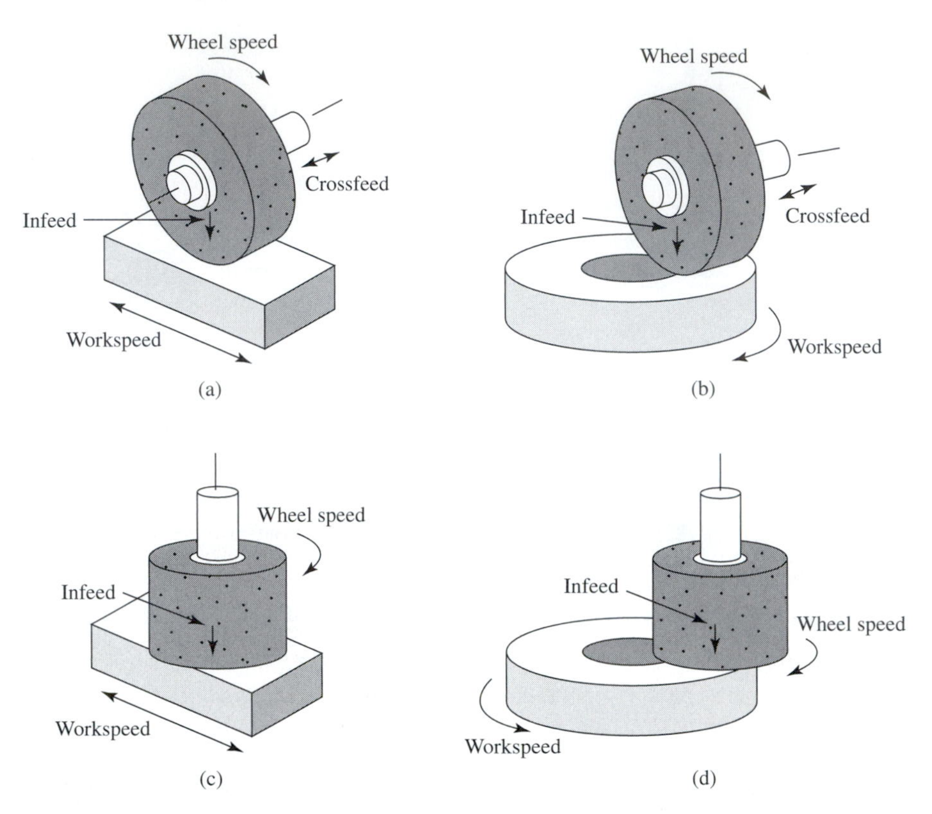

FIGURE 27.13 Four types of surface grinders commonly used in industry: (a) horizontal spindle/reciprocating table, (b) horizontal spindle/rotary table, (c) vertical spindle/reciprocating table, (d) vertical spindle/rotary table.

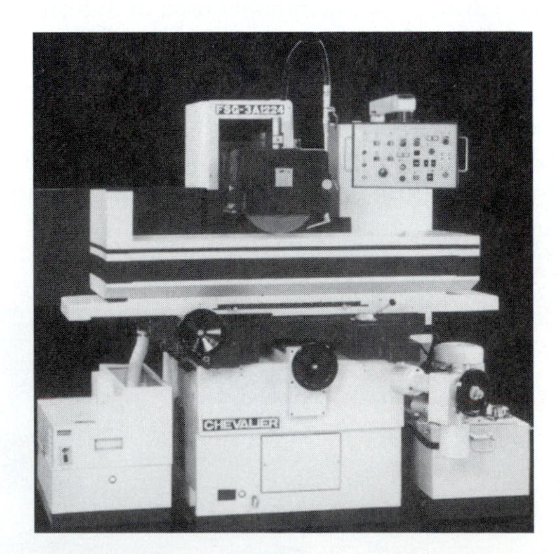

FIGURE 27.14 Automated surface grinder with coolant system. (Courtesy Chevalier Machinery, Inc.)

forth under the grinding wheel. The grinding wheel is mounted on a horizontal spindle and cuts on its periphery as it contacts the workpiece. The worktable is mounted on a saddle, which provides cross-feed movement of the workpiece. The wheelhead assembly moves vertically on a column to control the depth of cut required.

Horizontal Spindle/Rotary Table This surface grinder also has a horizontally mounted grinding wheel that cuts on its periphery. The workpiece rotates 360° on a rotary table underneath the wheelhead. The wheelhead moves across the workpiece to provide the necessary cross-feed movements. The metal removal rate is controlled by the amount of downfeed of the wheelhead assembly.

Vertical Spindle/Reciprocating Table This type of grinding machine is particularly suited for grinding long and narrow castings such as the bedways of an engine lathe. It removes metal with the face of the grinder wheel while the work reciprocates under the wheel. The wheelhead assembly, as on most other types of surface grinders, moves vertically to control the depth of cut. Cross feed is accomplished by the table moving laterally. The table is mounted on a saddle unit.

Vertical Spindle/Rotary Table This type of grinding machine (Fig. 27.15) is capable of heavy cuts and high metal removal rates. Vertical-spindle machines use cup, cylinder, or segmented wheels. Many are equipped with multiple spindles to successively rough, semifinish, and finish large castings, forgings, and welded fabrications. These grinding machines are available in various sizes and have up to 225-HP motors to drive the spindle.

Work-Holding Devices Almost any work-holding device used on a milling machine or drill press can be used on surface grinders. Vises, rotary tables, index centers, and other fixtures are used for special setups. However, the most common work-holding device on surface grinders is the **magnetic chuck**.

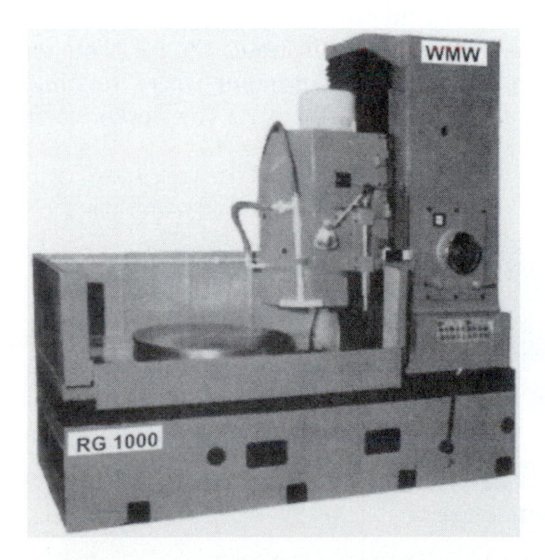

FIGURE 27.15 Vertical-spindle grinder with rotary table. (Courtesy WMW Machinery Co. Inc.)

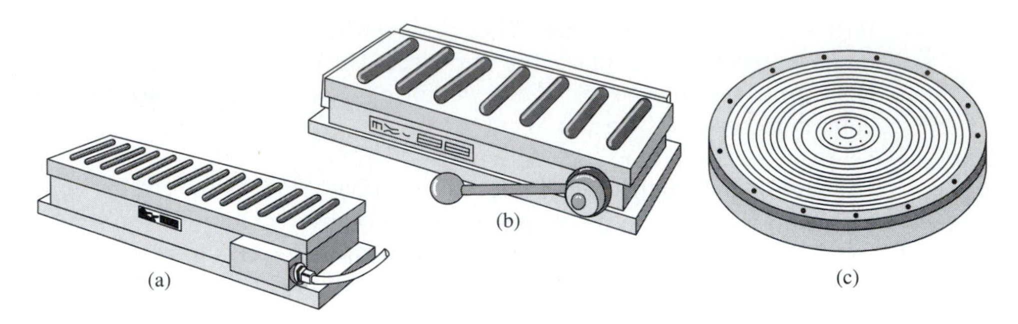

FIGURE 27.16 Three magnetic chucks: (a) electromagnetic chuck, (b) permanent magnet chuck, (c) rotary electromagnetic chuck.

Magnetic chucks hold the workpiece by exerting a magnetic attraction on the part. Only magnetic materials such as iron and steel may be mounted directly on the chuck. Two types of magnetic chucks are available for surface grinders: the permanent magnet and the electromagnetic chucks. Three types of magnetic chucks are shown in Figure 27.16.

On permanent magnet chucks, the holding power comes from permanent magnets. The work is placed onto the chuck, and a hand lever is moved to energize the magnets. The electromagnetic chuck operates on 110 or 220 volts and is energized by a switch. This type of chuck has two advantages. First, the holding power may be adjusted to suit the area of contact of the workpiece; small amounts of current are used with smaller parts, large amounts with larger parts. A second advantage is the demagnetizer switch. It reverses the current flow momentarily and neutralizes the residual magnetism from the chuck and workpiece.

27.4.2 Cylindrical Grinding

Cylindrical grinding is the process of grinding the outside surfaces of a cylinder. These surfaces may be straight, tapered, or contoured. Cylindrical grinding operations resemble lathe turning operations. They replace the lathe when the workpiece is hardened or when extreme accuracy and superior finish are required. Figure 27.17 illustrates the basic motion of the cylindrical grinding machine. As the workpiece revolves, the grinding wheel, rotating much faster in the opposite direction, is brought into contact with the part. The workpiece and table reciprocate

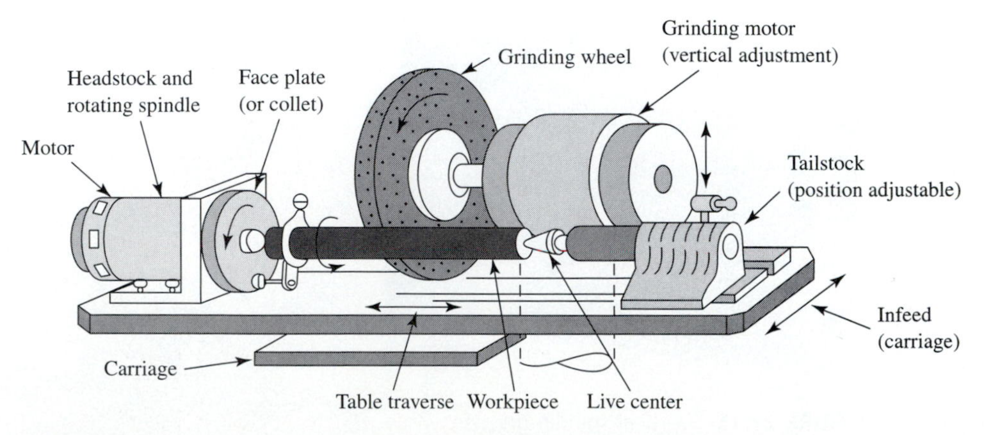

FIGURE 27.17 Schematic illustration of the basic components and motions of a cylindrical grinder.

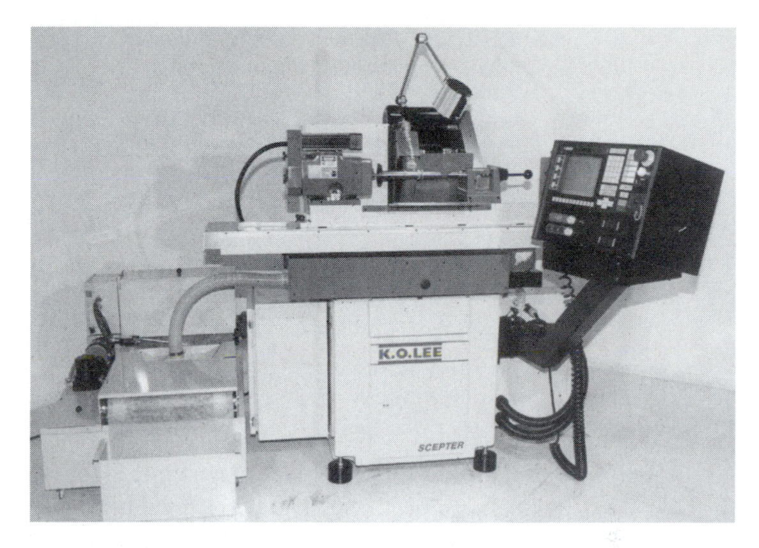

FIGURE 27.18 CNC cylindrical grinder with coolant system. (Courtesy K. O. Lee Co.)

while in contact with the grinding wheel to remove material. A CNC cylindrical grinder is shown in Figure 27.18.

Work-Holding Devices Work-holding devices and accessories used on center-type cylindrical grinders are similar to those used on engine lathes.

The primary method of holding work is between centers as shown in Figure 27.17. The points on these centers may be high-speed steel or tungsten carbide (Fig. 13.15). A lubricant is used with either type and is applied between the point of the center and the center hole in the work.

Independent, universal, and collet chucks can be used on cylindrical grinders when the work is odd-shaped or contains no center hole.

27.4.3 Centerless Grinding

Centerless grinding machines eliminate the need to have center holes for the work or to use work-holding devices. In centerless grinding, the workpiece rests on a work rest blade and is backed up by a second wheel, called the regulating wheel (Fig. 27.19). The rotation of the grinding

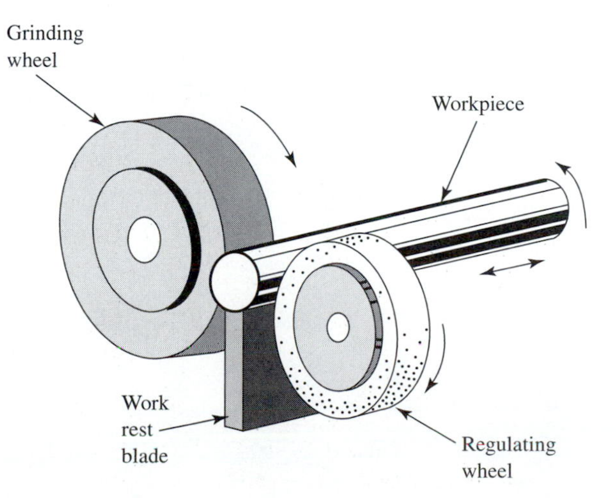

FIGURE 27.19 Operating principle of a centerless grinder.

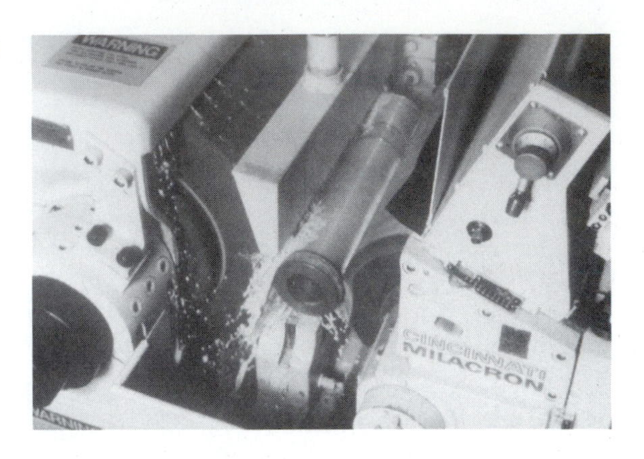

FIGURE 27.20 Typical centerless grinding operation. (Courtesy Cincinnati Machine)

wheel pushes the workpiece down on the work rest blade and against the regulating wheel. The regulating wheel, usually made of a rubber bonded abrasive, rotates in the same direction as the grinding wheel and controls the longitudinal feed of the work when set at a slight angle. By changing this angle and the speed of the wheel, the workpiece feedrate can be changed. The diameter of the workpiece is controlled by two factors: the distance between the grinding wheel and regulating wheel and the height of the work rest blade. A typical centerless grinding operation is shown in Figure 27.20.

27.4.4 Special Grinding Processes

Special types of grinders are grinding machines made for specific types of work and operations. A brief description of the more commonly used special types follows.

Internal Grinders Internal grinders are used to accurately finish straight, tapered, or formed holes. The most popular internal grinder is similar in operation to a boring operation in a lathe. The workpiece is held by a work-holding device, usually a chuck or collet, and revolved by a motorized headstock. The grinding wheel is revolved by a separate motor head in the same direction as the workpiece. It can be fed in and out of the work and also adjusted for depth of cut. An internal grinding operation with a steady-rest is shown in Figure 27.21.

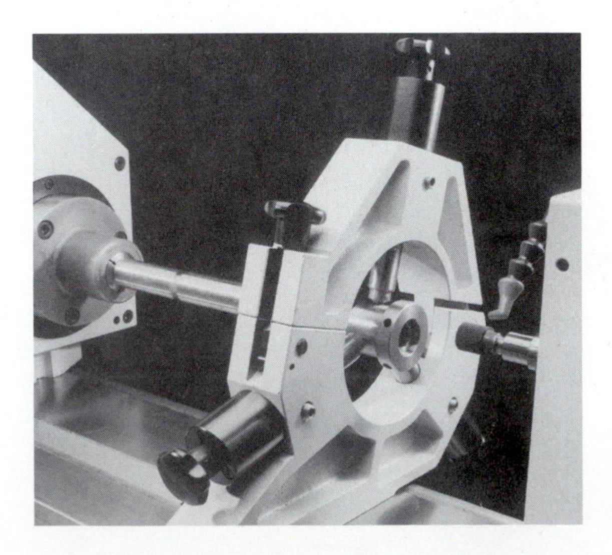

FIGURE 27.21 Internal grinding operation; the workpiece is held by a collet and supported in a steady-rest. (Courtesy Kellenberger, A Hardinge Co.)

FIGURE 27.22 Table top tool and cutter grinder is shown sharpening an end milling cutter. (Courtesy Chevalier Machinery, Inc.)

Tool and Cutter Grinders A tool and cutter grinder was introduced in Chapter 15, "Drilling Operations" (Fig. 15.12). These grinding machines are designed to sharpen milling cutters, reamers, taps, and other machine-tool cutters. A table top tool and cutter grinder is shown in Figure 27.22.

The general-purpose cutter grinder is the most popular and versatile tool-grinding machine. Various attachments are available for sharpening most types of cutting tools. Sharpening of a tap is shown in Fig. 27.23(a) and grinding of a milling cutter is shown in Fig. 27.23(b). Figure 27.24 shows sharpening of a carbide milling cutter with a diamond cup grinding wheel.

(a)

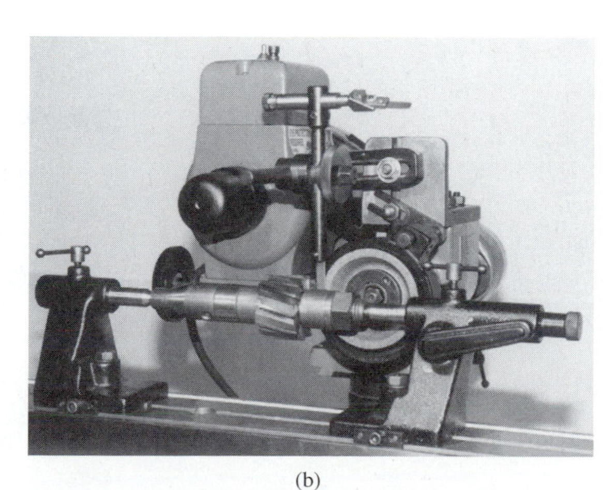

(b)

FIGURE 27.23 Tool and cutter grinder setups: (a) sharpening of a tap, (b) sharpening of a milling cutter. (Courtesy K. O. Lee Co.)

FIGURE 27.24 Sharpening of a carbide milling cutter with a diamond cup grinding wheel. (Courtesy Norton Company)

Jig Grinding Machines Jig grinders were developed to locate and accurately grind tapered or straight holes. Jig grinders are equipped with a high-speed vertical spindle for holding and driving the grinding wheel. They utilize the same precision locating system as do jig borers. A 5-axis continuous path jig grinder is shown in Figure 27.25.

Thread Grinding Machines These are special grinders that resemble the cylindrical grinder. They must have a precision lead screw to produce the correct pitch, or **lead**, on a threaded part. Thread grinding machines also have a means of dressing or **truing** the cutting periphery of the grinding wheel so that it will produce a precise thread form on the part.

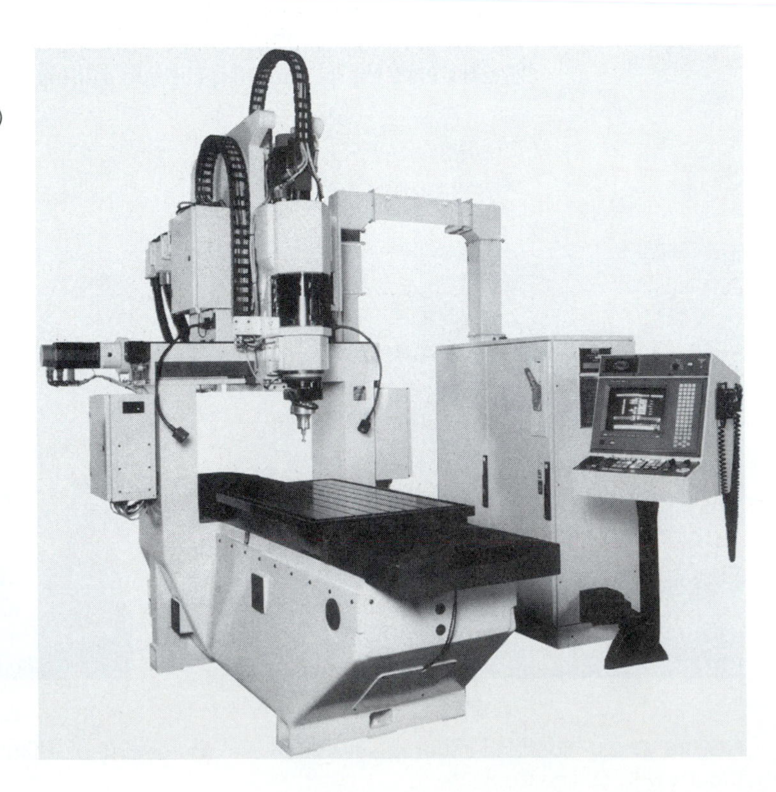

FIGURE 27.25 Continuous path 5-axis jig grinder. (Courtesy Moore Tool Co., Inc.)

27.5 GRINDING WHEEL SELECTION

Before attempting to select a grinding wheel for a particular operation, the operator should consider the following six factors for maximum and safe results.

Material to Be Ground If the material to be ground is carbon steel or alloy steel, aluminum oxide wheels are usually selected. Extremely hard steels and exotic alloys should be ground with Cubic Boron Nitride (CBN) diamond. Nonferrous metals, most cast irons, nonmetallics, and cemented carbides require a silicon carbide wheel. A general rule on grain size is to use a fine-grain wheel for hard materials and a coarse-grain wheel for soft and ductile materials. Close grain spacing and soft wheels should be used on harder materials; open structure and harder wheels are preferable on soft materials.

Nature of the Grinding Operation The finish required, accuracy, and amount of metal to be removed must be considered when selecting a wheel. Fine and accurate finishes are best obtained with small grain size and grinding wheels with resinoid, rubber, or shellac bonds. Heavy metal removal is obtained with coarse wheels with vitrified bonds.

Area of Contact The area of contact between the wheel and workpiece is also important. Close grain spacing, hard wheels, and small grain sizes are used when the area of contact is small, On the other hand, open structures, softer wheels, and larger grain sizes are recommended when the area of contact is large.

Condition of the Machine Vibration influences the finish obtained on the part as well as wheel performance. Vibration is generally due to loose or worn spindle bearings, worn parts, out-of-balance wheels, or insecure foundations.

Grinding Wheel Speed Wheel speed affects the bond and grade selected for a given wheel. Wheel speeds are measured in surface feet per minute (SFPM). Vitrified bonds are commonly used to 6500 SFPM or in selected operations up to 12,000 SFPM. Resinoid-bonded wheels may be used for speeds up to 16,500 SFPM.

Grinding Pressure Grinding pressure is the rate of in-feed used during a grinding operation; it affects the grade of wheel. A general rule to follow is that as grinding pressures increase, harder wheels must be used.

27.6 WHEEL BALANCING, DRESSING, AND TRUING

All grinding wheels are breakable, and some are extremely fragile. Great care should be taken in handling grinding wheels. New wheels should be closely inspected immediately after receipt to make sure they were not damaged during transit. Grinding wheels should also be inspected prior to being mounted on a machine.

To test for damage, suspend the wheel with a finger and gently tap the side with a screwdriver handle for small wheels and a wooden mallet for larger wheels. An undamaged wheel will produce a clear ringing sound; a cracked wheel will not ring at all.

Balancing It is important to balance wheels over 10 in. before they are mounted on a machine. The larger the grinding wheel, the more critical balancing becomes. Grinding wheel balance also becomes more critical as speed is increased. Out-of-balance wheels cause excessive vibration, produce faster wheel wear, chatter, poor finishes, and damage to spindle bearings, and can be dangerous.

The proper procedure for balancing wheels is to first statically balance the wheel. Next, mount the wheel on the grinding machine and dress. Then remove the wheel and rebalance it. Remount the wheel and dress slightly a second time.

Balancing of wheels is done by shifting weights on the wheel mount. The wheel is installed on a balancing arbor and placed on a balancing fixture. The weights are then shifted in a position to remove all heavy points on the wheel assembly.

Dressing and Truing Dressing is a process used to clean and restore a dulled or loaded grinding wheel's cutting surface to its original sharpness. In dressing, swarf is removed, as well as dulled abrasive grains and excess bonding material. In addition, dressing is used to customize a wheel face, so that it will give desired grinding results.

Truing is the process of removing material from the face of the wheel so that the resultant cutting surface runs absolutely true. This is very important in precision grinding, because an out-of-truth wheel will produce objectionable chatter marks on the workpiece. A new wheel should always be trued before being put to work. It is also a good idea to true the wheel if it is being remounted on a machine.

Dressing and truing conventional grinding wheels are two separate and distinct operations, although they may sometimes be done with the same tool. The tools used for conventional grinding wheel dressing include the following:

Mechanical dressers, commonly called star dressers, are held against the wheel while it is running. The picking action of the points of the star-shaped wheels in the tool remove dull grains, bond, and other bits of swarf. Star dressers are used for relatively coarse-grained conventional wheels, generally in offhand grinding jobs where grinding accuracy is not the main consideration.

Dressing sticks are used for offhand dressing of smaller conventional wheels, especially cup and saucer shapes. Some of these sticks are made of an extremely hard abrasive called **boron carbide.** In use, a boron carbide stick is held against the wheel face to shear the dull abrasive grains and remove excess bond. Other dressing sticks contain coarse Crystolon or Alundum grains in a hard vitrified bond. Various dressing sticks are shown in Figure 27.26.

Diamond dressing tools utilize the unsurpassed hardness of a diamond point to clean and restore the wheel's grinding face. Although single-point diamond tools were once the only products available for this kind of dressing, the increasing scarcity of diamonds has led to the development of multipoint diamond tools.

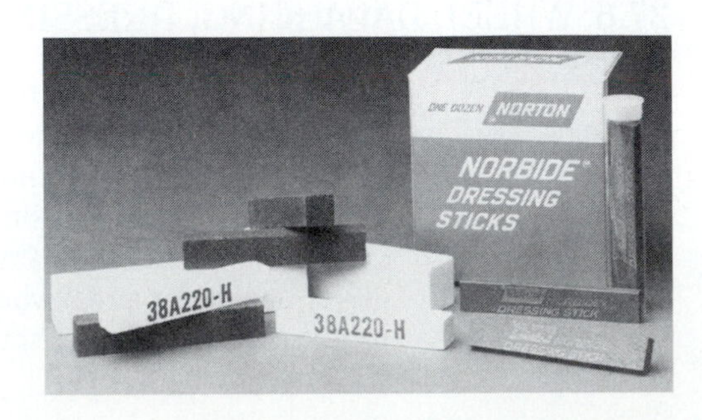

FIGURE 27.26 Various dressing sticks are shown. (Courtesy Norton Company)

FIGURE 27.27 Single- and multipoint diamond dressing tools. (Courtesy Norton Company)

Multipoint diamond dressing tools use a number of small diamonds held in a matrix. In use, the tool is held securely in the toolholder and held flat against the face of the running wheel. As it dresses, the tool is traversed across the wheel face until the job is done. As diamonds on the surface of the tool wear away, fresh new diamond points are exposed to offer extended life and use. This type of tool produces a very consistent wheel face from dress to dress.

Multipoint diamond dressing tools are available in a wide range of shank diameters and face shapes, to meet the requirements of a broad variety of grinding machines. Typical diamond tools used to dress grinding wheels are shown in Figure 27.27.

27.7 GRINDING WHEEL WEAR

The wear of a grinding wheel can be caused by three actions:

- Attrition or wearing down
- Shattering of the grains
- Breaking of the bond

In most grinding processes, all three mechanisms are active to some extent. Attritious wear is not desirable because the dulled grains reduce the efficiency of the process, resulting in increased power consumption, higher surface temperatures, and surface damage. However, attrition must go on to some extent, with the forces on the grit being increased until they are high enough to shatter the grit or break the bond posts holding the dulled grit. The action of particles breaking away from the grains serves to keep the wheel sharp without excessive wear. However, the grains must eventually break from the bond or the wheel will have to be dressed. Rupturing the bond post that holds the grit allows dull grains to be sloughed off, exposing new sharp edges. If this occurs too readily, the wheel diameter wears down too fast. This raises wheel costs and prohibits close sizing on consecutive parts.

Attritious Wear Attritious wear is responsible for the "glazed" wheel, which occurs when flat areas are worn on the abrasive grains but the forces are not high enough to break the dull grains out of the wheel face. Effective grinding ceases with a glazed wheel when the radial force becomes so high that the grit can no longer penetrate the workpiece surface to form chips. Attritious wear of the wheel occurs most often when fine cuts are taken on hard abrasive materials. It can often be avoided by taking heavier cuts or using a softer wheel, which will allow the grains to break out.

Grain Fracture The forces that cause the grain to shatter may arise from the cutting forces acting on the wheel, thermal conditions, shock loading, welding action between the grit and the chip, or combinations of these factors. In finish grinding, this type of wheel wear is desirable, because it keeps sharp edges exposed and still results in a low rate of wheel wear. In time, the wheel may become **loaded** and noisy, and require dressing. A loaded wheel should be dressed by taking a few deep cuts with the diamond so that the metal-charged layer is removed and the chips are not just pushed further into the wheel. Then it should be finish dressed according to the application requirements.

Bond Fracture It is desirable to have worn grit break out of the wheel so that new cutting edges will be exposed. This breaking down of the bond should progress fast enough that heat generation is sufficiently low to avoid surface damage. On the other hand, bond breakdown should be slow enough that wheel costs are not prohibitive. Normally, this means choosing the proper wheel grade for the job. A certain bond hardness is required to hold the grain in place. Softer wheels crumble too fast; harder wheels hold the dull grit too long.

27.8 GRINDABILITY

Grindability, in a like manner as machinability, may be thought of as the ease with which material can be removed from the workpiece by the action of the grinding wheel. Surface finish, power consumption, and tool (wheel) life can be considered to be fundamental criteria of the grindability of metals. In addition, there are the important factors of chip formation and susceptibility to damage of the workpiece. Chip formation, which leads to a loaded wheel, is detrimental.

The most important machine setting affecting machinability, the cutting speed, is not as important an influence on grindability because grinding is done at more or less constant speed. Instead, the important factor becomes the nature of the grinding wheel. The type of grit, grit size, bond material, hardness, and structure of the wheel all influence the grindability of the workpiece. The problems of tool material and configuration variables were discussed in connection with machinability.

In grinding operations like snagging and cutoff work, the surface finish, and even the metallurgical damage to the workpiece, may become relatively unimportant. Wheel life and the rate of cut obtainable then become the criteria of grindability.

The best way to determine grindability is to start with the selection of the proper wheel. This is done by beginning with the manufacturer's recommended grade for the conditions of the job and then trying wheels on each side of this grade. Any improvement or deterioration in the grinding action, as evidenced by wheel wear, surface finish, or damage to the workpiece, can be noted. After the proper wheel has been chosen, wheel life data may be obtained. Usually, this can be done during a production run.

Some of the factors to consider in establishing grindability ratings are discussed in the following examples of the grinding performance of metals.

Cemented Carbide This material cannot be ground with aluminum oxide grit wheels. Although cemented carbide can be ground with pure silicon carbide wheels, the grinding ratio is very low and the material is easily damaged. Carbide is easily ground with diamond wheels if light cuts are taken to prevent damage to the workpiece material. However, diamond grit wheels are quite expensive. The overall grindability of this material is very low.

High-Speed Steel Hardened high-speed steel can be ground quite successfully with aluminum oxide grit wheels. The grinding ratio is low, the relative power consumption high, and the possibility of damage to the workpiece is always present. Overall grindability is quite low.

Hardened Steel Medium hard alloy or plain carbon steels are easily ground with aluminum oxide wheels. The grinding ratio is good, and damage to the workpiece is not a serious problem. Relative power consumption is moderate. The grindability rating is good.

Soft Steel Annealed plain carbon steels grind with relatively low power consumption. Aluminum oxide wheels are satisfactory. The grinding ratio is quite high, but surface damage may be encountered. As a group, these materials are rated as having good grindability.

Aluminum Alloys These soft alloys grind with quite low power consumption, but they tend to load the wheel quickly, so wheels with a very open structure are needed. Grinding ratios are good. Silicon carbide grit works well, and belt grinding outperforms wheel grinding in many cases.

27.9 REVIEW QUESTIONS AND PROBLEMS

1. Define the grinding process.
2. Explain how a grinding wheel cuts metal.
3. Compare aluminum oxide and silicon carbide wheels.
4. Compare CBN and diamond wheels.
5. Explain the function of bonding materials.
6. Compare vitrified and resinoid bonds.
7. Discuss the applications of other bonds.
8. Discuss the various grain sizes and their applications.
9. Explain the concept of grinding wheel hardness.
10. What is meant by grinding wheel structure?
11. Describe the surface grinding process.
12. Discuss the advantages of magnetic chucks.
13. Describe the cylindrical grinding process.
14. Compare the work-holding methods of surface and cylindrical grinders.
15. Describe the centerless grinding process.
16. Discuss some of the special-purpose grinders.
17. Name and describe six factors to consider before selecting a grinding wheel.
18. Describe inspection methods for grinding wheels.
19. Why do grinding wheels need to be balanced?
20. Explain the difference between dressing and truing a grinding wheel.
21. Describe the dressing of a grinding wheel, using a multipoint diamond dressing tool.
22. Name and describe the three actions that cause grinding wheel wear.
23. Define grindability.
24. Compare the grindability of cemented carbide and various steels.

CHAPTER TWENTY-EIGHT

LAPPING AND HONING

CHAPTER OVERVIEW

28.1 INTRODUCTION

Lapping is a final abrasive finishing operation that produces extreme dimensional accuracy, corrects minor imperfections of shape, refines surface finish, and produces close fit between mating surfaces. Most lapping is done with a tooling plate or wheel (the lap) and fine-grained loose abrasive particles suspended in a viscous or liquid vehicle such as soluble oil, mineral oil, or grease. A typical lapping operation is shown in Figure 28.1.

Honing is a low-velocity abrading process. Material removal is accomplished at lower cutting speeds than in grinding. Therefore, heat and pressure are minimized, resulting in excellent size and geometry control. The most common application of honing is on internal cylindrical surfaces. The cutting action is obtained using abrasive sticks mounted on a metal **mandrel**. Since the work is fixed in such a way as to allow floating, and no clamping or chucking, there is no distortion.

28.2 LAPPING PROCESSES

The principal use of the lapping process is to obtain surfaces that are truly flat and smooth. Lapping is also used to finish round work, such as precision **plug gages**, to tolerances of .0005 to .00002 in.

FIGURE 28.1 Typical lapping machine. (Courtesy Engis Corporation)

Work that is to be lapped should be previously finished close to the final size. Although rough lapping can remove considerable metal, it is customary to leave only .0005 to .005 in. of stock to be removed.

Though it is an abrasive process, lapping differs from grinding or honing because it uses a "loose" abrasive instead of bonded abrasives as grinding wheels do (Fig. 28.2). These abrasives are often purchased ready mixed in a vehicle, often made with an oil–soap or grease base. These vehicles hold the abrasive in suspension before and during use. The paste abrasives are generally used in hand-lapping operations. For machine lapping, a light oil is mixed with dry abrasive so that it can be pumped onto the lapping surface during the lapping operation.

28.2.1 Lapping Machines

These machines are fairly simple pieces of equipment consisting of a rotating table, called a lapping plate, and three or four conditioning rings. Standard machines have lapping plates from 12 to 48 in. in diameter. Large machines up to 144 in. are made. These tables are run by 1- to 20-HP motors. A typical lapping machine is shown in Figure 28.3.

The lapping plate is most frequently made of high-quality soft cast iron, though some are made of copper or other soft metals. This plate must be kept perfectly flat. The work is held in the conditioning rings. These rings rotate as shown in Figure 28.4. This rotation performs two jobs. First, it "conditions" the plate, that is, it distributes the wear so that the lapping plate stays flat for a longer time. Second, it holds the workpiece in place. The speed at which the plate turns is determined by the job being done. In doing very critical parts, 10 to 15 RPM is used, and when polishing, up to 150 RPM is used.

A pressure of about 3 pounds per square inch (PSI) must be applied to the workpieces. Sometimes their own weight is sufficient. If not, a round, heavy pressure plate is placed in the

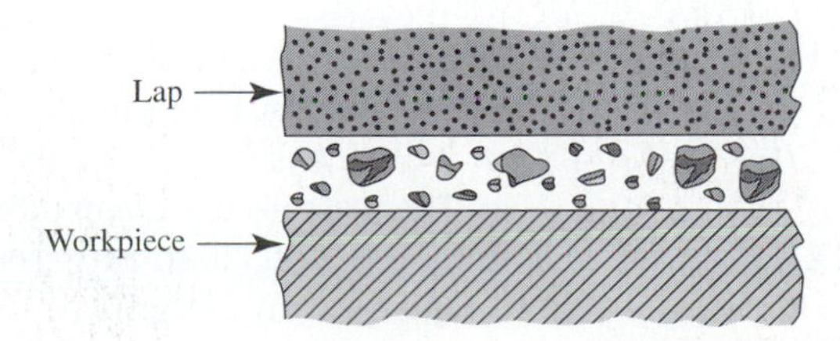

FIGURE 28.2 Abrasive grit must be uniformly graded to be effective in lapping.

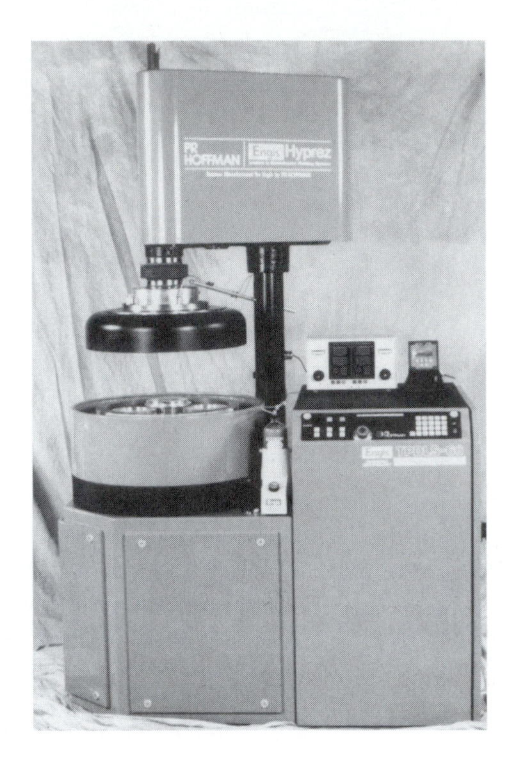

FIGURE 28.3 Typical dual plate lapping machine. (Courtesy Engis Corporation)

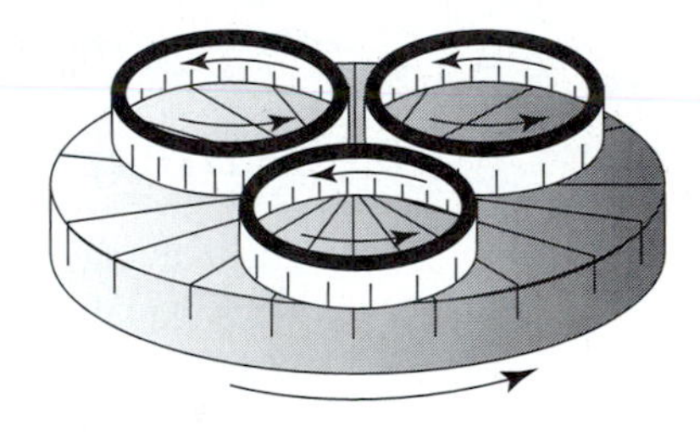

FIGURE 28.4 Conditioning rings used in lapping operations.

conditioning ring. The larger machines use pneumatic or hydraulic lifts to place and remove the pressure plates. Figure 28.5 shows various lapping plates.

The workpiece must be at least as hard as the lapping plate, or the abrasive will be charged into the work. It will take from 1 to 20 minutes to complete the machining cycle. Time depends on the amount of stock removed, the abrasive used, and the quality required. Figure 28.6 shows a production lapping machine.

28.2.2 Grit and Plate Selection

Flatness, surface finish, and a polished surface are not necessarily achieved at the same time or in equal quality. For example, silicon carbide compound will cut fast and give good surface finish, but will always leave a "frosty" or matte surface.

The grits used for lapping may occasionally be as coarse as 100 to 280 mesh. More often the "flour" sizes of 320 to 800 mesh are used. The grits, mixed in a slurry, are flowed onto the plate to replace worn-out grits as the machining process continues.

FIGURE 28.5 Typical lapping plates. (Courtesy Engis Corporation)

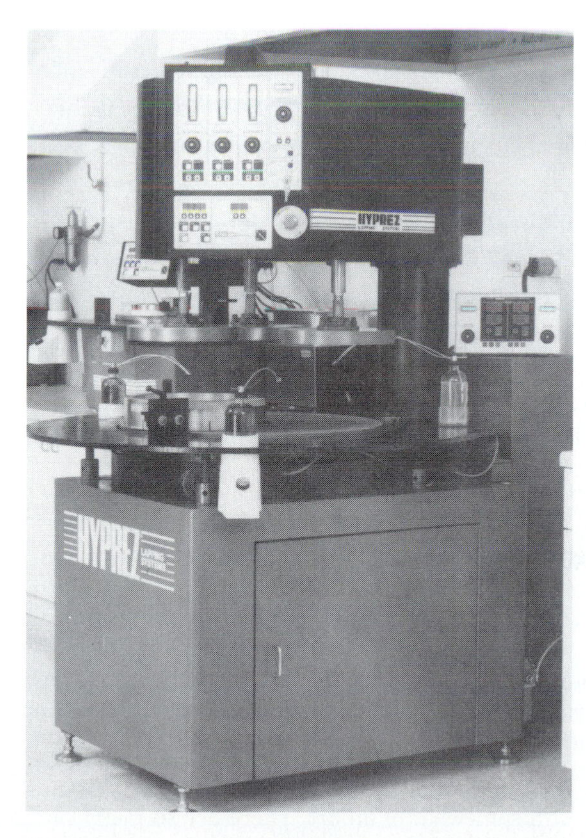

FIGURE 28.6 Single plate lapping production machine equipped for diamond abrasive slurry use. (Courtesy Engis Corporation)

The case for using diamond super abrasives rather than conventional abrasives such as aluminum oxide or silicon carbide, can be summed up in three words: Diamonds are *faster, cleaner,* and more *cost effective.*

With diamond slurries, the lapping and polishing phases of a finishing operation can often be combined into one step. Also, less time is required for cleaning parts and processing waste; throughput, along with overall productivity, is increased.

Lapping plates are manufactured from various materials as described here, and are available in standard sizes from 6 to 48 in. in diameter. Plates are supplied with square, spiral, concentric, and radial grooves as shown in Figure 28.5.

Iron: Aggressive Stock Removal

- Excellent primary/roughing lap plate, with long service life
- Often used as an alternative to cast-iron plates
- Produces a good surface finish on most materials, especially metals and ceramics

Copper: Moderate to Aggressive Stock Removal

- Most widely used, universal composite lap plate
- Excellent when primary and finishing lap are combined in a one-step operation
- Suitable for virtually any solid material: metal, ceramic, glass, carbon, plastic, etc.

Ceramic: Moderate Stock Removal

- Generally used to lap/polish ceramic parts and other stain-sensitive materials
- Used in applications where metallic-type contamination cannot be tolerated
- Affordable, more machinable alternative to “natural” ceramic plates

Tin/Lead: Fine Stock Removal

- Most widely used finishing lap/polishing plate
- Often used in place of polishing pads
- Suitable for metal, ceramic, and other materials

Tin: Fine Stock Removal

- Often used where lead-type contamination cannot be tolerated
- Suitable for charging of extra-fine particulates

28.2.3 Advantages and Limitations

Advantages Any material, hard or soft, can be lapped, as well as any shape, as long as the surface is flat. There is no warping, because the parts are not clamped and very little heat is generated. No burrs are created. In fact, the process removes light burrs. Any size, diameter, and thickness from a few thousandths thick up to any height the machine will handle can be lapped. Various sizes and shapes of lapped parts are shown in Figure 28.7.

Limitations Lapping is still somewhat of an art. There are so many variables that starting a new job requires experience and skill. Even though there are general recommendations and assistance from the manufacturers, and past experience is useful, trial and error may still be needed to get the optimum results.

FIGURE 28.7 Various sizes and shapes of lapped parts. (Courtesy Engis Corporation)

28.3 HONING PROCESSES

As stated earlier, honing is a low-velocity abrading process. Material removal is accomplished at lower cutting speeds than in grinding. Therefore, heat and pressures are minimized, resulting in excellent size and geometry control. The most common application of honing is on internal cylindrical surfaces. A typical honing operation is shown in Figure 28.8.

Machining a hole to within less than .001 in. in diameter and maintaining true roundness and straightness with finishes under 20 μin. is one of the more difficult jobs in manufacturing.

Finish boring or internal grinding may do the job, but spindle deflection, variation in hardness of the material, and difficulties in precise work holding make the work slow and the results uncertain. Because it uses rectangular grinding stones instead of circular grinding wheels, as shown in Figure 28.9(a) and (b), honing can correct these irregularities.

Honing can consistently produce finishes as fine as 4 μin., and even finer finishes are possible. It can remove as little as .0001 in. of stock or as much as .125 in. of stock. However, usually only .002 to .020 in. of stock is left on the diameter for honing. As shown in Figure 28.10, honing can correct a number of conditions or irregularities left by previous operations.

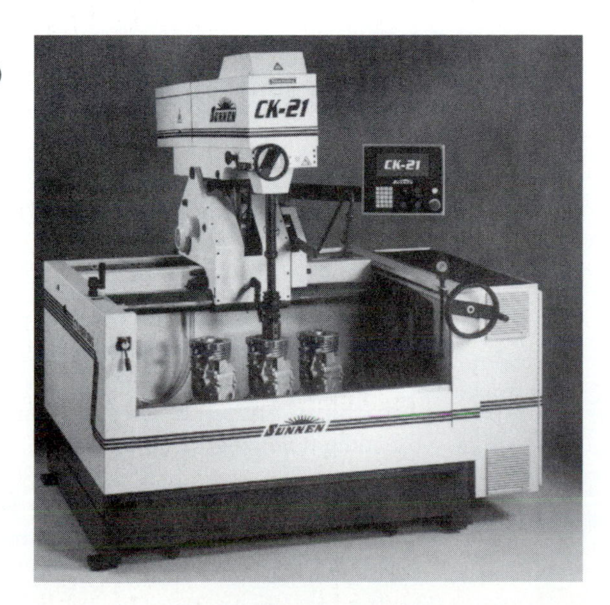

FIGURE 28.8 Typical vertical honing operation. (Courtesy Sunnen Products Co.)

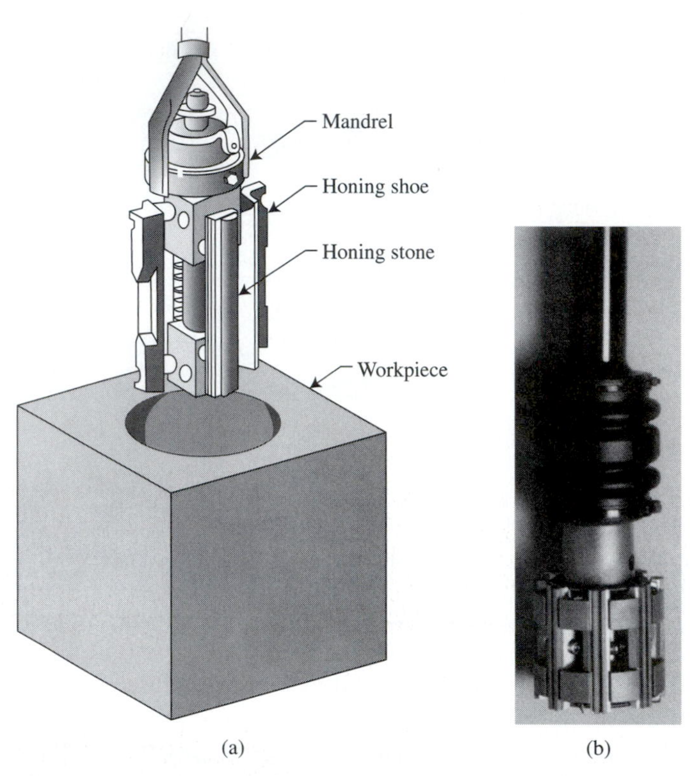

FIGURE 28.9 (a) Schematic illustration of the components of an internal hone. (b) Typical internal abrasive honing tool. (Courtesy Sunnen Products Co.)

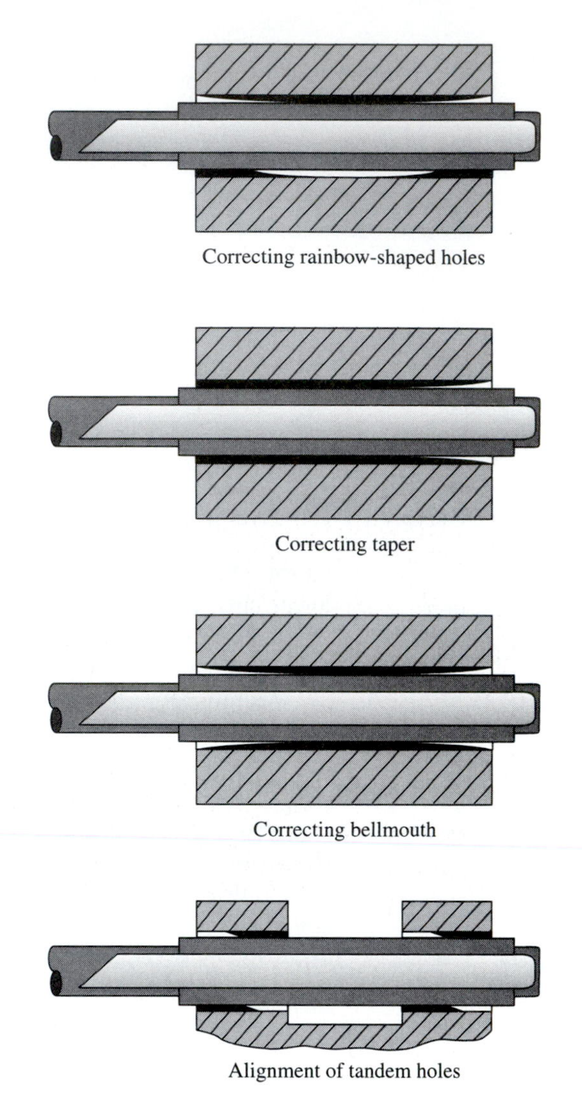

FIGURE 28.10 Undesirable conditions that can be corrected by honing.

28.3.1 Honing Machines

For most work, honing machines are quite simple. The most frequently used honing machines are made for machining internal diameters from .060 to 6 in. However, large honing machines are made for diameters up to 48 in. Larger machines are sometimes made for special jobs.

The length of the hole that can be honed may be anything from $\frac{1}{2}$ in. to 6 or 8 in. on smaller machines, and up to 24 in. on larger machines. Special honing machines are made that will handle hole lengths up to 144 in.

Horizontal-Spindle Machines Horizontal-spindle honing machines, for handheld work with bores up to 6 in., are among the most widely used. The machine rotates the hone at from 100 to 250 FPM.

The machine operator moves the work back and forth (strokes it) over the rotating hone. The operator must "float" the work, that is, not press it against the hone or the hole will be slightly oval. Sometimes the workpiece must be rotated.

Horizontal-spindle honing machines are also made with "power stroking." In these, the work is held in a self-aligning fixture and the speed and length of the stroke are regulated by controls on the machine.

As a hone is being used, it is expanded by hydraulic or mechanical means until the desired hole diameter is achieved. Various mechanical and electrical devices can be attached to the honing machine to control the rate of expansion and stop it when final size is reached.

On the simplest handheld machines, the operator may check the bore size with an **air gage**, continue honing, recheck, and so on until the size is correct. A horizontal-spindle honing machine is shown in Figure 28.11.

Vertical-Spindle Machines Vertical-spindle honing machines are used especially for larger, heavier work. These all have power stroking at speeds from 20 to 120 FPM. The length of the stroke is also machine controlled by stops set up by the operator. Vertical honing machines are also made with multiple spindles so that several holes may be machined at once, as in automobile cylinders (Figure 28.8).

Hone Body. The hone body is made in several styles, using a single stone for small holes and two to eight stones as sizes get larger (Fig. 28.9b). The stones come in a wide variety of sizes and shapes. Frequently there are hardened metal guides between the stones to help start the hone cutting in a straight line.

Cutting Fluid. A fluid must be used with honing. This has several purposes: to clean the small chips from the stones and the workpiece, to cool the work and the hone, and to lubricate the cutting action. A fine mesh filtering system must be used, because recirculated metal can spoil the finish. A vertical honing operation was shown in Figure 28.8. A few of the parts honed on such a machine are shown in Figure 28.12.

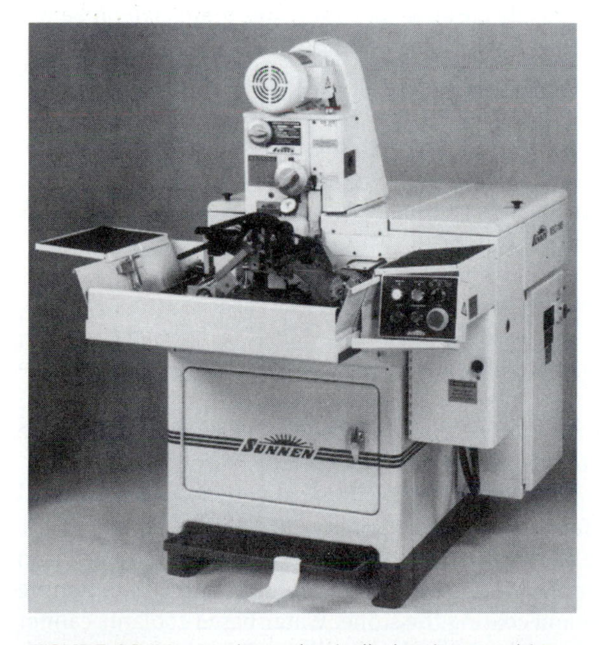

FIGURE 28.11 Horizontal-spindle honing machine. (Courtesy Sunnen Products Co.)

FIGURE 28.12 Parts honed on a vertical honing machine. (Courtesy Sunnen Products Co.)

28.3.2 Abrasive Tool Selection

The abrasive honing stone must be selected for the proper abrasive type, bond hardness, and grit size to deliver the fastest stock removal and desired surface finish. This selection is simple if done in the following three steps.

Step 1 Select the abrasive type with respect to the material composition of the bore. There are four different types of abrasives: aluminum oxide, silicon carbide, diamond, and CBN. All four of these were discussed in the previous chapter. Each type has its own individual characteristics that make it best for honing certain materials. Some simplified guidelines for their use are:

- Mild steel hones best with aluminum oxide.
- Cast iron, brass, and aluminum hone best with silicon carbide.
- Glass, ceramic, and carbide hone best with diamond
- High-speed tool steels and superalloys hone best with CBN.

Diamond and CBN are considered super abrasives because they are much harder than conventional abrasives. They cut easily and dull slowly, therefore allowing them to hone certain materials much faster and more efficiently than conventional abrasives. However, as shown above, super abrasives are not suited to honing all materials. For instance, diamond does not hone steel very well and CBN may not be as economical as using aluminum oxide to hone soft steel.

Step 2 Use the stone hardness suggested in the manufacturer's catalog. If the stone does not cut, select the next softer stone; if the stone wears too fast, select the next harder stone. Stone hardness does not refer to the hardness of the abrasive grain, but to the strength of the bonding material holding the abrasive grains together, as discussed in the previous chapter. A bond must be strong enough to hold sharp abrasive grains in position to cut, but weak enough to allow dulled grains to be sloughed off to expose underlying sharp grains. If the bond is too hard, the dulled abrasive grains will not be allowed to fall off and the stock removal rate will be reduced. If the bond is too soft, the stone will wear excessively because sharp abrasive grains fall off before they are fully used.

Diamond and CBN abrasive grains dull so slowly that standard ceramic or resin bonds may not be strong enough when honing rough out-of-round bores in hard materials or when CBN is used to hone soft steel. Metal bonds are best suited for these applications because the grains are held in a sintered metal matrix that is much stronger than standard bonds. As with choosing abrasive type, stone bond hardness must be matched to the application to maximize life and stock removal rates.

Step 3 Select the largest abrasive grit size that will still produce the desired surface finish. Surface finish is a function of the height of microscopic peaks and valleys on the bore surface, and honing can produce almost any degree of roughness or smoothness through the use of different abrasive grit sizes.

Honing oil can improve stock removal rates by helping the cutting action of the abrasive grains. It prevents pickup (spot welding of tool to bore) and loading (chips coating the stone). Honing oil does this, not by acting as a coolant, but through chemical activity. This chemical activity is produced by the ingredients in the oil. Whenever the temperature rises at one of the microscopic cutting points, the sulfur in the oil combines with the iron in the steel to form iron sulfide, an unweldable compound, and welding is prevented. The antiwelding property of honing oil also prevents chips from sticking together and coating the stone. Water-based coolants cannot produce this type of chemical activity. Use of water-based coolants will result in welding of metallic guide shoes to the part and loading of vitrified abrasive honing stones.

28.3.3 Production Honing

Honing will not only remove stock rapidly, it can also bring the bore to finish diameter within tight tolerances. This is especially true if the honing machine is equipped with automatic size control. With every stroke, the workpiece is pushed against a sensing tip that has been adjusted to the finish diameter of the bore. When the bore is to size, the sensing tip enters the bore and the machine stops honing. Size repetition from bore to bore is .0001 in. to .0002 in. The operator simply loads and unloads the fixture and presses a button; everything else is automatic.

Single-Stroke Honing A still faster and more accurate method of honing a bore to final size is single-stroke honing. The single-stroke tool (Fig. 28.13) is an expandable diamond-plated sleeve on a tapered arbor. The sleeve is expanded only during setup, and no adjustments are necessary during honing. Unlike conventional honing, where the workpiece is stroked back and forth over the tool, in single-stroke honing the rotating tool is pushed through the bore one time, bringing the bore to size. The return stroke does nothing to the bore except get the workpiece off the tool. Single-stroke honing is so accurate and consistent that honed bores do not require gaging.

Although single-stroke honing has many advantages, it is limited in the types and volumes of material that can be removed. The size and overall volume of chip produced in one pass must be no more than the space between the diamond grits or the tool will seize in the bore.

Workpieces are best suited for single-stroke honing when they are made of materials that produce small chips, such as cast iron, and that have interruptions that allow chips to be washed from the tool as the bore is being honed. Conventional honing should be used whenever the material to be honed produces long, stringy chips or the amount of stock to be removed is large.

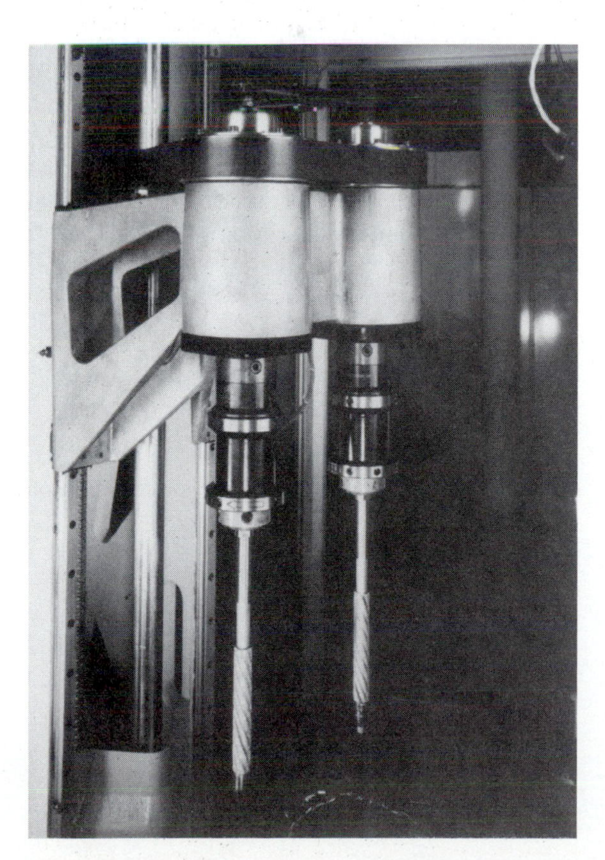

FIGURE 28.13 Single-stroke honing tools use expandable diamond-plated sleeves on a tapered arbor. (Courtesy Sunnen Products Co.)

28.3.4 Advantages and Limitations

Advantages The workpiece need not be rotated by power, and no chucks, faceplates, or rotating tables are needed, so there are no chucking or locating errors. The hone is driven from a central shaft, so bending of the shaft cannot cause tapered holes as it does when boring. The result is a truly round hole, with no taper or high or low spots, provided that the previous operations left enough stock so that the hone can clean up all the irregularities.

Honing uses a large contact area at slow speed compared with grinding or fine boring, which use a small contact area at high speed. Because of the combined rotating and reciprocating motion used, a crosshatched pattern is created, which is excellent for holding lubrication. Diameters with .001- to .0001-in. and closer accuracies can be repeatedly obtained in production work. Honing can be done on most materials from aluminum or brass to hardened steel. Carbides, ceramics, and glass can be honed by using diamond stones similar to diamond wheels.

Limitations Honing is thought of as a slow process. However, new machines and stones have shortened honing times considerably. Horizontal honing may create oval holes unless the work is rotated or supported. If the workpiece is thin, even hand pressure may cause a slightly oval hole.

28.4 REVIEW QUESTIONS AND PROBLEMS

1. Define and describe the lapping process.
2. Explain how a lapping machine works.
3. Describe the function and purpose of the conditioning rings.
4. Why is diamond slurry more economical despite its higher cost?
5. Name and describe the various lapping plate materials.
6. Define and describe the honing process.
7. Explain how a honing machine works.
8. Compare horizontal and vertical honing machines.
9. Why is cutting fluid necessary for honing?
10. Discuss the uses and limitations of super abrasives in honing.
11. Describe the chemical function of honing oils.
12. How does single-stroke honing differ from conventional honing?

CHAPTER TWENTY-NINE

NONTRADITIONAL MACHINING

CHAPTER OVERVIEW

- Introduction
- Chemical Processes
  - Chemical Milling
  - Photochemical Machining
- Electrical Processes
  - Electrochemical Machining (ECM)
  - Electrochemical Grinding (ECG)
- Mechanical Processes
  - Hydrodynamic Machining (HDM)
  - Ultrasonic Machining (USM)
  - Abrasive Jet Machining (AJM)
- Thermal Processes
  - Electron Beam Machining (EBM)
  - Electrical Discharge Machining (EDM)
  - Laser Beam Machining (LBM)
  - Plasma Arc Machining (PAM)
- Review Questions and Problems

29.1 INTRODUCTION

The term "nontraditional machining" is applied to a wide range of unique processes that remove material by chemical, electrical, mechanical, or thermal means. These processes were developed to meet the specific need of a production problem.

Some confusion exists regarding precisely which processes are considered to be nontraditional because many of the processes originated as long ago as the 1940s and have been in common use since. Other processes are relatively new. At the present time, all processes shown in Figure 29.1 are considered nontraditional. Other processes are constantly in development.

Nontraditional processes have certain general characteristics in common:

- They are used to answer specific, unique manufacturing problems.
- They usually improve surface quality or integrity over that produced by traditional processes.

NONTRADITIONAL MACHINING PROCESSES

Nontraditional Processes	Method of Metal Removal	Common Applications Applications
Chemical milling	Chemical removal of metal	• most metals • remove decarb • remove surface cracks • thinning & shaping • improved finish
Photochemical machining	Chemical removal of metal in masked predetermined pattern	• thin metal only • hard & brittle material • complex shapes
Electrochemical machining (ECM)	Reverse electroplating of metal	• hard metals • cavities • profiles
Electrochemical grinding (ECG)	Conventional grinding with ECM assist	• hard & tough metals • turbine blade notches • honeycombs • thin wall sections
Hydrodynamic machining (HDM)	Extremely high-velocity high-pressure water jet	• all nonmetallics • some very thin metals
Ultrasonic machining (USM)	Abrasive grit slurry cuts metal with vibrating tool	• hard materials Rc > 40 • blind holes • holes & slots
Abrasive jet machining (AJM)	Abrasive grit forced through nozzle by gas impacts & cuts metal	• narrow kerfs • trim & deburr • etching • remove corrosion
Electron beam machining (EBM)	High-energy electron beam melts & vaporizes material in a vacuum	• any known material • holes & slots • small cuts in thin parts
Electrical discharge machining (EDM)	Spark erosion of conductive metal by local heating, melting and vaporizing	• molds, tools & dies • prototypes • no distortion • any hardness
Laser beam machining (LBM)	High-energy laser melts & vaporizes metal	• any known material • holes, trimming & cutting • thin metal best
Plasma arc machining (PAM)	Ionic plasma extremely high temperature jets melt metal	• plate cutting up to 8″ • plate edge preparation • hole piercing

FIGURE 29.1 Overview of nontraditional machining processes.

- Most have higher power consumption rates for the same amount of metal removed than a traditional process.
- Most have lower stock removal rates than traditional processes.
- They do not create a chip.

Nontraditional machines are used to replace or supplement conventional machines that prove incapable or impractical for a process.

This chapter describes nontraditional process principles and highlights their use in processing. The Applications, Advantages, and Disadvantages sections are to aid as a general guide in selecting equipment for original processing or when selecting equipment to replace or supplement conventional machines that prove incapable or unproductive.

29.2 CHEMICAL PROCESSES

Chemical processes fall into two major categories: chemical milling and photochemical machining.

29.2.1 Chemical Milling

The chemical milling process shapes metal by the chemical removal of metal. The length of time the metal is immersed in the etching solution (etchant) determines the amount of metal removed. Areas of the workpiece that are not to be etched are protected by masking. Chemical milling is shown in Figure 29.2.

Operating Parameters The process consist of four main steps: cleaning, masking, etching, and demasking.

Cleaning. The workpiece must be properly cleaned to ensure a uniform adhesion of masking (maskant). Conventional cleaning processes are used, which include washing, vapor degreasing, and alkaline cleaning. Parts must be thoroughly dried before applying maskant.

Masking. The workpiece is masked by dipping, flow coating, or spraying. Two or more coats are usually applied and subsequently air dried or baked. Removal of portions of the mask exposes the desired metal portions to be etched.

Knives are used to cut at the required boundaries (scribing). Care must be taken not to cut or scratch the metal because this may intensify the action of the etchant solution in that localized area. After scribing, the cut portions of the mask are removed by hand.

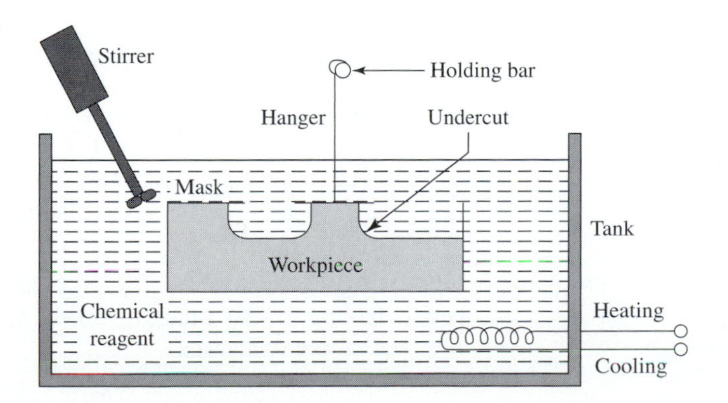

FIGURE 29.2 Schematic drawing of a chemical milling process.

Etching. Chemical milling is accomplished on the exposed areas of masked parts, which are then racked and submerged in the etchant solution for the required length of time. Etchants are the acids or mix of acids in proper concentrations required for each metal.

The etching process produces highly corrosive fumes that attack surrounding equipment and structures. Precautions and preventive measures must be employed such as shielding, covers, scrubbers, and so on for the safety of personnel and for the protection of equipment. Antipolluting laws must also be adhered to. Workpieces must be washed, flushed, and neutralized following etching.

Demasking. Masks are removed from etched workpieces by hand or by submerging in a demasking solution followed by washing and neutralizing.

Applications The major applications of chemical milling include

- Removal of material from all metal
- Removal of decarburized layers from steel
- Removal of surface cracks and defects
- Thinning and shaping of workpieces
- Improvement of surface finishes
- Removal of burrs

This process produces excellent results for all applications and especially improves surface finishes. Manufacturing Engineering must, however, observe the disadvantages and compensate, in this case, with sufficient capacity and washing operations.

Advantages

- No skilled labor required
- Simple, low-cost tooling required
- Processes any metal
- Processes any shape
- Produces no stress in the workpiece
- Processes large areas simultaneously

Disadvantages

- Slow
- Corrosive

29.2.2 Photochemical Machining

Photochemical machining is basically the fixing of a chemical-resistant image (resist) of a part design on a thin sheet or strip of metal that is to be machined to shape. Both are then subjected to chemical action by means of immersion, flow, or spray. The chemical action removes all metal not protected, leaving the desired part.

Operating Parameters The process consists of six main steps: cleaning, coating metal with resist, baking resist, image printing and removal of unexposed areas of resist, etching, and removal of exposed resist.

Cleaning. The metal must be clinically clean. Methods of cleaning include vapor degreasing and chemically washing and rinsing.

Coating Metal with Resist. The metal is coated with photographic material by dipping, spraying, roller coating, or flow processes. This material will act as a resist to the chemical action.

Baking Resist. The coated metal is baked prior to exposure to remove all solvents from the coating process.

Image Printing and Removal of Unexposed Areas of Resist. Art work that depicts the part design many times the size of the part is photographically reduced to actual size. This practice is more feasible for drafting (some parts' designs have exceedingly small, fine details) and also increases the accuracy of the image as the design is reduced.

Exposure, or printing the image on the resist, is generally accomplished by clamping the coated metal between two negatives (the designed images) and exposing to light rays. An image of the part will result on both the top and bottom surfaces, eventually allowing chemical action on both surfaces simultaneously. Developing is accomplished in a photographic solution that is then washed off. This wash also removes the unexposed portions of resist, exposing the metal areas to be removed in etching.

Etching. Etching removes the metal not protected by the photoresists and is done by immersion, spray, or flow of the proper etchant and followed by a wash or rinse.

Resist Removal. Many parts do not require the removal of resist, but where required, it can be removed in a solution, by mechanical, or by chemical action. Figure 29.3 illustrates steps in this process.

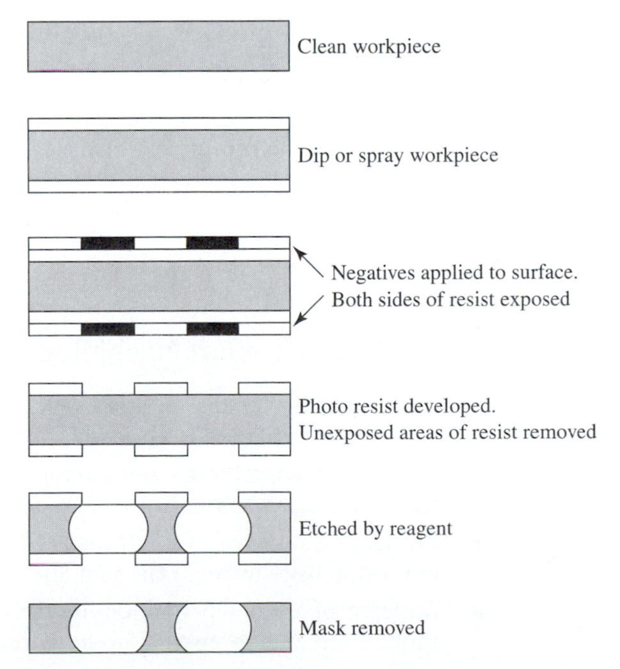

FIGURE 29.3 Steps to perform a photochemical machining process.

Applications The common applications of photochemical machining include

- Extremely thin metals
- Thin, hard, and brittle metals
- Complex parts (eliminates high die costs)
- Short runs

This process is especially usable for machining (etching) very thin, hard, and brittle metals. Very low tooling costs are involved. Care must be taken to counteract the disadvantages with sufficient capacity and rinses or washing operations.

Advantages

- Low tooling costs
- Excellent for extremely thin parts that would cause difficulty with die tooling
- Can process any metal
- Changes of designs easily handled with virtually no tool change costs

Disadvantages

- Slow
- Corrosive

29.3 ELECTRICAL PROCESSES

Electrical processes fall into two major categories: Electrochemical Machining (ECM) and Electrochemical Grinding (ECG).

29.3.1 Electrochemical Machining (ECM)

The electrochemical machining process is actually the reverse process of electroplating; the workpiece, in effect, is "deplated." High-amperage, low-voltage, DC electrical energy is passed from the tool (**cathode**) through electrolyte to the workpiece (**anode**). Disassociated (or deplated) material is removed from the gap between the tool and workpiece by a constant pressurized flow of electrolyte. The basic process is shown in Figure 29.4.

The Process The electrochemical machining process has basically five requirements:

- A cathode (the tool) machined to the configuration that is to be machined into the workpiece.
- An anode (the workpiece) and a work-holding device to locate and maintain the workpiece accurately in relation to the tool.
- A system to supply and maintain a pressurized flow of electrolyte through the gap between the tool and workpiece.
- An adequate source of controlled DC electrical energy capable of accurately maintaining current density between the tool and workpiece.
- A means of accurately maintaining an optimum gap between the tool and workpiece. Gaps range from .003 in. to .030 in. for various applications.

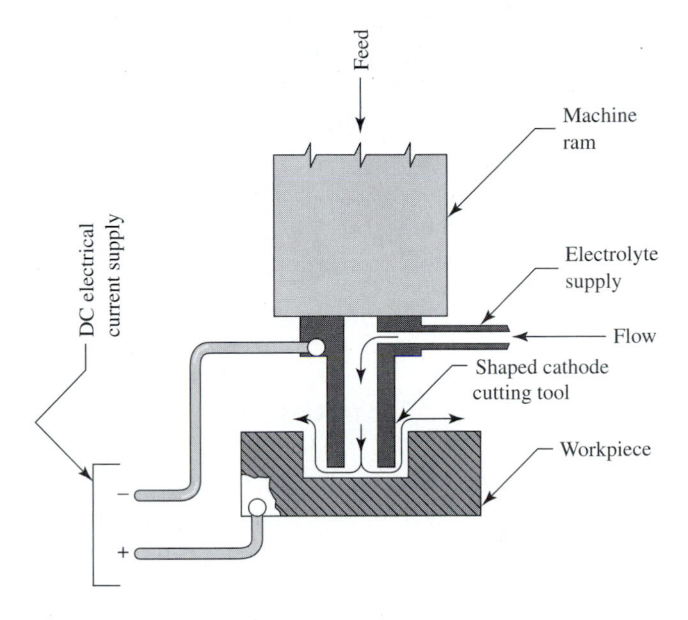

FIGURE 29.4 Schematic diagram of the electrochemical machining (ECM) process.

Electrolytes. Electrolytes are highly conductive solutions of water and salts such as NACl, $NANO_2$, NAOH, and others that are pumped through the gap between tool and workpiece at pressures up to 300 pounds per square inch.

Working temperatures of electrolyte range from 90°F to 125°F. By-products of the process, metal oxides and hydroxides, must be continually removed from the electrolyte to maintain maximum production rates.

Material is removed from the system by such typical means as

- Clarifiers and large settling tanks
- Continuous vacuum filtration
- Centrifugal separators

Applications The major applications of electrochemical machining of any electrically conductive metal include

- Generating almost any internal shape
- Any hardness
- Performing particular operations
    - Cutoff
    - Deburring
    - Drilling
    - Machined surfaces
    - Trepanning
    - Wire cutting

Electrochemical machining is increasingly used to generate internal and external shapes and for burr removal. Due to its high rate of metal removal, adequate means of continuous removal of by-products must be provided. Also, because the medium (electrolyte) is corrosive, rinses and washes must be provided.

Surface Finish and Accuracy Electrochemical machining produces burr-free and stress-free surfaces. The process can and does, however, release stresses in the workpiece resulting from previous operations. This factor affects final tolerances. Surface finishes range typically as follows:

Steels	10 μin.–30 μin.
Stainless steels	5 μin.–15 μin.
Cast iron	25 μin.–60 μin.

The process can hold a tolerance of ±.002 in. and in some instances as close as ±.0005 in.

Equipment and Tooling Machines are available for this process in steps of 500 amps with a total range of 500 to 5000 amps. Machine components must resist corrosion from saline solutions and are usually made of bronze or stainless steel. All machines have the very highest-quality electrical components and insulation not only to provide the required flux but also to enhance operator safety.

Machines are equipped with pumps, valves, and piping to provide the required pressurized electrolyte system. Also, all machines are provided with a system to maintain the critical gap between tool and workpiece. Filtration facilities, normally not part of the basic machine, must be provided separately.

ECM Tools The cathode or tool provided must be manufactured to close tolerances with an allowance for overcut in operation. It is common practice to try out a machine tool for an application, then tailor and modify the tool size based on tryout results. A very high quality of surface finish must also be provided for the tool.

It is economically sound to provide the highest-quality tool possible, because high-quality parts directly depend on the high-quality sizes and finishes of the tool; also, the tool does not wear as quickly. It may, however, be damaged if it is allowed to touch the workpiece and create an electrical arc.

Materials used for tools must be highly conductive and must be noncorrosive. They usually are left soft. Materials used for tools include copper, brass, bronze, stainless steel, and tungsten.

Advantages

- Removes burrs
- Versatile—produces any shape
- High metal removal rates
- Good to excellent surface finishes

Due to the high rate of metal removal, adequate means of continuous removal of by-products must be provided. Also, because the medium or electrolyte is corrosive, rinses or washes must be used.

Disadvantages

- Corrosive

29.3.2 Electrochemical Grinding (ECG)

The electrochemical grinding process combines the removal of metal by electrochemical machining process with the conventional abrasive removal process of a grinding wheel. Figure 29.5 illustrates the principle:

- A saline solution is pumped between the tool (the grinding wheel) and the workpiece.
- An electrical current is passed between the wheel and the workpiece, deplating the workpiece.
- Abrasive particles in the rotating grinding wheel remove the electrochemical oxidation formed on the workpiece surface by the electrochemical action.

Operating Parameters Operating parameters include power requirements, metal removal rates, abrasive wheels, and electrolytes.

Power. The power requirements for electrochemical grinding range from 4 to 15 volts direct current, with 8 volts typical. Current requirements range from 50 to 3000 amps, depending on the application.

Metal Removal. Metal removal rates at 1000 amps for various materials are

Aluminum	74 lb/hr
Copper	5.22 lb/hr
Chromium	2.14 lb/hr
Cobalt	2.42 lb/hr
Titanium	1.31 lb/hr
Tungsten	2.52 lb/hr

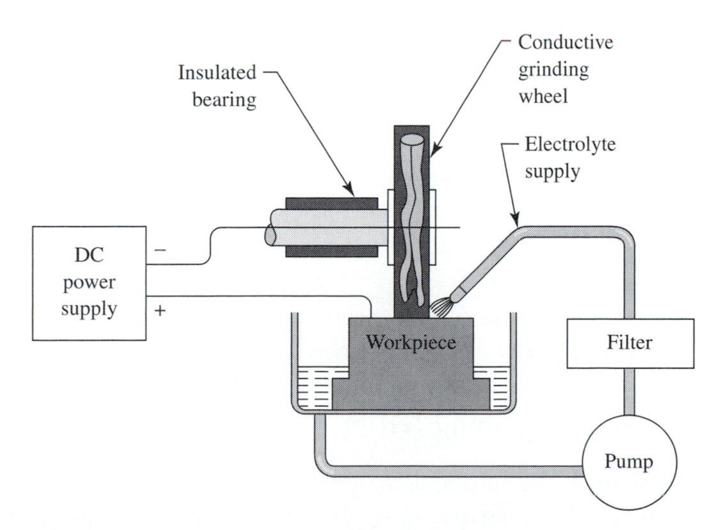

FIGURE 29.5 Schematic diagram of the electrochemical grinding (ECG) process.

As a rule of thumb, metal removal rates for material harder than Rc 45 can be considered to be 10 times faster than with conventional grinding.

ECG Wheels. Several types of abrasive wheels are used for electrochemical grinding including diamond wheels, but all must be current carrying. Wheels are resin bonded with copper mixed in the bond to provide electrical conductivity. Wheel speed is important because the rim acts as a pump to help provide an even, constant flow of electrolyte between the wheel and the workpiece. Speeds of 4000 to 7000 FPM are used.

Electrolytes. The most effective electrolyte for the process is sodium chloride solution, but it is also the most corrosive. Other electrolytes used include

- Sodium carbonate
- Sodium hydroxide
- Sodium nitrite

Operating temperatures range from 90°F to 110°F. Electrolytes must be filtered to maintain the operation at normal production levels.

Applications

- All hard, tough materials
- Carbide tools
- Titanium alloys

Manufacturing engineers specify this process for one or all of the advantages, but must also provide rinses or washes to control corrosion.

Advantages

- Much faster than conventional grinding
- Wheel wear negligible
- No stress from heat as is possible in conventional grinding

Disadvantages

- Corrosive
- Equipment cost higher than conventional grinding

29.4 MECHANICAL PROCESSES

Mechanical processes fall into three major categories, namely Hydrodynamic Machining (HDM), Ultrasonic Machining (USM), and Abrasive Jet Machining (AJM).

29.4.1 Hydrodynamic Machining (HDM)

Hydrodynamic machining is the cutting of material by means of an extremely fine, high-pressure, high-velocity stream of water. Pressures of up to 60,000 pounds per square inch (PSI) are used, produced by water pumps, hydraulic intensifiers, and accumulators. A schematic illustration of the process is shown in Figure 29.6.

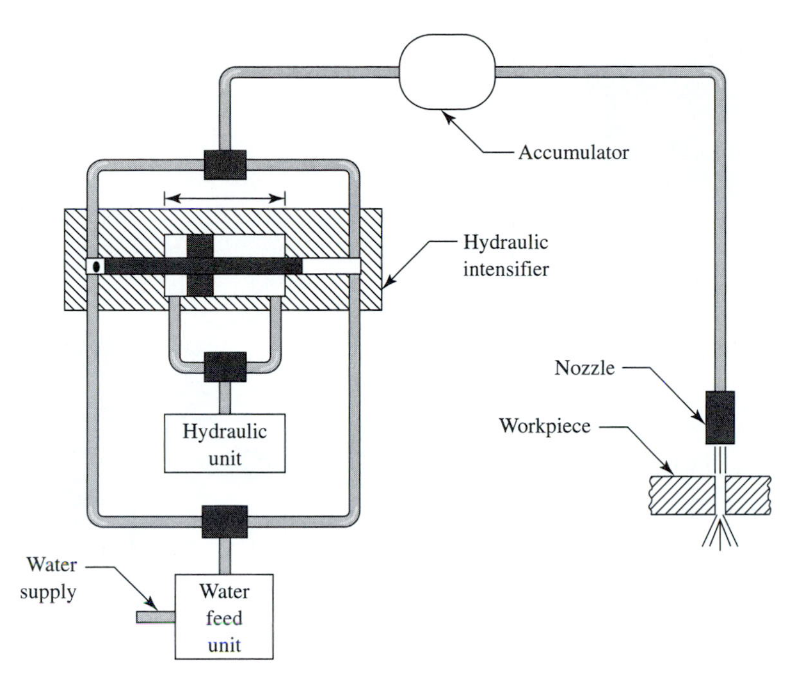

FIGURE 29.6 Schematic diagram of the hydrodynamic machining (HDM) process, also called the water jet machining (WJM) process.

A similar version of the process using lower pressure of 250 to 2000 PSI and known as Water Jet Machining (WJM) is employed principally for deburring. A Water Jet Machining (WJM) operation with two robots mounted upside-down is shown in Figure 29.7(a). Figure 29.7(b) shows a complete robotic water jet machining system.

Machining Equipment HDMs are normally custom designed and built to suit a specific application. The machine consists of six major components and systems:

- Base
- Movable, powered table
- Tracing system
- Fluid pressure system
- Tooling (nozzles)
- Fluid filter

Material to be machined is clamped to the table, which moves it under the cutting nozzles in a prescribed path controlled by the tracing system. The cutting fluid must be kept clean by filtering to maintain production rates and prolong nozzle life.

Nozzles. Nozzles are made of several long-lasting materials including

- Tungsten carbide
- Hardened steel alloys
- Synthetic sapphires
- Diamonds (the most costly)

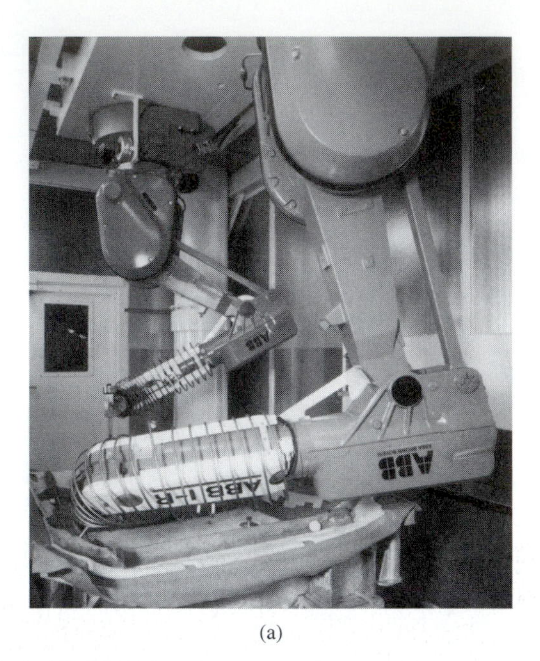

(a)

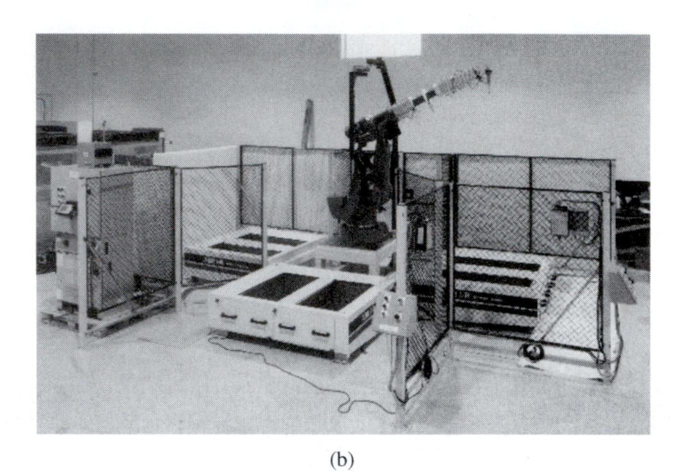

(b)

Figure 29.7 (a) Water jet machining (WJM) operation with two upside-down mounted robots. (b) Complete robotic water jet machining (WJM) system. (Courtesy: ABB I-R Waterjet Systems L.L.C.)

Operating Parameters

Operating parameters include water pressure, operating temperatures, nozzle orifice sizes, and fluids.

Water Pressure. Hydrodynamic machining operates over a pressure range of 10,000 PSI to 60,000 PSI and can go as high as 100,000 PSI for certain applications.

Operating Temperatures. Some heat or friction raises the temperature of the cutting fluid slightly, but is considered to have no effect on the process.

Nozzle Orifice Sizes. Orifice diameters in nozzles range from .003 to .015 in. depending on the application. Harder materials require the smaller orifice diameters.

Fluids. Water is the normal cutting medium, but a polymer such as glycerine is added to improve the cutting action where sharp corners are required.

Applications

Hydrodynamic machining is capable of cutting most soft materials. It is also capable of cutting unhardened, thin steel up to .005 in. thick and untempered soft aluminum up to .020 in. thick.

Final decision on the use of HDM applications or an alternate process normally depend on a cost study. Brittle materials tend to crack in the HDM process and are usually not a candidate. Some applications have an abrasive added to the water jet stream to cut metal.

Advantages

- Burr free and stress free
- Clean parts—usually require no washing

Disadvantages

- Limited to soft, thin metals
- Cannot be used where wetting the product is unacceptable
- Potential rust
- Equipment relatively expensive

29.4.2 Ultrasonic Machining (USM)

Ultrasonic machining accomplishes the abrasion of material from a workpiece with an ultrasonic vibrating tool that impels abrasive grit particles against the desired surface. Almost all metals, as well as many other nonmetallic materials, can be machined by USM. The process is especially effective on materials harder than Rc 40.

An abrasive grit is circulated through the gap between the tool end face and the workpiece. The ultrasonic unit, vibrating at high frequency normal to the surface to be eroded, propels grit particles against the workpiece, chipping and eroding the material to conform to the shape of the tool. A schematic view of the USM process is shown in Figure 29.8.

Machining Equipment Ultrasonic machines are constructed in sizes ranging from 200 to 2400 watts, with some built up to 4000 watts for special applications. The principal components or systems of these machines include

- An electronic oscillator that converts 60-cycle current to 20,000 cycles (Hz).
- An electrosonic unit with amplitude of vibrations adjustable from .0005 to .004 in.
- An abrasive slurry system

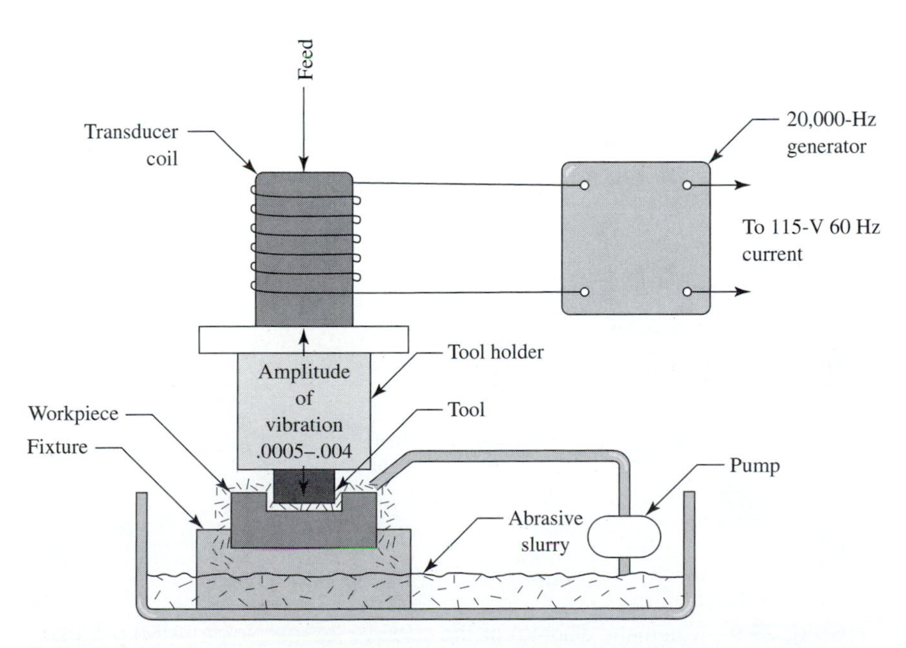

FIGURE 29.8 Schematic diagram of the ultrasonic machining (USM) process.

USM Tools Cutting tools are custom designed and made to suit the size, shape, and so on of the cavity to be produced. Cutting tools must be tough, ductile, and light in weight. Some metals used are SAE 52100 steel, stainless steel, and **molybdenum.**

Surface Finish and Accuracy Ultrasonic machining can hold a tolerance of plus or minus .001 in. with plus or minus .0002 in. possible in some cases. Surface finishes possible range from 20 μin. to 30 μin.

Application

- Almost any material: conductive, nonconductive, metallic, ceramic, composition
- Blind and through holes
- Slots and cavities
- Irregular shapes

Advantages

- Applicable to wide range of material
- Low-cost cutting tools

Disadvantages

- Produces tapered sides on deep cuts
- Not for heavy cuts

29.4.3 Abrasive Jet Machining (AJM)

Abrasive jet machining functions by propelling fine abrasive particles through a nozzle with a high pressure gas to cut many different materials. The action will cut a narrow kerf in a material, or can be used to clean or frost a metallic surface. A schematic view of the process is shown in Figure 29.9.

Operating Parameters Gas pressures used range from 75 pounds per square inch (PSI) to 125 PSI depending on the application. Dry, filtered compressed air, carbon dioxide, and nitrogen gases are used.

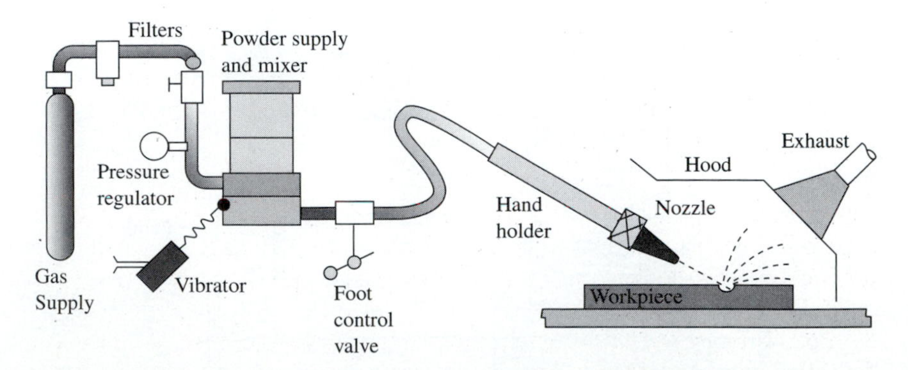

FIGURE 29.9 Schematic diagram of the abrasive jet machining (AJM) process.

Abrasive powders used range from 10 to 50 microns in size and include such varying substances as aluminum oxide, dolomite, sodium bicarbonate, and silicon carbide. Nozzle orifice sizes range from .003 to .030 in. depending on the application.

Applications

- Most materials, including diamonds
- Etching slots and holes
- Cleaning
- Frosting metallic surfaces
- Corrosion removal

Advantages

- Almost no heat generated
- Stress free and safe
- Can handle thin, brittle materials

Disadvantages

- Low material removal rates
- Dusty, requires dust collector unless abrasive medium is mixed and delivered to nozzle with water

29.5 THERMAL PROCESSES

Thermal processes fall into four major categories: Electron Beam Machining (EBM), Electrical Discharge Machining (EDM), Laser Beam Machining (LBM), and Plasma Arc Machining (PAM).

29.5.1 Electron Beam Machining (EBM)

The electron beam machining process uses electrical energy transformed to thermal energy for removing metal by melting or vaporizing. A stream of high-speed, pulsating electrons is produced by a generator and focused on the workpiece by means of a coil, known as a magnetic lens. The subsequent beam is moved by means of deflector coils.

The process, including the workpiece, is operated in a vacuum. Lead shielding is required to protect the operator from the x-ray radiation produced by the electron beam. Types of metal and hardness have no effect on the process. A schematic view of the electron beam machining process is shown in Figure 29.10.

Equipment Electron beam machines are available in a wide range of chamber sizes. Custom-built machines have chambers as large as 20 × 30 feet. Larger machines translate into slower production rates, because the chamber must be pumped down to create a vacuum for each piece or lead.

Both NC systems and digital computers are used to control workpiece movement, beam deflection, and beam focus. Fixturing is simple, usually requiring only locators and clamps. Most workpieces used in this process are flat.

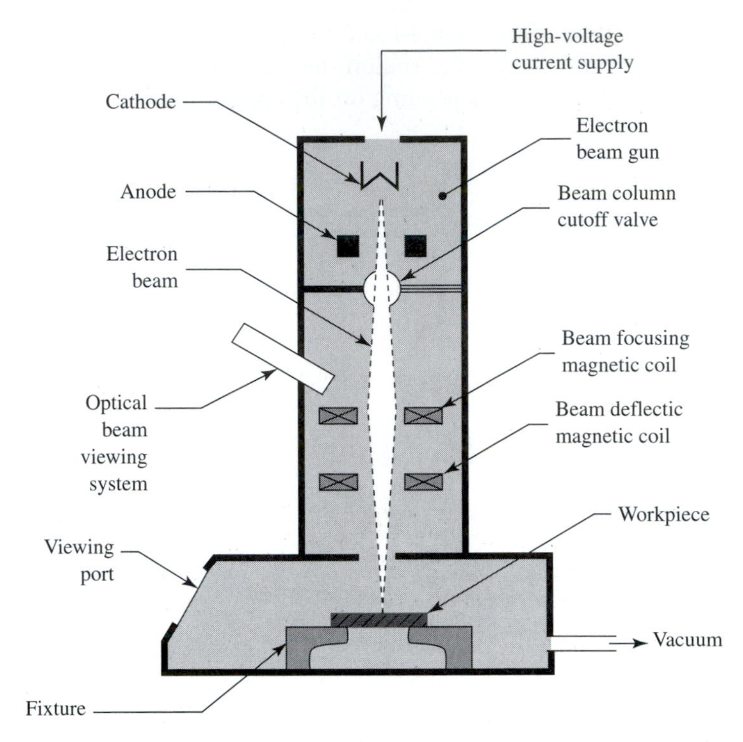

FIGURE 29.10 Schematic diagram of the electron beam machining (EBM) process.

Applications

- Almost any known material and metal (not suitable for diamonds)
- Slots and holes (both circular and irregular)

Surface Finishes and Tolerances

- Normal finish, 40 μin. (can hold in certain instances as low as 5 μin.)
- Normal tolerance, $\pm .001$ in. (can hold in certain instances as low as $\pm .0002$ in.)

Advantages

- Fast
- Clean

Disadvantages

- Parts may distort in heat
- Inconvenient loading into vacuum chamber

Safety Safeguards must be provided for operator safety with the use of electron beam machining:

Electrical	Insulation, covers, interlocks, and maintenance lockout system
Beta and x-rays	Shielding
Noise	Acoustic material for capacitor discharge and blowers
Viewing area	Closed circuit TV, light-absorbing materials, goggles

29.5.2 Electrical Discharge Machining (EDM)

The electrical discharge machining process is a method of removing metal by a constant stream (thousands of sparks per second) of discharges across a gap, melting, vaporizing, and eroding the shape of the tool into the workpiece in the presence of a dielectric fluid.

In practice, the workpiece is loaded in a fixture in contact with the machine base, making it the anode (+) polarity. The tool located in close proximity to the workpiece, but not in contact, becomes the cathode (−) polarity. Some applications reverse the polarity.

Electrical energy is applied to the tool, causing discharges across the gap between tool and work, eroding metal particles which are removed from the gap by the flow of dielectric fluid. A schematic view of the process is shown in Figure 29.11(a); and a typical EDM machine is shown in Figure 29.11(b).

Equipment and Tools A wire EDM machine and some of the parts produced with such a machine are shown in Figures 29.12 and 29.13 respectively. Electrical discharge machines are comprised of six major components, usually built in a C configuration of base, column, and head.

Base: Bases are either castings or weldments and normally form the dielectric fluid reservoir.

Workpiece Holder: The workpiece holder, or fixture, accurately locates and retains the workpiece in relation to the electrode.

Electrode Holder: This unit holds the cutting tool and the specially shaped electrode, and is mounted to the machine column in relation to the path the electrode is to follow.

Power Supply: This unit, normally separate from the machine, supplies the operating electrical current required for the process and includes the controls.

Feed Unit: This device measures the gap between the workpiece and the electrode end face, and feeds the electrode into the workpiece at a rate that maintains the gap.

Dielectrical System: The dielectrical system pumps the dielectrical fluid through the process gap to remove debris and metal particles as they are formed, and to carry them to a filter where they are removed. The system may be included in the machine base or may be a separate unit.

Electrodes Graphite is the most widely used material for electrodes. It is highly conductive, easily machined, and has a good working life. Fine-particle, high-density graphite is used for work requiring fine detail, but it is costly. Coarse-particle graphite, not capable of fine detail, is much less costly and therefore is used wherever feasible. Copper is used for electrodes where very smooth finishes are required. Other material such as brass, tellurium copper, copper tungsten, and tungsten carbide are used for special applications.

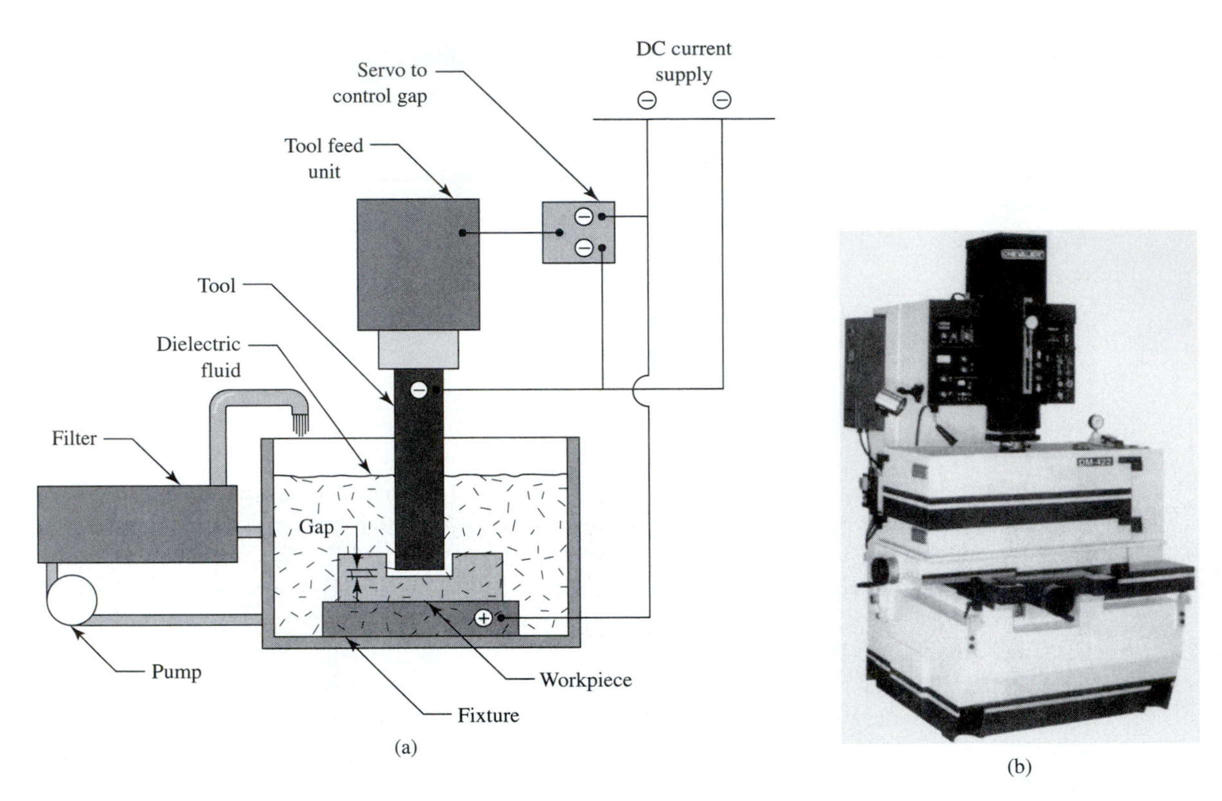

FIGURE 29.11 (a) Schematic diagram of the electric discharge machining (EDM) process. (b) Typical EDM machine. (Courtesy Chevalier Machinery, Inc.)

Dielectric Fluids Dielectric fluids must perform three functions:

- The fluid must act as an insulator up to the point where voltage applied across the gap between electrode and workpiece causes it to ionize, permitting a spark to flow to the workpiece.
- It must cool and solidify vaporized workpiece material into particles and also maintain the workpiece and electrode at working levels of temperature.
- It must wash, flush, and remove particles of metal and debris from the spark gap.

Most electrical discharge machining applications use petroleum-based dielectric fluids. Other types of fluids include glycol and water, silicone fluids, and mixtures. Straight kerosene is often used.

Applications

- Conductive materials only
- Dies—stamping—major user
- Dies—die cast—drawing, extrusion, forging, header
- Molds—permanent mold—aluminum and plastic
- Prototype parts
- Short-run production parts

FIGURE 29.12 Five-axis wire EDM machine. (Courtesy Hansvedt, A Hardinge Co.)

FIGURE 29.13 Typical parts produced on a wire EDM machine. (Courtesy Hansvedt, A Hardinge Co.)

Advantages

- Any hardness and any shape
- Fragile parts—no mechanical force to hold or machine

Disadvantages

- Slow
- Leaves remelt on cut surfaces

29.5.3 Laser Beam Machining (LBM)

The **laser** beam machining process is applicable to any material, hardness, and condition of electrical conductance. It is especially capable of producing very small holes (.001 in.) in steel up to .100 in. thick but is not practical for heavy metal cuts over .200 in. thick. A schematic diagram of the Laser Beam Machining (LBM) process is shown in Figure 29.14. Laser beam machining has potentially hazardous areas that must be controlled by proper safeguards and techniques for the safety of the operator.

Equipment Of the many types of lasers developed, the most practical for metal cutting include ruby lens, and CO_2. The solid state ruby lens laser produces more powerful output, but is more costly. CO_2 gas lasers are more efficient. Lasers can be programmed to cut, slot, drill, and **scribe.** Numerical Control is used on many applications to position and move material in relation to the beam.

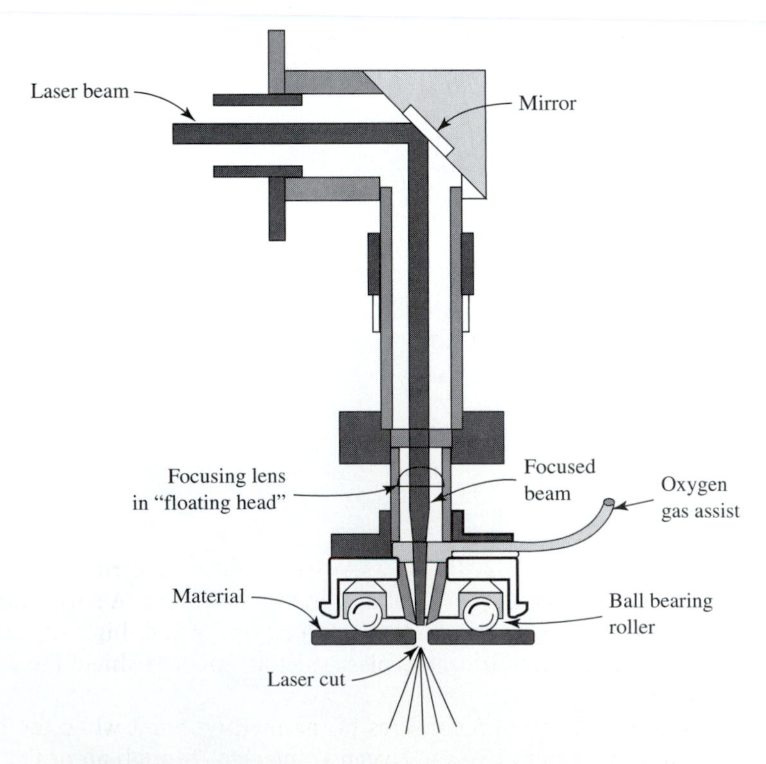

FIGURE 29.14 Schematic diagram of the laser beam machining (LBM) process.

Applications

- Any material—any hardness
- Thickness of cut to .400 in.
- Drilling, cutting, scribing

Laser Safety The laser beam machining process has several inherent conditions that can be dangerous and affect the safety of the operator or bystander.

High Electrical Power. High voltage and high currents (120 V and 1000 A) combine to create a potentially hazardous condition for operators. Proper insulation, covers, interlocks, and correct operating techniques must be provided and monitored for compliance with safe operating conditions.

X-Rays. Lead shielding must be provided and, in some instances, operators must be issued film badges to monitor exposure to x-rays.

Vision. Light-absorbing shields must be provided; in addition, operators must be issued eye protection goggles to prevent any accidental exposure during setup, part loading, and so on. Closed circuit TV systems are available and are in use for the operator to safely view and conduct such tasks as focusing the laser beam.

Advantages

- Versatile—processes any material, any hardness
- Nondistorting
- Can produce sharp corners
- No cutting tools required
- Simple fixturing required
- Adaptable to any type of control system

Disadvantages

- Limited ability to cut thick metal
- Potentially dangerous
- Equipment relatively costly.

29.5.4 Plasma Arc Machining (PAM)

Plasma arc machining is accomplished by constricting an electrical arc with a nozzle to increase and intensify the ensuing jets' voltage and temperature. A surrounding primary gas of nitrogen and argon is ionized, which develops a concentrated, high-velocity, intensely hot plasma jet. A secondary gas surrounds the plasma jet and acts to shield the jet from impingement of air molecules.

The plasma jet raises metal temperatures to the melting point while the high-velocity gas stream blows the molten material away. Oxygen is injected through an orifice of the nozzle in some applications to add another source of energy and further speed the process.

Water injection is used to constrict the arc, in some applications, to produce a square-edged cut, narrower kerf, and increased nozzle life at the expense of cutting speed. A schematic view of a conventional plasma arc torch is shown in Figure 29.15.

Machining Equipment Plasma arc torches are used with specially designed machines that resemble turret presses, having a rotating table for the workpiece over which is mounted the torch. Many conventional punch presses are also fitted with plasma torches.

PAM Tools Plasma arc torches, which are the cutting tools of the process, are available in several sizes. One development provides one size of torch head that is used with a variety of nozzles and electrodes to suit the application. Portable, handheld plasma arc torches are also available.

Applications

- Any metal—any hardness
- Cutting process capabilities
  - Straight lines, circles
  - Irregular shapes, grooves
- Hole piercing
  - Repetitive and good quality
- Stack cutting
  - Aluminum and stainless steel sheets (carbon steels tend to weld)
- Edge preparation—bevel edges before welding.

Advantages

- Fast—highly productive
- Good quality

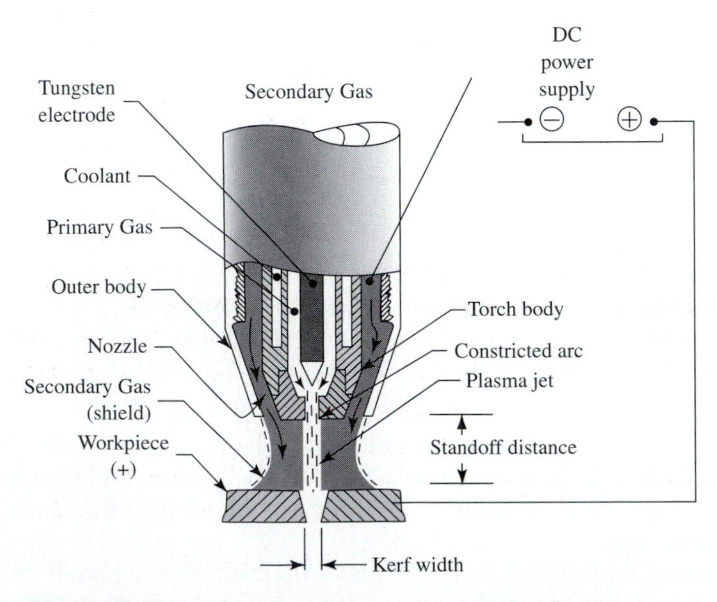

FIGURE 29.15 Schematic diagram of a conventional plasma arc torch.

- Low operating costs
- Easily automated

Disadvantages

- Metallurgical changes of material from heat (hardening of material and uneven top surface)

29.6 REVIEW QUESTIONS AND PROBLEMS

1. What is a nontraditional machine operation?
2. In what two ways are nontraditional machines used in processing?
3. Describe the principal points of a chemical milling process.
4. Name some parts that would suit a photochemical machining operation.
5. Describe the operation of an electrochemical machining process.
6. What operations are most suitable for the electrochemical process?
7. What is electrochemical grinding?
8. What is electrochemical grinding especially suited for?
9. Describe hydrodynamic machining.
10. What is hydrodynamic machining most suited for?
11. Describe ultrasonic machining.
12. Where can ultrasonic machining be used?
13. Where is abrasive jet machining used?
14. What material is not suitable for the electron beam machining process?
15. What safety measures must be employed with the electron beam machining process?
16. What is EDM?
17. What are the applications for EDM?
18. What are the advantages of the EDM process?
19. What material may be processed on a laser beam machining operation?
20. What safety precautions must be employed with laser beam machining?
21. Describe the operation of a plasma arc torch.
22. What are some applications of the plasma arc process?

Glossary

Abrasion Resistance The ability of material to withstand a change in dimensions due to a rubbing action by another material.

Abrasive, Artificial Silicon carbide, aluminum oxide, and other man-made materials.

Abrasive, Natural Diamond, emery, garnet, flint, pumice, and other natural materials.

Abrasive Wear The wear that occurs on a carbide part or tool in use, due to the rubbing action.

Acme Thread Screw thread with an included angle of 29°.

Actual A dimension that exists at the conclusion of the part's processing.

Air Gage Comparison device that senses variations in airflow rates, and converts them into dimensional scale.

Alternate Locators Duplicate locators used to position the workpiece from either of two acceptable surfaces.

Aluminum Silvery white metal that is one-third the weight of steel.

Aluminum Oxide Ceramic Also alumina (Al_2O_3). Cutting tools made of aluminum oxide.

Ancillary Those systems that are required by and used with, or in conjunction with, a production machine during manufacture.

Anneal A heat treatment in which metals are heated and then cooled very slowly for the purpose of decreasing hardness.

Anode Positive pole of an electrolytic cell.

Approximate Locators Used to facilitate loading suitable workpieces into a workpiece holder, where they aid positioning until clamping actions press the workpiece against the actual locator.

Arbor A rotating shaft upon which a cutting tool is fastened.

Arbors and Mandrels Spindles upon which a workpiece is mounted or pressed for purposes of machining or grinding; they may be run between centers or chucked.

Auxiliary Those systems that perform work, known as direct labor, on a part during manufacture but do not significantly change the part dimensionally.

Axis Centerline that is a reference point for mechanical or geometric relationships.

Backlash A condition created due to clearance between a thread and nut. The amount of thread turned before a component begins to move.

Backward Extrusion Metal flowing in the opposite direction to the motion of a ram; see Figure 21.15.

Back Rake The angle of inclination of the face of a tool away from the end cutting edge.

Baseline A system of dimensions in which dimensions are drawn from a common surface to the various related surfaces.

Bed-Lathe Main horizontal part of a lathe, which carries the headstock carriage and tailstock.

Bed-Press The base of a frame, usually open in the center to permit parts or slug ejection, or to enable the mounting of cushion devices below.

Bellmouth A condition in a machined hole in which the end is flared out in a bell shape to a dimension larger than the nominal size of the hole.

Bending, Edge Method Performed by clamping a workpiece to a die block with the portion to be bent extending cantilever fashion; as the punch moves down, it contacts the extended metal and wipes it down the side of the die block. See Figure 20.7.

Bending, V Method Accomplished by forcing a metal strip or blank into a die block that has the required V-angle by a wedge-shaped die with the same angle.

Billet Green slab or block of pressed material, often presintered, for slicing into smaller sizes also a cutoff length of steel bar or plate.

Binder The metallic component in cemented tungsten carbide that holds the carbide grains together.

Blank Holder Used to locate and apply pressure and friction to a blank, a blank holder controls the flow of metal as it is drawn into a cup by a punch.

Blank Holder Force The force required by a drawing operation, developed by trial and error; the force ranges from zero for certain thick sections to half the die force for thinner metal.

Blanking A die that creates a blank for succeeding operations by cutting or shearing a shape from strip or sheet stock; the die may be single purpose or a station in a progressive die.

Blind Hole A hole that does not go completely through an object.

Bolster Plate A flat steel plate bolted to the top face of the bed and upon which tooling and accessories are mounted.

Boring A machining process in which internal diameters are made in true relationship to the centerline of the spindle.

Boring Mill Upright machine for removing metal from a hole by using a single-point tool.

Boron Carbide Fine powder that is almost as hard as diamond.

Brass Alloy of copper and zinc; in some cases other materials are added.

Brinell Hardness The hardness of a metal or alloy measured by pressing a hard ball with a standard load into the specimen.

Brittleness The property of a material such as glass, ceramic, and cast iron, that causes it to suddenly break at a given stress without bending or distortion.

Broach A multitoothed cutting tool with progressively larger teeth, which is designed to finish a surface in a single stroke.

Broaching The process of removing unwanted material by pulling or pushing a tool, on which cutting teeth project, through or along the surface of a workpiece.

Bronze Alloy of copper and tin. In some cases other materials are added.

Built-Up Edge An adhering deposit of work material on the tooth face adjacent to the cutting edge.

Burnish To polish a metallic surface by rubbing it with a harder metallic surface.

Burr Sharp or ragged edge left from a machining operation.

Bushing A hollow cylinder used as a spacer; a reducer for a bore size or for a bearing.

CAD Computer-aided design; computer-aided drafting.

CAM Computer-aided manufacture.

Cam A device mounted on a rotating shaft that transmits rotary motion into uniform or nonuniform reciprocating motion.

Carburize The process of chemically combining carbon and a metal, by mixing and heating fine powders, to form a metallic compound such as carbide (WC) in the presence of heat.

Carriage The part of a lathe that moves along the ways, including the apron, cross slide, and other parts.

Cast Iron Alloys of iron, carbon, and other materials, which are primarily used for making castings.

Casting A metal object or workpiece produced by pouring or injecting molten metal into mold.

Cathode Negative pole of an electrolytic cell.

Cemented Carbide General term for all sintered carbide materials.

Cementite Iron carbide, a compound of iron and carbon (Fe_3C) found in steel and cast iron.

Center-Column Drilling Machine A drilling machine with a revolving, indexable turntable upon which holding fixtures are located and around which drilling heads are mounted in a column in the center, and around which the table revolves; the machine also mounts drilling heads.

Center Drill Short drill combined with a countersink, used to prepare the end of a shaft for a center.

Centerless Grinding Grinding operation in which the work is supported by a blade or rest between the grinding and the regulating wheel.

Centerline A reference line on a drawing or part layout from which all dimensions are located.

Center of Pressure The point where all bending forces of a blank cancel out and where the pressure at the center axis of a press can act vertically with no deflection; the point where all shearing forces balance.

Cermet A material composed of a nonmetallic compound and a metal; from the words "ceramic" and "metal."

Chamfer A bevel cut on a sharp edge of a part to improve resistance to damage and as a safety measure to prevent cuts.

Chart A graphical display of manufacturing dimensions, tolerances, and amounts of stock removed at each operation of a manufacturing sequence.

Chatter Vibration between tool and work, sufficient to cause an irregularity in the tool marks of the finished surface.

Chemical Cleaning A method of removing debris from a workpiece surface by use of chemical action; cleaning is done by spraying, flooding, or immersing.

Chemical Surface Treatment (CST) Applied by brush, spraying, or immersion; a treated surface is protected from corrosion and attack by various hostile elements.

Chip A fragment of the work material that has been separated in the machining operation.

Chipbreaker An irregularity in the face of the tool, or a separate piece fastened to the tool or toolholder, to cause the chip to break into short sections or curls.

Chipping The breakdown of cutting edges by loss of fragments broken away during the cutting action.

Chuck A mechanical device attached to the spindle of a machine to hold a workpiece by means of pressure and friction from jaws and jaw inserts.

Clamp A mechanical detail of a workholder that contacts a workpiece to press it against locators.

Clamping Force Also known as holding force, the clamping force is the pressure or force developed by the clamp against the workpiece.

Clean Up A surface is said to clean up when the entire surface from a preceding operation is removed by a succeeding operation.

Clearance The angle below or behind the cutting edge, which allows the tool to be forced into the work.
Climb Milling A metal-cutting operation in which the table moves in the same direction as the part of the cutter below the arbor, where the cutter attempts to "climb" the workpiece tool forces are down.
Clutch A component, usually found in a mechanical drive, that permits a driven component and a driving component to be mechanically connected or disconnected at will.
Cold Drawing Finishing metal bars by drawing them through a die while they are cold.
Cold Extrusion Metal is extruded without heating the billet.
Cold Rolled Steel bar stock produced by the rolling process without heat; cold rolling produces a good surface finish and close tolerances but has inherent stresses, which are relieved by heat application and produce distortion.
Cold Rolling Finishing metal sheets or flat bars by rolling them between polished rolls while they are cold.
Cold Sawing Any sawing process in which the chips are not heated to a softened state.
Collet A type of chuck, used for bar stock or tubular stock, to hold metal or tools while machining, a collet is operated by the closing action of a tapered ring.
Compacting The forming of an object from powder by compression of the powder, which is generally confined in a die or mold.
Contouring Machining an uneven but continuous path on a workpiece in two or three dimensions.
Conventional Milling The opposite of climb milling; chips are cut starting with a zero thickness and increase to a maximum thickness tool forces are up.
Coolant A cutting fluid used to cool the tool and workpiece, usually water based.
Copper A reddish soft ductile metal, with very good heat and electrical conductivity; also the basic element in brass and bronze.
Counterbore A cutting tool, fitted with a pilot, that engages a hole for part of its length.
Cradle Devices used to uncoil heavy coils of steel prior to use in a press.
Cratering The action whereby a tool face is eroded by chip contact and a depression is formed.
Critical A surface that has a critical relationship with other areas of a part and as such usually serves as a locating surface.
Crystalline The appearance of a coarse powder with a shape and structure like salt. Most cemented carbide fractures have a fine crystalline appearance.
Cushion An accessory that provides resistant force and motion required for such operations as blank holding, drawing, and redrawing, also used for knocking out and for stripping a part from a die.
Cutting Fluid Any of several materials used in cutting metals: cutting oils, synthetics, soluble or emulsified oils (water based), and sulfurized oils.
Cutting Tools Also known as perishable tooling, these are the tooling details incorporating cutting edges that do the actual machining.
Deburring To remove a sharp edge or corner caused by a machining or a manual process.
Deep Drawing A term used to define a drawn cup that has a depth greater than its radius.
Deflection Bending or displacement of a tool, spindle, or workpiece from its true position, deflection can be caused by clamping force or tool pressure.
Deformation The permanent change in shape due to cutting forces and temperature.
Deionization Removing ions from water by ion exchange.
Dial-Type Drilling Machine The same basic machine as a center-column drilling machine but without the center column.
Diamond Crystalline form of carbon (the hardest known mineral) used as a cutting and grinding tool and to dress grinding wheels.

Die A device that is mounted on a press for cutting and forming.

Die Block A block from which dies are cut; with some exceptions a die block is generally mounted on the lower shoe of a die set.

Die Casting Molten nonferrous metal is forced under pressure into a closed metal die.

Die Sets Dies are assembled to die sets, which keep them in alignment during operations with heavy guide posts. Die sets are commercially available in many sizes and configurations.

Dieing Machine A press with the drive mechanism located under the bed and the upper die fastened to the ram, which is pulled; the lower die is mounted on the press bed.

Dimensional Tolerances that are applicable to the specific design dimension.

Dimensional Control Control relating to the maintenance of physical dimensions as specified by the part design.

Direction The direction of a working or machining dimension from the locator, expressed "from the locator to the surface."

Distillation A process of heating a liquid to vapor form, condensing the vapor to form a liquid without impurities.

Dividing Head A mechanical device that spaces the perimeter of a workpiece into equal parts, also called an indexing head.

Double-Action Press A press with two independent rams or slides moving within one another that is used for deep drawing of large parts.

Double-End Machine A machine with machining units mounted at either end and with a work table possessing left and right movement capabilities. Workpieces held in fixtures located at the center of the table are moved alternately to each end of the machine for processing.

Dovetail An angular shape used on many types of interlocking slide components, especially on machine tools.

Drawing A process in which a punch, under pressure, causes flat metal to flow into a die cavity to assume a cuplike shape.

Drawing Force The force required to draw a specific part, calculated from an empirical formula that part design and specified metal dictate.

Drill Jig A device that accurately guides one or more drills into an object by means of accurately positioned hardened bushings.

Drop Forging Parts are so named because they are formed by the weighted upper die of a closed die set falling (dropping) onto a heated billet of metal located in the lower die and forcing metal into the die cavity.

Drop Forging Process A weighted upper die of a closed set that falls (drops) into a heated billet of metal located in the lower die, forcing metal into the die cavity; most dies have several consecutive cavities into which the billet is transferred by being struck a blow, each cavity forcing metal progressively into its final configuration.

Ductility The property of a metal to be deformed permanently without rupture, while under tension.

Electrochemical Grinding A form of electrochemical machining that is done on a grinding machine, which also removes metal by abrasion.

Electrochemical Machining Actually a reverse plating action in which high-amperage, low-voltage electrical current is passed through the workpiece in an electrolyte.

Electrolytic Grinding To abrasively machine with a diamond wheel, assisted by an electrolytic fluid.

End Cutting-Edge Angle The angle between the cutting edge and the end of the tool, perpendicular to the centerline of the toolholder.

End Mill A milling cutter, with cutting edges on both the face and the periphery, that may be used for either face cutting or peripheral cuts or both.

Equilibrium The state of balance between opposing forces.

Extreme-Pressure Lubricants Lubricants that contain sulfur, chlorine, or other compounds that react at high temperatures in a cutting or die zone, giving the lubricants or cutting fluids antiweld properties.

Extruding A process consisting of placing a billet of metal in a closed chamber and forcing it with pressure from a press through an opening in a die. A finite workpiece.

Extruding A die into which a billet of metal is placed, usually cold, after which a punch applies pressure, forcing the metal through an opening in the die.

Extrusion Forming metal bars, structural shapes, and tubes by forcing the metal through a die. A continuous process.

Face Milling Cutter A milling cutter used to generate a flat surface on the workpiece.

Faceplate Circular plate for holding workpieces, which may be attached to the nose of a lathe spindle.

Facing A cutting operation, usually at 90° to the spindle axis of the lathe.

Feed The relative amount of motion of the tool into the work for each revolution, stroke, or unit of time.

Ferrous From the Latin word "ferrum" meaning "iron"; an alloy containing a significant amount of iron.

Fillet Concave curve connecting two surfaces that meet at an angle.

Fine Blanking A die in which the blank is locked in with a V-shaped stinger ring to prevent metal flow while the punch is advanced, resulting in a blank with smooth full-length edges and no burrs.

Finish The quality of a surface in terms of roughness, waviness, and lay.

Fixture Workpiece holders that locate and clamp a part in position but do not have features to guide the cutting tool; they may have setup blocks or other means of aligning with a machine and cutting tools.

Flame Hardening A surface-hardening process in which ferrous metals are heated with a flame to the desired temperature and are then water quenched.

Flatness The permissible surface deviation from a plane.

Flaws Unintentional irregularities that occur at one place or at relatively infrequent intervals on the surface; flaws include cracks, checks, ridges, scratches, and blow holes.

Flexible Manufacturing Systems Systems that are composed of workstations, material-handling facilities, and controls, and are capable of performing all machining processes; each is special to its purpose—there is no standard.

Floating Free to move about over a given area; for example, a floating die holder or a floating reaming holder.

Floor Conveyor A conveyor designed to operate with its top surface at floor level to reduce lifting or lowering loads.

Flute The groove in a drill, tap, reamer, or milling cutter.

Fly Cutter A cutter body for facing operations on milling machines, in which one or more cutters can be used.

Foolproofing A workholder designed with a pin, block, or other device such that a workpiece cannot be loaded incorrectly is said to be foolproof.

Forging A workpiece produced by deforming to desired shape by presses, hammers, rolls, and upsetters.

Formability Previously known as ductility, a metal's formability is its comparative degree of satisfactory forming, determined by actual drawing tests of sample material.

Forming A method of working sheet metal into useful shapes by pressing or bending.

Forming Die A die that changes the shape of a metal piece but does not intentionally change the metal thickness.

Forward Extrusion Metal flows in the same direction as the motion of a ram; see Figure 21.14.

Frame The main structure of a press, a frame may be casting or weldment.

Gap Press A C-frame press that is not inclinable; see Figure 24.3.

Gear Train Two or more gears that drive machine parts at a specific ratio of speed.

Geometric Control Control relating to the stability of the workpiece.

Geometric Dimensioning and Tolerancing GD&T, as it is termed, is a method of specifying product design equivalents with respect to actual function and relationship of part features. The authoritative document governing its use in the United States is ANSI Y14.5M-198.

Geometry and Configuration Terms used to describe the shape of a product.

Gib A part of a slide mechanism used to adjust the clearance between two sliding parts.

Gibs Retainers that position and guide a ram through its stroke.

Glazing A dull grinding wheel whose surface grains have worn flat, causing the workpiece to be overheated.

Grain In metal, a single crystal consisting of parallel rows of atoms called a space lattice.

Grinding The machining process of removing metal by the cutting action of grains of abrasives in a wheel or disk.

Grit Any small, hard particle such as sand or grinding compound.

Group Technology A plan of manufacture for groups of parts or families of parts that have similar design and/or manufacturing sequences.

Gullet The bottom of the space between teeth on saws, broaches, and circular milling cutters.

Gun Drill A single-fluted drill for drilling deep and accurate holes.

Hardness Ability to withstand deformation, generally measured by the depth or area of penetration under a fixed load, using a diamond indentor.

Heat Treated Metal whose structure has been altered or modified by the application of heat.

Helical The geometry of a helix, where a point both rotates and moves parallel to the axis of a cylinder. Examples are threads, springs, and drill flutes.

Helix The path described by a point rotating about a cylinder while being moved along the cylinder.

Helix Angle Angle of a thread or gear tooth, with a line perpendicular to the longitudinal axis.

High-Carbon Steel Steel that has more than .60% carbon.

High-Speed Tool Steel (HSS) Steel containing tungsten or molybdenum as the major alloying element, which makes it a heat resistant cutting-tool material.

Honing Finishing process that uses abrasives to smooth and straighten holes.

Horning and Adjustable Bed Press Suitable for prototype work and low-volume production, this press has a horn for internal support of tubular parts and a bed that can be raised or lowered to suit different die heights. See Figure 24.4.

Hot Rolled Steel bar stock produced by hot rolling and allowed to cool; its surface finish is not as smooth as that of cold-rolled steel, and tolerances cannot be held as close. Sheet stock is also hot rolled. No internal stresses are set up and no distortion occurs with application of heat.

Hydraulic Press A press with a ram that is powered by hydraulic pressure, giving it flexibility in working pressures, lengths of stroke, and ram speeds; extremely high-tonnage machines are available. See Figure 24.12.

Hydrostatic Extrusion A ram applies pressure to a fluid surrounding a metal billet, forcing metal to flow through an opening in the die; see Figure 21.17.

Impact Extrusion Cold extrusion that is performed with a high-speed motion of the ram on an aluminum billet to produce such products as beverage cans; see Figure 21.23.

Indexable Tool A cutting tool that has several cutting edges and is used in milling cutters and single-point operations.

Induction Hardening A surface-hardening process in which ferrous metals are heated by electromagnetic induction to the desired temperature and then water quenched.

In-Line Banks Material in a process line and ahead of various metal-cutting and metal-working operations.

In-Line Transfer Machine This machine consists of workstations arranged in the required sequence in a straight line to perform machining operations progressively. Workpieces are usually located and clamped in pallet-type fixtures that are moved from station to station by powered transfer devices.

Inscribed Circle The circle that can be constructed internal to any closed figure or shape, such that all sides of the figure are tangent to the circle.

Internal Thread Threads cut on the inside of a hole, tube, or pipe.

IPM Inches per minute, a term used to express the cutting feedrate on straight line cuts.

IPR Inches per revolution, a term used to express the cutting feedrate on revolving cuts.

Isostatic Pressing Compacting or bonding using a gas or liquid as the multidirectional force.

Jig A device that locates and holds a workpiece and guides the cutting tools.

Kerf The width of a cut produced by a saw.

Keyway Grooves in a shaft or hub, in which a driving key is located.

Knockout A mechanism that operates on the press up-stroke to eject a part from a die; not used on all dies.

Knuckle Press Sometimes known as a knuckle joint press or toggle press; very high tonnage can be attained with the use of this feature. See Figure 24.10.

Lapping Finishing external or internal surfaces accurately by using very fine suspended abrasives.

Laser An intense source of coherent light energy, which may be used as cutting tool.

Laser Beam Cutting A process in which an intense, coherent beam of single-wavelength light is focused on a small area, melting and vaporizing it along the cutting line.

Lathe Machine tool for turning, facing, boring, and other similar operations.

Lay The direction of the predominant surface pattern, ordinarily determined by the production method used.

Lead The distance a thread or nut advances along a threaded rod in one revolution.

Lead Angle The angle between the side cutting edge and the projected side of the tool shank or holder, which leads the tool into the work.

Lead Screw Screw on a lathe that is used to move the carriage for thread cutting.

Lead Time In manufacturing, the time interval between a product release for manufacture and required start-up date for its manufacture.

Light Screen A press guard that prevents press motion if any part of a press operator's body comes within the hazardous area; see Figure 24.26.

Limits The maximum allowable dimensions of a part.

Line Boring A process in which several diameters arranged along a common centerline are bored simultaneously with a boring bar containing a boring cutting tool for each diameter.

Line Reaming A process in which several diameters arranged along a common centerline are reamed simultaneously with a reamer that has a reaming section for each diameter.

Loaders and Unloaders Devices used to load workpieces into and out of presses; may be mechanical aids, slides, chutes, or robots.

Loading A grinding wheel whose voids are being filled with metal, causing the cutting action of the wheel to be diminished.

Locating The surface upon which a part rests while a manufacturing operation is performed and from which working dimensions originate.

Locator The surface of the workholder upon which the workpiece rests, a locator may be a pin or surface.

Lockout System A safety system that requires turning off power and locking electrical controls during the time when maintenance personnel work on a press or other machine.

Machinability The relative ease of machining, which is related in part to the hardness of the material to be cut.

Machine (Equipment) Terms for the base machine, which is also called a machine tool.

Machining Center A machine, usually numerically controlled, that can automatically drill, ream, tap, mill, and bore workpieces, and is usually equipped with an automatic tool-changing system.

Magnetic Chuck A device that uses permanent magnets or electromagnets to hold iron or steel workpieces.

Mandrel The shaft onto which an object to be machined may be pressed or held.

Manufacturing Engineering A profession requiring planning, designing, and managing tooling and equipment, plants, and people to produce quality parts economically, safely, and with due consideration for the environment.

Mechanical Control Control relating to the proper application of forces on the workpiece.

Microinch The unit of measure for surface roughness.

Milling Machining surfaces by using a rotating cutter with one or more teeth.

Milling Cutter A rotating cutting tool provided with one or more teeth, which intermittently engage the workpiece and remove material by movement of both the cutter and the workpiece.

Modulus of Elasticity Ratio of stress per unit of area to the strain per unit of length, with the strain being below the elastic limit of the material.

Molybdenum Metal used as an element in steel, including high-speed steel.

Multiple-Spindle Automatic Bar and Chucking Machine A machine with multiple (six or eight) spindles arranged revolver-fashion in a carrier that indexes the spindles through machining stations from material load to workpiece cutoff or ejection; the machine is arranged horizontally. Feed tubes feed bar or tubular stock; chucks handle individual castings or forgings.

Multiple-Spindle Vertical Automatic Chucker A machine with vertical spindles that revolve and index horizontally carousel fashion; workpieces are chucked on the top face of the spindles and machined incrementally at each progressive station. Loading and unloading of parts can be manual or automated.

Negative Rake An angle of rake that is less keen or more blunt than 0° rake.

Neutral Rake A rake angle of 0°. This angle is perpendicular to the surface of the work.

Nickel Strong noncorrosive white metal, used as an alloying element in steels and for plating.

Nomenclature The names of individual parts of machines or tools.

Nominal Refers to basic sizes without tolerance, such as "3/8 drill."

Nonferrous Any metal that contains no iron.

Nose Radius The radius on a tool between the end and side cutting edges.

Numerical Control (NC) Control of a machine or a process by command instructions in symbolic form.

O.B.I Press Open-backed inclinable presses have a frame designed with a C configuration that is open at the back for material handling (dropping out parts), etc. The frame has a pivot permitting it to be inclined for various setups. See Figure 24.2.

Open Die Forging A die with a cavity into which heated metal can be placed by hammer blows.

Pallet Decoiler A tool that enables a wide range of metal widths, thicknesses, and forms to be used directly from the shipping pallet; see Figure 24.19.

Part Design Dimension A dimension on a part design, usually including the allowable deviation or tolerance.

Periphery Milling Cutter A milling cutter that cuts mainly on the perimeter or external boundary, to produce a finished surface.

Permanent Mold Casting A process in which molten nonferrous metal is poured by gravity into a metal die (mold) composed of hinged and sliding sections, which open and/or collapse to permit removal of the casting.

Pick and Place Manipulators Mechanical devices designed to load and unload workpieces into and out of a machine.

Piercing A die that cuts holes, openings, and slots in sheet metal strip or blanks; similar in design to a blanking die, this die can be used for secondary operations.

Pilot A pin that registers in pilot holes pierced along the edge of strip stock on a progressive die to maintain position in relation to each stroke of the press.

Pinion The smaller of a set of two gears.

Pitch Distance between the crests of adjacent threads or gear teeth.

Planer A machine tool with a reciprocating table, used to machine flat and angular surfaces.

Plasma Arc Cutting A process in which an electric arc created between a torch and the workpiece heats and ionizes a gas, such as nitrogen, and is forced through the torch, creating a plasma with temperatures as high as 50,000°F to melt and vaporize workpiece material.

Plug Gage An accurate fixed gage used to check the size of holes.

Pocket The space that has been machined out of a toolholder or milling cutter, to house the insert or insert package.

Polar Lubricant A tenacious lubricant that resists removal from a surface and then has properties capable of reducing the coefficient of friction between two metals.

Positioned A workpiece properly loaded, chucked, or placed in a workholder is positioned.

Positive Rake An angle of rake that is keener or more acute than 0° rake.

Powder Dry particles that make up a fine non-gritty material.

Powder Metallurgy Forming parts out of powdered metal by compacting the powder into a die or mold under great pressure and then heating it.

Power and Free Conveyor A conveyor designed with sections of power conveyor and high-powered interconnections with track, switches, controls, and usually elevating and lowering devices connected to idle or non-powered sections.

Power Press Brake A long, narrow press with bending capabilities used to bend angles on long workpieces or to produce several shorter workpieces simultaneously or progressively; see Figure 24.5.

Press Forging A forging produced by the squeezing action of a press forcing heated metal into a closed die cavity; see Figure 21.6.

Pressure Pad A die component that utilizes pressure from hydraulic, air, compressed die springs, or compressed rubber to function as a hold down, part stripper, blank holder, or component of a pad-type form die.

Producibility The ease and efficiency with which a product may be produced or manufactured; design parameters affecting production include specified complexity, hardness, material, surface finishes, and tolerances.

Product Engineering The department responsible for product design and testing, and in some instances, research and experiment.

Production Control The department of a manufacturing company that is responsible for production material, including ordering and receiving, movement, shipping, and storage.

Profile The contour of the surface in a plane perpendicular to the surface.

Programmable Robot Defined by the Robot Institute of America as "a programmable multifunctional manipulator designed to move material, parts, tools and special devices through variable programmed motions for the performance of various tasks."

Progressive Die This die, sometimes called a follower, gang, or cut and carry die, is one that performs several individual operations of a part sequentially as strip stock is advanced station by station through the die.

Pulley A flat-faced wheel used to transmit power by means of a flat belt or with grooved pulleys called V-Belt pulleys.

Punch A detail positioned in a punch holder (the upper portion of die set) that forms metal and/or pierces round, geometrically shaped, or irregularly shaped holes.

Quill The nonrotating but retracting and extending portion of a drill press or milling machine containing the bearings and the machine spindle.

Rack A gear with a straight pitch line.

Radial Drill Press A drilling machine with an arm that moves up or down and around a vertical column.

Rake Angle The angle between the top face of a cutting tool and a line at a 90° angle to the workpiece.

Ram A movable member of a press to which tooling and accessories are mounted.

Reaming Machining process to finish the interior of holes.

Rectangular Coordinates Dimensions from two right-angle surfaces to a point.

Redrawing The second and subsequent operations in which a cuplike shape is deepened.

Reduction of Diameter The method of calculating the amount of draw required for a part as compared with empirical recommended allowances that can be done in one drawing operation.

Reel Used to uncoil coils of metal, in particular where surfaces must not be marked or scratched; see Figure 24.17.

Relief Angle An angle that provides cutting-edge clearance for the cutting tool.

Resultant A dimension that is the calculated difference between two working dimensions or between a working dimension and another resultant dimension.

Resultant Clamping A single clamp, or clamping force, that is used in place of three separate clamps to push a workpiece against all three planes of location.

Reverse Osmosis The flow of clear or fresh water through a semipermeable membrane when pressure is applied to liquid solutions containing impurities on one side of the membrane.

Rockwell A hardness test that uses a penetrator and known weights. The Rockwell C scale is most commonly used for steel.

Roll Forming A process that shapes flat strip stock as it advances through a series of rolls; the cross-sectional area does not change. See Figure 20.24.

Rotary Transfer Machine This machine, similar to in-line machines in many respects, transfers workpieces around a circle and back to the starting point.

Roughness The fine irregularities of a surface texture that result from the production process.

RPM Revolutions per minute.

Runout An off-center rotation, such as a cylindrical part held in a lathe chuck and running off center as it rotates.

Saddle Assembly on a lathe that moves back and forth on the ways.

Sand Casting Molten metal poured into a shaped cavity made in sand by a pattern and allowed to cool to produce the desired object.

Scrap Cutter A device used at a press following die operation to cut scrap strips into convenient disposable lengths.

Scrap Strip Layouts Layouts of the scrap strip of a progressive die as it will appear at completion of all stations, showing processing as per the formal plan; efficient use of material; manual stops to aid starting new strips; automated stop for production run; pilot required; stock strippers where required; special conditions, flatness, tolerances, etc.; part disposal; and scrap disposal.

Scribe Making scratch marks on metal for the purpose of layout.

Seat A removable part of a tool holder or milling cutter, designed to support the cutting insert.

Setup The arrangement by which the machinist fastens the workpiece to a machine table or work-holding device, and aligns the cutting tools for metal removal.

Severity of Draw Another term for reduction of diameter expressed in percent.

SFM, SFPM Surface feet per minute.

Shaper A machine with a reciprocating tool attached to a ram, used to machine small parts.

Shear Angle The angle between the shear plane and the face of the work.

Shear Plane Metal is separated in the machining operation by a shearing action, which takes place in a plane passing through the cutting edge of the tool and the surface of the work directly above the chip.

Shedder Similar in design to a knockout, a shedder strips blanks that are held to face of punch by oily film.

Sheet A metal produced in thin sheets by the rolling process is the primary material for metalworking processes. Sheets are available in all commercially used metals.

Shell End Mill An end mill without an integral shank. It is designed to mount on the machine spindle by means of a stub arbor or adaptor.

Shim A thin metal spacer used in the adjustment of machine parts.

Shuttle-Type Machine A machine in which a workpiece is loaded into a pallet fixture and indexed to two or more machining stations; after machining operations, workpieces are returned (shuttled) back to the starting position.

Side Cutting-Edge Angle The angle between the side cutting edge and the projected side of the shank or holder.

Side Rake The inclination of the face of the tool away from the side cutting edge, measured in a plane perpendicular to the top plane of the tool and the side cutting edge.

Silicon Carbide A refractory and abrasive material made with sand, coke, and sawdust in an electric arc furnace.

Simultaneous Engineering A program composed of teams or groups in which Product Engineering, Manufacturing Engineering, and other involved areas work together simultaneously on new product designs and manufacturing planning.

Single-Spindle Automatic Lathes The majority of these lathes machine workpieces between centers; some have chucks, collets, or fixtures. Most have horizontal spindles.

Single-Spindle Automatic Screw Machines These machines have horizontal hollow spindles aligned with stock feeding tubes. Most are CAM-controlled but N/C or CNC versions are more flexible and quickly set up.

Sintering The operation of heating a powder compact so that it shrinks and fuses to a near void-free condition.

Smith Forging Also known as open die forging, the process consists of hammer blows to compress and form heated metal against an anvil or open die. Manual or power hammers are used.

Snagging Rough grinding to remove unwanted metal from castings and other products.

Solid, or from the Solid An expression used when a groove, hole, or slot is machined in virgin material where no previous operation has been performed.

Specific Tolerances Tolerances applied to a specific dimension.

Speeds Machining speeds are expressed in revolutions per minute (RPM); cutting speeds are expressed in surface feet per minute (SFPM).

Spindle A part of a machine that rotates and carries either the workpiece or cutters.

Spline Shaft A shaft on which teeth have been machined parallel to the shaft axis, which will engage similar internal teeth on a matching part to prevent rotation.

Springback Stresses created in metal parts by such operations as bending cause metal to move back toward its previous state after release from die.

Spur A gear with the centerline of the teeth parallel to the centerline of the shaft on which the gear is mounted.

Stacking Tolerance stacks are the sum of tolerances of two or more dimensional tolerances. The stack may be generated by the addition of the overall length tolerance of *two or more* parts, or it may come from the addition of two or more dimensions of the *same* part.

Stainless Steel An alloy that contains iron, carbon, chromium, and sometimes nickel.

Stamping A general term covering press operations and end product.

Stock Feeder A device consisting of powered rollers or ratchet feed gripping units used to move coiled strip stock through the die process; see Figure 24.20.

Stock Guides Also called gages, guides strip stock into and through a die.

Stock Removal As the name implies, the amount of material to be removed from a surface.

Stock Stop A device used to stop the motion of strip stock during the downstroke of a press.

Stock Stripper A device used to strip parts from a punch or die.

Storage Conveyor Any conveyor that is used to store finished or in-process material and parts; roller and monorail conveyors are much used.

Straight-Side Press This press is designed with columns or uprights across each end to form a boxlike structure. Windows, or openings, are provided in the ends to permit feeding stock to the die as well as for unloading, ejecting, and handling workpieces and scrap strips. See Figure 24.6.

Strip Layout Also known as a tool layout, a strip layout provides a progressive pictorial view of processing and cutting tools required to produce a part on a multispindle machine; see Figure 6.21.

Support Those systems that do not perform any work on a part but provide a function or service required in the manufacture of a part.

Supports Fixture components that resist deflection of an area of the workpiece, supports may be fixed or adjustable. Adjustable supports may be manually or automatically adjusted.

Surface Finish Specifications are shown on a product design using the standard symbols and expressed in microinches. See Figure 5.9.

Surface Grinder A machine that grinds flat surfaces with the workpiece held on a reciprocating or rotating table.

Swaging A process that reduces the diameter of rod, tubing, or wire by radial impacts.

Tantalum Carbide (TaC) A compound of equal parts by atomic weight of tantalum and carbon.

Tap A tool that cuts threads in a hole, with cutting edges on the flutes.

Template A metal, cardboard, or wooden form used to transfer a shape or layout when it must be repeated many times.

Tensile Strength The maximum unit load that can be applied to a material before ultimate failure occurs.

Thermal Deburring Also known as thermal energy deburring, thermal deburring removes burrs and flash of uniform thickness by placing parts in a chamber filled with a mixture of oxygen and natural gas and igniting. The resulting intense heat burns away burrs and flash.

Thermal Expansion The increase in size caused by heating.

Thread Chasing Making successive cuts in the same groove with a single-point tool to generate a thread.

Titanium A strong grayish metal that weighs 44% that of steel with same strength.

Titanium Carbide (TiC) An intermetallic compound of equal parts of titanium and carbon.

Tolerance The total amount of variation permitted from a specified dimension; it may be expressed as a plus, minus, or both.

Tool Force The amount of force applied to a workpiece by a cutting tool or grinding wheel.

Tool Geometry The proper shape of a cutting tool, which makes it work efficiently for a particular application.

Toolholder A tool that mechanically holds one or more indexable inserts.

Torque A force that tends to produce rotation or torsion.

Toughness The ability to withstand breakage or impact.

Transfer Device A unit used to move workpieces from one press to a succeeding press in tandem or in-line arrangements; see Figure 24.25.

Transfer Press Line Straight-sided presses are used for high production of large workpieces such as automobile body panels, doors, tops, and floorpans by arranging the presses in tandem (in-line) so that a workpiece can be transferred out of the first press into the front of the second press and so on until the finished part emerges from the final press. See Figure 24.8.

Transverse Rupture Strength (TRS) Breaking strength in a standard bending test.

Triple-Action Press This press has the same inner and outer ram construction and action as a double-action press, but also has a third ram in the press bed that moves up, enabling a reverse draw to be made during the press cycle. Triple-action presses are not common.

Truing In grinding operations, making a wheel concentric to the spindle with a dressing tool.

Trunnion Machine This machine can be thought of as a double-end machine that has several workpiece-holding fixtures mounted on an index unit located between the end machining units.

T-slot The slot in a machine table shaped like a T and used to hold T-nuts and studs for various clamping setups.

Tumbling Also known as barrel finishing, tumbling removes burrs and improves the finish of workpiece surfaces by revolving workpieces, abrasive slurry, and media together in a hexagon- or octagon-shaped barrel.

Tungsten Carbide An iron-gray powder composed of carbon and tungsten and used in sintered form as a cutting-tool material.

Turning A machine operation in which the work is rotated against a single-point tool.

Turret Lathe A lathe with a revolving turret head mounted either on the saddle or a special tailstock.

Turret Press Also known as a punching machine, a turret press has a work table for the workpiece (sheet) with *X*- and *Y*-axis movement capabilities and a turret or tool storage facility with appropriate sizes of punches. The press is programmed to position the hole location of the workpiece at the punching station as a correct punch is brought to the station. Hole punching is performed automatically in the planned sequence. See Figure 24.27a.

Upsetting Also known as upset forging and cold heading, a process that shapes the end of a blank or length of metal wire up to one inch in diameter, usually done cold; see Figure 21.28.

Unilateral Tolerances expressed in one direction only, such as $2.000^{-000}_{+.005}$

Vibratory Finishing A mass finishing process that combines rubbing and impacting of abrasive media against workpieces by vibrating in a tub- or bowl-shaped container mounted on coil springs.

Vise A workholder device consisting of two jaws and a means of bringing them together to grip a workpiece.

Water Jet and Abrasive Jet Cutting Processes that utilize extremely high pressures to cut all materials; pressure on the order of 40,000 psi is used for most applications but can go to 200,000 psi. Water jets deburr and also cut all nonmetallic materials. With the addition of abrasives, all metals are cut with this process.

Waviness An irregularity that occurs at greater spacing than roughness, waviness may result from such factors as machine or work deflections, vibrations, chatter, and heat treatment (roughness may be considered as superimposed on a wavy surface).

Ways Longitudinal surfaces on a machine on which components such as carriages and tables move.

Wearland A flat land worn on the relieved or backed off portion of a tool behind the cutting edge.

Wedge A locking device used to secure inserts in the pocket of a milling cutter.

White Cast Iron An extremely hard cast iron, which results from pouring the hot metal into a mold containing a chill plate.

Wire Produced by the drawing process using steel, copper, aluminum, and other metals, wire is drawn in diameters up to 1.000 inch and coiled for subsequent use.

Wire Drawing Steel rod is pointed and drawn through a die, cold, to a wire size and may be redrawn to succeedingly smaller wire sizes. See Figure 21.29.

Work Hardening Hardening that occurs when certain metals are hammered or rolled at room temperatures.

Workholder A term that is used interchangeably with the terms "workpiece holder" or "fixture."

Working A processing dimension and its tolerance used in a manufacturing operation; sometimes known as a machining dimension, it is the distance from the locator to the surface to be machined and its tolerance.

Workpiece A partially completed part or part in process.

Workpiece Controls The degree of accuracy within which the workpiece is held to specifications, dimensions, and tolerances regardless of material, machine, or tool variances; total workpiece control comprises dimensional, geometric, and mechanical controls.

Worm (Gear Set) A reduction gear composed of a worm and wheel with nonintersecting axes.

Index

Millimeters to Inches					
mm	*in.*	*mm*	*in.*	*mm*	*in.*
0.01	0.0004	0.35	0.0138	0.68	0.0268
0.02	0.0008	0.36	0.0142	0.69	0.0272
0.03	0.0012	0.37	0.0146	0.70	0.0276
0.04	0.0016	0.38	0.0150	0.71	0.0280
0.05	0.0020	0.39	0.0154	0.72	0.0283
0.06	0.0024	0.40	0.0157	0.73	0.0287
0.07	0.0028	0.41	0.0161	0.74	0.0291
0.08	0.0031	0.42	0.0165	0.75	0.0295
0.09	0.0035	0.43	0.0169	0.76	0.0299
0.10	0.0039	0.44	0.0173	0.77	0.0303
0.11	0.0043	0.45	0.0177	0.78	0.0307
0.12	0.0047	0.46	0.0181	0.79	0.0311
0.13	0.0051	0.47	0.0185	0.80	0.0315
0.14	0.0055	0.48	0.0189	0.81	0.0319
0.15	0.0059	0.49	0.0193	0.82	0.0323
0.16	0.0063	0.50	0.0197	0.83	0.0327
0.17	0.0067	0.51	0.0201	0.84	0.0331
0.18	0.0071	0.52	0.0205	0.85	0.0335
0.19	0.0075	0.53	0.0209	0.86	0.0339
0.20	0.0079	0.54	0.0213	0.87	0.0343
0.21	0.0083	0.55	0.0217	0.88	0.0346
0.22	0.0087	0.56	0.0220	0.89	0.0350
0.23	0.0091	0.57	0.0224	0.90	0.0354
0.24	0.0094	0.58	0.0228	0.91	0.0358
0.25	0.0098	0.59	0.0232	0.92	0.0362
0.26	0.0102	0.60	0.0236	0.93	0.0366
0.27	0.0106	0.61	0.0240	0.94	0.0370
0.28	0.0110	0.62	0.0244	0.95	0.0374
0.29	0.0114	0.63	0.0248	0.96	0.0378
0.30	0.0118	0.64	0.0252	0.97	0.0382
0.31	0.0122	0.65	0.0256	0.98	0.0386
0.32	0.0126	0.66	0.0260	0.99	0.0390
0.33	0.0130	0.67	0.0264	1.00	0.0394
0.34	0.0134	—	—	—	—

Inches to Millimeters					
in.	*mm*	*in.*	*mm*	*in.*	*mm*
0.001	0.025	0.290	7.37	0.660	16.76
0.002	0.051	0.300	7.62	0.670	17.02
0.003	0.076	0.310	7.87	0.680	17.27
0.004	0.102	0.320	8.13	0.690	17.53
0.005	0.127	0.330	8.38	0.700	17.78
0.006	0.152	0.340	8.64	0.710	18.03
0.007	0.178	0.350	8.89	0.720	18.29
0.008	0.203	0.360	9.14	0.730	18.54
0.009	0.229	0.370	9.40	0.740	18.80
0.010	0.254	0.380	9.65	0.750	19.05
0.020	0.508	0.390	9.91	0.760	19.30
0.030	0.762	0.400	10.16	0.770	19.56
0.040	1.016	0.410	10.41	0.780	19.81
0.050	1.270	0.420	10.67	0.790	20.07
0.060	1.524	0.430	10.92	0.800	20.32
0.070	1.778	0.440	11.18	0.810	20.57
0.080	2.032	0.450	11.43	0.820	20.83
0.090	2.286	0.460	11.68	0.830	21.08
0.100	2.540	0.470	11.94	0.840	21.34
0.110	2.794	0.480	12.19	0.850	21.59
0.120	3.048	0.490	12.45	0.860	21.84
0.130	3.302	0.500	12.70	0.870	22.10
0.140	3.56	0.510	12.95	0.880	22.35
0.150	3.81	0.520	13.21	0.890	22.61
0.160	4.06	0.530	13.46	0.900	22.86
0.170	4.32	0.540	13.72	0.910	23.11
0.180	4.57	0.550	13.97	0.920	23.37
0.190	4.83	0.560	14.22	0.930	23.62
0.200	5.08	0.570	14.48	0.940	23.88
0.210	5.33	0.580	14.73	0.950	24.13
0.220	5.59	0.590	14.99	0.960	24.38
0.230	5.84	0.600	15.24	0.970	24.64
0.240	6.10	0.610	15.49	0.980	24.89
0.250	6.35	0.620	15.75	0.990	25.15
0.260	6.60	0.630	16.00	1.000	25.40
0.270	6.86	0.640	16.26	—	—
0.280	7.11	0.650	16.51		

Inch–Millimeter Equivalents